SELECTED PAPERS ON ALGEBRA

THE
RAYMOND W. BRINK SELECTED MATHEMATICAL PAPERS

Published by
THE MATHEMATICAL ASSOCIATION OF AMERICA

The Raymond W. Brink Selected Mathematical Papers

VOLUME THREE

SELECTED PAPERS ON ALGEBRA

Reprinted from the

AMERICAN MATHEMATICAL MONTHLY
(Volumes 1–80)

and from the

MATHEMATICS MAGAZINE
(Volumes 1–45)

Selected and arranged by an editorial committee consisting of

SUSAN MONTGOMERY AND ELIZABETH W. RALSTON, Co-Chairmen
University of Southern California and *Fordham University*

S. ROBERT GORDON
University of California, Riverside

GERALD J. JANUSZ
University of Illinois

MURRAY M. SCHACHER
University of California, Los Angeles

MARTHA K. SMITH
University of Texas

Published and distributed by
THE MATHEMATICAL ASSOCIATION OF AMERICA

Library of Congress Catalog Card Number 77-792-80

Complete Set ISBN 0-88385-200-4
Vol. 3 ISBN 0-88385-203-9

Printed in the United States of America

Current printing (last digit):

10 9 8 7 6 5 4 3 2 1

FOREWORD

The RAYMOND W. BRINK SELECTED MATHEMATICAL PAPERS series of the Mathematical Association of America was established through a generous gift to the Association from Mrs. Carol Ryrie Brink in honor of her late husband, Professor Raymond W. Brink. The series provides a fitting and lasting memorial to Professor Brink, who served the Association in many significant ways including terms as Governor (1934–39 and 1943–48), as Vice-President (1940–41), and as President (1941–42).

Articles for inclusion in the SELECTED PAPERS volumes are selected from past issues of the Association's journals.

In expressing its deep appreciation to Mrs. Brink, the Board of Governors was particularly pleased to note that the gift was made to the Association at a time when our organization has become increasingly dependent on private sources for support. The Board therefore felt grateful to Mrs. Brink not only for her generosity but also for her wish that the gift will stimulate others to contribute to the projects of the Association.

The following RAYMOND W. BRINK SELECTED MATHEMATICAL PAPERS have been published:

Volume 1: SELECTED PAPERS ON PRECALCULUS

Volume 2: SELECTED PAPERS ON CALCULUS

Volume 3: SELECTED PAPERS ON ALGEBRA

PREFACE

In this volume the reader will find 108 articles on various topics in algebra which have been reprinted from the pages of the AMERICAN MATHEMATICAL MONTHLY and the MATHEMATICS MAGAZINE. The articles have been arranged in sections according to subject matter; each section ends with a bibliography, listing further articles on the same topic which for one reason or another it did not seem advisable to reprint, but which nevertheless contain interesting material.

We have on the whole selected articles we felt to be of interest to students and teachers of undergraduate and beginning graduate courses in algebra. Because abstract algebra is not taught to beginning students in most of our college and university curricula, this has naturally meant an emphasis on advanced undergraduate and graduate level mathematics. We have particularly looked for articles which extend familiar material past the point usually possible in university courses, or treat such material from an unfamiliar point of view (for example, by giving new proofs of old results), or cover material suitable for such courses but not usually included in them. We have included some very elementary articles which could be used in making up exercises for such courses (the Classroom Notes section of the MONTHLY has specialized in this kind of article). We have also chosen some articles for their historical comment. The MONTHLY in particular has a long history of publishing readable and authoritative expository articles, and it has been a pleasure to make some of these articles more available by reprinting them (a few especially fine expository articles unfortunately have proved simply too long to print and hence appear only in the bibliography). On the other hand, we have tried to avoid highly technical articles, or articles which seem solely intended as research. Finally, we have occasionally used our editorial license in reprinting articles which appealed to us for one or another idiosyncratic reason, but which fall outside the framework outlined above. We hope any resulting inconsistency justifies itself in the eyes of our readers.

From the above it will be evident that this volume contains material covering a huge range of difficulty; consequently the reader should not expect to find everything here both accessible and interesting. We are confident, however, that a wide circle of students and teachers with an interest in algebra should be able to find in this volume articles which can be read with both pleasure and profit.

THE EDITORIAL COMMITTEE

CONTENTS

(b) COMMUTATIVE RINGS

(c) MODULES

(d) DIVISION ALGEBRAS

(e) NON-COMMUTATIVE RINGS

(f) NON-ASSOCIATIVE RINGS AND ALGEBRAS

3. FIELD THEORY AND ALGEBRAIC NUMBER THEORY

(a) FIELD EXTENSIONS AND GALOIS THEORY

(b) FINITE FIELDS

(c) REDUCIBILITY OF POLYNOMIALS

(d) VALUATION THEORY

(e) ALGEBRAIC NUMBER THEORY

4. LINEAR ALGEBRA

(a) EQUATIONS, DETERMINANTS, AND RELATED TOPICS

(b) CHARACTERISTIC POLYNOMIALS AND EIGENVALUES

(c) INNER PRODUCTS AND QUADRATIC FORMS

5. HISTORY

6. ADDITIONAL TOPICS

1

GROUP THEORY

(a)

GROUP AXIOMS AND EXAMPLES

A SINGLE POSTULATE FOR GROUPS*

MICHAEL SLATER, Magdalen College, Oxford, England

Marlow Sholander [1] shows that if we make certain natural presuppositions we may characterize an abelian group by a single postulate. This note provides a similar result for an arbitrary group.

Let G be a nonempty set on which two operations are defined: a binary *multiplication* and a singulary *inversion*. For each $a, b \in G$ we write (ab) for the product in that order of a and b, and a' for the inverse of a.

THEOREM. *If* $(ab)c=(ad)f$ *implies* $b=d(fc')$ *for every* $a, b, c, d, f \in G$, *then* G *is a group relative to the operations described above.*

Proof. 1. Since $(ab)c=(ab)c$ we have $b=b(cc')$. Let us write a^* for aa'. Then $(ab^*)c=(ad^*)c$ so that

$$b^* = d^*c^* = d^*.$$

Thus for all $a \in G$, $aa'=e_r$ (a right identity).

2. If $ab=ad$ then $(ab)c=(ad)c$, so that

$$b = d(cc') = d.$$

We thus have left cancellation.

3. We have $(ae_r)b=(ab)e_r$, so that $e_r=b(e_rb')$. But $e_r=bb'$, so that, by left cancellation, $b'=e_rb'$. In particular,

$$e_r' = e_re_r' = e_r.$$

Next, $(ab)e_r=(ae_r)b$, so that $b=e_r(be_r')=e_rb$.

4. We have $(e_rc)c'=(e_rd)d'$ so that $c=d(d'c'')$. If we take $d=e_r$, we find, in

* From AMERICAN MATHEMATICAL MONTHLY, vol. 68 (1961), pp. 346–347.

particular, that $c=c''$. So

$$c = d(d'c),$$

and the equation $a=bx$ has a solution $x=b'a$.

5.† Given a, b, c let us choose d so that $[a(bc)]d=(ab)c$. Then $bc=b(cd')$, $ce_r=c=cd'$, $e_r=d'$, and $d=e_r$. We thus have associativity, and the rest is easy.

Added in proof. Since writing the above, I have come across an article [2] in which a single axiom for groups is given. The axiom is written in terms of the operator δ, where $ab\delta=ab'$.

More generally, let $W(x_1,\ldots,x_n,\delta)$ be a ("meaningful") word in the symbols $x_1,\ldots,x_n,\delta$. Suppose a class $\mathcal{C}$ of groups is such that a given group G with unit element e is a member of $\mathcal{C}$ if and only if $W(a_1,\ldots,a_n,\delta)=e$ for all $a_1,\ldots,a_n\in G$. For example, if $\mathcal{C}$ is the class of *all* groups, we may take $W=xx\delta$; if $\mathcal{C}$ is the class of abelian groups, we may take $W=xyyx\delta\delta\delta$.

It is then shown that the class $\mathcal{C}$ is characterized by the single axiom

$$xxx\delta W\delta y\delta z\delta xx\delta x\delta z\delta\delta\delta = y, \qquad (\alpha)$$

where $W=W(x_1,\cdots,x_n,\delta)$. That is, if G is a set on which a binary operation δ is defined, then G is a member of $\mathcal{C}$ if and only if (α) is satisfied for all x, y, z, $x_1,\cdots,x_n\in G$.

References

1. Marlow Sholander, Postulates for commutative groups, this MONTHLY, vol. 66, 1959, pp. 93–95.
2. Graham Higman and B. H. Neumann, Groups as groupoids with one law, Publ. Math. Debrecen, vol. 2, 1952, pp. 215–221.

THE EXISTENCE OF FREE GROUPS*

MICHAEL BARR, McGill University

Let X be a set. A free group generated by X consists of a group F and a function $f: X \to F$ satisfying the following condition:

For any group K and function $k: X \to K$, there is a unique group homomorphism $\phi: F \to K$ for which $\phi \circ f = k$.

It is a well-known theorem that *given any set X there is always a free group generated by X.* (It is an easy and instructive exercise for the reader to use the above definition to show that up to isomorphism, there is only one; also, the set $f(X)$ generates F, as explained below.) The usual proofs are by the construction of a semigroup of "words," multiplied by juxtaposition, that become a group modulo a rather complicated equivalence relation. Here we present a proof which never leads us out of the category of groups.

† I am indebted to the referee for pointing out a simplification in my treatment of 5.

* From AMERICAN MATHEMATICAL MONTHLY, vol. 79 (1972), pp. 364–367.

The proof is modeled after that of the general adjoint functor theorem of category theory and, as such, is readily adapted to solving any universal mapping problem in the category of groups, such as the existence of free products. It also works in any category consisting of all the algebras and algebra homomorphisms of any algebraic theory. (However, one must not foolishly exclude the empty set from such a category if it otherwise satisfies its axioms, for then the empty set might not generate a free algebra.) Thus included are all such categories as sets, sets with a basepoint (and base-point preserving functions), groups, abelian groups, rings, commutative rings, Lie rings, Jordan rings, algebras of these types, etc., each considered as a category with the evident definition of homomorphism.

1. Preliminaries. Let Γ be an index set, and let $\{G_\alpha\}$ be a family of groups indexed by Γ. The **product**

$$G = \prod_{\alpha \in \Gamma} G_\alpha$$

is the usual cartesian product with coordinate-wise multiplication. Each element of G is a Γ-tuple $\{x_\alpha\}$, where $x_\alpha \in G_\alpha$ for all $\alpha \in \Gamma$ and

$$\{x_\alpha\}\{y_\alpha\} = \{x_\alpha y_\alpha\}.$$

It is easily checked that G with this operation is a group. Also, each projection π_β defined by

$$\pi_\beta\{x_\alpha\} = x_\beta$$

is a group homomorphism, $\pi_\beta: G \to G_\beta$.

Let G be a group and A a sub-*set* of G. We say that A **generates** G if no proper sub-*group* of G contains A.

If $g: X \to G$ is a map, we say that g **generates** G if the image $g(X) = \{a \in G \mid a = g(x)$ for some $x \in X\}$ generates G.

Suppose $g: X \to G$ and g does not necessarily generate G. Then g does, in a sense, generate a subgroup of G. Precisely:

PROPOSITION 1. *Let $g: X \to G$ be a map of a set X into a group G. Then there is a subgroup H of G and a map $h: X \to H$ such that h* **generates** *H and $g = j \circ h$, where j is the inclusion map.*

Proof. Let H be the intersection of all subgroups of G which contain $g(X)$. Clearly no proper subgroup of H contains $g(X)$, so $g(X)$ generates H. The rest is clear.

In this situation, we call H the subgroup of G **generated** by $g: X \to G$.

PROPOSITION 2. *Let X be a set. Then there exists a collection C of pairs (G_α, g_α), indexed by some set Γ such that*

(1) *each G is a group and $g_\alpha: X \to G_\alpha$ generates G_α;*

(2) *If K is any group and $k: X \to K$ generates K, then for some α there is an isomorphism ψ on G_α onto K such that $\psi \circ g_\alpha = k$.*

It is tempting to say "but this is obvious; simply take the collection of *all* pairs (G, g), where G is a group and $g: X \to G$ generates G." The sticky point is "all," which leads into the usual logical paradoxes. The proof of Proposition 2 will be postponed until the last section.

2. The proof. We state precisely the main result of this paper.

THEOREM. *Let X be a set. Then there exists a free group generated by X.*

Proof. Take a collection C given by Proposition 2. Form

$$G = \prod_{\alpha \in \Gamma} G_\alpha \text{ and } g = \prod_{\alpha \in \Gamma} g_\alpha.$$

Then G is a group and $g: X \to G$. Note that

$$\pi_\alpha \circ g = g_\alpha.$$

Let F be the subgroup of G generated by g. By Proposition 1, there is a map $f: X \to F$ such that f generates F and $g = j \circ f$, where j is the inclusion map of F in G. Note that

$$\pi_\alpha \circ j \circ f = \pi_\alpha \circ g = g.$$

Now let $k: X \to K$, where K is a group. We must prove that a unique homomorphism $\phi: F \to K$ exists such that $\phi \circ f = k$.

First we prove uniqueness. Suppose that we have two homomorphisms $\phi_i: F \to K$, where $\phi_i \circ f = k$ for $i = 1, 2$. Let F_0 be the set of x in F such that $\phi_1(x) = \phi_2(x)$. Obviously F_0 is a subgroup of F. Since $\phi_1 \circ f = \phi_2 \circ f$, we have $f(X) \subseteq F_0$. But $f(X)$ generates F, hence $F_0 = F$, so $\phi_1 = \phi_2$. We pass to the existence proof.

First assume k generates K. Then Proposition 2 gives us an α and an isomorphism ψ on G_α onto K such that $k = \psi \circ g_\alpha$. Consider

$$F \xrightarrow[j]{} G = \prod G_\alpha \xrightarrow[\pi_\alpha]{} G_\alpha \xrightarrow[\psi]{} K.$$

Let ϕ be the composite map, $\phi = \psi \circ \pi_\alpha \circ j$. Then ϕ is a homomorphism, and

$$\phi \circ f = \psi \circ \pi_\alpha \circ j \circ f = \psi \circ g_\alpha = k.$$

Thus such an f exists in this case.

Now suppose k does not generate K. By Proposition 1, there is a subgroup K' of K and a map $k': X \to K'$ which generates K' such that $j' \circ k' = k$, where $j': K' \to K$ is the inclusion map.

We know there is a homomorphism $\phi'\colon F \to K'$ such that $\phi' \circ f = k'$. We set $\phi = j' \circ \phi'$; then $\phi\colon F \to K$ is a homomorphism, and

$$\phi \circ f = j' \circ \phi' \circ f = j' \circ k' = k.$$

3. The construction of C. In this section we wind up matters by proving Proposition 2. For any set S, let $|S|$ denote the cardinality of S.

LEMMA 1. *Let $k\colon X \to K$ generate K. Then $|K| \leqq \max(|X|, \aleph_0)$.*

Proof. Let $A = k(X)$. Then $|A| \leqq |X|$, so it is sufficient to show that $|K| \leqq \max(|A|, \aleph_0)$, where A is a set of generators of K. (Of course, if A is infinite, this is equivalent to the assertion that $|K| = |A|$.)

Let $A^{-1} = \{a^{-1} \mid a \in A\}$ and for B and C subsets of K, define $BC = \{bc \mid b \in B, c \in C\}$. Now let $A_1 = A \cup \{e\} \cup A^{-1}$, where e is the identity of K. Then define $A_2 = A_1A_1$, $A_3 = A_1A_2, \cdots, A_{n+1} = AA_n, \cdots$. Finally let $\bar{A} = \bigcup_{n=1}^{\infty} A_n$. By an easy induction, one sees that $A_n \subseteq A_n^{-1}$ and that $A_nA_m \subseteq A_{n+m}$. Thus $x, y \in \bar{A}$ implies $x^{-1} \in \bar{A}$ and $xy \in \bar{A}$, hence $\bar{A}$ is a subgroup of K. Since $A \subseteq \bar{A}$, this implies that $\bar{A} = K$. For any subsets B and C of K, the set BC is the image under the multiplication of $B \times C$, which implies that $|BC| \leqq |B| \cdot |C|$.

First we suppose that A is infinite. Then $|A_1| \leqq 1 + |A| + |A^{-1}|$, while $|A^{-1}| = |A|$, which gives $|A| = |A_1|$. Next $|A_2| \leqq |A_1| \cdot |A_1| = |A_1|$, while $A_1 \subseteq A_2$ (since $e \in A_1$, and $A_1 = eA_1 \subseteq A_1A_1$) implies $|A_1| \leqq |A_2|$, and then $|A_2| = |A|$. By induction, $|A_n| = |A|$ for all n, and thus $|\bar{A}| \leqq \Sigma |A_n| \leqq \aleph_0 \cdot |A| = |A|$.

When A is finite, the same argument shows instead that each A_n is finite, and the countable union $\bar{A}$ is at most countable.

LEMMA 2. *Let Y be any set. Then the collection of groups G whose underlying set is Y has cardinality at most*

$$c = |Y|^{|Y|^2}.$$

For a group is determined by its multiplication table, a function on $Y \times Y$ to Y. But c is the cardinality of the set of all such functions.

Proof of Proposition 2. For each cardinal s with $s \leqq \max(|X|, \aleph_0)$, choose a set Y_s with $|Y_s| = s$. Consider the set of all groups G such that the underlying set of G is some Y_s; by Lemma 2, this set exists. Then consider the collection C of all pairs (G_α, g_α), where G_α is one of these groups G_α and $g_\alpha\colon X \to G_\alpha$ generates G_α. The (2) of Proposition 2 is an immediate consequence of this construction and Lemma 1. The proof is complete.

NON-ABELIAN GROUPS OF ORDER pq*

S. K. BERBERIAN, Southern Illinois University

This note is concerned with the construction of non-abelian groups of order pq. The point of departure is the theorem characterizing such groups:†

THEOREM: *If G is a non-abelian group of order pq, p and q primes, $p<q$, then:*

(1) $q\equiv 1 \pmod p$;
(2) *G is generated by elements P and Q, of orders p and q, respectively, such that*
(3) $P^{-1}QP=Q^{\beta}$, where
(4) $\beta\not\equiv 1$ *is a root of the congruence*
$x^{p}\equiv 1 \pmod q$.

It is evident that as soon as two elements P and Q satisfying (2), (3), and (4) are located in a known group, they generate a subgroup satisfying the requirements for G.

We therefore assume property (1) to hold, and attempt to locate suitable elements P and Q in a known group. For embedding purposes, we utilize the symmetric permutation group of degree q.

First, there are numbers β satisfying the congruence (4); for, by assumption (1), the prime p is a divisor of $q-1$, the order of the multiplicative group of the field of integers modulo q. Since such an element β has order p, the full list of distinct solutions modulo q of the congruence (4) is:

$$1, \beta, \beta^2, \cdots, \beta^{p-1}.$$

For Q, we choose the q-cycle

$$Q=(1, 2, \cdots, q).$$

Modulo q, we may write

$$Q^{\beta}=(1, 1+\beta, 1+2\beta, \cdots, 1+(q-1)\beta).$$

P will necessarily be the product of disjoint p-cycles. In assumption (1), suppose that $q=1+mp$; we will choose P to be the product of m disjoint p-cycles. With such a structure, P will have to leave one symbol of Q unmoved; if we agree to let this symbol be 1 (this choice is arbitrary), then by (3), and the properties of conjugacy for permutations, P must act on the remaining symbols of Q according to the scheme:

$$\begin{array}{lccccccc} Q=(1, & 2, & 3, & \cdots, & i, & \cdots, & q) \\ (5)\qquad & \downarrow & \downarrow & & \downarrow & & \downarrow \\ Q^{\beta}=(1, & 1+\beta, & 1+2\beta, & \cdots, & 1+(i-1)\beta, & \cdots, & 1+(q-1)\beta). \end{array}$$

Thus, P must send i to $1+(i-1)\beta$, which in turn must be sent to

* From AMERICAN MATHEMATICAL MONTHLY, vol. 60 (1953), pp. 37–40.

† Burnside, W.: Theory of Groups of Finite Order; p. 48.

$1+[1+(i-1)\beta-1]\beta=1+(i-1)\beta^2$. A typical cycle of P would then be

$$(6) \qquad \pi = (1 + (i-1),\ 1 + (i-1)\beta,\ 1 + (i-1)\beta^2, \cdots, 1 + (i-1)\beta^{p-1}).$$

The cycle π terminates after p entries, since $\beta^p \equiv 1 \pmod q$, and hence

$$1 + (i-1)\beta^{p-1} \to 1 + (i-1)\beta^p \equiv i \pmod q.$$

The application of this principle is very direct.

Example 1: $p=3$, $q=7$

A primitive solution of the congruence $x^3 \equiv 1 \pmod 7$ is $\beta=2$. Choose

$$Q = (1\ 2\ 3\ 4\ 5\ 6\ 7).$$

Then,

$$Q^\beta = (1\ 3\ 5\ 7\ 2\ 4\ 6).$$

Assuming again that P leaves the symbol 1 fixed, we simply force P to meet the requirements:

$$P = (2\ 3\ 5)(4\ 7\ 6).$$

Then,

$$P^{-1}QP = Q^2$$

and P and Q generate the required group G.

We next show that the construction of a P, satisfying (3), is always possible. Example 1 suggests a method of proof. For the first cycle of P, write

$$\pi_1 = (1 + 1,\ 1 + \beta,\ 1 + \beta^2, \cdots, 1 + \beta^{p-1}).$$

Since $1, \beta, \beta^2, \cdots, \beta^{p-1}$ are incongruent modulo q, so are the entries of π_1. The cycling is proper, since the symbol $1+\beta^{p-1}$ of Q must be sent to

$$1 + \beta^p \equiv 1 + 1 \pmod q.$$

At this point, if $q=3$, then $p=2$ and the construction is ended. Otherwise, there is a symbol $i\neq 1$ of Q which does not appear in π_1; for the next factor of P, construct

$$\pi_2 = (1 + (i-1),\ 1 + (i-1)\beta,\ 1 + (i-1)\beta^2, \cdots, 1 + (i-1)\beta^{p-1}).$$

It has been already noted in connection with (6) that the cycling of π_2 is proper. The entries of π_2 are distinct modulo q, for, an equality

$$1 + (i-1)\beta^r \equiv 1 + (i-1)\beta^s \pmod q$$

would imply

$$\beta^{r-s} \equiv 1 \pmod q,$$

and since $0 \leq r,\ s < p$, and β has order p, it follows that $r=s$. Also, π_2 is disjoint from π_1; for suppose there were an equality

$$1 + (i - 1)\beta^r \equiv 1 + \beta^s \pmod{q}.$$

This would imply

$$i \equiv 1 + \beta^{s-r} \pmod{q}$$

contrary to the choice of i.

If there is a third step in the construction,

$$\pi_3 = (1 + (j - 1), 1 + (j - 1)\beta, \cdots, 1 + (j - 1)\beta^{p-1})$$

where $j \neq 1$ is neither in π_1 nor π_2, the preceding argument shows that π_3 is disjoint from π_1. To show that π_3 is disjoint from π_2, suppose there were an equality

$$1 + (j - 1)\beta^r \equiv 1 + (i - 1)\beta^s \pmod{q}.$$

Then,

$$j \equiv 1 + (i - 1)\beta^{s-r} \pmod{q}$$

whereas j is not in π_2.

Finally, $P = \pi_1\pi_2 \cdots \pi_m$, the construction terminating in exactly m steps, since there are $q-1$ symbols of Q to be moved, and $q-1=mp$. (An explicit induction step has been omitted; the argument surrounding π_3 is nearly adequate.) We have done little more than construct the system of prime residues modulo q.*

Example 2: The case $p=2$ is especially transparent, since we are then forced to choose $\beta=q-1$. If then $Q=(1, 2, \cdots, q)$, Q^β will be the inverse of Q:

$$Q^\beta = (1, q, q-1, \cdots, 2),$$

and if 1 is the symbol unmoved by P, then

$$P = (2, q)(3, q-1)(4, q-2) \cdots .$$

For example, if $q=7$, then

$$Q = (1\ 2\ 3\ 4\ 5\ 6\ 7)$$
$$Q^\beta = Q^6 = (1\ 7\ 6\ 5\ 4\ 3\ 2)$$
$$P = (27)(36)(45).$$

This representation of the non-abelian group of order $2q$ naturally suggests the dihedral group. Let the vertices of a regular q-gon be numbered in cyclic order. In the dihedral group for this polygon, Q may be interpreted as a rotation of $360°/q$, and P as a reflection in the axis of symmetry passing through vertex 1. Since q is odd, every reflection symmetry leaves fixed exactly one vertex; our choice of 1, as the symbol unmoved by P, may therefore be interpreted as a choice of vertex.

---

* Uspensky, J. V., and Heaslet, M. A.: Elementary Number Theory; p. 225.

BIBLIOGRAPHIC ENTRIES: GROUP AXIOMS AND EXAMPLES

The references below are to the AMERICAN MATHEMATICAL MONTHLY.

1. Paul Sally, Jr., The complete independence of some common axiom systems, 1962, 794–797.

Elementary examples showing that the axioms in the definition of a group and in the definition of an equivalence relation are all independent.

2. J. V. Whittaker, On the postulates defining a group, 1955, 636–640.

Shows that one can define a group in terms of an operation of subtraction instead of addition, still using only three postulates.

3. N. C. Meyer, Jr., A new proof of the existence of the free group, 1970, 870–873.

A variation on the "word" proof.

4. Trevor Evans, A condition for a cancellation semigroup to be a group, 1966, 1104–1108.

The condition: it satisfies an identity not a consequence of the commutative law.

5. R. A. Rosenbaum, Some simple examples of groups, 1959, 902–905.

Discusses "artificial groups" which can be used as examples in beginning algebra class. Groups are obtained by isomorphisms with known groups.

6. E. J. Finan, On groups of subtraction and division, 1941, 3–7.
7. J. S. Frame, Note on groups of subtraction and division, 1941, 468–469.

Discusses means of realizing dihedral groups as groups of functions generated by functions of the form $R_1(z)=r/z$, $R_2(z)=s-z$.

(b)

DECOMPOSITION OF GROUPS

ON CANCELLATION IN GROUPS*

R. HIRSHON, Polytechnic Institute of Brooklyn

Let $A \times B$ represent the direct product of the groups A and B. We shall say that B may be cancelled in direct products if

$$A \times B \approx A_1 \times B_1, \qquad B \approx B_1$$

imply $A \approx A_1$ for any A.

It seems natural to inquire about those groups which may be cancelled in direct products. We will show in this paper that a finite group B may be cancelled in direct products. As far as we can determine, this result does not appear

* From AMERICAN MATHEMATICAL MONTHLY, vol. 76 (1969), pp. 1037–1039.

in any standard text in group theory or algebra, perhaps because it appears to have been discovered as recently as 1947 ([4], introduction), and apparently is still not well known. Good use of it might be made, for example, in proving that the decomposition of a finite group as a direct product of indecomposable groups is unique up to isomorphism.

We present a proof of the cancellation theorem which we feel is the simplest available and is suitable for undergraduates. We also present in this paper an outline of a proof that an infinite cyclic group may not, in general, be cancelled in direct products, thus giving an example of the "simplest" type of group which may not be cancelled.

CANCELLATION THEOREM. *If B is a finite group, B may be cancelled in direct products.*

Proof. We observe first that it suffices to show

$$G = D \times B = D_1 \times B_1, \qquad B \approx B_1, \text{ imply } D \approx D_1. \tag{1}$$

We prove (1) by induction on $|B|$, the order of B.

Clearly (1) is true if $|B| = 1$. Assume (1) is true for groups B, with $|B| < k$. We prove (1) is true if $|B| = k$. First observe that if $B \cap D_1 = 1$ then $G = B \times D_1$, so that $D \approx G/B \approx D_1$. Hence, without loss of generality, we may assume $B \cap D_1 \neq 1$. Also by symmetry we may assume

$$F = B \cap D_1 \neq 1, \qquad K = B_1 \cap D \neq 1.$$

Now from (1), we may see

$$G/(F \times K) = (B \times D)/(F \times K) = (B_1 \times D_1)/(K \times F). \tag{2}$$

By a standard isomorphism theorem, we see from (2)

$$(B/F) \times (D/K) \approx (B_1/K) \times (D_1/F).$$

Hence, since $B \approx B_1$, we may write

$$B \times (B/F) \times (D/K) \approx B_1 \times (B_1/K) \times (D_1/F). \tag{3}$$

However,

$$\begin{aligned} B \times (B/F) \times (D/K) &\approx [B \times (D/K)] \times B/F \approx [(B \times D)/K] \times B/F \\ &= [(B_1 \times D_1)/K] \times B/F \approx (B_1/K) \times D_1 \times B/F. \end{aligned}$$

In summary, we have

$$B \times (B/F) \times D/K \approx (B_1/K) \times D_1 \times B/F. \tag{4}$$

Note that our hypothesis is symmetrical in B and B_1 and D and D_1, so if we interchange B and B_1 and D and D_1 (and hence F and K), we see from (4)

$$B_1 \times (B_1/K) \times D_1/F \approx (B/F) \times D \times B_1/K. \tag{5}$$

Now note from (3) that the groups on the left hand sides of (4) and (5) are isomorphic. Consequently, the groups on the right hand sides of (4) and (5) are isomorphic; that is,

$$(6) \qquad L_1 = D_1 \times (B/F) \times (B_1/K) \approx D \times (B/F) \times (B_1/K) = L_2.$$

Hence we may apply our inductive assumption twice in (6); that is, first cancel B_1/K in (6) and then cancel B/F. (To be quite precise, by using an isomorphism of L_1 onto L_2 obtained from (6), write (6) over as an equality between decompositions of L_2, and then apply the inductive assumption once, and then repeat this procedure again.) The result is $D_1 \approx D$, and the theorem is complete.

Kaplansky (in [5] p. 13) posed the following problem:

If B and B_1 are infinite cyclic abelian groups, and A is abelian and $A \times B \approx A_1 \times B_1$, is $A \approx A_1$? The question is answered affirmatively in [1], p. 55. It is surprising to discover that an infinite cyclic group may not be cancelled in general.

One can see the essential reason for this by considering a group H with the following properties:

(a) $H = \langle a \rangle L$, $L \cap \langle a \rangle = 1$, $L \Delta H$,

where $\langle a \rangle$ is an infinite cyclic group generated by a.

(b) There exists d, $d > 1$, such that a^d is in the centralizer of L.

(c) $K = \langle a^u \rangle L$ is not isomorphic to H, where u is an integer for which there exist integers s and e such that

(d) $eu - sd = \pm 1$.

Then if $\langle z \rangle$ is an infinite cyclic group and $G = \langle z \rangle \times H$ and if we set $w = z^e a^d$, $M = \langle z^s a^u \rangle L$, one can show $M \approx K$ and

$$G = \langle z \rangle \times H = \langle w \rangle \times M.$$

Since M and H are not isomorphic, this shows that an infinite cyclic group may not be cancelled in general. An example of such a group H is a group with two generators a and y, with defining relations $a^{-1}ya = y^4$, $y^{1024} = y$. One may take $L = \langle y \rangle$, $d = 5$, $u = 2$, $s = 1$, $e = 2$. We omit the proof that this group has the desired properties.

In closing, we point out that a group with a principal series, that is, one which obeys the ascending and descending chain condition for normal subgroups, may be cancelled in direct products. The proof is essentially the same as the one we have given for finite groups except that one uses induction on the length of a principal series. Some applications of this cancellation result appear in [2]. A sufficient condition for the cancellation of infinite groups which obey the maximal condition for normal subgroups, is given in [3].

References

1. L. Fuchs, Abelian Groups, Pergamon Press, New York, 1967.

2. R. Hirshon, Some Theorems on Hopficity. To appear. Transactions of the American Mathematical Society.

3. ——, Cancellation of groups with maximal condition. To appear in Proc. Amer. Math Soc.
4. B. Jónsson and A. Tarski, Direct Decompositions of Finite Algebraic Systems, Notre Dame Mathematical Lectures, Notre Dame, Indiana, 1947.
5. I. Kaplansky, Infinite Abelian Groups, The University of Michigan Press, Ann Arbor, 1962.

GROUPS AS UNIONS OF PROPER SUBGROUPS*

SEYMOUR HABER, Polytechnic Institute of Brooklyn, and
AZRIEL ROSENFELD, Ford Instrument Company and Yeshiva University

1. Introduction. It is evident that any group G which is not monogenic (generated by a single element) is expressible as a union of proper subgroups; for example, such a G is the union of its monogenic subgroups, which by hypothesis are all proper. Conversely, if a group is a union of proper subgroups of itself, it clearly cannot be monogenic.

It is not so easy to characterize groups which are *finite* unions of their proper subgroups. That there exist nonmonogenic groups which are not such finite unions is seen from the example of the additive group Q^+ of rational numbers. Indeed, suppose Q^+ were the union of its proper subgroups $H_1, \cdots, H_n$. Since there are only a finite number of these H's, it can evidently be assumed that this union is irredundant—that is, that none of the H's is contained in the union of all the others. It follows that any of the H's, for example H_1, contains a rational $r=m/n$ which is in no other H_i. This implies that for all integers h, r/h can be in no H_i but H_1, and hence must be in H_1. This is true in particular for h a multiple of m, say $h=dm$, which makes $r/h=m/ndm=1/nd$. But if this is in H_1, so is any integral multiple of it, and in particular so is any cn-tuple of it. Since $cn\cdot 1/nd=c/d$ is a completely arbitrary rational, this means that H_1 is the whole of Q^+, a contradiction. (This proof shows, incidentally, that Q^+ can never be an irredundant union of even infinitely many of its proper subgroups.)

Groups which *are* finite unions of their proper subgroups are the chief subject of the present note. Specifically, some conditions are derived which restrict the minimum number of proper subgroups into which a group can be decomposed.

2. Unions of two and three subgroups. We first prove

THEOREM 1. *No group is the union of two of its proper subgroups.*

Proof. If G is the union of its proper subgroups A and B, the union must be irredundant; that is, neither A nor B can contain the other. This being the case, let x be an element of A not contained in B, and y an element of B not contained in A. But if xy is in A then $x^{-1}xy=y$ is in A; similarly, if xy is in B, x must be in B. Thus in any case we have a contradiction.

The argument used to prove Theorem 1 can be applied to the proof of the following more general

* From AMERICAN MATHEMATICAL MONTHLY, vol. 66 (1959), pp. 491–494.

LEMMA. *Let G be the irredundant union of the subgroups H_i. Then for each i, H_i contains the intersection of all the remaining H's.*

Proof. Since the union is irredundant, H_i cannot be contained in the union H of the remaining H's. Let x be an element of H_i which is contained in none of the other H's, and let y be an element contained in all the other H's (and so in their intersection). If xy is in H, it must be in some H_j; but since y is in every H_j, it follows that x is also in H_j; a contradiction. On the other hand, if xy is not in H, it must be in H_i; but since x is in H_i, this implies that y is in H_i, which proves the Lemma.

That a group may be a union of three proper subgroups is shown by the example of the Klein 4-group ($\{e, a, b, c\}$ with the relations $a^2=b^2=c^2=e$, $ab=ba=c$, $bc=cb=a$, $ca=ac=b$), which is the union of the three subgroups $\{e, a\}$, $\{e, b\}$, $\{e, c\}$. In a sense, in fact, the 4-group is characteristic of all groups which are unions of three proper subgroups. Specifically, we have

THEOREM 2. *A group G is the union of three proper subgroups if and only if the Klein 4-group is a homomorphic image of G.*

Proof. Since the inverse image of a proper subgroup under a homomorphism is a proper subgroup, the "if" part is clear by the preceding paragraph. To prove the "only if," suppose that G is the union of the proper subgroups A, B and C. It follows from Theorem 1 that this union must be irredundant; hence the sets $A'=A-(B\cup C)$, $B'=B-(C\cup A)$, $C'=C-(A\cup B)$ are all nonempty. On the other hand, by the Lemma, $A\cap B\subset C$, $B\cap C\subset A$, and $C\cap A\subset B$, so that $A\cap B\cap C=A\cap B=B\cap C=C\cap A$; call it H. Thus evidently G is the disjoint union of H, A', B' and C'. By arguments similar to those used above, it can be shown that, in fact, H is normal; A', B' and C' are cosets of H; and G/H is the 4-group.

3. An indecomposability criterion. In setting bounds on the smallness of the number of proper subgroups of which a group G can be the union, a useful tool is the following

LEMMA. *Let the group G be the irredundant union of subgroups A_i ($i=1, \cdots, n>2$) and set $M=A_2\cup A_3\cup \cdots \cup A_n$. Then if x is not in M, we have x^k in M for some $k=1, \cdots, n-1$.*

Proof. Clearly x is in A_1. Choose y in A_2 and not in A_1. Then x^jy is in M for $j=1, \cdots, n-1$ since x^jy in A_1 would yield y in A_1. If x^jy is in A_2 for some $j=1, \cdots, n-1$, then x^j is in $A_2\subset M$, as desired. If we have $x^jy=x^my$ for some $j, m=1, \cdots, n-1$ with $j>m$, then $x^{j-m}=e$ is in M, as desired. Hence we may assume that the $n-1$ elements $xy, x^2y, \cdots, x^{n-1}y$ are distinct and in $A_3\cup \cdots \cup A_n$. It follows that x^jy, x^my are in A_q for some $q=3, \cdots, n$ and some $j, m=1, \cdots, n-1$ with $j>m$. Then $x^{j-m}=(x^jy)(x^my)^{-1}$ is in $A_q\subset M$, and the lemma is proved.

On the basis of this lemma we can now prove

THEOREM 3. *Suppose that kth roots can be taken in the group G for every positive integer k less than a certain n. Then G is not the irredundant union of n (or fewer!) of its proper subgroups.*

Proof. Since the hypothesis for the given n implies the analogous hypothesis for any m smaller than n, the "or fewer" part is an obvious consequence of the rest of the Theorem. Suppose, then, that G is the irredundant union of exactly n subgroups, and adopt the notation of the Lemma. If the element x is not in M, clearly no root y of x can be in M. This is true, for example, for $y=$ the $(n-1)!$th root of x, which exists in G by hypothesis. On the other hand, by the Lemma, y^k is in M for some k less than n. Since $x=(y^k)^r$, where $r=(n-1)!/k$, x too must be in M; a contradiction.

If G is a finite group of order N, the hypothesis of Theorem 3 is equivalent to the requirement that $(n-1)!$ be prime to N. (See, for example, this MONTHLY, vol. 60, 1953, pp. 185–6.) This gives us the immediate

COROLLARY. *Let G be a finite group of order N, p the smallest prime dividing N. Then G is not the union of p or fewer of its proper subgroups.*

The criterion of Theorem 3 cannot be strengthened. Indeed, let G be the abelian group generated by two elements x, y with the relations $x^p=y^p=e$; then G is the union of the $p+1$ proper subgroups generated by the elements x, y, xy, x^2y, x^3y, $\cdots$, $x^{p-1}y$, respectively.

For finite groups, the case in which (as in the example just given) the minimum imposed by Theorem 3 is actually assumed may be partially characterized by

THEOREM 4. *With notation as in the Corollary to Theorem 3, suppose that G is the union of exactly $p+1$ proper subgroups S_i; then at least one of the S's, say S_j, has index p. If, moreover, this S_j is normal, then all the S_i have index p and p^2 divides N.*

Proof. If none of the S's has index p, they must all have indexes greater than p; hence for each i, $o(S_i)\leqq N/p+1$. Then we have $N<\sum o(S_i)<(p+1)\cdot N/(p+1)=N$; a contradiction. Hence some S_j must have index p; assume now that this S_j is normal. Then for $i\neq j$, S_iS_j is a subgroup of G, which cannot be proper because G is the union of S_iS_j and the $p-1$ proper subgroups S_k for $k=1,\cdots,p+1$, $k\neq i$, $k\neq j$. Thus $S_iS_j=G$ for $i\neq j$, and, since S_j is normal, we have $o(G)o(S_i\cap S_j)=o(S_i)o(S_j)$, or $p(o(S_i\cap S_j))=o(S_i)$ for $i\neq j$. Set $o(S_i)=N/q_i$ for $i\neq j$, so that $q_i\geqq p$. Suppose that $q_i>p$ for some $i\neq j$. Then

$$
\begin{aligned}
N = o(G) &\leqq o(S_j) + \sum_{i\neq j} [o(S_i) - o(S_i \cap S_j)] \\
&\leqq (N/p) + \sum_{i\neq j} [(N/q_i) - (N/pq_i)] \\
&< (N/p) + p((N/p) - (N/p^2)) = N;
\end{aligned}
$$

a contradiction. Hence $q_i = p$ for every $i \neq j$, and $o(S_i \cap S_j) = N/p^2$ shows that p^2 divides N.

BIBLIOGRAPHIC ENTRY: DECOMPOSITION OF GROUPS

1. M. Brychheimer, A. C. Bryan, and A. Muir, Groups which are the union of three subgroups, AMERICAN MATHEMATICAL MONTHLY, 1970, 52–57.

(c)

HOMOMORPHISMS, AUTOMORPHISMS, AND MAPPINGS OF GROUPS

BIBLIOGRAPHIC ENTRIES: HOMOMORPHISMS, AUTOMORPHISMS, AND MAPPINGS OF GROUPS

The references below are to the AMERICAN MATHEMATICAL MONTHLY.

1. Kyle D. Wallace, Extension of mappings in finite abelian groups, 1972, 622–624.

Let G be a finite abelian group, A, B subgroups of G. Assume ϕ is an isomorphism of A with B and $G/A \cong G/B$. Article gives condition under which ϕ may be extended to an automorphism of G.

2. I. N. Herstein and J. E. Adney, A note on the automorphism group of a finite group, 1952, 309–310.

Proves that if p is a prime, $p^2 \| G|$, then $p \| \mathrm{Aut}(G)|$.

3. Roy Dubisch, A chain of cyclic groups, 1959, 384–386.

Let G_0 be a cyclic group, $G_i = \mathrm{Aut}\,(G_{i-1})$. Article gives conditions on $|G_0|$ for all G_i to be cyclic.

4. Eugene Schenkman and L. I. Wade, The mapping which takes each element of a group into its nth power, 1958, 33–34.

If the mapping is a homomorphism, when is G abelian?

(d)

ABELIAN GROUPS

THE BASIS THEOREM FOR FINITELY GENERATED ABELIAN GROUPS*

EUGENE SCHENKMAN, University of Wisconsin and Louisiana State University

The following proof of the basis theorem for finitely generated Abelian groups may be of interest because of its simplicity and because of the theorem presented here on which it is based.

If a_1 and a_2 are integers with greatest common divisor 1 then there are numbers r_1 and r_2 so that $a_1r_1+a_2r_2=1$. This fact is extended in the lemma below.

LEMMA. *If $a_1, \cdots, a_n (n>1)$ are integers whose greatest common divisor is* 1, *then there is an $n\times n$ matrix with integral elements whose determinant is* 1 *in which the $a_1, \cdots, a_n$ appear as the elements of the first row.*

Proof. For $n=2$ this follows from the statement of the paragraph preceding the lemma. We suppose that for $i=2, \cdots, n-1$, $a_i=b_id$ where d is the greatest common divisor of the a_i and hence the b_i have greatest common divisor 1; then by the induction assumption the b_i are elements of the first row of an $n-1$ square matrix of determinant 1.

Since a_n and d have greatest common divisor 1, there are integers s and t so that $sa_n+dt=1$. If e is the proper choice of ± 1 and if the $n-1$ square matrix of the previous paragraph is bordered as follows:

$$\begin{pmatrix} b_1d & \cdots & b_{n-1}d & a_n \\ * & \cdots & * & 0 \\ \vdots & & \vdots & \vdots \\ * & \cdots & * & 0 \\ esb_1 & \cdots & esb_{n-1} & t \end{pmatrix},$$

the resulting matrix has determinant 1 and the lemma is proved.

THEOREM. *If $x_1, \cdots, x_n$ are generators of an additively written Abelian group and $a_1, \cdots, a_n$ are integers with greatest common divisor* 1, *then $a_1x_1+\cdots+a_nx_n$ may be chosen as one of a set of n generators of the group.*

Proof. Let A be the matrix of the lemma. Then $A^{-1}=$Adjoint of A has integral entries also. If X is the column vector on the symbols $x_1, \cdots, x_n$ then the elements of the group corresponding to the rows of the column vector AX

* From AMERICAN MATHEMATICAL MONTHLY, vol. 67 (1960), pp. 770–771.

are a set of generators for the group. For it is clear that $X=A^{-1}AX$; hence the generators $x_1, \cdots, x_n$ may be expressed as integral combinations of the new generators.

BASIS THEOREM. *If G is an Abelian group generated by a finite number of elements, then G is the direct product of cyclic groups.*

Proof. Suppose every set of generators of G contains n or more elements. Then choose a set of n generators so that one of them, x_n, has minimal order k in the sense that no other set of n generators has an element of smaller order. The other $n-1$ generators, $x_1, \cdots, x_{n-1}$, generate a proper subgroup H of G. By an induction argument on n we may assert that H is a direct product of $n-1$ cyclic subgroups. The theorem will be proved when it is shown that the intersection of H with the cyclic group C generated by x_n is zero. If this were not so then for some integers $a_n<k$, and $a_1, \cdots, a_{n-1}$, $a_nx_n=a_1x_1+ \cdots +a_{n-1}x_{n-1}$ or $a_1x_1 + \cdots + a_{n-1}x_{n-1} - a_nx_n = 0$. If the greatest common divisor of $a_1, \cdots, a_n$ is d, then by the above theorem the element $g=(a_1/d)x_1+ \cdots +(a_{n-1}/d)x_{n-1}-(a_n/d)x_n$ is a member of a set of n generators of G of order a divisor of $d<a_n<k$. This is contrary to the choice of x_n. Accordingly we conclude that $H\cap C=0$ as desired.

A PROOF OF THE STRUCTURE THEOREM OF FINITE ABELIAN GROUPS*

E. R. GENTILE, Universidad de Buenos Aires

Let A be a finite abelian group. We denote its composition law multiplicatively. By using elementary arithmetic we know that A can be represented as a direct product of its p-primary components. So we restrict ourselves to the case where A is a finite abelian p-group of order p^k, p denoting a prime. For any $i\in N$, the set of nonnegative integers, we denote by G_i the group of p^ith roots of 1 as contained in the complex numbers C. With G_∞ we denote the union of G_i, $i\in N$, also in C. We will use the following elementary properties of G_∞:

(s) G_i and G_∞ are the only subgroups of G_∞,

(d) G_∞ is a divisible group, that is, for any $x\in G_\infty$ and any $m\in N$ there is $z\in G_\infty$ satisfying $x=z^m$.

* From AMERICAN MATHEMATICAL MONTHLY, vol. 76 (1969), pp. 60–61.

Property (d) follows easily from property (s).

In this note we give a very elementary proof of the following well-known result:

THEOREM. *A is a direct product of cyclic groups.*

We start with the following:

LEMMA. *Let A be a finite abelian group and H a subgroup. Then every homomorphism $\mathcal{G}: H \to G_\infty$ can be extended to A.*

Proof. Assume $H \neq A$ and let $x \in A - H$. Let $h_0 = \min \{h / x^h \in H\}$. Clearly $x^h \in H$, $h \in N \Rightarrow h_0 | h$. Since $\mathcal{G}(H)$ is a finite subgroup of G_∞ we have $\mathcal{G}(H) = G_m$ for some $m \in N$. Let $y \in H$ satisfy $\mathcal{G}(y) = w_m =$ a primitive p^mth root of 1. Let $w \in G_\infty$ satisfy $w^{h_0} = w_m^s = \mathcal{G}(x^{h_0})$; we need here property (d). Let $(x) \cdot H$ be the subgroup of A generated by x and H. We set

$$x^i \cdot h \to w^i \cdot \mathcal{G}(h) \qquad i \in Z,\ h \in H$$

and we claim that this gives a well defined homomorphism of $(x) \cdot H$ into G_∞. In fact, if $x^i \in H$ we have that $h_0 | i$, say $i = t \cdot h_0$. Therefore

$$w^i = (w^{h_0})^t = (w_m^s)^t = (\mathcal{G}(x^{h_0}))^t = \mathcal{G}(x^i)$$

and consequently if $e = x^i \cdot h =$ the identity element of A, we have

$$w^i \cdot \mathcal{G}(h) = \mathcal{G}(x^i) \cdot \mathcal{G}(x^{-i}) = \mathcal{G}(x^i) \cdot \mathcal{G}(x^i)^{-1} = 1.$$

From the finiteness of A the Lemma follows immediately.

Proof of theorem. We proceed to prove the theorem by induction on the exponent k of the order p^k of A. For $k = 1$, A is isomorphic to G_1 and nothing has to be proved. Let $1 < k$. Choose an element $a \in A$ with maximal order p^h in A. Then we have an isomorphism $(a) \to G_h$ of the subgroup (a) generated by a in A, into G_h. This isomorphism can be extended to a homomorphism $\mathcal{G}: A \to G_\infty$. We have $\mathcal{G}(A) = G_t$ for some t and clearly $G_h \subset G_t$. Let $b \in A$ satisfy $\mathcal{G}(b) = w_t =$ a primitive p^tth root of 1. Then b has order at least p^t, therefore we must have $t = h$ and so $\mathcal{G}(A) = G_h = \mathcal{G}((a))$. Let $x \in A$, then $\mathcal{G}(x) = w_h^m$ and hence $x \cdot a^{-m} \in \ker(\mathcal{G})$ which proves that $A = (a) \cdot \ker(\mathcal{G})$. Since the restriction of $\mathcal{G}$ to (a) is an isomorphism we can conclude that $A = (a) \times \ker(\mathcal{G})$. On applying the inductive argument to $\ker(\mathcal{G})$, the theorem follows.

REMARK. Incidentally we have proved that every element of maximal order in A generates a direct summand of A.

BIBLIOGRAPHIC ENTRIES: ABELIAN GROUPS

The references below are to the AMERICAN MATHEMATICAL MONTHLY.

1. Granville McCormick, A theorem on finite abelian groups, 1960, 670.

Shows every abelian group is isomorphic to a subgroup of G_m for some m (G_m is the multiplicative group of integers mod m).

2. Steven Bryant, Groups, graphs, and Fermat's last theorem, 1967, 152–156.

Proves that if G is an abelian group, G is determined by its order and the main diagonal of its multiplication table provided $p^2 \nmid |G|$, for $2^{p-1} \equiv 1 \pmod{p^2}$.

(e)

LINEAR GROUPS AND THE SYMMETRIC GROUPS

NOTE ON THE ALTERNATING GROUP*

E. L. SPITZNAGEL, JR., Northwestern University

The distinction between even and odd permutations is established in nearly every introductory modern algebra text. However, the most common proof leaves something to be desired, as Herstein remarks in *Topics in Algebra*, because it introduces a polynomial which seems completely extraneous to the subject. On the other hand, alternative proofs use permutation calculations which are fairly involved.

Following is a proof with more group-theoretic flavor than these other meth-

* From AMERICAN MATHEMATICAL MONTHLY, vol. 75 (1968), pp. 68–69.

ods. Perhaps it will be found aesthetically more pleasing.

Define $\tau \in S_n$ to be even if τ can be represented as the product of an even number of transpositions. Let A_n be the set of all even $\tau \in S_n$. Then A_n is a subgroup of S_n, of index at most 2, and all we need do is show that some element of S_n is not contained in A_n.

PROPOSITION. $(1, 2) \notin A_n$, *so* $[S_n : A_n] = 2$.

Proof. Suppose (1, 2) can be written as a product of an even number of transpositions. Then the identity permutation ϵ can be written as a product of an odd number of transpositions. Let $\epsilon = (a, b) \cdots$ be a product of the smallest odd number k of transpositions which gives ϵ, and among all such products beginning with $(a, \ )$ let it be one containing the smallest number of a's.

Since $a\epsilon = a$, there must be another transposition in this product which moves a. Let (a, c) be the one closest to the left. Since $(d, e)(a, c) = (a, c)(d, e)$ and $(c, d)(a, c) = (a, d)(c, d)$, we have $\epsilon = (a, b)(a, f) \cdots$ is a product of length k containing the same minimum number of a's.

But then if $f \neq b$, $\epsilon = (a, f)(b, f) \cdots$ is a product of length k with fewer a's, while if $f = b$, then ϵ is a product of $k - 2$ transpositions, contradicting minimality of k.

COROLLARY. *If τ is even, then every time it is written as a product of transpositions, that product contains an even number of transpositions.*

Proof. If not, then we would have $(1, 2)\tau \in A_n$, and $\tau \in A_n$, so $(1, 2) \in A_n$.

AN HISTORICAL NOTE ON THE PARITY OF PERMUTATIONS*

T. L. BARTLOW, Villanova University

Every beginning algebra student learns that the number of transpositions into which a given permutation can be decomposed is either always even or always odd. Many students find the traditional proof, involving the function

$$P = (x_2 - x_1)(x_3 - x_1) \cdots (x_n - x_1)(x_3 - x_2) \cdots (x_n - x_{n-1}),$$

unsatisfactory, because, as Herstein remarks, the polynomial "seems extraneous to the matter at hand" [7, p. 67]. Many alternative proofs have been offered, so many that we wonder why the traditional proof maintains its place in textbooks [**1**, **5**, **6**, **8** (p. 36), **10**, **11**, **12**, **14**]. Here we offer two more alternatives which derive from the origins of the subject.

I. The first is to explain that early studies of permutations occurred in a context in which the polynomial P is quite natural. Mathematicians of the sixteenth century knew that the coefficients of a polynomial could be expressed as elementary sym-

* From AMERICAN MATHEMATICAL MONTHLY, vol. 79 (1972), pp. 766–769.

metric functions of the roots of the polynomial. In 1770–1771, Lagrange [9] and Vandermonde [13] made the first efforts to exploit this fact to discuss the question of solvability of polynomials of degree greater than four. They recognized that any formula solving a general polynomial of degree n in terms of the coefficients of the polynomial must be a symmetric function in the n roots. This realization suggested to them the importance of studying the effect of permutations on functions of n variables. In 1815 Cauchy published the results of a careful study of this question [2, 3]. Cauchy credits Vandermonde with observing that the function P is a typical example of an alternating function, although Vandermonde appears to have made this observation only for $n=3$. Moreover, Cauchy proves that every alternating function of n variables is divisible by P. The first extensive study of permutations and permutation groups appeared in 1844–45 in several papers by Cauchy.

Thus early work on permutation groups was largely motivated and informed by investigations of the effect of permutations on a function of several variables, investigations in which the function P had a prominent role. An old text on Galois Theory takes this point of view [4].

II. The other alternative is to use what appears to be the original proof of the theorem in question, which did not involve P. In [3, pp. 98–104] Cauchy gives a proof which relates the parity of a permutation to the number of cycles which it involves.

LEMMA. *Every permutation is uniquely a product of disjoint cycles.*

Cauchy's proof is the one in use today. It will be important to count the number of cycles in a given permutation. For this purpose Cauchy counts one-cycles. Thus the identity permutation is, in modern notation, $(1)(2)\cdots(n)$ and $(1,3,4)$ is properly designated by $(1,3,4)(2)\cdots(n)$.

LEMMA. *If a product of g disjoint cycles is multiplied by a transposition, the result involves $g \pm 1$ disjoint cycles.*

Proof. Let α be a permutation involving g cycles, and let (a, b) be a transposition. If a and b belong to the same cycle of α, the product has the form

$$(a, c, \cdots, d, b, f, \cdots, h) \cdots (a, b) = (a, c, \cdots, d)(b, f, \cdots, h) \cdots,$$

which contains $g+1$ cycles. If a and b are in different cycles of α we have

$$(a, c, \cdots, f)(b, d, \cdots, h) \cdots (a, b) = (a, c, \cdots, f, b, d, \cdots, h) \cdots,$$

which involves $g-1$ cycles. The argument applies even if a or b stands alone in a cycle.

LEMMA. *Let the permutations of S_n be partitioned into classes according as they involve an even or an odd number of cycles. If a permutation is multiplied by a* *sequence of transpositions, it does or does not change classes according as the number of transpositions is odd or even.*

Proof. This is immediate from the preceding lemma.

THEOREM. *No permutation can be the product both of an even and of an odd number of permutations.*

Proof. Let α be any permutation and consider the partition of the preceding lemma. Let the identity permutation be multiplied by α, regarded as a product of transpositions. The number of transpositions is even if and only if α is in the same class as the identity and odd if any only if α is in the other class. Of course, α cannot be in both classes.

It follows that the partition of the preceding lemma is, in fact, the partition of S_n into even and odd permutations. Since the identity permutation involves n cycles even permutations involve an even number of cycles if and only if n is even. Thus, if $c(\pi)$ is the number of cycles involved in π, then π is even if and only if $n-c(\pi)$ is even. This observation of Cauchy's is the starting point for Phillips [**11**].

References

1. J. L. Brenner, A new proof that no permutation is both even and odd, this MONTHLY, 64 (1957) 499–500.

2. A. -L. Cauchy, Mémoire sur le nombre des valeurs qu'une fonction peut acquérir lorsqu'on y permute de toutes les manières possibles les quantités qu'elle renferme, J. l'École Polytechnique, 10 (1815) 1–28; Oeuvres Complètes, ser. II, vol. 1, pp. 64–90.

3. A. -L. Cauchy, Mémoire sur les fonctions qui ne peuvent obtenir que deux valeurs égales et de signes contraires par suite des transpositions opérées entre les variables qu'elles renferment, J. de l'École Polytechnique, 10 (1815) 29–112; Oeuvres Complètes, ser II. vol. pp. 91–169.

4. E. Dehn, Algebraic Equations: An Introduction to the Theories of Lagrange and Galois, Dover, New York, 1960.

5. E. L. Gray, An alternative proof for the invariance of parity of a permutation written as a product of transpositions, this MONTHLY, 70 (1963) 995.

6. I. Halperin, Odd and even permutations, Canadian Math. Bull., 3 (1960) 185–186.

7. I. N. Herstein, Topics in Algebra, Blaisdell, Waltham, Mass., 1964.

8. N. Jacobson, Lectures in Abstract Algebra, vol. I. Van Nostrand, Princeton, N. J., 1951.

9. J. L. Lagrange, Réflexions sur la résolution algébrique des équations, Nouveaux Mémoires de l'Académie Royale des Sciences et Belles-Lettres de Berlin (1770–1771); Oeuvres de Lagrange 3, 205–421.

10. Hans Liebeck, Even and odd permutations, this MONTHLY, 76 (1969) 668.

11. W. Phillips, On the definition of even and odd permutations, this MONTHLY, 74 (1967) 1249–1251.

12. E. L. Spitznagel, Note on the alternating group, this MONTHLY, 75 (1968) 68–69.

13. A. Vandermonde, Mémoire sur la résolution des équations, Histoire de l'Académie Royale des Sciences, 88 (1771) 365–416.

14. C. E. Weil, Another approach to the alternating subgroup of the symmetric group, this MONTHLY, 71 (1964) 545–546.

A NEW METHOD FOR THE DETERMINATION OF THE GROUP OF ISOMORPHISMS OF THE SYMMETRIC GROUP OF DEGREE N*

H. A. BENDER, University of Illinois

Our leading text books on group theory which determine the group of isomorphisms of the symmetric group of degree n make use of the following two theorems. First, the symmetric group of degree n ($n \neq 6$) contains n and only n subgroups of order $(n-1)!$ forming a single conjugate set. The symmetric group of degree 6 contains 12 subgroups of order 5!, which are simply isomorphic with one another and form two sets of conjugates of 6 each. Second, if G is a transitive substitution group of degree n and index n, then the group of isomorphisms of G can be represented as a transitive substitution group of degree n which contains G as an invariant subgroup.

In this article we shall study the group of isomorphisms of the symmetric group of degree n from the standpoint of independent generators of the symmetric group.†

It is known that the symmetric group G of degree n can be generated by a cyclic substitution s of degree $n-1$ and a transposition t which connects any one of the $n-1$ letters in s with the remaining letter.‡ The cyclic substitution s can be selected in $\dfrac{n!}{n-1}$ different ways, and for each cyclic substitution s the transposition t can be selected in $n-1$ different ways, hence the symmetric group of degree n can be generated in $n!$ different ways. Thus it follows that the symmetric group G can be made isomorphic with itself in $n!$ different ways such that all these isomorphisms are inner isomorphisms.§ Moreover, these isomorphisms constitute the group of inner isomorphisms.

Since the central of a symmetric group is the identity, it follows that the group of inner isomorphisms is simply isomorphic with this symmetric group. If this is not the group of isomorphisms, it is an invariant subgroup of the group of isomorphisms and the remaining isomorphisms are outer, or contragredient, isomorphisms.

We shall now consider the possibility of automorphisms of G in which t corresponds to substitutions of order 2 which are composed of transpositions. It is evident that the number of conjugates under G of such substitutions must be equal to the number of transpositions in G.

If we equate and simplify the number of transpositions in G to the number of substitutions of order two and degree four, six and $2r$ respectively, we have

$$(n-2)(n-3)=2\cdot 2!, \qquad (n-2)(n-3)(n-4)(n-5)=2^2\cdot 3!,$$
$$(n-2)(n-3)\cdots(n-2r+1)=2^{r-1}\cdot r!.$$

---

* From American Mathematical Monthly, vol. 31 (1924), pp. 287–289.

† O. Hölder, *Mathematische Annalen*, vol. 46 (1895), p. 345.

‡ *Cf.* R. D. Carmichael, *Quarterly Journal*, vol. 49 (1922), p. 226.

§ This may also be shown by using other sets of independent generators.

The first equation is not satisfied for real values of n. The second equation is satisfied for $n=6$ and for no other integral value of n. In general, suppose $n=2r$, then the third equation reduces to

$$(2r-2)(2r-3)\cdots(r+1)=2^{r-1},$$

and for $r>3$,

$$(2r-2)(2r-3)\cdots(r+1)>(r+1)^{r-2}>2^{r-1}.$$

Thus we have shown that in all possible automorphisms of the symmetric group of degree n $(n\neq 6)$ transpositions must correspond to transpositions.

We shall next consider the possibility of an automorphism of $G\,(n\neq 6)$ in which the cyclic substitution s corresponds to a non-cyclic substitution s_1 of order $n-1$. It is evident that s_1 can not be of degree less than n. If the substitution s_1 is of degree n, then the product of this substitution and a transposition will unite two of the cycles of s_1 into one cycle or decompose one of its cycles into two cycles according as the letters of the transposition appear in two cycles or in the same cycle respectively. Thus we have shown that outer isomorphisms of the symmetric group are possible only in the case $n=6$.

It is evident that in the case $n=6$ the substitution s of order 5 must correspond to cyclic substitutions of order 5 in every automorphism of G.

We shall now consider some of the conditions necessary for a cyclic substitution s of degree 5 and a substitution t_1 of order 2 and degree 6 to generate the symmetric group of degree 6.

The product of s and t_1 will omit two letters if the two substitutions contain two pairs of adjacent letters, and hence will not be of order 6. That is, the product

$$abcde\cdot ab\cdot cd\cdot ef=bdfe$$

will omit a and c. Likewise, t_1 and the powers of s can not have a pair of adjacent letters in common. Hence, for a given s the substitution t_1 can be selected in but five ways. For $abcde$ the five substitutions are

$$ab\cdot ce\cdot df,\qquad ac\cdot bf\cdot de,\qquad ad\cdot bc\cdot ef,\qquad ae\cdot bd\cdot cf,\qquad af\cdot be\cdot cd.$$

That s and t_1 so defined will generate the symmetric group of degree 6 is shown in the *Quarterly Journal*, vol. 49 (1922), p. 235.

Thus we have shown that *the group of isomorphisms I of the symmetric group of degree n* $(n\neq 6)$ *is the group of inner isomorphisms and is the symmetric group of this degree. For* $n=6$ *the order of the group of isomorphisms is* $\frac{6!}{5}(5+5)$, *or twice the order of the symmetric group of degree six, and exactly one half of the isomorphisms are outer isomorphisms.*

It can be shown that the group of isomorphisms of the symmetric group G of degree n $(n\neq 6)$ can be generated by the two isomorphisms S and T satisfying the following conditions:

$$S^{-1}sS=s,\qquad S^{-1}tS=s^{-1}ts,\qquad T^{-1}sT=t^{-1}st,\qquad T^{-1}tT=t.$$

For $n=6$ let us consider the automorphisms of G in which t corresponds to t_1.

Since all the substitutions of the same type as t or t_1 are conjugate under G, it follows that t_1 in turn must correspond to a substitution which is one of the conjugates of t under G. This being true for the remaining sets of conjugate substitutions, it follows that the square of any outer isomorphism, as well as the product of any two outer isomorphisms, is an inner isomorphism.

Furthermore, the square of any outer isomorphism T_1 which leaves s invariant is an inner isomorphism which may be obtained by transforming G by s or by a power of s. Since s is in the alternating group of G and all the isomorphisms brought about by transforming G by the operators of the alternating group form a subgroup in the I of G simply isomorphic with the alternating group, it follows that the square of T_1 is in the subgroup simply isomorphic with the alternating group. Since the commutators of an inner and an outer isomorphism are in the subgroup simply isomorphic with the alternating group, it follows that the square of every outer isomorphism is in the subgroup simply isomorphic with the alternating group, and hence this subgroup and one half of the outer isomorphisms constitute an invariant subgroup in the I of G. *Thus the group of isomorphisms of the symmetric group of degree* 6 *contains* 3 *invariant subgroups of order* 6!, *each having the common subgroup of order* 360.

THE GROUP OF THE PYTHAGOREAN NUMBERS*

J. MARIANI, Rutgers University

The complete solution of the equation

$$x^2 + y^2 - z^2 = 0 \tag{1}$$

in integral numbers x, y, z was given by Euclid, Book X, Lemma 1 to Proposition 29. He found the following well-known result: One of the numbers x, y must be even. Let x be even. Then every solution of (1) with $z > 0$ can be written in the form

$$x = 2ab, \qquad y = \pm (a^2 - b^2), \qquad z = a^2 + b^2, \tag{2}$$

where a and b are arbitrary integers. Vice versa, every triplet (2) gives us a solution of (1).

The left-hand side of (1) is an indefinite quadratic form with integral coefficients. The pseudo-orthogonal linear substitutions, with integral coefficients of the variables (in this case, x, y, z) which carry the quadratic form into itself are of interest, e.g., in the theory of automorphic functions. An account of the earlier part of the theory can be found in Fricke [1]. Clearly, these linear substitutions form a group the elements of which map integral solutions of (1) onto integral solutions.

In order to formulate our results, we shall use the following notations:

Three integers x, y, z will be called a *proper Pythagorean triplet* if they do not

* From AMERICAN MATHEMATICAL MONTHLY, vol. 69 (1962), pp. 125–128.

have a common divisor different from ± 1, if x is even, $z>0$, and if (1) is satisfied.

The group of all homogeneous linear transformations of x, y, z with integral coefficients, which leave $x^2+y^2-z^2$ invariant, will be denoted by $\bar{G}$. Its subgroup consisting of those transformations which map proper Pythagorean triplets onto proper Pythagorean triplets and have the determinant ± 1 will be called G.

THEOREM 1. *G acts transitively on the proper Pythagorean triplets. Every triplet can be mapped onto any other by a transformation belonging to G.*

THEOREM 2. *G is isomorphic to the group of all 2×2 matrices with integral coefficients and determinant ± 1.*

THEOREM 3. *The subgroup of substitutions of G which keeps a particular proper Pythagorean triplet fixed is the direct product of an infinite cyclic group and a group of order 2.*

THEOREM 4. *G is of finite index in $\bar{G}$.*

The proof of these results is simple. We observe first that, if x is even, it must be divisible by 4 since y^2-z^2 is divisible by 8 if y and z are odd. Therefore, we may assume that a in (2) is even. Consider the linear substitution c_2

$$a' = \lambda a + \mu b, \qquad b' = \nu a + \rho b \tag{3}$$

where λ, ρ are odd, ν, μ are even, and $\lambda\rho-\nu\mu=1$. By (3), a pair (a, b) is mapped onto a pair (a', b'), where a' is even and b' is odd and where a' and b' are coprime if a and b are coprime.

Now we have

LEMMA 1. *By choosing λ, μ, ν, ρ properly, we can map any coprime pair (a, b) onto the pair $a'=0$, $b'=1$.*

In fact, Euclid's algorithm guarantees the existence of two integers ν^*, ρ^* such that $\nu^*a+\rho^*b=1$. These integers are not uniquely determined; if we put $\nu=\nu^*+kb$, $\rho=\rho^*-ka$, $(k=0, \pm 1, \cdots)$ we also have $\nu a+\rho b=1$, and we can choose k in such a manner that ν is even ($k=0$ if ν^* is even, $k=1$ if ν^* is odd). Automatically, ρ will then be odd. Having determined ν and ρ, we determine λ and μ by $\lambda=b$, $\mu=-a$. Obviously, λ is odd and μ is even, and $\lambda\rho-\nu\mu=1$. This proves Lemma 1.

Now we observe that (2) and (3) define the substitution

$$\begin{aligned} x' &= 2a'b' = (\lambda\rho + \mu\nu)x + (\lambda\nu - \mu\rho)y + (\lambda\nu + \mu\rho)z, \\ y' &= a'^2 - b'^2 = (\lambda\mu - \nu\rho)x + \tfrac{1}{2}(\lambda^2 - \nu^2 - \mu^2 + \rho^2)y + \tfrac{1}{2}(\lambda^2 - \nu^2 + \mu^2 + \rho^2)z, \\ z' &= a'^2 + b'^2 = (\lambda\mu + \nu\rho)x + (\lambda^2 + \nu^2 - \mu^2 - \rho^2)y + \tfrac{1}{2}(\lambda^2 + \nu^2 + \mu^2 + \rho^2)z. \end{aligned} \tag{4}$$

Obviously, (4) is a pseudo-orthogonal transformation that maps proper Pythagorean triplets on proper Pythagorean triplets and carries the left hand side of (1) onto itself. Therefore, (4) is a substitution of G. According to Lemma 1, there exists a substitution (3) which carries a preassigned pair (a, b) (a even) into the pair $(0, 1)$. Therefore, every proper Pythagorean triplet can be carried into one of the triplets $(0, 1, 1)$ or $(0, -1, 1)$. These two triplets can be exchanged by the substitution

$$a' = b, \qquad b' = -a, \tag{5}$$

which induces the substitution $x'=x$, $y'=-y$, $z'=z$. To these substitutions, we may add the substitution

$$a' = -a, \qquad b' = b, \tag{6}$$

which induces the transformation $x'=-x$, $y'=y$, $z'=z$. Together the substitutions (3), (5) and (6) generate the extended modular group c_2', the quadratic representation of which, as described by (4), is at least a part G' of the group G described in Theorem 2, so that obviously G is transitive on the proper Pythagorean triplets. Now, no pseudo-orthogonal substitution outside of G' will map proper Pythagorean triplets onto triplets of the same type so that G' coincides with G. Consider any substitution of G

$$x' = P_2z + Q_2x + R_2y, \quad y' = P_3z + Q_3x + R_3y, \quad z' = P_1z + Q_1x + R_1y \tag{7}$$

satisfying the pseudo-orthogonality conditions (and the transpose):

$$\begin{aligned} P_1^2 + Q_1^2 - R_1^2 &= 1, & P_1P_2 - Q_1Q_2 - R_1R_2 &= 0;\\ P_2^2 - Q_2^2 - R_2^2 &= -1, & P_1P_3 - Q_1Q_3 - R_1R_3 &= 0;\\ P_3^2 - Q_3^2 - R_3^2 &= -1, & P_2P_3 - Q_2Q_3 - R_2R_3 &= 0. \end{aligned} \tag{8}$$

Using (2) and analogous formulas for x', y', z', we get from (7)

$$\begin{aligned} (a'+b')^2 &= \tfrac{1}{2}(P_1+P_2+Q_1+Q_2)(a+b)^2\\ &\quad + \tfrac{1}{2}(P_1+P_2-Q_1-Q_2)(a-b)^2 + \tfrac{1}{2}(R_1+R_2)\cdot 2(a-b)(a+b),\\ (a'-b')^2 &= \tfrac{1}{2}(P_1-P_2+Q_1-Q_2)(a+b)^2\\ &\quad + \tfrac{1}{2}(P_1-P_2-Q_1+Q_2)(a-b)^2 + \tfrac{1}{2}(R_1+R_2)\cdot 2(a-b)(a+b), \end{aligned} \tag{9}$$

the coefficients being integers. From (8) we get

$$\begin{aligned} (P_1+P_2)^2 - (Q_1+Q_2)^2 &= (R_1+R_2)^2,\\ (P_1-P_2)^2 - (Q_1-Q_2)^2 &= (R_1-R_2)^2. \end{aligned} \tag{10}$$

$\frac{1}{2}(R_1+R_2)$ and $\frac{1}{2}(R_1-R_2)$ being integers and relatively prime, we deduce from (10) that

$$\begin{aligned} \tfrac{1}{2}(P_1+P_2+Q_1+Q_2) &= m^2, & \tfrac{1}{2}(P_1+P_2-Q_1-Q_2) &= n^2,\\ \tfrac{1}{2}(P_1-P_2+Q_1-Q_2) &= r^2 & \tfrac{1}{2}(P_1-P_2-Q_1+Q_2) &= s^2, \end{aligned} \tag{11}$$

where m, n, r, s are integers. Using

$$P_1^2 - Q_1^2 - R_1^2 = P_1^2 - P_2^2 - P_3^2 = -(Q_1^2 - Q_2^2 - Q_3^2)$$
$$= -(R_1^2 - R_2^2 - R_3^2) = (ms - nr)^2,$$

we can solve (11) with respect to $P_1, P_2, \cdots, R_3$. Taking the square root in (9), we get a linear substitution from which we easily deduce (3). Moreover, the only substitutions of x, y, z with integral coefficients and with the properties of keeping the left-hand side of (1) fixed, map the triplet $(0, -1, 1)$ onto itself and are those induced by $a' = \pm a$, $b' = 2ka + b$, where k is an arbitrary integer.

Taking in (1), x, y, z positive and relatively prime, replacing them by $x' = x + \Delta$, $y' = y + \Delta$, $z' = z + \Delta$, and solving for Δ we get another proper solution, x', y', z' deriving from the transformation

$$\begin{bmatrix} -2 & -1 & 2 \\ -1 & -2 & 2 \\ -2 & -2 & 3 \end{bmatrix}$$

induced by matrix

$$\begin{pmatrix} 1 & 0 \\ 2 & 1 \end{pmatrix},$$

and $z' < z$. Taking x', y', z' positive and performing the same operation we get $z'' < z'$ and so on until we arrive by the method of infinite descent at $z^{(n)} = 1$, $x^{(n)} = 0$, $y^{(n)} = -1$.

This proves all our theorems, except Theorem 4. Now it is easily seen that a substitution of $\overline{G}$ must map a solution of (1) for which x, y, z have as a common divisor only ± 1 onto a solution x', y', z' of (1) with the same property. However, x', y', z' may not form a proper Pythagorean triplet since x' need not be even and z' need not be positive. Therefore, we must extend the group G by ad oining the substitutions

$$x' = y, \quad y' = x, \quad z' = z, \tag{12}$$

$$x' = x, \quad y' = y, \quad z' = -z, \tag{13}$$

in order to obtain $\overline{G}$. It is easily confirmed that G (group of substitutions of x, y, z) forms a normal divisor of index 4 in the group $\overline{G}$ arising from an adjunction of (12) and (13). This also proves Theorem 4.

Reference

1. R. Fricke and F. Klein, Vorlesungen über die Theorie der automorphen Funktionen, vol. 1, Leipzig, 1897.

THE CLASSICAL GROUPS AS A SOURCE OF ALGEBRAIC PROBLEMS*

C. W. CURTIS, University of Oregon

Introduction. Rather than attempting a general survey of what has been accomplished in algebra, I shall try to describe the influence on the development of algebra of one of the great sources of algebraic problems and ideas—the classical linear groups. I intend to show that these groups have provided an experimental, empirical basis for large parts of algebra. They are a family of examples which serve as raw material from which problems, conjectures, abstract ideas and general theorems have emerged, and which have provided direction and continuity to the research based on them. Surveys of recent developments and problems of current interest have been given by Brauer [5], [6], Carter [11], Dieudonné [24], M. Hall [26], Kaplansky [35], and MacLane [36].

It is impossible in this lecture to give an account of all the major results which involve the classical groups. In making a selection, I have simply included the material that has been most helpful to me.

1. Definitions and background. The story begins with Galois, who proved that a polynomial equation was solvable by radicals if and only if a certain finite group of automorphisms of a field was solvable. Thus he raised the problem of investigating groups in terms of their structure, and observed, as an example, that the set of all 2 by 2 invertible matrices with coefficients in a finite field of more than 3 elements, was a finite group that was not solvable.

Let us first give an up-to-date version of Galois' example. Let K be an arbitrary field, and V a finite dimensional vector space over K. The set of all linear transformations T of V which possess multiplicative inverses (and which we shall call invertible transformations) forms a group with respect to multiplication, called the general linear group $GL(V)$ of V. Hermann Weyl invented the term *the classical groups* to stand for $GL(V)$ and certain subgroups of $GL(V)$, which include the following ones.

(a) $SL(V)$, the unimodular group, consisting of all invertible transformations of determinant one.

(b) $O(V, f)$, the orthogonal group of a non-degenerate symmetric bilinear form f on V, consisting of all invertible transformations T that leave f invariant:

$$f(Tx, Ty) = f(x, y),\ x, y \in V.$$

For example, if V is n-dimensional euclidean space and $f(x, y)$ the usual inner product, $O(V, f)$ is the familiar group of linear transformations T that preserve lengths: $\|Tx\| = \|x\|$, where $\|x\| = f(x, x)^{1/2}$.

(c) $Sp(V, F)$, the symplectic group, consisting of all invertible transformations leaving invariant a non-degenerate skew-symmetric $(F(x, y) = -F(y, x))$ bilinear form F on V.

* From AMERICAN MATHEMATICAL MONTHLY, vol. 74 (1967), pp. 80–91 (part II).

In order to understand why the list of classical groups does not continue, we have to return to Galois' observation about the structure of the classical groups.

Suppose we have, more generally, a family of algebraic systems $\{G_\alpha\}$ for which the fundamental theorem of homomorphisms is valid, such as groups, rings, algebras, etc. Whenever we have a homomorphism

$$\phi: G \xrightarrow[\text{onto}]{} H$$

with kernel L, G is called an extension of H with kernel L. G is called *simple* if whenever G is an extension of H with kernel L, then either G is isomorphic to H, or G is isomorphic to L. Given a family $\{G_\alpha\}$ of simple systems, the *extension problem* asks for the construction of all systems G for which there is a sequence of homomorphisms

$$\phi_1: G \xrightarrow[\text{onto}]{} G_{\alpha_1} \in \{G_\alpha\}$$

with kernel G_1,

$$\phi_2: G_1 \xrightarrow[\text{onto}]{} G_{\alpha_*} \in \{G_\alpha\}$$

with kernel G_2, etc., and such that for some k, the kernel G_k belongs to the family $\{G_\alpha\}$. The systems G determined in this way are built up from the simple systems $\{G_\alpha\}$ by solving the extension problem.

Solvable groups are the result of applying this process to the family $\{G_\alpha\}$ of cyclic groups of prime order. Galois' observation that the classical groups are not solvable shows that the classical groups are built up from non-abelian simple groups by solutions of the extension problem.

The problem of which simple groups are derived from the classical groups was investigated first by Jordan and Dickson ([34], [21]) with Dickson's book in 1901 containing fairly complete results for the classical groups defined over finite fields. Dickson's proofs, while satisfactory at the time, involved exhausting calculations with matrices which are a tribute to his persistence, but often left obscure the essential ideas. The subject was taken up again by Dieudonné and Artin ([22], [23], [3]) who identified the simple groups derived from the classical groups over arbitrary fields by elegant methods using the geometric properties of linear transformations and subspaces of the underlying vector space.

The result of their investigations, which extended and clarified Dickson's work, can be summarized as follows. First of all consider all homomorphisms f_α of a group G onto abelian groups A_α with kernels G_α. It turns out that there is a unique minimal such kernel G' called the *commutator group* of G, which is generated by all commutators $(x, y) = xyx^{-1}y^{-1}$, $x, y, \in G$. Thus an arbitrary group is an extension of an abelian group with kernel equal to the commutator group. The first step in investigating the structure of classical groups was to find the commutator groups. In the case of $GL(V)$ the multiplication theorem for determinants gives a homomorphism of $GL(V)$ onto the abelian multiplicative group of the field. With one exception (when V is the two dimensional space over the field of two elements), the commutator group turns out simply to be the kernel of the determinant homomorphism, $SL(V)$. The symplectic group is its own commutator group. The commutator group of the orthogonal group is more complicated. When isotropic vectors $v \neq 0$ such that $f(v, v)=0$ are present, the commutator group of the orthogonal group is the kernel of a homomorphism called the spinor norm, whose construction depends on a certain associative algebra called the Clifford algebra of the form f, (see [23], [3]).

The main structure theorem on the classical groups, proved in special cases by Dickson, and in full generality by Dieudonné, is that, with a manageable number of exceptions, the commutator group G' of a classical group G is an extension of a non-abelian simple group with kernel Center(G'). In other words $G'/\text{Center}(G')$ is a simple group, in most cases.

Dickson accumulated many other facts about the classical groups that proved to be useful for experimental purposes: he found the orders of the finite simple groups he had obtained, and partial information on their conjugacy classes and subgroups. All these results provided evidence that there should be general methods for studying these groups, without considering them one family at a time, as is the case in the approach both of Dickson and Dieudonné-Artin.

2. First attempts at a classification; Lie algebras. With the examples of simple groups furnished by the classical groups, and the alternating and Mathieu groups in the finite case, the problem arises to classify the non-abelian simple groups in some reasonable way, and in particular, to ask whether there are others. This problem, for arbitrary groups, still seems well beyond our reach, but is basic because the simple groups are the building blocks of arbitrary groups.

The first successful classification of simple algebraic systems which are relevant to our problem was contained in E. Cartan's thesis [9] in 1894, in which the simple Lie algebras over the field of complex numbers were classified. Since then Lie algebras have developed a life of their own (see [38], [32], [35]). The link between Cartan's work and group theory, however, can only be understood by plunging into the deep waters of Lie group theory, invented by Cartan's predecessor, Sophus Lie. For an up-to-date account of Lie group theory see Chevalley [13], Helgason [27], or Hochschild [29]. Thus the first successful classification of some kind of simple groups required heavy use of nonalgebraic methods, and shows that it may be shortsighted to expect a problem stated in algebraic terms to have a purely algebraic solution.

Without giving details, a complex Lie group G is a group which is at the same time a complex analytic manifold, such that the group operations are given locally by analytic functions. The classical matrix groups, with the operation of matrix multiplication, are the prototypes of Lie groups. From the manifold structure, a Lie group G has a tangent space $\mathfrak{g}$ at the identity, which is a finite dimensional vector space over the complex numbers. It is possible to define on $\mathfrak{g}$

a bilinear multiplication $[X, Y]$, satisfying the axioms of a Lie algebra:

$$[XX] = 0, \qquad X \in \mathfrak{g}$$
$$[[XY]Z] + [[YZ]X] + [[ZX]Y] = 0, X, Y, Z \in \mathfrak{g}.$$

It can then be shown, by an interesting combination of algebraic and analytical reasoning, that the group multiplication in a certain neighborhood of the identity is completely determined by the multiplication in the Lie algebra. Simple Lie groups have simple Lie algebras (where the concept of a simple Lie algebra is defined by the general remarks on the extension problem in section 1). Then the purely algebraic problem of classifying the simple Lie algebras implies a local classification of simple Lie groups.

The classical groups over the field of complex numbers are Lie groups, and to the simple groups derived from the classical groups correspond simple Lie algebras. Cartan's great achievement was to prove, completing a program already begun by Killing, that besides the simple Lie algebras corresponding to the classical groups, there are only 5 additional simple ones, called the exceptional Lie algebras, of dimensions 14, 52, 78, 133, and 248. There is a simple Lie group belonging to each of the exceptional simple Lie algebras, and the actual construction of the exceptional algebras and groups, has been one of the most striking applications of the theory of alternative and Jordan algebras. (See e.g. [30], [31], [32], [16], [47].)

The importance of Cartan's classification is that questions about properties of simple Lie algebras and Lie groups can be answered by checking against each type of algebra (or group) in the classification. Many interesting general problems and conjectures have been made in this experimental way, and later general solutions, independent of the classification, have been found for those that survived the case-by-case examination.

One difficulty in studying Lie groups via Lie algebras is that different Lie groups can have the same Lie algebra, and although homomorphisms of Lie groups always generate homomorphisms of the Lie algebra, the reverse statement is not necessarily true. But the most fundamental criticism from the point of view of algebra is that the classical groups are defined for arbitrary fields, while the Lie group machinery works smoothly only for groups which are real or complex analytic manifolds.

3. Representation theory. The next milestone after E. Cartan's work was Hermann Weyl's series of papers in 1925–1926 [50] on the representation theory of simple Lie algebras and Lie groups.

There were two problems, partly suggested by physical applications, that motivated the work on the representations.

One arises from the observation that a classical group G on a vector space V admits a homomorphism into the general linear group of the space of tensors

$$V^{(m)} = V \otimes \cdots \otimes V, \ (m \text{ factors})$$

given by

$$A \to \underbrace{A \otimes \cdots \otimes A}_{m} = A^{(m)}.$$

The problem was to decompose the tensor space $V^{(m)}$ into minimal invariant subspaces relative to the set of transformations $A^{(m)}$, $A \in GL(V)$.

A second problem, also suggested by physics, was to split up the space of continuous functions on the 2-sphere into minimal invariant subspaces relative to the action of the rotation group in 3-dimensional space.

These problems are both special cases of the following general problem. A group G is said to have a representation on a vector space M over a field K if there is a homomorphism $T: G \to GL(M)$. A subspace N is called a G-subspace if $T(g)N \subset N$ for all $g \in G$. The representation on M is called *irreducible* if the only G-subspaces are the trivial subspace $\{0\}$ and the whole space M, and *completely reducible* if M is a direct sum of irreducible G-subspaces.

At the time of Weyl's papers, there was already an extensive theory of representations of finite groups, due mainly to Frobenius, Schur and Burnside. In particular it was known that every representation of a finite group in a field of characteristic zero was completely reducible. (See [20], p. 41. The bibliography of [20] contains references to the original papers on representations of finite groups.)

Weyl, completing work begun by Cartan, classified explicitly the irreducible representations of the simple Lie algebras over the complex field, and proved also the theorem of complete reducibility for representations of simple Lie algebras, by using integration over a compact group associated with the Lie algebra. The Lie group—Lie algebra correspondence then solves the corresponding problems for representations of Lie groups. The failure of homomorphisms of the Lie algebra to generate homomorphisms of the group is strikingly illustrated by the spin representation of the rotation group. In this case, a representation of the orthogonal Lie algebra gives rise to a representation of a two-sheeted covering group of the rotation group, which cannot be viewed as a representation of the rotation group itself ([7], [13], Chapter 2, Section XI).

Weyl's work led first to a purely algebraic proof of complete reducibility of representations of simple Lie algebras by van der Waerden and the physicist Casimir [12]. Then J. H. C. Whitehead gave a new and profoundly original proof of the theorem on complete reducibility ([52], [53]) that can be viewed as the beginning of the cohomology theory of Lie algebras, and contains the motivation for at least a chapter of homological algebra ([38], [10]).

In 1939, Weyl returned to the subject, and in his book on the classical groups [51], he gave a new determination of the irreducible tensor representations of the classical groups over fields of characteristic zero, using the Wedderburn structure theory of associative algebras in an ingenious manner due to Schur, Brauer and himself. This method avoided the Lie algebras and their accompanying analysis, and gave one of the first solutions since Dickson's book of a deep problem involving the classical groups by purely algebraic methods.

Weyl's paper ([50] p. 358–359) also contained an explicit formula for the characters of the irreducible representations of the simple Lie groups, which has inspired other combinatorial investigations on the representations of Lie groups and Lie algebras by Freudenthal, Kostant, Steinberg and others (see [32], Chapter VIII, for a complete account). This work includes a generalization of the Clebsch-Gordan formula for splitting up the tensor product of two irreducible representations of the unimodular group on a two dimensional space, into irreducible components (see [49], pp. 127–131).

4. Algebraic groups. The key to the discovery of a uniform approach to the classical groups was contained in Weyl's 1925–26 papers, where he proved (see [50] pp. 338–342) that a simple Lie algebra over the complex numbers was determined up to isomorphism by a certain finite group of permutations of one-dimensional subspaces of the Lie algebra. Weyl proved that these groups were all isomorphic to finite groups generated by reflections in euclidean space E^n.

In 1934–35, Coxeter proved ([17], [18]) that a finite group was isomorphic to a group generated by reflections in E^n if and only if as an abstract group, it had generators $S_1, \cdots, S_n$ which satisfied the defining relations

$$S_i^2 = 1,\ 1 \leqq i \leqq n, \text{ and } (S_iS_j)^{n_{ij}} = 1,\ i \neq j.$$

He determined all such groups, and proved that the ones corresponding to simple Lie algebras were precisely those groups generated by reflections satisfying a certain crystallographic condition on the angles between the reflecting hyperplanes.

In 1941, Stiefel proved that the Weyl-Coxeter group of a Lie algebra could be constructed within a compact Lie group belonging to the Lie algebra [41]. Specifically, the Weyl-Coxeter group of G is isomorphic to $N(T)/T$ where T is a maximal torus in G, and $N(T)$ its normalizer. Using Lie theory, it followed that the structure of G was determined, at least locally, by the finite group $N(T)/T$.

The stage was now set for attempting to classify simple groups of Lie type over arbitrary fields, now that an internal key to their structure had been found. To replace the analytical machinery of Lie groups, algebraic geometry was used, since by that time the foundations of algebraic geometry over arbitrary ground fields had been laid by Zariski and Andre Weil.

Hermann Weyl had already observed ([51], p. 147) that the classical groups could be viewed as the intersection of an algebraic variety in the space $M_n(K)$ of all n by n matrices over K with the general linear matrix group $GL(n, K)$. Such a group is called an *algebraic group*, and again the prototypes are the classical groups. For example the unimodular group is defined by the polynomial relation

$$\det(X) - 1 = \sum \pm x_{i_1 1}x_{i_2 2} \cdots x_{i_n n} - 1 = 0.$$

In 1956–58, Chevalley succeeded in classifying all simple algebraic groups over algebraically closed fields of arbitrary characteristic [15]. He proved that such a group was determined up to isomorphism by a Weyl-Coxeter group exactly as in the case of compact Lie groups, and that the Weyl-Coxeter groups which appeared satisfied the crystallographic restriction. Since the groups with these Weyl-Coxeter groups were known, it followed that the only simple algebraic groups were the classical groups and the five types of exceptional groups. The proof of Chevalley's result makes heavy use of the techniques of algebraic geometry. It depends on some basic work of A. Borel on solvable algebraic groups.

5. Finite groups of Lie type. Until 1955 no simple finite groups were discovered that were not already known to Dickson. In his famous Tôhoku Journal paper [14] of 1955, Chevalley constructed a simple group associated with every simple Lie algebra over the complex field and every field K, with some exceptions when K is the field of 2 or 3 elements. His construction yielded finite simple groups corresponding to the exceptional Lie algebras, and gave a uniform method for investigating the groups, which is based on new structural properties of the groups.

Chevalley proved first that a simple Lie algebra $\mathfrak{g}$ over the complex numbers has a basis $X_1, \cdots, X_n$ such that

$$[X_iX_j] = \sum c_{ijk}X_k,$$

where the $\{c_{ijk}\}$ are integers. This multiplication table serves to define a Lie algebra $\mathfrak{g}_K$ over an arbitrary field K, since the structure constants $\{c_{ijk}\}$ can be viewed as elements of K. The Chevalley group G associated with the Lie algebra $\mathfrak{g}$ and the field K is defined to be a certain subgroup of the automorphism group of $\mathfrak{g}_K$. Chevalley then proved by a long argument, but one that broke with tradition by treating all the groups at once, that with a few exceptions, the groups G were all simple.

For a Lie algebra $\mathfrak{g}$ associated with one of the classical groups, the Chevalley group of $\mathfrak{g}$ and K coincides with the simple group derived from the corresponding classical group [37]. In case K is the complex field, the Chevalley groups over K yield a complete set of simple Lie groups with complex parameters. If K is algebraically closed, of arbitrary characteristic, then the Chevalley groups give a complete set of examples of simple algebraic groups defined over K, by the classification theorem for algebraic groups. If K is a finite field, then the Chevalley groups include the finite simple groups investigated by Dickson, Dieudonné and Artin, and those associated with the exceptional Lie algebras give infinite families of finite simple groups not isomorphic to those in Dickson's original list.

The proof of this last fact requires formulas for the orders of the Chevalley groups defined over finite fields. Chevalley [14] derived these formulas using some topological properties of the Lie group of the same type, and recently L. Solomon [40] has given another derivation of the formulas using some topological properties of the Weyl-Coxeter group of the Lie algebra. No purely algebraic derivation of the formulas is known. A knowledge of the formulas is important because of a number theoretical study by Artin ([1], [2]) of the orders of the known simple groups which, among other things, gives a method of showing that different Chevalley groups in general cannot be isomorphic.

6. Further results on finite groups. The excitement over Chevalley's paper had hardly died down before new infinite families of finite simple groups were discovered, by Hertzig, Ree, Steinberg, Suzuki, and Tits (see [11]). Some of these were defined by analogy with the classification of simple algebraic or Lie groups, where it was known that among the groups parametrized by fields which were not algebraically closed, certain twisted versions of the Chevalley groups could occur. All are obtained from a Chevalley group by the following sort of construction. Certain Chevalley groups G, defined for certain fields K, admit automorphisms of period 2 or 3 such that the set of elements in G left fixed under the automorphism contains a new simple group. Of course we do not know whether we have seen the last of these constructions, especially in view of a new simple group defined by Janko [33], of order $11(11+1)(11^3-1)$, which is a subgroup of a Chevalley group defined over the field of 11 elements, but is not obtained by the preceding general method.

Nevertheless the situation has crystallized enough for us at least to hope for uniform description of all the known finite simple non-abelian groups. For this we are indebted above all to J. Tits, whose work is still in the process of publication. From a close inspection of the construction of the Chevalley groups, Tits developed the following very simple but still rather mysterious set of axioms which are satisfied by all the known finite simple groups except possibly the Janko group, and are suggested by the structural properties of the groups derived in Chevalley's Tôhoku Journal paper. (See [44], [45], [46].)

A group G (not necessarily finite) is said to have a BN-pair if G is generated by subgroups B and N such that

$$B \cap N \text{ is a normal subgroup of } N\text{; let } H = B \cap N.$$

$$N/H = \langle w_1, \cdots, w_n \rangle, \; w_i^2 = 1,$$

and for all cosets $w \in N/H$, and generators w_i of N/H,

$$w_i B w \subset BwB \cup Bw_i wB$$

and for each i

$$w_i B w_i \neq B.$$

The group $W = N/H$ is called the Weyl group of G, for the reason that the group N/H of an arbitrary finite group with a BN-pair is isomorphic to a group generated by reflections in euclidean space. Abstract groups with BN-pairs can be expressed as a union of (B, B) double cosets, $G = \bigcup_{n \in N} BnB$, with the double cosets in 1–1 correspondence with the elements of the Weyl group, a decomposition discovered first for the classical groups by Bruhat [8]. All the Chevalley groups, as well as the twisted types, have BN-pairs. Among other things Tits has found an elegant proof of simplicity for certain types of groups with BN-pairs [46], that provides in almost all cases a much shorter and easier proof of simplicity of the Chevalley groups and the twisted types than the one given in Chevalley's paper.

It has also been possible, using the methods of Chevalley and Tits in combination with Lie algebra methods similar to those in Weyl's 1925–26 papers, to classify all the irreducible representations of the finite Chevalley groups and the twisted types, in an algebraically closed field of the same characteristic as the field by which the groups are parametrized, completing the work begun by Weyl for the classical groups over fields of characteristic zero ([42], [19]).

The complex representations and characters of the finite Chevalley groups have not yet been classified, partly because of our lack of knowledge about the conjugacy classes.

In this connection, the Jordan normal form theorem tells us that two elements X and Y of $GL(V)$ are conjugate, i.e., $X=ZYZ^{-1}$, if and only if they have the same Jordan normal form. The Bruhat decomposition $G=\cup BnB$ fails to give a normal form up to conjugacy for elements of a general Chevalley group. R. Steinberg [43] has recently given a normal form for certain conjugacy classes of elements in "simply connected" versions of all the Chevalley groups over algebraically closed fields, that includes the Jordan normal form as a special case. This powerful result has already had applications to the classification of algebraic groups over non-algebraically closed fields and may suggest a method of determining the conjugacy classes in finite Chevalley groups.

Behind all the fascinating special results and problems suggested by these developments, lies the austere and immensely difficult problem of classifying the finite simple groups. The classical groups on 2- and 3-dimensional spaces were determined in terms of the structure of their subgroups by Zassenhaus, Suzuki, and Brauer and his students. See Section 92 of [20] for a survey of this work up to 1962. Then came the great achievement for which the Cole prize in algebra was awarded at the Denver meeting of the American Mathematical Society last winter—Walter Feit and John Thompson's proof, published in 1963, of Burnside's conjecture that a non-abelian simple group must have even order [25]. Since then the pace in finite groups has picked up, with Thompson's proof that the only finite simple groups, all of whose proper subgroups are solvable, are the known ones, and related works of Suzuki and Gorenstein and Walter. This work brings together all that is known about finite groups because the inductive method of proof used in these questions presents one with a group whose structure is to be determined, given fairly complete information on the structure of its subgroups which may be solvable groups, p-groups, etc.

If we continue to believe, as of course we must, in the possibility of communicating mathematical proofs as well as statements of results, then we are faced with the difficult and important task of simplifying some of these Herculean arguments, which often run to two and three hundred pages, to the point where they can meaningfully be presented in lectures and text books.

A bright ray of light in this direction is Tits' classification of finite groups with BN-pairs. It was known for some time that in a BN-pair G whose Weyl group was the symmetric group S_3, a coset geometry could be defined that satisfied the axioms for a projective plane. D. G. Higman and McLaughlin [28] proved that if the group G was finite, the plane was Desarguesian, and that G was an extension of the simple classical group $PSL(3, K)$ for some finite field K. Tits has proved that there is a geometry associated with every finite group with a BN-pair, and has succeeded in classifying the possible geometries in most cases [45]. The difficulty in applying his work to arbitrary finite groups is that it seems to be very difficult to tell whether a group has a BN-pair or not.

7. Conclusion. There are many other important directions I have not been able to take up. One basic experimental fact we have observed is the strong dependence of the methods used for studying the simple groups and their representations on the field over which the groups are defined. To make precise and formalize this kind of dependence is one of the tasks of homological algebra. In this direction we have, for example, applications of Galois cohomology [39] to the study of algebraic groups defined over nonalgebraically closed fields, and important new constructions in the category of modules over a ring—the algebraic K-theory of Bass [4]—which are at least partly suggested by the study of linear groups over the ring.

I have been able, nevertheless, to tell enough of the story to show that the classical groups have successfully resisted the best efforts of three generations of mathematicians to subdue them, and continue to suggest interesting problems for further research. The methods that have been used represent strong improvements and refinements over those available to Jordan and Dickson and have found their way into our elementary teaching. Their continual interaction with other branches of mathematics, and physics, would have heartened Hermann Weyl, who expressed in the preface of his book [51] his concern over the dangers of a too thorough specialization of mathematical research. The permanent feature of all this work has been the source of the problems—the groups themselves.

I hope that what I have said makes it clear that in our capacity as teachers we should introduce our students to the empirical sources, the physical facts of our science, as well as to the general ideas and methods we use in our work.

Other sources of algebraic ideas that have had a similar influence are the theory of algebraic numbers, representations of finite groups, and algebraic geometry; but these I leave to another time and other speakers.

References

1. E. Artin, The orders of the linear groups, Comm. Pure Appl. Math., 8 (1955) 355–366.

2. ———, The orders of the classical simple groups, Comm. Pure Appl. Math., 8 (1955) 455–472.

3. ———, Geometric Algebra, Interscience, New York, 1957.

4. H. Bass, K-Theory and stable algebra, Publ. Math., (I. H. E. S.), Paris, 22 (1964) 5–60.

5. R. Brauer, On finite groups and their characters, Bull. Amer. Math. Soc., 69 (1963) 125–130.

6. ———, Representations of finite groups, Lectures on Modern Mathematics, Vol. I, T. Saaty, Editor, Wiley, New York, 1963.

7. R. Brauer and H. Weyl, Spinors in n dimensions, Amer. J. Math., 57 (1935) 425–449; Selecta Hermann Weyl, Birkhäuser, Basel, 1956, pp. 431–454.

8. F. Bruhat, Représentations induites des groupes de Lie semi-simples connexes, C. R. Acad. Sci., Paris, 238 (1954) 437–439.

9. E. Cartan, Sur la structure des groupes de transformations finis et continus, Paris, Thèse, 1894; 2nd ed., Vuibert, Paris, 1933. Oeuvres complètes, Partie 1, Vol. 1, Paris, 1952.

10. H. Cartan and S. Eilenberg, Homological Algebra, Princeton, 1956.

11. R. W. Carter, Simple groups and simple Lie algebras, J. London Math. Soc., 40 (1965) 193–240.

12. H. Casimir and B. L. van der Waerden, Algebraischer Beweis der vollständigen Reduzibilität der Darstellungen halbeinfacher Liescher Gruppen, Math. Ann., 111 (1935) 1–12.

13. C. Chevalley, Theory of Lie groups I, Princeton, 1946.

14. ———, Sur certains groupes simples, Tôhoku Math. J., 7 (1955) 14–66.

15. ———, Séminaire Chevalley, Vols. I and II, École Normale Supérieure, Paris, 1958.

16. C. Chevalley, and R. D. Schafer, The exceptional simple Lie algebras F_4 and E_6, Proc. Nat. Acad. Sci. U. S. A., 36 (1950), 137–141.

17. H. S. M. Coxeter, Discrete groups generated by reflections, Ann. of Math., 35 (1934), 588–621.

18. ———, The complete enumeration of finite groups of the form $R_i^2 = (R_iR_j)^{k_{ij}} = 1$, J. London Math. Soc., 10 (1935) 21–25.

19. C. W. Curtis, Irreducible representations of finite groups of Lie type, Journal für Math., 219 (1965), 180–199.

20. C. W. Curtis, and I. Reiner, Representation Theory of Finite Groups and Associative Algebras, Wiley (Interscience), New York, 1962.

21. L. E. Dickson, Linear Groups, Leipzig, 1901; reprint, Dover, New York, 1958.

22. J. Dieudonné, Sur les groupes classiques, Hermann, Paris, 1948.

23. ———, La géométrie des groupes classiques, Ergebnisse der Math. N. F., No. 5, Springer, Berlin, 1955.

24. ———, Recent developments in mathematics, Amer. Math. Monthly, 71 (1964) 239–247.

25. W. Feit and J. Thompson, Solvability of groups of odd order, Pacific J. Math., 13 (1963) 775–1029.

26. M. Hall, Generators and relations in groups—the Burnside Problem, Lectures in Modern Mathematics, vol. 2, T. Saaty, Editor, Wiley, New York, 1964.

27. S. Helgason, Differential Geometry and Symmetric Spaces, Academic Press, New York, 1962.

28. D. Higman and J. McLaughlin, Geometric ABA-groups, Illinois J. Math., 5 (1961) 382–397.

29. G. Hochschild, The structure of Lie groups, Holden Day, San Francisco, 1965.

30. N. Jacobson, Cayley numbers and simple Lie algebras of type G, Duke Math. J., 5 (1939) 775–783.

31. N. Jacobson, Some groups of transformations defined by Jordan algebras, I, J. für Math., 201 (1959) 178–195, II, J. für Math., 204 (1960), 74–98; III, J. für Math., 207 (1961) 61–85.

32. ———, Lie Algebras, Interscience, New York, 1962.

33. Z. Janko, A new finite simple group . . . , J. Algebra, 3 (1966) 147–186.

34. C. Jordan, Traité des Substitutions, Paris, 1870.

35. I. Kaplansky, Lie algebras, Lectures on Modern Mathematics, V. 1, T. Saaty, Editor, Wiley, New York, 1963.

36. S. MacLane, Some recent advances in algebra, Studies in Mathematics, Vol. 2, Math. Association of America, (1963) 35–58.

37. R. Ree, On some simple groups defined by C. Chevalley, Trans. Amer. Math. Soc., 84 (1957) 392–400.

38. Séminaire "Sophus Lie" Théorie des algèbres de Lie, École Normale Supérieure, Paris, 1955.

39. J. P. Serre, Cohomologie Galoisienne des groupes algébriques linéaires, Colloque sur la Théorie des Groupes Algébriques, Gauthier-Villars, Paris, 1962.

40. L. Solomon, On the orders of the finite Chevalley groups, J. Algebra 3 (1966) 376–393.

41. E. Stiefel, Über eine Beziehung zwischen geschlossenes Lieschen Gruppen und diskontinuierlichen Bewegungsgruppen euklidischer Räume und ihre Anwendung auf der Aufzählung der einfachen Lieschen Gruppen, Comment, Math. Helv., 14 (1941) 350–380.

42. R. Steinberg, Representations of algebraic groups, Nagoya Math. J., 22 (1963) 33–56.

43. ———, Regular elements of semisimple algebraic groups, Publ. Math. (I. H. E. S.), Paris, 25 (1965) 49–80.

44. J. Tits, Théorème de Bruhat et sous groupes paraboliques, C. R. Acad. Sci., Paris, 254 (1962) 2910–2912.

45. ———, Groupes simples et géométries associées, Proc. Int. Congress Math., (1962) 197–221.

46. ———, Algebraic and abstract simple groups, Ann. of Math., 2, 80 (1964) 313–329.

47. M. L. Tomber, Lie algebras of type F, Proc. Am. Math. Soc., 4 (1953) 759–768.

48. B. L. van der Waerden, Gruppen von linearen Transformationen, Ergebnisse der Math., Bd. 4, Heft 2, Springer, Berlin, 1935.

49. H. Weyl, Theory of Groups and Quantum Mechanics, Dover reprint, New York, 1950.

50. ———, Theorie der Darstellung kontinuierlicher half-einfacher Gruppen durch lineare Transformationen (Teil I, II, III und Nachtrag), Selecta Hermann Weyl, Birkhäuser, Basel, 1956, pp. 262–366.

51. ———, The Classical Groups, 2nd edition, Princeton, 1946 (first edition, 1939).

52. J. H. C. Whitehead, On the decomposition of an infinitesimal group, Proc. Cambridge Philos. Soc., 32 (1936) 229–236.

53. ———, Certain equations in the algebra of an infinitesimal semi-simple group, Quart. J. Math. Oxford Ser., 8 (1937) 220–237.

BIBLIOGRAPHIC ENTRIES: LINEAR GROUPS AND THE SYMMETRIC GROUPS

The references below are to the AMERICAN MATHEMATICAL MONTHLY.

1. J. L. Brenner, A new proof that no permutation is both even and odd, 1957, 499–500.

2. Edwin L. Gray, An alternate proof for the invariance of parity of a permutation written as a product of transpositions, 1963, 995.

3. Clifford E. Weil, Another approach to the alternating subgroup of the symmetric group, 1964, 545–546.

4. W. Phillips, On the definition of even and odd permutations, 1967, 1249–1251.

Each of the above references provides a proof that the parity of a permutation is well-defined.

5. P. J. Lorimer, The outer automorphisms of S_6, 1966, 642–643.

6. H. H. Mitchell, Linear groups and finite geometries, 1935, 592–603.

7. George K. White, On generators and defining relations for the unimodular group M_2, 1964, 743–748.

8. Jessie MacWilliams, Orthogonal matrices over finite fields, 1969, 152–164.

Computes the order of the orthogonal groups over fields of characteristic 2.

(f)

SUBGROUPS AND ELEMENTS OF FINITE GROUPS

ANOTHER PROOF OF CAUCHY'S GROUP THEOREM*

JAMES H. McKAY, Seattle University

Since $ab=1$ implies $ba=b(ab)b^{-1}=1$, the identities are symmetrically placed in the group table of a finite group. Each row of a group table contains exactly one identity and thus if the group has even order, there are an even number of identities on the main diagonal. Therefore, $x^2=1$ has an even number of solutions.

Generalizing this observation, we obtain a simple proof of Cauchy's theorem. For another proof see [1].

CAUCHY'S THEOREM. *If the prime p divides the order of a finite group G, then G has kp solutions to the equation $x^p=1$.*

Let G have order n and denote the identity of G by 1. The set

$$S = \{(a_1, \cdots, a_p) \mid a_i \in G,\ a_1a_2 \cdots a_p = 1\}$$

has n^{p-1} members. Define an equivalence relation on S by saying two p-tuples are equivalent if one is a cyclic permutation of the other.

If all components of a p-tuple are equal then its equivalence class contains only one member. Otherwise, if two components of a p-tuple are distinct, there are p members in the equivalence class.

Let r denote the number of solutions to the equation $x^p=1$. Then r equals the number of equivalence classes with only one member. Let s denote the number of equivalence classes with p members. Then $r+sp=n^{p-1}$ and thus $p|r$.

Reference

1. G. A. Miller, On an extension of Sylow's theorem, Bull. Amer. Math. Soc., vol. 4, 1898, pp. 323–327.

* From AMERICAN MATHEMATICAL MONTHLY, vol. 66 (1959), p. 119.

ON A THEOREM OF FROBENIUS*

RICHARD BRAUER, Harvard University

1. A well-known theorem of Frobenius states that if G is a finite group of order g and if n is a divisor of g, the number of solutions β of the equation $\beta^n = 1$ in G is a multiple of n. This can be generalized in various ways. A number of proofs have been given [1–7]. I shall give here still another proof. I found this proof a very long time ago when I was still a student. I did not publish it, but I have used it for many years as material for problems in courses, giving a number of hints. I present the proof here in the hope that other teachers may find it interesting and useful for the same purpose.

2. We start with a lemma.

LEMMA. *Let H be a group. Let A be a normal subgroup of finite order a. If $\sigma \in H$ and if $\alpha \in A$, then σ^a and $(\sigma\alpha)^a$ are conjugate in H.*

Proof. Set $\tau = \sigma\alpha$. For each $\beta \in A$, consider the set S_β of distinct elements $\sigma^{-j}\beta\tau^j$ with $j = 0, \pm 1, \pm 2, \cdots$. Since

$$\sigma^{-j}\beta\tau^j \in \sigma^{-j}\beta(\sigma\alpha)^j A = \sigma^{-j}\sigma^j A = A,$$

S_β is a subset of A. In particular, S_β is a finite set. For two integers i and j, we have

$$\sigma^{-i}\beta\tau^i = \sigma^{-j}\beta\tau^j \tag{1}$$

if and only if $\beta^{-1}\sigma^{i-j}\beta = \tau^{i-j}$. It follows that the number N_β of distinct elements in S_β is the least positive exponent N for which

$$\beta^{-1}\sigma^N\beta = \tau^N \tag{2}$$

and that (1) holds if and only if N divides $i - j$. It is also clear that if β and γ are elements of A, the sets S_β and S_γ are equal or disjoint. Hence A is a disjoint union of sets S_β, say

$$A = S_{\beta_1} \cup S_{\beta_2} \cup \cdots \cup S_{\beta_r} \tag{3}$$

with $\beta_i \in A$. Let m denote the minimal value of N_{β_i} with $1 \leq i \leq r$. Then by (2), m is the least positive exponent for which conjugation by an element β of A carries σ^m into τ^m. Moreover, if exactly k of the r numbers N_{β_i} are equal to m,

---

* From AMERICAN MATHEMATICAL MONTHLY, vol. 76 (1969), pp. 12–15.

Professor Brauer did his doctoral work under I. Schur at the University of Berlin. He has been on the faculties of the Univ. of Koenigsberg, Univ. of Kentucky, Institute for Advanced Study, Univ. of Toronto, and Univ. of Michigan. Presently he holds the Perkins Professorship at Harvard. On leaves of absence he held positions at the Tata Institute, Nagoya Univ., the Akademie der Wissenschaften Göttingen (Gauss Professor), and the E. T. H. Zürich. He is best known for his contributions to the representation theory of groups and algebras and was awarded the AMS Cole Prize in Algebra in 1949. *Editor* the km elements of the corresponding sets S_{β_i} are exactly the elements β of A for which (2) holds with $N=m$. On the other hand, the number of these elements is the order of the centralizer of σ^m in A and hence divides a. Thus km divides a. Since σ^m and τ^m are conjugate and since m divides a, we see that σ^a and τ^a are conjugate as stated.

REMARK. The proof shows that there exist elements $\beta \in A$ which carry σ^a into τ^a.

3. Let G be a group and let H be a subgroup. We shall say that two elements β, γ of G are *equivalent with regard to* H, if $\beta^{-r}\gamma^r \in H$ for all integers r. Clearly this is an equivalence relation. If β is an element of G, denote by F_β the set of all elements $\sigma \in G$ for which $\beta^{-r}\sigma\beta^r \in H$ for all integers r.

PROPOSITION 1. *Let G be a group, H a subgroup, and β an element of G. Then F_β is a subgroup of H and $\beta^{-1}F_\beta\beta = F_\beta$. An element γ of G is equivalent to β with regard to H if and only if $\gamma \in \beta F_\beta$.*

Proof. The first statement is obvious. If $\gamma = \beta\phi$ with $\phi \in F_\beta$, then since F_β is normal in the group generated by β and F_β, for any integer r

$$\beta^{-r}\gamma^r = \beta^{-r}(\beta\phi)^r \in \beta^{-r}\beta^r F_\beta = F_\beta \subseteq H$$

and β and γ are equivalent.

Conversely, if β and γ are equivalent, set

$$\delta_r = \beta^{-r}\gamma^r \in H.$$

Then, for integral s

$$\beta^{-s}\delta_1\beta^s = \beta^{-s}\beta^{-1}\gamma\beta^s = \beta^{-s-1}\gamma^{s+1}\gamma^{-s}\beta^s = \delta_{s+1}\delta_s^{-1} \in H.$$

This shows that $\delta_1 \in F_\beta$. Hence $\gamma = \beta\delta_1 \in \beta F_\beta$ as stated.

PROPOSITION 2. *If H in Proposition 1 has finite order h and if β and γ are equivalent with regard to H, then β^h and γ^h are conjugate.*

Proof. By Proposition 1, $\gamma = \beta\phi$ with $\phi \in F_\beta$. Apply the lemma to the group $H = \langle\beta, F_\beta\rangle$ and its normal subgroup $A = F_\beta$, noting that the order of F_β divides h.

4. As before, let G be a group and H a subgroup. We shall say that two elements β, γ of G are *weakly equivalent with regard to* H, if there exists an element $\tau \in H$ such that $\beta^{-r}\tau\gamma^r \in H$ for all integers r. Again this is an equivalence relation. Since our condition can be written in the form

$$\beta^{-r}(\tau\gamma\tau^{-1})^r \in H,$$

we have

PROPOSITION 3. *Two elements β, γ of G are weakly equivalent with regard to H if and only if β is equivalent with regard to H to an H-conjugate $\tau\gamma\tau^{-1}$ of γ, $(\tau \in H)$.*

It follows from Propositions 3 and 1 that the class of β with regard to weak equivalence consists of the elements of the form

$$\gamma = \tau^{-1}\beta\delta\tau, \qquad \delta \in F_\beta, \qquad \tau \in H. \tag{4}$$

It is clear that it suffices to let τ range over a set T of representatives for right cosets in H modulo F_β. We claim that the corresponding representation (4) of γ is unique. This will be shown, if we can prove the following statement:

PROPOSITION 4. *Let G, H, β, F_β be as before. If $\sigma \in H$ and if for some δ and δ' in F_β*

$$\sigma^{-1}\beta\delta\sigma = \beta\delta' \tag{5}$$

then $\sigma \in F_\beta$.

Proof. It follows from (5) that

$$\beta^{-1}\sigma\beta \in F_\beta\sigma F_\beta.$$

Since β normalizes F_β, then for integral r,

$$\beta^{-r-1}\sigma\beta^{r+1} \in F_\beta\beta^{-r}\sigma\beta^r F_\beta.$$

Since $\sigma \in H$, it follows by induction on r, (both for $r>0$ and for $r<0$) that $\beta^{-r}\sigma\beta^r \in H$. Hence $\sigma \in F_\beta$ as claimed.

The next proposition is a consequence of the remark following (4), of Proposition 4, and of Proposition 2.

PROPOSITION 5. *Let G be a group and H a subgroup of finite order h. If $\beta \in G$, the class of β with regard to weak equivalence consists of exactly h elements γ and for all these elements γ, the elements β^h and γ^h are conjugate in G.*

5. One form of the Frobenius' Theorem reads

THEOREM. *Let G be a group of finite order g and let K be a conjugate class of G. If n is a positive integer and $(g, n) = d$, the number N of elements $\beta \in G$ with $\beta^n \in K$ is divisible by d.*

(It is well known that a slightly better result is obtained by applying the theorem to the centralizer in G of an element of K. If K consists of k elements, it follows that N is divisible by (g, nk).)

Using Sylow's Theorem, we can obtain Frobenius's Theorem as an immediate consequence of Proposition 5. Indeed, if p^r is any prime power dividing $d = (g, n)$, we choose H as a subgroup of order $h = p^r$ of G. Since the set of elements β with $\beta^n \in K$ consists of full weak equivalence classes with regard to H, the number N is divisible by p^r and hence by d.

It is clear that the same method still applies to infinite groups G, provided that G possesses suitable finite subgroups. We have

THEOREM. *Let G be a group and let K be a conjugate class of G. Let n be an integer and assume that the number N of elements β of G with $\beta^n \in K$ is finite. Then*

N is divisible by every prime power p^r dividing n for which there exist subgroups H of order p^r of G.

Far-reaching generalizations of Frobenius's Theorem have been given by P. Hall [5]. These cannot be obtained here.

6. We mention another consequence of the lemma in Section 2.

PROPOSITION 6. *Let G be a finite group of order g which has a normal subgroup H of order h. If $\beta\in G$ has order relatively prime to h and if C is the centralizer of β in G, then the centralizer $\overline{C}$ of βH in G/H is CH/H.*

Proof. The reciprocal image C^* of $\overline{C}$ in G consists of the elements $\gamma\in G$ for which

$$\gamma^{-1}\beta\gamma \in \beta H. \tag{6}$$

It is clear that $CH\subseteq C^*$. Conversely, if $\gamma\in C^*$ then by the lemma and the remark following it, there exists an element $\sigma\in H$ such that

$$\sigma^{-1}\beta^h\sigma = \gamma^{-1}\beta^h\gamma.$$

Since h and the order of β are coprime, then

$$\sigma^{-1}\beta\sigma = \gamma^{-1}\beta\gamma.$$

Hence $\gamma\sigma^{-1}\in C$, $\gamma\in CH$.

References

1. W. Burnside, The Theory of Groups of Finite Order, 2nd ed., Cambridge University Press, 1907, page 49.

2. G. Frobenius, Verallgemeinerung des Sylowschen Satzes, Sitzungsberichte der Preussischen Akademie, Berlin, 1895, pp. 437–449.

3. ———, Über einen Fundamentalsatz der Gruppentheorie I, II, Sitzungsberichte der Preussischen Akademie, Berlin, 1903, 987–991, 1907, 428–437.

4. M. Hall, Jr., The Theory of Groups, Macmillan, New York, 1959, p. 139.

5. P. Hall, On a theorem of Frobenius, Proc. London Math. Soc., 40 (1936) 468–501.

6. B. Huppert, Endliche Gruppen, vol. 1, Springer, Berlin, Heidelberg, New York, 1967, p. 44.

7. H. Zassenhaus, The Theory of Groups, 2nd ed., Vandenhoeck and Ruprecht, Göttingen, 1956.

A GROUP-THEORETIC PROOF OF WILSON'S THEOREM*

WALTER FEIT, Cornell University

Wilson's theorem states that if p is a prime, then p divides $(p-1)!+1$. We give a proof based on some elementary group theory results.

Let n be the number of elements in the symmetric group S_p on p letters, whose order divides p. A well-known result of Frobenius* states that p divides n. The number of p-cycles in S_p is easily seen to be $(p-1)!$. Since every element in S_p whose order divides p is either a p-cycle or the identity, $n=(p-1)!+1$. The proof is complete.

BIBLIOGRAPHIC ENTRIES: SUBGROUPS AND ELEMENTS OF FINITE GROUPS

1. C. D. H. Cooper, Subgroups of a supersolvable group, AMERICAN MATHEMATICAL MONTHLY, 1971, 1007.

Gives an elementary proof that if G is a finite supersolvable group, d a divisor of $|G|$, then G has a subgroup of order d.

2. Louis W. Shapiro, Finite groups acting on sets with applications, MATHEMATICS MAGAZINE, 1973, 136–147.

Presents a series of exercises whose aim is to derive a number of well-known theorems in group theory, including Sylow's Theorems.

(g)

HISTORY

THE FOUNDATION PERIOD IN THE HISTORY OF GROUP THEORY†

JOSEPHINE E. BURNS, University of Illinois

The earliest group notions. Henri Poincaré has pointed out that the fundamental conception of a group is evident in Euclid's work; in fact that the foundation of Euclid's demonstrations is the group idea. Poincaré establishes this assertion by showing that such operations as successive superposition and rotation about a fixed axis presuppose the displacements of a group. However much the fundamental group notions were unconsciously used in the work of early mathematicians, it was not until the latter part of the eighteenth century that these notions began to take life and develop.

* From AMERICAN MATHEMATICAL MONTHLY, vol. 65 (1958), p. 120.

† From AMERICAN MATHEMATICAL MONTHLY, vol. 20 (1913), pp. 141–148.

The foundation period. The period of foundation of group theory as a distinct science extends from Lagrange (1770) to Cauchy (1844–1846), a period of seventy-five years. We find Lagrange considering the number of values a rational function can assume when the variables are permuted in every possible way. With this beginning the development may be traced down through the contributions of Vandermonde, Ruffini, Abbati, Abel, Galois, Bertrand and Hermite, to Cauchy's period of active production (1844–1846). At the beginning of this period group theory was a discovery useful in the theory of equations; at the end it existed as a distinct science, not yet, to be sure, entirely free but so nearly so that this may be called the close of the foundation period.

LAGRANGE, RUFFINI, GALOIS.

Lagrange. The contributions of Lagrange are included in his memoir, *Réflexions sur la résolution algébrique des equations*, published in the Mémoires of the Academy of Science at Berlin in 1770–1771.† In this paper Lagrange first applies what he calls the "calcul des combinaisons" to the solution of algebraic equations. This is practically the theory of substitutions, and he uses it to show wherein the efforts of his predecessors, Cardan, Ferrari, Descartes, Tschirnhaus, Euler and Bezout fail in the case of equations of degree higher than the fourth. He studies the number of values a rational function can assume when its variables are permuted in every possible way. The theorem that the order of the subgroup divides the order of the group is implied but not explicitly proved. The theorem that the order of a group of degree n divides $n!$ is however explicitly stated. Lagrange does not use at all the notation or the terminology of group theory but confines himself entirely to direct applications to the theory of equations. He mentions the symmetric group of degree 4, and the four-group and the cyclic group of order 4 as subgroups of this group. This embodies practically all of the work of Lagrange in the theory of substitution groups.

Ruffini. The first man to follow Lagrange and to make any signal progress was Paolo Ruffini, an Italian, who published in 1799 in his *Teoria generale delle equazioni* a number of important theorems in the theory of substitutions. Burkhardt says* that several fundamental concepts are implied, if not explicitly stated. Ruffini's "permutation" corresponds to the later accepted term of group and to Cauchy's "system of conjugate substitutions." These permutations he classifies as "simple" and "complex." The first are of two sorts, of one cycle or of more than one cycle. The second he divides into three classes which correspond to the modern notions of (1) intransitive, (2) transitive imprimitive, and (3) transitive primitive. He enunciates the following theorems:

1. *If a substitution of n letters leaves the value of a rational function invariant, the*

† *Œuvres de Lagrange*, vol. 3, pp. 204–420.

* *Zeitschrift für Math. u. Phys.* (1892), 37, p. 119.

result of applying the substitution any number of times is that the function is left invariant.

2. *The order of a substitution is the least common multiple of the orders of the cycles.*

3. *There is not necessarily a subgroup corresponding to every arbitrary divisor of the order of the group.* (This theorem he established by showing that there is no eight-valued, four-valued or three-valued function on five letters.)

4. *A group of degree five that contains no cycle of degree five cannot have its order divisible by five.*

Abbati, Abel. From Ruffini to Galois, only two or three contributions of merit were made. Abbati gave the first complete proof that the order of the subgroup divides the order of the group. This he did by putting the substitutions of the group in rectangular array. Abbati also proved that there is no three or four valued function on n letters when n is greater than five. Cauchy published in 1815 a relatively unimportant memoir. Abel, by using the theory of substitutions to prove that it is impossible to solve algebraically equations of degree higher than the fourth, called attention to this useful instrument.

Galois. Up to the time of Cauchy, Galois had done the most for group theory. His work was written in 1831 and 1832, but not made public until 1846. To Galois is due the credit for the conception of the invariant subgroup, for the notion of the simple group and the extension of the idea of primitivity. Galois first used the term group in its present technical sense. Several important theorems are due to him. Among these are the following:

1. *The lowest possible composite order of a simple group is* 60.*
2. *The substitutions common to two groups form a group.*†
3. *If the order of a group is divisible by* p, *a prime, there is at least one substitution of order* p.
4. *The substitutions of a group that omit a given letter form a group.*

CAUCHY, THE FOUNDER OF GROUP THEORY.

Terminology. Cauchy's work on groups is found in his *Exercices d'analyse et de physique mathématique* published in 1844, and in the series of articles published in the Paris *Comptes Rendus* in 1845–1846.‡ Cauchy defines permutation and substitution in the same way, but uses the latter term almost entirely. He uses several devices to denote a substitution, the most common being (x, y, z, u, v, w) where each letter is replaced by the one which follows it and the last by the first. He defines

* *Œuvres Mathématiques d'Évariste Galois*, p. 26.

† *Manuscrits d'Évariste Galois*, p. 39.

‡ *Œuvres de A. Cauchy*, First series, vols. 9 and 10.

this as a cyclic substitution. In the article written in 1815* he defines the degree of a substitution as the first power of the substitution that reduces to identity, but later he also defines order in this way and uses the word order rather than degree in his subsequent work. He uses both the term unity and identical substitution. A transposition is defined and the terms similar, regular, inverse and permutable substitutions are found with their present day significance. A group he calls a "system of conjugate substitutions," which may be transitive or intransitive. A transitive imprimitive group is "transitive complex." The order of a system of conjugate substitutions is the number of substitutions that it contains. The term "divisor indicatif" is also found for the order of a group. Cauchy frequently transfers his definition of order to the theory of equations and speaks of the number of equal values a rational function can assume when the variables are permuted in every possible way. The number of distinct values of such a function is the index. In the early article the order of operation is from left to right, but in all his later work he reverses the order and operates from right to left.

The Memoir of 1815. In the paper published in 1815 the one theorem which is of special interest is that the number of distinct values of a non-symmetric function of degree n cannot be less than the largest prime that divides n, without becoming equal to 2. This theorem is proved. He states the special cases, that if the degree of a function is a prime number greater than 2, the number of distinct values cannot be less than the degree; and that if the degree is 6 the number of distinct values cannot be less than 6. He makes special reference to the functions belonging to (1) the intransitive group of degree 6 and order 36; (2) the transitive imprimitive group of degree 6 and order 72; (3) the intransitive group of degree 6 and order 48; (4) and the symmetric group of degree 5 considered as an intransitive group of degree 6. The functions are as follows:

(1) $a_1a_2a_3+2a_4a_5a_6$, (3) $a_1a_2a_3a_4+a_5a_6$,
(2) $a_1a_2a_3+a_4a_5a_6$, (4) $a_1a_2a_3a_4a_5+a_6$.

The Memoir of 1844. In the memoir published in 1844 in his *Exercices d'Analyse* there is much more of importance. The first theorem proved is that every substitution similar to a given substitution P is the product of three factors, the extremes of which are the inverse of each other and the middle term of which is P. Conversely, every product of three factors, the first and last of which are the inverse of each other, is similar to the middle term P. An important formula developed is for the number of substitutions similar to a given substitution. If ω is the number required, n the total number of letters and P the given substitution, composed of f cycles of order a, g cycles of order b, h cycles of order c, etc., and r

* *Journal de l'Ecole polytechnique*, 10 (1815), p. 1.

the number of letters fixed in P, then

$$\omega = \frac{n!}{(f!)(g!)(h!)\cdots(r!)a^f b^g c^h \cdots}.$$

Among other theorems that he proves in this article are the following:

1. *If P is a substitution of order i, h any number and θ the highest common factor of h and i, then P^h is of order i/θ.*

2. *If P is a substitution of order i, the substitutions among the powers of P that are of order i are the powers of P whose indices are prime to i. These substitutions are likewise similar to P, and hence the number of substitutions similar to P among the powers of P are in number equal to the number of numbers less than i and prime to it.*

3. *Let P be any substitution, regular or irregular; let i be its order and p any prime factor of i. Then a value for i can always be found such that P^i is a substitution of order p.*

4. *A substitution and its inverse are always similar.*

5. *The powers of a cycle constitute the totality of substitutions that transform the given cycle into itself.*

6. *The order of a system of conjugate substitutions is divisible by the order of each substitution.*

7. *Two permutable substitutions with no common power but identity, together generate a system of conjugate substitutions whose order is the product of the orders of the two substitutions.*

8. *If $1, P_1, P_2, \cdots P_{a-1}$ and $1, Q_1, Q_2, \cdots Q_{b-1}$ are two groups, one of order a and the other of order b, which are permutable and have no common terms but identity, then the group generated by these groups is of order ab.*

9. *The converse of* (8).

10. *If P and Q are two substitutions, one of order a and the other of order b, and if the two series $Q, PQ, P^2Q, \cdots P^{a-1}Q$ and $Q, QP, QP^2, \cdots QP^{a-1}$ are made up of the same terms in the same or different order, then the cyclic group generated by P is permutable with the cyclic group generated by Q.*

A most important element in establishing Cauchy's claim as the founder of group theory is the proof which he gives of the fundamental theorem that if m is the order of a group and p any prime which divides m, there is at least one substitution of order p. Galois had stated the theorem but had not proved it. Its importance is due to the fact that it is the first step toward Sylow's theorem which appeared nearly thirty years later. In this memoir there are a number of specific groups mentioned although no enumeration of groups of special degrees is attempted. The substitutions of several groups are written out. Among them are (1) the octic, (2) the four-group, (3) the holomorph of the cyclic group of order five, (4) the intransitive group of degree 6 and order 9 and (5) the intransitive group of degree 6 and order 16. In addition to these groups the substitutions of which are given, several other groups are mentioned, among them (1) the holomorph of order

42, (2) the group of degree 7 and order 21, (3) the holomorph of the cyclic group of order 9.

There is also mentioned a group of degree 9 and order 27, generated by the two substitutions

$$P \equiv x_0x_3x_6 \cdot x_1x_4x_7 \cdot x_2x_5x_8 \qquad \text{and} \qquad Q \equiv x_1x_2x_4x_8x_5 \cdot x_3x_6.$$

This is obviously impossible but a little study of his method reveals that he should have used $Q \equiv x_1x_3x_4x_8x_7x_5 \cdot x_3x_6$, and that in order to get the order of the group he has incorrectly applied the theorems he was announcing and illustrating. The order of the group should be 18. The two theorems which he was illustrating may be stated as follows:

1. *If P is a substitution on n letters,* $P = x_0x_1x_2 \cdots x_{n-1}$, *and if r is any number prime to n and if Q is a substitution derived from P by replacing each letter by another whose subscript is r times its own, then for any values of h and k*

$$Q^kP^h = P^{r^kh}Q^k.$$

2. *If in the above theorem we take v any divisor of n, distinct from unity, and let R be any substitution that replaces any letter* x_l *by the letter with the subscript* $l+v$, *then* Q^kR^h *generates a group of order vi, where i is the smallest value of k that satisfies the congruence* $r^k \equiv i \pmod{n}$.

Since in the present instance r is 2, i must be 6. The order of the group is then $vi = 3 \cdot 6 = 18$, whereas Cauchy seems to take $i = 9$ and derives the order of the group $vi = 3 \cdot 9 = 27$.

The Memoirs of 1845–1846. Some of the material in the memoirs* published in the *Comptes Rendus* had already appeared in the *Exercices*, but is treated more extensively here. There is much, however, in the later articles that is not touched in the *Exercices*. It is in the later memoirs that he first defines what he means by a system of conjugate substitutions. The definition is as follows:

I shall call derived substitutions all that can arise from the given substitutions by multiplying them one or more times by each other or by themselves; and the given substitutions together with all the derived substitutions form what I shall call a system of conjugate substitutions.

Then follow the general theorems:

1. *If i is the order of a substitution P and* $a, b, c, \cdots$ *are the prime factors of i, then the substitution P and its powers form a group generated by the substitutions* $P^{i/a}, P^{i/b}, P^{i/c} \cdots$.

2. *If* $P, Q, R, S, \cdots$ *are the substitutions that leave a given function* Ω *invariant, they form a group whose order is the number of equal values of* Ω *when the variables are permuted in every possible way.*

---

* *Œuvres de A. Cauchy*, First series, volumes 9 and 10.

3. *The order of a transitive group of degree n is n times the order of the subgroup that leaves one letter fixed.*

4. *If Ω is a transitive function on n letters, if m is the index of the corresponding transitive group under the symmetric group of degree n, then m will likewise be the index of the subgroup that leaves one letter fixed under the symmetric group of degree $n-1$.*

Intransitive groups. The general subject which Cauchy treats next is that of intransitive groups. He implicitly divides intransitive groups into two classes. The first class includes those in which the systems are independent, that is, in our terminology, those in which the systems of transitivity are united by direct product. The second class includes those in which the systems are dependent, that is, those in which the systems are united by some sort of isomorphism. The concept of isomorphism is however only implied. He states first that an intransitive group is formed by the combination in some way of transitive constituents. He then develops some interesting formulas for the order of intransitive groups. If the systems are independent, then the order M is the direct product of $A, B, C, \cdots$ the orders of the transitive constituents and the index will be $n!/ABC\cdots$ where n is the total number of letters. This same formula holds also when the systems are not independent if $A, B, C\cdots$ are defined in a special way. These definitions are as follows:

Let A be the order of the first transitive constituent;

Let B be the number of substitutions involving letters in the second system without involving any of the first;

Let C be the number of substitutions involving letters in the third system, without involving any of the first or second, and so on.

With these definitions of $A, B, C, D\cdots$ the order of the intransitive group is $M = ABC\cdots$.

Imprimitive groups. Cauchy next turns to imprimitive groups. Here he confines himself almost entirely to the consideration of those imprimitive groups which are simply transitive and which have for the subgroup that leaves one letter fixed the direct product of the transitive constituents. In regard to this particular type of imprimitive groups, he gives several theorems which are of interest, even though he does not touch upon the broader and more general principles.

If in a simply transitive group the subgroup that leaves one letter fixed is an intransitive group formed by the direct product of its transitive constituents, the group is imprimitive. He considers the special cases $a > n/2$, $a < n/2$, $a = n/2$ in which a, the number of letters in the largest transitive constituent and n is the degree of the group.

The only theorems which he enunciates that apply to imprimitive groups in general are:

1. *The number of letters in the systems of imprimitivity must be a divisor of the degree.*

2. *If A is the number of substitutions that permute the variables within the systems and K the number of ways the k systems can be permuted, then the order of the group is KA^k.*

Symmetric groups. Cauchy then considers briefly symmetric groups. One theorem which he states with its corollaries is as follows:

If a transitive group of degree n has a symmetric subgroup of degree a, where $a > n/2$, then the group is symmetric on n letters.

For $n > 2$, a transitive group of degree n is symmetric if it contains a symmetric subgroup of degree $n-1$.

If $n > 3$, a transitive group of degree n is symmetric if it contains a symmetric subgroup of degree $n-2$. The special case $n=4$ is excepted; for the octic group, belonging to the function $\Omega = xy + zu$ is symmetric on two letters, yet not symmetric on all four.

If $n > 4$, a transitive group of degree n is symmetric if it contains a symmetric subgroup of degree $n-3$. The case $n=6$ is excluded. A group requiring this exception is the imprimitive group of degree 6 *and order* 72 *which contains the symmetric group of order* 6.

If $n > 5$, a transitive group of degree n is symmetric if it contains a symmetric subgroup of degree $n-4$. If $n=6$ or $n=8$, this does not hold. Cauchy gives as examples of the case $n=6$ the imprimitive groups of degree 6 *and orders* 72 *and* 48.

Some implied theorems. One of the most important and interesting parts of Cauchy's work on group theory is that in which he develops some complicated formulas from which we may easily deduce the theorem that the average number of letters in the substitutions of a transitive group is $n-1$. Cauchy himself does not enunciate the theorem, although he brings the proof to the point where only the final statement is necessary to complete it. The formula for the case where the group is simply transitive is as follows:

$M = 2H_{n-2} + 3H_{n-3} + \cdots + (n-2)H_2 + n$ *where M is the order of the group, n its degree and H_{n-r} the number of substitutions involving $n-r$ letters.*

Further development of these formulas leads to the two conclusions:

1. *If G is l-fold transitive of degree n, the number of substitutions involving $n-l+1$ letters is equal to or greater than $n(n-1)\cdots(n-l+2)/(n-l)l!$;*

2. *If G is simply transitive, of degree n, the number of substitutions involving n letters is equal to or greater than $n-1$.*

We find here the assertion that if in an l-fold transitive group $l+1$ letters are left fixed, all are left fixed. That this is false is demonstrated by considering the imprimitive group of degree 6 and order 72. It is simply transitive and hence $l=1$; but $l+1=2$ letters may be left fixed without all being so.

The theorem that the order of the holomorph of a cyclic group is the product of the orders of the cyclic group and its group of isomorphisms is found implicitly in Cauchy. He gives two theorems relative to this point.

1. *Let* $P = x_0x_1x_2\cdots x_{n-1}$ *be a substitution of order n. Let r be a primitive root of the modulus n, and I the smallest of the indices of unity belonging to the base r. Then let Q be the substitution that replaces* x_t *by* x_{rt}. *The order of Q will be I and the order of the group will be nI.*

This I is nothing else than the order of the group of isomorphisms of the cyclic group. If n is a power of a prime p, then $I = n(1 - 1/p)$. *When n is a prime then* $I = n - 1$.

2. *With the same hypothesis as in* (1), P^a *and* Q^b, *where a and b are divisors of n and I respectively, generate a group of order* nI/ab.

Enumeration of orders. Cauchy was the first to attempt an enumeration of the possible orders of groups. This he did with a fair degree of accuracy up to and including the sixth degree. The enumeration including degree 5 is correct and complete, but several errors occur in the enumeration of those of degree 6. For instance, 150 is given as the index of a group of degree 6 under the symmetric group, although factorial six is not a multiple of 150.

Cauchy goes back to his original distinction between imprimitive groups with heads which are direct products and those with heads formed by isomorphisms. He gives as the possible orders of groups of degree 6 with heads the direct products of the transitive constituents, 72, 48, 24, 18, 16, and 8. The last two numbers are clearly impossible for there is no transitive group of degree 6 and order 8 or 16. The orders of the groups possible when the head is formed with isomorphisms are given as 6, 12, 4, while there is no imprimitive group of degree 6 and order 4. According to his classification there should also be included in this last enumeration 24 and 18. A complete omission occurs in the list of imprimitive groups since the groups of order 36 with two systems of imprimitivity are not mentioned. Otherwise the enumeration through degree 6 is correct and complete.

Errors of Cauchy. The errors in Cauchy's work on group theory may be divided into two classes. The first and smaller class includes a number of serious errors in logic. The second class includes a large number of minor errors, some typographical, others arising through careless statement. All of the serious errors found have been noted in this paper with the exception of one to which attention has already been called.* He states an erroneous theorem on imprimitive groups in the following form:

If G_1 *is a subgroup composed of all the substitutions that omit a given letter in a simply transitive group then G is imprimitive unless all the transitive constituents of* G_1 *are of the same degree.*

Conclusion. In conclusion we may say that the foundation period in the history of group theory includes the time from Lagrange to Cauchy inclusive; that at the beginning of this period group theory was a means to an end and not an end in

* See G. A. Miller in *Bibliotheca Mathematica* (1910), series 3, vol. 10, p. 321.

itself. Lagrange and Ruffini thought of substitution groups only in so far as they led to practical results in the theory of equations. Galois, while broadening and deepening the application to the theory of equations may be considered as taking the initial step toward abstract group theory. In Cauchy while a group is still spoken of as the substitutions that leave a given function invariant, and the order of a group is still thought of as the number of equal values which the function can assume when the variables are permuted in every possible way, nevertheless quite as often a group is a system of conjugate substitutions and its relation to any function is entirely ignored.

In Cauchy's work it is to be noted that use is made of all the concepts originated by the earlier writers in substitution theory except those of Galois, a number of which would have proved powerful instruments in his hands, notably the idea of the invariant subgroup of which he makes no explicit use.

Because of the important theorems Cauchy proved, because of the break which he made in separating the theory of substitutions from the theory of equations and because of the importance that he attached to the theory itself, he deserves the credit as the founder of group theory.

ON THE HISTORY OF SEVERAL FUNDAMENTAL THEOREMS IN THE THEORY OF GROUPS OF FINITE ORDER*†

G. A. MILLER, Leland Stanford Jr. University, Palo Alto

The three most prominent sources of the theory of groups of finite order are: geometric transformations, theory of numbers, and theory of algebraic equations. The group properties of the totality of the rotations through a submultiple of 360°, or through 180° around three concurrent lines, each perpendicular to the plane determined by the other two, must have been observed very early. The prominent place which the five regular polyhedra—the Platonic bodies—occupy in the history of thought make it appear probable that the group properties of the totality of the rotations into themselves of these solids were observed quite early.‡

The elementary theory of congruences and especially the combinatory laws of the n roots of unity when they are multiplied together exhibit group properties which cannot have escaped the notice of the early students of these questions. As Cayley aptly says (*Philosophical Magazine*, 1854, page 40) "A number of elementary group concepts have been employed by mathematicians for a long time but have not been especially noted on account of their great simplicity." Some of the theorems by Gauss and Schering along this line are of sufficient interest in themselves to attract attention. In particular the theorem by Gauss to the effect that an Abelian group can be resolved in only one way into factor groups whose orders are prime to each other, and the theorem by Schering to the effect that every Abelian group can be resolved in more than one way into factor groups such that

---

* From AMERICAN MATHEMATICAL MONTHLY, vol. 8 (1901), pp. 213–216.

† Read before the American Association for the Advancement of Science, August, 1901.

‡ Bravais, *Liouville's Journal*, Vol. 14, 1849, page 167.

the order of each is divisible by the orders of all those which follow it, have a strong claim to being considered the beginning of the theory of abstract groups of finite order.

The early developments in substitution groups were made in connection with the theory of algebraic equations, starting with the theories of Lagrange and becoming prominent in the development and extension of these theories by Ruffini, Abel, and Galois.* Cauchy seems to have been the first to develop the theory of substitution groups independently of any direct application to the theory of equations, and he was also the first to prove any remarkable theorems in this theory. The earlier theorems were almost self-evident, but this cannot be said of the theorem that a group must involve a substitution of a prime order (p) whenever its order is divisible by p.† This theorem, known in the theory of groups as Cauchy's theorem, is especially interesting since it is independent of the particular notation by means of which a group is represented and hence applies directly to the theory of abstract groups.

In proving this theorem, Cauchy employed another interesting theorem, due to himself, viz., that the symmetric group of degree n contains a subgroup of order p^m, p^m being the highest power of p that divides $n!$ About thirty years later Sylow extended very materially these results due to Cauchy by proving that a group (G) of order g must always contain $1+kp$ subgroups of order p^a, p^a being the highest power of p that divides g, and that all of these subgroups are conjugate under G. This fundamental theorem is known as *Sylow's theorem*. It is sometimes called the Cauchy-Sylow theorem, and was first published in the *Mathematische Annalen*, Vol. 5.

About twenty years later Frobenius published, in the *Berliner Sitzungsberichte*, a very important extension of Sylow's theorem in which it is proved that the number of subgroups of order p^a which are contained in any group is $\equiv 1 \bmod p$. These subgroups do not necessarily form a single set of conjugates unless p^a is the highest power of p that is contained in the order of the group. Frobenius deserves credit also for a simple proof of Sylow's theorem by means of a more prominent use of the concept of complete sets of conjugates. This concept is prominent in a large number of the publications by Frobenius.

From what precedes it may be observed that a very important part of Sylow's theorem was constructed on French soil. A Norwegian greatly enlarged and beautified this structure. Finally a German enhanced its usefulness by extending it very materially in certain directions. Hence this fundamental theorem stands before us as a truly international structure, which has required more than fifty years for its erection. One of its main contributors (Sylow) has published comparatively little along the line of group theory, while the other two have published extensively along this line.

The article in which Sylow made his famous theorem known contains another

* Cauchy, *Physique Mathématique*, Vol. 3, 1844, page 250.

† Pierpont, *Bulletin of the American Mathematical Society*, Vol. 1, 1895, page 196.

of not much less importance, viz., a group of order p^a contains at least p invariant operators. This furnishes the starting point of a large number of the theorems relating to groups whose order is a power of a prime. On account of its fundamental importance it has been proved in several different ways. The most recent of these proofs is deduced from the known fact that the subgroup which omits a given element of a transitive substitution group of order p^m omits a power of p of its elements.*

Closely related to this theorem is the question in regard to the largest Abelian subgroup in a group of order p^a. The first step towards a theorem along this line was the observation that every group of order 16 contains an Abelian subgroup of order 8. A little later it was observed (Comptes Rendus, February, 1896) that every group of order p^4 contains an Abelian subgroup of order p^3. This fact was then proved independently of the list of all the possible groups of order p^4.† Finally, it was proved in the *Messenger of Mathematics*, 1897, that every group of order p^a contains an Abelian group of order p^m, whenever

$$\alpha > \frac{m(m-1)}{2}.$$

A more special theorem which has a somewhat singular history is the one which states that there are just fifty-one groups of order 32. In 1896 a Frenchman stated in *Comptes Rendus de l'Académie des Sciences* that he had found seventy-five groups of this order and had not yet finished his enumeration. Shortly after this an American stated in the same journal that he also had investigated this problem and that he had proved that there are only fifty-one such groups.

About two years later an Italian published in *Annali di Matematica* his investigations in regard to all the groups of order p^5 and stated therein that both of these results were incorrect inasmuch as the correct number of these groups was just 50. Very shortly after this the said American reaffirmed his former results and called attention to several errors in the enumeration of the Italian. Finally, the latter published a separate investigation of this subject in which he agreed with the American and stated that he considered the determination of all the groups of order 32 settled beyond a doubt.

In tracing the history of a concept or theorem the most difficult part is that in which the concept seems implied, for one is always in danger of reading things into a paper, which the author did not have in mind. This is perhaps especially true of the concepts of isomorphisms and group of isomorphisms. Both of these are implied in the theory of intransitive substitution groups, but it is difficult to say whether any of the earliest workers along this line had them distinctly in mind.

It appears that Hölder and Moore were the first to call explicit attention to the fact that the totality of the simple isomorphisms of a group with itself constitute a group. Hölder remarked that the group of cogredient isomorphisms is an invariant

* *American Journal of Mathematics*, Vol. 23, page 173.

† *Quarterly Journal of Mathematics*, Vol. 28, page 233.

subgroup of the group of isomorphisms. Recently the group of isomorphisms of a group (G) has been studied from various standpoints. It has been observed that it is the largest subgroup of the holomorph of G, which does not include one of the elements. When G is cyclic it is Abelian, but when G is any non-cyclic Abelian group its group of isomorphisms is non-Abelian. In the latter case its invariant operators are those which transform each of the operators of G into the same power.

The second volume of Weber's Algebra is one of the best and most popular works on the theory of groups of finite order. This may perhaps justify our noting a very singular error which occurs on page 54 of the first edition. The author states at this place that the natural numbers when combined by multiplication furnish the most important example of a commutative group. It is very easy to see that these numbers, combined in the said manner, do not form any group at all. So that what is called the most important example has no existence.

In his second edition Weber recognizes this fact and replaces the given example by another "most important example." It would probably be difficult to prove that numbers combined by multiplication furnish a more important example of Abelian groups than when they are combined by addition. The greatness of this work may perhaps justify the noting of another slight defect, viz., in the treatment of the new subject of commutator groups, the author gives reference to the man who first published the name of these groups but he does not give any reference to the one who first published their properties.

BIBLIOGRAPHIC ENTRY: HISTORY

1. G. A. Miller, The founder of group theory, AMERICAN MATHEMATICAL MONTHLY, 1910, 162–165.

Discusses which mathematician can be called the founder of group theory.

(h)

APPLICATIONS

CAMPANOLOGICAL GROUPS*

T. J. FLETCHER, Sir John Cass College

The art of change ringing has always been considered highly mathematical, but little has been said about the precise mathematical notions involved. This may be because the art has been restricted almost entirely to English-speaking

* From AMERICAN MATHEMATICAL MONTHLY, vol. 63 (1956), pp. 619–626.

countries, and so has not attracted the attentions of algebraists on the continent of Europe. This paper contains examples of some of the methods used by bell-ringers, and shows that they are decompositions of a symmetric group into Lagrange cosets. In consequence they provide useful practical illustrations of a topic which most textbooks treat in an entirely abstract fashion. We also see that a knowledge of the elementary notions of group theory might have enabled campanologists to construct peals without the tedious empirical methods which seem to have been the practice in the past.

The problem which a team of ringers undertakes is to produce all the permutations (changes) on a set of bells, proceeding according to certain rules. The highest bell, number one, is called the *Treble*, and the lowest is called the *Tenor*. When the bells ring down the scale, from Treble to Tenor, they are said to be in *rounds*. The rules to which a peal must conform are:

i) the peal must begin and end in rounds,

ii) no bell may move more than one place between successive changes,

iii) no bell may lie still for more than two successive changes.

(The last rule is relaxed occasionally.)

A peal may be composed of any number of bells up to twelve, and as it takes about 28 hours to ring all the changes on eight bells alone, the ringers cannot hope to ring all the possible changes if the number of bells is larger. Then they restrict attention to ringing those changes which are regarded as the most musical.

The six changes on three bells may be rung as follows:

1 2 3
2 1 3
2 3 1
3 2 1
3 1 2
1 3 2
1 2 3

They could equally well be rung in the reverse order, but only these two ways conform to the rules. The course of the Treble is particularly simple—it *hunts up* and then *hunts down*. The other bells follow a similar *hunting course* but naturally the phase is different.

Notation. We will denote changes as ringers do by numbering the bells from Treble to Tenor and giving the order in which they ring in the change. The symbols which result could be regarded as the elements of a symmetric group, but it is more strictly logical to regard the transitions between changes as constituting the group, and we will denote these by the familiar cycle notation. Thus the transition

$$\begin{bmatrix} 213654 \\ 514263 \end{bmatrix}$$

will be denoted by the symbol (145)(36). (Rule ii of course prohibits the carrying out of this transition in one move.) The notation means that the bell in first place moves to fourth place, and the bell in fourth place moves to fifth place, and the bell in fifth place moves to first place; while the bells in third and sixth place change over. The cycle symbol is an operator which transforms the first change to the second, and such operators multiply together according to well-known rules. We will use capital letters to denote these transitions. Initially they will denote transitions between adjacent changes, but later on they will be used to denote transitions from one change to another which may occur elsewhere in the peal. Because of Rule ii, capital letters denoting transitions between adjacent changes are operators of period two. That is to say they are their own inverses. It is important to avoid confusing the change 1342 with the change generated by applying the operator (1342) to rounds, because this is 2413.

Singles. A peal on four bells is called *Singles*. If four bells hunt they produce the eight changes in the first block of Table I. A further move of a similar kind would restore rounds, so, in ringers' language, "the bell that the treble takes from lead makes second place and leads again; and the other bells dodge at the back-stroke lead of the treble." The mysteries of "hand-stroke" and "back-stroke" do not concern mathematicians; they arise because a bell swings two ways and successive strokes have to be rung with a different action. The meaning of "dodge" should be sufficiently clear from the table.

TABLE I

SINGLES		
1 2 3 4	1 3 4 2	1 4 2 3
2 1 4 3	3 1 2 4	4 1 3 2
2 4 1 3	3 2 1 4	4 3 1 2
4 2 3 1	2 3 4 1	3 4 2 1
4 3 2 1	2 4 3 1	3 2 4 1
3 4 1 2	4 2 1 3	2 3 1 4
3 1 4 2	4 1 2 3	2 1 3 4
1 3 2 4	1 4 3 2	1 2 4 3

The process of hunting consists of employing the two operators $A=(12)(34)$ and $B=(23)$ alternately. These generate a group of order eight, the first eight changes. In general hunting on n bells generates a group of order $2n$. $AB=(1342)$ and this element has period four. It is convenient to call these eight elements the *hunting sub-group* and to denote it by $\mathcal{H}$. The last of these changes is 1324, and the peal continues by following this with the irregular move $C=(34)$, instead of employing B which would produce rounds. The second block therefore consists of the hunting sub-group premultiplied by $B^{-1}C=(243)$. This can be denoted by $(243)\mathcal{H}$. After it has been rung the irregular move C has to be performed again, and the third block consists of $(243)^2\mathcal{H}$. When the last change is followed once more by C we return to rounds because $(243)^3=I$.

This method of ringing therefore displays the decomposition into cosets:

$$\mathcal{G} = \mathcal{H} + (243)\mathcal{H} + (243)^2\mathcal{H}.$$

Other methods are possible on four bells but their structure is not quite so clear as in this example.

Doubles. A peal on five bells is called *Doubles.* For five bells or more it is usually necessary to have a conductor who calls *bobs* and *singles* as required. That is to say he does the multiplications which change from one coset to the next—or to put the matter more correctly he does some of them, because the ringers engaged on any particular peal proceed under a set of standing orders which ensure that a number of the transitions from one coset to the next are done automatically. The first five bell method which we will consider is called *Grandsire Doubles.* The hunting sub-group of ten changes is generated by the two operations $A=(23)(45)$ and $B=(12)(34)$. In Grandsire Doubles the number of changes which the ringers can ring without further instructions is increased to 30 by using the further operation $C=(12)(45)$. In practice this irregular move is performed at the very beginning, before hunting commences; and the peal begins as shown in Table II.

TABLE II

GRANDSIRE DOUBLES

1 2 3 4 5		
2 1 3 5 4	2 1 5 4 3	2 1 4 3 5
2 3 1 4 5	2 5 1 3 4	2 4 1 5 3
3 2 4 1 5	5 2 3 1 4	4 2 5 1 3
3 4 2 5 1	5 3 2 4 1	4 5 2 3 1
4 3 5 2 1	3 5 4 2 1	5 4 3 2 1
4 5 3 1 2	3 4 5 1 2	5 3 4 1 2
5 4 1 3 2	4 3 1 5 2	3 5 1 4 2
5 1 4 2 3	4 1 3 2 5	3 1 5 2 4
1 5 2 4 3	1 4 2 3 5	1 3 2 5 4
1 2 5 3 4	1 2 4 5 3	1 2 3 4 5

These 30 changes are called *plain course.* It is easy to see why the ringers prefer to begin with the operation C. If this is done the return of the Treble to the lead can be taken as the cue to perform the irregular operation C again, whereas if they commence with hunting no such clear indication is given them. It is now necessary to introduce some means whereby the ringers may pass from one plain course to another; and to do this the conductor calls bobs and singles. A *single* is a change which interrupts the regular work of an even number of bells, and a *bob* is a change which interrupts the regular work of an odd number. The distinction is again not one which a mathematician would have chosen. In some cases a bob or a single alters the parity of the changes which are being rung, whereas the other does not, but this is not the case in general.

Grandsire uses the single

$$\begin{bmatrix} 15432 \\ 15423 \end{bmatrix} = (45),$$

and Rule iii is relaxed at this point in this method of ringing. To ring a full peal of Grandsire, four bobs and two singles are necessary. Grandsire Doubles is a decomposition of the symmetric group on five symbols into twelve cosets of ten, masked by a pre-multiplication by C. Without the pre-multiplication plain course may be written:

$$\mathfrak{H} + (354)\mathfrak{H} + (354)^2\mathfrak{H}.$$

$(354) = B^{-1}C$ and is obviously of period three. This leads one to suspect that it is possible to generate a longer plain course by employing some move different from C. This can be done, and Plain Bob employs $D = (23)$ instead of C. $B^{-1}D = (1342)$, and so plain course consists of 40 changes, and the entire peal can be rung with fewer calls.

One of the most pleasing methods from a mathematical point of view is *Stedman's Doubles*. This method was invented round about 1640, and it displays a striking knowledge of decomposition into cosets very nearly one hundred years before Lagrange was born. Troyte [1] describes the method as follows: "Three bells go through the three bell changes while the other bells dodge behind; at the completion of each six changes one bell coming down from behind to take its part in the changes, and one going up behind to take its part in the dodging." In the Grandsire the work of the Treble is different from that of the other bells, as he pursues a hunting course the whole of the time, whereas the other bells modify their hunting course now and again. In Stedman's method the work of every bell is the same, and the symmetric group is decomposed into cosets of six. By tradition the method begins in the middle of a set, so the decomposition into cosets is pre-multiplied by three operations, or alternatively it is carried out from the reference order 32415 instead of 12345. The peal is shown in Table III. It will be noticed that the sets of six changes are rung as in our very first example, but taken forwards and backwards alternately.

The method requires two singles, and in this case they change the parity of the permutations being rung. The singles are indicated on the table by the letter S. They interchange some of the members of the cosets in which they occur, and they also decompose the whole group in another way—into two cosets of 60. All the members of the alternating sub-group occur in the peal before and after the singles, while the members of the other coset occur between them.

TABLE III

STEDMAN'S DOUBLES

1 2 3 4 5			
2 1 3 5 4			
2 3 1 4 5			
3 2 4 1 5	5 2 3 4 1	3 1 4 2 5	5 1 3 4 2
2 3 4 5 1	5 3 2 · ·	1 3 4 · ·	5 3 1 · ·
2 4 3 1 5	3 5 2 · ·	1 4 3 · ·	3 5 1 · ·
4 2 3 5 1	3 2 5 · ·	4 1 3 · ·	3 1 5 · ·
4 3 2 1 5	2 3 5 · ·	4 3 1 · ·	1 3 5 · ·
3 4 2 5 1	2 5 3 1 4	3 4 1 5 2	1 5 3 2 4
4 3 5 2 1	5 2 1 3 4	4 3 5 1 2	5 1 2 3 4
S 4 5 3 1 2	2 5 1 · ·	*S* 4 5 3 2 1	1 5 2 · ·
5 4 3 1 2	2 1 5 · ·	5 4 3 2 1	1 2 5 · ·
5 3 4 2 1	1 2 5 · ·	5 3 4 1 2	2 1 5 · ·
3 5 4 1 2	1 5 2 · ·	3 5 4 2 1	2 5 1 · ·
3 4 5 2 1	5 1 2 4 3	3 4 5 1 2	5 2 1 4 3
4 3 2 5 1	1 5 4 2 3	4 3 1 5 2	2 5 4 1 3
3 4 2 1 5	1 4 5 · ·	3 4 1 · ·	2 4 5 · ·
3 2 4 5 1	4 1 5 · ·	3 1 4 · ·	4 2 5 · ·
2 3 4 1 5	4 5 1 · ·	1 3 4 · ·	4 5 2 · ·
2 4 3 5 1	5 4 1 · ·	1 4 3 · ·	5 4 2 · ·
4 2 3 1 5	5 1 4 3 2	4 1 3 2 5	5 2 4 3 1
2 4 1 3 5	1 5 3 4 2	1 4 2 3 5	2 5 3 4 1
2 1 4 5 3	5 1 3 · ·	1 2 4 · ·	5 2 3 · ·
1 2 4 3 5	5 3 1 · ·	2 1 4 · ·	5 3 2 · ·
1 4 2 5 3	3 5 1 · ·	2 4 1 · ·	3 5 2 · ·
4 1 2 3 5	3 1 5 · ·	4 2 1 · ·	3 2 5 · ·
4 2 1 5 3	1 3 5 2 4	4 1 2 5 3	2 3 5 1 4
2 4 5 1 3	3 1 2 5 4	1 4 5 2 3	3 2 1 5 4
4 2 5 3 1	3 2 1 · ·	4 1 5 · ·	3 1 2 · ·
4 5 2 1 3	2 3 1 · ·	4 5 1 · ·	1 3 2 · ·
5 4 2 3 1	2 1 3 · ·	5 4 1 · ·	1 2 3 4 5
5 2 4 1 3	1 2 3 · ·	5 1 4 · ·	
2 5 4 3 1	1 3 2 4 5	1 5 4 3 2	

The number of changes generated by some particular method can be calculated quite easily. Consider for example *Plain Bob Major*. (A peal on eight bells is called *Major*.) The complete 40,320 changes on eight bells were first rung at Leeds, Kent, in 1761, by relays of men who took 27 hours to complete the task. Plain Bob Major generates the hunting sub-group by means of $A=(12)(34)(56)(78)$ and $B=(23)(45)(67)$, and when the Treble returns to the lead, rounds are avoided by using $C=(34)(56)(78)$ which means the second coset commences with $B^{-1}C=(2468753)$ which is of period 7. Plain course therefore consists of 7 cosets of 16, which is 116 changes. This method uses the bob $(23)(56)(78)$ and the single (56)(78), and it is interesting to work out how many more changes these will generate and where they must be used.

Treble Bob. The *Treble Bob* methods of ringing are more complicated than the plain methods because the bells do not follow a simple hunting course, but dodge on their way up and down. This means that a "lead" (the set of changes between successive appearances of the Treble as the first bell) is twice as long as in the plain method, and that the changes of a lead do not possess the group property, being usually a complex generated by three operations employed according to some scheme such as $ABACABAC \cdots$. Even here group properties are very apparent in the tables which are given in books on campanology. The move at the end of a lead in Treble Bob is arranged so that when the Treble returns to the front the other $n-1$ bells are cyclically permuted; in this way plain course consists of as many leads as possible.

We have so far pointed out only that the three blocks of Table I are a decomposition into left-hand cosets. If we read the table by lines we see a decomposition into right-hand cosets using a sub-group of order three. This can also be seen in Table II although it is a little obscured by the irregular way in which the peal begins, and it is a general feature of Plain and Treble Bob methods.

As we have indicated earlier most of the methods of campanology are empirical, *ad hoc* methods devised some hundreds of years ago long before the invention of formal group theory, but the subject has not been without its great theorists. One of the finest of these was W. H. Thompson, who was not a practicing ringer but a civil servant in India who made a hobby of solving change ringing problems. As far as can be ascertained he was not a mathematician and his writings [**3**] do not suggest that he realized that the calculations which he was performing were in fact pieces of group theory. He was concerned in particular with the following problem: "Is it possible to ring all the changes on seven bells (Triples) using the Grandsire method with only plain and bob leads?" He devised his own notation for this problem, and it bears striking analogies to the notation invented by the celebrated Irish mathematician Hamilton for another problem [**4**] a few years earlier.

Because the work in each lead is perfectly regular (being the hunting subgroup), the analysis of a complicated peal is greatly simplified by considering only the "lead-ends." Putting this in a way which seems more natural to mathematicians we need only consider the first member of each coset. The effect of a plain lead in Grandsire Triples is to send bells which have just rung a lead starting with rounds into the next lead in the order 1253746. We may therefore write $P=(34675)$. The effect of a bob lead is to send bells which have just rung a lead starting with rounds into the next lead in the order 1752634, and so we put $B=(247)(365)$.

There are 7! changes in Grandsire Triples and each lead contains 14. The total number of leads is therefore $6!/2$ and the initial members of the cosets form the alternating group on six symbols. Denoting these six symbols by 2, 3, 4,

5, 6, 7 (because bell number 1 is always in front at the beginning and end of a lead) the question which Thompson was discussing was: "Will P and B generate the alternating group on six symbols, and furthermore generate it unicursally?"

The answer was "No," but the interest is in Thompson's methods. Using the notion of lead-ends he drew up a chart in such a way that the unicursal generation of the group became a matter of the unicursal description of a polytope whose edges represented the leads of the peal. The problem is remarkably like the problem of the game of visiting the vertices of an icosahedron which Hamilton had discussed a few years before [4]. Hamilton's problem is by far the simpler of the two, and it is not exactly the same as he is concerned with visiting the vertices of the polytope and Thompson is concerned with traversing the edges, but the two employ an algebraic notation which is almost identical. The solution of Hamilton's game presumably gives a peal on six bells (or rather a *touch*, because the 20 vertices of the icosahedron correspond to only 40 leads, whereas 60 leads are necessary to ring a full peal in Minor).

After constructing the polytope Thompson argues as follows. If the result of following x with a plain ending is the same as the result of following y with a bob, then clearly the leads x and y must be treated in the same way, that is both must be plain or both must be bob. Now if $xP=yB$ then $y=xPB^{-1}$. Therefore if x is plain or bob, $x(PB^{-1})$ must be the same. But the operator PB^{-1} is of period five, and so all the leads may be split up into sets of five which must all be treated the same way. Thompson called these Q-sets, and they are of course cosets in the strict meaning of the term.

If all the 360 leads are plain they are split up into 72 *round blocks* since P is of period five. On the other hand if they are all bobs they are split up into 120 round blocks since B is of period three. The peal can be rung if we may start at one of these extremes and by bobbing or plaining suitable Q-sets convert the whole 360 leads into one round block. This would be denoted by a chain of letters of the form *PPBPBPPPB etc*, where the whole product was equal to the identity but no sub-section was. Thompson showed this to be impossible by proving that bobbing or plaining any Q-set always results in the loss or gain of an even number of round blocks. Hence the number of round blocks, which is 72 or 120 to start with, can never be reduced to one. The beauty of the proof is marred by the fact that the stage showing that the number of round blocks lost or gained is always even is carried out by a long and tedious process of enumeration. But it is very difficult to see any means by which this could have been avoided. The enumeration of the cosets of a group of large order is inevitably tedious, and modern processes do not seem to offer any way of reducing Thompson's labours to any marked extent.

The study of campanology therefore shows how bell-ringers acquired a considerable amount of empirical knowledge about groups long before the theoretical basis was established by mathematicians, and it also provides instructive examples of the abstract theorems. The main problem of campanology is a problem of Group Theory to which no practical solution seems to have been found even yet: "Given a particular set of elements of a group, will they generate the whole group, and will they generate it unicursally?"

Bibliography

1. Troyte, C. A. W. Change Ringing, London, 1869.
2. Morris, E. History and Art of Change Ringing, 1931.
3. Thompson, W. H. A Note on Grandsire Triples, London, 1886. (This is reprinted in Grandsire, J. W. Snowden, London, 1905.)
4. Hamilton, W. R. Quarterly Journal of Mathematics, London, 1862, vol. V, p. 305; or Philosophical Magazine, January, 1884, series 5, vol. XVII, p. 42. (This is also described in W. W. Rouse Ball, Mathematical Recreations and Essays, Eleventh Edition, London, 1939.)

THE GROUP GENERATED BY CENTRAL SYMMETRIES, WITH APPLICATIONS TO POLYGONS*

EDWARD KASNER, Columbia University

The object of this article is to generalize the following well known theorem of elementary geometry: If the mid-points of the consecutive sides of any quadrilateral are joined the resulting figure is a parallelogram. In the case of a triangle the corresponding construction gives a triangle not having any peculiar property. The question therefore arises as to whether or not in case of polygons of more than four sides there is any theorem analogous to that concerning the quadrilateral. It will be shown that, in this respect, there is an essential distinction between polygons of an even number of sides, and those of an odd number of sides. This depends fundamentally upon the character of the group which is discussed in § 1.

§ 1. The Group.

1. Any translation T of the plane may be written

$$(T) \qquad \begin{aligned} x' &= x + h \\ y' &= y + k, \end{aligned}$$

where h and k are the components, in the direction of the coördinate axes, of the vector corresponding to the translation. It is obvious that the totality of translations form a group; for the combination of any two, say T_1 and T_2 whose vector components are h_1, k_1 and h_2, k_2 respectively, gives

$$\begin{aligned} x' &= x + h_1 + h_2 \\ y' &= y + k_1 + k_2, \end{aligned}$$

which is itself a translation.

2. Consider now the transformation termed *central or point symmetries*. Such a

* From American Mathematical Monthly, vol. 10 (1903), pp. 57–63.

symmetry is defined by

$$(S) \qquad \begin{aligned} x' &= -x+2a \\ y' &= -y+2b, \end{aligned}$$

where a, b are the coördinates of the center of the symmetry, *i.e.*, the fixed point P with respect to which corresponding points x, y and x', y' are symmetric. The symmetries themselves do not form a group, but we now show that

The translation T and the central symmetries S form a group.

In the first place, *the product of two symmetries is a translation*. For, if the center of S_1 is P_1, with coördinates (a_1, b_1), and if the center of S_2 is P_2, with coördinates (a_2, b_2), the combination S_1S_2 gives

$$\begin{aligned} x' &= x+2(a_2-a_1) \\ y' &= y+2(b_2-b_1). \end{aligned}$$

The vector of this translation is double the vector P_1P_2. Similarly, the product S_2S_1 is the translation whose vector is twice P_2P_1. In the second place, *the product of a symmetry and a translation is a symmetry*. For, the transformation ST is

$$\begin{aligned} x' &= -x+2a+h \\ y' &= -y+2b+k; \end{aligned}$$

this is the symmetry whose center is obtained from the center of S by applying the vector of T. Similarly, the product in the reverse order, *i.e.*, TS is the symmetry whose center is obtained from the center of S by applying the vector opposite to that of T.

It follows then that any combination of transformations T and S is itself either a T or an S, so that the group property is proved. In Lie's terminology the group considered is a mixed two-parameter group consisting of two continuous systems of transformations. The translations constitute a self-conjugate sub-group.

3. Since the product of two symmetries is a translation, and since the translations constitute a group, it follows that *the product of an even number of symmetries is a translation*. The product of the $2k$ symmetries $S_1, S_2, \ldots, S_{2k}$ is in fact

$$\begin{aligned} x' &= x+2(a_{2k}-a_{2k-1}+\cdots+a_2-a_1) \\ y' &= y+2(b_{2k}-b_{2k-1}+\cdots+b_2-b_1), \end{aligned}$$

where a_i, b_i denote the coördinates of P_i the center of S_i. The formulae may be interpreted geometrically by observing that the differences a_2-a_1, b_2-b_1, for example, are the components of the vector P_1P_2; therefore *the vector of the resulting translation is twice the vector sum*

$$P_1P_2+P_3P_4+\cdots+P_{2k-1}P_{2k}.$$

4. An odd number of symmetries may be combined by combining the first with the product of all the remaining, which by 3 is a translation. Therefore, *the product of an odd number of symmetries is a symmetry*. If the symmetries are $S_1, S_2, \ldots, S_{2k+1}$

their product is

$$x' = -x + 2(a_{2k+1} - a_{2k} + \cdots + a_3 - a_2 + a_1)$$
$$y' = -y + 2(b_{2k+1} - b_{2k} + \cdots + b_3 - b_2 + b_1).$$

The center of the resulting symmetry is obtained from the center P of the first symmetry by applying the vector sum

$$P_2P_3 + P_4P_5 + \cdots + P_{2k}P_{2k+1}.$$

5. The application to be made depends essentially upon the *fixed points* of the transformations, *i.e.*, the points which are transformed into themselves. Excluding points at infinity from consideration, we observe in the first place that in case of a symmetry there is one and but one fixed point, namely, the center of the symmetry. On the other hand, in case of a translation, there are no fixed points, except when the translation reduces to the identical transformation, in which case all the points of the plane are fixed.

§ 2. Mid-Point polygons.

6. Consider any polygon whose vertices in order may be denoted by $Q_1, Q_2, \ldots, Q_n$. If the middle points of the sides are connected in order we derive a new polygon of the same number of sides which for brevity may be termed the *inscribed polygon*; the original polygon in its relation to the derived is then termed the *circumscribed polygon*. For every polygon there is then a definite inscribed polygon. The question now to be considered concerns the converse problem: *Given an arbitrary polygon $P_1, \ldots, P_n$, is it possible to construct a circumscribed polygon, i.e.*, is it possible to find n points $Q_1, Q_2, \ldots, Q_n$ such that P_1 shall be mid-way between Q_1 and Q_2, P_2 mid-way between Q_2 and Q_3, and so on until finally P_n shall be mid-way between P_n and P_1?

To answer this question, take tentatively any point Q in the plane; construct with respect to P_1 the symmetric point; then with respect to P_2 construct the point symmetric to the one just obtained; and so in order until finally by symmetry with respect to P_n a point Q' is obtained. The original polygon $P_1, \ldots, P_n$ thus defines a definite transformation by which to any point Q corresponds an unique point Q'. This transformation is simply the product of n symmetries and therefore, by the previous section, is either a translation or a symmetry according as n is even or odd.

7. Consider first the case of a polygon with an odd number of sides $n = 2k+1$. The transformation from Q to Q' is then a symmetry. There is, therefore, by § 5 a single point which remains invariant under the transformation. Denoting this point by Q_1, we obtain from it, by successive symmetry with respect to $P_1, P_2, \ldots, P_{2k}$, the points $Q_2, Q_3, \ldots, Q_{2k+1}$; Q_{2k+1} and Q_1 are then necessarily symmetric with respect to the last vertex P_{2k+1}, so that $Q_1, \ldots, Q_{2k+1}$ are in fact the vertices of the unique circumscribed polygon.

Any polygon with an odd number of sides can be obtained as an inscribed polygon; there exists one and only one circumscribed polygon.

The circumscribed polygon may be constructed by applying the result stated at the end of 4. The first vertex Q_1 is obtained from the first vertex P_1 of the original polygon by applying the vector sum $P_2P_3 + P_4P_5 + \cdots + P_{2k}P_{2k+1}$; then the remaining vertices $Q_2, \ldots, Q_{2k+1}$, are obtained by successive symmetry as described above.

8. If the polygon has an even number of sides $n = 2k$, then from 3 the transformation from Q to Q' is a translation which in general does not reduce to the identical transformation. In this case, from 5, there exists no fixed point, and therefore no circumscribed polygon.

In the case of an arbitrary polygon of an even number of sides no circumscribed polygon exists, i.e., not all such polygons can be obtained as inscribed polygons.

9. The construction will, however, be possible in the exceptional case where the resulting translation reduces to identity. If then we term a $2k$-gon *special* when it is possible to circumscribe a polygon about it, the result may be stated:

Any special $2k$-gon is characterized by the fact that the product of the symmetries having for centers the vertices of the polygon is identity.

The class of polygons considered may be defined otherwise as follows: From the formulae in 3, we have as the conditions for reducing to identity,

$$a_{2k} - a_{2k-1} + \cdots + a_2 - a_1 = 0$$
$$b_{2k} - b_{2k-1} + \cdots + b_2 - b_1 = 0;$$

which together express the vanishing of the vector sum

$$P_1P_2 + P_3P_4 + \cdots + P_{2k-1}P_{2k}.$$

The vanishing of this sum necessitates the vanishing of

$$P_2P_3 + P_4P_5 + \cdots + P_{2k}P_1,$$

since for any polygon the vector sum of all the sides is zero. Therefore

In any special $2k$-gon the vector sum of the alternate sides vanish; this condition is also sufficient.

The equation of condition above may also be written

$$\frac{a_1 + a_3 + \cdots + a_{2k-1}}{k} = \frac{a_2 + a_4 + \cdots + a_{2k}}{k}$$
$$\frac{b_1 + b_3 + \cdots + b_{2k-1}}{k} = \frac{b_2 + b_4 + \cdots + b_{2k}}{k},$$

which may be interpreted as follows:

In any special $2k$-gon with vertices $P_1, P_2, \ldots, P_{2k}$, the mean point (or center of gravity) of the alternate vertices $P_1, P_3, \ldots, P_{2k-1}$ coincides with the mean point of the remaining vertices $P_2, P_4, \ldots, P_{2k}$.

Both points obviously coincide with the mean point of all the vertices of the $2k$-gon.

10. For a special $2k$-gon the transformation from Q to Q' described in 6 reduces to identity, so that every point of the plane is an invariant point. Therefore in constructing the circumscribed polygon any point may be assumed for the first vertex Q_1, the other vertices then being determined by successive symmetry with respect to $P_1, P_2, \ldots, P_{2k-1}$.

About a special $2k$-gon a double-infinity of circumscribed $2k$-gons may be constructed; otherwise stated, if it is possible to circumscribe one polygon about a given $2k$-gon, it is possible to circumscribe a double infinity.

We shall now prove that among this double infinity of $2k$-gons there is one which is itself special, so that *about any special $2k$-gon it is possible to circumscribe one and only one special $2k$-gon.* Let the first vertex Q_1 of any circumscribed polygon be x, y; then the next vertex Q_2, obtained by symmetry with respect to P_1, is $-x+2a_1$, $-y+2b_1$; similarly Q_2 is $x-2a_1+2a_2$, $y-2b_1+2b_2$; finally, Q_{2k} is $-x+2a_1-2a_2+\cdots+2a_{2k-1}$, $-y+2b_1-2b_2+\cdots+2b_{2k-1}$. If now the circumscribed polygon is to be special we must have

$$Q_1Q_2+Q_3Q_4+\cdots+Q_{2k-1}Q_{2k}=0,$$

which is equivalent to

$$kx-(2k-1)a_1+(2k-2)a_2-\cdots-a_{2k-1}=0$$
$$ky-(2k-1)b_1+(2k-2)b_2-\cdots-b_{2k-1}=0.$$

These equations determine x and y, that is, the first vertex Q_1, uniquely, which proves the theorem announced.

The special quadrilaterals are simply parellelograms. About any parallelogram a double infinity of quadrilaterals may be circumscribed, of which one is itself a parallelogram; about this in turn a parallelogram may be circumscribed and so on indefinitely. So for any special $2k$-gon, one can not only inscribe special $2k$-gons indefinitely, but also circumscribe them. Again, just as the quadrilateral inscribed in a general parallelogram is itself an arbitrary parallelogram, so the $2k$-gon inscribed in a special $2k$-gon is not further specialized, but is an arbitrary special $2k$-gon.

11. The second characteristic given in 9 gives the following *construction for special $2k$-gons*. Take an arbitrary k-gon $D', D'', \ldots, D^{(k)}$; on each side construct a parallelogram, thus on $D'D''$ construct $P_1D_1D_2P_2$, on $D''D'''$ construct $P_3D''D'''P_4$, finally construct $P_{2k-1}D^{(k-1)}D^{(k)}P_{2k}$; then $P_1, P_2, \ldots, P_{2k}$ will constitute a special $2k$-gon. To prove this we need merely observe that since the alternate sides $P_1P_2, P_3P_4, \ldots,$ are equal and parallel to $D'D'', D''D'''$, respectively, the vector sum of the former sides is equal to the vector sum of all the sides of the auxiliary k-gon and therefore vanishes.

It is seen from this that a *special $2k$-gon is completely determined by $2k-1$ of its vertices*. For if $P_1, P_2, \ldots, P_{2k-1}$ are given we can construct the auxiliary k-gon by starting at an arbitrary point D', drawing the vector $D'D''$ equal to P_1P_2, then $D''D'''$ equal to $P_3P_4, \ldots$, finally, $D^{(k-1)}D^{(k)}$ equal to $P_{2k-3}P_{2k-2}$; P_{2k} is then found by drawing from P_{2k-1} a vector equal to $D^{(k)}D'$. This is the generalization of the fact that a parallelogram is determined by three of its vertices (given of course in order).

After the case $k=2$ of the parallelogram, the first case deserving particular attention is the case $k=3$, *i.e.*, the *special hexagon.* Such a hexagon may be obtained, in accordance with the result above, by constructing parallelograms on the sides of an arbitrary triangle. Another construction is as follows: Take any two parallelograms $ABCO$, $ODEF$ having a vertex in common; the remaining vertices $ABCDEF$ constitute a special hexagon. The same hexagon may be obtained in this way by means of three distinct pairs of parallelograms. This may be generalized so as to apply to $2k$-gons.

The third characteristic stated in 9, in the case of a special hexagon $ABCDEF$, shows that the median point of the triangle ACE coincides with that of the triangle BDF; therefore the six lines obtained by joining each vertex to the mid-point of the opposite diagonal of the hexagon are concurrent, the point of concurrence being the mean point of the hexagon.

§ 3. Extension to space.

12. The preceding results admit of immediate extension to space of three dimensions, and to higher spaces. In fact, the translations and central symmetries still constitute a group, and results for polygons in space follow in a manner entirely analogous to that employed above. One difference as to the character of point symmetries may be noticed: In the plane a point symmetry is identical with a rotation of the plane in itself through 180°; in space however point symmetry is not equivalent to a rotation since in fact corresponding figures are not congruent but differ in the order of arrangement of their parts. If we consider a central symmetry in the plane as a rotation, the analogue in space would be a line symmetry, which is in fact rotation about an axis through 180°. However, in the application to space polygons only the symmetries of the former type, *i.e.*, point symmetries with respect to the vertices, are considered.

The results hold also for one dimension, that is, for sets of points in a line. Thus, for any set $P_1, P_2, \ldots, P_n$ there is a derived "inscribed" set consisting of the mid-points of the segments, $P_1P_2, P_2P_3, \ldots, P_nP_1$. This set is entirely arbitrary if n is odd, but not if n is even; the characteristics stated in 9 apply almost literally to these special sets of points of the latter type.

BIBLIOGRAPHIC ENTRY: APPLICATIONS

1. D. J. Dickinson, On Fletcher's paper 'Campanological groups', AMERICAN MATHEMATICAL MONTHLY, 1957, 331–332.

2

RING THEORY

(a)

EUCLIDEAN RINGS AND UNIQUE FACTORIZATION DOMAINS

UNIQUE FACTORIZATION OF GAUSSIAN INTEGERS*

WALTER RUDIN, University of Wisconsin

The usual proof of the unique factorization theorem in $R[i]$, the ring of all complex numbers of the form $m+ni$, where m and n are integers, depends on the existence of a Euclid algorithm in $R[i]$.† In the present note an elementary fact from plane geometry is exploited to yield a very simple and short proof of the theorem.

LEMMA. *Suppose C is a circle of radius r and Q is a square whose center lies on C and whose diagonal is not longer than $2r$. Then at least one vertex of Q lies in the interior of C.*

To see this, let t be the radius of Γ, the circle which passes through the vertices of Q. Let Γ' be the intersection of Γ with the interior of C. Since $r \geqq t$, the length of Γ' is at least one third of the circumference of Γ, and hence one or two vertices of Q lie on Γ'.

The *units* of $R[i]$ are the numbers i^n ($n=0, 1, 2, 3$). For θ in $R[i]$, the four numbers $i^n\theta$ are the *associates* of θ, and θ is a *prime* if θ is not a unit and if θ is not the product of any two members of $R[i]$ neither of which is a unit. If α is in $R[i]$, $|\alpha| > 1$, and α is not a prime, these definitions imply that $\alpha = \alpha'\alpha''$, where $|\alpha'| < |\alpha|$ and $|\alpha''| < |\alpha|$; finitely many repetitions of this process lead to a factorization of α into primes. This factorization is unique in the following sense:

THEOREM. *If $\theta_1, \cdots, \theta_r$ and $\phi_1, \cdots, \phi_s$ are primes in $R[i]$ and if $\theta_1 \cdots \theta_r = \phi_1 \cdots \phi_s$, then $r = s$ and the numbers ϕ_j can be so ordered that ϕ_j is an associate of θ_j $(1 \leqq j \leqq r)$.*

---

* From AMERICAN MATHEMATICAL MONTHLY, vol. 68 (1961), pp. 907–908.

† See, for instance, W. J. LeVeque, Topics in Number Theory, vol. I, Reading, Mass., 1956.

Proof. Suppose the theorem is false. Since there are only finitely many elements of $R[i]$ in every bounded region of the complex plane, there exists α in $R[i]$ such that

(A) α has two *distinct* factorizations into primes

$$\alpha = \theta_1 \cdots \theta_r = \phi_1 \cdots \phi_s, \tag{1}$$

(B) no β in $R[i]$ with $|\beta| < |\alpha|$ has property (A).

Note that no θ_j in (1) is an associate of any ϕ_k, for otherwise α/θ_j would satisfy (A), in contradiction to (B). We may assume that $|\theta_1| \geqq |\phi_1|$. Applying the lemma to the square whose vertices are $\theta_1 + i^n\phi_1$ ($n=0, 1, 2, 3$), we see that ϕ_1 has an associate, say ϕ_1^*, such that

$$|\theta_1 - \phi_1^*| < |\theta_1|. \tag{2}$$

Put

$$\beta = (\theta_1 - \phi_1^*)\theta_2 \cdots \theta_r. \tag{3}$$

By (2), $|\beta| < |\alpha|$. Since $\beta = \alpha - \phi_1^* \cdot \theta_2 \cdots \theta_r$, we see that ϕ_1 divides β. Since ϕ_1 is not an associate of any of the primes $\theta_2, \cdots, \theta_r$, (3) and (B) imply that ϕ_1 divides $\theta_1 - \phi_1^*$. Hence ϕ_1 divides θ_1, and we have a contradiction.

It may be of interest to students to pinpoint just where the above proof breaks down in the ring of all numbers of the form $m + ni\sqrt{3}$ (to give just one example); in this ring, 4 has two distinct factorizations into primes:

$$4 = 2\cdot 2 = (1 + i\sqrt{3})(1 - i\sqrt{3}).$$

Note, however, that the unique factorization theorem does hold in the ring of all numbers of the form $m + n\theta$ if $2\theta = 1 + i\sqrt{3}$, and that it can be proved in the above manner (with regular hexagons in place of squares).

A CHARACTERIZATION OF POLYNOMIAL DOMAINS OVER A FIELD*

TONG-SHIENG RHAI, Taiwan University, Formosa

Usually a Euclidean domain (D, φ) is defined to be an integral domain D on which there is a nonnegative integral valued function φ satisfying the following conditions:

ED1. $\varphi(a) = 0$ if $a = 0$, the zero of D.

ED2. $\varphi(a) \leqq \varphi(b)$ if a divides b.

ED3. For any pair of elements a, $b \neq 0$ in D, there exist elements q and r in D such that

$$a = bq + r \quad \text{and} \quad \varphi(r) < \varphi(b).$$

For example, see Zariski-Samuel [1] p. 23; later we shall make use of results stated there.

In general, the existence of q and r is not assumed to be unique. The purpose of this note is to point out the fact that if their uniqueness is assumed then D is

* From AMERICAN MATHEMATICAL MONTHLY, vol. 69 (1962), pp. 984–986.

either a field (commutative, always) or a polynomial domain over a field. We shall prove

THEOREM 1. *A Euclidean domain* (D, φ) *is either a field or a polynomial domain over a field if* (D, φ) *satisfies the condition*:

(U). *For any pair of elements* $a, b \neq 0$ *in* D, *there is a unique pair of elements* q, r *in* D *such that*

$$a = bq + r \quad \text{and} \quad \varphi(r) < \varphi(b).$$

N. Jacobson proved the same result (see [2]), due to the lemma given below. It is based on the slightly different definition of Euclidean domain which requires conditions ED1, ED3 and

ED2′. $\varphi(ab) = \varphi(a) \cdot \varphi(b)$ for all a, b, in D,

in place of ED2. This note shows, therefore, that Jacobson's result still holds in wider sense.

To prove the theorem, we need a lemma:

LEMMA. *For a Euclidean domain* (D, φ), *the condition* (U) *is equivalent to condition* (V):

(V). $\varphi(a+b) \leqq \max(\varphi(a), \varphi(b))$ *for all* a, b *in* D.

Proof of the lemma. (V)⇒(U). Suppose, for $a, b \neq 0$ in D, that there are two pairs of elements q_1, r_1, and q_2, r_2, such that

$$\begin{aligned} a &= bq_1 + r_1 \quad \text{and} \quad \varphi(r_1) < \varphi(b), \\ &= bq_2 + r_2 \quad \text{and} \quad \varphi(r_2) < \varphi(b). \end{aligned}$$

Then it follows that

$$b(q_1 - q_2) = -(r_1 - r_2) \neq 0,$$

i.e., b is a divisor of $r_1 - r_2$, hence

$$\varphi(b) \leqq \varphi(r_1 - r_2).$$

Now by (V), and since $\varphi(-a) = \varphi(a)$ for any a in D, we have

$$\varphi(b) \leqq \text{Max.}(\varphi(r_1), \varphi(r_2)),$$

which is a contradiction.

(U)⇒(V). Suppose that there are elements a and b in D such that

$$\varphi(a + b) > \varphi(a), \varphi(b).$$

Then, for the pair of elements $a^2 - b^2 + b$ and $a + b$, the uniqueness of quotient and remainder fails. In fact

$$\begin{aligned} a^2 - b^2 + b &= (a + b)(a - b) + b, & \varphi(b) &< \varphi(a + b), \\ &= (a + b)(a - b + 1) - a, & \varphi(a) &< \varphi(a + b). \end{aligned}$$

Proof of Theorem 1. It follows from ED2 that, for any nonzero element a in D, $\varphi(1) \leqq \varphi(a)$. Now define the set F by

$$F = \{a \in D \mid \varphi(a) \leqq \varphi(1)\}.$$

Then there are two cases to be considered:

i) $F=D$. In this case, all nonzero elements in D are units (i.e., every nonzero elements has an inverse); hence D is a field.

ii) $F \neq D$. In this case there is at least one element a in D such that $\varphi(a)>\varphi(1)$, and we can choose an element x in D-F with the property that $\varphi(1)<\varphi(x) \leqq \varphi(a)$, for all a in D-F.

It is to be noticed that F is a subfield of D, since F is closed under subtraction because of condition (V).

Now, by assumption, for any a in D-F there exist unique q and r in D such that

$$a=xq+r, \qquad \varphi(r)<\varphi(x).$$

Clearly, r is in F, and it can be verified that $\varphi(q)<\varphi(a)$. If $\varphi(q) \geqq \varphi(x)$, i.e., q is in D-F, we can get, in similar manner, a unique pair of elements q_1, r_1 such that

$$q = xq_1 + r_1, \qquad r_1 \in F, \qquad \varphi(q_1) < \varphi(q).$$

We continue this process. After a finite number of steps it has to break off, since

$$\varphi(a) > \varphi(q) > \varphi(q_1) > \cdots$$

is a decreasing sequence of nonnegative integers. Thus any a in D can be uniquely expressed in the form

$$a = x^n q_{n-1} + x^{n-1} r_{n-1} + \cdots + x r_1 + r,$$

where $q_{n-1}, r_{n-1}, \cdots, r_1, r$ are in F.

Hence D is a polynomial domain over F.

Conversely, if a polynomial domain D is endowed with standard norm function φ, i.e. $\varphi(a)=\deg a+1$, $\deg 0=-1$, then (D, φ) is a Euclidean domain satisfying the condition (U). Furthermore we obtain

THEOREM 2. *For a polynomial domain $D=F[x]$ over a field F, if there is a nonnegative integral valued function φ defined on D such that (D, φ) is a Euclidean domain satisfying the condition* (U), *then φ must be a monotone increasing function of the degree.*

Proof. First we shall establish a relation:

$$\text{(A)} \qquad \varphi(a) = \varphi(x^n) \quad \text{if } \deg a = n.$$

If $n=0$ this is true, so we go on with the induction. Suppose (A) is true for every integer $m<n$. Now any a in D of degree n with leading coefficient 1 can be written

$$a = x^n + b, \quad \text{where} \quad \deg b = m < n.$$

Then, by (V) and the inductive hypothesis, we have

$$\varphi(a) \leqq \text{Max.}\,(\varphi(x^n), \varphi(x^m)) = \varphi(x^n).$$

On the other hand, $x^n = a - b$ gives

$$\varphi(x^n) \leqq \text{Max.}\,(\varphi(a), \varphi(x^m)) = \varphi(a).$$

Hence $\varphi(x^n) = \varphi(a)$. Because $\varphi(fa) = \varphi(a)$ holds for any nonzero element f in F, this completes the proof of (A).

Let a, b in D have deg $a = m$, deg $b = n$ and $m < n$. Then by (A)

$$\varphi(a) = \varphi(x^m) < \varphi(x^n) = \varphi(b)$$

which shows that φ is a monotone increasing function of the degree.

References

1. O. Zariski and P. Samuel, Commutative algebra, vol. 1, Van Nostrand, Princeton, 1958.
2. N. Jacobson, A note on non-commutative polynomials, Ann. of Math., 35 (1934) 209–210.

A NOTE ON QUADRATIC EUCLIDEAN DOMAINS*

P. J. ARPAIA, C. W. Post College

To begin with, by an integral domain we shall always mean a commutative ring *with identity* such that $ab = 0$ implies $a = 0$ or $b = 0$. By a subdomain of an integral domain we will mean a subset of an integral domain that forms an integral domain under the given operations—hence *must possess an identity* element. By a Euclidean domain we mean an integral domain that is Euclidean. We now pose our question. Do there exist Euclidean domains having subdomains that are not Euclidean? We propose to answer this question in the affirmative.

It is a well-known fact that the integers under addition and multiplication form a Euclidean domain. Furthermore, there are many subrings of the integers that are *not* Euclidean. For example $2Z = \{n \in Z\colon n = 2k,\ k \in Z\}$. We have $36 = 2 \cdot 18 = 6 \cdot 6$. Since factorization into irreducible elements in a Euclidean ring must be unique, $2Z$ is not Euclidean. However, $2Z$ does not contain an identity element hence is not an integral domain. Furthermore, since Z is a cyclic group under addition and since 1 generates Z, Z can possess no *subdomain* that is not Euclidean. Hence we must look elsewhere for Euclidean domains possessing subdomains that are not Euclidean.

In [1], LeVeque states that $R[\sqrt{d}]$ is a quadratic Euclidean domain if and only if d has one of the 21 values:

$$-11, -7, -3, -2, -1, 2, 3, 5, 6, 7, 11, 13, 17, 19, 21, 29, 33, 37, 41, 57 \quad \text{or} \quad 73.$$

We will now show that each of these domains possesses a subdomain that is not Euclidean.

* From AMERICAN MATHEMATICAL MONTHLY, vol. 75 (1968), pp. 864–865.

Before we proceed recall that

$$N(a+b\sqrt{d}) = a^2 - db^2 \quad \text{and} \quad N((a+b\sqrt{d})(e+f\sqrt{d})) = N(a+b\sqrt{d})N(e+f\sqrt{d}).$$

We will need the following lemmas:

LEMMA 1. *Let $R[\sqrt{d}]$ be any quadratic domain and define D by $D = \{a+b\sqrt{d}\in R[\sqrt{d}]: a, b\in Z, b\equiv 0 \pmod{p}, p$ prime$\}$. Then D is a subdomain of $R[\sqrt{d}]$.*

Proof. For $a+b\sqrt{d}$, $e+f\sqrt{d}\in D$ we have $(a+b\sqrt{d})+(e+f\sqrt{d})\in D$. Now

$$(a+b\sqrt{d})(e+f\sqrt{d}) = (ae+dbf)+(af+be)\sqrt{d}.$$

Hence $(a+b\sqrt{d})(e+f\sqrt{d})\in D$. Therefore D is a subdomain of $R[\sqrt{d}]$. Further, $1=1+0\sqrt{d}\in D$.

LEMMA 2. *Let D be defined as in Lemma 1. If $a+b\sqrt{d}\in R[\sqrt{d}]$ and $N(a+b\sqrt{d})=p$, p a prime; then $a+b\sqrt{d}\notin D$.*

Proof. Suppose $a+b\sqrt{d}\in D$. Now $a^2-db^2=p$. Since $b\equiv 0 \pmod{p}$, $b^2=p^2n^2$. Hence $a^2=p(1+dpn^2)$. But then $a^2\equiv 0 \pmod{p}$. Hence $a\equiv 0 \pmod{p}$. Therefore $a^2=p^2m^2$. But then $p^2(m^2-dn^2)=p$. This implies that p divides 1. Hence $a+b\sqrt{d}\notin D$.

LEMMA 3. *If $p=a^2-db^2$, then p^3 has the two distinct factorizations ppp and $(pa+pb\sqrt{d})(pa-pb\sqrt{d})$ in D.*

Proof. Since $N(p)=p^2$, p is a prime in D by Lemma 2. Furthermore, since $N(pa+pb\sqrt{d})=N(pa-pb\sqrt{d})=p^3$, each of $pa+pb\sqrt{d}$ and $pa-pb\sqrt{d}$ is prime in D by Lemma 2. Finally, since $pa\pm pb\sqrt{d}=p(a\pm b\sqrt{d})$; p is not an associate of $pa+pb\sqrt{d}$ or of $pa-pb\sqrt{d}$.

THEOREM. *Each of the Quadratic Euclidean domains $R[\sqrt{d}]$ possesses a subdomain that is not Euclidean.*

Proof. In view of Lemma 3, all we need show is that for each of the 21 values of d stated in the first paragraph, there exists a solution to $a^2-db^2=p$, p a prime. The following table suffices:

d	$p=a^2-db^2$
-11	$47=36-(-11)(1)$
-7	$11=4-(-7)(1)$
-3	$7=4-(-3)(1)$
-2	$3=1-(-2)(1)$
-1	$2=1-(-1)(1)$
2	$-7=1-(2)(4)$
3	$-2=1-(3)(1)$
5	$11=16-(5)(1)$
6	$-5=1-(6)(1)$
7	$-3=4-(7)(1)$
11	$-7=4-(11)(1)$

d	$p=a^2-db^2$
13	$3=16-(13)(1)$
17	$-13=4-(17)(1)$
19	$-3=16-(19)(1)$
21	$-17=4-(21)(1)$
29	$-13=16-(29)(1)$
33	$-29=4-(33)(1)$
37	$107=144-(37)(1)$
41	$-37=4-(41)(1)$
57	$-53=4-(57)(1)$
73	$-37=36-(73)(1)$

Reference

1. W. J. LeVeque, Topics in Number Theory, Vol. 2, Addison-Wesley, Reading, Mass., 1956.

SOME APPLICATIONS OF A MORPHISM*

RICHARD SINGER, Webster College, St. Louis

The beginning student in abstract algebra is introduced to a wide variety of concepts and terminology which he may have difficulty in relating to his previous mathematical knowledge. This paper uses some of these concepts to prove two theorems about polynomials with integer coefficients. While these theorems were originally proved without these concepts, more concise proofs can be given by using them. Both proofs use the natural morphism from $Z[X]$ onto $Z_p[X]$ to eliminate extraneous information. The proofs also use the fact that when p is prime, $Z_p[X]$ is an integral domain and in particular a unique factorization domain.

The system of integers will be denoted by Z, the system of integers modulo p by Z_p.

THEOREM (Gauss). *Let $f, g \in Z[X]$. If f and g are primitive then fg is primitive.*

Proof. If fg is not primitive, then there is a prime p in Z such that $p|fg$. Letting α be the morphism from $Z[X]$ onto $Z_p[X]$ with kernel (p) it follows that

$$(\alpha f)(\alpha g) = \alpha(fg) = 0,$$

and since $Z_p[X]$ is an integral domain either $\alpha f=0$ or $\alpha g=0$. Thus either $p|f$ or $p|g$, contradicting the hypothesis that f and g are primitive.

One application of this theorem, which should relate to the student's previous mathematical experience, is to show that if a polynomial with integral coefficients cannot be factored into polynomials of lower degree with integer coefficients then a factorization into polynomials of lower degree with rational coefficients is also impossible.

THEOREM (Eisenstein). *Let $f=a_0+\cdots+a_nX^n\in Z[X]$. If there exists a prime $p\in Z$ such that p is not a factor of a_n, $p|a_i$ for all $i<n$, and p^2 is not a factor of a_0, then f is irreducible in $Z[X]$.*

Proof. Suppose $f=gh$, where $g=b_0+\cdots+b_jX^j$ and $h=c_0+\cdots+c_kX^k$, with j and k greater than 0. Let α be the morphism from $Z[X]$ to $Z_p[X]$ with kernel (p) then αa_n, αb_j and αc_k are not zero and

$$(\alpha g)(\alpha h) = \alpha f = (\alpha a_n)X^n.$$

Since $Z_p[X]$ is a *UFD*, $X|\alpha(g)$ and $X|\alpha(h)$. Thus $\alpha(b_0)=0=\alpha(c_0)$, and we have $p^2|b_0\cdot c_0=a_0$.

* From AMERICAN MATHEMATICAL MONTHLY, vol. 76 (1969), pp. 1131–1132.

A PRINCIPAL IDEAL RING THAT IS NOT A EUCLIDEAN RING*

JACK C. WILSON, University of North Carolina at Asheville.

1. Introduction. In introductory algebra texts it is commonly proved that every Euclidean ring is a principal ideal ring. It is also usually stated that the converse is false, and the student is often referred to a paper by T. Motzkin [**1**]. Unfortunately, this reference does not contain all of the details of the counterexample, and it is not easy to find the remaining details from the references given in Motzkin's paper. The object of this article is to present the counterexample in complete detail and in a form that is accessible to students in an undergraduate algebra class.

Not all authors use precisely the same definitions for these two types of rings. Throughout this paper the following definitions will hold.

DEFINITION 1. *An integral domain R is said to be a Euclidean ring if for every $x \neq 0$ in R there is defined a nonnegative integer $d(x)$ such that:*

(i) *For all x and y in R, both nonzero, $d(x) \leqq d(xy)$.*

(ii) *For any x and y in R, both nonzero, there exist z and w in R such that $x = zy + w$ where either $w = 0$ or $d(w) < d(y)$.*

DEFINITION 2. *An integral domain R with unit element is a principal ideal ring if every ideal in R is a principal ideal; i.e., if every ideal A is of the form $A = (x)$ for some x in R.*

The ring, R, to be considered is a subset of the complex numbers with the usual operations of addition and multiplication:

$$R = \{a + b(1 + \sqrt{-19})/2 \mid a \text{ and } b \text{ are integers}\}.$$

It is elementary to show that R is an integral domain with unit element. The purpose of this article then is to show that R is a principal ideal ring, but that it is impossible to define a Euclidean norm on R so that with respect to that norm R is a Euclidean ring.

2. The ring is a principal ideal ring. In R there is the usual norm, $N(a + bi) = a^2 + b^2$, which has the property that $N(xy) = N(x)N(y)$ for all complex numbers x and y. In R this norm is always a nonnegative integer. The essential theorem for this part of the example is due to Dedekind and Hasse, and the proof is taken from [**2**, p. 100].

THEOREM 1. *If for all pairs of nonzero elements x and y in R with $N(x) \geqq N(y)$, either $y \mid x$ or there exist z and w in R with $0 < N(xz - yw) < N(y)$, then R is a principal ideal ring.*

* From MATHEMATICS MAGAZINE, vol. 46 (1973), pp. 34–38.

Proof. Let $A \neq (0)$ be an ideal in R. Let y be an element of A with minimal nonzero norm, and let x be any other element of A. For all z and w in R, $xz - yw$ is in A so that either $xz - yw = 0$ or $N(xz - yw) \geqq N(y)$. Hence the assumed conditions on R require that $y \mid x$; i.e., $A = (y)$.

The ring R under consideration will now be shown to satisfy the hypotheses of Theorem 1. Observe that $0 < N(xz - yw) < N(y)$ if and only if $0 < N[(x/y)z - w] < 1$. Given x and y in R, both nonzero and $y \nmid x$, write x/y in the form $(a + b\sqrt{-19})/c$ where a, b, c are integers, $(a, b, c) = 1$, and $c > 1$. First of all, assume that $c \geqq 5$. Choose integers d, e, f, q, r such that $ae + bd + cf = 1$, $ad - 19\,be = cq + r$, and $|r| \leqq c/2$. Set $z = d + e\sqrt{-19}$ and $w = q - f\sqrt{-19}$. Thus,

$$\begin{aligned}(x/y)z - w &= (a + b\sqrt{-19})(d + e\sqrt{-19})/c - (q - f\sqrt{-19})\\ &= r/c + \sqrt{-19}/c.\end{aligned}$$

This complex number is not zero and has norm $(r^2 + 19)/c^2$, which is less than 1 since $|r| \leqq c/2$ and $c \geqq 5$. The only case that is not immediately obvious is $c = 5$, but then $|r| \leqq 2$ so that $r^2 + 19 \leqq 23 < c^2$.

The remaining possibilities are $c = 2, 3$, or 4. Consider these in order:

(i) If $c = 2$, $y \nmid x$ and $(a, b, c) = 1$ imply that a and b are of opposite parity. Set $z = 1$ and $w = [(a-1) + b\sqrt{-19}]/2$ which are elements of R. Thus, $(x/y)z - w = 1/2 \neq 0$ and has norm less than 1.

(ii) If $c = 3$, $(a, b, c) = 1$ implies that $a^2 + 19b^2 \equiv a^2 + b^2 \not\equiv 0 \pmod 3$. Let $z = a - b\sqrt{-19}$ and $w = q$ where $a^2 + 19b^2 = 3q + r$ with $r = 1$ or 2. Thus, $(x/y)z - w = r/3 \neq 0$ and has norm less than 1.

(iii) If $c = 4$, a and b are not both even. If they are of opposite parity, $a^2 + 19b^2 \equiv a^2 - b^2 \not\equiv 0 \pmod 4$. Let $z = a - b\sqrt{-19}$ and $w = q$, where $a^2 + 19b^2 = 4q + r$ with $0 < r < 4$. Thus, $(x/y)z - w = r/4 \neq 0$ and has norm less than 1. If a and b are both odd, $a^2 + 19b^2 \equiv a^2 + 3b^2 \not\equiv 0 \pmod 8$. Let $z = (a - b\sqrt{-19})/2$ and $w = q$, where $a^2 + 19b^2 = 8q + r$ with $0 < r < 8$. Thus, $(x/y)z - w = r/8 \neq 0$ and has norm less than 1.

This completes the proof that R is a principal ideal ring.

3. The ring is not a Euclidean ring. This part of the counterexample is taken from [1]. The material is repeated and slightly elaborated here in order to give a self-contained result accessible to an undergraduate class. As with the previous section the results are stated within the context of the ring R under consideration, but the theorem applies to more general integral domains. Throughout this section R_0 will denote the set of nonzero elements of R.

DEFINITION 3. *A subset P of R_0 with the property $PR_0 \subset P$; i.e., xy is an element of P for all x in P and y in R_0, is called a product ideal of R. (Notice that R_0 is a product ideal.)*

DEFINITION 4. *If S is a subset of R, the derived set of S, denoted by S', is defined by $S' = \{x \in S \mid y + xR \subset S$, for some y in $R\}$.*

LEMMA 1. *If S is a product ideal, then S' is a product ideal.*

Proof. If x is in S', then x is in S and there exists y in R such that $y + xR \subset S$. Let z be in R_0. Since S is a product ideal and x is in S, xz is in S. Further, $y + (xz)R \subset y + xR \subset S$. This shows that $S'R_0 \subset S'$; i.e., S' is a product ideal.

LEMMA 2. *If $S \subset T$, then $S' \subset T'$.*

Proof. If x is in S', then x is in S and hence in T, and there exists a y in R such that $y + xR \subset S \subset T$. Therefore, x is in T', and $S' \subset T'$.

THEOREM 2. *If R is a Euclidean ring, then there exists a sequence, $\{P_n\}$, of product ideals with the following properties:*
(i) $R_0 = P_0 \supset P_1 \supset P_2 \supset \cdots \supset P_n \supset \cdots$,
(ii) $\cap P_n = \varnothing$,
(iii) $P'_n \subset P_{n+1}$, *for each* n, *and*
(iv) *For each* n, $R_0^{(n)}$, *the nth derived set of* R_0, *is a subset of* P_n.

Proof. Let the Euclidean norm in R be symbolized by $d(x)$ for x in R_0. For each nonnegative integer n, define $P_n = \{x \in R_0 \mid d(x) \geqq n\}$. This defines the sequence which obviously has properties (i) and (ii). Suppose that x is in P_n and y is in R_0. $d(xy) \geqq d(x) \geqq n$ which implies that xy is in P_n. This shows that $P_n R_0 \subset P_n$; i.e., for each n, P_n is a product ideal.

For property (iii) let x be in P'_n; i.e., x is in P_n and there exists a y in R such that $y + xR \subset P_n$. Applying the Euclidean algorithm, there exist elements q and r in R with $y = xq + r$ and $r = 0$ or $d(r) < d(x)$. Hence, $r = y + x(-q)$ is in $y + xR \subset P_n$, which implies that $d(r) \geqq n$, and in turn, $d(x) > d(r) \geqq n$, so that $d(x) \geqq n + 1$ and x is in P_{n+1}. This proves property (iii) $P'_n \subset P_{n+1}$.

For property (iv), clearly $R_0 = P_0$ and application of (ii) gives $R'_0 = P'_0 \subset P_1$. Assuming that $R_0^{(n)} \subset P_n$, Lemma 2 and (iii) yield $R_0^{(n+1)} \subset P'_n \subset P_{n+1}$. By induction, (iv) is proved.

COROLLARY. *If $R'_0 = R''_0 \neq \varnothing$, then R is not a Euclidean ring.*

Proof. The hypotheses of the corollary imply that for all n, $R_0^{(n)} = R'_0$. If R is a Euclidean ring, the theorem would require $R'_0 = \cap R_0^{(n)} \subset \cap P_n = \varnothing$.

This corollary is now used to show that R is not a Euclidean ring. First R'_0 is determined. If x is a unit in R, say $xy = 1$, and z is an element of R, $z + x(-yz) = 0$ is not in R_0. This shows that units are not in R'_0. If x is not a unit in R, then using $z = -1$, $z + xy \neq 0$ for all y in R, which shows that if x is not zero and not a unit, x is in R'_0. Altogether, R'_0 is precisely the set of elements of R that are neither units nor zero. Notice that the only units of our example R are 1 and -1. Next, in order to determine the elements of R_0'', it is convenient to use the following terminology:

DEFINITION 5. *An element x of R_0' is said to be a side divisor of y in R provided there is a z in R that is not in R_0' such that $x \mid (y+z)$. An element x of R_0 is a universal side divisor provided that it is a side divisor of every element of R.*

If x is in R_0'', then x is in R_0' and there is a y in R such that $y + xR \subset R_0'$; i.e., x never divides $y + z$ if z is zero or a unit. Thus, x is not a side divisor of y, and therefore, not a universal side divisor. Conversely, if x is not in R_0'', and is in R_0', then for every y in R there exists a w in R with $y + xw$ not in R_0'; i.e., $y + xw$ is zero or a unit, and therefore, x is a side divisor of y. Since this holds for every y in R, x is a universal side divisor. Together, these two arguments show that R_0'' is the set R_0' exclusive of the universal side divisors. If it can now be shown that R has no universal side divisors, this will show that $R_0' = R_0'' \neq \varnothing$, and the corollary will complete the proof that R is not a Euclidean ring.

A side divisor of 2 in R must be a nonunit divisor of 2 or 3. In R, 2 and 3 are irreducible, and therefore, the only side divisors of 2 are 2, -2, 3, and -3. On the other hand, a side divisor of $(1+\sqrt{-19})/2$ must be a nonunit divisor of $(1+\sqrt{-19})/2$, $(3+\sqrt{-19})/2$, or $(-1+\sqrt{-19})/2$. These elements of R have norms of 5, 7, and 5, respectively, while the norms of 2 and 3 and their associates are 4 and 9, respectively. As a result, no side divisor of 2 is also a side divisor of $(1+\sqrt{-19})/2$, and there are no universal side divisors in R. All of the details of the counterexample are complete.

References

1. T. Motzkin, The Euclidean algorithm, Bull. Amer. Math. Soc., 55 (1949) 1142–1146.
2. H. Pollard, The Theory of Algebraic Numbers, Carus Monograph 9, MAA, Wiley, New York, 1950.

UNIQUE FACTORIZATION*

PIERRE SAMUEL, Institut Henri Poincaré, Paris

1. Introduction. It is well known that every ordinary integer is, in a unique way, a product of prime numbers. With an eye to generalizations it is better to state this unique factorization property in the *ring* Z of rational (i.e., >0 or <0) integers. Thus, if we denote by P the set of all prime numbers, every nonzero element x of Z may be written, in a unique way, as

$$x = \pm 1 \prod_{p \in P} p^{v_p(x)}, \tag{1}$$

where the exponents $v_p(x)$ are positive integers, almost all 0 (i.e., equal to 0

* From AMERICAN MATHEMATICAL MONTHLY, vol. 75 (1968), pp. 945–952.

except for a finite number of them) in order that formula (1) makes sense. The somewhat abstract formulation given by (1), with its seemingly infinite product, has the great advantage of indicating how the exponents $\nu_p(x)$ depend on x. If we allow negative exponents, we see that (1) holds also for all *nonzero* rational numbers x. Furthermore, for any pair x,y of nonzero rational numbers, we see that we have

$$(2) \qquad \nu_p(xy) = \nu_p(x) + \nu_p(y), \qquad \nu_p(x + y) \geqq \inf(\nu_p(x), \nu_p(y)).$$

Algebraists express formulae (2) by saying that the mapping $\nu_p: Q^* \to Z$ is a *discrete valuation* of the field of rational numbers.

More generally, we define a *factorial ring* (or a "unique factorization domain," U.F.D.) to be an integral domain A for which there exists a subset P of A such that every nonzero element x of A may be written, in a *unique* way, as

$$(1') \qquad x = u \prod_{p \in P} p^{\nu_p(x)}$$

where u is a *unit* (i.e., an invertible element) in A, and where the exponents $\nu_p(x)$ are positive integers, almost all 0. It can easily be proved that the subset P is uniquely determined up to units; more precisely the set $(Ap)_{p \in P}$ of principal ideals is uniquely determined, and coincides with the set of all maximal principal ideals distinct from A. Let us notice that a principal ideal Ab of a domain A is maximal (among principal ideals distinct from A) iff every divisor d of b is either a unit or is such that db^{-1} is a unit; such an element b is called an *irreducible* element of A.

For a ring A, factoriality is a very useful property. At least for multiplicative questions, the *arithmetic* in a factorial ring A is as nice as in the ring Z of ordinary integers. It may be recalled that, in the 19th century, arithmeticians like Kummer and Dedekind noticed that some rings of algebraic integers failed to be factorial; e.g., the formulae

$$2 \cdot 3 = (1 + \sqrt{-5})(1 - \sqrt{-5}), \qquad 3 \cdot 3 = (\sqrt{10} + 1)(\sqrt{10} - 1)$$

show that the rings $Z[\sqrt{-5}]$, $Z[\sqrt{10}]$ are not factorial; this led Kummer and Dedekind to introduce the important notion of an *ideal*, and to replace the unique factorization of elements by the unique factorization of ideals, thus inaugurating the theory of rings which we now call "Dedekind rings." Lack of time prevents me from talking more about this important and beautiful theory.

The interest of factorial rings does not come only from arithmetical reasons. Factoriality has also a very simple *geometric* interpretation. In geometry, more precisely in the study of algebraic, analytic or formal varieties, a ring A occurs as a ring of functions (algebraic or analytic, as the case may be) on some variety V, or in the neighborhood of some point of V. To say that A is a domain means that V is irreducible. Denoting by n the dimension of V, the factoriality of A then means, roughly speaking, that every subvariety of W of dimension $n-1$ of V may be defined *by a single equation*; more precisely the functions $f \in A$ which vanish on W form an ideal $p(W)$ in A, and factoriality means that these ideals $p(W)$ (for dim $W = n-1$) are principal.

2. How to prove factoriality. We have just seen that factoriality is a desirable property for a ring. On the other hand proving that a ring is factorial is rarely trivial, so it is useful to have at hand as many characterizations of factorial rings as possible.

As we have seen in Section 1, factoriality of A means that every nonzero element of A admits a decomposition as a product of irreducible ones, and that this decomposition is unique up to units. The existence of such a decomposition is usually easy to check; it follows from this "chain condition" for principal ideals (valid in any factorial ring):

(3) *Every strictly increasing sequence of principal ideals of A is finite*, which is itself equivalent to the "maximal condition":

(3′) *Every nonempty family of principal ideals of A admits a maximal element.*

For example (3) (or (3′)) holds when the ring A is noetherian, and most rings that are encountered in arithmetic or in geometry are noetherian. Furthermore, with proper caution, property (3) may pass to the direct limits. We henceforth assume that (3) holds.

As to *uniqueness*, things are not so easy. Unique factorization in a ring A implies that any irreducible element p of A enjoys the stronger property that

(4) *If p divides a product ab, then it divides a or b.*

Conversely, assuming (3), a well-known proof copied from elementary number-theory shows that (4) implies the uniqueness of the decomposition into irreducible factors. An element p which enjoys property (4) is called a *prime* element of A; this means that the principal ideal Ap is a prime ideal ($ab \in Ap \Rightarrow a \in Ap$ or $b \in Ap$), or, equivalently, that the factor ring A/Ap is a domain. As shown in elementary number-theory, property (4) is equivalent to

(4′) *Any two elements of A admit a greatest common divisor*, and also to:

(4″) *Any two elements of A admit a least common multiple.*

A rather handy form of (4″) is

(4‴) *The intersection of any two principal ideals of A is a principal ideal.*

If we deal with a *noetherian* domain A, it can be proved that every nonunit in A is contained in a prime ideal of height 1 (i.e., a prime ideal which is minimal among nonzero prime ideals). From this one easily deduces that a noetherian domain A is factorial iff

(5) *Every prime ideal of height* 1 *of A is principal.*

This condition has already been met at the end of Section 1, when we discussed the geometric meaning of factoriality.

More technical characterizations of factorial rings may be given in the framework of the theory of *Krull rings*, for which we refer to Bourbaki, "Algèbre

Commutative," Chap. VII, "Diviseurs" [4]. Let us only say that the class of Krull rings contains the class of factorial rings and is more stable under various ring-theoretic operations. Furthermore, it is in general easy to test whether a given ring is a Krull ring or not. The problem is therefore to test whether a given Krull ring is factorial, and, if not, to measure its "nonfactoriality."

3. Properties stronger than factoriality. For proving that Z is factorial, one usually first proves that Z is *principal* (i.e., that every ideal of Z is principal). Then the chain condition (3) is very easy, and condition (4‴) is obvious. The example of a polynomial ring in several variables over a field shows that being principal is a stronger property than being factorial; thus it could seem to be dangerous to concentrate on this stronger property for proving factoriality. However, we have a reliable touchstone for telling us whether the danger exists or not. In fact the commutative algebraists have developed an extensive theory of the *dimension* of a ring, and many methods for computing the dimension of a ring are available. Moreover the principal rings are characterized as being the factorial rings of *dimension* 0 *or* 1. Thus the dimension of the ring A we are studying will tell us whether it is reasonable to attempt to prove that A is principal.

In most geometric cases, principality is proved by proving separately factoriality and one-dimensionality. But, in *algebraic number theory*, there are methods for proving directly that a ring is principal. For instance, let K be a number-field of finite degree n over the rationals, let A be the ring of algebraic integers of K, d the absolute discriminant of A, and $2r_2$ the number of nonreal conjugates of K in C. Then, by using Minkowski's theory of lattice points in convex sets, one can prove that every nonzero ideal $\mathfrak{A}$ of A may be written as $\mathfrak{A}=xb$, where x is an element of K^* and where b is an ideal in A for which

$$\text{card}(A/b) < \left(\frac{4}{\pi}\right)^{r_2} \frac{n!}{n^n} (|d|^{1/2}). \tag{6}$$

Now the right hand side of (6) can be computed by standard methods, whereas the ideals b for which A/b has a given cardinal c are finite in number, and are easy to determine if c is not too large. Thus, if it happens that all the ideals b for which (6) is satisfied are principal, then the ring A is principal. The reader may apply the method to the ring $A=Z[i]$ of Gaussian integers (here $r=2$, $r_2=1$, $|d|=4$, whence the right hand side of (6) is <2, and (6) thus implies $b=A$); he may then feel that this is a very sophisticated method for proving that $Z[i]$ is principal! In fact the usual proof for $Z[i]$, as well as for Z or for a polynomial ring $k[X]$ over a field k, uses the fact that these rings are *euclidean*. Let us recall that an integral domain A is said to be euclidean if there exists a mapping $\phi: A \to N$ (the positive integers) such that, for every nonzero b in A, every class modulo Ab admits a representative r such that $\phi(r) < \phi(b)$ (i.e., every a in A may be written $a=bq+r$ with $\phi(r) < \phi(b)$). A euclidean ring A is principal for, given a nonzero ideal b in A, we choose a nonzero element x of b for which

$\phi(x)$ is minimal, and see that x generates b. For this proof, it is not necessary to assume that ϕ takes its values in N; any well ordered set W would work as well. A mapping $\phi: A \to W$ satisfying the above property is called an algorithm on A. If we consider a given ring A and a large-enough well ordered set W (e.g., such that card $(W) >$ card (A)), the theory of well ordered sets shows that every algorithm on A is isomorphic (in an obvious sense) with an algorithm with values in W. Furthermore, if $\phi_\alpha: A \to W$ is a family of algorithms on A, then $\phi = \inf_\alpha \phi_\alpha$ is also an algorithm, so that A (if euclidean) admits a smallest algorithm. If the residue fields of A are finite, this smallest algorithm ϕ_0 actually takes its values in N (the general case is still open). But it is not necessarily the usual algorithm: in the case of Z, $\phi_0(n)$ is the number of binary digits of the integer $|n|$ $(n \in Z)$; however, for polynomials in X over a field k, $\phi_0(P(X))$ is the degree of the polynomial $P(X)$.

Much work has been done by arithmeticians for determining the number fields for which the ring A of integers is euclidean; most of them studied the more restricted problem of finding out whether the usual "norm-function" (i.e., $\phi(x) = \text{card}\ (A/Ax)$ for $x \neq 0$) is an algorithm or not. For imaginary quadratic fields, the five fields $Q(\sqrt{-d})$ for $d = 1, 2, 3, 7, 11$ are the only ones for which the norm is an algorithm, and are also the only euclidean ones. But there are four principal noneuclidean rings of integers in imaginary quadratic fields for $d = 19$, 47, 67 and 163 (the problematic existence of a fifth one has recently been disproved). As to real quadratic fields $Q(\sqrt{m})$ $(m > 0)$, the list of those which are euclidean for the norm is known:

$$m = 2, 3, 5, 6, 7, 11, 13, 17, 19, 21, 29, 33, 37, 41, 57, 73.$$

Many others are known to be principal, but we do not know whether their number is finite or not. Also we do not know whether some of them might not be euclidean for another algorithm than the norm; a bit of evidence induces the writer to think that $Q(\sqrt{14})$ deserves to be studied in this respect (see [5], [6]). Summarizing, one might say that the theory of euclidean rings has a quite different flavor from that of factoriality.

4. Nagata's Theorem. Masayoshi Nagata has proved a theorem which is very useful for showing that a ring is factorial. We recall that, if A is an integral domain with quotient field K and if S is a multiplicatively closed subset of A $(0 \notin S)$, then the fractions a/s with $a \in A$ and $s \in S$ form a subring of K, denoted by $S^{-1}A$, and called the *ring of quotients* of A with respect to S. Now suppose that A satisfies the finiteness condition (3) (see Section 2), that S is generated by *prime* elements (Section 2), and that $S^{-1}A$ is factorial; then Nagata's theorem states that A itself is *factorial*. If S is generated by a finite number of prime elements, one can dispense with condition (3). A very easy converse of Nagata's theorem is that any ring of quotients of a factorial ring is factorial.

Gauss's lemma about *polynomial rings* is an easy consequence of Nagata's theorem. In fact let R be a factorial ring, L its quotient field, and $S = R - \{0\}$.

Since a prime element p of R remains prime in the polynomial ring $A=R[X]$ (for $A/pA=(R/pR)[X]$ is a domain), S is generated by prime elements of A. But $S^{-1}A=L[X]$ is a polynomial ring in one variable over a field, whence is euclidean and factorial. Hence $A=R[X]$ is factorial by Nagata. By induction the same holds for polynomial rings in several variables over a factorial ring.

Let us sketch three other *examples* of application of Nagata's theorem (complete proofs are left to the reader):

(a) Let k be an algebraically closed field of characteristic $\neq 2$, and $F(X_1, \cdots, X_n)$ a nondegenerate quadratic form over k, with $n\geqq 5$. Then $A=k[X_1, \cdots, X_n]/(F)$ is factorial. (By a change of variables, write $F=X_1X_2+G(X_3, \cdots, X_n)$; denote by x_j the image of X_j in A; then x_1 is prime since G is irreducible (for $n\geqq 5$); taking $S=\{1, x_1, \cdots, x_1^j, \cdots\}$, we see that $S^{-1}A=k[x_1, x_3, \cdots, x_n][1/x_1]$ is factorial as a ring of quotients of a polynomial ring.)

(b) Let k be a field in which -1 is not a square, and $A=k[X, Y, Z]/(X^2+Y^2+Z^2-1)$ ("the ring of the 2-sphere"); then A is factorial. (Denote by x, y, z the images of X, Y, Z in A; then $x^2+y^2=(1+z)(1-z)$; take S generated by $1-z$, which is prime; now $S^{-1}A$ is factorial as in (*a*).)

(c) Let k be a field and $A=k[X, Y, Z]/(X^r+Y^s+Z^t)$ where the exponents r, s, t are pairwise relatively prime; then A is factorial. (Denote by x, y, z the images of X, Y, Z in A, so that $z^t=-(x^r+y^s)$; suppose first that $t\equiv 1 \pmod{rs}$, i.e., $t=1+drs$; take S generated by z (which is prime), and set $x'=x/z^{ds}$, $y'=y/z^{dr}$; then $z=-(x'^r+y'^s)$ and $S^{-1}A=k[x', y'][1/z]$ is factorial; in the general case, one chooses an integer j such that $jt\equiv 1 \pmod{rs}$, and replaces z by some jth root w of z.)

5. Further Results. The theory of factorial rings is nowadays much more developed than has been sketched above. For example, in [2] of the bibliography, we find about 80 pages of lecture notes entirely devoted to factoriality with sizeable prerequisites from commutative and homological algebra; moreover these notes did not contain everything known on the subject when they were written (1963), and the theory has progressed since that time. We will thus briefly sketch some highlights of this theory, without defining some of the terms we use; for detailed definitions, proofs and connected results, we refer the reader to the bibliography.

(1) *Power Series.* In Section 4 we have stated Gauss's lemma about polynomial rings. It is a particular instance of the "transfer" of some property from a ring A to the polynomial ring $A[X]$. Many similar transfers are known, and also transfers of properties from a ring A to the formal power series ring $A[[X]]$. Thus it was reasonable to conjecture that, if A is factorial, so is $A[[X]]$. This conjecture has been disproved (see [7]). In the first counter-examples given, the ground ring A was a noncomplete local ring, and taking formal power series over a noncomplete local ring could be deemed, by some mathematicians, to be an unnatural (or even immoral) operation. Doubts were settled very recently by

P. Salmon [13], who constructed a complete local factorial ring A such that $A[[X]]$ is not factorial.

(2) *Regular Rings.* The notion of a regular ring is defined in commutative algebra; in the geometric case, it corresponds to the notion of a nonsingular variety. In 1957, M. Auslander and D. Buchsbaum proved, by homological methods, that any regular local ring is factorial. Their proof has been streamlined by I. Kaplansky [1], [2], and by N. Bourbaki [4].

On the other hand, if A is a regular and factorial ring (not necessarily local), then both $A[X]$ and $A[[X]]$ are factorial.

(3) *Galoisian going-down.* Let A be a factorial ring, and G a finite group of automorphisms of A; the elements of A which are invariant by G form a subring of A, traditionally denoted by A^G. Let A^* be the multiplicative group of units A. Then if the cohomology group $H^1(G, A^*)$ vanishes, the ring A^G is factorial ([2], [3], [17]).

Here the writer cannot resist giving an amusing example. We take for A a polynomial ring $A = k[X_1, \cdots, X_n]$ (k: a field, $n \geqq 5$), and for G the *alternating group* A_n, acting on A by permutations of the variables. Here the ring of invariants A^G is generated over k by the elementary symmetric functions $s_1, \cdots, s_n$ and by the "discriminant" $d = \Pi_{i<j}(X_i - X_j)$; it is known that $d^2 = P(s_1, \cdots, s_n)$ where P is a polynomial over k. Furthermore $A^* = k^*$ is trivially operated by G, so that $H^1(G, A^*) = \mathrm{Hom}(G, k^*)$ (classical formula in the cohomology of groups). Since $G = A_n$ is a simple group ($r \geqq 5$) and since k^* is commutative, we have $\mathrm{Hom}(G, k^*) = 0$ and A^G is factorial.

The same method has given an example of a factorial ring which is not a Macaulay ring [18]. Notice that P. Murthy has proved that a factorial Macaulay ring is necessarily a Gorenstein ring.

In characteristic $p \neq 0$, there is a parallel theory in which automorphisms are replaced by derivatives ([2], [9], [16]). As above the proofs of factoriality are partly computational, and (especially in characteristic 2) the complete performance of these computations is sometimes more accessible than in the case of automorphisms.

(4) *Complete Intersections.* A local ring A is called a "complete intersection" if it is isomorphic to some R/I, where R is a regular local ring and I an ideal generated by a regular R-sequence (this means that I may be generated by $\dim(R) - \dim(R/I)$ elements). By using powerful methods of his theory of schemes (the latest version of algebraic geometry), A. Grothendieck proved that a complete intersection A, such that A_p is factorial whenever $\dim(A_p) \leqq 3$, is itself factorial ([19]). This generalizes older geometric theorems of F. Severi, S. Lefshetz and A. Andreotti. No purely ring-theoretic proof of Grothendieck's theorem is known.

(5) *Two-dimensional Factorial Rings.* We have already said that the factorial rings of dimension one are the principal rings; among them, the local ones are the discrete valuation rings and are considered as well known. In dimension 2, we have already seen a good number of examples of factorial rings: e.g., the rings

of the surfaces $x^i+y^j+z^k=0$ (i, j, k pairwise relatively prime) and of the sphere $x^2+y^2+z^2-1=0$; localizing the first ones at the origin, we obtain many non-regular local factorial rings of dimension 2. These local rings are not complete and, moreover, the factoriality of their completions $C=K[[x, y, z]]$ was in doubt. First G. Scheja and D. Mumford proved that the complete ring C of the surface $x^2+y^3+z^5=0$ is factorial. Then P. Salmon, for the counterexample alluded to in (1), proved the same for the surface $x^2+y^3+tz^6=0$ over a field K of the form $K=k(t)$ with t transcendental over k.

In this last example the ground field K is not algebraically closed. Now E. Brieskorn, by using techniques from algebraic geometry, has proved that, among the complete two-dimensional local rings over an algebraically closed field K, only two are factorial: the regular ring $K[[X, Y]]$ (formal power series), and the ring $K[[x, y, z]]$ with $x^2+y^3+z^5=0$ (cf. [13]). It can be noted that the latter is the ring of invariants of an icosahedral group acting on the former [1].

A bibliography of factorial rings

An elementary exposition can be found in:

1. P. Samuel, Anneaux Factoriels (red. A. Micali), Bol. Soc. Mat., São Paulo, 1964.

More complete results in:

2. P. Samuel, Lectures on unique factorization domains, (notes by Pavman Murthy) Tata Institute for Fundamental Research lectures, No. 30, Bombay, 1964.

3. ———, Lectures in commutative algebra (notes by M. Bridger), mimeographed by Brandeis University, Waltham, Mass., 1964–65 (write to Brandeis).

For a treatment of factorial rings, in the framework of Krull rings, see:

4. N. Bourbaki, Algèbre Commutative, Chap. VII "Diviseurs," Hermann, Paris, 1966.

For the case of number-fields, see:

5. Hardy-Wright, An introduction to the theory of numbers, Clarendon, Oxford, 1960, and also the tables in

6. Borovič-Safarevič, Théorie des nombres, Gauthier-Villars, Paris, 1966. (German and English translations also available.)

Most results, up to 1964, about factorial rings are given in [1], [2], [3]. For the reader's convenience, we however quote:

7. P. Samuel, On unique factorization domains, Ill. J. Math., 5 (1961) 1–17.

8. ———, Sur les anneaux factoriels, Bull. SMF, 89 (1961) 155–173.

9. ———, Classes de diviseurs et dérivées logarithmiques, Topology, 3, Supp. 1 (1964) 81–96.

10. ———, Modules réflexifs et anneaux factoriels, In Colloque International de Clermont-Ferrand, ed. CNRS, Paris, 1965.

11. P. Samuel, Sur les séries formelles restreintes, C.R. Acad. Sci., Paris, 1962.

The ring of the surface $x^2+y^3+z^5=0$ is studied in

12. F. Klein, Lectures on the icosahedron, Dover, New York, 1956, Chap. 2, Sections 12 and 13.

13. E. Brieskorn, Local rings which are UFD's, (preprint, MIT, Oct. 1966), and in articles of G. Scheja (Math. Ann., 1965), and D. Mumford (Publ. I.H.E.S., 9 (1961)).

The first example of a complete local ring A for which $A[[t]]$ is not factorial was given in:

14. P. Salmon, Sulla non-factorialita . . . , Rend. Lincei, June 1966.

A further discussion of this example is in:

15. N. Zinn-Justin, Dérivations des corps et anneaux de caractéristique p, (Thèse Paris 1967); in print in Mémoires Soc. Math. France, 1967.

For the theory of the "purely inseparable going-down," see:

16. N. Hallier, Utilisation des groupes de cohomologie dans la théorie de la descente p-

radicielle, C.R. Acad. Sci. Paris, 261 (1965) 3922–3924 and also [15]. (NB: Hallier is the maiden-name of Mrs. Zinn-Justin.)

For examples of "galoisian going-down," see:

17. M. J. Dumas, Sous anneaux d'invariants d'anneaux de polynômes, C.R. Acad. Sci. Paris, 260 (1965) 5655–5658.

18. M. J. Bertin, Sous groupes cycliques d'ordre $p^n \cdots$, C.R. Acad. Sci. Paris, April 1967. (NB: Dumas is the maiden-name of Mrs. Bertin.)

A proof of Grothendieck's theorem on the factoriality of some complete intersections is in

19. A. Grothendieck, Séminaire de Géometrie Algébrique 1961–1962, exposé XI, mimeographed by the Institut des Hautes Etudes Scientifiques, 35 route de Chartres, 92-Bures sur Yvette. France.

UNIQUE FACTORIZATION DOMAINS*

P. M. COHN, Bedford College, University of London

1. Introduction. One of the most fascinating developments in ring theory in recent years is the way in which large parts of algebraic geometry can now be stated entirely in terms of commutative Noetherian rings [**23**]. In the other direction this has led to new tools for classifying and investigating these rings; furthermore, these applications are no longer confined to Noetherian rings, and although commutativity is assumed as a rule, one suspects that even this is not always essential. There is an extensive and rapidly growing literature on the subject, and it would be difficult to do justice to it in a brief article. Instead, we have singled out a special class of rings: unique factorization domains. They provide a good example of how ring-theoretical properties can be illustrated by geometrical ideas. The non-commutative case is described separately; this is less well developed, and the connection with geometry is less clear, but eventually any geometric ring theory must also comprehend the non-commutative case.

We begin with the definition of a commutative unique factorization domain (UFD) in section 2, and its relation to such basic notions as Dedekind and Krull domains. In section 3 we analyse the definition and reduce it to a statement about primes. Nagata's theorem and its application to Felix Klein's theorem on line complexes is discussed in section 4.

The remainder of the article is concerned with the non-commutative case. In

---

* From AMERICAN MATHEMATICAL MONTHLY, vol. 80 (1973), pp. 1–18.

Professor Cohn did his Cambridge Ph. D. under Philip Hall, and he has held positions at the Univ. de Nancy, Manchester University, Queen Mary College London, and (presently) Bedford College, London. He has spent leaves-of-absence at Yale Univ., Univ. of Chicago, and Rutgers. He has published extensively in universal algebra and in many branches of algebra; in 1965–67 he was the Secretary of the London Mathematical Society. His Books are *Lie Groups* (Cambridge Univ. Press 1957), *Linear Equations* (Routledge and Kegan Paul 1958), *Solid Geometry* (Routledge and Kegan Paul 1961), *Universal Algebra* (Harper and Row 1965), and *Free Rings and Their Relations*, LMS Monograph 2 (Academic Press 1971). He received a MAA Lester R. Ford Award in 1972. *Editor.*

section 5 we discuss the lattice-method of defining and recognizing UFD's and give some non-commutative examples. All are consequences of the fact that an atomic 2-fir is a UFD. If we drop atomicity, we are left with the Schreier refinement property, which so far has mainly been studied in the commutative case (section 6). Section 7 describes two special cases of interest in the non-commutative theory: the lattice of factors is distributive, respectively a chain. In section 8 we examine the shortcomings of the definition of non-commutative UFD and describe some remedies that have been proposed, and section 9 notes the problems of factorizing zero-divisors.

Throughout, some of the easier proofs have been sketched and others omitted, with a reference where full proofs can be found. When no convenient reference was available, proofs are given in more detail. The article is based on a lecture delivered at the British Mathematical Colloquium at Leicester on April 1, 1964.

2. Commutative unique factorization domains. All rings are understood to be associative, with a unit-element 1, which is inherited by subrings and preserved by homomorphisms. Moreover, all modules are unital. Usually our ring R will be an **integral domain,** i.e., the set R^* of non-zero elements is non-empty and closed under multiplication. This terminology will be used even for non-commutative rings (though at first our rings will be commutative). An element u of a ring R is called a **unit,** if there exists $v \in R$ such that $uv = vu = 1$; any non-unit which cannot be written as a product of two non-units is said to be **irreducible** or an **atom**. An integral domain is said to be **atomic** if every element, not zero or a unit, is a product of atoms.

DEFINITION. A commutative integral domain R is said to be **factorial** or a **unique factorization domain** (UFD) if it satisfies the following conditions:

A. *R is atomic,*

U. *Any two factorizations of an element into atoms differ only in the order of the factors, and by unit factors.*

Thus if $c = a_1 \cdots a_r = b_1 \cdots b_s$, where a_i, b_j are atoms, then $r = s$ and after a suitable renumbering of the b's, a_i is associated to b_i, i.e., $a_i = b_i u_i$, where u_i is a unit.

The best-known example of a UFD is the ring $\mathbf{Z}$ of integers. There are two basic ways of proving that $\mathbf{Z}$ is a UFD, which we shall call the **prime method** and the **lattice method**. Both are capable of generalization and we shall deal with each in turn (sections 3 and 5).

UFD's are important for several reasons: In the first place, their characteristic property makes them more amenable; secondly, to impose unique factorization often singles out a significant class of rings, while thirdly, the methods used to prove factoriality have often given rise to other notions important in their own right.

The unique factorization property of the integers can also be shown to hold for the ring of Gaussian integers $a + b\sqrt{-1}$ $(a, b \in \mathbf{Z})$, and this led to efforts to prove the same for the ring of integers in any finite algebraic extension of $\mathbf{Z}$. These efforts, though doomed to failure, led Kummer to introduce 'ideal numbers' in an attempt to restore unique factorization. Dedekind [**18**] gave a general definition of ideals and showed that rings of algebraic integers possess unique factorization for their ideals. Thus if $\mathfrak{p}_i$ $(i \in I)$ are the different prime ideals, any non-zero ideal $\mathfrak{a}$ has a unique representation

$$\mathfrak{a} = \prod \mathfrak{p}_i^{\nu_i}, \tag{1}$$

where the integers $\nu_i = \nu_i(\mathfrak{a})$ are non-negative and all but a finite number of them are zero. A commutative integral domain with unique factorization of ideals is called a *Dedekind domain*; such a ring is necessarily Noetherian, i.e., it satisfies the ascending chain condition, briefly ACC, for ideals. There are UFD's that are not Noetherian, and hence not Dedekind, e.g., the polynomial ring in infinitely many indeterminates over a field.

To get a common generalization of UFD and Dedekind domain, we observe that in both cases we have a family of integer-valued functions on R^* satisfying the familiar conditions for an exponential valuation:

V.1. $v(a) \geqq 0$,

V.2. $v(a - b) \geqq \min\{v(a), v(b)\}$,

V.3. $v(ab) = v(a) + v(b)$.

Such a valuation extends in a unique way to the field of fractions K of R. Now Krull [**30**] considered more generally, integral domains with a family $(v_i)_{i \in I}$ of such valuations, where for any $c \in K^*$, $v_i(c) = 0$ for almost all i, and $c \in R$ if and only if $v_i(c) \geqq 0$. Such rings are called *Krull domains* (Krull called them "endliche diskrete Hauptordnungen"); clearly they include both UFD's and Dedekind domains as special cases, e.g., a Noetherian integral domain is a Krull domain if and only if it is integrally closed (in its field of fractions). Krull domains retain at least some of the useful features of UFD's; moreover, unlike UFD's, the class of Krull domains is closed under integrally closed integral algebraic extensions [**6**]. For any Krull domain its departure from factoriality is measured by the *divisor class group*, i.e. the group of all divisors (formal products $\prod \mathfrak{p}_i^{\nu_i}$) modulo the principal divisors [**36**]. Its vanishing characterizes UFD's; it is unchanged under adjunction of indeterminates, but may change under algebraic extension.

The relation between the different types of ring becomes clearer if we adopt a slightly different point of view. Let K be any commutative integral domain and K its field of fractions. On K^* we define the relation of divisibility: a divides b, in symbols: $a \mid b$, if $ba^{-1} \in R$. Clearly this relation is reflexive and transitive, i.e., it is a *preordering* of K^*. Moreover, it is compatible with multiplication: if $a \mid b$, then $ac \mid bc$ for all $c \in K^*$. In this way K^* becomes a preordered group; if U is the group of units in R, then $D = K^*/U$ is the partially ordered group associated with the preordered group K^*; it is called the *divisibility group* of R. Now various classes of rings can be described entirely in terms of the order type of their divisibility group. For comparison we shall need ${}^{I}\mathbf{Z}$, the direct sum of card (I) copies of $\mathbf{Z}$, with the componentwise ordering: $(x_i) \leqq (y_i)$ if and only if $x_i \leqq y_i$ for all $i \in I$. Then

(i) R is a UFD if and only if D is order-isomorphic to ${}^{I}\mathbf{Z}$, for some I,

(ii) if R is a Krull domain then D is order-isomorphic to a subgroup of ${}^{I}\mathbf{Z}$, for some I,

(iii) R is a valuation ring if and only if D is totally ordered,

(iv) R is a discrete valuation ring if and only if $D \cong \mathbf{Z}$.

3. The relation of UFD's to primes. Let us analyse the notion of UFD more closely. An element p of a commutative integral domain R is said to be *prime* if p is not zero or a unit and $p \mid ab$ implies $p \mid a$ or $p \mid b$. From this definition it is easy to see that each prime must be an atom. The converse need not hold; in fact with a finiteness condition it is equivalent to unique factorization:

THEOREM 1. *A commutative ring is a UFD if and only if it is an atomic integral domain and every atom is prime.*

This is a sort of localization of the condition U. It is easily proved; in fact the usual proof that $\mathbf{Z}$ is a UFD consists in verifying that every atom is prime, using the Euclidean algorithm, and then carrying out what is in effect a proof of Theorem 1 [**42**].

Examples of atomic integral domains that are not UFD's are well known, e.g., the integers in the field $Q(\sqrt{-5})$. To give an example of a different kind, take R to be the ring generated by x_0, x_1, x_2, x_3 over a field k, with the defining relation

$$x_0 x_1 = x_2 x_3. \tag{2}$$

Here $x_1 \mid x_2 x_3$, but $x_1 \nmid x_2$, $x_1 \nmid x_3$, so x_1 is not prime, but it is clearly an atom.

Regarding atomicity, this is a finiteness condition which clearly holds in every Noetherian domain. More generally, it holds in every integral domain with ascending chain condition on principal ideals, ACC_1 for short. Conversely, every UFD satisfies ACC_1, but there are atomic domains not possessing ACC_1 (notwithstanding an assertion to the contrary in Proposition 1.1 of [**12**]); for a counter-example, see Anne Grams, Proc. Cambridge Phil. Soc. 75 (1974) 321–329. On the other hand, the Noetherian condition is not necessary in a UFD, as we have seen, but as a rule this is the important case for algebraic geometry.

A less obvious factoriality criterion is obtained by using prime ideals. An ideal $\mathfrak{p}$ in a ring R is said to be *prime* if $R/\mathfrak{p}$ is an integral domain. E.g., in an integral domain a principal ideal (p) is prime precisely when p is 0 or a prime element. Another way of describing a prime ideal is as an ideal whose complement is multiplicative, i.e., a nonempty multiplicatively closed set.

We recall that given any subset S of a ring R, and any ideal $\mathfrak{a}$ disjoint from S, the standard application of Zorn's lemma produces an ideal $\mathfrak{p}$ containing $\mathfrak{a}$ and maximal subject to the condition $\mathfrak{p} \cap S = \varnothing$. Moreover, if S is multiplicative, $\mathfrak{p}$ is prime, as is easily checked (and well known). Then we have the following characterization of UFD's in terms of prime ideals [**28**]:

THEOREM 2. *An integral domain is a UFD if and only if every nonzero prime ideal contains a prime element.*

We recall the essence of the proof. If R is a UFD and $\mathfrak{p} \neq 0$ a prime ideal, let $\mathfrak{p}$ contain $a = p_1^{\alpha_1} \cdots p_r^{\alpha_r}$, where the p_i are primes, then $p_i \in \mathfrak{p}$ for some $i = 1, \cdots, r$. Conversely, assume the condition and let S be the set of all products of primes. Then S is multiplicative and moreover, it is **saturated**, i.e., any factor of an element of S is itself in S. If there is a non-unit c not in S, then $(c) \cap S = \varnothing$, so a maximal ideal $\mathfrak{p}$ containing c and disjoint from S exists; it must be prime and by hypothesis contains a prime element, which contradicts the fact that all these elements lie in S. Now the usual proof of Theorem 1 shows that a product of primes is necessarily unique.

In particular we have the

COROLLARY 1. *In a UFD, every minimal non-zero prime ideal is principal.*

In a Noetherian domain the converse holds, i.e., a Noetherian domain in which every minimal non-zero prime ideal is principal is a UFD; this follows from Theorem 2, because in this case every non-zero prime ideal contains a minimal non-zero prime ideal. But this is a non-trivial result, the consequence of Krull's 'principal ideal theorem' (cf. [**28**], where the latter is described as "probably the most important single theorem in the theory of Noetherian rings").

To give a geometrical illustration, consider a twisted cubic. This cannot be obtained as the complete intersection of two surfaces in 3-space, for if the surfaces had degrees m and n, then $mn = 3$ and so m or n is 1 and the cubic would be plane. In fact, as is well known, a twisted cubic can be obtained as the intersection of three suitable quadrics, or as the intersection of two quadrics with a common generator, if we ignore the generator.

Geometrically, any non-degenerate quadric in complex projective 3-space can be brought to the form $x_0x_1 = x_2x_3$. The ring of functions on this quadric is the ring $A = \mathbf{C}[X_0, X_1, X_2, X_3]/(f)$, where $f = X_0X_1 - X_2X_3$. This ring is an integral domain, but not a UFD, as we saw earlier. Now a minimal non-zero prime ideal of A corresponds to a maximal proper subvariety of the quadric, i.e., a curve, and by Th. 2, we cannot always expect this to be given by a single equation; the twisted cubic is a case in point. (This is a slight oversimplification, because subvarieties are actually defined by homogeneous ideals in this case.)

Generally, if k is an algebraically closed field, an algebraic set over k is given by the zeros in k^n of a set of polynomials in $X_1, \cdots, X_n$ and the coordinate ring of this algebraic set is

$$A = k[X_1, \cdots, X_n]/\mathfrak{a},$$

where $\mathfrak{a}$ is the ideal generated by the given set of polynomials. We have a variety (= irreducible algebraic set) if and only if $\mathfrak{a}$ can be taken to be a prime ideal, and then A is an integral domain. If we take for granted the fact that every maximal subvariety of an n-dimensional variety has codimension 1 (i.e., is $(n-1)$-dimensional, cf. [**32**], p. 36), Th. 2, Cor. 1 gives the necessity of the next result; the sufficiency follows by the remark following Th. 2, Cor. 1, because the coordinate ring of a variety is Noetherian.

COROLLARY 2. *The coordinate ring of a variety is a UFD if and only if every subvariety of codimension* 1 *determines a principal ideal (thus the subvariety is a complete intersection).*

If V is any variety, the set $\mathfrak{o}_x$ of functions defined at a given point x of V is a **local ring**, i.e., a ring whose non-units form an ideal (the ideal of functions vanishing at x). If x is a simple point of V, $\mathfrak{o}_x$ is what is called a **regular** local ring, and this is necessarily a UFD. There are several proofs of this fact; for a thorough analysis of the algebraic background, see [**28**], and for a history of the problem, see [**35**].

4. Nagata's Theorem. How does one prove that a given ring is a UFD? In the case of $\boldsymbol{Z}$ we needed the Euclidean algorithm ([**19**], Book VII, Prop. 1–2) to prove that every atom is prime. For polynomial rings in one variable over a field one can use the same method (introduced by Stevin [**40**] to find the greatest common divisor of two polynomials), but for more than one variable this method is no longer available (in the next section we shall see why). However, it is still true that a polynomial ring in any number of variables over a field is a UFD. More generally, if R is a UFD, then so is $R[X]$; the proof depends on forming rings of fractions.

Let R be an integral domain with field of fractions K. Given a multiplicative subset S of R^*, write

$$R_S = \{a/s \mid a \in R, s \in S\}.$$

This is again a ring, in fact a subring of K, e.g., $R_{R^*} = K$. Any prime p in R either becomes a unit in R_S or it remains prime, depending on whether or not p divides an element of S. Moreover, any atom in R_S comes from an atom in R. Thus by Th. 1 we obtain:

THEOREM 3. *If R is a UFD and S any multiplicative subset of R^*, then R_S is also a UFD.*

Conversely, we have the following result, first proved by Nagata [**34**] (for Noetherian domains):

THEOREM 4. *Let R be an atomic integral domain and S a multiplicative subset of R consisting of products of primes. Then if R_S is a UFD, so is R.*

The proof consists roughly in this: every atom of R either divides an atom of S and is then shown to be prime (using the fact that S consists of prime products), or if it divides no element of S it stays an atom in R_S and is then prime because R_S is a UFD (cf. [17], p. 116).

With the help of Th. 4 it is very easy to show that factoriality is preserved by adjunction of indeterminates.

COROLLARY 1. *If R is a UFD, then so is the polynomial ring $R[X]$.*

For if $K = R_{R^*}$ is the field of fractions, then by the Euclidean algorithm, $K[X]$ is a UFD. Now $K[X] = R[X]_{R^*}$ and R^* consists of prime products, because R is a UFD, while Gauss's lemma ensures that any prime of R is still a prime in $R[X]$. Further, $R[X]$ satisfies ACC_1: one gets a bound on the number of non-unit factors by considering leading terms. Hence by Th. 4, $R[X]$ is a UFD.

Cor. 1 shows (by induction) that the polynomial ring in any finite number of indeterminates over a field is a UFD. To extend the result to infinitely many indeterminates one can proceed as follows.

Let A, B be any rings, such that A is a subring of B. The inclusion $A \subseteq B$ is said to be *inert* if for any $c \in A$ such that $c = ab$, where $a, b \in B$, there exists a unit $u \in B$ such that au, $u^{-1}b \in A$. E.g., any integral domain R is inert in the polynomial ring $R[X]$. Now let R be a ring which is a union of a directed system of subrings R_λ (i.e. any two subrings of the system are contained in a third). If all the R_λ are UFD's and all the inclusions $R_\lambda \subseteq R_\mu$ are inert, then R is a UFD. For any $c \in R$ lies in some R_λ and so has a unique factorization into atoms in R_λ; moreover any factorization of c in a bigger ring R_μ can by inertia be pulled down to R_λ and so must agree with the factorization already found.

We note that the inertia condition cannot be omitted: the semigroup algebra (over a field) of the additive semigroup of positive rational numbers is not a UFD, but it can be written as the union of a directed system of UFD's.

Now let $\mathfrak{X} = (X_i)_{i \in I}$ be an infinite family of indeterminates and (R_λ) the family of rings obtained by adjoining finitely many of the X's to a given UFD R. The R_λ form a directed system of subrings of $R[\mathfrak{X}]$ with inert inclusions, and each is a UFD (by Cor. 1), hence $R[\mathfrak{X}]$ is again a UFD.

As a second application of Th. 4, due to Nagata [34, 37] we show that the coordinate ring of a quadric in more than three dimensions is a UFD.

COROLLARY 2. *Let k be an algebraically closed field of characteristic not two, and $Q(X_1, \ldots, X_n)$ a non-degenerate quadratic form in $n \geqq 5$ variables. Then $A = k[X_1, \ldots, X_n]/(Q)$ is a UFD.*

Proof. We can always write $Q = X_1 X_2 + Q_1(X_3, \cdots, X_n)$; here Q_1 is irreducible because $n \geqq 5$. Writing x_i for the residue class of X_i (mod Q), we have $A = k[x_1, \cdots, x_n]$. Let S be the multiplicative set generated by X_2 in $k[X_1, \cdots, X_n]$ and S' the multiplicative set generated by x_2 in A, then

$$A_{S'} = k[x_2, \cdots, x_n][x_2^{-1}] = k[x_2, \cdots, x_n]_{S'} = k[X_2, \cdots, X_n]_S.$$

Since $k[X_2,\cdots,X_n]$ is a UFD, the ring on the right is a UFD by Th. 3. Now x_2 is prime in A because Q_1 is irreducible, so A is a UFD by Th. 4.

Cor. 2 has an interesting geometrical consequence due to Klein. The lines in projective 3-space are described by Plücker coordinates π_{ij} $(i,j = 0,\cdots,3)$, subject to the relation (cf. [**41**], p. 22):

$$\pi_{01}\pi_{23} + \pi_{02}\pi_{31} + \pi_{03}\pi_{12} = 0. \tag{3}$$

Each set of ratios (π_{ij}) satisfying (3) defines a line, so that the set of lines in 3-space may be viewed as a quadric (clearly non-degenerate) in projective 5-space, the **Klein quadric.** Thus the lines in 3-space form an algebraic variety; an algebraic subset of codimension 1 on this variety is called a **line complex**. Now by Th. 4, Cor. 2, the coordinate ring of the Klein quadric is a UFD and so (by Th. 2, Cor. 2) its subvarieties of codimension 1 are complete intersections. Thus we get:

KLEIN'S THEOREM. *Every irreducible line complex in projective* 3*-space is given by a single equation in Plücker coordinates* (*besides* (3)).

5. The lattice method. We now turn to the second method of studying UFD's, the **lattice method.** This starts from quite a different definition of UFD, though of course equivalent to the one given earlier. It leads to other generalizations, and in particular, it does not require the ring to be commutative.

Thus let R be an integral domain (not necessarily commutative). If $c \in R^*$ and

$$c = a_1 \cdots a_r, \tag{4}$$

we consider the sequence of right ideals from R to cR:

$$R \supseteq a_1R \supseteq a_1a_2R \supseteq \cdots \supseteq a_1\cdots a_rR = cR;$$

with it we associate the corresponding quotients

$$R/a_1R,\, a_1R/a_1a_2R \cong R/a_2R, \cdots, R/a_rR. \tag{5}$$

If we have a second factorization of c:

$$c = b_1 \cdots b_s \tag{6}$$

with quotients $R/b_1R,\cdots,R/b_sR$, we say that (4) and (6) are **isomorphic** if $r = s$ and there is a permutation $i \mapsto i'$ of $\{1,\cdots,r\}$ such that $R/a_iR \cong R/b_{i'}R$. Now we define a (general) UFD as an atomic integral domain in which any two complete factorizations of a given element are isomorphic.

This provides a definition of UFD for the non-commutative case. Although stated in terms of right modules, it turns out to be left-right symmetric. Further, it reduces to the previous definition in the commutative case; to see this we note that if in a commutative integral domain, $R/aR \cong R/bR$, then aR is the annihilator of R/aR and so $aR = bR$, from which it follows that a and b are associated.

Let R be any ring; a right R-module M is said to be **strictly cyclic** if it can be written as R/cR, where c is a non-zero divisor. We denote by $\mathscr{C}_R$ the category whose objects are all the strictly cyclic right R-modules, while the morphisms are all the homomorphisms between them. The category ${}_R\mathscr{C}$ of strictly cyclic left R-modules is defined correspondingly. Any homomorphism $f: R/aR \to R/bR$ in $\mathscr{C}_R$ is given by an equation

$$ca = bc',$$

and based on this fact one shows that there is a duality (i.e., a category anti-isomorphism) between $\mathscr{C}_R$ and ${}_R\mathscr{C}$, for any ring R, [**13, 17**]. We shall call it the **factorial duality** in R; in particular this shows that

$$(7) \qquad R/aR \cong R/bR \text{ if and only if } R/Ra \cong R/Rb,$$

from which the left-right symmetry of the above notion of UFD follows immediately. Let us call two non-zerodivisors a, b of a ring R **similar** if $R/aR \cong R/bR$ [**20, 24**]. By (7) this notion is left-right symmetric; the corresponding notions for zero-divisors are distinct, as examples by Fitting [**20**] show.

Earlier we saw that in a commutative ring two elements are similar precisely when they are associated; in general there is no such simple criterion. Some equivalent conditions are given in

THEOREM 5. *Let a,a' be two non-zerodivisors in a ring R. Then the following three conditions are equivalent*:

(i) *a and a' are similar,*

(ii) *the matrices $\begin{pmatrix} 1 & 0 \\ 0 & a \end{pmatrix}$ and $\begin{pmatrix} 1 & 0 \\ 0 & a' \end{pmatrix}$ are associated in R_2,*

(iii) *there exist mutually inverse 2×2 matrices μ with a in the $(1,1)$-position and μ^{-1} with a' in the $(2,2)$-position.*

For a proof see [**17**], p. 124 f. ((i) $\Leftrightarrow$ (ii) was proved in [**20**] and (i) $\Leftrightarrow$ (iii) in [**9**]). Here are some examples of non-commutative UFD's.

1. The ring of integral quaternions (a rational quaternion is said to be **integral** if its coefficients are integers or halves of odd integers).

2. The ring of linear differential operators. Let $k = \mathbf{C}(t)$ be the field of rational functions in a single variable and D an indeterminate over k with the commutation rule

$$(8) \qquad Df = fD + f', \text{ where } f' = df/dt.$$

The skew polynomials $\Sigma f_i D^i$ $(f_i \in k)$ with multiplication according to (8) form a UFD. This, probably one of the first examples, was established by Landau [**31**] and Loewy [**33**]. Two polynomials in D which are similar in the sense explained above define differential equations which are equivalent in the sense of Poincaré.

3. Free associative algebras [**13, 17**]. Every free associative algebra $k\langle x_1, \cdots, x_n \rangle$ over a field is a UFD; as an example of a non-trivial factorization in the free

algebra $k\langle x, y\rangle$ we have

$$xyx + x = x(yx + 1) = (xy + 1)x,$$

and as is easily seen, $xy + 1$ and $yx + 1$ are similar atoms.

4. Group algebras of free groups [**13**, **17**].

5. Free products of skew fields [**13**].

The definition given at the beginning of this section suggests a way of proving a ring to be a UFD:

THEOREM 6. *An integral domain R is a UFD whenever for each $c \in R^*$ the set $L(cR, R)$ of principal right ideals between R and cR forms a modular lattice of finite length of the lattice of all right ideals of R.*

For then we can apply the Jordan-Hölder theorem for modular lattices [**4**, **10**]. An example of a modular lattice is the lattice of all right ideals of a ring R; this is the lattice of all submodules of R regarded as a right R-module. The lattice of all submodules of any module is modular (hence the name). Therefore in a principal ideal domain (i.e., an integral domain in which every left or right ideal is principal) the principal right ideals between R and cR form a modular lattice; the *ACC* holds because every right ideal is finitely generated, while the *DCC* follows by the factorial duality, using the fact that R has *ACC* for left ideals. So we obtain

COROLLARY 1. *Every principal ideal domain is a UFD.*

Both the integral quaternions and the ring of linear differential operators are principal ideal domains and therefore are UFD's. The free associative algebra (on more than one free generator) is clearly not a principal ideal domain; so we look for weaker hypotheses from which to deduce Th. 6. Whenever the principal right ideals form a sublattice of the lattice of all right ideals, they form a modular lattice; this is so provided that for any $a, b \in R$ there exist $d, m \in R$ such that

$$aR + bR = dR, \quad aR \cap bR = mR.$$

The first equation leads to the Bezout identity $au + bv = d$ for the greatest common divisor d of a and b, and these rings are called **right Bezout domains**. They are just the integral domains in which every finitely generated right ideal is principal, thus they are not much more general than principal right ideal domains. To get a wider class, let us look at free algebras. Any free associative algebra over a field has the following property [**13**]:

Every right ideal is free as right R-module, of unique rank.

A ring with this property is called a **free right ideal ring** or **right fir** for short. Left firs are defined similarly and a left and right fir is called a **fir**. E.g., free algebras (over a field), group algebras of free groups, and free products of skew fields are all firs. This is proved by the **weak algorithm**, a generalization of the Euclidean algorithm (to which it reduces in the commutative case). From this point of view the polynomial ring in one variable $k[X]$ is just the free associative algebra on one generator. This explains why the Euclidean algorithm for polynomials in one variable does not extend to more variables: it only applies to free algebras.

Any fir satisfies left and right ACC_1 [**13, 17**]. Moreover, the mapping $(x, y) \leftrightarrow ax - by$ from R^2 to $aR + bR$ defines an exact sequence

$$0 \to aR \cap bR \to R^2 \to aR + bR \to 0, \tag{9}$$

which necessarily splits (because $aR + bR$ is free); hence $aR + bR$ and $aR \cap bR$ are principal whenever $aR \cap bR \neq 0$. Thus all the conditions of Th. 6 hold and we find [**13, 17**]:

COROLLARY 2. *Any (left and right) fir is a UFD.*

Let us define, for any $n \geqq 1$, an **n-fir** as a ring in which every right ideal on at most n generators is free, of unique rank. The notion so defined is left-right symmetric and for larger n we get smaller classes, until we get to **semifirs**, the rings that are n-firs for all n. The 1-firs are just the integral domains, and a 2-fir is a ring in which each 2-generator right ideal is free, of unique rank. Looking more closely to see what was needed to prove Cor. 2, we obtain [**9, 17**]:

COROLLARY 3. *Every atomic 2-fir is a UFD.*

This generalizes Cor. 2, because every fir is clearly an atomic 2-fir. In the commutative case, every atomic 2-fir is a principal ideal domain, so Cor. 2 and 3 tell us nothing new for commutative rings. But we have seen examples of non-principal firs (free algebras) and there are also atomic 2-firs that are not firs [**3, 11**]. For example, to obtain a ring R such R^* can be embedded in a group but R cannot be embedded in a (skew) field (Malcev's problem), Bowtell [**7**] constructs an atomic 2-fir; this ring cannot be a fir because every fir is embeddable in a field [**16**]. We remark in passing that if R is any atomic 2-fir, then R^* is embeddable in a group [**16**]; it is not known whether this property is shared by all 2-firs, or by all UFD's.

The problem of unique factorization has also been studied in rings with a set of defining relations of the form $ab = cd$, where a, b, c, d are atoms in the free algebra, by Bokut' [**5**].

6. The Schreier refinement property. A look at the lattice method of defining UFD's immediately suggests the generalization obtained by giving up atomicity. Let us define an *S-ring* (for Schreier) as an integral domain in which any two factorizations of any non-zero element have isomorphic refinements.

For the moment let us return to the commutative case; in the presence of ACC_1, S-rings reduce of course to UFD's, but in general these classes are distinct. In fact there is an intermediate class, the HCF-rings: an **HCF-ring** is an integral domain in which any two elements have a highest common factor. In terms of the divisibility group D of a ring (section 2) we can say that R is an HCF-ring if and only if D is lattice-ordered, while R is an S-ring precisely if D has the (m,n)-interpolation property, for all m,n: given $x_1,\cdots,x_m,y_1,\cdots,y_n \in D$, if $x_i \leqq y_j$ $(i = 1,\cdots,m,$ $j = 1,\cdots,n)$, then there exists $z \in D$ such that $x_i \leqq z \leqq y_j$ (all i,j). This is actually a consequence of the (2,2)-interpolation property [**4, 12**].

Clearly we have the implications

$$\text{UFD} \Rightarrow HCF\text{-ring} \Rightarrow S\text{-ring},$$

and neither of these arrows can be reversed [**12**]. Moreover, any HCF-ring is integrally closed (in its field of fractions), but this need not be true of S-rings, as is shown by the following example, due (independently) to G. M. Bergman and M. Kneser (unpublished):

Let F be a field with a proper algebraic extension E, and consider the ring of all formal power series $a_0 + \sum_1^\infty a_i x^{\lambda_i}$, where $a_0 \in F$, $a_i \in E$ $(i > 0)$ and (λ_i) is a sequence of positive rational numbers tending to infinity. Then R is an S-ring, but not integrally closed.

By an argument somewhat analogous to the proof of Th. 4 one shows that if R is an integrally closed S-ring, then so is $R[X]$ (cf. [**12**]). Here the hypothesis of integral closure cannot be omitted; in fact if $R[X]$ is an S-ring, it is easy to see that R must be an integrally closed S-ring. Thus it is more natural to confine attention to integrally closed S-rings. These rings are studied in [**12**], where they are called **Schreier rings**.

Turning to the non-commutative case, we note that every 2-fir is an S-ring. It is not difficult to give examples of non-atomic 2-firs: apart from the commutative examples there is the group algebra of a free product of copies of the additive group of rational numbers; this is a non-Ore semifir which is non-atomic. It seems more difficult to produce examples of non-Ore semifirs (or even 2-firs) that are atomless.

The commutative case suggests that there should be a condition analogous to integral closure which plays a part in the study of general S-rings, but it is not clear what form this condition should take, or indeed, what its precise role would be.

For other studies of infinite factorizations see [**1, 3, 25, 27**].

7. Special cases in the non-commutative theory. In a commutative UFD the principal ideals form a lattice; more generally this is so (by definition) in a commutative integral domain with highest common factors (HCF) and least common multiples (LCM). These are the HCF-rings we met in section 6; in fact it is enough to assume the existence of a HCF for each pair of elements, or equivalently assume the existence of a LCM for each pair [**12**]. Curiously enough, this symmetry disappears when we consider individual pairs: in an integral domain, any pair of elements having an LCM also has an HCF, but the converse need not hold (consider the elements 2 and $2x$ in the subring of the polynomial ring $\mathbf{Z}[X]$ consisting of all polynomials with even coefficient of X).

If R is an HCF-ring and K its field of fractions, then the principal fractional ideals in K form a group under multiplication, and this group structure is compatible with the ordering by inclusion. Thus we have a lattice-ordered group; it is well known that such a group is distributive as a lattice ([4], p. 292). In particular, in a Bezout domain R, for any $c \in R^*$ the set $L(cR, R)$ of principal ideals between R and cR is a distributive lattice. For non-commutative rings this need not be so, even in the case of 2-firs, where $L(cR, R)$ is a sublattice of the lattice of all right ideals. Let us say that an integral domain R has a **distributive factor lattice** if for each $c \in R^*$ the set $L(cR, R)$ is a distributive sublattice of the lattice of all right ideals of R. By the factorial duality the notion so defined is left-right symmetric; moreover, any ring with a distributive factor lattice is a 2-fir.

A principal ideal domain has a distributive factor lattice if and only if every (left or right) ideal is two-sided [17] and this is a fairly stringent requirement. It is therefore of interest that among general 2-firs quite a wide class of rings have a distributive factor lattice, e.g., free associative algebras and group algebras of free groups. This follows from some technical results of G. M. Bergman proved in [3, 17]. These results show more generally that a 2-fir defined over a field k, which remains a 2-fir under all field extensions, has a distributive factor lattice.

In an atomic 2-fir with distributive factor lattice, the factors of a given element form a distributive lattice of finite length. These lattices have been described in terms of partially ordered sets [3, 4, 17]; to be precise, the categories of finite distributive lattices and homomorphisms, and finite partially ordered sets and order-preserving mappings are dual to each other via the functor $\mathrm{Hom}(-, 2)$, where 2 is the 2-element lattice resp. ordered set. This description has been used by Bergman to study the possible factorizations that can occur. For example, in a commutative UFD, the only distributive lattices which can be realized in this way are direct products of chains; the corresponding partially ordered sets are disjoint unions of chains. However, in a free associative algebra, every finite distributive lattice can be realized as a lattice of factors in this way. The simplest case not occcurring in commutative rings is the partially ordered set 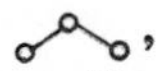 , with the corresponding lattice . It is the factor lattice of the element $x(x+1)y$ in the free algebra on x and y (cf. [3, 17]).

We can specialize 2-firs still further by requiring the set $L(cR, R)$ of principal right ideals to be a chain. Any element c with this property is said to be **rigid**, and an integral domain in which all non-zero elements are rigid is called a **rigid domain**. A commutative domain is rigid precisely if it is a valuation ring (by definition of the latter), and a rigid commutative UFD is a discrete valuation ring.

Among non-commutative rings a typical example is the ring of formal power series in several non-commuting indeterminates over a field: $k \langle\langle x_1, \cdots, x_n \rangle\rangle$ [17]. More generally, an integral domain is rigid if and only if it is a 2-fir and a local ring. An element c in an atomic 2-fir is rigid whenever all the factors of c in a complete factorization generate a proper ideal [29, 17].

A right discrete valuation ring may be defined as an integral domain R with an atom p such that every non-zero right ideal has the form $p^n R$ and $\cap p^n R = 0$. Then a rigid UFD is a right discrete valuation ring if and only if it contains a non-unit c such that cR meets every non-zero right ideal of R non-trivially [17].

8. Remarks on the definition of non-commutative UFD. In some respects the definition of non-commutative UFD given in section 5, is not entirely satisfactory: there is no analogue to Nagata's theorem (Th. 4). Any reasonable analogue should enable one to prove that the free algebra $Z\langle x_1, \cdots, x_n \rangle$ over the integers is a UFD, but this is not the case according the above definition, as the factorizations

$$xyx + 2x = x(yx+2) = (xy+2)x$$

show. They are complete factorizations of $xyx + 2x$, but $xy + 2$ is not similar to $yx + 2$.

In order to describe the various possibilities that can arise, let us assume that we have an equivalence relation $\mathfrak{q}$ defined on the set of non-zerodivisors in each ring such that

E. 1. *If* $a \mathfrak{q} a'$ *and* a *is an atom, then so is* a',

E. 2. *In a commutative integral domain,* $a \mathfrak{q} a'$ *if and only if* a *is associated to* a'.

We shall call R a $\mathfrak{q}$-UFD if every element not zero or a unit has a complete factorization into atoms, and given any two such factorizations of the same element:

$$c = a_1 \cdots a_r = b_1 \cdots b_s,$$

we have $r = s$ and there exists a permutation $i \mapsto i'$ of $\{1, \cdots, r\}$ such that $a_i \mathfrak{q} b_{i'}$. For example, the class of UFD's defined in section 5 may now be described more accurately as **similarity-UFD's.** As we remarked earlier, $Z\langle x, y \rangle$ is not a similarity-UFD, and we therefore try to find a wider equivalence $\mathfrak{q}$ than similarity, for which this ring is a $\mathfrak{q}$-UFD. A number of different choices for $\mathfrak{q}$ have been proposed; all are wider than similarity and in addition to E. 1–2 satisfy

E. 3. *In an atomic 2-fir,* $\mathfrak{q}$ *reduces to similarity.*

Their usefulness depends on the ease with which $\mathfrak{q}$ can be checked; apart from this, the main requirement is that the property of being a $\mathfrak{q}$-UFD should be reflected by taking rings of fractions (i.e., Nagata's theorem). To describe this property we must first define primes in non-commutative rings.

An element c of a ring R is said to be **invariant** if c is a non-zerodivisor such that $cR = Rc$. E.g., in a commutative integral domain every non-zero element is invariant. In general the condition on c just states that the left multiples of c are the same as the right multiples; for an invariant element c we can therefore write $c \mid b$ without ambiguity to indicate that b is a multiple of c. Now a **prime** is defined as an invariant non-unit p in R, such that

$$p \mid ab \text{ implies } p \mid a \text{ or } p \mid b.$$

Clearly any product of invariant elements is invariant; thus if S is a multiplicative set consisting of prime products, then every $s \in S$ is invariant and hence, for each $a \in R$, there exists $a' \in R$ satisfying $as = sa'$. This shows that the pair R, S satisfies the Ore right multiple condition, and so S is a right denominator set [**15**], which can be used to form the ring of fractions R_S. We note that since S consists of non-zerodivisors, the natural homomorphism $\lambda: R \to R_S$ is injective, so we may take R to be embedded as subring in R_S.

Suppose now that we have an equivalence $\mathfrak{q}$ defined on each ring R, satisfying E. 1–3 and moreover,

E. 4. *Let R be a ring and S a multiplicative subset consisting of prime products. If $a, a' \in R^*$ are such that $a \,\mathfrak{q}\, a'$ in R_S, then $a \,\mathfrak{q}\, a'$ in R.*

With the help of this condition it is possible to establish an analogue of Nagata's theorem. Thus let R be an atomic integral domain and S a multiplicative set consisting of prime products, such that R_S is a $\mathfrak{q}$-UFD, where $\mathfrak{q}$ is an equivalence relation satisfying E. 1–4. Then R is also a $\mathfrak{q}$-UFD; for given two atomic factorizations

$$c = a_1 \cdots a_r = b_1 \cdots b_s, \tag{10}$$

if one of $a_1, \cdots, a_r, b_1, \cdots, b_s$ is a prime, it is a left factor of c and so may be taken to be a_1 say. Thus $a_1 \mid b_1 \cdots b_s$ and by primeness, $a_1 \mid b_i$ for some i; since b_i is an atom, it must be associated to a_1 and so b_i is also prime. But then we can divide (10) by a_1 and use induction on $r + s$ to complete the proof that every element has a unique factorization into atoms. We may therefore assume that no a_i or b_j is prime; then it follows that the a_i and b_j cannot divide any element of S. Going over to R_S we find that each a_i, b_j is still an atom in R_S. Since R_S is a $\mathfrak{q}$-UFD, $r = s$ and there is a permutation $i \mapsto i'$ of $1, \cdots, r$ such that $a_i \,\mathfrak{q}\, b_{i'}$ in R_S, and by E. 4, $a_i \,\mathfrak{q}\, b_{i'}$ in R. Thus we have proved:

THEOREM 7. *Let $\mathfrak{q}$ be an equivalence on rings satisfying E. 1–4. If R is an atomic integral domain and S a multiplicative set consisting of prime products such that R_S is a $\mathfrak{q}$-UFD, then R is a $\mathfrak{q}$-UFD.*

What was said earlier shows that similarity is an equivalence relation satisfying E. 1–3 but not E. 4. Since we are looking for an equivalence $\mathfrak{q}$ wider than similarity, two elements a, b defining isomorphic modules R/aR, R/bR will lie in the same $\mathfrak{q}$-class, so that $\mathfrak{q}$ can be described in terms of the category $\mathscr{C}_R$ of strictly cyclic modules. Brungs in [**8**] defines a preordering '$\prec$' on R by putting $a \prec b$ whenever there is an injective homomorphism $R/aR \to R/bR$. The associated equivalence: '$a \prec b$ and $b \prec a$' is called **(right) subsimilarity**. It satisfies E. 1–3 and in place of E. 4 satisfies an analogous condition (with a rather more complicated definition of prime), which enables one to prove, e.g., that any free algebra over $\mathbf{Z}$ is a subsimilarity-UFD.

A still wider notion of equivalence was introduced in [**14**]: We again define a preordering by putting a before b whenever there is a monomorphism $R/aR \to R/bR$ (i.e., a right cancellative map in $\mathscr{C}_R$); the associated equivalence is called **right monosimilarity**. This satisfies E. 1–3 and also E. 4 if we limit ourselves to multiplicative sets S in the centre of the ring. In this way we reach the notion of a **right monosimilarity-UFD**; dually one can define **right episimilarity-UFD's,** on replacing mono- by epimorphisms, but by the factorial duality this is the same as a left monosimilarity-UFD [**14**].

Still wider notions of equivalence are possible. Let us call two elements a, b of a ring R **left coprime** if they have no common left factor apart from units, and define right coprime similarly. A relation

$$ab' = ba' \tag{11}$$

is said to be **coprime** if a, b are left coprime and a', b' right coprime. It can be shown ([**17**] p. 126) that in a 2-fir two elements a, a' are similar if and only if they can be put in a coprime relation (11). This shows that the relation between a and a', expressed by a coprime equation (11) in a 2-fir, is an equivalence. In general rings this is not so, but we can construct an equivalence as follows. Let us say that a, a' (in that order) are **perspective** if they can be put in a coprime relation (11); two elements a, a' will be called **projective** if there is a chain $a_0 = a, a_1, \cdots, a_n = a'$ such that for $i = 1, \cdots, n$ either a_{i-1}, a_i or a_i, a_{i-1} are perspective. Then projectivity is an equivalence, in fact it is the equivalence 'generated' by perspectivity, and by what has been said, it reduces to similarity in 2-firs. This relation has been studied by Beauregard [**2**] for certain classes of rings. Let us define a **weak HCF-ring** as an integral domain R such that for each $c \in R^*$ the set $L(cR, R)$ is a modular lattice, relative to the ordering by inclusion. By the factorial duality this notion is left-right symmetric, and in the commutative case, it reduces to the notion of HCF-ring considered in section 6. Beauregard [**2**] shows that every atomic weak HCF-ring is a projectivity-UFD. Now it is not hard to verify that projectivity is an equivalence satisfying E. 1–4, therefore Th. 7 applies in this case.

To give an example of a ring which definitely falls outside all these definitions of UFD, let us take the Weyl algebra, i.e., the ring A generated by elements x, y with the defining relation $xy - yx = 1$ over a field k of characteristic not 2. This ring is a Noetherian domain and hence has a skew field of fractions. Let S be the set of all non-zero polynomials in y, then S is a right denominator set and the ring A_S of fractions is a principal ideal domain and hence a (similarity-)UFD. However, we have the following factorizations in A:

$$xyx + x = (xy + 1)x = x^2y.$$

It is easily checked that $x, y, xy + 1$ are atoms, so not even the number of factors in a complete factorization is constant.

9. Factorizing zerodivisors. So far we have confined ourselves almost entirely to integral domains. A very similar theory is possible for the factorization of non-zero-divisors in general rings, but this is of less interest and so has not received as much attention. For 'full' matrices over firs there is a fairly satisfactory theory (cf. [17], ch. 5, and, for an application, [16]). The corresponding theory for rectangular matrices still faces difficulties, in that not all complete factorizations of a given matrix have the same number of factors [17].

Finally there is the problem of factorizing zero-divisors. A definition of commutative unique factorization ring (with zero-divisors) has been given by Fletcher [21, 22], who shows that the unique factorization rings so defined are just the finite direct products of UFD's and 'special' principal ideal rings (i.e., homomorphic images of discrete valuation rings). The main difficulty in factorizing zero-divisors is that one cannot expect uniqueness; nevertheless legitimate questions can be asked, as is shown by the theorem on the diagonal reduction of matrices over a principal ideal domain, which may be regarded as the prototype of a unique factorization theorem for this case.

References

1. R. A. Beauregard, Infinite primes and unique factorization in a principal right ideal domain, Trans. Amer. Math. Soc., 141 (1969) 245–254.

2. ———, Right LCM-domains, Proc. Amer. Math. Soc., 30 (1971) 1–7.

3. G. M. Bergman, Commuting elements in free algebras and related topics in ring theory, Thesis, Harvard University, 1967.

4. G. Birkhoff, Lattice theory, 3rd ed. (AMS, Providence 1967).

5. L. A. Bokut', Factorization theorems for certain classes of rings without zero-divisors (Russian) I, Algebra i Logika 4, No. 4 (1965) 25–52; II. *ibid.* No. 5 (1965) 17–46.

6. N. Bourbaki, Algèbre commutative, ch. 7 (Hermann, Paris 1965).

7. A. J. Bowtell, On a question of Malcev, J. Algebra, 6 (1967) 126–139.

8. H. H. Brungs, Ringe mit eindeutiger Faktorzerlegung, J. Reine Angew. Math., 236 (1969) 43–66.

9. P. M. Cohn, Noncommutative unique factorization domains, Trans. Amer. Math. Soc., 109 (1963) 313–331; correction 119 (1965) 552.

10. ———, Universal algebra, Harper & Row, New York, London, Tokyo, 1965.

11. ———, Some remarks on the invariant basis property, Topology, 5 (1966) 215–228.

12. ———, Bezout rings and their subrings, Proc. Cambridge Phil. Soc., 64 (1968) 251–264.

13. ———, Free associative algebras, Bull. London Math. Soc., 1(1969) 1–39.

14. ———, Factorization in general rings and strictly cyclic modules, J. Reine Angew. Math., 239/40 (1970) 185–200.

15. ———, Rings of fractions, this MONTHLY, 78 (1971) 596–615.

16. ———, The embedding of firs in skew fields, Proc. London Math. Soc., (3) 23 (1971) 193–213.

17. ———, Free Rings and their relations, Academic Press, London & New York, 1971.

18. R. Dedekind, Über die Theorie der ganzen algebraischen Zahlen, Vieweg, Braunschweig, 1964.

19. Euclid, Elements (–300).

20. H. Fitting, Über den Zusammenhang zwischen dem Begriff der Gleichartigkeit zweier Ideale und dem Äquivalenzbegriff der Elementarteilertheorie, Math. Ann., 122 (1936) 572–582.

21. C. R. Fletcher, Unique factorization rings, Proc. Cambridge Phil. Soc., 65 (1969) 579–583.

22. ———, The structure of unique factorization rings, Proc. Cambridge Phil. Soc., 67 (1970) 535–540.

23. A. Grothendieck, Éléments de géométrie algébrique (PUF, Paris 1960).

24. N. Jacobson, Theory of rings, AMS, Providence, 1943.

25. A. V. Jategaonkar, A counter-example in homological algebra and ring theory, J. Algebra, 12 (1969) 418–440.

26. R. E. Johnson, Unique factorization in a principal right ideal domian, Proc. Amer. Math. Soc., 16 (1965) 526–528.

27. ———, Unique factorization monoids and domains, Proc. Amer. Math. Soc., 28 (1971) 397–404.

28. I. Kaplansky, Commutative rings, Allyn & Bacon, Boston, 1970.

29. E. G. Koševoi, On the multiplicative semigroup of a class of rings without zero-divisors (Russian), Algebra i Logika 5, No. 5 (1966) 49–54.

30. W. Krull, Idealtheorie, Ergeb. d. Math. vol. 4, 3, Springer, Berlin, 1935.

31. E. Landau, Ein Satz über die Zerlegung homogener linearer Differentialausdrücke in irreduzible Faktoren, J. Reine Angew. Math., 124 (1902) 115–120.

32. S. Lang, Introduction to algebraic geometry, Interscience, New York, 1958.

33. A. Loewy, Über reduzible homogene Differentialausdrücke, Math. Ann., 56 (1903) 549–584.

34. M. Nagata, A remark on the unique factorization theorem, J. Math. Soc. Japan, 9 (1957) 143–145.

35. ———, Local rings, Interscience, New York — London, 1962.

36. P. Samuel, Sur les anneaux factoriels, Bull. Soc. Math. France, 89 (1961) 155–178.

37. ———, Anneaux factoriels, Sao Paulo, 1963.

38. ———, Lectures on unique factorization domains, TIFR, Bombay, 1964.

39. ———, Unique factorization, this MONTHLY, 75 (1968) 945–952.

40. S. Stevin, Arithmétique (1585, new ed. 1958).

41. B. L. van der Waerden, Moderne algebraische Geometrie, Springer, Berlin, 1939.

42. O. Zariski and P. Samuel, Commutative algebra I, Van Nostrand, Princeton, 1968.

CORRECTION TO "UNIQUE FACTORIZATION DOMAINS"

P. M. COHN, Bedford College, University of London

The statement "Any Noetherian UFD is a Dedekind domain" (this MONTHLY, 80 (1973) 1–18) should be omitted.

The assertion is of course well known to be false; a correct statement would be: A Dedekind domain is a UFD if and only if it is a principal ideal domain.

I am indebted to Professor J. H. Hays for drawing my attention to this error.

BIBLIOGRAPHIC ENTRIES: EUCLIDEAN RINGS AND UNIQUE FACTORIZATION DOMAINS

The references below are to the AMERICAN MATHEMATICAL MONTHLY.

1. G. B. Birkhoff, Note on certain quadratic number systems for which factorization is unique, 1906, 156–159.

Uses a geometric argument in the complex plane to show that certain rings of algebraic integers are Euclidean.

2. Elizabeth R. Bennett, Factoring in a domain of rationality, 1908, 222–226.

After some introductory historical remarks, considers the factorization of rational integers in the ring of integers in $Q(\sqrt{-6})$.

3. G. Szekeres, A canonical basis for the ideals of a polynomial domain, 1952, 379–386.

Enumerates the ideals in $R[x]$, where R is a principal ideal domain, by finding a "canonical basis" which characterizes the ideal.

4. W. S. Brown, Reducibility properties of polynomials over the rationals, 1963, 965–969.

Proves that "almost all" nth degree polynomials over the rationals, for a fixed positive integer n, are irreducible.

5. E. A. Maier, A proof of the fundamental theorem of arithmetic, 1964, 1116–1117.

The proof avoids greatest common divisors. The method is valid for any Euclidean domain.

6. D. R. Hayes, A Goldbach theorem for polynomials with integral coefficients, 1965, 45–46.

Proves that every polynomial in $Z[x]$ of positive degree is a sum of two irreducible polynomials of the same degree, by using Eisenstein's irreducibility criterion.

7. Bernard Jacobson and Robert J. Wisner, Matrix number theory: an example of non-unique factorization, 1965, 399–402.

(b)

COMMUTATIVE RINGS

DIVISORS OF ZERO IN POLYNOMIAL RINGS*

W. R. SCOTT, The University of Kansas

The purpose of this note is to give a brief proof of the following theorem, proved by Alexandra Forsythe in a note with the above title, this MONTHLY, vol. 50, 1943, pp. 7–8. (*Cf.* also, N. H. McCoy, *Rings and Ideals*, p. 34, Th. 4.)

THEOREM. *Let R be a commutative ring, and let $R[x]$ be the ring of polynomials over R. If f in $R[x]$ is a divisor of zero, then there is a c in R such that $c\neq 0$, $cf=0$.*

Proof. Deny the theorem, and let g be a non-zero polynomial of smallest degree such that $fg=0$. Let

$$f = a_0 + a_1x + \cdots + a_mx^m,$$
$$g = b_0 + b_1x + \cdots + b_nx^n,$$

where $b_n\neq 0$ and $n\geqq 1$. Since $b_nf\neq 0$, $a_ib_n\neq 0$ for some i, and therefore $a_ig\neq 0$. Let

* From AMERICAN MATHEMATICAL MONTHLY, vol. 61 (1954), p. 336.

r be the largest integer such that $a_r g \neq 0$. Then

$$fg = (a_0 + \cdots + a_r x^r)(b_0 + \cdots + b_n x^n) = 0.$$

Hence $a_r b_n = 0$ and deg $(a_r g) < n$. However, $(a_r g)f = 0$, which is a contradiction.

UNIMODULAR COMPLEMENTS*

IRVING REINER, Institute for Advanced Study and University of Illinois

A vector with rational integral components is called *primitive* if the greatest common divisor of its components is unity. A well-known theorem* (valid when the components lie in any principal ideal ring) states that any primitive vector can be completed to a unimodular matrix, that is, given any primitive row vector $\mathfrak{a}$, there exists a matrix of determinant ± 1 whose first row is $\mathfrak{a}$.

Due to the many applications of this theorem, it may be of interest to point out that the result holds in situations more general than that of a principal ideal ring, although the standard proof† depends strongly on the hypothesis that all ideals are principal. In 1911, Steinitz‡ developed the theory of elementary divisors for matrices over the ring R of algebraic integers in an algebraic number field. Even though R is not in general a principal ideal ring, his results imply that if $a_1, \cdots, a_n \in R$ generate the unit ideal in R, then there exists a matrix with first row $(a_1 \cdots a_n)$ whose determinant is a unit in R. The purpose of this note is to give a simple proof of this theorem, valid for any ring R in which classical ideal theory holds (see van der Waerden, *Modern Algebra* II, p. 83. New York, 1950, Ungar), and in particular valid for the case considered by Steinitz.

The theorem clearly holds for $n=1$. When $n=2$, we note that the ideal (a_1, a_2) generated by a_1 and a_2 in R consists of all sums $x_1a_1+x_2a_2$, $x_1 \in R$, $x_2 \in R$. Hence if $(a_1, a_2) = (1)$, there exist $b_1, b_2 \in R$ for which

$$\begin{vmatrix} a_1 & a_2 \\ b_1 & b_2 \end{vmatrix} = 1,$$

so the result holds for $n=2$.

Now let $n>2$, and assume the result established for a primitive k-component row vector, where $k<n$. Observe that if U is unimodular, then $\mathfrak{a} = (a_1 \cdots a_n)$ is completable if and only if $\mathfrak{a}U$ is completable; for if $\mathfrak{a}$ is the first row of the unimodular matrix T, then $\mathfrak{a}U$ is the first row of the unimodular matrix TU. Since U has an inverse, the argument also goes in the other direction.

Next we show there exists $b \in R$ for which $(a_1, \cdots, a_{n-2}, a_{n-1} - ba_n) = (1)$. For let $(a_1, \cdots, a_{n-2}) = \mathfrak{p}_1^{\alpha_1} \cdots \mathfrak{p}_r^{\alpha_r}$ be the factorization of $(a_1, \cdots, a_{n-2})$

* From AMERICAN MATHEMATICAL MONTHLY, vol. 63 (1956), pp. 246–247.

† MacDuffee, Theory of Matrices, Th. 21.1, p. 31. Berlin 1933, Springer.

‡ Math. Ann., Vol. 71, 1911, pp. 328–354.

into powers of distinct prime ideals. We need only choose b so that

$$a_{n-1} \not\equiv ba_n \pmod{\mathfrak{p}_i} \qquad (i = 1, \cdots, r).$$

For any $\mathfrak{p}_i$ dividing a_n, this holds for all b, since such a $\mathfrak{p}_i$ does not divide a_{n-1}. For the $\mathfrak{p}_i$ not dividing a_n, the congruence $a_n x \equiv a_{n-1} \pmod{\mathfrak{p}_i}$ has a unique solution x_i. For these $\mathfrak{p}_i$, we impose the condition

$$b \equiv x_i + 1 \pmod{\mathfrak{p}_i}.$$

These congruences can be solved for b by the Chinese Remainder Theorem, which is certainly valid under the hypotheses on R (see van der Waerden, *op. cit.*, II, p. 41).

Now we have

$$(a_1 \cdots a_{n-1} \quad a_n)\begin{pmatrix} I^{(n-2)} & & 0 \\ & 1 & 0 \\ 0 & & \\ & -b & 1 \end{pmatrix} = (a_1 \cdots a_{n-2} \quad a_{n-1} - ba_n \quad a_n),$$

with the above matrix unimodular. Furthermore, by the induction hypothesis, $(a_1 \cdots a_{n-2}\ a_{n-1} - ba_n)$ is completable to a unimodular matrix V. Hence $(a_1 \cdots a_{n-2}\ a_{n-1} - ba_n\ a_n)$ is completable to

$$\begin{pmatrix} & & a_n \\ & V & 0 \\ & & \vdots \\ 0 & \cdots\ 0 & 1 \end{pmatrix}.$$

This completes the proof.

IF $R[X]$ IS NOETHERIAN, R CONTAINS AN IDENTITY*

R. W. GILMER, JR., Florida State University

The well-known Hilbert basis theorem shows that if R is a commutative ring with identity and if R is Noetherian, then $R[X]$, the polynomial ring in one indeterminate over R, is also Noetherian. In this note we establish the converse of this result.

THEOREM. *If R is a commutative ring such that $R[X]$ is Noetherian, then R contains an identity.*

Proof. The mapping $f(X) \to f(0)$ is a homomorphism of $R[X]$ onto R. Hence R is Noetherian. If now $r \in R$, then the chain of ideals $(r) \subseteq (r, rX) \subseteq \cdots \subseteq (r, rX, \cdots, rX^n) \subseteq \cdots$ contains only finitely many distinct ideals. Thus, for

* From AMERICAN MATHEMATICAL MONTHLY, vol. 74 (1967), p. 700.

some n, $rX^{n+1} \in (r, rX, \cdots, rX^n)$, say

$$(*) \qquad rX^{n+1} = \sum_{i=0}^{n} f_i(X) \cdot rX^i + \sum_{i=0}^{n} n_i \cdot rX^i,$$

where $f_i(X) = \sum_{j=0}^{m_i} f_j^{(i)} X^j$, $f_j^{(i)} \in R$, and each n_i is an integer. Equating coefficients of X^{n+1} in (*) we have $r = \sum_{i=0}^{n} (f_{n-i}^{(i)} \cdot r) = g \cdot r$ where $g = \sum_{i=0}^{n} f_{n-i}^{(i)} \in R$. We then note that for any elements $r_1, r_2, g_1, g_2 \in R$ such that $r_1 = g_1 r_1$ and $r_2 = g_2 r_2$, we have $r_1 = (g_1 + g_2 - g_1 g_2) r_1$ and $r_2 = (g_1 + g_2 - g_1 g_2) r_2$. By induction it follows that if $\{r_i\}_{i=1}^{k}$ is a finite set of elements of R then there exists an element $u \in R$ such that $r_i = u r_i$ for $i = 1, \ldots, k$. Since R is Noetherian, there is a finite subset T of R such that T generates R as an ideal of R. If $e \in R$ is such that $et = t$ for each $t \in T$, then it follows that e is an identity of R.

AN EXISTENCE THEOREM FOR NON-NOETHERIAN RINGS*

ROBERT GILMER, Florida State University

In developing the theory of Noetherian rings, it is desirable to have at hand some examples of non-Noetherian rings. One such example is the ring of polynomials in infinitely many indeterminates over any nonzero ring. A second example is $R[X]$, where R is any nonzero commutative ring without identity [1], but in this case, $R[X]$ is also a ring without identity. We present here a theorem which provides a method for constructing, as subrings of a polynomial ring in finitely many indeterminates, a wide class of commutative rings with identity which are not Noetherian. It should be noted that the proof of the theorem given uses only one result (Lemma 1) outside the basic theory of Noetherian rings, namely, that a finitely generated idempotent ideal of a commutative ring is principal and is generated by an idempotent element. An examination of the proof of this result reveals that even it is obtained as a direct application of Cramer's Rule for determinants over a commutative ring with identity.

THEOREM 1. *Suppose that R is a nonzero commutative ring, that A is a nonzero ideal of R distinct from R, and that $\{X_\lambda\}_{\lambda \in \Lambda}$ is a set of indeterminates over R. The subring $S = R + A[\{X_\lambda\}]$ of $R[\{X_\lambda\}]$, consisting of those polynomials over R having each of their nonconstant coefficients in A, is Noetherian if and only if these three conditions hold*: (1) *Λ is finite*, (2) *R is Noetherian, and* (3) *the ideal A is idempotent.*

Proof. Suppose that S is Noetherian. It is clear that Λ must be finite, for if not, the ideal $A[\{X_\lambda\}]$ of S would not be finitely generated. The mapping on S which sends each polynomial $f \in S$ onto its constant term is a homomorphism from S onto R, so that R is Noetherian. If σ is a fixed element of Λ and if $a \in A$, the ideal $(aX_\sigma, aX_\sigma^2, \cdots, aX_\sigma^m, \cdots)$ of S is finitely generated, so that aX_σ^{m+1}

* From AMERICAN MATHEMATICAL MONTHLY, vol. 77 (1970), pp. 621–623.

$\in(aX_\sigma, \cdots, aX_\sigma^m)$ for some positive integer m:

$$aX_\sigma^{m+1} = \sum_{i=1}^{m} (s_i aX_\sigma^i + n_i aX_\sigma^i)$$

for some $s_i \in S$, $n_i \in Z$. Equating coefficients of X_σ^{m+1} from each side of this equality, we conclude that $a = ba$ for some element b in A. Thus $a \in A^2$, $A \subseteq A^2$, and therefore $A = A^2$.

To prove the other half of Theorem 1, we need to prove a lemma. This result is known (for example, a proof appears in [2], p. 58); we include its proof here for the sake of completeness.

LEMMA 1. *If B is a finitely generated idempotent ideal of a commutative ring T, then B is principal and is generated by an idempotent element.*

Proof. We first assume that T has an identity and we let $\{b_i\}_{i=1}^n$ be a finite set of generators for B. Then $B = B^2 = \sum_{i=1}^n Bb_i$ so that we obtain a system of equations

$$b_k = \sum_{i=1}^{n} s_{ki} b_i,$$

where $s_{ki} \in B$ and where $1 \leqq k \leqq n$. This gives rise to the system of equations

$$\sum_{i=1}^{n} (\delta_{ki} - s_{ki}) b_i = 0 \qquad 1 \leqq k \leqq n,$$

where δ_{ki} is the Kronecker delta.

By Cramer's rule, $db_i = 0$ for $1 \leqq i \leqq n$, where d is the determinant of the matrix $[\delta_{ki} - s_{ki}]$. It is easy to see, however, that d is of the form $1 - b$ for some element b in B. Since $0 = db_i = b_i - bb_i$ for each i, B is the principal ideal generated by b. And since $1 - b$ annihilates B, $(1-b)b = 0$, or $b = b^2$.

If T contains no identity element, we consider a commutative ring T^* obtained by adjoining an identity element to T. Then B is a finitely generated idempotent ideal of T^*, and hence is principal as an ideal of T^*, generated by an idempotent element v. Since T^* is obtained by adjoining an identity element to T, it follows that v also generates B as an ideal of T.

To complete the proof of Theorem 1, we assume that Λ is finite—say $\Lambda = \{1, 2, \ldots, t\}$, that R is Noetherian, and that A is idempotent. Then A is principal and is generated by an idempotent element e. Moreover, R is the direct sum of A and the ideal $B = \{x - ex \mid x \in R\}$. Since R is Noetherian, $A \cong R/B$ and $B \cong R/A$ are Noetherian rings, and A is a ring with identity. Hence, $A[X_1, \ldots, X_t]$ and B are ideals of S which are Noetherian rings. It then follows that

$$S = R + A[X_1, \cdots, X_t] = B \oplus A[X_1, \cdots, X_t]$$

is a Noetherian ring.

COROLLARY. *Suppose that R is a commutative ring with identity containing no idempotent elements other than* 0 *and* 1. *If A is a nonzero ideal of R distinct from R, and if $\{X_\lambda\}_{\lambda\in\Lambda}$ is any set of indeterminates over R, then $S=R+A[\{X_\lambda\}]$ is a commutative ring with identity which is not Noetherian.*

References

1. R. Gilmer, If $R[X]$ is Noetherian, R contains an identity, this MONTHLY, 74 (1967) 700.
2. ———, Multiplicative Ideal Theory, Queen's University, Kingston, Ontario, 1968.

BIBLIOGRAPHIC ENTRIES: COMMUTATIVE RINGS

The references below are to the AMERICAN MATHEMATICAL MONTHLY.

1. N. H. McCoy, Remarks on divisors of zero, 1942, 286–295.

   An expository article on zero divisors and on rings with no nilpotent elements.

2. H. S. Butts and L. I. Wade, Two criteria for Dedekind domains, 1966, 14–21.

   The two criteria are a multiplicative norm or an additive length.

3. J. A. Eagon, Finitely generated domains over Jacobson semi-simple rings are Jacobson semi-simple, 1967, 1091–1092.

   Gives a simple proof of this, with consequences about Hilbert rings.

4. Herbert Elliot, Jr., Transcendentals and generators in commutative polynomial rings, 1969, 267–270.

   Characterizes those elements y in the polynomial ring $R[x]$ such that $R[y]\cong R[x]$.

(c)

MODULES

FINITELY GENERATED MODULES OVER PRINCIPAL IDEAL DOMAINS*

W. J. WONG, University of Otago, Dunedin, New Zealand

It is well known that every finitely generated module over a principal ideal domain is a direct sum of cyclic submodules. The object of this note is to provide a short simple proof of this result. The ideas are similar to those of Rado's proof of the corresponding theorem for finitely generated Abelian groups [1].

Let M denote a finitely generated module over a principal ideal domain R. Let the letters x, y denote elements of M, and the letters a, b, c elements of R. If $a\neq 0$, let $f(a)$ denote the number of prime factors occurring in a factorization

* From AMERICAN MATHEMATICAL MONTHLY, vol. 69 (1962), pp. 398–400.

of a (counting multiplicities); put $f(0)=\infty$. With the understanding that every integer is less than ∞, we see that, if a divides b, then $f(a)\leqq f(b)$, equality holding if and only if a and b are associates. In particular, $f(a)=0$ if and only if a is a unit. For an element x of M, the set of all a for which $ax=0$ is an ideal of R; we call a generator of this ideal the *order* $o(x)$ of x. $o(x)$ is determined to within associates.

The symbol $\{y_1, \cdots, y_m\}$ denotes the submodule generated by $y_1, \cdots, y_m$. If $\{y_1, \cdots, y_m\}=M$, and the relation $a_1y_1+\cdots+a_my_m=0$ implies that $a_iy_i=0$ for every i, the y_i are said to constitute a *basis* of M.

LEMMA. *Let* $M=\{x_1, \cdots, x_m\}$, *and* $a_1x_1+\cdots+a_mx_m=0$, *where not all the* a_i *are* 0. *Then there are elements* $y_1, \cdots, y_m$ *of* M *such that* $M=\{y_1, \cdots, y_m\}$, *and the order of* y_1 *divides every* a_i.

Proof. We notice first the truth of the result if one of the a_i is a unit. For example, if a_1 is a unit, x_1 is a linear combination of the other x_i, so that we may take $y_1=0$, $y_i=x_i$, $i=2, \cdots, m$.

Let $s=\sum f(a_i)$, the sum being taken over those i for which a_i is nonzero; then s is a nonnegative integer. If $s=0$, at least one a_i is a unit; we therefore proceed by induction on s. If only one a_i is nonzero, the result is trivial, so we may assume that a_1 and a_2 are nonzero, and are not units. Let b be their greatest common divisor. There exist elements b_1, b_2, c_1, c_2 of R such that $a_1=bb_1$, $a_2=bb_2$, $b=c_1a_1+c_2a_2$. Then, $b_1c_1+b_2c_2=1$, so that

$$x_1 = b_2(c_2x_1 - c_1x_2) + c_1(b_1x_1 + b_2x_2),$$
$$x_2 = - b_1(c_2x_1 - c_1x_2) + c_2(b_1x_1 + b_2x_2).$$

Hence,

$$M = \{c_2x_1 - c_1x_2, b_1x_1 + b_2x_2, x_3, \cdots, x_m\},$$
$$b(b_1x_1 + b_2x_2) + a_3x_3 + \cdots + a_mx_m = 0.$$

Since $f(b)\leqq f(a_1)<f(a_1)+f(a_2)$, the induction hypothesis applies, and provides us with elements $y_1, \cdots, y_m$ which generate M, such that $o(y_1)$ divides $b, a_3, \cdots, a_m$. Since b divides a_1 and a_2, $o(y_1)$ divides all a_i.

THEOREM. *M has a basis.*

Proof. The result being trivial if M can be generated by one element, we may suppose that M can be generated by a set of m elements, where $m>1$, and that the result is true for all modules which can be generated by $m-1$ elements. There exists a number n (possibly ∞) which is the least possible value of $f(o(x))$ for elements x which can occur in generating sets of M containing m elements. Let $M=\{x_1, \cdots, x_m\}$, where $f(o(x_1))=n$.

If $M=\{x_1\}+\{x_2, \cdots, x_m\}$ (direct sum), the induction hypothesis gives a basis for $\{x_2, \cdots, x_m\}$ which together with x_1 forms a basis for M. In the contrary case, we have $a_1x_1+\cdots+a_mx_m=0$, where $a_1x_1\neq 0$. Let b be the greatest common divisor of a_1 and $o(x_1)$. Then, $f(b) < f(o(x_1))$, for else b and $o(x_1)$ would be associates, so that $o(x_1)$ would divide a_1, contradicting the fact that $a_1x_1 \neq 0$. There exist c_1, c_2 such that $b = c_1a_1 + c_2o(x_1)$. Then,

$$bx_1 + c_1a_2x_2 + \cdots + c_1a_mx_m = 0.$$

By the lemma, $M = \{y_1, \cdots, y_m\}$, where $o(y_1)$ divides b, so that $f(o(y_1)) \leqq f(b) < n$, an impossibility.

Remark. The usual definition of a basis requires that it contain only nonzero elements. Such a basis may be obtained from a basis as defined above by dropping any zero elements; hence the existence theorem is still valid with the more restricted definition.

Reference

1. R. Rado, A proof of the basis theorem for finitely generated Abelian groups, J. London Math. Soc., 26 (1951) 74–75.

MODULES OVER COMMUTATIVE RINGS*

W. G. LEAVITT, University of Nebraska

The following is another short proof of the fact that for a commutative ring with unit R, any finitely based R-module is "dimensional" in the sense that all of its bases have the same number of elements.

THEOREM. *Let R be a commutative ring with unit. If M is a unitary R-module with a basis of n elements, then all bases of M contain exactly n elements.*

Proof. (The method is that of [1], p. 115.) Let $\{\alpha_i\}$ $(i=1, \cdots, n)$ be a basis for M. It is easy to see that M cannot have an infinite basis. (See [2], p. 241–2. Applied to modules, the method shows that for a module with an infinite basis all bases have the same cardinality.) Thus let $\{\beta_j\}$ $(j=1, \cdots, m)$ be another basis of M. Write $\alpha_i = \sum_{j=1}^{m} a_{ij}\beta_j$ $(i = 1, \cdots, n)$ and $\beta_j = \sum_{k=1}^{n} b_{jk}\alpha_k$ $(j=1, \cdots, m)$. If $A = [a_{ij}]$ and $B = [b_{ij}]$, it follows from the independence of the α_i's and the β_j's that

$$(1)\ AB = I_n \quad \text{and} \quad (2)\ BA = I_m,$$

where I_n and I_m are unit matrices. Conversely, the existence of relations (1) and (2) in a ring R implies the existence of an R-module with bases of lengths m and n, namely the module of all m-tuples. This module has, of course, the rows of I_m as a basis, but also has as an alternative basis the rows of A. This is clear, since from (2) each row of I_m is a linear combination of the rows of A, while from (1), $XA = 0$ implies $XI_n = X = 0$, so the rows of A are independent.

Now any homomorphism of R preserves the relations (1) and (2), and so any nonzero homomorphic image of R also admits a module with bases of lengths

* From AMERICAN MATHEMATICAL MONTHLY, vol. 71 (1964), pp. 1112–1113.

m and n. But if we apply Zorn's lemma in the usual way (relative to ideals not containing the unit, partially ordered by set inclusion) we obtain a maximal ideal I of R. Since R/I is a field, its modules are vector spaces all of whose bases are of the same length. Thus since R/I is a homomorphic image of R, we must conclude that $m=n$.

References

1. W. G. Leavitt, The module type of a ring, Trans. Am. Math. Soc., 103 (1962) 113–130.
2. N. Jacobson, Lectures in abstract algebra, vol. II, Van Nostrand, New York, 1953.

NOTE ON MODULES*

ROLANDO E. PEINADO, State University of Iowa

It is well known and not too difficult to show (see [3]) that for a left (right) vector space over a division ring, there exists a basis, and all bases have the same cardinality. The same fact about bases is no longer true in a general module over an arbitrary ring. Even if it has bases, in case these bases are finite they do not necessarily have the same cardinality. Everett in [1] and Leavitt in [4] have constructed examples. Consider the following "simpler" example.

Let **V** be a left vector space over a division ring D, with a countable infinite basis $\{a_i\}$. Let R be the ring of endomorphisms of **V** into **V**, where addition is the obvious one and product is defined as composition of mappings. R is a left module over itself, and has a basis of a single element namely the identity endomorphism. But $\{u_1, u_2\}$ also form a basis, where we define for all $n=1, 2, \cdots$

$$\begin{aligned} a_{2n-1}v_1 &= 0 & a_{2n}v_1 &= a_n \\ a_{2n-1}v_2 &= a_n & a_{2n}v_2 &= 0 \\ a_nu_1 &= a_{2n} & a_nu_2 &= a_{2n-1} \end{aligned}$$

where u_i and v_i $(i=1, 2)$ are members of R. Then $e=v_1u_1+v_2u_2$ is clearly the identity endomorphism, and hence $\{u_1, u_2\}$ are generators for R. The R-modules Ru_1 and Ru_2 form a direct sum, since if $z\in Ru_1\cap Ru_2$ then $z=du_1=bu_2$. But since $u_1v_2=u_2v_1=0$, this implies $z=ze=0$. Thus $\{u_1, u_2\}$ is an independent set of generators, and therefore a basis. Similarly we can also construct a basis of n elements.

Nevertheless if an R-module M, for arbitrary ring R, has an infinite basis, all bases will be infinite and with the same cardinality. This fact is known (follows from results in [2]). The following is an elementary and straightforward proof:

THEOREM. *If an R-module has an infinite basis, then all its bases have the same cardinality.*

Proof. Let M be an R-module with bases $\{u_i\}$, $i\in I$ and $\{v_j\}$, $j\in J$, where I

* From MATHEMATICS MAGAZINE, vol. 37 (1964), pp. 266–267.

and J are index sets. Assume that the cardinality I is infinite. Write

$$v_j = \sum c_{ij} u_i \qquad \text{for all } j \text{ in } J. \tag{1}$$

This immediately shows that every u_i appears in the right hand side of (1) for if not we can write the one not appearing, say u_k, in terms of the v_j's and hence in terms of the remaining u_i's, contradicting the independence of the u_i's. Now associate with each v_j the subset S_j of the $\{u_i\}$ given by its representation. Each u_i appears in at least one of the S_j, as remarked above. Hence cardinality J is infinite, for otherwise the collection of all sets S_j would be finite, since every S_j is finite. But each u_i is in at least one S_j, contradicting cardinality I infinite.

Let c be the cardinality of the set of subsets S_j. Then $c \leqq$ cardinality J, since some of these sets may be repeated. From set theory we have that the cardinality of an infinite union of disjoint, nonempty finite sets is equal to the cardinality of the set of sets. Let S be the disjoint union of the above sets (any repeated u_i is regarded as a separate object); then cardinality $S = c$. But $\{u_i\} \subseteq S$, so cardinality $I \leqq$ cardinality S. Thus cardinality $I \leqq$ cardinality J and, by symmetry, cardinality $J \leqq$ cardinality I. Hence cardinality $I =$ cardinality J.

References

1. C. J. Everett, Jr., Vector spaces over rings, Bull. Amer. Math. Soc., 48 (1942) 312–316.
2. T. Fujiwara, Note on the isomorphism problem for free algebraic systems, Proc. Japan Acad., 31 (1955) 135–6.
3. N. Jacobson, Lectures in abstract algebra, vol. 2, Van Nostrand, New York, 1951.
4. W. G. Leavitt, Modules over rings of words, Proc. Amer. Math. Soc., 7 (1956) 188–193.

BIBLIOGRAPHIC ENTRIES: MODULES

The references below are to the AMERICAN MATHEMATICAL MONTHLY.

1. V. Dlab, The concept of a torsion module, 1968, 973–976.

   Discusses torsion theories for modules over non-commutative rings.

2. Morris Orzech, Onto endomorphisms are isomorphisms, 1971, 357–362.

   Generalizes a result of Vasoncelos on modules over commutative rings, and uses this to give unified proofs of a number of results on modules.

(d)

DIVISION ALGEBRAS

THE CLASSIFICATION OF REAL DIVISION ALGEBRAS*

R. S. PALAIS, Brandeis University

Let D be a finite dimensional division algebra over the field $\mathbf{R}$ of real numbers. One way of stating the fundamental theorem of algebra is to say that if D is commutative (i.e. a field) then D is isomorphic over $\mathbf{R}$ to either $\mathbf{R}$ or the field $\mathbf{C}$ of complex numbers. A famous theorem of Frobenius asserts that if we allow D to be noncommutative then there is only one new possibility: D can be isomorphic over $\mathbf{R}$ to the quaternion algebra of Hamilton. This is an algebra $\mathbf{H}$ of dimension four generated as a vector space by basis elements $1, i, j, k$ which satisfy the multiplication table

$$i^2 = j^2 = k^2 = -1; \qquad ij = -ji = k; \qquad jk = -kj = i; \qquad ki = -ik = j.$$

The proofs of Frobenius' theorem in the literature seem to be of two types. Either they are elementary, but rather computational, e.g. [2], or else they deduce the theorem from sophisticated general results about division algebras, e.g. [1]. We wish to give here a short, self-contained proof which seems both elementary and conceptual. Besides the inevitable use of the fundamental theorem of algebra we use only the simplest facts about the eigenvalues of linear transformations.

Before starting the proof we note that the two-dimensional subspace of $\mathbf{H}$ generated by 1 and i is isomorphic to the complex numbers. If we denote it by $\mathbf{C}$ then $\mathbf{H}$ becomes a vector space over $\mathbf{C}$ (using left multiplication for the scalar operations). Moreover $\mathbf{C}$ is clearly $\{x \in D \mid ix = xi\}$, while the complementary two-dimensional space spanned by j and k is just $\{x \in D \mid ix = -xi\}$. It is this observation which motivates the proof.

Let 1 denote the unit of D. As usual we can think of $\mathbf{R}$ as embedded in D via the map $x \to x \cdot 1$. We may assume $D \neq \mathbf{R}$. Let d be any element of D not in $\mathbf{R}$ and let $\mathbf{R}\langle d \rangle$ denote the two-dimensional subspace $\mathbf{R} + \mathbf{R}d$ spanned by 1 and d. We claim:

(1) $\mathbf{R}\langle d \rangle$ *is a maximal commutative subset of* D, *consisting of all the elements of* D *which commute with* d. *Moreover it is a field isomorphic to* $\mathbf{C}$.

Proof. Choose a subspace F of D of maximal dimension which includes $\mathbf{R}\langle d \rangle$ and is commutative. If $x \in D$ commutes with everything in F then $F + \mathbf{R}x$ is commutative and so must equal F, so $x \in F$ proving that F is a maximal com-

* From American Mathematical Monthly, vol. 75 (1968), pp. 366–368.

mutative subset of D. In particular, if $x \neq 0$ is in F then x^{-1} commutes with everything in F (because $xy = yx \Rightarrow yx^{-1} = x^{-1}y$) so $x^{-1} \in F$ and F is a field. By the fundamental theorem of algebra F is isomorphic over $\mathbf{R}$ to $\mathbf{C}$. In particular, F has dimension two so that $F = \mathbf{R}\langle d \rangle$. Finally, if $x \in D$ commutes with d it commutes with everything in $\mathbf{R}\langle d \rangle = F$, hence belongs to F.

According to (1) we can select an element $i \in D$ such that $i^2 = -1$ and we may identify $\mathbf{R}\langle i \rangle$ with $\mathbf{C}$. We can now view D not merely as a vector space over $\mathbf{R}$, but also as a vector space (of half the dimension) over $\mathbf{C}$ as well, the scalar operations of $\mathbf{C}$ on D being given by multiplication on the left. On the other hand multiplication on the *right* by i can then be interpreted as a (complex) linear transformation T on the (complex) vector space D; i.e., we define

(2) $Tx \equiv xi$.

Since $T^2 = -$(identity), the only possible eigenvalues of T are $+i$ and $-i$; denote by D^+ and D^- the corresponding eigenspaces:

(3) $D^+ = \{x \in D \mid xi = ix\}$, $D^- = \{x \in D \mid xi = -ix\}$.

Of course $D^+ \cap D^- = \{0\}$. We claim moreover

(4) $D = D^+ \oplus D^-$.

This follows immediately from the decomposition $x = \frac{1}{2}(x - ixi) + \frac{1}{2}(x + ixi)$ for all $x \in D$, the two summands being respectively in D^+ and D^- as one checks by (3). We next note that

(5) $D^+ = \mathbf{C}$ and $x, y \in D^- \Rightarrow xy \in D^+$.

The first statement is immediate from (1), the second from (3).

If $D^- = 0$ then by (4) and (5) we have $D = \mathbf{C}$, so let us assume $D^- \neq 0$ and show that D must be isomorphic to $\mathbf{H}$. First of all the real dimension of D must be four, i.e., its complex dimension must be two. This follows from (4), (5) and

(6) $\dim_{\mathbf{C}} D^- = 1$.

Proof. Select any nonzero $\alpha \in D^-$. Then right multiplication by α gives a complex linear transformation on D which is nonsingular (its inverse is right multiplication by α^{-1}), and it interchanges D^+ and D^- by (5) so $\dim_{\mathbf{C}} D^- = \dim_{\mathbf{C}} D^+ = 1$. Moreover

(7) $\alpha^2 \in \mathbf{R}$ and $\alpha^2 < 0$.

Proof. Since by (1) $\mathbf{R}\langle \alpha \rangle$ is a field it contains α^2. But also $\alpha^2 \in \mathbf{C}$ by (5) and therefore $\alpha^2 \in \mathbf{C} \cap \mathbf{R}\langle \alpha \rangle = \mathbf{R}$. If $\alpha^2 > 0$ it would have two square roots in $\mathbf{R}$ hence three square roots in the field $\mathbf{R}\langle \alpha \rangle$ which is impossible by field theory (or more concretely here, because $\mathbf{R}\langle \alpha \rangle \simeq \mathbf{C}$).

By (7) a suitable positive multiple of α is an element j of D^- satisfying $j^2 = -1$. Define $k = ij$, so that by (6), j and k form a basis for D^- over $\mathbf{R}$, and hence by (4) the elements $1, i, j, k$ form a basis for D over $\mathbf{R}$. Since $j, k \in D^-$, they anticommute with i. This together with $i^2 = j^2 = -1$ and $k = ij$ show that $1, i, j, k$ satisfy the multiplication table given above for the quaternions.

References

1. A. Albert, The Structure of Algebras, AMS Colloquium Publication, 1939.
2. I. N. Herstein, Topics in Algebra, Blaisdell, New York, 1964 p. 326.

THE INVOLUTORY ANTI AUTOMORPHISMS OF THE QUATERNION ALGEBRA*

RABE VON RANDOW, Otago University, Dunedin, New Zealand

Let H denote the algebra of quaternions and A^* the set of automorphisms and antiautomorphisms of H. Clearly A^* is a group and, since the square of any antiautomorphism is an automorphism, the automorphisms of H form a subgroup A of index 2 in A^*. Furthermore, as H is a vector space of dimension 4 over the field R of real numbers, A and A^* are clearly subgroups of GL (4, R). The object of this note is to determine the involutory antiautomorphisms of H. If E is an involutory antiautomorphism of H and $P \in A^*$, then obviously $P^{-1}EP$ is again an involutory antiautomorphism of H, and we will say that E and $P^{-1}EP$ are equivalent modulo A^*. We will show that, up to equivalence mod A^*, there are only two involutory antiautomorphisms of H, namely

$$(1)\qquad \begin{aligned} a + bi + cj + dk &\to a - bi - cj - dk, \\ a + bi + cj + dk &\to a - bi + cj + dk. \end{aligned}$$

The first of these is of course the usual conjugation of quaternions. In [1] we consider Hermitian forms over H, utilising the second involutory antiautomorphism. This gives rise to a theory of Hermitian forms completely analogous to that of quadratic forms over C, whereas the usual theory of Hermitian forms over H is analogous to that of quadratic forms over R or Hermitian forms over C.

Determination of A and A^*. Suppose that $S \in A^*$, $S \notin A$, and express S as a real 4×4 matrix with respect to the usual base of H. Then, as

$$(2)\qquad \begin{gathered} S(1) = 1, \qquad (S(i))^2 = (S(j))^2 = (S(k))^2 = -1, \\ S(j) \cdot S(i) = S(k), \qquad S(k) \cdot S(j) = S(i), \qquad S(i) \cdot S(k) = S(j), \end{gathered}$$

an easy calculation shows that S is of the form

$$S = \begin{pmatrix} 1 & 0 \\ 0 & T \end{pmatrix},$$

where T is orthogonal of determinant -1.

Similarly, suppose that $S \in A$. Then the relations (2) and $S(i) \cdot S(j) = S(k)$, $S(j) \cdot S(k) = S(i)$, $S(k) \cdot S(i) = S(j)$ show that S is of the form

$$S = \begin{pmatrix} 1 & 0 \\ 0 & T \end{pmatrix},$$

where T is orthogonal of determinant 1.

* From AMERICAN MATHEMATICAL MONTHLY, vol. 74 (1967), pp. 699–700.

Determination of the involutory antiautomorphisms of H. Let E be an involutory antiautomorphism of H. Then E has the form

$$E = \begin{pmatrix} 1 & 0 \\ 0 & F \end{pmatrix},$$

where F is orthogonal of determinant -1 and involutory, hence symmetric and orthogonal of determinant -1. The eigenvalues of E can therefore only be 1, 1, 1, -1, and 1, -1, -1, -1. Hence E has, with respect to some base in H, one of the following standard forms:

$$\begin{pmatrix} 1 & 0 & 0 & 0 \\ 0 & -1 & 0 & 0 \\ 0 & 0 & 1 & 0 \\ 0 & 0 & 0 & 1 \end{pmatrix}, \quad \begin{pmatrix} 1 & 0 & 0 & 0 \\ 0 & 1 & 0 & 0 \\ 0 & 0 & -1 & 0 \\ 0 & 0 & 0 & 1 \end{pmatrix}, \quad \begin{pmatrix} 1 & 0 & 0 & 0 \\ 0 & 1 & 0 & 0 \\ 0 & 0 & 1 & 0 \\ 0 & 0 & 0 & -1 \end{pmatrix}, \quad \begin{pmatrix} 1 & 0 & 0 & 0 \\ 0 & -1 & 0 & 0 \\ 0 & 0 & -1 & 0 \\ 0 & 0 & 0 & -1 \end{pmatrix}.$$

The first three are equivalent mod A^*, however, and not equivalent to the last one, as they have differing eigenvalues. This leaves the two involutory antiautomorphisms (1), which are clearly the only ones mod A^*, as equivalence mod A^* is nothing but equivalence under orthogonal reduction.

The involutory automorphisms of H are obtained similarly.

Reference

1. R. von Randow, A note on Hermitian forms, this MONTHLY 74 (1967) 575–577.

EQUATIONS IN QUATERNIONS*

IVAN NIVEN, University of Illinois

1. Introduction. We prove the existence of a quaternion root of the equation

$$a(x) = x^m + a_1x^{m-1} + a_2x^{m-2} + \cdots + a_m = 0, \qquad a_m \neq 0, \tag{1}$$

with coefficients from the algebra of real quaternions. The writer had proved this result when m is odd, but the proof was rendered obsolete when Nathan Jacobson pointed out that the result (without restriction on m) can be obtained as a simple consequence of some work of Ore [1]. This is given in detail in §2.

In §3 we give a method for obtaining the roots of (1), which is not very practical in the sense that it involves the simultaneous solving of two real equations of degree $2m-1$. The method used is a generalization of Sylvester's treatment [2] of the quadratic equation corresponding to (1). Sylvester's conclusion that a quadratic equation has six roots is incorrect because he neglects to show that they exist, and also overlooks the possibility of an infinite number of roots; a complete analysis is given in §4 (Theorem 2). The number of roots of (1) is

* From AMERICAN MATHEMATICAL MONTHLY, vol. 48 (1941), pp. 654–661.

discussed in §5 (Theorem 3), necessary and sufficient conditions being given for an infinite number of roots.

The proof given here of the existence of a root of (1) is stated for the general case where the coefficients of the equation are quaternions over any real-closed field R (*i.e.*, no sum of squares in R is equal to -1, and no algebraic extension of R has this property).

Reinhold Baer, on hearing of this existence proof, proved the converse, so that we have the following strong result:

THEOREM 1. *Let D be a non-commutative division algebra with centrum C. Then every equation* (1) *with coefficients from D has a solution in D if and only if C is a real-closed field, and D is the algebra of real quaternions over C.*

The necessity of these conditions is shown in §6; the proof gives a slightly stronger result than stated in the theorem above, since only those equations with coefficients from C are used. The writer is indebted to Jacobson and Baer for permission to give their proofs here.

Note that (1) is a special equation. The most general quadratic term, for example, would have the form $bxcxd$, involving three coefficients. However, the results are valid for equations similar to (1) having all coefficients to the right of the powers of the unknown.

2. The existence of a root. Let the coefficients of (1) be quaternions over a real-closed field R. By replacing the quaternions a_r by their conjugates $\bar{a}_r$, we obtain a polynomial $\bar{a}(x)$. We multiply this on the right by $a(x)$, and allow x to be commutative with the coefficients. Thus we obtain a polynomial $\alpha(x)$ with coefficients in R, which is, by the fundamental theorem of algebra, factorable into linear factors in $R(i, x)$, and hence *a fortiori* in $R(i, j, x)$. Theorem 1 on p. 494 of Ore's paper [1] states that any other factorization of $\alpha(x)$ in $R(i, j, x)$ must also have linear factors. Now $a(x)$ can be factored into irreducible factors, and these are factors of $\alpha(x)$. Hence by Ore's theorem they are linear. Taking $x-c$ as the right linear factor, we can write

$$a(x) = (x^{m-1} + b_1x^{m-2} + \cdots + b_{m-1})(x - c).$$

That $x=c$ is a root of $a(x)=0$ is not apparent from this equation, since we have assumed that x commutes with the coefficients. However, upon rewriting the above equation in a form analogous to (1),

$$a(x) = x^m + (b_1 - c)x^{m-1} + (b_2 - b_1c)x^{m-2} + \cdots + (b_{m-1} - b_{m-2}c)x - b_{m-1}c,$$

we verify immediately that c is a root.

3. A right-division algorithm. The norm n of any quaternion x is defined as the product of x and its conjugate $\bar{x}$; and the addition of x and $\bar{x}$ gives t, the trace of x. It is well known that x satisfies the equation

$$x^2 - tx + n = 0. \tag{2}$$

We now divide $a(x)$ on the right by the expression in this equation, and obtain the algorithm

$$a(x) = q(x^2 - tx + n) + f(a_r, t, n)x + g(a_r, t, n), \tag{3}$$

the remainder being comprised of the last two terms, polynomials in t, n and the coefficients of $a(x)$. The nature of the quotient q does not interest us. Note that the remainder vanishes for any root of equation (1), and conversely.

When $f \neq 0$, the vanishing of this remainder can be expressed in the form

$$x = -\frac{1}{f}\cdot g = -\frac{1}{\bar{f}f}(\bar{f}g), \tag{4}$$

where $\bar{f}$, the conjugate of f, is obtained by replacing each a_r in f by its conjugate $\bar{a}_r$. Since the conjugate of a product equals the product of the conjugates in reverse order, we have

$$\bar{x} = -\frac{1}{\bar{f}f}(\bar{g}f).$$

By multiplication and addition of the last two equations we get the norm and trace of x; thus

$$n = \frac{1}{\bar{f}f}(\bar{g}g), \qquad t = -\frac{1}{\bar{f}f}(\bar{f}g + \bar{g}f), \tag{5}$$

since $\bar{f}f$ and $\bar{g}g$ have real coefficients and are commutative with the other polynomials. These equations may be written in the forms

$$N(t, n) = n\bar{f}f - \bar{g}g = 0, \qquad T(t, n) = t\bar{f}f + \bar{f}g + \bar{g}f = 0, \tag{6}$$

where $N(t, n)$ and $T(t, n)$ are polynomials in t and n with real coefficients.

First we note that any root x_0 of (1) has a trace t_0 and a norm n_0 which satisfy equations (6). This is apparent except when $f(a_r, t_0, n_0) = 0$, in which case equation (4) is meaningless. But in this case equation (3) implies that $g(a_r, t_0, n_0)$ vanishes, and equations (6) are satisfied.

Conversely, any simultaneous real solution (t_0, n_0) of (6) gives one or more roots of (1). First suppose that $f(a_r, t_0, n_0) \neq 0$. The values t_0 and n_0 can be substituted in (4) to give a quaternion x_0, and since these quantities satisfy (2), equation (3) indicates that x_0 is a root of $a(x) = 0$. It is important to note that in this case one solution of equations (6) gives exactly one solution of (1).

On the other hand, if $f(a_r, t_0, n_0) = 0$, then the first equation (6) gives

$$\bar{g}(a_r, t_0, n_0)g(a_r, t_0, n_0) = 0.$$

But the product of a quaternion and its conjugate is zero only if the quaternion is zero, and hence $g(a_r, t_0, n_0) = 0$. Returning to (3), we see that any solution of

$$x^2 - t_0x + n_0 = 0 \tag{7}$$

is also a solution of (1). The above analysis enables us to inquire into the number of roots of (1), but first we must know the number of solutions in quaternions of equation (7).

4. Quadratic equations. Consider the equation

$$x^2 + bx + c = 0, \qquad c \neq 0, \tag{8}$$

b and c being real quaternions. We assume that $t(b)$, the trace of b, is zero, for otherwise the substitution $x = y - \frac{1}{4}t(b)$ gives a quadratic equation with the required property. For example, the substitution $x = y + \frac{1}{2}t_0$ in equation (7) gives

$$y^2 - d = 0, \qquad d = \tfrac{1}{4}t_0^2 - n_0. \tag{9}$$

We shall need the treatment of this equation to complete the discussion of (8). Suppose that $y = y_0 + y_1 i + y_2 j + y_3 ij$, each y with a subscript being real. We substitute in (9) and separate the result with respect to the linearly independent units 1, i, j, and ij, to obtain

$$y_0^2 - y_1^2 - y_2^2 - y_3^2 = d, \qquad y_0 y_1 = y_0 y_2 = y_0 y_3 = 0.$$

If $d \geqq 0$, then $y_1 = y_2 = y_3 = 0$, and the roots of (9) are $\pm\sqrt{d}$. If $d < 0$, then $y_0 = 0$, and we obtain an infinitude of quaternion solutions of (9), corresponding to the real solutions of $y_1^2 + y_2^2 + y_3^2 = -d$. Henceforth we take b and c to be not both real.

Applying the division algorithm of §3 to (8), we obtain the following values for the functions f and g:

$$f = b + t, \qquad g = c - n. \tag{10}$$

Hence equations (6) become

$$nt^2 + nb\bar{b} - c\bar{c} + n(c + \bar{c}) - n^2 = 0, \tag{11}$$

and

$$t^3 + tb\bar{b} - 2nt + t(c + \bar{c}) + \bar{b}c + \bar{c}b = 0, \tag{12}$$

since $b + \bar{b}$ vanishes. Following the theory of §3, we see that if a real solution $t = t_0$, $n = n_0$ of these equations satisfies $f = 0$ and $g = 0$, then we have $b = -t_0$ and $c = n_0$. But t_0 is real, and b has zero trace, so that both are zero; also, c must be real. Hence equation (8) reduces to one of type (9), contrary to hypothesis. Consequently the solutions of (11) and (12) do not satisfy $f = 0$, and this, by §3, implies that each of these real solutions gives exactly one solution of (8); the solution is given by the substitution of the functions (10) in equation (4).

We introduce the notation

$$B = b\bar{b} + c + \bar{c}, \qquad C = c\bar{c}, \qquad D = \bar{b}c + \bar{c}b, \tag{13}$$

noting that B, C, and D are real. First we consider solutions of (11) and (12) with $t = 0$, so that $D = 0$, by (12). Then the possible values of n are given by

(11), which reduces to

$$n^2 - Bn + C = 0. \tag{14}$$

We want real roots; any real root will be positive because of the manner in which equations (11) and (12) were set up. Thus we obtain 0, 1, or 2 roots of (8) according as B^2-4C is negative, zero, or positive.

Finally, we search for solutions of (11) and (12) with $t\neq 0$. We solve (12) for n; thus

$$n = (t^3 + tB + D)/2t, \tag{15}$$

and we substitute this value in (11) to obtain

$$t^6 + 2Bt^4 + (B^2 - 4C)t^2 - D^2 = 0. \tag{16}$$

Each distinct real root of this equation gives us, by use of (15), a root of (8). In order to find the number of real roots of (16), we prove the following:

LEMMA 1. *If B is negative, so is B^2-4C.*

Proof. Since $b\bar{b}$ is not negative, $c+\bar{c}$ must be negative by the hypothesis. We can write

$$B^2 - 4C = b\bar{b}B + b\bar{b}(c + \bar{c}) + (c - \bar{c})^2.$$

If c has the form $c_0+c_1i+c_2j+c_3ij$, then the last term on the right side of this equation equals $-4(c_1^2+c_2^2+c_3^2)$. Hence the three terms on the right are real, and none of them is positive. They cannot all be zero, for that would imply that $b=0$ and $c=\bar{c}$, contrary to hypothesis, and this proves the lemma.

We consider (16) as a cubic in t^2, and look for positive roots. If $D\neq 0$, the number of positive roots is one by Descartes' rule of signs and Lemma 1. Thus equation (16), considered as a sextic, has two real roots when $D\neq 0$.

If $D=0$, we divide the obvious zero roots out of (16), and have

$$t^4 + 2Bt^2 + B^2 - 4C = 0. \tag{17}$$

Considering this as a quadratic equation, we see that the discriminant is not negative, so that the roots are real. If B^2-4C is positive, B is positive by Lemma 1, and the quadratic (17) has negative roots. Hence the quartic (17) has no real root. Similarly, if B^2-4C is zero, we find that the quartic (17) has no real roots other than zeros. In both these cases, all roots of (8) are obtained from (14). Finally, if B^2-4C is negative, the quartic (17) has exactly two real roots, giving two solutions of (8). Note that in this case no solutions of (8) result from (14).

We summarize these results in the following:

THEOREM 2. *Consider the quaternion equation* (8), *the trace of b being zero. If b and c are real (so that $b=0$), the equation has an infinite number of roots or just two roots according as c is positive or negative. Otherwise, the equation has one or two roots according as the quantities defined in* (13) *satisfy the relations $D=B^2-4C=0$ or not.*

5. The number of roots of (1). We suppose first that no real solution of equations (6) satisfies $f=g=0$, so that there is a one-to-one correspondence between the roots of (1) and the real solutions of (6). We now need some information about the nature of the functions f and g of equation (3).

LEMMA 2. *The functions f and g of equation* (3) *are of degree $m-1$ in n and t; moreover, f has only one term of this degree, namely t^{m-1}. Also, every term of g is divisible by n, with the exception of a_m.*

PROOF. The proof is by induction on m. Equations (10) indicate the truth of the lemma in case $m=2$. We now obtain recurrence relations for f and g. The polynomial of degree $m+1$ analogous to $a(x)$ can be written in the form $a(x)\cdot x+a_{m+1}$, and corresponding to the algorithm (3) we have

$$\begin{aligned} a(x)\cdot x + a_{m+1} &= qx(x^2 - tx + n) + fx^2 + gx + a_{m+1} \\ &= (qx + f)(x^2 - tx + n) + ftx - fn + gx + a_{m+1}. \end{aligned}$$

Calling the remainder in the last expression above $Fx+G$, we have the relations $F=ft+g$ and $G=-fn+a_{m+1}$. The induction is completed by noting that if the functions f and g have the properties stated in the lemma, so do F and G, m being replaced by $m+1$.

It is a consequence of the above lemma that equations (6) are of degree $2m-1$ in n and t. Now it is known [3] that the curves represented by (6) cannot have more than $(2m-1)^2$ intersections provided that the polynomials N and T are relatively prime; and this is the case when neither of the two resultants of N and T vanishes identically. We now show that this is the case.

By Lemma 2, equations (6) can be written in the forms

$$\begin{aligned} N(t, n) &= c_0 + c_1 t + \cdots + c_{2m-3}t^{2m-3} + c_{2m-2}t^{2m-2}, \\ T(t, n) &= d_0 + d_1 t + \cdots + d_{2m-2}t^{2m-2} + t^{2m-1}, \end{aligned}$$

the coefficients c_r and d_r being polynomials in n. Then the resultant obtained by eliminating t is

$$\left|\begin{array}{ccccccccccc} c_0 & c_1 & \cdot & \cdot & \cdot & c_{2m-2} & 0 & \cdot & \cdot & \cdot & 0 \\ 0 & c_0 & c_1 & \cdot & \cdot & c_{2m-3} & c_{2m-2} & 0 & \cdot & \cdot & 0 \\ \cdot & \cdot & \cdot & \cdot & \cdot & \cdot & \cdot & \cdot & \cdot & \cdot & \cdot \\ 0 & \cdot & \cdot & \cdot & 0 & c_0 & c_1 & \cdot & \cdot & \cdot & c_{2m-2} \\ d_0 & d_1 & \cdot & \cdot & \cdot & d_{2m-2} & 1 & 0 & \cdot & \cdot & 0 \\ 0 & d_0 & d_1 & \cdot & \cdot & \cdot & d_{2m-2} & 1 & \cdot & \cdot & 0 \\ \cdot & \cdot & \cdot & \cdot & \cdot & \cdot & \cdot & \cdot & \cdot & \cdot & \cdot \\ 0 & \cdot & \cdot & \cdot & d_0 & d_1 & \cdot & \cdot & \cdot & d_{2m-2} & 1 \end{array}\right| \begin{array}{l} \left.\vphantom{\begin{array}{c}1\\1\\1\\1\end{array}}\right\} 2m-1 \text{ rows} \\ \left.\vphantom{\begin{array}{c}1\\1\\1\\1\end{array}}\right\} 2m-2 \text{ rows} \end{array}$$

This is a polynomial in n; in order to show that it does not vanish identically, we prove that it has a non-zero constant term. When we set $n=0$, Lemma 2 and equation (6) show that each $c_r=0$ for $r=1, 2, \cdots, 2m-2$, whereas c_0 assumes the value $-\bar{a}_m a_m$; that is, all elements of the determinant above the principal diagonal vanish. Hence the value of the constant term of this resultant is $(-\bar{a}_m a_m)^{2m-1}$, which is not zero.

Having shown that the polynomials f and g have no common factor involving t, we now eliminate the possibility that they have a polynomial in n alone as a common factor. We could show that this is not possible by proving that the resultant which eliminates n does not vanish identically. But it is easier to proceed directly. If a polynomial in n divides T, it must divide the coefficient of the highest power of t. But this coefficient is unity.

Having shown that equation (1) cannot have more than $(2m-1)^2$ roots when no real solution of (6) satisfies $f=g=0$, we turn to the case where these equations are satisfied by certain solutions of (6). Let there be s such real solutions. Then we have s equations of type (7), each having either two roots or an infinite number of roots (by Theorem 2). If the number of roots is finite, that is, if these equations have two roots each, we divide them out of equation (1). Thus we obtain an equation of degree $m-2s$, which has no factors of the form (7), and hence has at most $(2m-4s-1)^2$ quaternion roots. Adding $2s$ to account for the roots of the quadratic equations, we note that

$$2s + (2m - 4s - 1)^2 < (2m - 1)^2,$$

when $m \geqq 2s$ and $s>0$. Hence we have shown that if the number of roots of (1) is finite, it cannot exceed $(2m-1)^2$. Theorem 2, and in particular equation (9), can be used now to give the following result:

THEOREM 3. *Equation* (1) *has an infinite number of quaternion roots if and only if* $a(x)$ *is divisible by an expression of type* (7), *with the real values* t_0 *and* n_0 *satisfying the inequality* $t_0^2<4n_0$. *If the number is finite, it cannot exceed* $(2m-1)^2$.

6. The necessity of the conditions of Theorem 1. Denote by u the order of D over its centrum C. Thus there exist u elements in D which are linearly independent over C, but any $u+1$ elements in D are linearly dependent over C. Given any element x in D, there exist therefore elements c_i in C such that $x^u+\sum_{i=0}^{u-1} c_i x^i=0$. Since D is a division algebra, it follows now that the sub-field $C(x)$ of D which is generated by adjoining the element x of D to C, is a commutative field, finite over C, and the irreducible equation in C which is satisfied by x has a degree not exceeding u. Since every equation in C has a solution in D, this implies that the degrees of irreducible equations in C do not exceed u.

LEMMA 3. *Let A denote the essentially uniquely determined algebraically closed commutative field which contains C and is algebraic over C. Then A is finite over C.*

Proof. Suppose first that C is of characteristic $p\neq 0$. We prove that there is no element t in C such that the equation $z^p-t=0$ has no solution in C. For, if there were such an element, then none of the equations $z^{p^i}-t=0$ would have a solution in C. Since this last equation has the form $(z-t_i)^{p^i}=0$ in the field A, it has one and only one solution in A; since the p^{i-1}th power of this solution is a solution of $z^p - t = 0$, it follows that each of these equations is irreducible in C. But this is impossible, since some p^i is larger than u. Hence every element in C is the pth power of an element in C. Consequently [4], A is separable over C; and this result is also true when the characteristic of C is zero.

If B is some field between A and C, and if B is finite over C, then B is a simple extension of C since it is separable over C. Thus the degree of B over C is equal to the degree of the irreducible equation in C whose solution generates B over C. Since the degrees of irreducible equations in C do not exceed u, it follows that the degrees of finite extensions of C do not exceed u. Consequently, there exists a field M between A and C which is finite of maximal degree over C. If w is any element of A, then $M(w)$ is finite over C. Since the degree of M over C is as large as possible, it follows that M and $M(w)$ have the same degree over C. Hence w is in M, and $M = A$, and the lemma is proved.

Now it follows from a theorem by Artin-Schreier [5] that either C is a real-closed field, or A equals C. The latter is impossible since C, the centrum of a non-commutative division algebra, cannot be algebraically closed. Hence C is a real-closed field; but the only non-commutative division algebra over C is the algebra of real quaternions [6], and this completes the proof.

References

1. Oystein Ore, Theory of non-commutative polynomials, Annals of Math. (II), vol. 34, 1933, pp. 480–508.

2. Outlined by Cayley in P. G. Tait's Quaternions, third edition, Cambridge Press, 1890, pp. 157–159.

3. *Cf.*, Bôcher's Introduction to Higher Algebra, Macmillan, 1927, problem 4 on p. 239, and the theorem on p. 202. The theorem on p. 202 is not precisely what we want here, since we need a proposition about *polynomials* in two variables. Two polynomials $f(x, y)$, $g(x, y)$ have two resultants, R_x obtained by eliminating x, and R_y by eliminating y. Bôcher's proof shows that the vanishing of R_x identically is a necessary and sufficient condition for the two polynomials to have a common factor involving x. But it is possible for $f(x, y)$ and $g(x, y)$ to have a common factor in y alone, with R_x not identically zero; in such a case R vanishes (*e.g.*, consider the polynomials $xy - 4x - 3y + 12$, $xy + y^2 - 4x - 4y$).

4. See E. Steinitz, Algebraische Theorie der Körper, 1930, p. 50.

5. Artin-Schreier, Hamburger Abhandlung, vol. 5, 1927, p. 230, Theorem 4.

6. See A. A. Albert, Structure of Algebras, A.M.S. Coll. Publ. 24, Chap. 9, Theorem 28, p. 146.

WEDDERBURN'S THEOREM AND A THEOREM OF JACOBSON*

I. N. HERSTEIN, Cornell University

In teaching an undergraduate class in modern algebra (whose students, although very bright, were mostly sophomores and so had little algebraic knowledge or technique) the author was faced with the problem of presenting to the class two theorems which long had been among his favorites, namely, Wedderburn's theorem on finite division rings and (in the division ring case) Jacobson's theorem that a ring in which $x^{m(x)}=x$ for all x is commutative. In order to do so he devised the proofs presented here; these proofs may be of interest to others confronted with similar problems.

The only facts needed to follow the proofs and which are not absolutely trivial are:

1. The multiplicative group of a finite field is cyclic.
2. If F is a finite field, $\alpha\neq 0\in F$ then there exist λ, $\mu\in F$ so that $1+\lambda^2-\alpha\mu^2=0$.

Of course, many proofs of Wedderburn's theorem exist. In fact, we presented Wedderburn's original proof [2] and the slight twist, in its "punch-line," introduced by Witt [3]. Other proofs, to name but a few, are those of Artin [1] and Zassenhaus [4]. The proof here is closest in spirit to that of Artin, but seems to be both shorter and more elementary. Moreover, no use is made either of counting or of nontrivial number theory. It is of interest that the proof finally hinges on the fact that the quaternions over a finite field do not form a division ring. This is equivalent to making use of fact (2) in the introduction.

We begin with

LEMMA 1. *Let D be a division ring of characteristic $p>0$ with center Z, and P the prime field with p elements contained in Z. Suppose $a\in D$, $a\notin Z$ is such that $a^{p^n}=a$ for some $n>0$. Then there exists an $x\in D$ such that*

(1) $xax^{-1}\neq a$,

(2) $xax^{-1}\in P(a)$, *the field obtained by adjoining a to P.*

Proof. Define the mapping $\delta: D\to D$ by $\delta(x)=xa-ax$ for all $x\in D$. $P(a)$ is a finite field, and has, say, p^m elements. These all satisfy the equation $u^{p^m}=u$. By a trivial verification we immediately have that $\delta^p(x)=xa^p-a^px$, $\delta^{p^m}(x)=xa^{p^m}-a^{p^m}x=xa-ax=\delta(x)$ for all $x\in D$. Thus $\delta^{p^m}=\delta$.

Now if $\lambda\in P(a)$, $\delta(\lambda x)=(\lambda x)a-a(\lambda x)=\lambda(xa-ax)=\lambda\delta(x)$ since λ commutes with a. Thus if I denotes the identity map on D, $\delta\lambda I=(\lambda I)\delta$ for all $\lambda\in P(a)$. Now the polynomial $u^{p^m}-u$ considered over $P(a)$ has all its p^m roots as the elements of $P(a)$. Thus $u^{p^m}-u=\prod_{\lambda\in P(a)}(u-\lambda)$. Thus since δ commutes with all λI, $\lambda\in P(a)$,

$$0=\delta^{p^m}-\delta=\prod_{\lambda\in P(a)}(\delta-\lambda I).$$

* From AMERICAN MATHEMATICAL MONTHLY, vol. 68 (1961), pp. 249–251.

If for every $\lambda\neq 0\in P(a)$, $\delta-\lambda I$ annihilates no nonzero element in D, then since $\delta(\delta-\lambda_1 I)\cdots(\delta-\lambda_k I)=0$ we would get that $\delta=0$, that is that $xa-ax=0$ for all $x\in D$, forcing $a\in Z$ contrary to hypothesis. Thus there is a $\lambda\neq 0\in P(a)$ and an $x\neq 0\in D$ so that $(\delta-\lambda I)x=0$; that is $xa-ax=\lambda x$ and so $xax^{-1}=a+\lambda\neq a$ $\in P(a)$, proving the lemma.

COROLLARY. *In Lemma* 1 $xax^{-1}=a^i\neq a$ *for some integer* i.

Proof. Let a be of order s, then in the field $P(a)$ all the roots of the polynomial u^s-1 are $1, a, a^2, \cdots, a^{s-1}$. Since xax^{-1} is in $P(a)$ and is a root of this polynomial, $xax^{-1}=a^i$ follows.

We first prove the

THEOREM (Wedderburn). *A finite division ring is a field.*

Proof. Let D be a finite division ring. We assume the theorem to be true for division rings with fewer elements than D.

We first remark that if $a, b\in D$ are such that $b^ta=ab^t$ but $ab\neq ba$, then $b^t\in Z$. For consider $N(b^t)=\{x\in D \mid xb^t=b^tx\}$. $N(b^t)$ is a subdivision ring of D, so if it were not D it would be commutative. Since $a, b\in N(b^t)$ and these do not commute, it must then be that $N(b^t)=D$.

Pick $a\in D$, $a\notin Z$ such that a minimal positive power of a falls in Z. Clearly this minimal power is a prime, r. By the corollary to Lemma 1, there is an $x\in D$ such that $xax^{-1}=a^i\neq a$. Since r is a prime, using the little Fermat theorem, $x^{r-1}ax^{-(r-1)}=a^{i^{r-1}}=a^{1+ru}=aa^{ru}=\lambda a$ where $a^{ru}=\lambda\in Z$. Since $x\notin Z$, $x^{r-1}\notin Z$ by the minimal nature of r; thus by the remark in the paragraph above, $x^{r-1}a$ $\neq ax^{r-1}$, so that $\lambda\neq 1$. Let $b=x^{r-1}$. Thus $bab^{-1}=\lambda a$; consequently $a^r=ba^rb^{-1}$ $=(bab^{-1})^r=\lambda^r a^r$, forcing $\lambda^r=1$. We claim that if $y\in D$ is such that $y^r=1$ then $y=\lambda^i$, for in the field $Z(y)$ there are at most r roots of the polynomial t^r-1 and these are already given by the powers of λ. Now $b^r=\lambda^r b^r=(a^{-1}ba)^r=a^{-1}b^ra$; thus $b^ra=ab^r$, $ba\neq ab$ which implies that $b^r\in Z$. The multiplicative group of Z is cyclic and is generated by an element γ. Thus $a^r=\gamma^n$, $b^r=\gamma^m$. If $n=kr$ then $(a/\gamma k)^r=1$, which would make $a/\gamma k=\lambda^i$ and would lead to $a\in Z$. Thus $r\nmid n$; similarly $r\nmid m$. Let $a_1=a^m$, $b_1=b^n$. Thus $a_1^r=a^{mr}=\gamma^{mn}=b^{nr}=b^r$ and

$$a_1b_1=\mu b_1a_1(\mu\neq 1\in Z) \tag{1}$$

($\mu\neq 1$ since $r\nmid n, m$ so $a_1, b_1\notin Z$) and $\lambda^r=1$. Computing $(b_1^{-1}a_1)^r$ using (1) we arrive at

$$(b_1^{-1}a_1)^r=\mu^{-(1+2+\cdots+(r-1))}b_1^{-r}a_1^r=\mu^{-r(r-1)/2}.$$

If r is odd then since $\mu^r=1$, we have that $(b_1^{-1}a_1)^r=1$. But then $b_1^{-1}a_1=\lambda^i$; and this implies that $a_1b_1=b_1a_1$, a contradiction. Hence if r is odd the theorem is proved.

If $r=2$, then since $\mu^2=1$, $\mu\neq 1$ we have $\mu=-1$. The characteristic must also then be different from 2. Also $\alpha=a_1^2=b_1^2\in Z$, $a_1b_1=-b_1a_1\neq b_1a_1$. In Z there are elements ξ, η so that $1+\xi^2-\alpha\eta^2=0$. But then $(a_1+\xi b_1+\eta a_1b_1)^2=\alpha(1+\xi^2-\alpha\eta^2)$

$=0$. Being in a division ring this yields $a_1+\xi b_1+\eta a_1 b_1=0$. Thus

$$0 \neq 2a_1^2 = a_1(a_1 + \xi b_1 + \eta a_1 b_1) + (a_1 + \xi b_1 + \eta a_1 b_1)a_1 = 0.$$

This contradiction finishes the proof.

We now proceed to prove the

THEOREM (Jacobson). *Let D be a division ring such that for every $x \in D$ there exists an integer $n(x)>1$ so that $x^{n(x)}=x$. Then D is commutative.*

Proof. For $a \neq 0 \in D$, $a^n=a$, $(2a)^m=2a$. Putting $q=(n-1)(m-1)+1$ we have $q>1$, $a^q=a$, $(2a)^q=2a$, so that $(2^q-2)a=0$. Thus D has characteristic $p>0$. If P is the prime field with p elements contained in Z, then $P(a)$ has p^h elements, so that $a^{p^h}=a$. So if $a \notin Z$ the conditions of Lemma 1 are satisfied and so there exists a $b \in D$ such that (1) $bab^{-1}=a^u \neq a$. Suppose $b^{p^k}=b$ and consider

$$W = \left\{x \in D \mid x = \sum_{i,j}^{p^h, p^k} p_{ij} a^k b^j, p_{ij} \in P\right\}.$$

W is finite, is closed under addition, and by virtue of (1) is closed under multiplication. Thus W is a finite ring and being a subring of the division ring D the two cancellation laws hold so it is a finite division ring. But then it is commutative. Since $a, b \in W$ this forces $ab=ba$, contrary to $a^u b=ba$. This proves the theorem.

References

1. Emil Artin, Über einen Satz von Herrn J. H. MacLaglan-Wedderburn, Abh. Math. Sem. Univ. Hamburg, vol. 5, 1927, pp. 245–250.

2. J. H. M. Wedderburn, A theorem on finite algebras, Trans. Amer. Math. Soc., vol. 6, 1905, pp. 349–352.

3. Ernst Witt, Über die Kommutativität endlicher Schiefkörper, Abh. Math. Sem. Univ. Hamburg, vol. 8, 1931, p. 413.

4. Hans Zassenhaus, A group-theoretic proof of a theorem of MacLaglan-Wedderburn, Proc. Glasgow Math. Assoc., vol. 1, 1952, pp. 53–63.

ANOTHER PROOF OF WEDDERBURN'S THEOREM*

T. J. KACZYNSKI, Evergreen Park, Illinois

In 1905 Wedderburn proved that every finite skew field is commutative. At least seven proofs of this theorem (not counting the present one) are known. See [1], [2], [5] (Part Two, p. 206 and Exercise 4 on p. 219), [6] (two proofs), and [7]. Unlike these proofs, the proof to be given here is group-theoretic, in the sense that the only non-group-theoretic concepts employed are of an elementary nature.

LEMMA. *Let q be a prime. Then the congruence $t^2+r^2 \equiv -1 \pmod q$ has a solution t, r with $t \not\equiv 0 \pmod q$.*

---

* From AMERICAN MATHEMATICAL MONTHLY, vol. 71 (1964), pp. 652–653.

Proof. If -1 is a quadratic residue, take $r=0$ and choose t appropriately. Assume -1 is a nonresidue. Then any nonresidue can be written in the form $-s^2 \pmod q$ with $s \not\equiv 0$. If t^2+r^2 is ever a nonresidue for some t, r, set $t^2+r^2 \equiv -s^2$, and we have $(ts^{-1})^2+(rs^{-1})^2 \equiv -1$. (Throughout this note, x^{-1} denotes that integer for which $xx^{-1} \equiv 1 \pmod q$.) On the other hand, if t^2+r^2 is always a residue, then the sum of any two residues is a residue, so $-1 \equiv q-1 = 1+1+ \cdots +1$ is a residue, contradicting our assumption.

Proof of the theorem. Let F be our finite skew field, F^* its multiplicative group. Let S be any Sylow subgroup of F^*, of order, say, p^α. Choose an element g of order p in the center of S. If some $h \in S$ generates a subgroup of order p different from that generated by g, then g and h generate a commutative field containing more than p roots of the equation $x^p=1$, an impossibility. Thus S contains only one subgroup of order p and hence is either a cyclic or a generalized quaternion group ([3] p. 189).

If S is a generalized quaternion group, then S contains a quaternion subgroup generated by two elements a and b, both of order 4, where $ba=a^{-1}b$. Now a^2 generates a commutative field in which the only roots of the equation $x^2=1$ or $(x+1)(x-1)=0$ are ± 1, so since $(a^2)^2=1$, we have

$$a^2 = -1. \tag{1}$$

Hence $a^{-1}=a^3=-a$, so

$$ba = -ab. \tag{2}$$

Similarly,

$$b^2 = -1. \tag{3}$$

Taking $q=$characteristic of F $(q \cdot 1=0)$, choose t and r as specified in the lemma. Using relations (1), (2), (3), we have

$$(t+ra+b)(r^2+1+rta+tb) = r(t^2+r^2+1)a+(t^2+r^2+1)b = 0.$$

One of the factors on the left must be 0, so for some numbers $u, v, w, u \not\equiv 0 \pmod q$, we have $w+va+ub=0$, or $b=-u^{-1}va-u^{-1}w$. So b commutes with a, a contradiction. We conclude that S is not a generalized quaternion group, so S is cyclic.

Thus every Sylow subgroup of F^* is cyclic, and F^* is solvable ([4], pp. 181–182). Let Z be the center of F^* and assume $Z \neq F^*$. Then F^*/Z is solvable, and its Sylow subgroups are cyclic. Let A/Z (with $Z \subset A$) be a minimal normal subgroup of F^*/Z. A/Z is an elementary abelian group of order p^k (p prime), so since the Sylow subgroups of F^*/Z are cyclic, A/Z is cyclic. Any group which is cyclic modulo its center is abelian, so A is abelian. Let x be any element of F^*, y any element of A. Since A is normal, $xyx^{-1} \in A$, and $(1+x)y=z(1+x)$ for some $z \in A$. An easy manipulation shows that $y-z=zx-xy=(z-xyx^{-1})x$.

If $y-z=z-xyx^{-1}=0$, then $y=z=xyx^{-1}$, so x and y commute. Otherwise, $x=(z-xyx^{-1})^{-1}(y-z)$. But A is abelian, and $z, y, xyx^{-1} \in A$, so x commutes with y. Thus we have proven that A is contained in the center of F^*, a contradiction.

References

1. E. Artin, Über einen Satz von Herrn J. H. M. Wedderburn, Abh. Math. Sem. Hamburg, 5 (1927) 245.
2. L. E. Dickson, On finite algebras, Göttingen Nachr., 1905, p. 379.
3. M. Hall, The theory of groups, Macmillan, New York, 1961.
4. Miller, Blichfeldt and Dickson, Theory and applications of finite groups, Wiley, New York, 1916.
5. B. L. van der Waerden, Moderne Algebra, Ungar, New York, 1943.
6. J. H. M. Wedderburn, A theorem on finite algebras, Trans. Amer. Math. Soc., 6 (1905) 349.
7. E. Witt, Über die Kommutativität endlicher Schiefkörper, Abh. Math. Sem. Hamburg, 8 (1931) 413.

BIBLIOGRAPHIC ENTRIES: DIVISION ALGEBRAS

The references below are to the AMERICAN MATHEMATICAL MONTHLY.

1. M. Leum and M. F. Smiley, A matric proof of the fundamental theorem of algebra for real quaternions, 1953, 99–100.
2. K. Rogers, Skew fields of prime characteristic, 1962, 287–288.

   Gives examples.

3. Kenneth O. May, The impossibility of a division algebra of vectors in three-dimensional space, 1966, 289–291.

   Shows that the triples (x,y,z) over the real numbers cannot form a division algebra, when the pairs $(x,y)=(x,y,0)$ behave like the complex numbers. Includes historical remarks.

(e)

NON-COMMUTATIVE RINGS

THE ALGEBRA OF SEMI-MAGIC SQUARES*

L. M. WEINER, DePaul University

1. Introduction. A semi-magic square is a square array of n^2 numbers for which the sum of the n numbers in any row or column is a constant S. If the sum of each diagonal is also equal to S, the array is referred to as a magic square. The objects referred to as numbers will, for the purposes of this paper, be taken to be the elements of a field F whose characteristic is prime to n. Any

* From AMERICAN MATHEMATICAL MONTHLY, vol. 62 (1955), pp. 237–239.

such semi-magic square may be considered as an n by n matrix and will hereafter be called an S-matrix of order n according to the following

DEFINITION. *An S-matrix* $A=(a_{ij})$ *of order n is an n by n matrix for which* $\sum_{i=1}^{n} a_{ij}=\sum_{i=1}^{n} a_{ji}=S(A)$ *for every j.*

THEOREM 1. *The set of all S-matrices of order n forms a subalgebra* R_n *of the total matric algebra of degree n which satisfies the following conditions:*

(i) $$S(A+B) = S(A)+S(B),$$

(ii) $$S(kA) = kS(A) \qquad \text{for } k \text{ in } F,$$

(iii) $$S(AB) = S(A)S(B).$$

Proof. Let $A=(a_{ij})$ and $B=(b_{ij})$ be S-matrices of order n, and $C=(c_{ij})=A+B$. Then $c_{ij}=a_{ij}+b_{ij}$, and

$$\sum_{i=1}^{n} c_{ij} = \sum_{i=1}^{n} (a_{ij}+b_{ij}) = \sum_{i=1}^{n} a_{ij} + \sum_{i=1}^{n} b_{ij} = S(A)+S(B).$$

The same holds for the rows of C. The matrix kA has elements ka_{ij}, and

$$\sum_{i=1}^{n} ka_{ij} = k\sum_{i=1}^{n} a_{ij} = k\sum_{i=1}^{n} a_{ji} = kS(A)$$

for every j. Next let $D=(d_{ij})=AB$. Then

$$d_{ij} = \sum_{k=1}^{n} a_{ik}b_{kj},$$

and

$$\sum_{i=1}^{n} d_{ij} = \sum_{i=1}^{n}\sum_{k=1}^{n} a_{ik}b_{kj} = \sum_{k=1}^{n}\sum_{i=1}^{n} a_{ik}b_{kj} = \sum_{k=1}^{n} b_{kj}\sum_{i=1}^{n} a_{ik} = \sum_{k=1}^{n} b_{kj}S(A) = S(B)S(A).$$

A similar computation holds for the rows of D.

Unfortunately, this result cannot be extended to magic squares as is shown by the following counterexample:

$$A = \begin{pmatrix} 8 & 1 & 6 \\ 3 & 5 & 7 \\ 4 & 9 & 2 \end{pmatrix}, \qquad B = \begin{pmatrix} 6 & 1 & 8 \\ 7 & 5 & 3 \\ 2 & 9 & 4 \end{pmatrix}, \qquad AB = \begin{pmatrix} 67 & 67 & 91 \\ 67 & 91 & 67 \\ 91 & 67 & 67 \end{pmatrix}.$$

The set of all magic squares of order n, however, does form a linear subspace of the algebra R_n.

The remainder of this paper will be devoted to the determination of the structure of the algebra R_n.

2. The decomposition theorem. Unless otherwise stated, the order of the matrices comprising the algebra R_n will be considered fixed throughout this section, and so the subscript n will be dropped in this and the following section.

We shall first exhibit two ideals of the algebra R. Let M be the set of all S-matrices A for which the entries a_{ij} are all equal to a fixed number a, and let N be the set of all S-matrices A for which $S(A)=0$. It is easy to see that M and N are both ideals of R. If $A=(a_{ij})=(a)$ belongs to M and $B=(b_{ij})$ belongs to R, then the element in the ith row and jth column of $AB=a\sum_{k=1}^{n} b_{kj}=aS(B)$ which indicates that AB belongs to M. Similarly, BA belongs to M. If, on the other hand, A belongs to N and B belongs to R, then $S(AB)=S(BA)=S(A)S(B)=0$.

Next let A be any matrix belonging to R, and let $B=A-C$ where C is the matrix all of whose entries are equal to $S(A)/n$. Then C belongs to M, $S(C)=nS(A)/n=S(A)$, $S(B)=S(A)-S(C)=0$, and B belongs to N. Since $A=B+C$, we see that any element of R is expressible as the sum of an element of M and an element of N; that is, the algebra R is the sum of the algebras M and N. This sum is actually a direct sum; for if A belongs to both M and N, A has a fixed entry a in each place, $S(A)=na=0$, $a=0$, and A is the zero matrix of R. We have proved

THEOREM 2. *The algebra R is the direct sum of the one dimensional ideal M and the ideal N.*

3. The structure of the algebra N. Define A_{ij} to be the matrix which has the number 1 in the first row and first column, the number -1 in the first row and jth column, the number -1 in the ith row and first column, the number 1 in the ith row and jth column, and zeros elsewhere for $i, j=2, 3, \cdots, n$.

THEOREM 3. *Let the S-matrices A_{ij} be as defined above for $n>1$, and $i, j=2, 3, \cdots, n$. Then the algebra N has dimension $(n-1)^2$, and the matrices A_{ij} form a basis for N.*

Proof. Assume first that we have a relationship of the form $\sum_{i,j=2}^{n} a_{ij}A_{ij}=0$. When $2\leqq p, q\leqq n$, $a_{pq}A_{pq}$ is the only component of this sum which contributes anything to the entry in the pth row and qth column which is zero. It follows that $a_{pq}=0$ for $p, q=2, 3, \cdots, n$, and the A_{ij} are linearly independent.

Next let $B=(b_{ij})$ be any S-matrix belonging to N, and consider the S-matrix $\sum_{i,j=2}^{n} b_{ij}A_{ij}$. This matrix is equal to B since when $2\leqq p, q\leqq n$, the element in the pth row and qth column is b_{pq}, the element in the first row and qth column is $-\sum_{i=2}^{n} b_{iq}=b_{1q}$, the element in the pth row and first column is $-\sum_{j=2}^{n} b_{pj}=b_{p1}$, and the element in the first row and first column is $\sum_{j=2}^{n}\sum_{i=2}^{n} b_{ij}=-\sum_{j=2}^{n} b_{1j}=b_{11}$. This shows that the S-matrices A_{ij} span N and completes the proof of the theorem.

A RELATIONSHIP BETWEEN SEMI-MAGIC SQUARES AND PERMUTATION MATRICES*

ARTHUR A. SAGLE, University of Washington

A semi-magic square is an n by n matrix A over a field F for which the sum of the n elements in any row or column is a constant $a=S(A)$. It was shown in [1] that the semi-magic squares of order n under the usual matrix addition and multiplication form an algebra $\mathfrak{M}$ over F.

THEOREM. *If* Π *is the set of permutation matrices and* $\{\Pi\}$ *is the algebra generated by* Π *over* F, *then* $\mathfrak{M}=\{\Pi\}$.

Proof. Let $\mathfrak{R}$ be the set of matrices of the form aI where a is in F and let $\mathfrak{N}$ be the set of matrices for which $S(A)=0$. It was shown in [1] that the matrices

$$N_{ij} = i\begin{bmatrix} 1 & 0\cdots0 & \overset{j}{-1} & 0\cdots0\\ 0 & & 0 & \\ \vdots & & \vdots & \vdots\\ 0 & & 0 & \\ -1 & 0\cdots0 & 1 & 0\cdots0\\ 0 & & 0 & \\ \vdots & & \vdots & \\ 0 & \cdots & 0 & \cdots0\end{bmatrix}$$

form a basis of $\mathfrak{N}$ over F. Considering $\mathfrak{M}$ as a vector space over F, $\mathfrak{M}=\mathfrak{R}+\mathfrak{N}$: if A is in $\mathfrak{M}$ where $S(A)=a$, then $A=aI+(A-aI)$ and $A-aI$ is obviously in $\mathfrak{N}$.

Now suppose P is in Π, then $S(P)=1$. Thus P is in $\mathfrak{M}$ and therefore $\Pi\subset\mathfrak{M}$. Since $\mathfrak{M}$ is an algebra the sums and products of any matrices in $\mathfrak{M}$ are again in $\mathfrak{M}$, and since $\Pi\subset\mathfrak{M}$, $\{\Pi\}\subset\mathfrak{M}$. Conversely to show that $\mathfrak{M}\subset\{\Pi\}$ it will be sufficient to show that the basis matrices N_{ij} and I can be written as sums of products of permutation matrices. Since I is in $\{\Pi\}$, we need only show the N_{ij} are in $\{\Pi\}$. Now $N_{ij}=PN_{nn}Q$ for suitable permutation matrices P and Q *i.e.* P interchanges the nth row of N_{nn} and the ith row of N_{nn}, and Q interchanges the nth column of N_{nn} and the jth column of N_{nn}. However

$$N_{nn} = \begin{bmatrix} 1 & & 0\\ & \ddots & \\ 0 & & 1\end{bmatrix} - \begin{bmatrix} 0 & & \cdots 0 & 1\\ 0 & 1 & & 0\\ \vdots & & \ddots & \vdots\\ 0 & & 1 & \\ 1 & 0\cdots0 & & 0\end{bmatrix} = I - P_1$$

* From AMERICAN MATHEMATICAL MONTHLY, vol. 64 (1957), pp. 658–659.

is in $\{\Pi\}$ since I and P_1 are in Π. Therefore $N_{ij}=P(I-P_1)Q=PQ-PP_1Q$ is in $\{\Pi\}$ and hence $\mathfrak{M}\subset\{\Pi\}$.

Reference

1. L. M. Weiner, The algebra of semi-magic squares, this MONTHLY, vol. 62, 1955, pp. 237–239.

ANNIHILATORS IN POLYNOMIAL RINGS*

NEAL H. MCCOY, Smith College

It is known that if f is a divisor of zero in the polynomial ring $R[x]$, where R is a commutative ring, there exists a non-zero element c of R such that $cf=0$. Proofs of this theorem have been given by McCoy [3], Forsythe [2], Cohen [1], and Scott [4]. It is clear that the theorem as stated does not immediately generalize to polynomials in more than one indeterminate. Moreover, it has been pointed out in Problem 4419 of this MONTHLY (1950, p. 692 and 1952, p. 336) that the theorem itself is not necessarily true for noncommutative rings. The purpose of this note is to obtain a suitable generalization of this result to the case of an arbitrary polynomial ring in any finite number of indeterminates.

Let R be an arbitrary ring and $R[x_1, \cdots, x_k]$ the ring of polynomials in the indeterminates $x_1, \cdots, x_k$, with coefficients in R. If A is a right ideal in $R[x_1, \cdots, x_k]$, let us denote by A^r the set of right annihilators of A, that is, the ideal consisting of all elements h of the ring $R[x_1, \cdots, x_k]$ such that $Ah=0$. We shall prove the following theorem.

THEOREM. *If R is an arbitrary ring and A is a right ideal in $R[x_1, \cdots, x_k]$ such that $A^r\neq 0$, then $A^r\cap R\neq 0$.*

The proof is by induction on k, so we begin by considering the case in which $k=1$. Throughout this part of the proof we shall write x in place of x_1.

We assume that $A^r\cap R=0$, and shall obtain a contradiction. From this assumption it follows that if c is a non-zero element of R, then $fc\neq 0$ for some $f\in A$. Let g be a non-zero element of A^r of minimum degree m, and let us set

$$g = b_0x^m + \cdots + b_m \qquad (b_0\neq 0).$$

Then $m>0$, since $A^r\cap R=0$. Let f be an element of A such that $fb_0\neq 0$, and let

$$f = a_0x^n + \cdots + a_n \qquad (a_0 \neq 0).$$

Since $fb_0\neq 0$, we have $a_jg\neq 0$ for some $j=0, 1, \cdots, n$. Choose p as the smallest integer such that $a_pg\neq 0$. Then $fg=0$ yields $a_pb_0=0$, and it follows that the degree of a_pg is less than m. Moreover, $a_pg\in A^r$ since A^r is an ideal in $R[x]$. We therefore have a contradiction, and the theorem is established for the case in which $k=1$. This part of the proof was suggested by the referee and is a simplified version of our original proof.

---

* From AMERICAN MATHEMATICAL MONTHLY, vol. 64 (1957), pp. 28–29.

To complete the proof we must establish the theorem for $k>1$ indeterminates under the assumption that it holds for $k-1$ indeterminates. Any element f of $R[x_1, \cdots, x_k]$ may be considered as a polynomial in x_k with coefficients that are polynomials in $x_1, \cdots, x_{k-1}$. That is, we may write f in the form

$$f = h_0 x_k^n + \cdots + h_n,$$

where the h_i are elements of the ring $R[x_1, \cdots, x_{k-1}]$. For convenience, we may refer to the h_i as the *coefficients* of f. Since $A^r \neq 0$, the case already proved (with R replaced by $R[x_1, \cdots, x_{k-1}]$) shows that there exists a non-zero element g of $R[x_1, \cdots, x_{k-1}]$ such that $Ag=0$. However, since g does not contain x_k, this implies that $hg=0$ for *every* coefficient h of an element of A. Hence if B is the right ideal in the ring $R[x_1, \cdots, x_{k-1}]$ generated by all coefficients of elements of A, it follows that $g \in B^r$ and hence $B^r \neq 0$. By the induction hypothesis, we then have $B^r \cap R \neq 0$. But clearly $B^r \subseteq A^r$, and so $A^r \cap R \neq 0$, completing the proof.

COROLLARY. *If R is a commutative ring and f is a divisor of zero in the polynomial ring $R[x_1, \cdots, x_k]$, then there exists a non-zero element c of R such that $cf=0$.*

This follows at once from the observation that if R is commutative and A is the principal ideal generated by f in $R[x_1, \cdots, x_k]$, then $g \in A^r$ if and only if $fg=0$.

References

1. I. S. Cohen, On the structure and ideal theory of complete local rings, Trans. Amer. Math. Soc., vol. 59, 1946, pp. 54–106.
2. A. Forsythe, Divisors of zero in polynomial rings, this MONTHLY, vol. 50, 1943, pp. 7–8.
3. N. H. McCoy, Remarks on divisors of zero, this MONTHLY, vol. 49, 1942, pp. 286–295.
4. W. R. Scott, Divisors of zero in polynomial rings, this MONTHLY, vol. 61, 1954, p. 336.

ORDERS FOR FINITE NONCOMMUTATIVE RINGS*

D. B. ERICKSON, Concordia College, Moorhead, Minnesota

Introductory Comments. The solution to problem E 1529 (this MONTHLY, 70, 1963, p. 441) answers the question of the order for the smallest noncommutative ring. It is 4. This paper answers the more general question concerning a characterization of the orders for finite noncommutative rings in a series of three theorems. (In view of the first theorem the comment following the solution of E 1529, to the effect that there exists a noncommutative ring of each composite order N, must be incorrect.)

THEOREM 1. *If R is a finite ring of order $n>1$ and if n has square free factorization, then R is a commutative ring.*

* From AMERICAN MATHEMATICAL MONTHLY, vol. 73 (1966), pp. 376–377.

Proof. It is well known that if the additive group of a ring is cyclic then the ring is commutative. The additive group of R is a finite abelian group which, by the Basis Theorem for Finite Abelian Groups, may be expressed as the direct product of s cyclic subgroups of R, say $R_1, \cdots, R_s$, where the order of each R_i divides the order of R_{i+1} $(i=1, \cdots, s-1)$, and s is the number of elements in a minimal generating system (basis) for R. But by the Lagrange Theorem the order of each R_i divides the order of R. Since the order of R is square free, the additive group of R is the cyclic group of order n (abbr. C_n), and therefore R is a commutative ring.

In the next two theorems we construct a noncommutative ring of order m for each positive integer $m>1$ having square factors.

THEOREM 2. *If p is a prime integer, then there exists a noncommutative ring of order p^2.*

Proof. Let R be the direct product of C_p with itself. It is clear that R is an abelian group of order p^2 and that R is not cyclic. A minimal generating system for R is $\{(a, 0), (0, a)\}$, where a is a generator for C_p. We define multiplication on the basis elements by the requirement that the product of two basis elements is the left factor, and extend it to the whole system by the distributive law, whence

$$(j_1a, k_1a)\cdot(j_2a, k_2a) = (j_2 + k_2)(j_1a, k_1a),$$

for all (j_ia, k_ia) in R, where the j_i and k_i are integers. Closure, the associative law, and the distributive laws are easily verified; thus, R is a ring of order p^2. Since $(0, a)(a, 0)=(0, a)$ but $(a, 0)(0, a)=(a, 0)$ R is not commutative.

This last result is extended to all orders np^2 by Theorem 3.

THEOREM 3. *Let R_1 be a ring of order p^2 as constructed in Theorem 2, and let R_2 be any ring of order n, then the ring $R=R_1 \dotplus R_2$ (i.e. the direct sum of R_1 and R_2) is a noncommutative ring of order np^2.*

REMARK. The proof of this statement is obvious from the properties of direct sums of rings. We only remark that for R_2 we may use the trivial ring of order n which has as its additive group C_n and in which all products are zero; note also that R contains a subring isomorphic to R_1 and is therefore noncommutative.

We summarize in the following corollary.

COROLLARY. *If m is a positive integer, $m>1$, then there exists a noncommutative ring of order m if and only if m has square factors.*

Supported by NSF Grant GE-1082.

INVERSES IN RINGS WITH UNITY*

C. W. BITZER, U. S. Naval Ordnance Test Station, China Lake, California

It is believed that the following proof of a theorem by Kaplansky is noteworthy.

If u is an element of a ring R with unit 1 such that u has more than one right inverse, then u has infinitely many right inverses.

Define

$$S = \{x \mid ux = 1\},$$
$$T = \{xu - 1 + s \mid x \in S\},$$

where s is some fixed member of S. Note that by hypothesis S contains at least two distinct elements. Then $T \subseteq S$, since if $xu-1+s \in T$ then $x \in S$, and $u(xu-1+s) = uxu - u + us = 1u - u + 1 = 1$.

The mapping of S onto T given by $x \to xu - 1 + s$ is one-to-one, since if $xu-1+s = yu-1+s$, where $x, y \in S$, then $xu = yu$ and $x = y$ due to the fact that u has a right inverse. If S is finite then $T = S$. In particular $s \in T$ so that for some $x \in S$ we have $xu-1+s = s$ or $xu = 1$. Consequently, for $t \in S$, distinct from x, we arrive at the following contradiction: $x = x(ut) = (xu)t = t$.

Therefore S is infinite.

Reference

1. N. Jacobson, Lectures in Abstract Algebra, Vol. 1, Van Nostrand, Princeton, 1951.

ON THE MATRIX EQUATION $AB = I$†

PAUL HILL, University of Houston

In a course in linear algebra, and elsewhere, it is a great computational convenience to know that if A and B are n-square matrices such that $AB = I$, where I is the identity matrix of order n, then necessarily $BA = I$ and B is the inverse of A. A standard proof of this result is the identification of a matrix with its corresponding linear transformation and the proof that a linear transformation of a finite dimensional vector space is onto if and only if it is one-to-one. There is, however, an elementary direct proof of an even stronger result which is almost an immediate consequence of an interesting theorem about matrices that is not so well known, although its proof is quite simple.

Suppose that R is a ring and that n is a positive integer. Let R_n denote the ring of all n-square matrices over R. We emphasize that R need not be commutative.

THEOREM. *If R satisfies the ascending chain condition for right ideals, then so does R_n.*

---

* From AMERICAN MATHEMATICAL MONTHLY, vol. 70 (1963), p. 315.
† From AMERICAN MATHEMATICAL MONTHLY, vol. 74 (1967), pp. 848–849.

Proof. Denote by $f_{i,j}$ the function from R_n to R that sends a matrix onto its (i, j)-component. Observe that if S is a nonempty subset of R_n which is closed with respect to addition, subtraction, and scalar multiplication on the right, then $f_{i,j}(S)$ is a right ideal of R.

Assume that $\{E_k\}$ is a sequence of right ideals of R_n such that E_k is properly contained in E_{k+1} for each positive integer k. For each positive integer t not exceeding n^2 let $E_{k,t}$ denote the subset of E_k consisting of those matrices in E_k that have zero in each of the first t components—for definiteness, count the components by rows. Now suppose that t is less than n^2 and that we have already shown that $E_{k,t}$ is properly contained in $E_{k+1,t}$ for all but a finite number of k. We wish to show that the latter result must also hold for $t+1$. Since $f_{i,j}(E_{k,t})$ is a right ideal of R and since R satisfies the ascending chain condition, we know that $f_{i,j}(E_{k,t})=f_{i,j}(E_{k+1,t})$ for all but a finite number of k. If this equality holds, however, and if $E_{k,t}$ is proper in $E_{k+1,t}$, then $E_{k,t+1}$ is proper in $E_{k+1,t+1}$; for if A is in $E_{k+1,t}$ but not in $E_{k,t}$, then there exists B in $E_{k,t}$ having the same $(t+1)$-component as A and $B-A$ is in $E_{k+1,t+1}$ but not in $E_{k,t+1}$. We are led to the conclusion that the set consisting of only the zero matrix in R_n is a proper subset of itself. Thus our assumption must be denied, and R_n satisfies the ascending chain condition for right ideals.

The next theorem is a weak version of Theorem 4 in [1].

THEOREM. *If R is a ring with unit which satisfies the ascending chain condition for right ideals, then the equation $ab=1$ in R implies that $ba=1$.*

Proof. Define a mapping π from R into itself by: $x\rightarrow ax$. Since $\pi(bx)=x$ for each x in R, π is onto. Let $K_n=\{x\in R \mid \pi^n(x)=0\}$ denote the kernel of π^n for each positive integer n. Since K_n is a right ideal of R, there must be a positive integer n such that $K_n=K_{n+1}$. For any such n, we see that $\pi(\pi^n(x))=0$ implies that $\pi^n(x)=0$. Since π^n is onto and since π preserves addition, π must be one-to-one. Since $\pi(1)=\pi(ba)$, $ba=1$ and the theorem is proved.

COROLLARY. *If R is a field or division ring and if A and B are n-square matrices over R such that $AB=I$, then $BA=I$.*

Proof. R has only the two trivial right ideals.

Both of the above theorems can be proved, in much the same way, for the descending chain condition and also for left ideals, so we actually have

THEOREM. *If R is a ring with unit which satisfies the ascending (descending) chain condition for right (left) ideals and if A and B are n-square matrices over R such that $AB=I$, then $BA=I$.*

Reference

1. R. Baer, Inverses and zero divisors, Bull. Amer. Math. Soc., 48 (1942).

ON THE SYMMETRY OF SEMI-SIMPLE RINGS*

A. KERTÉSZ, University of Debrecen, and O. STEINFELD, Mathematical Institute of the Hungarian Academy of Sciences, Budapest

By a semisimple ring we mean a ring containing no nonzero nilpotent left ideal and satisfying the descending chain condition for left ideals. It is known that for semisimple rings a complete symmetry prevails between concepts involving "right-hand" notions and between their "left-hand" analogues. Thus to any characterization of these rings there exists a "dual" characterization which arises if for the "right-hand" and "left-hand" concepts occurring in the original characterization we substitute the corresponding "left-hand" and "right-hand" ones. For example, by the well-known theorem of Noether an arbitrary ring R is semisimple if and only if it has a right unit element and can be decomposed into the direct sum of minimal left ideals (see [2], or [3], Sec. 123), and by the above-mentioned duality we obtain that the ring R is semisimple if and only if R has a left unit element and can be decomposed into the direct sum of minimal right ideals.† To the best of our knowledge the demonstration of the equivalence of the two dual characterizations of semisimple rings has been effected so far only in an indirect way, mostly with the aid of the self-dual Wedderburn-Artin theorem. It is the purpose of this note to give a simple and straightforward proof for the equivalence of the two dual noetherian characterizations.

We are going to prove the following theorem:

If a ring R has a right unit element and admits a decomposition into a direct sum of a finite number of minimal left ideals, then R has also a left unit element and can be decomposed into a direct sum of finitely many minimal right ideals, and conversely.

It will clearly be sufficient to prove the first assertion of the theorem.

Let e be a right unit element of the ring R, and let the direct decomposition

$$R = L_1 + \cdots + L_m \tag{1}$$

hold, where the L_i $(i=1, \cdots, m)$ are mininal left ideals of R. In view of (1) we have for e the decomposition

$$e = e_1 + \cdots + e_m \qquad (e_i \in L_i;\ i = 1, \cdots, m). \tag{2}$$

First we show that $L_i = Re_i$ $(i=1, \cdots, m)$ and $e_1, \cdots, e_m$ are orthogonal idempotents.‡ In view of (2) we have for each component of e, $e_i = e_ie = e_ie_1 + \cdots$

* From AMERICAN MATHEMATICAL MONTHLY, vol. 67 (1960), pp. 450–452.

† In E. Noether's original formulation of the theorem there occurs a "two sided unit element." It was one of us who remarked that it is sufficient to postulate the existence of a one sided unit element [1].

‡ The proof of this assertion is essentially the same as that of Lemma 5 in §123 of [3]. We give here this proof only for completeness sake and in view of the circumstance that in our considerations we use only a right unit element instead of a two sided one.

$+e_i^2+\cdots+e_ie_m$, whence by the uniqueness of the decomposition the relations

(3) $$e_ie_k = \begin{cases} e_i & \text{for } i = k, \\ 0 & \text{for } i \neq k, \end{cases}$$

follow. Since by $e_i^2=e_i(\neq 0)$ we have the relation $Re_i\neq 0$ and by $e_i\in L_i$ the relation $Re_i\subseteq L_i$, the minimality of L_i implies $Re_i=L_i$.

Let us now show that e is a left unit element in R. Consider an arbitrary element r of R, which, by (2), admits the decomposition

(4) $$r = re = re_1 + \cdots + re_m.$$

For a fixed i $(1\leqq i\leqq m)$ the set of all elements of the form

(5) $$ere_i - re_i$$

is a left ideal $L(\subseteq Re_i)$ of R. By the minimality of Re_i we must have either $L=Re_i$ or $L=0$. The case $L=Re_i$ is, however, impossible since the product of two elements of (5) is always zero, and so L cannot contain the idempotent element e_i. Now $L=0$ implies

(6) $$ere_i = re_i, \qquad (i = 1, \cdots, m).$$

Hence, by (4), $er=ere_1+\cdots+ere_m=re_1+\cdots+re_m=re=r$ and this proves e to be a left unit element. Accordingly we have, by (2), for any element $r(\in R)$,

(7) $$r = e_1r + \cdots + e_mr.$$

On the other hand, since from the relation $e_1r_1+\cdots+e_mr_m=0$ $(r_1,\cdots,r_m\in R)$ it follows by left multiplication with e_i $(i=1,\cdots,m)$ in view of (3) that $e_ir_i=0$, and hence any element of the ring R admits a unique representation as a sum of elements belonging to the right ideals $e_1R,\cdots,e_mR$, respectively. Consequently we have the direct decomposition

(8) $$R = e_1R + \cdots + e_mR.$$

In order to prove the minimality of the right ideals e_iR $(i=1,\cdots,m)$, it will be sufficient to show that for an arbitrary fixed element $e_is(\neq 0; s\in R)$ there exists a $t(\in R)$ such that

(9) $$e_is\cdot t = e_i.$$

First of all we remark that, by (4), the element $e_is(\neq 0)$ can be written in the form $e_is=e_ise_1+\cdots+e_ise_m$, where we may consider the component e_ise_k to be different from zero. Let us now consider the left ideal $Re_i\cdot e_ise_k$. By the idempotency of e_i and by the minimality of Re_k we have for this left ideal,

(10) $$Re_i\cdot e_ise_k = Re_k.$$

From (10) there follows $e_kRe_i\cdot e_ise_k=e_kRe_k$, and this assures the existence of an element $s^*(\in R)$ for which

$$e_k s^* e_i \cdot e_i s e_k = e_k \tag{11}$$

holds. Now we are going to prove that

$$e_i s e_k \cdot e_k s^* e_i = e_i. \tag{12}$$

In order to shorten our notations, we put $e_i s e_k \cdot e_k s^* e_i = z$. Multiplying equation (11) from the left by $e_i s e_k$ we obtain with the aid of $e_i s e_k \neq 0$ the relation $z \neq 0$, and again by (11) the equality

$$z^2 = z. \tag{13}$$

Finally, by the definition of z and by (3) we have

$$e_i z = z. \tag{14}$$

On the other hand, by $Rz \neq 0$, $Rz \subseteq Re_i$ and by the minimality of Re_i we have $Rz = Re_i$. From this $yz = e_i$ follows for a suitable element $y (\in R)$. However, by (13) this implies

$$e_i z = yz^2 = yz = e_i, \tag{15}$$

and (14) and (15) together prove (12). Since by (12) the equation (9) has the solution $t = e_k s^* e_i$, the right ideals $e_i R$ $(i = 1, \cdots, m)$ are minimal right ideals in R, and this completes the proof of the theorem.

References

1. A. Kertész, Beiträge zur Theorie der Operatormoduln, Acta Math. Acad. Sci. Hungar., vol. 8, 1957, pp. 235–257.
2. E. Noether, Hyperkomplexe Grössen und Darstellungstheorie, Math. Z., vol. 30, 1929, pp. 641–692.
3. B. L. van der Waerden, Algebra II, Berlin, 1955.

A THEOREM OF LEVITZKI*

YUZO UTUMI, University of Rochester

THEOREM. *Let S be a ring with maximum condition for left ideals. Then every nil one-sided ideal of S is nilpotent.*

This theorem is well known. (See Levitzki's "On Multiplicative Systems," Comp. Math. 8 (1950), and N. Jacobson, "Structure of Rings," p. 199, and I. N. Herstein "A Theorem of Levitzki," Proc. AMS 13 (1962), where a proof by A. W. Goldie is also cited.) The following proof is believed to be simpler than any known proof.

We denote the left annihilator of an element x by $l(x)$, and the (two-sided) ideal generated by x by $(x)_t$.

LEMMA. *Let S be a semigroup with maximum condition for annihilator left*

* From AMERICAN MATHEMATICAL MONTHLY, vol. 70 (1963), p. 286.

ideals. If S has a nonzero nil left or right ideal, then S contains a nonzero nilpotent ideal.

Proof. Let A be a nonzero nil right ideal, and let $l(a)$, $0 \neq a \in A$, be a maximal one among $l(x)$ for all $0 \neq x \in A$. If $au \neq 0$, then $(au)^n = 0$, $(au)^{n-1} \neq 0$ for some $n > 1$. Hence $au \in l((au)^{n-1}) = l(a)$ and $aua = 0$. Thus, $aSa = 0$ and $(a)_l^3 = 0$.

In case S has a nonzero nil left ideal B, Sb is nil for any $b \in B$, and hence bS also is nil. Let $0 \neq b \in B$. If $bS = 0$, $(b)_l^2 = 0$. If $bS \neq 0$, we may apply the above argument to $A = bS$.

Proof of the theorem. Let N be the maximal nilpotent ideal of S, and C a nil right (left) ideal of S. If C were not contained in N, then $(C+N)/N$ would be a nonzero nil right (left) ideal of S/N, and hence S/N would contain a nonzero nilpotent ideal by the lemma, contradicting the maximality of N. Thus, $C \subset N$, whence any nil one-sided ideal C is nilpotent, as desired.

RINGS OF FRACTIONS*

P. M. COHN, Bedford College, London

Introduction. A well-known (and easily proved) theorem states that each integral domain can be embedded in a field [**37**]. This was generalized to certain noncommutative rings in a brilliant paper by Ore [**33**] in 1930; his results are essentially as simple as in the commutative case, and the proofs, though longer, are no harder. Beyond this, very little is known, so little that it can be set down in quite a brief article. I thought this was worth doing because some interesting problems on the embedding of rings in skew fields remain, and because it gives me the chance to mention some recent work which makes it seem that these embedding problems are not quite as hard as they appear at first sight.

The central problem, finding *fields* of fractions, is really part of the problem of constructing *rings* of fractions (i.e., inverting certain elements), which also has other important applications. We shall therefore arrange the discussion so as to include this more general case.

Conventions. Every ring has a unit-element, denoted by 1, which is preserved by homomorphisms, inherited by subrings and acts as the identity operator on modules. The same conventions apply to semigroups. Of course a ring may very well consist of 0 alone; this is the case precisely when $1 = 0$. We use 0 to denote

---

* From AMERICAN MATHEMATICAL MONTHLY, vol. 78 (1971), pp. 596–615.

Professor Cohn did his Cambridge Ph.D. under Philip Hall, and he has held positions at the Univ. de Nancy, Manchester University, Queen Mary College London, and (presently) Bedford College, London. He has spent leaves-of-absence at Yale Univ., Univ. of Chicago, and Rutgers. He has published extensively in universal algebra and in many branches of algebra; in 1965–67 he was the Secretary of the London Mathematical Society. His books are *Lie Groups* (Cambridge Univ. Press 1957), *Linear Equations* (Routledge and Kegan Paul 1958), *Solid Geometry* (Routledge and Kegan Paul 1961), and *Universal Algebra* (Harper and Row 1965). *Editor.*

both the zero element and the set consisting of the zero element; the context will always make clear which is intended.

In any ring R the set of nonzero elements is denoted by R^*. A ring R, not necessarily commutative, such that R^* is a group under multiplication will be called a *field*. In the current literature this is often called a *skew field* or *division ring*; we shall occasionally use the prefix 'skew' for emphasis. A ring R such that R^* is a semigroup under multiplication is said to be *entire*, and a commutative entire ring is called an *integral domain*. Note that in a field $1 \neq 0$; the same is true in entire rings, by our convention about semigroups.

An element u in a ring is *invertible* or a *unit* if it has an *inverse* u^{-1} satisfying $uu^{-1} = u^{-1}u = 1$; of course the inverse is unique if it exists at all. In an entire ring, if $uv = 1$, then $u(vu-1) = (uv-1)u = 0$, hence $vu = 1$ and v is the inverse of u. Thus all one-sided inverses are two-sided in this case. An element u is called a *zero-divisor* if $u \neq 0$ and if for some $v \neq 0$, either $uv = 0$ or $vu = 0$. A *nonzero-divisor* is a nonzero element which is not a zero-divisor. Thus by our convention 0 is neither a zero-divisor nor a nonzero-divisor.

If R is any ring, a *field of fractions* of R is a field containing R as a subring and generated, as a field, by R.

Outline. In Section 1 we review the commutative case; Section 2 introduces the obvious but rather useful notion of a 'universal S-inverting ring' and also gives Malcev's example of an entire ring not embeddable in a field. The Ore construction occupies Section 3, with applications in Section 4, including the theorems of Goldie and Posner. The remaining sections describe methods of constructing fields of fractions which go beyond Ore's theorem. In Section 5 we examine generalized inverses, and the relation to the Johnson-Utumi 'ring of quotients'; this turns out to be not very close in the case of chief interest to us, that of entire rings. The topological methods in Section 6 essentially generalize the notion of a decimal fraction. The final Section 7 reports briefly on the author's recent method of embedding rings in fields by inverting matrices rather than elements.

1. The commutative case. Let R be a commutative ring. The need for fractions arises when we try to enlarge R so as to ensure that equations of the form

$$xb = a \tag{1}$$

can be solved. At this stage our instinct should tell us to beware of the case $b = 0$, but we shall leave this question aside for the moment. If we denote the solution of (1) by ab^{-1} or a/b, we see immediately that $a/b = ac/bc$, or more generally,

$$a/b = a'/b' \quad \text{if and only if } ab' = ba', \tag{2}$$

under suitable restrictions to exclude division by 0. Further, if the solutions are to form a ring containing R, they must add and multiply according to the rules

$$\frac{a}{b}+\frac{a'}{b'}=\frac{ab'+a'b}{bb'},\qquad \frac{a}{b}\cdot\frac{a'}{b'}=\frac{aa'}{bb'}. \tag{3}$$

We learn at an early stage of our algebra course that if R is an integral domain, then we can find a field containing R as subring by taking all fractions a/b with $b\neq 0$ and combining them according to the rules (3), bearing in mind the cancellation rule (2).

To generalize this construction we observe that to form the new denominators in (3), we need only *multiply* the denominators b and b' together, not add them; this suggests taking a subsemigroup S of R as our stock of denominators. Now we shall in general no longer obtain a field; we may not even get a ring containing R as subring, but by following essentially the same construction we get a ring R_S say, with a homomorphism

$$\lambda: R \to R_S$$

which maps the elements of S to invertible elements, and it will be a simple matter to find out when λ is injective.

Thus we are given a subsemigroup S of a commutative ring R and we define a relation on the product set $R\times S$ by the rule:

$$(a, s)\sim(a', s') \text{ if and only if } as't=a'st \text{ for some } t\in S. \tag{4}$$

This reduces to (2) when S consists of nonzero-divisors; in general the form (4) is necessary to make sure that '$\sim$' is really an equivalence relation. Let us only check transitivity (reflexivity and symmetry are obvious): If $(a, s)\sim(a', s')$ and $(a', s')\sim(a'', s'')$, then $as't=a'st$ and $a's''t'=a''s't'$ for some $t, t'\in S$; hence

$$as''\cdot s'tt' = a'ss''tt' = a''s'stt' = a''s\cdot s'tt'.$$

Since $s'tt'\in S$, this shows that $(a, s)\sim(a'', s'')$ and the transitivity is proved. We denote the equivalence class containing (a, s) by a/s and define addition and multiplication by the formulae (3); of course we must verify that the definitions really only depend on the classes of a/s, a'/s' and not on the representatives of these classes used in the formulae. This is a routine verification, as is the proof that the set of classes a/s with these operations forms a ring, denoted by R_S say, with zero $0/1$ and unit-element $1/1$. The mapping

$$\lambda: a \mapsto a/1 \tag{5}$$

of R into R_S is clearly a homomorphism; it maps each element of S to an invertible element of R_S, because $(s/1)(1/s)=s/s=1/1$.

Let us say that a homomorphism $f: R\to R'$ is *S-inverting* if each element of S is mapped by f to an invertible element of R'. For example the mapping (5) is S-inverting, but we can say more than this. Let $f: R\to R'$ be any S-inverting homomorphism and define a mapping $f_1: R \times S \to R'$ by the rule

$$(a, s)^{f_1} = a^f(s^f)^{-1}.$$

This makes sense because f is S-inverting, by hypothesis. We observe now that f_1 takes the same value on pairs that are equivalent according to (4): If $as't = a'st$, then $a^f s'^f t^f = a'^f s^f t^f$ and hence $a^f(s^f)^{-1} = a'^f(s'^f)^{-1}$. This means that we obtain a well-defined mapping f' of R_S into R' by putting $(a/s)^{f'} = (a, s)^{f_1}$. This mapping f' has the property that for any $a \in R$,

$$(a/1)^{f'} = a^f, \tag{6}$$

an equation which may also be expressed by saying that the accompanying diagram commutes, i.e., $\lambda f' = f$. Moreover, f' is uniquely determined by (6), because

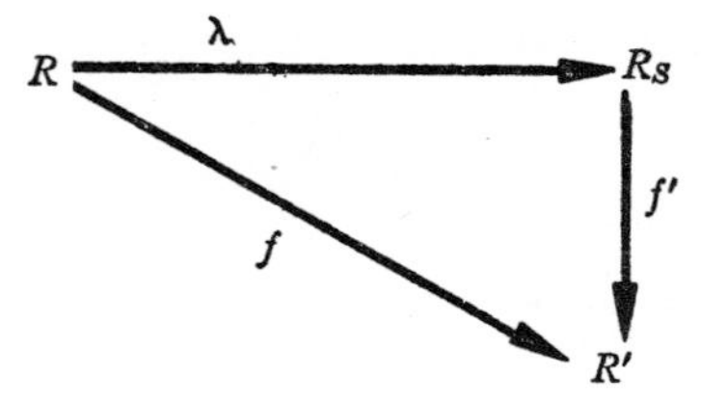

that equation determines its value on the elements $a/1$, and its value on $1/s$ must be the inverse of its value on $s/1$. Thus the mapping

$$\lambda: R \to R_S \tag{7}$$

is not just an S-inverting homomorphism, but the most general such homomorphism, in the sense that each S-inverting homomorphism can be obtained by taking a uniquely determined homomorphism from R_S. This property is expressed by saying that (7) is the *universal S-inverting homomorphism*, and R_S itself is called the *universal S-inverting ring*. This universal property in effect determines R_S up to isomorphism.

We shall wish to know when λ is injective; more generally, let us determine the kernel of λ. Clearly $a^\lambda = 0$ if and only if $a/1 = 0/1$, and by definition this means that $at = 0$ for some $t \in S$. We can now sum up our results:

THEOREM 1.1. *Let R be a commutative ring and S a subsemigroup of R. Then there is a universal S-inverting homomorphism $\lambda: R \to R_S$; the elements of R_S can be written as fractions a/s ($a \in R$, $s \in S$), where $a/s = a'/s'$ if and only if $as't = a'st$ for some $t \in S$. Further,*

$$\ker \lambda = \{a \in R \mid at = 0 \textit{ for some } t \in S\}.$$

REMARKS. 1. Note that R_S is again commutative.

2. The ring R_S reduces to 0 precisely when $0 \in S$; it is to avoid this trivial case that one usually assumes $0 \notin S$.

3. The mapping λ is injective if and only if S contains only nonzero-divisors. In that case R_S is called a *ring of fractions* of R; the largest such ring is obtained

by taking S to be the set of all nonzero-divisors, a set which is always a subsemigroup. The ring obtained by inverting all nonzero-divisors is called the *total ring of fractions* of R. In case R is an integral domain, this is the universal R^*-inverting ring of R, which is of course the field of fractions of R.

An important special case of the theorem is obtained by taking S to be the complement of a prime ideal $\mathfrak{p}$ in R (a *prime* ideal is an ideal of R whose complement is a semigroup under multiplication; note that this does not allow R as prime ideal). In that case one often writes $R_\mathfrak{p}$ instead of R_S, somewhat inconsistently, but without risk of confusion, because S never contains 0, whereas $\mathfrak{p}$ always does.

The problem of constructing fractions already arises in a semigroup, and should really be considered in that setting. We have nevertheless treated the case of rings first, on account of its importance; it also happens to coincide with the historical order of its development [16]. In any case we can easily extract the answer for semigroups from our conclusion:

THEOREM 1.2. *Let M be a commutative semigroup and S a subsemigroup of M. Then there is a universal S-inverting homomorphism*

$$\lambda: M \to M_S,$$

the elements of M_S can be written as fractions a/s ($a \in M$, $s \in S$) as in the ring case, and a, a' have the same image under λ if and only if $at = a't$ for some $t \in S$.

2. Some observations on the general case. Let us return to our basic problem, which is to construct a field of fractions for a given ring, when possible. If a ring R is to be embedded in a field, then whether commutative or not, R must be entire. But this necessary condition, which in the commutative case was sufficient, in general is no longer so. The first example of an entire ring not embeddable in a field was given by Malcev [28]. He takes the ring R generated by eight elements a, b, c, d, x, y, u, v, with defining relations

$$ax = by, \qquad cx = dy, \qquad cu = dv. \tag{1}$$

To show that this ring is entire, one uses a normal form for its elements. In outline the argument goes as follows. Each element of R can, by use of (1), be expressed as a noncommutative polynomial in the given generators, in which there are no occurrences of by, dy, dv (the right-hand sides of the equations (1)), and such an expression is unique. The verification that the product of nonzero elements is nonzero is fairly straightforward, though care is needed to ensure that all possibilities are considered at each stage. The normal form also shows that $au \neq bv$, but if R were embeddable in a field, or even in a ring in which a, c, y, v have inverses, we could deduce from (1) that $a^{-1}b = xy^{-1}$, $xy^{-1} = c^{-1}d$, $c^{-1}d = uv^{-1}$, hence $a^{-1}b = uv^{-1}$, and so

$$au = bv, \tag{2}$$

which is a contradiction.

Malcev obtained his example in the course of studying conditions under which a semigroup is embeddable in a group [29]. In fact he was able to write down an infinite series of 'quasi-identities', i.e., conditions of the form

$$A_1, \cdots, A_n \text{ imply } B,$$

(where $A_1, \cdots, A_n, B$ are equations in a semigroup) which he proved necessary and sufficient for a semigroup to be embeddable in a group. The simplest of these quasi-identities are left and right cancellation: '$xy=xz$ implies $y=z$' and '$xz=yz$ implies $x=y$.' The next condition is of the form 'the equations (1) imply (2)', and the example just given shows it to be independent of cancellation (more generally, the infinite set of quasi-identities cannot be replaced by any finite subset). A detailed account of Malcev's Theorem can be found in [11]; it seems likely that any corresponding criterion for the embeddability of rings in skew fields is rather more complicated. But it follows from results in general algebra that the embeddability of a nonzero ring in a field can be expressed by a (possibly infinite) set of quasi-identities ([11], p. 235).

Recently, in [41], A. A. Klein has found an infinite set of quasi-identities which are necessary for embeddability in a field and which he conjectures to be sufficient.* They express that R is entire and for all n, each nilpotent $n \times n$ matrix C satisfies $C^n=0$.

Malcev has asked whether rings exist whose nonzero elements are embeddable in groups, but which are not embeddable in fields. Such rings were found simultaneously and independently by three people in 1966 [4, 5, 24].

Let us now take the general situation and see whether anything found in the commutative case can be used here. If R is a ring and S any subset (not necessarily a subsemigroup), we can define an S-inverting homomorphism as before. Given an S-inverting homomorphism $f: R \to R'$, let $\bar{S}$ be the subset of R whose elements are mapped into invertible elements of R'. Clearly $\bar{S} \supseteq S$, but equality need not hold; in particular $\bar{S}$ contains all invertible elements of R, and if $u, v \in \bar{S}$, then $uv \in \bar{S}$, because $(uv)^f = u^f v^f$, and the latter is invertible when u^f, v^f are. This shows that $\bar{S}$ is always a subsemigroup, so nothing is lost by taking the set to be inverted as a semigroup.

We can again construct a universal S-inverting ring R_S: simply take a presentation of R (by generators and defining relations) and for each $s \in S$ adjoin a new generator s' and extra relations

$$ss' = s's = 1. \tag{3}$$

The ring R_S so obtained may no longer contain R as subring, e.g., if we apply this construction to Malcev's example, with $S = \{a, c, y, v\}$, we get a collapse because then $au = bv$. But we always have a natural homomorphism $R \to R_S$ which is S-inverting, and in fact this is the universal S-inverting homomorphism,

* Added in proof (March 29, 1971): A counterexample to this Conjecture has just been found by G. M. Bergman.

because the relations holding in R_S, namely (3) together with the relations of R itself, must hold in any image of R under an S-inverting homomorphism. (This is essentially an application of Dyck's Theorem, see, e.g., [11], p. 183.) In this way we obtain the following result:

THEOREM 2.1. *Let R be any ring and S any subset of R. Then there is a universal S-inverting homomorphism $\lambda: R \to R_S$, where R_S is unique up to isomorphism. Moreover, λ is injective if and only if R can be embedded in a ring in which all elements of S have inverses.*

The assertion is quite general and, because of its very generality, rather easy to prove. It is also not hard to see that the correspondence $(R, S) \mapsto R_S$ is a *functor* (from the category of pairs R, S to the category of rings). This means that to each homomorphism of rings $f: R \to R'$ such that $S^f \subseteq S'$ for subsets S, S' of R, R' respectively, there corresponds a homomorphism $\bar{f}: R_S \to R'_{S'}$ such that $\overline{fg} = \bar{f}\bar{g}$ and the identity mapping on R corresponds to the identity on $R_S: \bar{1} = 1$. Moreover, the natural mappings $\lambda: R \to R_S$ and $\lambda': R' \to R'_{S'}$ have the property of making the following diagram commute:

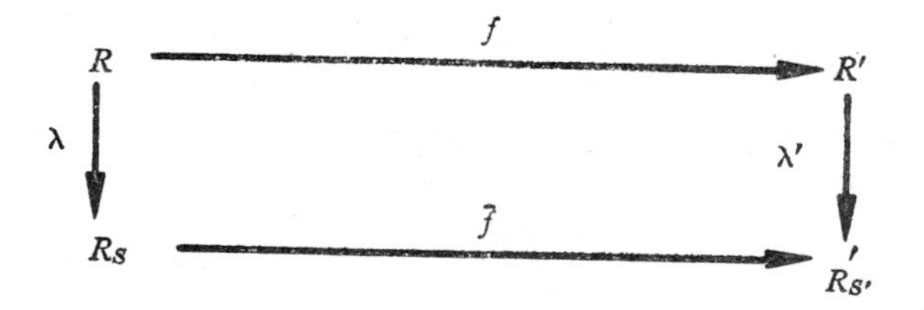

This is expressed by saying that λ is a *natural transformation* (for details cf. e.g., [27]).

At first sight Th.2.1 looks deceptively like Th.1.1, but it has the serious drawback that no normal form for the elements of R_S is given. This makes it hard to decide when λ is injective; also we cannot be sure, even when R is embeddable in a field, that R_{R^*} will be the whole field of fractions. The trouble is that after adjoining inverses of all the nonzero elements to R, there may still be elements without inverses, e.g., elements of the form $ab^{-1}c + de^{-1}f$. So the process of adjoining inverses may have to be repeated, perhaps infinitely often. The following observation is sometimes useful:

THEOREM 2.2. *Let R be any ring. If there is an R^*-inverting homomorphism of R into a field K, then R is embeddable in a field.*

For by hypothesis, we have a homomorphism

$$f: R \to K, \tag{4}$$

and since any nonzero element of R maps to an invertible element of K, it cannot map to 0, i.e., (4) is injective.

The necessity of having to repeat the process of adjoining inverses does not arise for semigroups: If a semigroup M is embeddable in a group G, then the subsemigroup of G generated by the elements of M and their inverses already forms a group. Neither does the problem arise in the special case treated by Ore, to which we now turn.

3. Ore's Construction. In an attempt to carry over the results of Section 1 to the noncommutative case, let us examine the situation where every element of the universal S-inverting ring R_S can be written in the form of a fraction a/s. If this is to be possible, we must be able to express $(1/s)(a/1)$ in this form, say

$$(1/s)(a/1) = a'/s'. \tag{1}$$

Multiplying both sides by $s/1$ on the left and by $s'/1$ on the right, we get

$$as'/1 = sa'/1. \tag{2}$$

This gives us a clue to the extra condition required now.

THEOREM 3.1. *Let R be any ring, S a subsemigroup, and assume further that*
(i) *for any $a \in R$ and $s \in S$, $aS \cap sR \neq \emptyset$,*
(ii) *for any $a \in R$ and $s \in S$, if $sa = 0$, then $at = 0$ for some $t \in S$.*
Then the elements of the universal S-inverting ring R_S can be constructed as fractions a/s $(a \in R,\ s \in S)$, where

$$a/s = a'/s' \Rightarrow au = a'u', \qquad su = s'u' \in S \text{ for some } u, u' \in R. \tag{3}$$

The kernel of the natural mapping is then

$$\ker \lambda = \{a \in R \mid at = 0 \text{ for some } t \in S\}.$$

Of course here we must distinguish carefully between as^{-1} and $s^{-1}a$; the expression a/s corresponds to the former.

A subsemigroup S of R satisfying the conditions (i), (ii) of this theorem will be called a *right denominator set* in R. The proof of this result is largely an exercise in patience, and the reader is recommended to verify at least some of the steps. The basic observation is that any two fractions can be brought to a common denominator in S, using (i), and they represent the same element of R_S if and only if, over a suitable common denominator, their numerators are equal (cf. (2)). Addition of fractions over the same denominator is straightforward, and the multiplication of fractions is based on the rule (1).

REMARKS: 1. Again $R_S = 0$ if and only if $0 \in S$; one usually excludes this case.

2. There is a left-right analogue of the theorem, obtained by switching sides; it shows how to form *left* fractions, starting from a *left* denominator set.

3. There is a corresponding theorem for constructing fractions in a semigroup. More generally, the construction can be performed in any category (cf. [20], p. 28).

4. Any subsemigroup S consisting of invertible elements of R is a right (and

left) denominator set, and the universal S-inverting homomorphism is then an isomorphism.

5. Any *central* subsemigroup S (i.e., satisfying $as=sa$ for all $a\in R$, $s\in S$) is a right (and left) denominator set.

6. The universal S-inverting mapping is injective if and only if S contains only nonzero-divisors. In that case condition (ii) becomes superfluous. This case is sufficiently important to be stated separately.

COROLLARY 1. *Let R be a ring and S a subsemigroup of R consisting of nonzerodivisors, such that $aS\cap sR\neq\varnothing$ for any $a\in R$, $s\in S$. Then the universal S-inverting homomorphism is injective.*

When S is as in Corollary 1, R_S is again called a *ring of fractions, total* in case S consists of all nonzero-divisors. But this time we cannot be sure that there is a total ring of fractions, because the set of all nonzero-divisors need not satisfy the hypothesis of Corollary 1.

If R is entire and R^* is a right denominator set, we get the case originally treated by Ore (cf. [**33**]; the generalizations were given by Asano [**3**] and others).

COROLLARY 2. *Let R be an entire ring such that*

$$aR\cap bR\neq 0 \qquad \textit{for any } a,\, b\in R^*. \tag{4}$$

Then R can be embedded in a skew field K, whose elements have the form of fractions a/b ($a\in R$, $b\in R^*$).

Condition (4) is called the *right Ore condition*, and an entire ring satisfying (4) is called a *right Ore ring*. We observe that (4) is necessary as well as sufficient for the conclusion to hold. For if an entire ring R can be embedded in a field K in such a way that each element of K has the form ab^{-1} (a, $b\in R$), then in particular, for any a, $b\in R^*$ we can find a', $b'\in R^*$ such that $b^{-1}a=a'b'^{-1}$; hence $ab'=ba'\neq 0$.

4. Applications of Ore's Construction. An important class of Ore rings (which Ore himself had studied and clearly had in mind when making his construction, cf. [**34**]) is formed by the *skew polynomial rings*. Given any field K, passing to the polynomial ring $K[x]$ in an indeterminate x is a familiar construction, which still works even when K is skew. In that case it is usual to require x to be *central*, i.e., to commute with the field elements. But we often need a more general case: we still assume that each element of our ring can be written as a polynomial in just one way:

$$f = a_0 + xa_1 + \cdots + x^n a_n \qquad (a_i\in K); \tag{1}$$

we no longer assume that x is central, but instead that for each $a\in K$ there exist $\bar{a}$, $a'\in K$ satisfying

$$ax = x\bar{a} + a'. \tag{2}$$

It is not too hard to show that the mapping $\alpha: a \mapsto \bar{a}$ is an endomorphism of K, and that $\delta: a \mapsto a'$ is an additive mapping satisfying

$$(ab)' = a'\bar{b} + ab'. \tag{3}$$

Any additive mapping δ of a field into itself, satisfying (3) (for some endomorphism $\alpha: a \mapsto \bar{a}$) is called an *$\alpha$-derivation*. E.g., on the field of rational functions $F(t)$ over some commutative field F, the usual derivative $f' = df/dt$ defines a 1-derivation (associated with the identity automorphism of $F(t)$).

Conversely, given any endomorphism α of a field K and any α-derivation δ, we can define a multiplication on the set of polynomials (1) by using the commutation rule (2). This leads to a ring denoted by $K[x; \alpha, \delta]$ and called a *skew polynomial ring*. This ring is entire and is in fact a right Ore ring; for the proof we can use the Euclidean algorithm, as in ordinary polynomial rings, but we must take care here to perform all divisions on the right. It follows that we can form the field of fractions, denoted by $K(x; \alpha, \delta)$. Of course for $\alpha = 1$, $\delta = 0$ the skew polynomial ring reduces to the ordinary polynomial ring $K[x]$ with a central indeterminate and its field of fractions $K(x)$.

It is important to note that the construction just given is unsymmetric, and $K[x; \alpha, \delta]$ will not in general be a *left* Ore ring. The condition for it to be one is that α should be an automorphism, for this is the condition which enables us to rewrite (2) as a commutator formula in the other direction:

$$xa = a_1x + a_2. \tag{4}$$

Explicitly, if $ax = xa^\alpha + a^\delta$ and $\alpha^{-1} = \beta$, then $a^\beta x = xa + a^{\beta\delta}$, hence (4) holds with $a_1 = a^\beta$, $a_2 = -a^{\beta\delta}$. Conversely, if α is not an automorphism, take a in K but not in the image under α. Then it is easily checked that x and xa have no common left multiple other than 0; so $K[x; \alpha, \delta]$ is a *left* Ore ring precisely when α is an automorphism.

We observe that the class of right Ore rings is closed under forming polynomial rings [15].

THEOREM 4.1. *If R is a right Ore ring, then so is the ring $R[x]$ of polynomials in a central indeterminate.*

Proof. Since R is right Ore, it has a field of fractions, K say, and $R[x]$ can be embedded in the field $K(x)$ of rational functions in x. Each element of $K[x]$ has the form fa^{-1}, where $f \in R[x]$ and $a \in R^*$ is a common denominator for the coefficients. Hence each element w of $K(x)$ can be written $w = fa^{-1}(gb^{-1})^{-1} = fa^{-1}bg^{-1}$, where $f, g \in R[x]$ and $a, b \in R^*$. Since R is right Ore, it contains a', b' such that $ab' = ba' \neq 0$; hence $a^{-1}b = b'a'^{-1}$, so $w = fb'a'^{-1}g^{-1} = fb'(ga')^{-1}$. This shows that each element of $K(x)$ is a fraction of elements of $R[x]$; therefore the latter is a right Ore ring, as asserted.

It is a remarkable fact, first observed by Goldie [18], that every right Noetherian entire ring is a right Ore ring. (A ring is *right Noetherian* if all its right ideals are finitely generated.) For if R is entire and right Noetherian, take

$x, y \neq 0$, and consider the right ideal generated by the elements $x^n y$ $(n=0, 1, \cdots)$. This must be finitely generated, say $x^n y = ya_0 + xya_1 + \cdots + x^{n-1}ya_{n-1}$. If $a_0 = 0$, we can cancel a power of x from the left, so we may assume that $a_0 \neq 0$, and then

$$x(x^{n-1}y - ya_1 - \cdots - x^{n-2}ya_{n-1}) = ya_0 \neq 0.$$

The result is sometimes called the "little Goldie theorem":

THEOREM 4.2. *Every right Noetherian entire ring is a right Ore ring.*

Similarly the total ring of fractions introduced earlier plays a role in the "big Goldie theorem." To state it we recall that a ring R is said to be *prime* if the product of nonzero ideals in R is nonzero and *semiprime* if the square of each nonzero ideal is nonzero. A ring is *right Artinian* if its right ideals satisfy the descending chain condition; this is a very much stronger condition than being right Noetherian, and much more is known about the Artinian rings (cf. e.g., [26]). Goldie's theorem provides a connection between the two; in one form it states:

If a right Noetherian ring R is prime (*respectively semiprime*), *then R has a total ring of fractions which is right Artinian and simple* (*respectively semisimple*).

For a brief proof see [19].

Let us return to Th.4.2 and consider an entire ring R which is not right Ore. For simplicity we take R to be an algebra over a commutative field F. By hypothesis we can find $x, y \in R^*$ such that $xR \cap yR = 0$. It follows that there is no polynomial in x and y (treated as noncommuting variables) which is zero, except the one whose coefficients are all 0. For each such polynomial is of the form $f = \alpha + xf_1 + yf_2$, where $\alpha \in F$ and f_1, f_2 are polynomials of lower degree than f. Suppose that f is the polynomial of least degree in two noncommuting indeterminates that vanishes for x and y. If $\alpha \neq 0$, then f_1, f_2 cannot both vanish; say $f_2 \neq 0$, hence $\alpha x + xf_1x + yf_2x = 0$, i.e., $x(f_1x + \alpha) = -yf_2x \neq 0$, a contradiction. Hence $\alpha = 0$, so $xf_1 = -yf_2$; by the choice of x and y this implies $f_1 = f_2 = 0$, which contradicts the choice of f. So we have proved that there is no polynomial f other than the zero polynomial such that $f(x, y) = 0$. In other words, the subalgebra generated by x and y is the *free associative algebra* on these generators. The restriction on the coefficients is easily lifted, so one has the following result ([23], [12], [25]):

THEOREM 4.3. *An entire ring is either a left and right Ore ring, or it contains a free algebra on two generators.*

By definition a *polynomial identity* is an identical relation not holding in all rings, in particular not in free rings. Thus Th.4.3 has the following immediate consequence [1]:

COROLLARY 1. *An entire ring with a polynomial identity is a* (*left and right*) *Ore ring.*

In analogy with Goldie's theorem, Posner [35] has generalized this result to show that any prime ring with a polynomial identity has a total ring of fractions which is a central simple algebra of finite dimension over its centre.

Jategaonkar who first proved Th.4.3 has also shown how to use it to embed the free algebra in a field [23]. Take a ring R which is a right but not left Ore ring. (E.g., $K[x; \alpha, 0]$ with a non-surjective endomorphism α, say $K=F(t)$ with $\alpha:f(t)\mapsto f(t^2)$.) By Th.4.3 R contains a free algebra and by Th.3.1, Cor.2 it has a field of (right) fractions. So the free algebra is embedded in a field. This is of interest because the free algebra is very far from being an Ore ring. However, this embedding is rather artificial; indeed most automorphisms of the free algebra cannot be extended to the field of fractions just constructed. Later, in Section 7, we shall meet fields of fractions which do not suffer from this defect.

The process of forming fractions can be applied to modules as well as to rings. If R is a ring with a subsemigroup S, and $\lambda:R\to R_S$ is the universal S-inverting homomorphism, then to each right R-module M there corresponds a right R_S-module M_S with an R-module homomorphism $\mu:M\to M_S$ (where M_S is regarded as R-module by means of $\lambda: x\cdot a = xa^\lambda$ for $x\in M_S$, $a\in R$), and μ is the universal mapping with this property, i.e., given any R-module homomorphism of M into an R_S-module N, there exists a unique R_S-module homomorphism of M_S into N such that the accompanying diagram commutes:

This much is general theory, proved in the same way as Th.2.1. There is even a formula for M_S if we are willing to use tensor products (cf. e.g., [27]):

$$M_S = M \otimes R_S,$$

but this formula makes it no easier to study M_S in detail. Now let us assume that S is a right denominator set in R; then the elements of M_S can be written as fractions m/s, where $m\in M$, $s\in S$, and two fractions represent the same element of M_S if and only if, over a suitable common denominator, they have the same numerator. The kernel of the natural mapping $\mu:M\to M_S$ is a submodule tM of M, called the *S-torsion submodule* of M. It consists of all $m\in M$ such that $ms=0$ for some $s\in S$. If $tM=0$, the module M is said to be *S-torsionfree*; e.g., the quotient M/tM is always S-torsionfree.

When R is the ring $\mathbf{Z}$ of integers and $S=\mathbf{Z}^*$ the set of all nonzero integers, tM reduces to the usual torsion subgroup of an abelian group.

5. Strongly regular rings. There have been many attempts to generalize the notion of 'inverse' of an element, to take account of zero-divisors, usually in the form of a 'relative inverse' a', satisfying $aa'a=a$. We shall present the part of this theory that is relevant to the embedding problem.

A ring R is said to be *regular* if to each $a \in R$ there corresponds $x \in R$ such that $axa=a$; if R is such that for each $a \in R$ there exists $x \in R$ satisfying $a^2x=a$, it is called *strongly regular*. In the commutative case this is the same as requiring R to be regular, but in general it is stronger. This is not apparent at first sight, but it will follow from the structure theorems given below. We shall need some of the standard theory of the Jacobson radical; this can be found in [22], to which we refer when necessary. Let us recall that from any family R_λ of rings we can form a *direct product* $P = \prod R_\lambda$ by taking the Cartesian (set-theoretical) product and performing all the operations componentwise. (In the older books this is also called the direct sum.) A subring R of the direct product P is said to be a *subdirect* product if the canonical projections ϵ_λ on the factors R_λ, when restricted to R, are still surjective. E.g., $\boldsymbol{Z}$ can be expressed as subdirect product of fields $\boldsymbol{Z}_p$, where p ranges over all primes. The projection of $\boldsymbol{Z}$ on $\boldsymbol{Z}_p$ maps each integer n to its residue class (mod p).

As an example of a strongly regular ring we have any field, or more generally, any direct product of fields. However, a subdirect product of fields need not be strongly regular, as the example of the integers shows. Nevertheless, these two notions are closely related:

THEOREM 5.1. *Every strongly regular ring is a subdirect product of fields.*

Proof. Let R be strongly regular. Then its Jacobson radical J is 0. For an element $a \in R$ lies in J precisely when $1-ax$ is invertible for all $x \in R$. By hypothesis we can find $x \in R$ such that $a(1-ax)=a-a^2x=0$ and $1-ax$ is invertible; hence $a=0$. It follows ([22], p. 14, [26], p. 58) that R is a subdirect product of primitive rings, each a homomorphic image of R and therefore again strongly regular, so it only remains to show that a strongly regular ring which is also primitive is a field. Now any primitive ring is a dense ring of linear transformations in a vector space V over a field ([22], p. 28, [26], p. 54), and we shall be done if we show that V is 1-dimensional. Assume that V contains two linearly independent elements v_1, v_2. By density there exists $a \in R$ such that $v_1a=v_2$, $v_2a=0$, and by strong regularity we can find $x \in R$ such that $a^2x=a$; hence $v_2=v_1a=v_1a^2x=v_2ax=0$, a contradiction. This completes the proof.

COROLLARY. *A ring is strongly regular if and only if it is regular and has no nilpotent elements other than* 0.

Proof. Let R be strongly regular; then it is a subdirect product of fields and therefore cannot have any nonzero nilpotent elements. Further, if $a^2x=a$, then for each projection ϵ_λ of R on a factor K_λ of the product, either $a\epsilon_\lambda=0$ or $a\epsilon_\lambda \cdot x\epsilon_\lambda = 1 = x\epsilon_\lambda \cdot a\epsilon_\lambda$. In all cases, $a\epsilon_\lambda \cdot x\epsilon_\lambda \cdot a\epsilon_\lambda = a\epsilon_\lambda$; hence $axa=a$ and R is regular.

Conversely, let R be regular without nonzero nilpotent elements, and take $x \in R$ to satisfy $axa=a$. Then

$$(a^2x - a)^2 = a^2xa^2x - a^2xa - a^3x + a^2 = a^3x - a^2 - a^3x + a^2 = 0,$$

hence $a^2x-a=0$ and R is strongly regular.

The connection with fields of fractions is provided by the following result [7]:

THEOREM 5.2. *A subring of a strongly regular ring is embeddable in a field if and only if it is entire.*

Proof. The condition is clearly necessary; so assume that R is entire and is a subring of a strongly regular ring. The latter is a subdirect product of fields, so R is itself a subring of a direct product of fields, say

$$R \subseteq P = \prod K_\lambda.$$

Let $\epsilon_\lambda: P \to K_\lambda$ be the canonical projection and define for each $x \in P$,

$$\Gamma_x = \{\lambda \in \Lambda \mid x\epsilon_\lambda = 0\}, \qquad I_x = \prod_{\lambda \in \Gamma_x} K_\lambda.$$

Each I_x is an ideal in P. Let I be the ideal generated by all the I_x such that $x \in R^*$. Then $R \cap I = 0$; for if $x \in R \cap I$, then $x \in I_{y_1} + \cdots + I_{y_n}$, where $y_i \in R^*$, and hence $xy_1y_2 \cdots y_n = 0$. Therefore $x = 0$.

Let $f: P \to P/I$ be the natural homomorphism. Then by the construction of I, f is R^*-inverting. Since P/I, like P, is strongly regular, there is a homomorphism g of P/I into a field; now fg is an R^*-inverting homomorphism of P into a field, and (by Th.2.2) this provides an embedding of R in a field.

The following consequence was first proved using ultraproducts [36]:

COROLLARY. *An entire subring of a direct product of fields is embeddable in a field.*

Th.5.2 shows that an entire ring can be embedded in a strongly regular ring if and only if it can be embedded in a field. By contrast, any entire ring can be embedded in a regular ring; for there is a construction which associates with any ring R its 'total (right) quotient ring' $Q(R)$, and when R is entire (more generally, for any ring with 'zero singular ideal') $Q(R)$ is regular (cf. [17, 26]). Moreover, this total quotient ring agrees with the total ring of fractions when the latter exists, i.e., by Th.3.1, when the nonzero-divisors of R form a right denominator set. But in general the total quotient ring of R gives no clue about the embeddability of R in a field. E.g., though for an entire ring $Q(R)$ is always regular, it is not strongly regular unless R is a right Ore ring.

For a very thorough survey of quotient rings, see [40].

6. Topological embedding methods. Besides the time-honoured way of forming fractions there is another method of embedding rings in fields, which is also taught at school and goes back to the 16th century [38]. This is the method of decimal fractions; it consists of taking all expressions of the form

$$\sum_{-n}^{\infty} a_\nu t^\nu, \tag{1}$$

where $a_\nu = 0, 1, \cdots, 9$ and $t = 1/10$, and adding and multiplying in the usual way. Division is possible because if in (1), $a_{-n} \neq 0$ say, the series (1) can be written as $t^{-n}a_{-n}(1 - \sum_1^\infty b_\nu t^\nu)$, where each factor is invertible. Of course all the series are convergent, as Laurent series, because $|t| < 1$. This method can be generalized; the general form is even simpler in some respects, because the usual absolute value is replaced by a non-Archimedean valuation.

Let R be a ring with a *valuation*, i.e., a function $v(x)$ taking the integers or $+\infty$ as values, such that

V.1. $v(x) = \infty$ if and only if $x = 0$,

V.2. $v(xy) = v(x) + v(y)$,

V.3. $v(x - y) \geqq \min\{v(x), v(y)\}$.

Such a valuation may be thought of as defining a topology on R; since our aim is to embed R in a field (if possible), we shall specify the topology by its neighbourhoods of 1. We shall limit ourselves to the case where the set

$$P = \{ab^{-1} \mid a \in R, b \in R^*\} \tag{2}$$

is dense in the field to be constructed, so we must say when ab^{-1} is close to 1. The natural condition for this is to require $v(ab^{-1} - 1)$ to be large. Of course v is only defined on R in the first instance, but if we assume that it can be extended to the field of fractions in such a way as to satisfy V.1–3, we have

$$v(a - b) = v([ab^{-1} - 1]b) = v(ab^{-1} - 1) + v(b);$$

hence ab^{-1} is close to 1 precisely when $v(a - b) - v(b)$ is large.

We now have a topology (in fact a uniformity, cf. [6]) on the set P of fractions ab^{-1}. What is still needed to make it into a ring? We shall not make P itself into a ring, but its completion in the given topology. To enable us to add and multiply we need an "asymptotic Ore condition":

A. For any a, b in R^* the function

$$f(x, y) = v(ax - by) - v(by)$$

is unbounded above.

This enables us to embed R in a field, by defining the ring operations in the completion of P, much in the same way as the usual Ore condition was used before. The details are somewhat technical, so we omit them (cf. [8]), but there is a simple way of restating the result in terms of graded rings.

Let us define

$$R_n = \{x \in R \mid v(x) \geqq n\}.$$

Then the R_n form a descending series of additive subgroups of R such that $\bigcap R_n = 0$, $\bigcup R_n = R$ and $R_iR_j \subseteq R_{i+j}$. In other words, we have a *filtered ring*. With each such filtered ring R one associates another ring, its *graded ring* gr R, as follows: the additive group of gr R is the direct sum of the terms

$$\mathrm{gr}_n R = R_n/R_{n+1}. \tag{3}$$

To define multiplication it is enough, by the distributive law, to specify the product of an element of $\mathrm{gr}_i R$ and one of $\mathrm{gr}_j R$. Let $\alpha \in \mathrm{gr}_i R$, $\beta \in \mathrm{gr}_j R$; according to (3), these are cosets, say $\alpha = a + R_{i+1}$, $\beta = b + R_{j+1}$. We define $\alpha\beta = ab + R_{i+j+1}$; it is easily verified that this definition does not depend on the choice of a, b within their cosets. Associativity is clear, so gr R becomes a ring in this way. Loosely speaking, it is the ring formed by taking 'leading terms' in R.

We shall get a graded ring even if the function v satisfies, instead of V.2, only $v(xy) \geqq v(x) + v(y)$. The stronger condition V.2 merely ensures that gr R is entire; further the asymptotic Ore condition can be shown to be equivalent to the Ore condition for gr R. The result may be summed up as follows [8]:

THEOREM 6.1. *Let R be a filtered ring whose associated graded ring is a right Ore ring. Then R can be embedded in a field K. In fact K can be taken as a complete topological field, with P given by* (2) *as a dense subset.*

The result can be used to embed the universal associative envelope of a Lie algebra (even infinite-dimensional) in a field. In particular it provides another embedding of the free algebra, because this can be regarded as the universal associative envelope of the free Lie algebra. (Cf. [8] and for other applications [9].) For a further generalization see [42].

Another 'topological' method of constructing fields of fractions consists of taking an ordered group and forming 'Laurent series': Given a totally ordered group G and a commutative field F, consider the direct power F^G of F indexed by G. With each $f \in F^G$ we associate a subset $D(f)$ of G, its *support*, defined as

$$D(f) = \{s \in G \mid f(s) \neq 0\}.$$

E.g., the group algebra FG of G may be identified with the set of elements of finite support. In general it will not be possible to define the multiplication of F^G in such a way as to extend the operation on FG, for this requires that

$$fg = (\Sigma f(s)s)(\Sigma g(t)t) = \sum_u \Big[\sum_{st=u} f(s)g(t) \Big] u, \tag{4}$$

and here the inner sum on the right will generally contain infinitely many non-zero terms to be added. However, if both f and g have a *well-ordered* support, then the equation $st = u$ has, for a given $u \in G$, only finitely many solutions (s, t) in $D(f) \times D(g)$. Moreover, the sum (4) itself will then have well-ordered support. We can therefore define a ring structure on the set A of all elements of F^G whose support is well-ordered, in such a way that the group algebra FG becomes a subalgebra of A.

Finally it can be shown that A is in fact a field, so that the group algebra of G has been embedded in a field. This result was obtained simultaneously and independently by Malcev [30] and Neumann [32]. Later Higman [21] proved a general result on ordered algebraic systems which includes the Malcev-Neumann construction as a special case. For a simplified presentation of Higman's result see [11], p. 123.

We now have another way of embedding free algebras in fields: Let G be the free group on a set X. Then G can be totally ordered (by writing elements as products of basic commutators and taking the lexicographic ordering of the exponents, cf. [31]). Hence its group algebra FG is embeddable in a field. Since FG clearly contains the free algebra $F\langle X\rangle$ as subalgebra, this provides an embedding of the latter in a field.

7. The matrix method. In Section 2 we observed that to embed a noncommutative ring R in a field it may not be enough to produce inverses of all nonzero elements of R. We can try to overcome this difficulty by adjoining inverses of suitable matrices. Given a set Σ of square matrices over R, we can formally adjoin inverses of these matrices as follows. For each $n\times n$ matrix $A=(a_{ij})$ in Σ, take a set of n^2 symbols $A'=(a'_{ij})$ and adjoin the a'_{ij} to R as extra generators with defining relations, in matrix form,

$$AA' = A'A = I.$$

The resulting ring is denoted by R_Σ, and we have a natural homomorphism $\lambda: R\to R_\Sigma$. This ring has properties entirely analogous to the ring R_S described in Th.2.1, to which it reduces when all the matrices in Σ are 1×1. So we again call R_Σ and λ the *universal Σ-inverting ring* and *homomorphism*, respectively.

It is of interest to note that under suitable conditions on Σ, each element of R_Σ is some a'_{ij}; thus the generating set of R_Σ described above is then the whole ring. To state the result, let us say that the set Σ of matrices is *admissible* if (i) $1\in\Sigma$, (ii) the result of applying elementary row (or column) transformations to any matrix of Σ again lies in Σ, and (iii) if $A, B\in\Sigma$, then $\left(\begin{smallmatrix}A & C\\ 0 & B\end{smallmatrix}\right)\in\Sigma$ for any matrix C and zero matrix 0 of the appropriate size.

THEOREM 7.1. *Let R be a ring, Σ an admissible set of matrices over R, and $f: R\to S$ any Σ-inverting homomorphism. Then the set $\bar{R}$ consisting of all components of inverses of matrices in Σ^f is a subring of S.*

When S is a field, this result applies in particular to the set Σ_1 of all matrices over R whose images are invertible in S, for then Σ_1 can easily be shown to be admissible. When $\Sigma=\Sigma_1$, the set $\bar{R}$ is also called the *rational closure* of R under the homomorphism f.

The question now is: Which matrices do we have to invert to get a field? In the commutative case the answer was easy: we had to invert all nonzero elements, and this ensured that all matrices that are nonzero-divisors also become invertible. But in the general case there may well be matrices that are nonzero-divisors and yet are not invertible in any larger ring. Thus let R be any entire ring that is neither a right nor a left Ore ring. Then there exist $a, b, c, d\in R^*$ such that $Ra\cap Rb=0$, $cR\cap dR=0$, and it follows easily that the matrix

$$\begin{pmatrix} ac & ad\\ bc & bd\end{pmatrix} = \begin{pmatrix} a\\ b\end{pmatrix}(c\ d) \tag{1}$$

is a nonzero-divisor. But this matrix cannot be invertible in any field; more generally, no homomorphism of R into a field can map (1) to an invertible matrix. This example suggests the following definition:

A matrix A over a ring R is said to be *full* if it is square, say $n\times n$, and it cannot be written as a product $A=PQ$, where P is $n\times r$, Q is $r\times n$, and $r<n$. Clearly, the most we can hope for, in mapping a ring R into a field, is to invert the full matrices. The next result goes some way towards saying when this can be done [13].

THEOREM 7.2. *Let R be a ring such that the set Φ of all full matrices over R is admissible. Then the universal Φ-inverting ring R_Φ is either* 0 *or a field; moreover, when R_Φ is a field, the universal Φ-inverting homomorphism $\lambda:R\to R_\Phi$ is injective.*

This is proved by exhibiting each element of R_Φ as a component of the solution of a matrix equation with a full matrix of coefficients, and showing that the inverse element satisfies a similar equation. The last part of the theorem follows from Th.2.2, because any nonzero element of a ring is full.

The hypotheses of Th.7.2 are satisfied if R_Φ is a nonzero ring in which each one-sided matrix inverse is two-sided (i.e., $AB=I$ implies $BA=I$, cf. [14]), but it is more difficult to find conditions in terms of R itself. Here is one case where this has been done.

A *free ideal ring*, or *fir* for short, is a ring R in which each right ideal (and each left ideal) is free as an R-module, and all bases of a free module have the same number of elements [10]. Examples of firs are: (i) free algebras over a commutative field (on any free generating set), (ii) group algebras of free groups, and (iii) free products of fields, over a common subfield, [10]. For a fir one can show that the set Φ of full matrices is admissible and that $\lambda:R\to R_\Phi$ is an embedding. Hence using Th.7.2 we see that each fir can be embedded in a field. Since the class of full matrices is preserved under automorphisms, each automorphism of the fir can be extended (in just one way) to an automorphism of its field of fractions.

A final point concerns the uniqueness. The field of fractions of an integral domain or, more generally, of a right Ore ring is unique up to isomorphism. For any ring isomorphism $a\mapsto a'$ extends to an isomorphism of the field of fractions by the formula $(a/s)'=a'/s'$. In the general case this is no longer so: there may be several nonisomorphic fields of fractions of a noncommutative ring. (E.g., for the free algebra, the field of fractions obtained from Th.7.2, using the fact that the free algebra is a fir, can be shown to be different from the field obtained by Jategaonkar's construction in Section 4.) Given two fields of fractions K_1, K_2 of a ring R, we define a *specialization* from K_1 to K_2 as a homomorphism f from a subring R_1 of K_1 to K_2 that reduces to the identity map on R and such that any nonunit of R_1 is mapped to 0 by f. Of course no unit can be mapped to 0, so ker f is the precise set of nonunits. This means that the nonunits of R_1 form an ideal, $\mathfrak{m}$ say; a ring with this property is said to be *local*. Clearly $R_1/\mathfrak{m}$ is a field, isomorphic to the image of R_1 under f. This image is a subfield of K_2 containing R, but since K_2 is generated (as a field) by R, the image is all of K_2, i.e., f is surjective. This then shows each specialization to be surjective.

A field of fractions K of R is said to be *universal* if for each field of fractions K' of R there is a unique specialization from K to K'. This property determines K up to isomorphism, and in looking for fields of fractions, we are naturally interested in finding the universal one, if it exists. Any free algebra (over a commutative field) has a universal field of fractions [2] and also has other non-universal ones. More generally one can show [14]:

THEOREM 7.3. *Let R be a ring such that the set Φ of full matrices is admissible. If $R_\Phi \neq 0$ (so that R_Φ is a field, by Th.7.2), then R_Φ is the universal field of fractions of R.*

In particular this shows that each fir has a universal field of fractions.

These notions will be useful when one tries to do noncommutative algebraic geometry, which might be defined as the study of zero-sets of rational functions in skew fields, just as the usual kind is the study of zero-sets of polynomials in commutative fields. In the commutative case polynomial zero-sets and rational zero-sets are the same, which is why we could confine ourselves to the former. In general this may not be so (cf. the examples in [39]), and some thought is needed even to construct rational functions. In [2] Amitsur classified rational function fields according to the (rational) identities they satisfy; this is taken up by Bergman [39] from a more general view-point; he also shows that affine space over a field with infinite center is irreducible and describes a universal field of functions.

Clearly many problems remain; we end by listing a few:

1. Find criteria for the existence of a homomorphism of a ring into a field (remember that the zero-mapping is not a homomorphism).
2. Which rings have fields of fractions?
3. Which rings have more than one field of fractions?
4. Which rings have a universal field of fractions?

Added in preparation of this volume:

The author reports that he has solved the problem of finding criteria for the embeddability of rings in skew fields (raised on p. 150, line 12). The complete solution, together with an answer to the above problems, appears in Chapter 7 of his book, Free rings and their relations, Academic Press, London, 1971.

References

1. S. A. Amitsur, On rings with identities, J. London Math. Soc., 30 (1955) 464–470.
2. ———, Rational identities and applications to algebra and geometry, J. Algebra, 3 (1966) 304–359.
3. K. Asano, Über die Quotientenbildung von Schiefringen, J. Math. Soc. Japan, 1 (1949) 73–78.
4. L. A. Bokut', On Malcev's problem, Sibirsk. Mat. Zh., 10 (1969) 965–1005.
5. A. J. Bowtell, On a question of Malcev, J. Algebra, 6 (1967) 126–139.
6. N. Bourbaki, Topologie générale, Ch. I–II, Hermann, Paris 1949.
7. I. E. Burmistrovič, On the embedding of rings in fields, Sibirsk. Mat. Zh., 4 (1963) 1235–1240.

8. P. M. Cohn, On the embedding of rings in skew fields, Proc. London Math. Soc., (3) 11 (1961) 511–530.

9. ———, Quadratic extensions of skew fields, Proc. London Math. Soc., (3) 11 (1961) 531–556.

10. ———, Free ideal rings, J. Algebra, 1 (1964) 47–69.

11. ———, Universal Algebra, Harper & Row, (New York, London, Tokyo 1965).

12. ———, Free associative algebras, Bull. London Math. Soc., 1 (1969) 1–39.

13. ———, The embedding of firs in skew fields, Proc. London Math. Soc., (3) 23 (1971) 193–213.

14. ———, Universal skew fields of fractions, Symposia Math., VIII (1972) 135–148.

15. C. W. Curtis, A note on noncommutative polynomials, Proc. Amer. Math. Soc., 3 (1952) 965–969.

16. M.-L. Dubreil-Jacotin, Sur l'immersion d'un semigroupe dans un groupe, C. R. Acad. Sci. Paris, 225 (1947) 787–788.

17. C. C. Faith, Lectures on injective modules and quotient rings, Springer Lecture notes in Math. No. 49 (Berlin 1967).

18. A. W. Goldie, The structure of prime rings under ascending chain conditions, Proc. London Math. Soc., (3) 8 (1958) 589–608.

19. ———, Some aspects of ring theory, Bull. London Math. Soc., 1 (1969) 129–154.

20. R. Hartshorne, Residues and duality, Springer Lecture notes in Math., No. 20 (Berlin 1966).

21. G. Higman, Ordering by divisibility in abstract algebras, Proc. London Math. Soc., (3) 2 (1952) 326–336.

22. N. Jacobson, Structure of rings, Providence, R. I., 1956, 1964.

23. A. V. Jategaonkar, Ore domains and free algebras, Bull. London Math. Soc., 1 (1969) 45–46.

24. A. A. Klein, Rings nonembeddable in fields with multiplicative semigroups embeddable in groups, J. Algebra, 6 (1967) 101–125.

25. E. G. Koševoi, On certain associative algebras with transcendental relations, Algebra i Logika, Sem. 9, No. 5 (1970).

26. J. Lambek, Lectures on rings and modules, Blaisdell, Waltham, Mass., 1966.

27. S. MacLane, Homology, Springer, Berlin-New York, 1963.

28. A. I. Malcev, On the immersion of an algebraic ring into a field, Math. Ann., 113 (1937) 686–691.

29. ———, Über die Einbettung von assoziativen Systemen in Gruppen I, Mat. Sbornik, 6 (48) (1939) 331–336, II. dto 8 (50) (1950) 251–264.

30. ———, On the embedding of group algebras in division algebras, Dokl. Akad. Nauk SSSR, 60 (1948) 1499–1501.

31. B. H. Neumann, On ordered groups, Amer. J. Math., 71 (1949) 1–18.

32. ———, On ordered division rings, Trans. Amer. Math. Soc., 66 (1949) 202–252.

33. O. Ore, Linear equations in noncommutative fields, Ann. of Math., 32 (1931) 463–477.

34. ———, Theory of noncommutative polynomials, Ann. of Math., 34 (1933) 480–508.

35. E. C. Posner, Prime rings satisfying a polynomial identity, Proc. Amer. Math. Soc., 11 (1960) 180–184.

36. A. Robinson, A note on embedding problems, Fund. Math., 50 (1961/62) 167–309; 455–461.

37. E. Steinitz, Algebraische Theorie der Körper, J. Reine Angew. Math., 137 (1910) new edition by R. Baer and H. Hasse, Berlin, 1930.

38. S. Stevin, Arithmétique (Amsterdam 1585).

39. G. M. Bergman, Skew fields of noncommutative rational functions, after Amitsur, to appear.

40. V. P. Elizarov, Rings of fractions, Algebra i Logika Sem., 8, no. 4 (1969) 381–424.

41. A. A. Klein, A conjecture concerning embeddability of rings into fields, J. Algebra, (to appear).

42. J. Dauns, Embeddings in division rings, Trans. Amer. Math. Soc., 150 (1970) 287–299.

NIL ALGEBRAS AND PERIODIC GROUPS*

IRWIN FISCHER AND RUTH R. STRUIK,
University of Colorado

Introduction. The purpose of this paper is to give an account of some recent work ([2], [3]) in groups and rings which the authors hope will be comprehensible to a person with the background of a standard first year graduate course in modern algebra.

We discuss two questions which have turned out to be more closely related than appeared originally. The first was originally asked by Burnside in 1902 [1], and can be stated as follows:

Let G be a finitely generated, periodic group. Is G finite? *Finitely generated* means that G contains a finite set of elements, $g_1, g_2, \cdots, g_r$ (called its generators) such that every element can be expressed as a finite product of the generators and their inverses. A group, G, is *periodic* if for every element $g \in G$, there exists some integer n (which may depend on g) such that $g^n = 1$. If $g^n = 1$ with n fixed, for all $g \in G$, and n is the smallest positive integer for which this is true, then n is called the *exponent* of G.

Suppose G is of exponent 3 and generated by two elements a and b, then $1, a, b, a^2, b^2, ab, ba, a^2b, ab^2, aba, bab, \cdots$ may or may not be distinct as elements of G. If an infinite number of them are distinct, then G would be an infinite group. In this particular case, G is finite, but this requires proof.

The other question was originally asked by Kuroš in 1941 [14]: Let A be a finitely generated, algebraic algebra. Is A finite-dimensional (as a vector space)? An *algebra*, A, is a ring which is simultaneously a vector space over a field K. In addition, the following holds:

$$\alpha(ab) = (\alpha a)b = a(\alpha b) \qquad \text{for all } \alpha \in K,\ a, b \in A.$$

A is *finitely-generated* if there is a finite subset $a_1, \cdots, a_r$ (called its generators) such that every element of A can be obtained from the generators by a finite number of additions, multiplications and/or scalar multiplications. An algebra A is *algebraic* if every $a \in A$ satisfies an equation of the form

$$k_n a^n + k_{n-1} a^{n-1} + \cdots + k_0 = 0, \tag{1}$$

where $k_i \in K$. The equation may differ for different $a \in A$.

Since each of these questions was originally proposed, a great many attempts to solve them have occurred. Most of the results obtained indicated that the answer is "yes" if other hypotheses were added. For example, Burnside [1] considered the following special cases:

(1) G of exponent 2,
(2) G of exponent 3,
(3) G of exponent 4, $r = 2$ (i.e., G with two generators).

---

* From American Mathematical Monthly, vol. 75 (1968), pp. 611–623.

The answer in these cases is "yes." Then in 1940, Sanov [19] obtained an affirmative answer for exponent 4 and an arbitrary (but finite)number of generators. Marshall Hall Jr. in 1958 [6] similarly disposed of groups of exponent 6. If $g^5=1$ for all $g\in G$, the answer is still unknown. Then Novikov in 1959 [18] announced that the answer is "no" if $n\geqq 72$, and $r\geqq 2$.

The work on the second question has had a similar history. Kuroš [14] discussed several special cases, all with affirmative answers. In [10], [15] and [16] Jacobson and Levitzki settled the question affirmatively for algebras of bounded degree. In an algebraic algebra A the *degree* of $a\in A$ is the least degree of the polynomials of which a is a root. If the degrees of all $a\in A$ have a maximum, then A is of *bounded degree.* Since these papers, many special cases have been studied. (See, e.g., [4], [7], [8], [11], [12], [20].)

Then in 1964, Golod [3] announced that the answer to both questions is "no." In particular, he constructed a finitely generated nil algebra which is infinite-dimensional. In a *nil algebra,* $a^n=0$ for all $a\in A$ (n might depend on a). (A nil algebra is obviously algebraic: put $k_n=1$; $k_i=0$, $1\leqq i\leqq n-1$ in (1).) Using this nil algebra, Golod constructed a finitely generated group which is periodic and infinite. This settles negatively both Burnside's problem and Kuroš' problem.

In view of the many previous attempts to solve these two problems, the Golod-Šafarevič construction is remarkable for its simplicity and the small amount of background needed to understand it. The details of their method will be given in subsequent sections of this exposition.

Even with Golod's results, much still remains unknown. For example, given Novikov's result, it is still unknown whether finitely generated groups of exponent 5, 7, 8, 9, $\cdots$, 71 are finite or infinite. Also, the details of Novikov's proof have not yet been published as of the writing of this paper, and the proof, when it appears, will very likely be long and difficult. Hence the Golod-Šafarevič results are still very significant.

In Section 1 we will give a sufficient condition for an algebra to be infinite-dimensional. In Section 2 we will apply this method to settle Burnside's and Kuroš' questions.

1. In this section we give a sufficient condition for an algebra A to be infinite-dimensional. Before the main theorem can be stated, a few terms will be defined.

Let K be any field and let $R_d=K[x_1, x_2, \cdots, x_d]$ be the polynomial ring over K in the noncommuting indeterminates $x_1, \cdots, x_d$. This means that the x_i do not commute with each other, but they do commute with the elements of K. Then

$$R_d = T_0 \oplus T_1 \oplus \cdots \oplus T_n \oplus \cdots, \tag{2}$$

where $T_0=K$ and T_n is the vector space over K spanned by the d^n monomials $x_{i_1}x_{i_2}\cdots x_{i_n}$. In (2), $\oplus$ means direct sum: that is, every $u\in R_d$ can be written uniquely as a sum

$$u = u_0 + u_1 + \cdots + u_k, \quad \text{where} \quad u_i \in T_i. \tag{3}$$

The elements of T_i are said to be *homogeneous* of degree i. For example, $x_1^3+x_1x_2x_3$ is homogeneous of degree 3, but $x_1^4+x_1x_2x_3$ is not homogeneous. If u is a homogeneous polynomial of degree i, we will denote by $\partial(u)=i$ the degree of u.

Let F be a set of nonzero homogeneous polynomials, $f_1, f_2, \cdots$ such that $2 \leqq \partial(f_i) \leqq \partial(f_{i+1})$ and such that the number, r_n, of polynomials of degree n is finite (possibly zero). We will be studying the ideal, I, generated by $f_1, f_2, \cdots$. This ideal is made up of polynomials whose summands are of the form

$$uf_jv, \qquad u, v \in R_d. \tag{4}$$

If u and v are written as sums of homogeneous polynomials, this means that (4) can be written as a sum of homogeneous polynomials of the form $u_if_jv_k$, where u_i and v_k are homogeneous polynomials of R_d. We now form the factor algebra

$$A = R_d/I. \tag{5}$$

If $r=u_2+u_3+\cdots+u_s \in I$, $u_i \in T_i$, then each $u_i \in I$, since I is generated by homogeneous polynomials. This means that if

$$v_1 + \cdots + v_s \equiv w_1 + \cdots + w_t \pmod{I}, \; v_i, w_i \in T_i,$$

then $v_i \equiv w_i \pmod{I}$ for each i. Hence,

$$A = A_0 \oplus A_1 \oplus \cdots \oplus A_n \oplus \cdots, \tag{6}$$

where $A_n=(T_n+I)/I$. In particular,

$$A_0 \simeq T_0 \simeq K \quad \text{and} \quad A_1 \simeq T_1,$$

since $\partial(f_i) \geqq 2$. Since A_n is a vector space over K, it makes sense to talk about $b_n = \dim A_n$. Obviously $\dim T_n = d^n$.

We will also be considering formal power series such as

$$1 - dt + \sum_{i=2}^{\infty} r_i t^i \tag{7}$$

and

$$\left(1 - dt + \sum_{i=2}^{\infty} r_i t^i\right)^{-1} = 1 + \sum_{i=1}^{\infty} x_i t^i. \tag{8}$$

Here a formal power series is simply a function from the positive integers into the real numbers, and the coefficient of t^n is merely the value of the function at n; convergence is irrelevant. The "inverse" (8) is obtained by solving the following equations:

$$
(9) \qquad
\begin{aligned}
x_1 &= d \\
x_2 &= dx_1 - r_2 = d^2 - r_2 \\
x_3 &= dx_2 - r_2x_1 - r_3 = d(d^2 - r_2) - r_2d - r_3 \\
&\cdots\cdots\cdots\cdots\cdots\cdots \\
x_n &= dx_{n-1} - \sum_{i=2}^{n} r_i x_{n-i}, \quad \text{where} \quad x_0 = 1.
\end{aligned}
$$

We can now state the Golod-Šafarevič theorem ([2] and [3]). The proof we give here is a modification of one appearing in [21].

THEOREM 1.1. *Let* $R_d = K[x_1, \cdots, x_d]$ *be the polynomial ring over* K *in the noncommuting indeterminates* $x_1, \cdots, x_d$. *Let* $f_1, f_2, \cdots, \in F$ *be a set of homogeneous polynomials of* R_d, *and let the number of polynomials of degree* i *be* r_i. *Let* $2 \leqq \partial(f_i) \leqq \partial(f_{i+1})$. *Let* I *be the ideal generated by* $f_i \in F$. *Let*

$$
(10) \qquad R_d/I = A = A_0 \oplus A_1 \oplus \cdots \oplus A_n \oplus \cdots.
$$

If all the coefficients of

$$
(11) \qquad \left(1 - dt + \sum_{i=2}^{\infty} r_i t^i\right)^{-1}
$$

are nonnegative, then A *is infinite-dimensional.*

We first prove a lemma:

LEMMA 1.2. *Under the conditions of Theorem* 1.1, *let* $b_n = \dim A_n$. *Then*

$$
(12) \qquad b_n \geqq db_{n-1} - \sum_{i=2}^{n} r_i b_{n-i}, \qquad n \geqq 2.
$$

(Note the similarity between (12) and the last equation of (9).)

Proof. Let $I_n = I \cap T_n$, i.e., I_n is the vector space spanned by the homogeneous polynomials of I of degree n. By an elementary theorem on vector spaces, I_n has a complementary subspace; call it S_n. Then

$$
T_n = I_n \oplus S_n.
$$

Obviously,

$$
(13) \qquad \dim T_n = \dim I_n + \dim S_n = \dim I_n + b_n.
$$

For the case $n = 2$, I_2 is generated by the $f_i \in F$ such that $\partial(f_i) = 2$. Hence $\dim I_2 \leqq r_2$. (If the f_i of degree 2 were linearly independent, then $\dim I_2 = r_2$.) Hence

$$
d^2 = \dim T_2 = \dim I_2 + b_2 \leqq r_2 + b_2
$$

i.e., $b_2 \geqq d^2 - r_2 = db_1 - r_2 b_0$, since $b_1 = d$ and $b_0 = 1$, which is (12) for $n = 2$.

Let now $T_{n-1} = S_{n-1} \oplus I_{n-1}$, and let $s_1, s_2, \cdots, s_{b_{n-1}}$ be a basis for S_{n-1} and $g_1, \cdots, g_m$ a basis for I_{n-1}, both considered as vector spaces. Then the elements

$$s_i x_j, \qquad (i = 1, 2, \cdots, b_{n-1} \text{ and } j = 1, \cdots, d)$$

together with the elements

$$g_k x_j, \qquad (k = 1, 2, \cdots, m \text{ and } j = 1, \cdots, d)$$

obviously form a basis for T_n. Let J be the vector space spanned by the $g_k x_j$. Let $v_1, v_2, \ldots$ be a set of homogeneous polynomials of degrees up to $n-1$, which forms a basis for $S_0 \oplus S_1 \oplus \cdots \oplus S_{n-1}$. Let L be the vector space spanned by all elements of degree n of the form $v_i f_j$, $f_j \in F$. We now show that if $u \in I_n$, then $u = v + w$, $v \in L$ and $w \in J$. Since $u \in I_n$, it can be written as a sum of polynomials of the form $u_i f_j u_k$, where u_i, u_k are homogeneous polynomials and $\partial(u_i f_j u_k) = n$.

Case I. $\partial(u_k) \geqq 1$. Then $u_i f_j u_k \in J$. Let us indicate the proof of this by an example. Let $u_i = x_1 x_2$, $u_k = x_3 x_4$. Then

$$x_1 x_2 f_j x_3 x_4 = (x_1 x_2 f_j x_3) x_4.$$

But $x_1 x_2 f_j x_3 \in I_{n-1}$ and therefore $u_i f_j u_k \in I_{n-1} x_4 \subseteq J$, since $J = I_{n-1} x_1 \oplus I_{n-1} x_2 \oplus \cdots \oplus I_{n-1} x_d$.

Case II. $\partial(u_k) = 0$, i.e., $u_i f_j u_k = u_i f_j$. If $u_i \in T_k$, then $u_i = v' + w'$, where $v' \in S_k$, $w' \in I_k$. Then $u_i f_j = v' f_j + w' f_j$. By the argument given for Case I, $w' f_j \in J$. By definition of $v_1, v_2, \ldots, v' f_j = \sum_k c_{i_k} v_{i_k} f_j \in L$, $c_{i_k} \in K$.

Hence $u \in I_n$ implies $u = v + w$, $v \in L$, $w \in J$. This means that $\dim I_n \leqq \dim L + \dim J$. (We cannot say $\dim I_n = \dim L + \dim J$, for it may be that nonzero members of I_n are contained in $L \cap J$.) Hence

$$\dim T_n = \dim I_n + \dim S_n \leqq \dim J + \dim L + b_n. \tag{14}$$

But $\dim L \leqq \sum_{i=2}^n b_{n-i} r_i$, i.e.,

$$\dim T_n \leqq \dim J + \sum_{i=2}^n b_{n-i} r_i + b_n. \tag{15}$$

Now the $s_i x_j$ along with $g_k x_j$ form a basis for T_n. This means

$$\dim T_n = d b_{n-1} + \dim J. \tag{16}$$

Putting (15) and (16) together gives (12).

Proof of Theorem. Consider

$$\left(1 + \sum_{i=1}^{\infty} b_i t^i\right)\left(1 - dt + \sum_{j=2}^{\infty} r_j t^j\right) = A(t). \tag{17}$$

These are just formal power series. If one multiplies the left hand side of (17) and compares with (12), one obtains that the coefficients of $A(t)$ are all nonnegative.

Multiplying on both sides by

$$\left(1 - dt + \sum_{j=2}^{\infty} r_j t^j\right)^{-1} = C(t),$$

one obtains

$$1 + \sum_{i=1}^{\infty} b_i t^i = A(t)C(t). \tag{18}$$

By hypothesis the coefficients of $C(t)$ are nonnegative. By definition of b_i, $b_i \geqq 0$ for $i \geqq 1$. We want to make sure that the left hand side of (18) is *not* a polynomial. Since $A(t)$ has nonnegative coefficients, it is sufficient to show that $C(t)$ is *not* a polynomial. Suppose $C(t)$ is a polynomial: then $(1 - dt + \sum_{j=2}^{\infty} r_j t^j)\, C(t) = 1$, and hence

$$\left(1 + \sum_{j=2}^{\infty} r_j t^j\right) C(t) = 1 + dtC(t). \tag{19}$$

Comparing the two sides of (19), it is impossible for $C(t)$ to be a polynomial. This completes the proof of Theorem 1.

COMMENT: In [21], Vinberg shows that if (11) is replaced by

$$(1 - t)\left(1 - dt + \sum_{i=2}^{\infty} r_i t^i\right)^{-1},$$

then the requirement that the f_i be homogeneous can be removed. In this case, the degree of the polynomial is defined to be the least degree of its homogeneous components. Recently, R. H. Bruck has shown that for $d \geqq 2$, $r_i \geqq 0$, any formal power series satisfying the requirements of Theorem 1.1 also satisfies Vinberg's requirements and vice-versa. R. H. Bruck has also obtained necessary and sufficient conditions for such formal power series to satisfy these requirements.

COROLLARY 1.3. *Under the hypotheses of Theorem* 1.1, *let* $r_i \leqq s_i$. *If*

$$\left(1 - dt + \sum_{i=2}^{\infty} s_i t^i\right)^{-1} \tag{20}$$

has all its coefficients nonnegative, then A is infinite dimensional.

Proof of Corollary 1.3. Let

$$H = 1 - dt + \sum_{n=2}^{\infty} r_n t^n,$$

$$G = 1 - dt + \sum_{n=2}^{\infty} s_n t^n,$$

$$U = \sum_{n=2}^{\infty} (s_n - r_n) t^n,$$

then $U+H=G$. We are given that G^{-1} has all its coefficients nonnegative, and we want to prove that the same is true for H^{-1}. Now $H=G-U$ and hence,

$$H^{-1} = [G(1 - UG^{-1})]^{-1} = G^{-1}(1 - UG^{-1})^{-1}.$$

Since U and G^{-1} have all their coefficients nonnegative, so does UG^{-1}. Then all the coefficients of

$$(1 - UG^{-1})^{-1} = 1 + \sum_{n=1}^{\infty} (UG^{-1})^n$$

are nonnegative, and hence the same is true for $G^{-1}(1-UG^{-1})^{-1}$. This shows that all the coefficients of H^{-1} are nonnegative, and we can now use Theorem 1.1.

In order to apply Theorem 1.1 or Corollary 1.3 in a particular case, we need some expressions for possible r_i:

THEOREM 1.4 (Golod [3]). *Let r_i and A be as in Theorem 1.1. If*

$$r_i \leqq \epsilon^2(d - 2\epsilon)^{i-2},$$

where ϵ is any positive number such that $(d-2\epsilon)>0$, then A is infinite-dimensional.

Proof. It is sufficient to examine the coefficients of

$$\left(1 - dt + \sum_{i=2}^{\infty} \epsilon^2(d - 2\epsilon)^{i-2}t^i\right)^{-1}. \tag{21}$$

To do this, we make use of the following manipulations with formal series:

$$\frac{1}{1-a} = 1 + a + a^2 + \cdots = \sum_{i=0}^{\infty} a^i; \tag{22}$$

$$(1 - a)^2(1 + 2a + 3a^2 + \cdots) = 1, \text{ i.e.,} \tag{23}$$

$$\frac{1}{(1-a)^2} = \sum_{n=1}^{\infty} na^{n-1}. \tag{24}$$

Hence

$$1 - dt + \sum_{i=2}^{\infty} \epsilon^2(d - 2\epsilon)^{i-2}t^i$$

$$= 1 - dt + \epsilon^2t^2[1 + (d - 2\epsilon)t + \cdots + (d - 2\epsilon)^nt^n + \cdots]$$

$$= 1 - dt + \frac{\epsilon^2t^2}{1 - (d - 2\epsilon)t} = \frac{1 + (2\epsilon - 2d)t + (d - \epsilon)^2t^2}{1 - (d - 2\epsilon)t}$$

$$= \frac{[1 - (d - \epsilon)t]^2}{1 - (d - 2\epsilon)t}.$$

Hence (21) becomes

$$\frac{1-(d-2\epsilon)t}{[1-(d-\epsilon)t]^2}. \tag{25}$$

Now use (24) with $a=(d-\epsilon)t$:

$$\begin{aligned}\frac{1-(d-2\epsilon)t}{[1-(d-\epsilon)t]^2} &= [1-(d-2\epsilon)t]\left[1+\sum_{n=1}^{\infty}(n+1)(d-\epsilon)^n t^n\right] \\ &= 1+\sum_{n=1}^{\infty}[(n+1)(d-\epsilon)^n-(d-2\epsilon)n(d-\epsilon)^{n-1}]t^n \qquad (26) \\ &= 1+\sum_{n=1}^{\infty}(d-\epsilon)^{n-1}[d+(n-1)\epsilon]t^n.\end{aligned}$$

Since $d-2\epsilon>0$, $d-\epsilon>\epsilon>0$, and all the coefficients of (26) are nonnegative.

COROLLARY 1.5. *In Theorem* 1.4, *let* $d=2$ *and* $r_i=0$ *for* $i=2, 3, \cdots, 9$ *and* $r_i=0$ *or* 1 *for* $i\geqq 10$. *Then A is infinite-dimensional.*

Proof. Use Theorem 1.4 with $\epsilon=1/4$ and $d=2$. Then

$$\frac{(2-1/2)^8}{16}=\frac{(1+1/2)^8}{16}.$$

Expand $(1+\frac{1}{2})^8$ using the binomial theorem. From the first four terms, $(1+\frac{1}{2})^8>16$. Hence $\epsilon^2(d-2\epsilon)^8>1$. Since $(d-2\epsilon)^i<(d-2\epsilon)^{i+1}$ if $d-2\epsilon>1$, this is sufficient to prove Corollary 1.5.

COMMENT. Actually $(1.5)^7>16$ and $r_i=0$, $i=1, 2, \cdots, 7, 8$; $r_i=0$ or 1, $i\geqq 9$ will be sufficient to make A infinite-dimensional.

2. We can now start constructing examples of nil algebras and periodic groups. The first two examples, simple ones, are modifications of those that appear in [8].

THEOREM 2.1. *Let K be any finite or countable field. Then there exists an infinite-dimensional nil algebra over K generated by* 2 *elements.*

Proof. Consider $K[x_1, x_2]$, the polynomial ring over K generated by the non-commuting variables x_1 and x_2. Consider all polynomials with constant term equal to zero. These can be enumerated: $u_1, u_2, \cdots$. Then

$$u_1^{10}=s_{11}+s_{12}+\cdots+s_{1k_1},$$

where s_{1j} are homogeneous polynomials and $10\leqq\partial(s_{1j})<\partial(s_{1,j+1})$. Let $\partial(s_{1k_1})=N_1$. Then

$$u_2^{N_1+1}=s_{21}+\cdots+s_{2k_2},$$

where $N_1+1 \leqq \partial(s_{2j}) < \partial(s_{2,j+1})$. Continue in this way. This gives a collection of homogeneous polynomials, s_{ij}, all of degree $\geqq 10$, and for each degree, there is at most one polynomial. We now use Corollary 1.5 and Theorem 1.1 to obtain an infinite-dimensional algebra,

$$A = A_0 \oplus A_1 \oplus \cdots \oplus A_n \oplus \cdots$$

Consider $B = A_1 \oplus A_2 \oplus \cdots \oplus A_n \oplus \cdots$ B is an algebra generated (as an algebra) by x_1+I and x_2+I, and every element $b \in B$ is such that $b^m=0$ for some m. This is our example.

THEOREM 2.2. *For every prime p, there is an infinite group generated by two elements in which every element has order a power of p.*

Proof. Let K be the field of integers modulo p. Let A be the algebra constructed in the proof of Theorem 2.1. Then if $1+u \in A$, with $u \in B$, there exists an m such that $u^m=0$. Let $p^j > m$. Then

$$(1 + u)^{p^j} = 1 + u^{p^j} = 1, \tag{27}$$

since if one uses the binomial theorem, the coefficients of u^i, $1 \leqq i \leqq p^j - 1$ will be divisible by p. Let

$$a = 1 + x_1 + I \quad \text{and} \quad b = 1 + x_2 + I.$$

Consider the subset G of A consisting of all finite power products of a and b, with nonnegative exponents. Since a and b are of finite multiplicative order, G is a group, the multiplication of A being the operation of G. By (27) every element of G has order a power of a prime.

We want to show that G is infinite. Suppose not. Each element of G is a coset of the form $1+u_1+u_2+\cdots+u_s+I$, u_s homogeneous polynomials. If G is finite, we can choose coset representatives so that there is a maximum M to the degrees of the homogeneous components u_i. Since A, the algebra of Theorem 2.1, is infinite-dimensional, there is a monomial $x_{i_1}x_{i_2} \cdots x_{i_{M+1}} \notin I$. Then

$$\begin{aligned} g &= (1 + x_{i_1})(1 + x_{i_2}) \cdots (1 + x_{i_{M+1}}) + I \\ &= 1 + v_1 + \cdots + v_M + x_{i_1}x_{i_2} \cdots x_{i_{M+1}} + I \neq 1 + I, \end{aligned}$$

v_i homogeneous of degree i. g cannot be any of the elements of G already enumerated. Hence G must be infinite.

Theorem 2.1 assumed that K was finite or countable. This is not necessary, and we now give Golod's construction which holds for an arbitrary field:

THEOREM 2.3. (Golod [3]). *Let K be any field. Then there exists a nil algebra over K with $d \geqq 2$ generators which is infinite-dimensional.*

Proof. Let $R_d = K[x_1, \cdots, x_d]$ as previously. We want to construct a set of homogeneous polynomials $f_1, f_2, \cdots \in F$ such that the number of polynomials of degree $i \leqq \epsilon^2(d-2\epsilon)^{i-2}$ for some $\epsilon>0$, and such that if $u \in R_d$ has its constant term 0, then $u^m \in$ ideal generated by $f_1, f_2, \cdots$ for some m. As usual, we denote the ideal generated by $f_1, f_2, \cdots$ by I. We will construct a sequence of ideals, $I_1, I_2, \cdots$ such that $I_j \subseteq I_{j+1}$ and such that if $u \in R_d$, with constant term 0, and the highest degree of the monomials appearing in u is j, then $u^m \in I_j$ for some m. Then we will choose

$$I = I_1 \cup I_2 \cup \cdots \cup I_n \cup \cdots \qquad \text{(set-theoretic union).}$$

We start with I_1. Consider an arbitrary monomial of degree 1 with constant term 0: $u = c_1x_1 + c_2x_2 + \cdots + c_dx_d$, $c_i \in K$. Then

$$u^M = (c_1x_1 + c_2x_2 + \cdots + c_dx_d)^M. \tag{28}$$

If one expands the right hand side of (28), one obtains a sum of monomials of degree M. Consider the c_i as "variables" with the x_j as "coefficients." For example, the "coefficient" of $c_1c_2 \cdots c_M$ (let's say $d \geqq M$) would be

$$\sum x_{i_1}x_{i_2} \cdots x_{i_M}$$

where the summation is over all permutations, $(i_1, i_2, \cdots, i_M)$ of $(1, 2, \cdots, M)$. Similarly the "coefficient" of c_3^M would be x_3^M. For any $c_{i_1}c_{i_2} \cdots c_{i_M}$, the "coefficient" is a homogeneous polynomial in $x_{i_1}, \cdots, x_{i_M}$. These "coefficients" will be our f_i. We have still to decide what M will be. Each of the "coefficients" is a homogeneous polynomial of degree M. How many are there? It is the number of *commutative* monomials in d variables of degree M. This is equal to the binomial coefficient

$$\binom{M+d-1}{d-1}. \tag{29}$$

(If this is not "well-known" to you, (29) can be obtained by the following argument: Put $M+d-1$ points on a horizontal line, and color $d-1$ of them blue. The number of points from the left end to the first blue point is the exponent of x_1, the number of points between the first and second blue points is the exponent of x_2, etc. This gives a monomial of degree M in $x_1, \cdots, x_d$. To each selection of $d-1$ points, there corresponds a unique monomial and vice-versa.) But

$$\binom{M+d-1}{d-1} \leqq (M+d-1)^{d-1}. \tag{30}$$

We want to choose an M such that

$$(M+d-1)^{d-1} \leqq \epsilon^2(d-2\epsilon)^{M-2}. \tag{31}$$

Pick any $\epsilon>0$, such that $d-2\epsilon>1$, say $\epsilon=\frac{1}{3}$. Then (31) becomes

$$(M+d-1)^{d-1} \leqq \frac{(d-2/3)^{M-2}}{9}. \tag{32}$$

Consider (32) as an inequality with M as the "variable" and d as a constant.

Considered as a function of M, $(M+d-1)^{d-1}$ is a polynomial, while the right-hand side of (32) is an exponential function of M. This means that there exists an M_1 (M_1 a positive integer) such that

$$(M_1 + d - 1)^{d-1} \leqq \frac{(d - 2/3)^{M_1-2}}{9},$$

and (32) holds for all $M \geqq M_1$. The corresponding

$$\binom{M_1 + d - 1}{d - 1}$$

homogeneous polynomials will be used to generate I_1. Evidently for every linear homogeneous polynomial, u, $u^{M_1} \in I_1$.

Suppose we now assume that $I_1, I_2, \cdots, I_{k-1}$ have been formed from the homogeneous polynomials $f_1, f_2, \cdots, f_{m_{k-1}}$, with $\partial(f_i) \leqq \partial(f_{i+1})$. Let $\partial(f_{m_{k-1}}) = M_{k-1}$. We now form a typical polynomial of degree k with constant term 0:

$$(33) \qquad u = c_1^{(1)} x_1 + \cdots + c_d^{(1)} x_d + c_1^{(2)} x_1^2 + \cdots + c_{d^k}^{(k)} x_d^k.$$

Again we look at the expansion of u^M, and consider the x_i as "coefficients" of the $c_i^{(j)}$. Repeating the argument used in obtaining I_1, we find that we will have to count the number of *commutative* monomials in $q = d + d^2 + \cdots + d^k$ variables. This is

$$\binom{M + q - 1}{q - 1} \leqq (M + q - 1)^{q-1}.$$

Again we look for an M such that

$$(34) \qquad (M + q - 1)^{q-1} \leqq \frac{(d - 2/3)^{M-2}}{9}.$$

The M we choose in addition to satisfying (34) must also satisfy the requirement

$$(35) \qquad M > M_{k-1} = \partial(f_{m_{k-1}}),$$

i.e., to form I_k, we use all the $f_1, \cdots, f_{m_{k-1}}$, and only add homogeneous polynomials whose degree is greater than that of any of the preceding f_i. Once an M is found satisfying (34) and (35), we can then add the corresponding homogeneous polynomials to $f_1, \cdots, f_{m_{k-1}}$ to generate I_k. This gives us $f_1, \cdots, f_{m_k}$. Evidently for all $u \in R_d$, u with zero constant term and of degree k, $u^m \in I_k$ for some m.

This gives a sequence of homogeneous polynomials, $f_1, f_2, \cdots$ and a sequence of ideals, $I_1 \subseteq I_2 \subseteq \cdots$ which have the desired properties.

COMMENT. Golod, in addition, shows that the group G of Theorem 2.2 has another property which he calls "finitely-approximable." This means that the intersection of all normal subgroups of G of finite index is the identity. Novikov's [18] infinite groups cannot have this property. It is beyond the scope of this exposition to prove these things. Some further interesting constructions are given in [17].

The authors wish to thank those who read a preliminary version of this paper for their helpful comments. In particular, the authors are grateful to J. T. McCall for his careful reading of that version and his many thoughtful suggestions. The second author wishes to thank the Council on Research and Creative Work at the University of Colorado for a grant which supported some of her work on this paper.

Added in proof. The details of the proof announced in [18] are now appearing: *On infinite periodic groups, I,* Izv. Akad. Nauk SSSR, Ser. Mat. 32 (1968) 212–244 by P. S. Novikov and S. I. Adyan begins this proof. $n \geqq 72$ has been replaced by odd $n \geqq 4381$.

References

1. W. Burnside, On an unsettled question in the theory of discontinuous groups, Quart. J. Pure and Applied Math., 33 (1902) 230–238.

2. E. S. Golod and I. R. Šafarevič, On class field towers, Izv. Akad. Nauk SSSR. Ser. Mat., 28(1964) 261–272. AMS Translations, Ser. 2, 48 (1965) 91–102.

3. E. S. Golod, On nil-algebras and finitely-approximable p-groups, Izv. Akad. Nauk SSSR. Ser. Mat., 28(1964) 273–276. AMS Translations. Ser. 2, 48 (1965) 103–106.

4. M. P. Drazin, A generalization of polynomial identities in rings, Proc. AMS, 8 (1957) 352–361.

5. Marshall Hall, Jr., The Theory of Groups, Macmillan, New York, 1959.

6. ———, Solution of the Burnside problem for exponent 6, Illinois J. of Math., 2 (1958) 764–786.

7. I. N. Herstein, Theory of Rings (1961), mimeographed notes obtainable from the University of Chicago Math Dept.

8. ———, Topics in Ring Theory (1965), mimeographed notes obtainable from the University of Chicago Math Dept.

9. Graham Higman, On a conjecture of Nagata, Proc. Camb. Phil. Soc., 52 (1956) 1–4.

10. N. Jacobson, Structure theory for algebraic algebras of bounded degree, Ann. Math., 46(1945) 695–707.

11. ———, Structure of Rings, Revised edition, 1964, American Mathematical Society.

12. Irving Kaplansky, Rings with a polynomial identity, Bull. AMS, 54 (1948) 575–580.

13. ———, Topological representation of algebras II, Trans. AMS, 68 (1950) 62–75.

14. A. G. Kuroš, Problems in ring theory, connected with the Burnside problem in periodic groups, Izv. Akad. Nauk SSSR Ser. Mat., 5 (1941) 233–241.

15. Jakob Levitzki, On the radical of a general ring, Bull. AMS, 49 (1943) 462–466.

16. ———, On a problem of A. Kurosch, Bull. AMS, 52 (1946) 1033–1035.

17. M. F. Newman, A theorem of Golod-Šafarevič and an application in group theory, preprint obtainable from author at Dept. of Pure Math., SGS, Australian National Univ, GPO Box 4, Canberra ACT, Australia.

18. P. S. Novikov, On periodic groups, Dokl. Akad. Nauk SSSR, 127 (1959) 4, 749–752. AMS Translations, Ser. 2, 45 (1965) 19–22.

19. I. N. Sanov, Solution of Burnside's problem for exponent 4. Leningrad State Univ. Ann., 10 (1940) 166–170.

20. A. I. Širšov, On a problem of Levitzki, Dokl. Akad. Nauk SSSR, 120 (1958) 41–42.

21. E. B. Vinberg, On the theorem concerning the infinite-dimensionality of an associative algebra, Izv. Akad. Nauk SSSR Ser. Mat., 29 (1965) 209–214.

BIBLIOGRAPHIC ENTRIES: NON-COMMUTATIVE RINGS

The references below are to the AMERICAN MATHEMATICAL MONTHLY.

1. Henryk Minc, Left and right ideals in the ring of 2×2 matrices, 1964, 72–75.

Calculates explicitly all the one-sided ideals in the ring of 2×2 matrices over a field and in the ring of 2×2 upper triangular matrices.

2. Neal H. McCoy, Certain systems of linear equations over a ring with an application to polynomial rings, 1962, 847–851.

Proves a theorem on right annihilators of right ideals, using a non-commutative determinant.

3. D. W. Henderson, A short proof of Wedderburn's theorem, 1965, 385–386.

4. Irving Kaplansky, "Problems in the theory of rings" revisited, 1970, 445–454.

Describes progress on the open questions proposed at the Ram's Head Conference, 1956. Extensive bibliography.

5. E. E. Bray, K. A. Byrd, and R. L. Bernhardt, The injective envelope of the upper triangular matrix ring, 1971, 883–886.

Gives an elementary proof of the fact that the injective envelope of the upper triangular matrix ring over a division ring D is the full ring of $n\times n$ matrices over D.

6. J. V. Brawley and L. Carlitz, A characterization of the $n\times n$ matrices over a finite field, 1973, 670–672.

Characterizes them as those rings R such that every function from R to R is a generalized polynomial over R.

(f)

NON-ASSOCIATIVE RINGS AND ALGEBRAS

TRACE FUNCTIONS ON ALGEBRAS WITH PRIME CHARACTERISTIC*

HANS ZASSENHAUS, McGill University

The natural trace function on the matrix algebra F_d of degree d over a field F is defined as

$$f((\alpha_{ik})) = \operatorname{tr}((\alpha_{ik})) = \sum_{i=1}^{d} \alpha_{ii}.$$

* From AMERICAN MATHEMATICAL MONTHLY, vol. 60 (1953), pp. 685–692.

It satisfies the rules

$$f(a+b) = f(a) + f(b) \tag{1}$$

$$f(\lambda a) = \lambda f(a), \tag{2}$$

$$f(ab) = f(ba), \tag{3}$$

for a, b in F_d, λ in F.

Introducing the Lie-multiplication in F_d by

$$a \bigcirc b = ab - ba, \tag{4}$$

the rules (1)–(3) become equivalent to (1), (2) and

$$f(a \bigcirc b) = 0 \tag{3a}$$

for $a, b \in F_d$.

In an associative ring A of prime characteristic p the identity

$$(a_1 + a_2)^p = a_1^p + a_2^p + \Lambda(a_1, a_2) \tag{5}$$

holds, with $\Lambda(a_1, a_2)$ a certain sum of Lie-products of the form

$$a_{i_1} \bigcirc (a_{i_2} \bigcirc (\cdots (a_{i_{p-1}} \bigcirc a_{i_p}) \cdots)).*$$

If A is commutative then (5) assumes the simpler form

$$(a_1 + a_2)^p = a_1^p + a_2^p. \tag{5a}$$

Let f be a function defined on an associative algebra A over a field F of prime characteristic p and assume that the values of f are in F satisfying (1)–(3). It follows from (1), (3a) and (5) that

$$f((a+b)^p) = f(a^p) + f(b^p) \tag{6}$$

and more generally

$$f\left(\left(\sum_{i=1}^{r} \lambda_i a_i\right)^p\right) = \sum_{i=1}^{r} \lambda_i^p f(a_i^p). \tag{7}$$

E.g. for $A = F_d$, $f = \text{tr}$ the matrices $e_{rs} = (\delta_{ir}\delta_{ks})$ form a basis of F_d over F with the multiplication rule $e_{rs}e_{uv} = \delta_{su}e_{rv}$ such that for $a = (\alpha_{ik})$,

$$a = \sum \alpha_{ik} e_{ik} \dagger$$

and, according to (7), (5a),

$$\text{tr}\,(a^p) = \sum \alpha_{ik}^p (\text{tr}\, e_{ik})^p = \sum \alpha_{ii}^p = \left(\sum \alpha_{ii}\right)^p = (\text{tr}\, a)^p,$$

a rule which may be obtained also by using ordinary tools of matrix theory.

---

* Hans Zassenhaus, Über Lie'sche Ringe mit Primzahl Charakteristik, Hamburger Abhandlungen, Bd. 13, p. 89, 1939.

† The mere summation symbol indicates summation over all indices occurring twice.

DEFINITION. *A function f defined on an associative algebra A over a field F of prime characteristic p with values in an algebraically closed extension Ω of F is called a trace function A over F if it satisfies the rules* (1)–(3) *and the rule*

(8) $$f(a^p) = f(a)^p \qquad \textit{for } a \textit{ contained in } A.$$

An example is given by the natural trace function on a matrix algebra of characteristic p. More generally, every representation Δ of an associative algebra A over a field F of characteristic $p>0$ by matrices of degree d with coefficients in Ω leads to the trace function tr $\Delta(a)$ on A over F. This follows from the conditions

(9) $$\begin{aligned} \Delta(a+b) &= \Delta a + \Delta b, \\ \Delta(\lambda a) &= \lambda \Delta a, \\ \Delta(ab) &= \Delta a \cdot \Delta b, \end{aligned}$$

for $a, b \in A$, $\lambda \in F$ which every representation must satisfy. It is the purpose of this paper to show that the only trace functions on A over F are the traces defined by representations, and to give an application to the theory of representations of finite groups for characteristic $p>0$.

From now on A will always be an associative algebra over a field F of characteristic $p>0$.

PROPOSITION 1. *The trace functions on A form a module $T(A/F)$ of characteristic p.*

Proof obvious.

Iterating (8) we obtain the rule

(8a) $$f(a^{p^j}) = f(a)^p \qquad (j = 0, 1, 2, \cdots)$$

for trace functions. In particular,

(8b) if an equation $a^{p^j} = 0$ holds then $f(a) = 0$.

This leads to

PROPOSITION 2. *Each trace function on A over F vanishes on the radical $R(A)$ of A.*

For any element a of A there must be a non-trivial linear relation between the infinitely many elements $a, a^p, a^{p^2}, \cdots$ say

$$\sum_{j=0}^{m} \lambda_j a^{p^j} = 0, \qquad \text{with } \lambda_j \in F, \lambda_m = 0.$$

For a trace function on A over F it follows that

$$0 = f\left(\sum_{j=0}^{m} \lambda_j a^{p^j}\right) = \sum_{j=0}^{m} \lambda_j f(a)^{p^j}.$$

Hence

PROPOSITION 3. *The values of a trace function on A over F are algebraic over F.*

If A is the ring sum of the algebras A_1 and A_2 over F then every trace function f on A induces a trace function f_i on $A_i (i=1, 2)$. The trace function f on A over F is uniquely determined by the pair f_1, f_2 according to the rule

$$f(a_1 + a_2) = f_1(a_1) + f_2(a_2) \qquad \text{for } a_1 \in A_1,\ a_2 \in A_2. \tag{9}$$

Conversely any pair of trace functions f_i on A_i $(i=1, 2)$ determines a trace function f on A according to (9). Hence

PROPOSITION 4. *The module formed by the trace functions on a ring sum of two algebras over F is direct sum of the modules formed by the trace functions on the summands.*

For the matrix algebra of degree d over F we obtain for each trace function f the equations

$$f(e_{rs}) = f(e_{rs} \circ e_{ss}) = 0, \qquad f(e_{rr} - e_{ss}) = f(e_{rs} \circ e_{sr}) = 0$$

in case of $r \neq s$. Hence for $a = \sum \alpha_{ik} e_{ik}$ it follows that

$$\begin{aligned} f(a) &= \sum \alpha_{ik} f(e_{ik}) = \sum \alpha_{ii} f(e_{ii}) = \sum \alpha_{ii}(f(e_{ii} - e_{11}) + f(e_{11})) \\ &= \sum \alpha_{ii} f(e_{11}) = \operatorname{tr} a \cdot f(e_{11}). \end{aligned}$$

Furthermore

$$f(e_{11})^p = f(e_{11}^p) = f(e_{11});$$

hence $f(e_{11})$ belongs to the prime field. We have obtained

PROPOSITION 5. *The module of the trace functions on a matrix algebra A of characteristic d is generated by the natural trace function so that there are exactly p trace functions on a matrix algebra.*

Let $b_1, b_2, \cdots, b_n$ be a basis of the associative algebra A over F. There are linear relations

$$b_i \circ b_k = b_i b_k - b_k b_i = \sum \gamma_{ikl} b_l, \tag{10}$$

$$b_i^p = \sum \gamma_{il} b_l \tag{11}$$

with $\gamma_{ikl}, \gamma_{il}$ contained in F. In view of the rules (1)–(3), (8) each trace function f on A over F satisfies the rules

$$f\left(\sum \lambda_l b_l\right) = \sum \lambda_l f(b_l), \tag{12}$$

$$\sum \gamma_{ikl} f(b_l) = 0, \tag{13}$$

$$f(b_i)^p = \sum \gamma_{il} f(b_l). \tag{14}$$

Conversely we have

PROPOSITION 6. *If a set of n constants* $f(b_1), f(b_2), \cdots, f(b_n)$ *belonging to the algebraically closed extension* Ω *of* F *satisfies* (13), (14), *then it defines a trace function on* A *over* F *according to* (12).

Proof: The rules (1)–(3) follow from (12), (13) by obvious computations. Using (1)–(3), (14) and (7) we obtain

$$\begin{aligned} f((\sum \lambda_i b_i)^p) &= \sum \lambda_i^p f(b_i^p) = \sum \lambda_i^p f(\sum \gamma_{il} b_l) \\ &= \sum \lambda_i^p \sum \gamma_{il} f(b_l) = \sum \lambda_i^p f(b_i)^p \\ &= (\sum \lambda_i f(b_i))^p = (f(\sum \lambda_i b_i))^p \end{aligned}$$

and hence (8).

From propositions 1–6 we have the following

THEOREM ON TRACE FUNCTIONS. *The trace functions on an associative algebra* A *over a field* F *of prime characteristic* p *with values in the algebraically closed extension* Ω *of* F *coincide with the trace functions belonging to the representations of* A *over* F *by matrices of finite degree with coefficients in* Ω. *The number of the trace functions on* A *over* F *is equal to* p^ρ *where* ρ *denotes the number of classes of equivalent absolutely irreducible representations of* A *over* F.

Proof: We recapitulate that a representation Δ of A over F by matrices of degree d with coefficients in Ω is called absolutely irreducible if there are d^2 matrices linearly independent over Ω among the matrices $\Delta(a)$ with a contained in A. Denoting by A_Ω the result of extending the associative algebra A with basis $b_1, b_2, \cdots, b_n$ over F to an associative algebra with basis $b_1, b_2, \cdots, b_n$ over Ω with the same rules of multiplication, we conclude from proposition 6 that for every trace function f on A over F by the formula

$$f_\Omega(\sum \Lambda_l b_l) = \sum \Lambda_l f(b_l) \qquad \text{with } \Lambda_l \in \Omega, \tag{15}$$

there is defined a trace function f_Ω on A_Ω over Ω. Conversely every trace function on A_Ω over the algebraically closed field Ω has its values, according to proposition 2, entirely in Ω. Hence it induces a trace function on A over F. And this correspondence between the trace functions on A_Ω over Ω and the trace functions on A over F provides an isomorphism between the modules $T(A_\Omega/\Omega)$ and $T(A/F)$. According to McLagan-Wedderburn the difference algebra of A_Ω over its radical decomposes into the ring sum of ρ matrix algebras over Ω where by known results ρ coincides with the number of classes of equivalent absolutely irreducible representations of A over F. Due to proposition 2 the number of trace functions of A_Ω over Ω coincides with the corresponding number for $A_\Omega - R(A_\Omega)$ over Ω. From propositions 4, 5 it follows that this number is p^ρ.

Since each trace function on a matrix algebra over F is obtained by multiplying the natural trace function with an element of the prime field, we conclude that it may be interpreted as the trace of a multiple of the natural representation by matrices. In view of the construction leading to proposition 4 each trace function on a ring sum of finitely many matrix algebras over F is the trace of a certain fully reducible representation. Hence any trace function on A_Ω over Ω is the trace of a sum of irreducible representations of A_Ω over Ω by matrices of finite degree with coefficients in Ω, or what amounts to the same, any trace function on A over F is the trace of the sum of absolutely irreducible representations of A over F, *q.e.d.*

By application of proposition 6 and the main theorem to the group algebra of a finite group over the prime field of characteristic $p>0$ we obtain

PROPOSITION 7. *Let ρ_p denote the number of classes of equivalent absolutely irreducible representations of a finite group G for characteristic $p>0$. There are p^{ρ_p} trace functions on G. They are characterized as functions f on G with values in the algebraic algebraically closed field Ω of characteristic p satisfying*

$$(16) \qquad f(xy) = f(yx)$$

$$(17) \qquad f(x^p) = f(x^p)$$

for $x, y \in G$.

Equivalent to (16), (17) are the rules

$$(16a) \qquad f(txt^{-1}) = f(x)$$

$$(17a) \qquad f(x^{p^j}) = f(x)^{p^j}$$

for $x, y \in G$, j a non-negative integer.

DEFINITION. *Two elements a, b of G are p-conjugate if there is an equation*

$$xa^{p^j}x^{-1} = yb^{p^k}y^{-1}$$

with $x, y \in G$ and j, k non-negative integers. We write

$$a \underset{p}{\sim} b$$

to indicate the p-conjugacy of a and b, whereas

$$a \sim b$$

indicates that a and b are conjugate under G.

By definition

$$b \underset{p}{\sim} a \qquad \text{if } a \underset{p}{\sim} b,$$

$$a \underset{p}{\sim} b \qquad \text{if } a \sim b,$$

$$a^{p^j} \underset{p}{\sim} a \qquad (j = 0, 1, 2, \cdots).$$

The relation $a \sim b$ has the 3 properties of an equivalence relation. Due to the identity

$$(xax^{-1})^m = xa^m x^{-1},$$

we have

$$a^m \sim b^m \qquad \text{if } a \sim b$$

and

$$a^m \underset{p}{\sim} b^m \qquad \text{if } a \underset{p}{\sim} b.$$

Furthermore $a \underset{p}{\sim} b$ if and only if $a^{p^j} \sim b^{p^k}$ holds for some non-negative integers j, k. Suppose that we also have $b^{p^l} \sim c^{p^m}$. Then

$$a^{p^{j+l}} = (a^{p^j})^{p^l} \sim (b^{p^k})^{p^l} = (b^{p^l})^{p^k} \sim (c^{p^m})^{p^k} = c^{p^{m+k}},$$

$$a \underset{p}{\sim} c.$$

Hence p-conjugacy satisfies the 3 requirements for an equivalence relation. The elements of G are distributed among classes of p-conjugate elements each of which consists of some classes of conjugate elements. Let the order of G be $n = p^\nu \cdot n'$ with n' prime to p. Then for $a \in G$ we have $a \underset{p}{\sim} a^{p^\nu}$ so that the order of a^{p^ν} divides n'. Hence each element of G is p-conjugate to an element with order prime to p, *i.e.*, to a p-regular element. Let a be a p-regular element, then from $a^n = 1$ follows $a^{n'} = 1$.

Among the classes of conjugate elements represented by $a, a^p, a^{p^2}, \cdots$ there must be repetitions. Let $a^{p^{d(a)}}$ be the first element conjugate to a previous element say

$$a^{p^{d(a)}} \sim a^{p^i} \qquad \text{with } 0 \leqq i < d(a)$$

and denote by p' a solution of the congruence $pp' \equiv 1(n')$; then in case $i > 0$ we have

$$a^{p^{d(a)-1}} = (a^{pp'})^{p^{d(a)-1}} = (a^{p^{d(a)}})^{p'} \sim (a^{p^i})^{p'} = a^{p^{i-1}}$$

contrary to the minimum property of d. Hence $i = 0$, $a^{p^{d(a)}} \geqq a$, $a^{p^{2d(a)}} \geqq a^{p^{d(a)}}, \cdots$

$$a^{p^{jd(a)}} \sim a \qquad (j = 0, 1, 2, \cdots)$$

$$a^{p^i} \sim a^{p^{i+jd(a)}} \qquad (i, j = 0, 1, 2, \cdots).$$

It follows that there are precisely $d(a)$ classes of conjugate elements containing an element a^{p^i} and these $d(a)$ classes are represented by the elements $a, a^p, \cdots, a^{p^{d(a)-1}}$. Every p-regular element which is p-conjugate to a must be in one of the $d(a)$ classes of conjugate elements represented by $a, a^p, \cdots, a^{p^{d(a)-1}}$.

Now we answer the following question: Let a be p-regular, b an element of order $p^\mu q$ where p does not divide q, and let

$$(18) \qquad a^{p^j} \sim b^{p^k}$$

for some non-negative integers j, k. Under which conditions is

$$(19) \qquad a^{p^l} \sim b^{p^m}, \qquad \text{with some non-negative integers } l, m?$$

Answer: The necessary and sufficient conditions are

$$m \geqq \mu \text{ and the congruence } l + k \equiv j + m(d(a)).$$

Proof: Assume (19). Since a is p-regular, the same is true for a^{p^l} and b^{p^m}, hence $m \geqq \mu$. Furthermore

$$a^{p^{l+k}} \sim b^{p^{m+k}} \sim a^{p^{j+m}}.$$

But due to the previous considerations two powers a^{p^r} and a^{p^s} are conjugate under G if and only if $r \equiv s(d(a))$. Hence $l+k \equiv j+m(d(a))$. Conversely let $m \geqq \mu$ and $l+k \equiv j+m(d(a))$; then

$$a^{p^{l+k}} \sim a^{p^{j+m}}$$

and b^{p^m} is p-regular,

$$b^{p^{m+k}} = (b^{p^k})^{p^m} \sim (a^{p^j})^{p^m} = a^{p^{j+m}} \sim a^{p^{j+k}},$$

$$b^{p^m} = (b^{p^m})^{(pp')^k} = (b^{p^{m+k}})^{p'^k} \sim (a^{p^{l+k}})^{p'^k} = (a^{(pp')^k})^{p^k} = a^{p^k},$$

which completes the proof.

For the construction of the trace functions on G we choose a representative system $a_1, a_2, \cdots, a_r$ of the classes of p-conjugate elements such that each representative is p-regular. Due to (16a) and (17a) and the relation

$$a_i^{p^{d(a_i)}} \sim a_i,$$

we obtain for each trace function f on G the equations

$$f(a_i)^{p^{d(a_i)}} = f(a_i)$$

which are equivalent to the statement that $f(a_i)$ belongs to the Galois field $GF(p^{d(a_i)})$ of $p^{d(a_i)}$ elements. Furthermore for x contained in G we have the relation

$$(20) \qquad x^{p^j} \sim a_i^{p^k} \qquad \text{with some } i, j, k,$$

from which it follows that

$$(21) \qquad f(x)^{p^j} = f(a_i)^{p^k}$$

which determines $f(x)$ uniquely once the values $f(a_1), f(a_2), \cdots, f(a_r)$ are known.

Conversely let us assign to each representative a_i one of the $p^{d(a_i)}$ elements of $GF(p^{d(a_i)})$ as the value $f(a_i)$ and let us define $f(x)$ according to (20) and (21) for every element x of G. Due to the answer which we gave to the question above it follows that the value of $f(x)$ is independent of the relation (20) connecting x with the representative a_i. Since $(x^{p^l})^{p^j} \sim a^{p^{k+1}}$ we find that $f(x^{p^l})^{p^j} = f(a_i)^{p^{k+1}}$ and hence $f(x^{p^l}) = f(x)^{p^l}$. If $y \geqq x$ then

$$y^{p^j} \sim x^{p^j} \sim a_i^{p^k},$$
$$f(y)^{p^j} = f(a_i)^{p^k},$$
$$f(y) = f(x).$$

Hence f is a trace function.

According to the previous construction, the number of trace functions on G is p^σ where

$$\sigma = \sum_{i=1}^{r} d(a_i).$$

The exponent of p coincides with the number of classes of conjugate p-regular elements.

Hence we have obtained another proof of the theorem of Richard Brauer* that the number ρ_p of classes of absolutely irreducible representations of a finite group G for characteristic $p>0$ coincides with the number of classes of conjugate p-regular elements.† In addition we have found an explicit construction of the p^{ρ_p} trace functions on G for characteristic p.

ON ASSOCIATORS IN JORDAN ALGEBRAS‡

C. E. TSAI, Michigan University

Jacobson proved in [1] that if A is an (associative) algebra over a field of characteristic zero and a, b are elements in A such that $[a, [a, b]] = 0$, then $[a, b]$ is a nilpotent element in A where $[a, b] = ab - ba$. Kuźmen [3] has generalized this result to certain nonassociative algebras.

In this note, we shall look at this problem from another angle and prove the following: If J is a finite dimensional Jordan algebra and if $(x, (x, a, y), y) = 0$, then (x, a, y) is a nilpotent, where $(x, a, y) = (xa)y - x(ay)$ is the associator.

LEMMA 1. *If J be a power associative algebra, D be a derivation of J and $D^2a = 0$ for some element a in J; then $D^m a^n = 0$ if m, n are positive integers and $m > n$.*

Proof (by induction). Let $n = 1$ and m be any integer greater than 1. Then $D^m a = 0$ is given. If the lemma holds for all $n' < n$, then

$$D^{n+1}a^n = D^{n+1}(a \cdot a^{n-1})$$

* Über die Darstellung von Gruppen in Galois'schen Feldern. Actualités scientifiques et industrielles 195, 1935.

† This application has been suggested to me by a lecture of Professor R. Brauer, in which he makes use of (1), (2), (3) in proving his theorem.

‡ From AMERICAN MATHEMATICAL MONTHLY, vol. 75 (1968), pp. 748–749.

$$= \sum_{i=0}^{n+1} \binom{n+1}{i} D^i(a) D^{n+1-i}(a^{n-1})$$

$$= aD^{n+1}(a^{n-1}) + (n+1)D(a)D^n(a^{n-1}) = 0.$$

Suppose $D^{n+k'}(a^n) = 0$ for every positive integer $k' < k$ and all n; then $D^{n+k}a^n = D[D^{n+(k-1)}a^n] = 0$.

LEMMA 2. *If J is a power associative algebra with derivation D and if a is an element of J such that $D^2a = 0$, then $D^n a^n = n!(Da)^n$ for all positive integer n.*

Proof (by induction). The lemma is obviously true if $n = 1$. If we assume that $D^{n-1}(a^{n-1}) = (n-1)!(Da)^{n-1}$, then

$$D^n(a^n) = D^n(a \cdot a^{n-1}) = \sum_{i=0}^{n} \binom{n}{i} D^i(a) D^{n-i}(a^{n-1}) = aD^n(a^{n-1})$$

$$+ nD(a)D^{n-1}(a^{n-1}) = nD(a) \cdot (n-1)!(Da)^{n-1} = n![D(a)]^n.$$

THEOREM. *Let J be a finite dimensional power associative algebra over a field K and D be a derivation on J. If a is an element of J and if the characteristic of K is either 0 or greater than the degree of the minimal polynomial of a, then $D^2a = 0$ implies $D(a)$ is nilpotent.*

Proof. Let $\phi(\lambda) = \alpha_0 + \alpha_1\lambda + \alpha_2\lambda^2 + \cdots + \alpha_t\lambda^t$ be the minimal polynomial of a. Then

$$0 = \phi(a) = \alpha_0 e + \alpha_1 a + \alpha_2 a^2 + \cdots + \alpha_t a^t.$$

Hence

$$0 = D^t(\phi(a)) = D^t[\alpha_0 e + \alpha_1 a + \alpha_2 a^2 + \cdots + \alpha_t a^t]$$

$$= \alpha_0 D^t(e) + \alpha_1 D^t(a) + \cdots + \alpha_t D^t(a^t)$$

$$= a_t D^t(a^t) = t! a_t [D(a)]^t.$$

Thus $D(a)$ is nilpotent by our assumption on the characteristics of the field K.

COROLLARY. *Let a, x, y be elements of a finite dimensional Jordan algebra over a field K. If the characteristic of K is either 0 or greater than the degree of the minimal polynomial of a, then $(x, (x, a, y), y) = 0$ implies $(x, a, y) = 0$.*

Proof. It is well known that the mapping $a \rightarrow (x, a, y) = (xa)y - x(ay)$ is a derivation on a Jordan algebra (see, for example, [4]).

References

1. N. Jacobson, Rational methods in the theory of Lie algebras, Ann. of Math., 36(1935) 875–881.
2. D. Kleinecke, On operator commutators, Proc. Amer. Math. Soc., 8(1957) 535–536.
3. E. N. Kuźmin, On commutators in flexible rings, Math. Reviews, 23-A 1682 (1962).
4. R. Schafer, An Introduction to Non-Associative Algebras, Academic Press, New York, 1966.

LINEARIZATION IN RINGS AND ALGEBRAS*

JERRY GOLDMAN, De Paul University, and SEYMOUR KASS,
Illinois Institute of Technology, Chicago

1. Introduction. If R is a ring in which every element satisfies the identity $x^2=0$, then it is easy to show that R is anticommutative, i.e., that $xy=-yx$ for all x, y in R. This is so because in $xy+yx=(x+y)^2-x^2-y^2$ each of the terms $(x+y)^2$, $-x^2$, and $-y^2$ is zero and, therefore, so is their sum. The method by which the anticommutativity of R is deduced from $x^2=0$ is a simple illustration of a process known as "linearization" or "polarization," a technique by means of which an identity of high degree in one variable is made to yield an identity of lower degree in several variables. Linearization is an important technique in associative and nonassociative rings and algebras, but it usually appears in the literature in a form that fails to reveal its general features. Although the linearization process has been studied in [4] and [7], to our knowledge an elementary exposition of this technique has never been made.

In this article we shall describe a technique for linearizing polynomial identities systematically. To demonstrate its utility, in Section 4 we use linearization to prove the new result that a strictly power-associative p-ring is commutative.

2. Preliminaries. Let A be an algebra over a field F. Let $\mathcal{P}$ be the algebra of all polynomials (not necessarily associative or commutative) in indeterminates $x_1, x_2, \cdots$ over F. Let $P(x_1, \cdots, x_n)\in\mathcal{P}$. We say that A satisfies P if P vanishes when the $x_1, \cdots, x_n$ are replaced by any $a_1, \cdots, a_n$ from A. Also, we identify the polynomial $P(x_1, \cdots, x_n)$ with the identity $P(x_1, \cdots, x_n)=0$ in order to say that A satisfies $P=0$.

We will need a special case of the class of linear operators studied by Gerstenhaber in [4]. Let α be a monomial in $\mathcal{P}$. Corresponding to each variable $x_i\in\mathcal{P}$ we define an operator $\delta_i(\alpha)$ from the monomials of $\mathcal{P}$ to $\mathcal{P}$ as follows: $m(x_1, \cdots, x_n)\delta_i(\alpha)=0$ if x_i does not appear in m; otherwise, $m(x_1, \cdots, x_n)\delta_i(\alpha)$ $=$ the polynomial that is obtained by making *all possible* replacements of the arguments x_i one at a time by α and summing the resulting monomials. The number of monomials obtained is the degree of x_i in $m(x_1, \cdots, x_n)$. For example:

$$[(xx)(yx)]\delta_x(\alpha) = (\alpha x)(yx) + (x\alpha)(yx) + (xx)(y\alpha).$$

The operator $\delta_i(\alpha)$ is then extended to all of $\mathcal{P}$ by linearity. We denote the rth iterate of $\delta_x(\alpha)$ by $\delta_x^r(\alpha)$. Thus

$$(*)\qquad [(xx)(yx)]\delta_x^2(\alpha) = 2(\alpha\alpha)(yx) + 2(\alpha x)(y\alpha) + 2(x\alpha)(y\alpha).$$

* From AMERICAN MATHEMATICAL MONTHLY, vol. 76 (1969), pp. 348–355.

In general, if m is a monomial of degree n in x and if $i \leqq n$, then $m\delta_x^i(\alpha)$ is a polynomial with $\binom{n}{i}$ *distinct* terms, each of which occurs with multiplicity $i!$. We define

$$m\frac{\delta_x^i(\alpha)}{i!}$$

to denote the sum of the $\binom{n}{i}$ distinct terms. Thus, the $i!$ occurring in the operator $(\delta_x^i(\alpha))/i!$ is not to be construed as a field element. In (*), if the characteristic of F is 2, then

$$[(xx)(yx)]\delta_x^2(\alpha) = 0,$$

but

$$[(xx)(yx)]\frac{\delta_x^2(\alpha)}{2!} = (\alpha\alpha)(yx) + (\alpha x)(y\alpha) + (x\alpha)(y\alpha).$$

The operator $\delta_x^i(\alpha)/i!$ is extended to all of $\mathcal{P}$ by linearity.

We will need to use the fact that $\delta_x(\alpha)$ is a *derivation* on $\mathcal{P}$.

PROPOSITION. *Let* $P_1, P_2 \in \mathcal{P}$. *Then*

$$(P_1P_2)\delta_x(\alpha) = P_1\delta_x(\alpha)P_2 + P_1\cdot P_2\delta_x(\alpha). \tag{1}$$

Proof. Without loss of generality we may assume that P_1 and P_2 are monomials. Suppose x occurs m times in P_1 and n times in P_2. There are $m+n$ occurrences of x in P_1P_2; therefore, $(P_1P_2)\delta_x(\alpha)$ is a sum of $m+n$ monomials. We divide this set of monomials into two disjoint classes. The first consists of all terms arising from a replacement of the x's in P_1 by α; the second consists of terms arising from a replacement of the x's in P_2 by α. Every monomial in $P_1\delta_x(\alpha)\cdot P_2$ belongs to the first class; every monomial in $P_1\cdot P_2\delta_x(\alpha)$ belongs to the second class. Since $P_1\delta_x(\alpha)\cdot P_2$ has m terms and $P_1\cdot P_2\delta_x(\alpha)$ has n terms, we must have equality in (1).

3. Linearization and identities. We define $P(x_1, \cdots, x_n)=0$ to be a *homogeneous* identity if P is homogeneous in each variable singly. That is, each x_i occurs the same number, k_i, of times in each monomial of P. We then say that the degree of x_i is k_i and assign the degree $k_1+ \cdots +k_n$ to P. For example, if $P(x_1, x_2) = (x_1^2x_2)x_1 - x_1(x_2x_1^2)$, then the degree of x_1 is 3, the degree of x_2 is 1, and P is homogeneous of degree 4.

It is known [5] that if A satisfies a multilinear identity (every monomial is of degree 1 in each variable) $P(x_1, \cdots, x_n)=0$, then A will satisfy a homogeneous identity regardless of the cardinality of F. For if x_1 does not appear in some monomial of P, then $P(0, x_2, \cdots, x_n)=0$ is a multilinear identity of degree not greater than P satisfied by A. Continuation of this process will yield a homogeneous identity satisfied by A.

Let A be an algebra over a field F which satisfies a homogeneous polynomial identity $P(x_1, \cdots, x_m)=0$. It is natural to ask what new identities also satisfied by A can be derived from P. It is clear that the set $\mathcal{S}$ of all identities satisfied by A is a subalgebra, in fact, an ideal of $\mathcal{P}$.

Let $\mathcal{P}_m$ be defined as the subalgebra of $\mathcal{P}$ of all polynomials in m indeterminates. Then, for any elements $a_1, \cdots, a_m$ of A, we can define a natural homomorphism, $P(x_1, \cdots, x_m) \to P(a_1, \cdots, a_m)$, of $\mathcal{P}_m$ onto A. The kernel $\mathcal{S}(a_1, \cdots, a_m)$ of this homomorphism is merely the collection of all polynomials of $\mathcal{P}_m$ which are satisfied by $a_1, \cdots, a_m$. Thus, the collection of all elements of $\mathcal{P}_m$ which are satisfied by A is

$$\bigcap_a \mathcal{S}(a_1, \cdots, a_m),$$

the intersection being taken over all sequences, $a=(a_1, \cdots, a_m)$, of m elements of A.

It is easy to see from the above that $\mathcal{S}=\bigcup_m \bigcap_a \mathcal{S}(a_1, \cdots, a_m)$.

For any indeterminates x, y in $\mathcal{P}$ and λ in F, we have the following

LEMMA.

$$(x+\lambda y)^n = \sum_{i=0}^{n} \lambda^i x^n \frac{\overset{i}{\delta_x}(y)}{i!}. \tag{2}$$

Proof. We proceed by induction on n. If $n=1$, (2) is clear. We suppose that (2) holds and will show that

$$(x+\lambda y)^{n+1} = \sum_{i=0}^{n+1} \lambda^i x^{n+1} \frac{\overset{i}{\delta_x}(y)}{i!}. \tag{3}$$

We have

$$\begin{aligned}(x+\lambda y)^{n+1} &= (x+\lambda y)^n(x+\lambda y) \\ &= x^{n+1} + \sum_{i=1}^{n} \lambda^i \left[x^n \frac{\overset{i}{\delta_x}(y)}{i!}\right] x + \sum_{i=0}^{n-1} \lambda^{i+1} \left[x^n \frac{\overset{i}{\delta_x}(y)}{i!}\right] y + \lambda^{n+1} y^{n+1}.\end{aligned} \tag{4}$$

By shifting the index in the third term of (4), we find that

$$(x+\lambda y)^{n+1} = x^{n+1} + \sum_{i=1}^{n} \lambda^i \left\{\left[x^n \frac{\overset{i}{\delta_x}(y)}{i!}\right] x + \left[x^n \frac{\overset{i-1}{\delta_x}(y)}{(i-1)!}\right] y\right\} + \lambda^{n+1} y^{n+1}.$$

The proof will be completed by showing that

$$x^n \frac{\overset{i}{\delta_x}(y)}{i!} \cdot x + x^n \frac{\overset{i-1}{\delta_x}(y)}{(i-1)!} \cdot y = x^{n+1} \frac{\overset{i}{\delta_x}(y)}{i!} \qquad \text{for } i = 1, \cdots, n. \tag{5}$$

However, (5) merely states that the $\binom{n+1}{i}$ possible distinct terms resulting by replacing y, i-times, in x^{n+1} and summing, can be alternatively gotten by adding the results of the following two operations:

(i) summing the $\binom{n}{i}$ terms resulting from replacing y, i-times, in the first n factors of x^{n+1},

(ii) summing the $\binom{n}{i-1}$ terms resulting from replacing y, $(i-1)$ times, in $x^n y$.

The equality of these two alternate procedures for computing $x^{n+1}\delta_x^i(y)/i!$ follows directly from the definition. Thus, (5) follows and the lemma is proved.

When we want to fix our attention upon some one variable, say $x_i = x$ of degree n, we write $P(x)$ instead of $P(x_1, \cdots, x_m)$. We must have

$$P(x + \lambda y) = \sum_{i=0}^{n} \lambda^i S_i(x, y)$$

for some polynomials $S_i(x, y)$, where y is chosen as an indeterminate independent of x and $\lambda \in F$.

THEOREM 1. $S_i(x, y) = P(x)[\delta_x^i(y)/i!]$.

Proof. It suffices to prove the theorem for P a monomial. We just note that since $S_i(x, y)$ is the "coefficient" of λ^i, we must have i occurrences of y. Thus, as in the lemma, $S_i(x, y)$ is equal to the sum of all $\binom{n}{i}$ possible distinct monomials formed from $P(x)$ by substituting y for x, i-times, in all possible ways. This sum is precisely $P(x)[\delta_x^i(y)/i!]$.

THEOREM 2. *Let A be an algebra over the field F. If the homogeneous polynomial $P(x_1, \cdots, x_m)$ belongs to $\mathcal{S}$ and F has at least n elements, where n is the degree of x in $P(x)$, then $P(x)[\delta_x^i(y)/i!]$ belongs to $\mathcal{S}$.*

That is, application of the δ-operator to a homogeneous identity $P = 0$, satisfied by such an A, yields a new identity, $P(x)[\delta_x^i(y)/i!] = 0$, also satisfied by A. Thus, the original identity of high degree in x yields a set of new identities, each in more indeterminates, but of lower degree in x. The process by means of which the S_i are obtained from P is called *partial linearization*. Of course, S_1 may be further linearized by repeating the process and introducing new variables, until finally a multilinear polynomial, called the *linearized form* of P, is obtained. If the new variables introduced are set equal to one another, the new result is a scalar multiple of P.

Proof of Theorem 2. $P(x) = 0$ is satisfied by A; thus, A satisfies $P(x + \lambda y) = 0$ for $\lambda \in F$. Therefore, $\sum_{i=0}^{n} \lambda^i S_i(x, y) = 0$. We note that $S_0(x, y) = P(x) = 0$ and that $S_n(x, y) = P(y) = 0$. Thus, $\sum_{i=1}^{n-1} \lambda^i S_i(x, y) = 0$.

Now successively replace λ in this last equation by the $n-1$ distinct nonzero scalars $\lambda_1, \cdots, \lambda_{n-1}$ to obtain the system:

$$\begin{array}{l} \lambda_1 S_1 + \lambda_1^2 S_2 + \cdots + \lambda_1^{n-1} S_{n-1} = 0, \\ \quad\vdots \\ \lambda_{n-1} S_1 + \lambda_{n-1}^2 S_2 + \cdots + \lambda_{n-1}^{n-1} S_{n-1} = 0. \end{array}$$

Let the (Vandermonde) determinant of the system be $V=\lambda_1\cdots\lambda_{n-1}\Pi(\lambda_r-\lambda_s)$ for $1\leqq s<r\leqq n-1$, and let V_j denote the cofactor of the element in column one and row j of the coefficient matrix. Multiply the jth equation by V_j for each $j=1,\ldots,n-1$, and add the resulting equations. By elementary determinant theory, $VS_1=0$, and because $V\neq 0$, we have $S_1=0$. Similarly, each $S_i=0$, and the theorem is proved.

Theorem 2 may be restated to say that $\mathcal{S}$ is invariant under $\delta_x^i(y)$ if F is of high enough characteristic. Certainly a field F of characteristic zero will do. We give some examples.

The ring R, mentioned at the beginning of Section 1, satisfies $P(x)=x^2=0$. Using Theorem 2, we see that $0=P(x)\delta_x(y)=xy+yx$. Thus, R is anticommutative.

Next, just as the commutator $[x,y]=xy-yx$ measures departure of a ring from commutativity, so the associator $(x,y,z)=(xy)z-x(yz)$ measures departure from associativity. If an algebra satisfies $P(x)=(x,x,x)=0$ (3rd power-associativity), we have

$$S_1(x,y) = (x,x,x)\delta_x(y) = (y,x,x)+(x,y,x)+(x,x,y) = 0.$$

We may linearize further to obtain the multilinear identity

$$\begin{aligned}P(x)\delta_x(y)\delta_x(z) &= S_1(x,y)\delta_x(z)\\ &= (y,z,x)+(y,x,z)+(z,y,x)+(x,y,z)+(z,x,y)+(x,z,y) = 0.\end{aligned}$$

We mention that an algebra A over F which satisfies a multilinear identity $P=0$ will also satisfy $P=0$ over any scalar extension of F. Thus, in matters involving an enlarging of a field of scalars, it is of obvious utility to try to reduce a given polynomial identity to an equivalent multilinear identity. An important illustration of this is given by power-associativity.

An algebra A over F is said to be *power-associative* if it satisfies $x^nx^m=x^{n+m}$ for all positive integers n and m, where powers are defined inductively by $x^1=x$ and $x^{n+1}=x^nx$ for $n=1,2,3,\cdots$.

An algebra A which is power-associative over every scalar extension of F is called *strictly* power-associative. Kokoris [6] has shown that a power-associative algebra need not be strictly power-associative, but that the two concepts coincide for characteristic $\neq 2,3,5$. One insures strict power-associativity in a power-associative algebra by requiring that all partial linearizations of identities equivalent to $x^nx^m-x^{n+m}=0$ also be identities satisfied by A.

The theory of power-associative algebras was created by A. A. Albert. Under the aegis of Albert and Jacobson the subject has been developed extensively in the past twenty-five years (see the bibliography in [8] for references to their work).

Jordan algebras are among the more important power-associative algebras. A Jordan algebra A is a commutative algebra over a field F which satisfies the identity

$$(x^2y)x - x^2(yx) = 0.$$

A successful structure theory exists for these algebras which depends upon a Peirce decomposition analogous to the classical Peirce decomposition for associative algebras. As another example of the use of linearization, we obtain the Peirce decomposition for Jordan algebras.

We assume here that the characteristic of F is not 2 or 3. Apply the operator $\delta_x(u)$ to the Jordan identity to get

$$[(ux)y]x + [(xu)y]x + (x^2y)u - [(ux)(yx) + (xu)(yx) + x^2(yu)] = 0.$$

Or, since A is commutative, $2[(ux)y]x+(x^2y)u-2(ux)(yx)-x^2(yu)=0$.

Now we apply $\delta_x(v)$ to this last equation and obtain the linearized form of the Jordan identity:

$$(6) \quad [(uv)y]x + [(ux)y]v + [(xv)y]u - (uv)(yx) - (ux)(yv) - (xv)(yu) = 0.$$

(Setting $u=v=x$, we see that this reduces to $3(x^2y)x-3x^2(yx)=0$. Thus, the Jordan identity and its linearized form are equivalent for our choice of F.)

For any $a\in A$ define the mapping $R_a:A\to A$ by $xR_a=xa$ for every $x\in A$. It is easy to see that each R_a is a linear transformation on A. If A has an idempotent e (an element e such that $e^2=e\neq 0$), we can set $x=y=v=e$ in (6) to get $u(2R_e^3-3R_e^2+R_e)=0$ for any $u\in A$. Thus,

$$2R_e^3 - 3R_e^2 + R_e = R_e(2R_e - 1)(R_e - 1) = 0.$$

Hence, the characteristic roots of R_e are in the set $\{0, \frac{1}{2}, 1\}$. A can now be decomposed as a vector space direct sum

$$A = A_e(0) + A_e(\tfrac{1}{2}) + A_e(1),$$

where $A_e(\lambda)=\{x\in A|xR_e=\lambda x\}$ are R_e-invariant subspaces for $\lambda=0, \frac{1}{2}$, or 1. Each $a\in A$ can be written $a=a_0+a_{1/2}+a_1$ where $a_0=a+2(ae)e-3ae$, $a_{1/2}=4[ae-(ae)e]$, and $a_1=2(ae)e-ae$, with $a_\lambda\in A_e(\lambda)$. This sum is known as the Peirce decomposition of A with respect to the idempotent e. Properties of this decomposition are fundamental in Albert's structure theory for Jordan algebras [**1,2**].

4. An application. A p-ring is a ring R with the property that for all $a\in R$, $a^p=a$ and $pa=0$, where p is a fixed prime. An elementary proof that associative p-rings are commutative was given by Forsythe and McCoy in [**3**]. We will now extend this result to power-associative p-rings. The proof is made by essentially rewriting the proof in [**3**] in terms of δ-operators and then using the fact that $\delta_x(y)$ is a derivation.

THEOREM 3. *A strictly power-associative p-ring is commutative.*

Proof. Let a and b be any two elements of R. We may regard R as an algebra over the field J_p, the integers modulo p. Let $\mathcal{P}_2$ be the algebra in two indeterminates x and y over J_p as defined in Section 3. The correspondence $x\to a$ and

$y \to b$ can be extended in a natural way to a homomorphism of $\mathcal{P}_2$ into R. Denote the kernel by K. Clearly, $z^p - z \in K$ for any z in $\mathcal{P}_2$. In particular, $(x+\lambda y)^p - (x+\lambda y) \in K$ for every $\lambda \in J_p$. Using (2) we see that

$$\sum_{i=1}^{p-1} \lambda^i x^p \frac{\delta_x^i(y)}{i!} \in K.$$

Recalling that $x^p(\delta_x^i(y)/i!)$ is just S_i of Theorem 2, and replacing λ by each of its nonzero values in turn, we obtain

$$\begin{array}{l} S_1 + S_2 + \cdots + S_{p-1} \in K, \\ 2S_1 + 2^2 S_2 + \cdots + 2^{p-1} S_{p-1} \in K, \\ \vdots \\ (p-1)S_1 + (p-1)^2 S_2 + \cdots + (p-1)^{p-1} S_{p-1} \in K. \end{array}$$

If we operate on each of these elements by cofactors of the coefficient matrix as in the proof of Theorem 2, we reach the conclusion that $VS_1 \in K$, where V is the (nonzero) determinant of the coefficient matrix. Thus, $S_1 \equiv x^p \delta_x(y) \in K$. Now $xx^p - x^p x \in K$, and because R is *strictly* power-associative, we know that $(xx^p - x^p x)\delta_x(y) \in K$. Since $\delta_x(y)$ is a derivation, we have

$$\begin{aligned} K \ni (xx^p - x^p x)\delta_x(y) &= x\delta_x(y)x^p + x(x^p\delta_x(y)) \\ - [x^p\delta_x(y)x + x^p(x\delta_x(y))] &= yx^p + x(x^p\delta_x(y)) - x^p\delta_x(y)x - x^p y. \end{aligned}$$

Because $x^p\delta_x(y) \in K$ and K is an ideal, $x(x^p\delta_x(y)) \in K$ and $x^p\delta_x(y)x \in K$. Thus, $yx^p - x^p y \in K$, so that $ba^p - a^p b = 0$, or $ba - ab = 0$ and Theorem 3 is proved.

References

1. A. A. Albert, On Jordan algebras of linear transformations, Trans. Amer. Math. Soc., 59 (1946) 524–555.

2. ———, A structure theory for Jordan algebras, Ann. of Math., 48 (1947) 546–567.

3. A. Forsythe and N. H. McCoy, On the commutativity of certain rings, Bull. Amer. Math. Soc., 52 (1946) 523–526.

4. M. Gerstenhaber, On nilalgebras and linear varieties of nilpotent matrices II, Duke Math. J., 27 (1960) 21–31.

5. I. Kaplansky, Notes on ring theory, University of Chicago Math. Lecture Notes, Chicago, 1965, pp. 72–76.

6. L. A. Kokoris, New results on power-associative algebras, Trans. Amer. Math. Soc., 77 (1954) 363–373.

7. J. M. Osborn, Identities of nonassociative algebras, Canad. J. Math., 17 (1965) 78–92.

8. R. D. Schafer, An Introduction to Nonassociative Algebras, Academic Press, New York, 1966.

BIBLIOGRAPHIC ENTRIES: NON-ASSOCIATIVE RINGS AND ALGEBRAS

Except for the entry labeled MATHEMATICS MAGAZINE, the references below are to the AMERICAN MATHEMATICAL MONTHLY.

1. S. I. Goldberg, On the Euler characteristic of a Lie algebra, 1955, 239–240.

Gives an elementary proof that the Euler characteristic of any Lie algebra over an arbitrary field vanishes.

2. Alvin Hausner, The alternative isotopes of division algebras, 1963, 730–733.

Proves that an alternative isotope of an associative division ring must be isomorphic to the original division ring (so in particular is associative).

3. E. W. Wallace, Two-dimensional power-associative algebras, MATHEMATICS MAGAZINE, 1970, 158–162.

Gives canonical forms for all the two-dimensional power associative algebras over the real and complex numbers.

4. Raymond Coughlin and Michael Rich, Some associativity conditions for algebras, 1971, 1107.

Proves that if every associator is a scalar, then the algebra is associative.

3

FIELD THEORY AND ALGEBRAIC NUMBER THEORY

(a)

FIELD EXTENSIONS AND GALOIS THEORY

ON THE FUNDAMENTAL THEOREM OF ALGEBRA*

HANS ZASSENHAUS, Ohio State University

Introduction. In the year 1799 C. F. Gauss [4] gave the first formal proof of the theorem that every nonconstant polynomial with real coefficients can be factored into a product of linear factors and quadratic factors. A constructive proof based on arguments of a purely algebraic nature and on assumptions about the real number field that were stated in purely algebraic terms (though the proof of these assumptions would require analytic methods) was first given by O. Perron [6]. In this paper Perron's method is further developed towards an algorithmic routine for solving algebraic equations with complex coefficients.

1. Really closed fields. The real number field has the following properties:

1.1. The negative of a nonsquare is a square.

1.2. The sum of squares is zero only if each summand is zero.

1.3. An algebraic equation of odd degree has a real solution.

The first two properties derive from the existence of an algebraic ordering of the real number field and a square root of every positive real number.

The last property follows from an application of the intermediate-value theorem of analysis to a polynomial: $f(x)=x^n+a_1x^{n-1}+\cdots+a_n$ of odd degree n with real coefficients, in view of the inequalities:

$$f\left(1+\sum_{i=1}^{n}|a_i|\right)>0>f\left(-\left(1+\sum_{i=1}^{n}|a_i|\right)\right). \tag{1.4}$$

* From American Mathematical Monthly, vol. 74 (1967), pp. 485–497.

Fields with the properties **1.1**, **1.2**, **1.3** are said to be *really closed.* E.g. the real number field is really closed. For a really closed field F an algebraic ordering is given by the rule:

1.5 The element a of F is greater than the element b of F if and only if the difference element $a-b$ is a nonzero square element of F. This implies the positivity concept:

1.6 An element of F is positive if and only if it is a nonzero square element of F.

In order to deduce the ordering properties of F from 1.5 we have to verify the positivity rules:

1.7 The negative of any nonpositive nonzero element is positive.

1.8 The sum and product of two positive elements are positive.

Indeed, if a is non positive and nonzero, then a is a non square of F. It follows from 1.1 that $-a$ is a square element of F so that $-a$ is positive.

If a and b are positive elements of F then the equations

$$a = \xi^2, \qquad b = \eta^2$$

are solvable by non zero elements ξ, η of F. From 1.2 it follows that $a+b$ does not vanish. If there holds an equation: $-(a+b)=\zeta^2$ in F then

$$\xi^2+\eta^2+\zeta^2=0$$

holds contrary to 1.2.

Thus $a+b$ is a square element of F so that the sum of two positive elements of F always is positive. Since,

$$ab = \xi^2\eta^2 = (\xi\eta)^2 \neq 0$$

the product of two positive elements is positive.

Clearly, for any really closed field only one positivity concept exists.

2. Ordered division rings. (See [1], [2], [3], [5], [7]). For a unital ring D with positivity concept satisfying 1.7, 1.8 we define an algebraic ordering by declaring the relation

$$a > b \text{ (or: } b < a)$$

to mean that the difference of the elements a, b of D is positive.

This definition implies the rules customarily demanded of an algebraic ordering relation:

2.1 (trichotomy). For any two elements a, b of R, there holds one and only one of the three relations:

$$a > b, \qquad a = b, \qquad b > a.$$

2.2 (transitivity). If $a>b$, $b>c$ then $a>c$.

2.3 If $a>b$, $c>d$ then $a+c>b+d$ and $ac+bd>ad+bc$.

Conversely, if a $>$relation is defined in D which satisfies 2.1, 2.3 then the relation $a>0$ defines a positivity concept satisfying 1.7 and 1.8, from which the $>$ relation can be derived as was done above.

The derived relation:

$$a \geqq b \text{ (or: } b \leqq a)$$

meaning that either a is greater than b or a is equal to b (or: not $a<b$) and the functions:

$$\operatorname{sign} a = \begin{cases} 1 & \text{if } a > 0 \\ 0 & \text{if } a = 0 \\ -1 & \text{if } a < 0 \end{cases}$$

$$|a| = a \cdot \operatorname{sign} a = (\operatorname{sign} a) \cdot a$$

satisfy the basic rules:

$a \geqq a, \quad a \geqq b$ and $b \geqq a \Rightarrow a = b, \quad a \geqq b$ and $b \geqq c \Rightarrow a \geqq c,$

$a \geqq b$ and $c \geqq d \Rightarrow a + c \geqq b + d, \quad ab + cd \geqq ad + cb,$

$\operatorname{sign}(ab) = \operatorname{sign} a \cdot \operatorname{sign} b,$

$|a| \cdot \operatorname{sign} a = (\operatorname{sign} a) \cdot |a| = a,$

$|a| = |-a| \geqq 0, \quad |a| = 0 \Leftrightarrow a = 0,$

$|a+b| \leqq |a| + |b|, \quad |a-b| \geqq ||a| - |b||, \quad |ab| = |a| \cdot |b|.$

The positive elements of an ordered division ring D form a halfring H (according to [8], p. 95, a halfring is a nonempty subset of a ring that is closed under addition and multiplication,) with the following properties:

2.4. The halfring H contains the square of every non zero element of the subdivision ring D_H generated by H.

2.5. The halfring H is a maximal subhalfring of D not containing the zero element of D.

Conversely, every halfring H of D satisfying 2.4 and 2.5 is associated with a positivity concept of D_H satisfying 1.7, 1.8.

Indeed, for any two non zero elements u, v of D we have the identity:

2.6 $$uvu^{-1}v^{-1} = (uv)^2(v^{-1}u^{-1}v)^2(v^{-1})^2$$

so that

2.7 $$uvu^{-1}v^{-1} \in H \quad \text{if } 0 \neq u \in D_H,\ 0 \neq v \in D_H.$$

Thus $uh = h(h^{-1}u\ hu^{-1})u \in Hu$ if $0 \neq u \in D_H$, $h \in H$; hence $hu \in uH$, $uH \subseteq Hu$. Similarly $Hu \subseteq uH$; hence $uH = Hu (u \in D_H)$.

For h of H we have

$$H = h^{-1}hH \subseteq h^{-1}H, \qquad h^{-1} = h^{-1}1^2 \in h^{-1}H,$$

$$h^{-1}H + h^{-1}H \subseteq h^{-1}H,$$

$$h^{-1}Hh^{-1}H = h^{-1}(Hh^{-1})H = h^{-1}h^{-1}HH \subseteq H \subseteq h^{-1}H,$$

hence $h^{-1}H$ is a halfring of D_H containing H, but not zero. Since $h^{-1}H$ contains every non zero square element of $D_H = D_{h^{-1}H}$ it follows from the maximal property of H that $h^{-1}H = H$, hence $h^{-1} \in H$.

If the non zero element c of D_H is not contained in H then $-c$ belongs to H. Indeed,

$$c = c \cdot 1^2 \in cH \subseteq \hat{H} = H \cup cH \cup (H + cH)$$
$$(H + cH) + (H + cH) \subseteq H + cH$$
$$HcH = cHH \subseteq cH,$$
$$cHcH = c(Hc)H = c(cH)H = c^2HH \subseteq H$$
$$(H + cH)(H + cH) \subseteq H + cH.$$

Hence $\hat{H}$ is a halfring of D_H properly containing H, so $\hat{H}$ contains zero because of the maximal property of H.

There are elements h_1, h_2 of H for which $h_1 + ch_2 = 0$. Hence,

$$-c = h_1 h_2^{-1} \in HH = H.$$

THEOREM 1. *A division ring D can be algebraically ordered if and only if the sum of finitely many finite products of square elements is zero only if all summands are zero.* (What happens if one makes the weaker assumption that the sum of finitely many squares can be zero only in the trivial way?)

Proof. That the condition is necessary was shown above. Let it be satisfied.

The set H_0 of all sums of finitely many finite products of non zero square elements of D does not contain zero. H_0 is a halfring. By Zorn's lemma [9] H_0 is contained in a maximal halfring H of D not containing zero. Hence, $2 = 1^2 + 1^2 \in H_0 \subseteq H$, $2 \neq 0$.

Since for any element u of D we have $u = \frac{1}{2}((u+1)^2 - u^2) \in D_H$ it follows that $D_H = D$ and D has indeed the algebraic ordering which is given by

2.8. $a > b$ if and only if $a - b$ is in H.

THEOREM 2. *If the field F is algebraically ordered and if F is contained in an algebraically closed field extension Ω, then there is a really closed subfield Φ of Ω such that*

2.9. *every positive element of F is a square element of Φ.*

2.10. *Ω is algebraic over Φ.*

Proof. By Zorn's lemma [9] there is a maximal halfring H of Ω containing all positive elements of F and every non zero square element of the subfield Φ generated by H, but not zero.

We have shown above that the given ordering of F can be extended to an algebraic ordering of the field Φ such that H is the halfring of the positive elements.

If there is an element ξ of Ω that is not algebraic over Φ then the halfring formed by the sums of finitely many nonzero square elements of $\Phi(\xi)$ with coefficients in H is larger than H and it contains all nonzero square elements of $\Phi(\xi)$. Because of the maximal property of H it contains zero; hence there holds an equation

$$0 = \sum_{i=1}^{n} h_i \left(\frac{P_i(\xi)}{N(\xi)}\right)^2 \quad (h_i \in H)$$

where $N(x), P_1(x), \cdots, P_n(x)$ are non zero polynomials of $\Phi[x]$. Let m be the maximum degree of the polynomials $P_1(x), \cdots, P_n(x)$ and let a_i be the coefficient of x^m in $P_i(x)$. Then

$$0 = \sum_{i=1}^{n} h_i P_i(\xi)^2 = \sum_{i=1}^{n} h_i P_i(x)^2 = \sum_{i=1}^{n} h_i a_i^2,$$

contrary to the construction of H. Hence Ω is algebraic over Φ.

For every element u of H the equation $u=\xi^2$ is solvable in Ω. If ξ does not belong to Φ then the elements $a+b\xi$ with a, b contained in Φ such that $a \geqq 0, b \geqq 0, a+b>0$ forms a halfring $\bar{H}$ of Ω larger than H; it will be contained in the halfring $\bar{H}\bar{H}^{-1}$ containing all square elements of $\Phi(\xi)$, but not zero, which is a contradiction. (It is clear that $\bar{H}$ is a halfring larger than H, not containing zero and that $\bar{H}\bar{H}^{-1}$ has the same properties. Moreover, if a, $b \in \Phi$, $a \geqq 0$, $b<0$, $a^2>b^2u$, then both $-bu+a\xi$ and $(a+b\xi)(-bu+a\xi)=(a^2-b^2u)\xi$ belong to $\bar{H}$; hence $a+b\xi$ belongs to $\bar{H}\bar{H}^{-1}$. Now, if c, $d \in \Phi$ and $c+d\xi \neq 0$, then $(c+d\xi)^2 \neq 0$, then $(c+d\xi)^2=a+b\xi$ when $a=c^2+d^2u \in H$, and $b=2cd$, $a^2-b^2u=(c^2-d^2u)>0$ so that indeed $(c+d\xi)^2$ belongs to $\bar{H}\bar{H}^{-1}$.) Hence every element of H is a square element of Φ and 1.1, 1.2 are satisfied by Φ.

If there is a polynomial $f(x)$ of odd degree with coefficients Φ for which the equation

2.11 $$-1 \equiv \sum_{i=1}^{s} f_i(x)^2 (f(x))$$

can be solved by polynomials $f_1(x), \ldots, f_s(x)$ with coefficients in Φ, then we can find among these $f(x)$ one of minimal degree $2n+1$. Upon substitution of the least remainders with regard to division by $f(x)$ it follows that a relation 2.11 obtains in which the maximum degree k of the polynomials $f_1(x), \ldots, f_s(x)$ is not greater than $2n$.

Moreover, since Φ has an algebraic ordering we conclude that $n>0$ and that the coefficient of x^{2k} in

$$\sum_{i=1}^{s} f_i(x)^2$$

does not vanish, therefore the polynomial $1+\sum_{i=1}^{s} f_i(x)^2 = f(x)g(x)$ is of degree $2k$ and consequently the polynomial $g(x)$ is of odd degree less than $2n+1$.

But this is impossible in view of the congruence:

$$-1 \equiv \sum_{i=1}^{s} f_i(x)^2(g(x))$$

and the minimal property of $f(x)$.

Hence, for a polynomial $f(x)$ of odd degree over Φ a relation of the form 2.11 never holds. On the other hand there must be a polynomial $g(x)$ of odd degree among the irreducible divisors of $f(x)$ in $\Phi[x]$. Since Ω is algebraically closed, there is a root ξ of $g(x)$ in Ω. As was shown above, no sum of square elements of the extension field $\Phi(\xi)$ can be equal to -1. This implies that the sums of finitely many nonzero square elements of $\Phi(\xi)$ form a halfring $\overline{H}$ containing H as well as all nonzero square elements of $\Phi(\xi)$. Because of the maximal property of H we conclude that $\overline{H}$ coincides with H, $\Phi(\xi)=\Phi$, $\xi\in\Phi$, Φ is really closed.

3. The ring extensions associated with an equation. Suppose that the polynomial

3.1. $f(x)=x^n+a_1x^{n-1}+\cdots+a_n$ with coefficients $a_0=1, a_1, a_2, \cdots, a_n$ in the commutative unital ring v has the root ξ in v then

$$\begin{aligned} f(x) &= f(x) - f(\xi) = \sum_{h=0}^{n} a_{n-h}(x^h - \xi^h) \\ &= (x-\xi)\sum_{h=1}^{n} a_{n-h}\left(\sum_{j=0}^{h-1} \xi^{h-j-1}x^j\right) \\ &= (x-\xi)\frac{f}{x-\xi}(x) \end{aligned}$$

where

$$\frac{f}{x-\xi}(x) = \sum_{j=0}^{n-1}\left(\sum_{h=0}^{n-1-j} a_{n-j-h-1}\xi^h\right)x^j. \tag{3.2}$$

Hence every root of the polynomial 3.2 is also a root of $f(x)$. Moreover, if v is an integral domain then every root of $f(x)$ that is not equal to ξ is also a root of 3.2. And if for some polynomial $g(x)$ of degree $n-1$ over v the equation $f(x)=(x-\xi)g(x)$ holds then $g(x)=f(x)/(x-\xi)$. This is because the equation for $g(x)$ allows us to determine recursively the coefficients of $g(x)$.

The preceding remarks motivate the following formal construction.

Let us denote by $v[\xi; f]$ the v-module with basis $1, \xi, \cdots, \xi^{n-1}$ over v. The rule

$$\begin{aligned} \xi_r(1) &= \xi \\ \xi_r(\xi^i) &= \xi^{i+1} \quad (0 < i < n-1) \\ \xi_r(\xi^{n-1}) &= -a_1\xi^{n-1} - a_2\xi^{n-2} - \cdots - a_n 1 \qquad (3.3) \\ \xi_r\left(\sum_{j=0}^{n-1} b_j\xi^j\right) &= \sum_{j=0}^{n-1} b_j\xi_r(\xi^j) \quad (b_j \in v, 0 \leqq j < n) \end{aligned}$$

establishes a v-endomorphism ξ_τ of $v[\xi; f]$ satisfying the equation:

3.4 $$f(\xi_\tau) = 0$$

which is an equation for ξ_τ over v of minimal degree.

We define a multiplication on $v[\xi; f]$ by the rule:

$$\left(\sum_{h=0}^{n-1} k_h \xi^h\right)\eta = \sum_{h=0}^{n-1} k_h \xi_\tau^h(\eta) \, (\eta \in v[\xi; f])$$

which turns the v-module $v[\xi; f]$ into a commutative unital ring extension of v when the mapping $b \to b1 + 0\xi^1 + \cdots + 0\xi^{n-1} (b \in v)$ provides the embedding isomorphism of v into $v[\xi; f]$. The unit element of v also is the unit element of $v[\xi; f]$. Together with

$$\xi = 01 + 1\xi^1 + 0\xi^2 + \cdots + 0\xi^{n-1}$$

the ring v generates the v-ring $v[\xi; f]$ with v-basis $1, \xi, \xi^2, \cdots, \xi^{n-1}$ such that $f(\xi) = 0$. If in a commutative unital ring extension v^* of v for some element η of v the equation $f(\eta) = 0$ holds, then the mapping

$$\sum_{j=0}^{n-1} b_j \xi^j \to \sum_{j=0}^{n-1} b_j \eta^j (b_j \in v; 0 \leqq j < n)$$

provides a v-homomorphism of $v[\xi; f]$ into v^*.

Extending the preceding construction we define the commutative unital ring extension $v[\xi_1, \cdots, \xi_n; f]$ of v with $n!$ basis elements

$$\xi_1^{\nu_1} \xi_2^{\nu_2} \cdots \xi_n^{\nu_n} \quad (0 \leqq \nu_j < n - j; j = 1, 2, \cdots, n)$$

over v as follows:

3.5 $$v[\xi_1, \cdots, \xi_n; f] = (v[\xi_1; f])[\xi_2, \cdots, \xi_n; f/(x - \xi_1)].$$

It follows that in $v[\xi_1, \cdots, \xi_n; f]$ there holds the factorization:

3.6 $$f(x) = (x - \xi_1)(x - \xi_2) \cdots (x - \xi_n)$$

of $f(x)$ into n linear factors.

THEOREM. 3 *The $n!$ permutations of the n distinct roots $\xi_1, \xi_2, \cdots, \xi_n$ of f determine automorphisms of $v[\xi_1, \cdots, \xi_n; f]$ such that the automorphism induced by the permutation π maps the polynomial expression $P(\xi_1, \xi_2, \cdots, \xi_n)$ on $P(\pi\xi_1, \cdots, \pi\xi_n)$ for any polynomial $P(x_1, \cdots, x_n)$. These $n!$ automorphisms form a group γ_n such that every element of v is fixed by every member of γ_n. The elements of v are the only elements of the ring $v[\xi_1, \cdots, \xi_n; f]$ that are fixed by γ_n.*

Proof. This is clear if $n = 1$. Apply induction over n. Let $n > 1$. The subset $\bar{v}$ of all elements of the ring extension that are fixed by γ_n is a subring containing v. Because of 3.5 and the induction assumption it follows that $\bar{v}$ is contained in $v[\xi_1]$. Applying a permutation automorphism interchanging ξ_1, ξ_2 to an element of $v[\xi_1]$ that is not contained in v, we obtain an element of $v[\xi_1]\ [\xi_2]$ which is not contained in $v[\xi_1]$. Hence the theorem.

COROLLARY. (Theorem on symmetric functions.) *Let* $R=v[x_1, x_2, \cdots, x_n]$ *be the polynomial ring in n commuting independent variables* $x_1, x_2, \cdots, x_n$ *over* v. Let $s_i=\sum x_{\alpha_1}x_{\alpha_2}\cdots x_{\alpha_i}\mid 1\leqq\alpha_1<\cdots<\alpha_i\leqq n$ *be the i-th basic symmetric function* $(1\leqq i\leqq n)$. *Then every polynomial in* $x_1, x_2, \cdots, x_n$ *over v that is fixed by all variable permutations (symmetric polynomial in* $x_1, x_2, \cdots, x_n$*) is equal to a polynomial in* $s_1, s_2, \cdots, s_n$ *over* v.

Proof. We note that, for the polynomial $f(x)=x^n-s_1x^{n-1}+s_2x^{n-2}-\cdots+(-1)^ns_n$,

3.7 $$f(x) = (x - x_1)(x - x_2)\cdots(x - x_n),$$

3.8 $$f(x_i) = 0 \quad (1 \leqq i \leqq n).$$

If v coincides with the rational integer ring $\mathbf{Z}$, then R is an integral domain. Hence

3.9 $$\frac{f}{x-x_1}(x_2)=0.$$

Using the homomorphism of $\mathbf{Z}[x_1, x_2, \cdots, x_n]$ into R which sends 1 into 1, x_i into $x_i(1\leqq i\leqq n)$ we find that 3.9 always holds. By induction over n we conclude that there is a homomorphic mapping ψ of the ring extension $T=S[\xi_1\cdots, \xi_n; f]$ over the subring S of R generated by v and $s_1, s_2, \cdots, s_n$ which maps ξ_i onto x_i. Clearly T is generated by v and $\xi_1, \xi_2, \cdots, \xi_n$, hence there is a homomorphism σ of R onto T over v mapping x_i onto $\xi_i(1\leqq i\leqq n)$. We have $\sigma\psi=1_T$, $\psi\sigma=1_R$, hence σ, ψ are isomorphisms over v. Now the corollary follows from Theorem 3.

Let $\bar{s}_1, \bar{s}_2, \cdots, \bar{s}_n$ be another set of n independent commuting variables over v and let $\bar{f}(x)=x^n-\bar{s}_1x^{n-1}+\bar{s}_2x^{n-2}-\cdots+(-1)^n\bar{s}_n$.

There is a homomorphism ϕ of the ring extension $\overline{T}=\overline{S}[\bar{\xi}_1, \cdots, \bar{\xi}_n; \bar{f}]$ of $\overline{S}=v[\bar{s}_1, \cdots, \bar{s}_n]$ onto T over v which maps $\bar{\xi}_i$ onto $\xi_i(1\leqq i\leqq n)$ and hence $\bar{s}_i$ onto s_i. On the other hand, there is a homomorphism κ of R onto $\overline{T}$ over v mapping x_i onto $\bar{\xi}_i(1\leqq i\leqq n)$ and we have: $\kappa\psi\phi=1_{\overline{T}}$, $\phi\kappa\psi=1_T$ therefore ϕ, $\kappa\psi$ are isomorphisms over v. It follows that the homomorphic mapping of $\overline{S}$ onto S over v which maps $\bar{s}_i$ onto $s_i(1\leqq i\leqq n)$is an isomorphism. In other words the basic symmetric functions are independent over v.

From the theorem on symmetric functions it follows that the coefficients of the polynomials

$$S_j(f)(x) = \prod_{1\leqq\alpha_1<\alpha_2<\cdots<\alpha_j\leqq n} (x - (\xi_{\alpha_1} + \xi_{\alpha_2} + \cdots + \xi_{\alpha_j}))$$

of degree $\binom{n}{j}$ are in v. Moreover, if v is an integral domain in which the factorization $f(x)=\prod_{j=1}^n (x-\eta_j)$ obtains then the mapping of ξ_j onto $\eta_j(1\leqq j\leqq n)$ can be extended in precisely one way to a homomorphism of $v[\xi_1, \cdots, \xi_n; f]$ onto v over v, and the equations $S_j(f)(\eta_{\gamma_1} + \cdots + \eta_{\gamma_j}) = 0$ hold whenever $1 \leqq \gamma_1 < \gamma_2 < \cdots < \gamma_j \leqq n$.

Assume that v is a field and that $S_2(f)$ has the root ξ in v. Let $d(x)$ be the greatest common divisor of the polynomials $f(x)$ and $f(\xi - x)$ in $v[x]$ with leading coefficient 1.

There holds an equation $d(x) = A(x)f(x) + B(x)f(\xi - x)$ with polynomials $A(x)$, $B(x)$ in $v[x]$. Hence in $v[\xi_1, \cdots, \xi_n; f]$:

$$d(x) = A(x)f(x) + B(x)f(\xi_1 + \xi_2 - x) + (\xi - \xi_1 - \xi_2)g(x, \xi_1 + \xi_2)$$

where $g(x, y)$ is a polynomial in two variables x, y over v. Upon substitution of ξ_2: $d(\xi_2) = (\xi - \xi_1 - \xi_2)g(\xi_2, \xi_1 + \xi_2)$.

If $d(x)$ is a nonzero constant then $d(x) = 1 = d(\xi_2)$, then $\xi - \xi_1 - \xi_2$ is invertible in $v[\xi_1, \cdots, \xi_n; f]$. The same applies to $\xi - \xi_i - \xi_k$ $(1 \leqq i < k \leqq n)$ and hence to the product.

But this product is equal to $S_2(f)(\xi)$ which is zero, a contradiction. Hence, $d(x)$ is not constant.

4. The fundamental theorem of algebra.

THEOREM 4. *Let F be a really closed field. The field extension E formed by the symbols $a + bi$ $(a, b \in F)$ with the operational rules*

4.1
$$a + bi = c + di \Leftrightarrow a = c,\ b = d$$
$$(a + bi) + (c + di) = (a + c) + (b + d)i$$
$$(a + bi)(c + di) = (ac - bd) + (ad + bc)i$$
$$(a, b, c, d \in F)$$

is algebraically closed.

Proof. The field property of E is shown in the customary manner. The zero element of E is the symbol $0 + 0i$, the unit element of E is the symbol $1 + 0i$, the inverse of $a + bi$ is the symbol

$$\frac{a}{a^2 + b^2} + \left(\frac{-b}{a^2 + b^2}\right)i$$

provided not both of the elements a, b of F are zero. The mapping of a onto $a + 0i$ provides an embedding isomorphism of F into E such that upon identification of a and $a + 0i$ the field F becomes a subfield of E. The symbol $0 + 1i$ generates the quadratic extension E over F. Since $a + bi = (a + 0i) + (b + 0i)(0 + 1i)$, we are entitled to denote the symbol $0 + 1i$ by i, so that $a + bi$ is the actual sum of the element a of F and the product of the element b of F by i. The element i is a root of the irreducible quadratic equation:

4.2
$$i^2 + 1 = 0$$

over F the other root being $-i$. Hence the complex conjugate mapping:

$$4.3 \qquad a + bi \to \overline{a + bi} = a - bi (a, b \in F)$$

establishes an involutoric automorphism of E over F. An element γ of E is in F if and only if its complex conjugate $\bar{\gamma}$ coincides with γ. The automorphism 4.3 is extended to an involutoric automorphism:

$$4.4 \qquad f(x) \to \bar{f}(x)$$

of $E[x]$ over $F[x]$ by setting

$$4.5 \qquad \bar{f}(x) = \bar{a}_0 x^n + \bar{a}_1 x^{n-1} + \cdots + \bar{a}_n$$

if

$$4.6 \qquad f(x) = a_0 x^n + a_1 x^{n-1} + \cdots + a_n (a_0, a_1, \cdots, a_n \in E).$$

The polynomial 4.6 lies in $F[x]$ if it coincides with its complex conjugate polynomial $\bar{f}(x)$, and conversely.

Let us further mention that for every element $a+bi$ of $E(a, b \in F)$ the equation

$$4.7 \qquad a + bi = (\xi + \eta i)^2 (\xi, \eta \in F)$$

is solved in E by:

$$4.8 \qquad \begin{aligned} \xi &= \left| \sqrt{1/2(a + |\sqrt{a^2 + b^2}|)} \right| \\ \eta &= (\operatorname{sign} b) \cdot \left| \sqrt{1/2(-a + |\sqrt{a^2 + b^2}|)} \right|. \end{aligned}$$

Our task is to find a root in E for every polynomial

$$4.9 \qquad f(x) = x^n + a_1 x^{n-1} + a_2 x^{n-2} + \cdots + a_n$$

with coefficients $a_1, a_2, \cdots, a_n$ in E.

Let $n = 2^s n'$ when $0 \leqq s$, $n' \equiv 1(2)$.

Suppose it would not always be possible to find a root in E for a given nonconstant polynomial, then there would be polynomials 4.9 with minimum value of s for which no root could be found in E and among those polynomials there would be a polynomial 4.9 with minimum value of n'.

If $s=0$ and if all coefficients $a_1, a_2, \cdots, a_n$ are real then by assumption about F we can find a root of $f(x)$ in F, a contradiction.

If there is a root η of $S_2(f)$ in E then the greatest common divisor $d(x)$ of $f(x)$ and of $f(\eta - x)$ can be formed in $E[x]$. It is not constant as we have seen previously.

There holds an equation $f(x) = d(x)e(x)$ in $E[x]$ so that every root of $d(x)$ or $e(x)$ also is a root of $f(x)$. Since there is no root of $f(x)$ in E the same applies to the polynomials $d(x)$, $e(x)$. Because of the minimal property of $f(x)$ it follows that $d(x)$ has the same degree as $f(x)$ whereas $e(x)$ is a nonzero constant. In other words $f(x)=f(\eta-x)$ if $s>0$. Or, setting $y=x-\eta/2$ we have

$$f(x) = g(y), \qquad g(y) = g(-y), \qquad g(y) = h(y^2), \qquad f(x) = h((x - \eta/2)^2)$$

when h is a polynomial of degree $n/2$ with coefficients in E.

It follows from this argument that $S_2(f)$ has no root in E if n is even. On the other hand, the degree of $S_2(f)$ is not divisible by 2^s, hence we can find a root of $S_2(f)(x)$ in E due to the minimal property of $f(x)$.

These arguments show that n is odd and that not all coefficients of $f(x)$ are real. But the degree of $f_1(x)=f(x)\bar{f}(x)$ is $2n$, the degree of $S_2(f_1)$ is the odd number $n(2n-1)$. Since $f_1(x)$ has real coefficients, also $S_2(f_1)$ has real coefficients. Therefore, $S_2(f_1)$ has a real root and by the argument given above $f_1(x)$ has a root ξ in E. Since $0=f_1(\xi)=f(\xi)\bar{f}(\xi)$, but by assumption $0\neq f(\xi)$ it follows that $0=\bar{f}(\xi)$. Forming the complex conjugate of this equation we obtain $0=f(\bar{\xi})$ so that $f(x)$ does have the root $\bar{\xi}$ in E which is a contradiction.

Thus we have established Theorem 4.

This existence proof can be reformulated so as to yield an algorithm:

I. If $s=0$, $f(x)\in F[x]$ then we know by assumption how to find a root of f in F.

II. If the algorithm is defined already for polynomials of odd degree less than $2n+1$ with coefficients in E and if $f(x)$ is a polynomial of degree $2n+1$ with coefficients in E, then solve $S_2(f\bar{f})(\eta)=0$ as in I. Form $d(x)$ as above. If $[d]\equiv 1$ (2) then $[d]<2n+1$; apply the algorithm to find a root of d in E. It turns out to be a root of f or of $\bar{f}$ too. If $[d]\equiv 0(2)$, then the algorithm can be applied to $e(x)=(f/d)(x)$ and provides a root of e which turns out to be also a root of f.

Thus, the algorithm is obtained for solving in E equations of odd degree over E.

III. If $s>0$ and if the algorithm is defined for solving in E equations with coefficients in E such that either the degree is not divisible by 2^s or it is smaller than the degree of f and not divisible by 2^{s+1}, then apply the algorithm to the equation $S_2(f)(\eta)=0$ and form $d(x)$ as above. Apply the algorithm to $d(x)$ in case $2^s\nmid[d]$ or $2^{s+1}\nmid[d]$ and $[d]<[f]$. Apply the algorithm to f/d in case $[d]<[f]$, $2^{s+1}\mid[d]$. In either case, a root of f is obtained. If $[d]=[f]$, then form h as above. Apply the algorithm to h, to find a root ξ of h. As above we obtain the root $\eta/2+\sqrt{\xi}$ of f.

However, the degrees of the auxiliary equations involved may become very large. E.g. for $n=8$ we may find the degrees 8,

$$\binom{8}{2} = 28, \qquad \binom{28}{2} = 378, \qquad \binom{378}{2} = 71253.$$

But let us note that the binomial coefficient

$$\binom{n}{2^i} = \frac{2^s n' \cdot (n-1) \cdots (n-2^i+1)}{2^i \cdot 1 \cdots (2^i-1)}$$

is divisible by 2^{s-i}, but not divisible by 2^{s-i+1} for $i=0, 1, 2, \cdots s$. Hence the same arguments also apply to the auxiliary equations:

$$f(x) = 0, \qquad S_2(f)(x) = 0, \qquad S_4(f)(x) = 0, \cdots, S_{2^{s-i}}(f)(x) = 0, \cdots$$
$$S_{2^s}(f)(x) = 0.$$

Note that the polynomial $S_{2^{s-1}}(f)$ is symmetric about $-(1/2)a_1$ in case $n=2^s$. Hence in that case it is of the form $g(x+(1/2)a_1)^2$ where $g(x)$ is of odd degree.

In any event we obtain considerable economy; e.g., for $n=8$ we have to form and to solve equations of degrees 8, 28, 2, 35.

Assuming that there exists an algorithm for finding the real roots of a polynomial 4.9 with real coefficients we extend it as follows to an algorithm for finding all the roots of $f(x)$.

(1) If $\gcd(f, df/dx)=d(x)$ is not constant, let $(f/d)(x)=e_0(x)$, $\gcd(e_0,(df/dx)) = e_1(x)$ not constant, $(e_0/e_1)(x)=f_1(x)$, $\gcd(e_{j-1},(d^jf/dx^j))=e_j(x)$ not constant, $(e_{j-1}/e_j)(x)=f_j(x)$ and

$$\gcd\left(e_j, \frac{d^{j+1}f}{dx^{j+1}}\right)=1, \qquad (j>0).$$

Hence $f(x)=f_1(x)f_2(x)^2\cdots f_j(x)^jf_{j+1}(x)^{j+1}$ when $f_1(x),f_2(x),\ldots,f_j(x),f_{j+1}(x)$ are mutually prime, separable polynomials.

The task is now to find the roots of these polynomials.

(2) $\gcd(f,df/dx)=1$, $f(x)$ has the distinct real roots $\alpha_1,\alpha_2,\ldots,\alpha_r$, $f(x)= (x-\alpha_1)(x-\alpha_2)\cdots(x-\alpha_r)g(x)$, $g(x)$ is separable and has no real root. The task is to find the roots of $g(x)$.

(3) $f(x)$ is separable and has no real root.

Find the real roots of $S_2(f(x))$ say $\beta_1<\beta_2<\cdots<\beta_\rho$. Normally speaking each of the real roots β_j is simple and in this case we have $\gcd(f(x), f(\beta_j-x)) = (x-\beta_j/2)^2+\gamma_j^2$ where γ_j is positive.

In this case the roots of $f(x)$ are the $n=2\rho$ complex numbers

$$\beta_j/2 \pm i\gamma_j \quad (1 \leqq j \leqq \rho).$$

If the real roots β_j are not all simple, a more elaborate algorithm must be carried out.

Let B_0 be the set of all roots β_j and let $\mu_0(\beta_j)$ be the multiplicity of the root β_j of $S_2(f)(x)$. Denote by A_1 the set of all elements of B_0 that are not the arithmetic mean of two distinct elements of B_0. We find that $\gcd(f(x), f(\beta-x)) =\phi_\beta(x)=\phi_\beta(\beta-x)$ is nonconstant for any β of A_1. Hence $\phi_\beta(x)=h_\beta((x-\beta/2)^2)$ where $\phi_\beta(x)$ has degree $2\mu_0(\beta)$ and h_β is a polynomial with half the degree of $\phi_\beta(x)$ such that all roots of h_β are negative, say, they are of the form: $-\gamma_{\beta_k}$ $(1 \leqq k \leqq \frac{1}{2}\mu_0(\beta))$ when γ_{β_k} is positive. Hence the roots of $\phi_\beta(x)$ are the $\mu_0(\beta)$ complex numbers $\beta/2\pm\gamma_{\beta_k}i$. If the total number of these roots is n then the task is completed.

If the total number is less than n then we determine for each member γ of B_0 that is not contained in A_1 the number of times, say $\nu_0(\gamma)$ that $\gamma=\frac{1}{2}(\beta+\beta')$

$(\beta, \beta' \in A_1, \beta < \beta')$ and

$$\gamma_{\beta_k} = \gamma_{\beta'_{k'}} (1 \leqq k \leqq \mu_0(\beta)/2, 1 \leqq k' \leqq \mu_0(\beta')/2).$$

In view of the connection between f and $S_2(f)$ we find that $\mu_1(\gamma) = \mu_0(\gamma) - 2\nu_0(\gamma) \geqq 0$ when $\mu_1(\gamma)$ is not always zero.

Let us form the set B_1 of all γ's for which $\mu_1(\gamma)$ is positive.

We proceed as above, substituting B_1 for B_0, μ_1 for μ_0, A_2 for A_1 when A_2 is the subset of all members of B_1 that are not an arithmetic mean of two distinct members of B_1. In this way some further roots of $f(x)$ will be obtained. If not yet all of them are found, proceed as before until all n roots are constructed.

References

1. E. Artin, Kennzeichnung des Körpers der reellen algebraischen Zahlen, Abhg. Math. Sem. Univ. Hamburg, 3 (1924) 170–175.
2. E. Artin und O. Schreier, Algebraische Konstruktion reeller Körper, Abh. Math. Sem. Univ. Hamburg, 5 (1927) 85–99.
3. E. Artin und O. Schreier, Eine Kennzeichnung der reell abgeschlossenen Körper, Abh. Math. Sem. Univ. Hamburg, 5 (1927) 225–231.
4. Carl Friedrich Gauss, Demonstratio nova theorematis omnem functionem algebraicam rationalem integram unius variabilis in factores reales primi vel secundi gradus resolvi posse, Inaugural dissertation, Göttingen, 1799.
5. Nathan Jacobson, Lectures in Abstract Algebra, Vol. III, Van Nostrand, Princeton, N. J., 1951.
6. Oskar Perron, Algebra, 3rd ed., Berlin, 1951.
7. Wolfgang Krull, Elementare und klassische Algebra vom modernen Standpunkt, Berlin 1963 (see last chapter).
8. Hans Zassenhaus, The Theory of Groups, 2nd ed., Chelsea, New York, 1958.
9. Max Zorn, A remark on a method in transfinite algebra, Bull. Amer. Math. Soc. 41 (1935) 667–670.

A NOTE ON THE ALGEBRAIC CLOSURE OF A FIELD*

ROBERT GILMER, Florida State University

In order to prove that a given field F has an algebraic closure, it is sufficient to prove the existence of an extension field K of F such that each nonconstant polynomial f over F splits into linear factors over K, for if such a field K exists, then the set of elements of K algebraic over F is an algebraic closure of F [3, p. 194]. To establish the existence of such a field K, Lang in his text *Algebra* [2] proceeds as follows. He first proves that if E is any field, then there is an extension field E_1 of E such that each nonconstant polynomial f over E has a root in E_1. Applying this result successively to $E_1, E_2, \cdots$, Lang obtains an ascending chain $E_1 \subset E_2 \subset \cdots$ of extension fields of E such that for each i, every nonconstant polynomial over E_i has a root in E_{i+1}. Then if $L = \bigcup_{i=1}^{\infty} E_i$, it is easy to show that each nonconstant polynomial over E splits into linear factors over L so that L contains an algebraic closure of E. (See [2] pp. 169–170.) The purpose

* From AMERICAN MATHEMATICAL MONTHLY, vol. 75 (1968), pp. 1101–1102.

of this note is to observe that in the above process, the field E_1 already contains an algebraic closure of E. This we do by establishing the following result:

THEOREM. *If K is a subfield of a field L and if each nonconstant polynomial with coefficients in K has a root in L, then each nonconstant polynomial with coefficients in K splits into linear factors in $L[X]$.*

Proof. It suffices to prove that for f a nonconstant irreducible monic polynomial with coefficients in K, f is a product of linear factors in $L[X]$. If K has characteristic 0, we define $p=1$; otherwise p denotes the characteristic of K. (In Bourbaki's terminology [1, p. 71], *p is the characteristic exponent of K.*) In a splitting field F of f over K, the roots of f all have the same multiplicity p^e for some nonnegative integer $e: f=\prod_{i=1}^{n}(X-\alpha_i)^{p^e}=\prod_{i=1}^{n}(X^{p^e}-\alpha_i^{p^e})$. Further, if $f=X^{np^e}+f_{n-1}X^{(n-1)p^e}+\cdots+f_1X^{p^e}+f_0$, and if $\beta_i=\alpha_i^{p^e}$ for each i, $\{\beta_1,\cdots,\beta_n\}$ is the set of roots of the irreducible, separable polynomial $g=x^n+f_{n-1}X^{n-1}+\cdots+f_0$ over K [3, p. 120]. The field $K(\beta_1,\cdots,\beta_n)$ is a finite, normal, separable extension of K. It is therefore a simple extension of K: $K(\beta_1,\cdots,\beta_n)=K(\gamma)$. If h is the minimal polynomial for γ over K, then $K(\gamma)$ is a splitting field for h over K and there is, by hypothesis, a root θ of h in L. The fields $K(\theta)$ and $K(\gamma)$ are each K-isomorphic to $K[X]/(h)$, and hence are K-isomorphic to each other. It follows that $k(\theta)$ is a splitting field of g over K. Therefore, if $p^e=1$—that is, if f is separable over K, then L contains a splitting field of f over K. Hence we have proved our theorem in case the algebraic closure of K in L is separable over K. In particular, our proof is complete if K is a perfect field.

In case K is not perfect, we let K_0 be the subfield of L consisting of those elements which are purely inseparable over K. We show that the field K_0 is perfect. Hence if $s\in K_0$ and if $s^{p^e}\in K$, then by hypothesis, $X^{p^{e+1}}-s^{p^e}$ has a root t in L. Therefore, $t^{p^{e+1}}\in K$, $t\in K_0$, and $(t^p-s)^{p^e}=0$, implying that $t^p=s$ so that K_0 is perfect [3, p. 124]. Further, if $q(x)=\sum_{i=0}^{r}q_iX^i$ is any nonconstant polynomial over K_0, then for some positive integer e, $[q(X)]^{p^e}=v(X)$ $=\sum q_i^{p^e}X^{ip^e}\in K[X]$ so that $v(X)$ has a root α in L. We have $0=v(\alpha)=[q(\alpha)]^{p^e}$, implying that $q(\alpha)=0$. We conclude that K_0 is a perfect field with the property that each nonconstant polynomial over K_0 has a root in the extension field L of K_0. By our previous proof, each nonconstant polynomial over K_0 splits into linear factors in $L[X]$. This property holds then, in particular, for nonconstant polynomials over K, and this completes the proof of our theorem.

References

1. N. Bourbaki, Eléments de Mathématique, Algèbre, Book II, Chap. 5, Hermann, Paris, 1950.
2. S. Lang, Algebra, Addison-Wesley, Reading, 1965.
3. B. L. van der Waerden, Modern Algebra, vol. I, English ed., Ungar, New York, 1948.

BALANCED FIELD EXTENSIONS*

JOSEPH LIPMAN, Purdue University

Let k be a field, and let K be an algebraic extension of k. K is then a purely inseparable extension of a separable extension of k; for reasons of symmetry, one might wonder when K will be a separable extension of a purely inseparable extension of k. (This is not always so: cf. example at the end of this note.) When this does happen, let us say that K/k is "balanced." We wish to set down some simple observations about such extensions.

For basic notions of field theory see O. Zariski and P. Samuel: *Commutative Algebra*, Van Nostrand, Princeton, N. J., 1958, Volume I, chapter 2; also chapter 3, section 15, for the definition and properties of free joins.

PROPOSITIONS: A. *The following are equivalent:*

1. *K/k is balanced.*
2. *There exists a separable algebraic extension of K which is normal over k.*
3. *If L is a field of algebraic functions over k, then the order of inseparability $[(L, K): K]_i$ is the same for all free joins (L, K) of L/k and K/k.*

In geometric language, 3. reads: If V/k is an irreducible algebraic variety, then all the irreducible components of V/K have the same order of inseparability.

B. *Let $\bar{k}$ be an algebraic closure of k, and let $\bar{k}_s$ (respectively $\bar{k}_i$) be the subfield of $\bar{k}$ consisting of all elements which are separable, (resp. purely inseparable) over k. We know that the subfields of $\bar{k}$ (resp. $\bar{k}_s$, $\bar{k}_i$) form a lattice Z (resp. Z_s, Z_i) under the operations of field composition and intersection.*

The balanced extensions of k in $\bar{k}$ form a sublattice of Z isomorphic with the direct product $Z_s \times Z_i$.

Proofs: A. We show that $1 \xrightarrow{a)} 3 \xrightarrow{b)} 2 \xrightarrow{c)} 1$.

a) Let $k \subseteq I \subseteq K$, I being a field such that I/k is purely inseparable, and K/I is separable. Any free join (L, K) contains a free join (L, I), and since K/I is separable, we have $[(L, K): K]_i = [(L, I): I]_i$. Thus we may assume that $K = I$. But then there is nothing to prove, since all free joins of L/k and I/k are equivalent.

b) Let $x \in K$, and let $L = k(x)$. The free joins of L/k, K/k, are all the fields of the form $K(\bar{x})$, where $\bar{x}$ is k-conjugate to x, in some fixed algebraic closure $\overline{K}$ of K. Since $[K(x): K]_i = [K: K]_i = 1$, we have $[K(\bar{x}): K]_i = 1$, i.e. *all k-conjugates of x* (in $\overline{K}$) *are separable over K*. From this it follows immediately that if N is the least extension of K normal over k, then N/K is separable.

c) Let $N \supseteq K$ be such that N/K is separable and N/k is normal. Let $I \subseteq N$ be the field of invariants of all automorphisms of N/k. Then I/k is purely inseparable, and N/I is separable. It will be sufficient to show that $I \subseteq K$. But

---

* From AMERICAN MATHEMATICAL MONTHLY, vol. 73 (1966), pp. 373–374.

any x in I is separable over K (since N/K is separable), and purely inseparable over K (since I/k is purely inseparable).

B. Let B be the set of balanced extensions of k in $\bar{k}$. Clearly $K \in B$ iff $K/(K \cap \bar{k}_i)$ is separable, and this latter condition may be expressed as follows:

(1′) If $x \in K$, if $f(X)$ is the minimum monic polynomial of x over k, and if $\bar{f}(X) \in \bar{k}[X]$ is *the* polynomial without multiple roots, of which $f(X)$ is a power, then $\bar{f}(X) \in (K \cap \bar{k}_i)[X]$.

It follows immediately that an arbitrary intersection of members of B is again a member of B.

Again, if $K \in B$, then K is a separable extension of $K \cap \bar{k}_i$; also K is purely inseparable over $K \cap \bar{k}_s$; if K' is the compositum $(K \cap \bar{k}_s, K \cap \bar{k}_i)$, then K/K' is both separable and purely inseparable, i.e., $K = K'$. We see then, that $K \in B$ *iff* K *is generated by a separable extension of* k *and a purely inseparable extension of* k.

It follows easily that the field generated by an arbitrary collection of members of B is itself a member of B.

We have shown, therefore, that B is a sublattice of Z (in fact, a complete sublattice). We have also given an order-preserving map F from $Z_s \times Z_i$ *onto* B; if $S \in Z_s$, $I \in Z_i$, then $F(S, I)$ is the composed field (S, I). Now if $x \in (S, I) \cap \bar{k}_i$, then x is separable over I and purely inseparable over I; hence $(S, I) \cap \bar{k}_i = I$. Similarly $(S, I) \cap \bar{k}_s = S$. Thus F is injective, and the proof is complete.

EXAMPLE. For an example of a nonbalanced extension, let L be a field of characteristic two, let Y, Z, be indeterminates over L, let $k = L(Y, Z)$, and let $K = k(x)$, x being a root of

$$f(X) = X^4 + YX^2 + Z = 0.$$

One checks that $f(X)$ is irreducible over k, that $[K:k]_i = 2$, and that, in the notation of (1′) above, $\bar{f}(X) = X^2 + \sqrt{Y}X + \sqrt{Z}$. According to (1′), K/k cannot be balanced unless $\sqrt{Y} \in K$ and $\sqrt{Z} \in K$. Since $[k(\sqrt{Y}, \sqrt{Z}) : k]_i = 4$, this is impossible.

ON EXTENSIONS OF Q BY SQUARE ROOTS*

R. L. ROTH, University of Colorado

In courses on field theory one often uses the field of all algebraic numbers as an example of an extension of Q, the field of rational numbers, which is both algebraic and of infinite degree over Q. A more intuitively grasped example is perhaps the field $Q(\sqrt{2}, \sqrt{3}, \sqrt{5}, \cdots, \sqrt{p}, \cdots)$, where we adjoin the square roots of all the positive primes to Q. However, to prove that each new adjunction of the square root of a prime is in fact a proper extension, it turns out to be easier to prove the following stronger statement by induction. (In this note, "prime" will always mean positive prime integer.)

THEOREM. *Let $p_1, \cdots, p_n$ be n distinct positive prime integers. Let $F=Q(\sqrt{p_1}, \cdots, \sqrt{p_n})$. Let $q_1, \cdots, q_r$ be any set of distinct positive primes none of which appear in the list $\{p_1, \cdots, p_n\}$. Then $\sqrt{q_1q_2\cdots q_r} \notin F$.*

Proof. By induction on n.

Let $n=0$. Then $F=Q$. If $q_1, \cdots, q_r$ are any distinct primes, then $x^2-q_1q_2\cdots q_r$ is irreducible over Q by Eisenstein's criterion and $\sqrt{q_1q_2\cdots q_r} \notin Q$. (This can also be shown by a proof analogous to the classical proof that $\sqrt{2} \notin Q$.)

Now assume the theorem is true for $n-1$ and suppose that

$$F = Q(\sqrt{p_1}, \cdots, \sqrt{p_{n-1}}, \sqrt{p_n}).$$

Set $F_0=Q(\sqrt{p_1}, \cdots, \sqrt{p_{n-1}})$. By induction the theorem holds for F_0, and $F=F_0(\sqrt{p_n})$ is an extension of F_0 of degree 2. Let $q_1, q_2, \cdots, q_r$ be any distinct primes not on the list $\{p_1, \cdots, p_n\}$. Suppose it were true that $\sqrt{q_1q_2\cdots q_r} \in F$. Then $\sqrt{q_1q_2\cdots q_r} = a+b\sqrt{p_n}$ where a and b belong to F_0. Then

$$q_1q_2\cdots q_r = a^2 + b^2p_n + 2ab\sqrt{p_n}. \tag{1}$$

(i) If a and b are both not 0, equation (1) shows that

$$\sqrt{p_n} = \frac{q_1q_2\cdots q_r - a^2 - p_nb^2}{2ab}$$

would lie in F_0.

(ii) If $b=0$, then $\sqrt{q_1\cdots q_r} = a \in F_0$.

(iii) If $a=0$, then $\sqrt{q_1\cdots q_r} = b\sqrt{p_n}$; hence $\sqrt{q_1\cdots q_rp_n} = p_nb \in F_0$. However, each of these three possibilities contradicts the inductive hypothesis (with respect to F_0). So $\sqrt{q_1\cdots q_r} \notin F$.

COROLLARY 1. *If q is not on the list of primes $\{p_1,\cdots,p_n\}$ and q is a prime, then $\sqrt{q} \notin Q(\sqrt{p_1},\cdots,\sqrt{p_n})$.*

* From AMERICAN MATHEMATICAL MONTHLY, vol. 78 (1971), pp. 392–393.

COROLLARY 2. *If $p_1, \cdots, p_n$ are any distinct primes then*

$$[Q(\sqrt{p_1}, \sqrt{p_2}, \cdots, \sqrt{p_n}):Q] = 2^n.$$

COROLLARY 3. *If $p_1, p_2, \cdots, p_i, \cdots$ is any infinite sequence of distinct primes then $Q(\sqrt{p_1}, \sqrt{p_2}, \cdots, \sqrt{p_i}, \cdots)$ is an infinite algebraic extension of Q. In particular $Q(\sqrt{2}, \sqrt{3}, \sqrt{5}, \cdots, \sqrt{p}, \cdots)$ where p runs over all primes has this property.*

(Note, the latter is the same as the field $Q(\sqrt{1}, \sqrt{2}, \sqrt{3}, \cdots, \sqrt{n}, \cdots)$ where n runs over all positive integers.)

The theorem may also be applied in special examples, for instance $[Q(\sqrt{2}, \sqrt{7}, \sqrt{15}):Q]=8$. $[Q(\sqrt{14}, \sqrt{15}):Q]=4$ since $\sqrt{15}\notin Q(\sqrt{14})\subset Q(\sqrt{2}, \sqrt{7})$. $[Q(\sqrt{14}, \sqrt{6}):Q]=4$, since if $\sqrt{14}\in Q(\sqrt{6})\subset Q(\sqrt{2}, \sqrt{3})$, then $\sqrt{7}\in Q(\sqrt{2}, \sqrt{3})$.

ON DEMOIVRE'S QUINTIC*

R. L. BORGER, University of Illinois

§1. For the domain of rational numbers, DeMoivre's quintic

$$x^5 + px^3 + \tfrac{1}{5}p^2x + r = 0, \tag{1}$$

for values of p and r making the discriminant

$$\Delta \equiv \left(\frac{r}{2}\right)^2 + \left(\frac{p}{5}\right)^5 \neq 0,$$

will be shown to have as its Galois group *either the metacyclic group G_{20} or a cyclic group C_4*. We may then readily deduce the following properties:

DeMoivre's quintic is solvable by radicals.

Either all the roots are real or only one root is real.

Not more than one root is rational; if the equation is reducible in $R(1)$, its left member is the product of a linear and an irreducible quartic factor.

If the equation is irreducible in $R(1)$, any root is a rational function of an arbitrary pair of roots.

To determine the Galois group of (1), we make use on the one hand of Cayley's resolvent sextic for any quintic, and on the other hand of the following lemma:†

If we know a rational function of the roots of an algebraic equation $f(x)=0$ having the properties:

(*i*) That it is formally invariant under the substitutions of a group G' and under no others.

(*ii*) That it has a value in the domain of rationality.

---

* From AMERICAN MATHEMATICAL MONTHLY, vol. 15 (1908), pp. 171–174.

† Dickson, *Algebraic Equations*, p. 59, §65.

(*iii*) That it is distinct from its conjugates under the substitutions of the symmetric group $G_n!$, then the Galois group G of $f(x)=0$ is a subgroup of G'.

§2. We exclude those values of p and r for which the discriminant $\triangle=0$. They give rise to equal roots and these may be removed by the process of highest common divisor. The function

$$\phi\equiv(x_1x_2+x_2x_3+x_3x_4+x_4x_5+x_5x_1)-(x_1x_3+x_2x_4+x_3x_5+x_4x_1+x_5x_2)$$

belongs to the group G_{10} consisting of the substitutions

$$1;\quad (12345);\quad (13524);\quad (14253)\quad (15432);$$
$$(12)(35);\quad (25)(34);\quad (15)(24);\quad (14)(32);\quad (13)(45).$$

Under the substitutions of G_{60}, ϕ takes six values, which are the roots of a resolvent sextic. For the general quintic Cayley has computed* this resolvent sextic, which becomes for equation (1):

$$\phi^6-7p^2\phi^4+11p^4\phi^2-\frac{32(5^5r^2+2^2p^5)}{5^2\sqrt{5}}\phi+4000pr^2+\tfrac{3}{25}p^6=0. \tag{2}$$

One root of (2) is $\phi=p\sqrt{5}$. By differentiating (2) with respect to ϕ we see that this root is simple unless

$$121p^5+5^5r^2=0. \tag{3}$$

We now divide the discussion into the two cases

I) p and r not satisfying (3).

II) p and r satisfying (3).

I) In this case, ϕ is distinct from its conjugates under G_{60}. Hence $\phi^2=5p^2$ belongs to G_{20}† and is distinct from its conjugates. Hence (§1), G_{20} contains the Galois group of (1). The Galois group G for the domain $R(1)$ may then be

$$G_{20},\ G_{10},\ C_5,\ C_4,\ G_2,\text{ or } G_1\equiv 1.$$

The groups G_{10}, G_5, G_2, G_1 may be at once excluded. By the definition of the Galois group of an equation, every rational function of the roots which remains invariant under the substitutions of G is rationally known. If G is G_{10} or a subgroup of it, then ϕ, belonging to G_{10}, would be rationally known. Since $\phi=p\sqrt{5}$ this is impossible unless $p=0$. Hence when $p\neq 0$, G is not contained in G_{10}.

If $p=0$ we know that (1) reduces to a binomial equation p and its group is metacyclic when r is not the fifth power of a rational number. If $r=k^5$ (k rational) the group G is then C_4. Hence when p and r do not satisfy (3), $G=C_4$ or G_{20}.

---

* Cayley, *Collected Mathematical Papers*, Vol. IV. p. 319.

† The substitutions of G_{20} are given by $\begin{pmatrix}\alpha \\ \alpha x+\beta\end{pmatrix}$, $\begin{pmatrix}\alpha=1\ 2\ 3\ 4 \\ \beta=0\ 1\ 2\ 3\ 4\end{pmatrix}$.

§**3.** Next, we consider the case in which p and r satisfy (3). By solving (3) we find

$$r=\frac{11p^2}{5^2}\sqrt{\frac{p}{5}}. \tag{4}$$

Since r must be a rational number, $\sqrt{\frac{p}{5}}=a$ (a rational),

$$p=5a^2, \quad r=11a^5. \tag{5}$$

Substituting these values in (1), we get

$$x^5+5a^2x^3+5a^4x+11a^5=0. \tag{6}$$

This equation has the root $x=-a$, and the depressed equation is

$$x^4-ax^3+6a^2x^2-6a^3x+11a^4=0. \tag{7}$$

Calling the roots of (7) x_1, x_2, x_3, x_5, and setting

$$y_1=x_1x_2+x_3x_5,$$
$$y_2=x_1x_3+x_2x_5,$$
$$y_3=x_1x_5+x_2x_3,$$

we obtain the cubic resolvent of (7),

$$y^3-6a^2y^2-38a^4y+217a^6=0. \tag{8}$$

The roots of (8) are:

$$y_1=7a^2; y_2=\frac{a^2}{2}(-1+5^2\sqrt{5}); y_3=\frac{a^2}{2}(-1-5^2\sqrt{5}); \tag{9}$$

$y_1=x_1x_2+x_3x_5$ belongs to the group

$G_8\equiv[1,(12); (35); (1325); (1523); (12)(35); (13)(25); (15)(23)]$.

And since y_1 is distinct from its conjugates under G_{24} the Galois group of (7) is G_8 or a subgroup of G_8. As y_2 and y_3 are irrational the group for the domain $R(1)$ cannot be contained in

$$G_4\equiv[1; (12)(35); (13)(25); (15)(23)].$$

The function $\psi\equiv(x_1+x_2)-(x_3+x_5)=a\sqrt{5}$ belongs to the group $H_4\equiv[1; (12); (35); (12)(35)]$. Since the value of ψ is irrational* G is not contained in H_4.

The function $\chi\equiv(x_1-x_2)(x_3-x_5)[x_1+x_2-(x_3+x_5)]\equiv a^3 5^5$ belongs to the group $C_4\equiv[1; (1325), (1523); (12)(35)]$, since χ is rational and takes two values under the substitutions of G_8.

Therefore, $G=C_4$ or one of its subgroups. G cannot be a subgroup of C_4 as the subgroups of C_4 are contained in H_4. Therefore, $G=C_4$.

Hence when p and r satisfy (3), the group of (1) is C_4.

---

* a is not equal to 0 because of (3) and not both p and $r=0$.

§4. We have now proved that *the Galois group of DeMoivre's quintic for the domain $R(1)$ is either the cyclic group C_4 or the metacyclic group G_{20}.* We therefore get the following results:

I. *DeMoivre's quintic is solvable by radicals.* The solution can be effected by the well known substitution $x = y - p/5y$.

Since the group may be C_4 the equation may be reducible. Hence

II. *If the equation is reducible it must reduce to the product of a linear factor and an irreducible quartic factor.*

As an equivalent form of II, we have,

III. *DeMoivre's quintic can never have more than one rational root.*

By means of a property of metacyclic equations* we may also conclude that

IV. *All the roots of DeMoivre's quintic are real or only one of them is real.*

If the group of the equation is G_{20} the equation is metacyclic† and,

V. *Each root is a rational function of an arbitrary pair of them.*

This problem was suggested to me by Prof. L. E. Dickson and I wish to thank him for criticisms and suggestions in connection with its solution.

BIBLIOGRAPHIC ENTRIES: FIELD EXTENSIONS AND GALOIS THEORY

Except for the entry labeled MATHEMATICS MAGAZINE, the references below are to the AMERICAN MATHEMATICAL MONTHLY.

1. D. H. Lehmer, Quasi-cyclotomic polynomials, 1932, 383–389.

Discusses polynomials all of whose roots have the same modulus and whose arguments are commensurable with 2π.

2. D. H. Lehmer, A note on trigonometric algebraic numbers, 1933, 165–166.

Determines the degree over the rational numbers of $\cos 2\pi r$ and $\sin 2\pi r$, when r is rational.

3. Howard Kleiman, The determination of a Galois polynomial from its root polynomials, 1968, 54–55.

Proves that over a formally real field a Galois polynomial is uniquely determined by its root polynomials.

4. Paul B. Yale, Automorphisms of the complex numbers, MATHEMATICS MAGAZINE, 1966, 135–141.

5. Kenneth Kalmanson, A familiar constructibility criterion, 1972, 277–278.

Gives an example of a non-constructible algebraic number whose minimal polynomial has degree 4 over the rationals.

---

* Weber, *Algebra*, I, p. 620, VIII.

† Weber, *Algebra*, I, p. 618, VI.

(b)

FINITE FIELDS

MODULAR FIELDS*†

SAUNDERS MAC LANE, Harvard University

1. Introduction. The general theory of modular fields, though elementary in its presuppositions, offers an instructive cross-section of modern algebraic methods. These fields exhibit the generality of subject-matter inherent in abstract algebra, and at the same time illustrate the intimate connection between algebraic and arithmetic problems.

Modular fields arise first in number theory in the consideration of congruences with a prime modulus p. For integers a and b the ordinary definition states that

$$a \equiv b \pmod{p} \qquad \text{means that} \qquad p \text{ divides } (a - b).$$

Any integer a on division by p yields a quotient q and a remainder r,

$$a = qp + r, \qquad 0 \leqq r < p;$$

hence $a \equiv r \pmod{p}$, where the remainder r is one of the integers

$$(1) \qquad F_p\colon\ 0, 1, 2, \cdots, p - 2, p - 1.$$

Any integer is congruent to one of those in this set of p numbers.

With these numbers alone one can still carry out algebraic operations, provided one adds and multiplies these numbers in the ordinary fashion, and then reduces the answer by congruence to one of the numbers (1). For example, if $p=5$, the product $2\cdot 3=6$ should really be $2\cdot 3\equiv 6-5=1$. In this fashion one can make multiplication and addition tables for $p=5$, as shown. It is strange

+	0	1	2	3	4
0	0	1	2	3	4
1	1	2	3	4	0
2	2	3	4	0	1
3	3	4	0	1	2
4	4	0	1	2	3

·	0	1	2	3	4
0	0	0	0	0	0
1	0	1	2	3	4
2	0	2	4	1	3
3	0	3	1	4	2
4	0	4	3	2	1

that this idea has not appeared more‡ in texts on number theory, for the idea

* From AMERICAN MATHEMATICAL MONTHLY, vol. 47 (1940), pp. 259–274.

† An address delivered before the Mathematical Association of America at Columbus, Ohio, December 30, 1939.

‡ *Cf.* remarks in Weiss [**26**].

is an essentially simple one. One can introduce it by the intuitively natural algebra of the words "even" and "odd," as

$$\text{even} \cdot \text{even} = \text{even}, \quad \text{even} \cdot \text{odd} = \text{even}, \quad \text{odd} \cdot \text{odd} = \text{odd},$$
$$\text{even} + \text{even} = \text{even}, \quad \text{even} + \text{odd} = \text{odd}, \quad \text{odd} + \text{odd} = \text{even}.$$

This is just the algebra of integers modulo $p=2$.

A congruence modulo p has all the properties of an equation; congruences can be added and multiplied term by term, and the relation of congruence is reflexive, symmetric, and transitive. If the modulus p is fixed, one might just as well dub congruence "equality." Every integer is then "equal" to one of the p symbols, $0, \cdots, p-1$, and the sums and products of these symbols, so identified, give exactly the algebra of the integers modulo p, as described above.

If one objects to rebaptizing "congruence" by fiat, one may adopt the more sophisticated procedure* of replacing each remainder r modulo p by the *class* r_p of all integers $r, r+p, r+2p, \cdots$ congruent to it. Such "congruence classes" are then added and multiplied according to the rules

$$(2) \qquad r_p + s_p = (r+s)_p, \qquad r_p \cdot s_p = (rs)_p.$$

Furthermore the congruence classes r_p and s_p will be equal (*i.e.*, will contain the same elements) if and only if the integers r and s are congruent, so the desired "equality" has now been properly introduced. In any event the *integers modulo p* form a finite set of objects (1) satisfying all rules of algebra.

The presence of such arithmetic objects, which are certainly not ordinary numbers but which still obey ordinary algebra, is the reason why modern algebra is abstract. To separately discuss the algebra of numbers, then the algebra of congruence classes, then the algebra of functions, and so on would be most inefficient. Instead, theorems are better proved for any (abstractly conceived) system of objects whatever to which the basic rules of algebra apply.

These laws of algebra for a set F of objects, such as the integers modulo p, are codified as follows: For a and b in F there is uniquely defined a *sum* $a+b$ and a *product* $a \cdot b$. This product is *commutative* $[ab=ba]$ and *associative* $[a(bc)=(ab)c]$, as is also the sum. The *distributive* law $a(b+c)=ab+ac$ holds for all a, b, and c. The set F contains a zero 0 and a unit 1, with the characteristic properties

$$a+0=a=0+a, \qquad 1 \cdot a = a = a \cdot 1,$$

respectively. Finally, *subtraction* and *division* are possible, which is to say that the equations $a+x=0$ and $b \cdot y=1$ have solutions x and y in F, except when $b=0$. Any set F of elements with all these properties is called a *field*. One may say that a field is any system of elements within which addition, subtraction, multiplication, and division (excluding division by zero) can be carried out in the usual fashion.

* *Cf.* Albert [1, p. 7]; van der Waerden [27, p. 13]; or Mac Lane [17, Chapter I].

Well known fields are: (a) the set of all rational numbers; (b) the set of all real numbers; (c) the set of all complex numbers. The field (1) composed of the integers modulo p is often called the *Galois field* $GF[p]$. A *modular field* is any field containing such a $GF[p]$.

These fields $GF[p]$ are not the only finite fields. One may construct larger fields by simply adjoining to a $GF[p]$ the roots of certain algebraic equations. The process resembles the construction of the complex numbers from the field R of real numbers. Here one adjoins to R a symbol i representing a root of the equation $x^2+1=0$; the field C of all complex numbers $a+bi$ then contains everything which can be expressed rationally in terms of i and real numbers. The fact that C is generated over R by adjoining i is symbolized by $C=R(i)$. Note in particular that the polynomial x^2+1 used to generate this extension is *irreducible* over R, because it cannot be factored into polynomials of smaller degree with coefficients in R.

In similar vein consider the polynomial $f(x)=x^2+x+1$ over the field F_2 with two elements (the integers modulo 2). Neither $f(1)$ nor $f(0)$ is zero, so this polynomial $f(x)$ has no roots in F_2, hence has no linear factors, hence is irreducible over F_2. Invent a symbol u to denote a root of $f(x)=0$, so that

$$u^2 + u + 1 = 0, \qquad u^2 = -u - 1 = u + 1.$$

(Recall that $-1=+1$, modulo 2.) All higher powers of u can thereby be successively reduced to linear expressions in u. Reciprocals can be similarly reduced, so that the field generated by u contains all told just four linear expressions: 0, 1, u, $u+1$. These combine under addition and multiplication

$+$	0	1	u	$u+1$
0	0	1	u	$u+1$
1	1	0	$u+1$	u
u	u	$u+1$	0	1
$u+1$	$u+1$	u	1	0

$\cdot$	0	1	u	$u+1$
0	0	0	0	0
1	0	1	u	$u+1$
u	0	u	$u+1$	1
$u+1$	0	$u+1$	1	u

as shown in the tables. The process of obtaining this field by adjoining to the original F_2 a root u of x^2+x+1 is known as *algebraic extension* of F_2, and the resulting field $F_2(u)$ is called a Galois field of 4 elements.

For each prime p and each integral exponent n one may analogously extend the field of integers modulo p to a field consisting of exactly* p^n elements. As E. H. Moore first showed, *any* two fields with p^n elements each are algebraically indistinguishable (isomorphic). The arithmetic origin of all these finite fields is the study of algebraic integers. If $\mathfrak{p}$ is a prime ideal in a field K of algebraic numbers, then the congruences modulo this ideal behave as do ordinary congruences, and yield like them a finite field with p^n elements, where p^n is the so-called "norm" of the ideal $\mathfrak{p}$. The properties of the resulting finite fields play

* See detailed discussion of finite fields in van der Waerden [27, §31]; or Albert [1, p. 166].

an essential rôle in the class field theory and in the study of rational division algebras (Albert [2, ch. 9]).

2. Characteristics. The integers modulo p have one peculiar property. The unit 1, added p times to itself, yields $p \equiv 0 \pmod{p}$ as answer; hence

$$(3) \qquad 1 + 1 + \cdots + 1 = 0, \qquad (p \text{ summands}).$$

On multiplying this equation by any integer a, one has

$$(4) \qquad a + a + \cdots + a = 0, \qquad (p \text{ summands}),$$

in the Galois field F_p. Any field F, all of whose elements a have the property (4), is called a field of *characteristic* p, or a *modular field*. It can be shown* that any non-modular field has an infinite characteristic, in the sense that $a \neq 0$ entails $a+a+ \cdots +a \neq 0$, for any number of summands. Any finite field of p^n elements essentially contains the integers modulo p, hence satisfies (3) and therefore (4). Thus any finite field is modular.

Watch the effect of (4) on the binomial expansion,

$$(a + b)^p = a^p + pa^{p-1}b + (p(p-1)/2)a^{p-2}b^2 + \cdots + pab^{p-1} + b^p.$$

According to the genesis of this expansion, the term $pa^{p-1}b$ second on the right really represents a sum of p products $a^{p-1}b+a^{p-1}b+ \cdots +a^{p-1}b$. In a field of characteristic p this sum is zero. The other intermediate terms of the binomial expansion suffer the same fate, for each binomial coefficient $p(p-1)/2, \cdots, p$ is a multiple of the characteristic p. One has left only

$$(5) \qquad (a + b)^p = a^p + b^p, \qquad (a, b \text{ in } F \text{ of characteristic } p).$$

As S. C. Kleene has remarked, a knowledge of the case $p=2$ of this equation would corrupt freshman students of algebra!

The pth power of a product is always a product of pth powers, so the rules

$$(6) \qquad (a \pm b)^p = a^p \pm b^p, \qquad (ab)^p = a^p b^p, \qquad (a/b)^p = a^p/b^p$$

hold in any field of characteristic p. These rules state that the process of raising to a pth power leaves the operations of addition, division, *etc.*, unchanged. This process yields a correspondence

$$(7) \qquad a \longleftrightarrow a^p, \qquad (\text{from } F \text{ to } F^p),$$

which carries the field F into the field F^p composed of all pth powers from F. The correspondence is one-to-one, for the equality of two pth powers $a^p = b^p$ would entail $0 = b^p - a^p = (b-a)^p$, and hence $b = a$. To summarize, the correspondence $a \longleftrightarrow a^p$ is an isomorphism, where an *isomorphism* between two fields is defined to be any one-to-one correspondence which preserves sums and products.

Repeated application of the rules in (6) shows that the pth power of any rational expression can be computed by applying the exponent p to each term or

* *Cf.* Albert [1, p. 30]; Mac Lane [17, §21]; van der Waerden [27, §25].

factor in the expression. In particular,

$$(1+1+\cdots+1)^p = 1^p + 1^p + \cdots + 1^p = 1+1+\cdots+1 \tag{8}$$

holds in the field of integers modulo p. If we use m summands here, this is $m^p = m$. In terms of congruences this is $m^p \equiv m \pmod p$, which is the little Fermat Theorem!

3. Algebraic and transcendental extensions. Our major concern is the structure of the general modular field, finite or infinite. In the analogous case of fields of numbers it is customary to distinguish the algebraic numbers, such as $\sqrt{3}$, which satisfy some polynomial equation with rational coefficients, from the transcendental numbers (e, π), which satisfy no such equation. In general, let a given field F be contained in any larger field K. An element u of K is *algebraic* over F if u is a root of a polynomial

$$f(x) = a_n x^n + a_{n-1}x^{n-1} + a_{n-2}x^{n-2} + \cdots + a_1 x + a_0 \tag{9}$$

with coefficients a_i in F. If this equation $f(x)=0$ for u be chosen with a degree n as small as possible, the polynomial $f(x)$ is *irreducible* over F. For, a reducible $f(x)$ would have factors $f(x)=f_1(x)f_2(x)$ with coefficients in F, and u would satisfy one of the equations $f_1(x)=0$, $f_2(x)=0$, of degree smaller than n. An element u in K not algebraic over F is called *transcendental*; for u transcendental, $f(u)=0$ implies that all the coefficients in $f(x)$ are zero.

Important is not the element u in K by itself, but the field $F(u)$ which it generates. The field consists of all rational combinations of u with coefficients in F, and is called a *simple extension* of F, "algebraic" or "transcendental" according as u is algebraic or transcendental over F. This dichotomy is the root of one of the basic results found by Steinitz in his pioneering investigations of fields (Steinitz [23]): *Any modular field can be obtained by successive transcendental and algebraic extensions of a field (isomorphic to the field) of integers modulo p.*

Such extensions can be used not only to build up a given field K from a subfield F, but also to manufacture new fields from old. Given a polynomial $f(x)$ irreducible over a field F, one can concoct a symbol u for a root of this polynomial and construct therewith an algebraic extension $F(u)$ generated by the root u. In point of fact, $F(u)$ consists of elements expressible as polynomials $b_0+b_1u+\cdots+b_{n-1}u^{n-1}$, with coefficients in F and of degree less than the degree n of the given $f(x)$.

Alternatively, a variable t over a modular field gives rise to rational functions

$$\frac{g(t)}{h(t)} = \frac{b_0 + b_1 t + \cdots + b_r t^r}{c_0 + c_1 t + \cdots + c_m t^m}, \qquad (c_i,\ b_j \text{ in } F, \text{ not all } c_i = 0). \tag{10}$$

Under the usual rules for adding and multiplying such expressions, the totality of these rational functions is a field $F(t)$ which is a simple transcendental extension of F. If F is a finite field, the resulting field $F(t)$ is the simplest instance of an infinite modular field.

4. Inseparable equations. Over the transcendental extension $F(t)$ there are in turn algebraic extensions, such as that generated by a root of the polynomial $f(x)=x^p-t$. This $f(x)$ is irreducible over $F(t)$, for if it could be factored, the denominators in t could be eliminated, and we could write $x^p-t=g(x, t)h(x, t)$, with factors which are polynomials in x and t. Since the product of these two polynomials is linear in t, one of them must be linear in t, while the other cannot involve t at all! This is absurd unless one of the factors is a constant; hence $f(x)$ is indeed irreducible.

But trouble arises with the introduction of a root u for this equation $x^p-t=0$. Since this u is a pth root of t, we have a factorization

$$x^p - t = x^p - u^p = (x - u)^p, \tag{11}$$

according to the rule (6) for the pth power of a difference. This means that u is a p-fold root of x^p-t, so this irreducible polynomial has all its roots equal, and t has only one pth root.

This differs drastically from the usual situation with ordinary complex nth roots, for an irreducible polynomial $f(x)$ with *rational* coefficients can never have a multiple root. Let us trace the proof of this fact. If $f(x)$ has a complex number r as m-fold root, then $f(x)=(x-r)^m g(x)$, with $m>1$. The derivative is

$$f'(x) = (x - r)^{m-1}[mg(x) + (x - r)g'(x)]. \tag{12}$$

Since $m>1$, this insures that $f(x)$ and $f'(x)$ have a common factor $(x-r)^{m-1}$, not a constant. But the highest common factor of $f(x)$ and $f'(x)$ can be found by the euclidean algorithm, using only rational operations. This highest common factor then has rational coefficients, and its degree is at most that of $f'(x)$. It must divide $f(x)$, counter to the assumed irreducibility of that polynomial.

Can this contradiction be deduced for a polynomial $f(x)$, irreducible not over the rationals but over some modular field, and having a multiple root r in a larger field? The derivative $f'(x)$ of calculus is no longer available, but for any polynomial $f(x)$ as in (9) a "formal" derivative can still be defined as

$$f'(x) = na_n x^{n-1} + (n - 1)a_{n-1}x^{n-2} + (n - 2)a_{n-2}x^{n-3} + \cdots + a_1. \tag{13}$$

Here the coefficient ia_i of the term x^{i-1} is to denote the sum

$$ia_i = a_i + a_i + \cdots + a_i, \qquad (i \text{ summands}). \tag{14}$$

Apply this derivative to the troublesome polynomial x^p-t of (11). We find

$$(x^p - t)' = px^{p-1} = x^{p-1} + \cdots + x^{p-1} = 0, \qquad (p \text{ summands}).$$

No wonder that an argument on the H. C. F. of x^p-t and 0 runs aground! Looking back, one sees that the argument following (12) about multiple roots will work, except in such cases when $f'(x)$ vanishes.

When do all coefficients ia_i of $f'(x)$ vanish? In a modular field $ia_i=0$ means either that a_i itself is zero, or that the number i of summands, in (14), is a multiple of the characteristic p. A coefficient a_i can thus differ from zero only for terms a_ix^i with exponent $i\equiv 0 \pmod p$. The vanishing of $f'(x)$ means therefore that $f(x)$ can involve x only as powers of x^p, so that $f(x)$ has the form

$$g(x) = b_m x^{mp} + b_{m-1}x^{(m-1)p} + \cdots + b_1 x^p + b_0. \tag{15}$$

An irreducible polynomial $g(x)$ of this form must always have p-fold roots. Such a polynomial is called *inseparable* (its roots cannot be "separated" into distinct roots). Many properties of ordinary equations fail for inseparable equations.

An element u algebraic over a modular field F is called *separable* over F if the irreducible equation for u is separable (*i.e.*, has no multiple roots). Of the inseparable algebraic elements the simplest examples are pth roots which satisfy inseparable equations $x^p = a$. Consider an arbitrary inseparable element u, root of an inseparable polynomial (15) of degree mp. This polynomial involves only pth powers of its variable, so u^p is a root of an equation

$$h(y) = b_m y^m + b_{m-1}y^{m-1} + \cdots + b_1 y + b_0, \tag{16}$$

of smaller degree m. The adjunction of the root u to our field F can then be effected in two stages

$$F \to F(u^p) \to F(u^p, \sqrt[p]{u^p}) = F(u).$$

The element u^p first adjoined may still belong to an inseparable equation $h(y) = 0$; in that event the process can be reapplied to get u^{p^2} satisfying an equation of still smaller degree. The adjunction of an inseparable algebraic element to a modular field can be accomplished by adjoining successive pth roots of a suitable separable algebraic element (Steinitz [**23**]). This reduction of algebraic extensions to separable extensions followed by extensions by pth roots, indicates that the novel properties are concerned chiefly with the latter type of extension.*

5. Perfect fields. There are no inseparable algebraic elements over the field of integers modulo p, for this field already contains the pth roots of all of its elements—indeed, the Fermat Theorem, $a^p = a$, asserts that every element is its own pth root. A *perfect* field F of characteristic p is a field in which each element a has a pth root. Over such a field each pth root equation $x^p = a$ is reducible, as $x^p - a = (x - \sqrt[p]{a})^p$. More generally *any inseparable polynomial* $g(x)$ *involving only pth powers of* x *must be reducible over a perfect field*. For, each coefficient b_i of the polynomial $g(x)$ in (15) has in F a pth root $b_i^{1/p}$; according to the simple behavior of pth powers this gives a factorization

$$g(x) = (b_m^{1/p} x^m + b_{m-1}^{1/p} x^{m-1} + \cdots + b_1^{1/p} x + b_0^{1/p})^p.$$

Every finite field F is perfect, hence has no inseparable algebraic extensions. To prove this, recall the correspondence $a \longleftrightarrow a^p$ of (7), which is a one-to-one

---

* Technically, the least power $q = p^e$ such that u^q is separable over F is known as the *exponent* of u over F. The *degree* of u over F is the degree of its irreducible equation, while the degree of u^q is known as the *reduced degree* of u.

correspondence between *all* elements of F and those elements a^p which are pth powers. Since there are but a finite number of elements in F, there must be the same number of pth powers. This means that every element is a pth power.

A simple transcendental extension $F(t)$ of a modular field can never be perfect. To verify this we need only produce an element with no pth root in the field. The variable t itself is such an element, for if t had as pth root some rational function $g(t)/h(t)$ in the field, t would equal $[g(t)/h(t)]^p$, a pth power which can be calculated by the rule (6). In the notation of (10), the result is

$$(c_0^p + c_1^p t^p + \cdots + c_m^p t^{mp})t = b_0^p + b_1^p t^p + \cdots + b_r^p t^{rp},$$

an identity which clearly cannot hold good. For similar reasons a multiple transcendental extension $F(t_1, t_2, \cdots, t_n)$, consisting of all rational functions of n independent variables t_i, cannot be a perfect field.

6. Galois theory. To what extent can one generalize to modular fields the ordinary properties of fields of rational and algebraic numbers? A major topic is the Galois theory, which analyzes the solvability of a polynomial equation $f(x)=0$ over a field F. The roots $r_1, \cdots, r_n$ of this equation generate over F a *root field*

$$(17)\quad K = F(r_1, r_2, \cdots, r_n), \qquad \text{where} \qquad f(x) = (x - r_1)(x - r_2)\cdots(x - r_n);$$

the Galois Theory studies K in terms of its group of automorphisms, each of which is an isomorphism of the field K with itself, induced by a permutation of the roots r_i. Should these roots all be equal, the only such permutation is the identity, and the theory breaks down. Only if one assumes that the roots are all distinct, *i.e.*, that $f(x)$ is separable, does the standard theory of root fields hold* over a modular F.

This straightforward generalization does not suffice for irreducible *in*separable polynomials. The first process to fail is the construction of a "Galois resolvent," which is an equation with a root u in K such that all the roots r_i can be rationally expressed in terms of this single quantity u. In terms of fields, this means that the multiple algebraic extension $K=F(r_1, \cdots, r_n)$ can be represented as a simple extension $F(u)$. Over an imperfect field F there may be multiple algebraic extensions which cannot be so represented. Consider for instance the rational function field,

$$(18)\qquad F_0 = P(t_1, t_2), \qquad P \text{ perfect},$$

in two independent variables t_1 and t_2. An adjunction of pth roots will yield an extended field

$$(19)\qquad K_0 = F_0(u_1, u_2); \qquad u_1^p = t_1, \qquad u_2^p = t_2,$$

which consists of all elements expressible as polynomials

* *Cf.* Albert [1, ch. VIII]; van der Waerden [27, ch. 7]; Mac Lane [17, §68].

$$w = \sum_{i,j} a_{ij} u_1^i u_2^j = h(u_1, u_2), \qquad (i, j = 0, \cdots, p-1), \tag{20}$$

with coefficients a_{ij} in F_0. This field K_0 is not a simple extension $K_0 = F_0(w)$ for any w. For, if there were a generator w, then by the rule for pth powers,

$$w^p = \sum_{i,j} a_{ij}^p u_1^{ip} u_2^{jp} = \sum_{i,j} a_{ij}^p t_1^i t_2^j$$

is in F_0, so w is a pth root of an element of F_0. That such a single pth root could generate the field K_0 containing two independent pth roots u_1 and u_2 is unreasonable. This hunch can be substantiated by an argument on the degree* of the extension K_0 of F_0.

If a multiple extension does not have one generator, what is then the *minimum* number of generators? Miriam Becker [6] has recently found the answer. Over the particular field $P(t_1, t_2)$ of (18) it appears that *any* multiple algebraic extension can be expressed by two generators, just as in the case of the special extension K_0 of (19). The underlying reason is the presence of just two independent pth roots, $\sqrt[p]{t_1}$ and $\sqrt[p]{t_2}$, not in the field $P(t_1, t_2)$; the pth root of any other rational function $g(t_1, t_2)$ in the field can be expressed by the rule (6) in terms of these two pth roots, together with pth roots of coefficients which already lie in the perfect base field P.

Over any modular field F one calls the r pth roots $a_1^{1/p}, a_2^{1/p}, \cdots, a_r^{1/p}$ *p-independent* if no one of them can be rationally expressed in terms of F and the others. Becker proves that *any multiple algebraic extension of an imperfect field F can be generated by m elements, where m is the maximum number of independent pth roots over F.* If $m=0$, F is perfect: if $m=1$, any multiple algebraic extension is simple, as shown by Steinitz.

7. Derivatives. The solution of an ordinary equation $f(x)=0$ by radicals (if possible) proceeds in successive stages which correspond to successive fields lying between the coefficient field F and the root field K. For a separable equation the whole array of possible intermediate fields is finite—but not so for some inseparable extensions. Between the fields F_0 and K_0 of (19) lie infinitely many distinct fields $F_0((t_1+t_2^m)^{1/p})$, with $m=1, p+1, 2p+1, \cdots$. For a separable equation the fields intermediate between K and F can be put into one-to-one correspondence with the sub-groups of the Galois group of automorphisms of K over F. This certainly fails for an inseparable extension like (19), for in that case the Galois group of K_0 over F_0 consists of the identity alone and so has no proper sub-groups to correspond to intermediate fields. Specifically, the Galois group consists of all isomorphisms of K_0 with itself which leave fixed each element in the base field F_0; but an isomorphism leaving fixed the elements t_1 and t_2 of F_0

* This degree is the maximum number of elements of K_0 "linearly independent" over F_0. This maximum is p^2, for any w is linearly dependent on the p^2 elements $u_1^i u_2^j$ of (20). For a simple extension $F_0(w)$ the degree would be only p. Hence $F_0(w)$ cannot equal K_0.

must likewise leave fixed their *unique* pth roots u_1 and u_2 and hence must leave all elements of K_0 fixed.

For this description of intermediate fields by the Galois group Jacobson has found a substitute, in the special case of extensions K obtained by adjoining any number of pth roots to a modular field F, as

$$K = F(a_1^{1/p}, a_2^{1/p}, \cdots, a_n^{1/p}), \qquad \text{each } a_i \text{ in } F. \tag{21}$$

By a piece of poetic justice, his solution depends on exploiting the very formal derivatives whose misbehavior (*cf.* §4) is at the root of inseparability. For example, in the field K_0 of (19) one has two "derivative" operators D_1 and D_2, defined for the arbitrary element $w=h(u_1, u_2)$ of (20) by

$$h(u_1, u_2)D_1 = \partial h(u_1, u_2)/\partial u_1, \qquad h(u_1, u_2)D_2 = \partial h(u_1, u_2)/\partial u_2. \tag{22}$$

This time the properties of pth powers are fortunate, for $u_1^pD_1=pu_1^{p-1}=0$, as it ought to be, for $u_1^p=t_1$ is in the base field and so should have derivative 0 according to the definition (22). These derivatives can be used to characterize sub-fields of K_0; for example, the sub-field $F_0(u_1)$ consists of everything annihilated by the operator D_2 (*i.e.*, of all w with $wD_2=0$).

In general, Jacobson considers [12] all *formal differentiation operators* D which map K into itself by a correspondence $w \to wD$ which carries elements of F into zero and which obeys the usual formal rules for differentiation:

$$(v + w)D = vD + wD, \qquad (vw)\, D = v(wD) + (vD)w.$$

From any two such operators D_1 and D_2 one may construct new differentiations $D_1 \pm D_2$, D_1^p, and D_1c, for c in F. Furthermore, the commutator $[D_1, D_2]=D_1D_2 - D_2D_1$ is again a formal differentiation. This commutator satisfies the identity

$$[[D_1, D_2], D_3] + [[D_2, D_3], D_1] + [[D_3, D_1], D_2] = 0,$$

which is one of the essential postulates for a Lie algebra. The set $\mathfrak{L}$ of all differentiations is in fact a Lie algebra over the base field F. This algebra acts as a substitute for the Galois group of a field K of type (21), in the sense that *there is a one-to-one correspondence between the fields intermediate between K and F and the restricted Lie sub-algebras of the algebra $\mathfrak{L}$ of all formal differentiations of K over F.* For this purpose a *restricted* sub-algebra of $\mathfrak{L}$ is a sub-set $\mathfrak{L}'$ of $\mathfrak{L}$ which is itself a Lie algebra and which is restricted to contain D^p for each D of $\mathfrak{L}'$.

8. Algebraic geometry. A skew curve can be represented as the intersection of two surfaces, which may often be taken as cylinders

$$f(x, y) = 0, \qquad g(x, z) = 0 \tag{23}$$

with axes parallel to the z and y coördinate axes, respectively. If f and g are polynomials, the intersection of these cylinders is an algebraic curve. Alternatively, x may be viewed as a quantity transcendental over the field C of complex numbers; the polynomial equations then make the quantities y and z algebraic over the field $C(x)$ of rational functions of x. All told they give a field $C(x, y, z)$

generated by "algebraic functions" y and z of x. This field is the algebraic invariant of the curve (23). The ordinary analytic theory of these algebraic function fields can be developed, without using the geometry of the Riemann surface, if the base field C of complex numbers is replaced by a perfect modular field P or even by an imperfect one.*

In an n-dimensional euclidean space an r-dimensional algebraic manifold can be described as the set of points common to $n-r$ suitable algebraic hypersurfaces. These hypersurfaces may be taken, as in (23), in the form of "cylinders"

$$(24)\quad f_1(y_1, \cdots, y_r, y_{r+1}) = f_2(y_1, \cdots, y_r, y_{r+2}) = \cdots = f_{n-r}(y_1, \cdots, y_r, y_n) = 0,$$

where each f_i is an irreducible polynomial actually containing y_{r+i}. As coefficients in (24) we use not complex numbers but elements from a perfect modular field P. If this field P is finite, this means that we are considering a manifold in some finite affine (or projective) geometry, consisting of a finite number of "points" specified by coördinates in P. Algebraically, the symbols $y_1, \cdots, y_n$ related by (24) generate a field $K=P(y_1, \cdots, y_r, y_{r+1}, \cdots, y_n)$, consisting of all rational functions of these quantities, subject only to the rules of algebra and the special conditions (24). This field is obtained from the base field P by r successive simple extensions by the transcendentals $y_1, \cdots, y_r$, followed by $n-r$ successive algebraic extensions by the roots $y_{r+1}, \cdots, y_n$ of the polynomial equations (24). In a sense, the geometry of the manifold depends on the structure of this field.

What of the presence of inseparable equations in the definition (24) of such a manifold? Suppose, for instance, that the equation $f_1=0$ is inseparable in y_{r+1}, so that this variable appears only as a pth power. Certainly this could not simultaneously be the case for all the variables $y_1, \cdots, y_r, y_{r+1}$ in f_1, for in that event we could extract the pth root of every term in the equation $f_1=0$, thus making $f_1=(g_1)^p$, counter to the assumed irreducibility of f_1 over the perfect field P. Suppose then that y_1 is one of the variables which does not appear in $f_1(y_1, \cdots, y_r, y_{r+1})$ only as a pth power. The equation $f_1(y_1, \cdots, y_{r+1})$, which originally defined y_{r+1} inseparably over the field $P(y_1, \cdots, y_r)$, can be turned about and viewed as a definition of y_1 as a quantity *separable* and algebraic over the field $P(y_2, \cdots, y_r, y_{r+1})$, generated by the r independent transcendentals $y_2, \cdots, y_{r+1}$. A further juggling of the independent variables can then be applied to any subsequent equations of (24) which may be inseparable. Hence the result: *If a field $K=P(y_1, \cdots, y_n)$ is obtained from a perfect field P by adjoining a finite number of elements $y_1, \cdots, y_n$, one can find for K a generation $K=P(t_1, \cdots, t_r; u_1, \cdots, u_{n-r})$ involving r simple transcendental extensions by variables t_i, followed by $n-r$ separable algebraic extensions.* Whenever independent transcendents t_i in K have this property, that every element in K is *separable*

* *Cf.* general discussion of these abstract algebraic functions in Mac Lane-Nilson [19] or Schilling [21]. Especially interesting is the introduction of a Riemann Zeta function when P is finite (Hasse [9]), the peculiar behavior of the Weierstrass points whenever P is modular (Schmidt [22]), and the generalizations of Abelian functions (Schilling [20]).

and algebraic over $P(t_1, \cdots, t_r)$, we say that the $t_1, \cdots, t_r$ form a *separating transcendence basis* for K over P.

This construction of separating transcendence bases was discovered independently for different purposes: by the author, in connection with Albert's theory of pure forms (Albert [4]); by van der Waerden [28], for a new proof of the theorem that two distinct irreducible algebraic manifolds M_r and M_{n-r} in projective n-space intersect in a finite number of points, and, moreover, that the "number" of points, properly counted, is the product of the degrees of M_r and M_{n-r}.

9. Preservation of independence. The troubles of inseparable equations can be avoided whenever we find a separating transcendence basis for the field under consideration. Unfortunately this cannot always be done. Suppose, for instance, that the base field is the field $F_0 = P(t_1, t_2)$ of all rational functions of two transcendents t_1 and t_2 over a perfect field P, and construct a larger field L by adjoining first a new transcendent z and then an algebraic element u, with

$$u^p = t_1 + t_2 z^p, \qquad L = F_0(z, u). \tag{25}$$

Since the pth root u is inseparable over $F_0(z)$, this z is surely not a *separating* transcendence basis for L over F_0. The order of adjunction might have been inverted, adding u first as a transcendent to L and then z, but the equation (25) indicates that z would then be a pth root. The same trouble would always arise: one can prove that L has over F_0 *no* separating transcendence basis.* The same troublesome example arises in Krull's general ideal theory [13].

To find the reason for this absence of separability one must look at the possible independent pth roots in the base field F_0. In §6 we saw that the pth roots $\sqrt[p]{t_1}$ and $\sqrt[p]{t_2}$ were p-independent there, because neither can be expressed in terms of F_0 and the other. These pth roots are no longer p-independent in the top field L, for the defining equation (25) of that field gives an expression $\sqrt[p]{t_1} = u - z\sqrt[p]{t_2}$. This suggests that we restrict attention to those extensions L over F which *preserve p-independence*, in the sense that any set of p-independent pth roots over F remains p-independent over L. The relevance of this concept is indicated by the following alternative description: *a field L preserves p-independence over F if and only if the adjunction to F of any finite set of elements $y_1, \cdots, y_n$ from L yields a field $F(y_1, \cdots, y_n)$ which has over F a separating transcendence basis.*

This concept also makes it possible to find explicit conditions that given extensions have separating transcendence bases (Mac Lane [16]). One simply stated result is this: *If a field K has a finite separating transcendence basis over a sub-field M, then any field L between K and M also has a finite separating transcendence basis over M.* In other words, one can find a set S of independent tran-

* Even though, according to the Theorem of §8, L has over the *original* perfect field P a separating transcendence basis consisting of u, z, and t_1.

scendents in L, such that every element of L satisfies over $M(S)$ an algebraic irreducible equation without multiple roots.

10. General field towers. What can be said of the structure of arbitrarily complicated modular fields? The fields $P(y_1, \cdots, y_n)$ associated with algebraic manifolds had separating transcendence bases over a perfect field P. Does every modular field have a separating transcendence basis T over a suitable perfect sub-field?

The answer is no. A simple counterexample may be built from the extension $P(t)$ of a finite field P by a transcendental t. We saw in §5 that $P(t)$ is imperfect because t has in it no pth root. If we try to embed $P(t)$ in a larger field P' which will be perfect, we must have in P' a pth root $t^{1/p}$ and hence the whole rational function field $P(t^{1/p})$ generated by this root. In this field $t^{1/p}$ has no pth root, so we add $t^{p^{-2}}$, and so on, till we have the "tower"

$$P(t) \subset P(t^{p^{-1}}) \subset P(t^{p^{-2}}) \subset P(t^{p^{-3}}) \subset \cdots. \tag{26}$$

The field enveloping everything in this tower may be called $P(t^{p^{-\infty}})$; it consists of all elements lying in any one of the fields (26). Furthermore this sum field $P(t^{p^{-\infty}})$ is perfect, for an element in any one of the fields of (26) does have a pth root in the next field of the tower.

This perfect field $P(t^{p^{-\infty}})$ can have over P no separating transcendence basis. Any such basis would consist of a single transcendent t', which must lie in some one of the fields $P(t^{p^{-e}})$ of the tower (26). The generating element $t^{p^{-(e+1)}}$ of the next field is then a quantity inseparable over $P(t')$, so t' cannot have been the desired separating basis.

The tower (26) as written shows $P(t^{p^{-\infty}})$ generated by a transcendental extension followed by successive (inseparable) extractions of pth roots. Nevertheless each field of this tower, considered by itself, is a simple transcendental extension of P by $t^{p^{-e}}$. The whole field is thereby approximated by a tower of fields, each of which has a separating transcendence basis over the base field P, and each of which consists of pth powers of elements in the next field. F. K. Schmidt has shown that any perfect field P' has a similar "separating tower" over any one of its perfect sub-fields. He also stated without proof an analogous tower theorem for an imperfect field, but it was later shown by examples* that this general theorem could not hold. Recently F. K. Schmidt and the author have jointly [18] found a modified tower theorem: *If a modular field K is generated from a perfect sub-field P by a denumerable number of elements, then there is a sub-field L with a separating transcendence basis over P and a tower of fields $L \subset M_0 \subset M_1 \subset \cdots$ which collectively exhaust K, such that each M_i has over L a separating transcendence basis and is generated over L by pth powers from M_{i+1}.* The non-denumerable cases can then be broken down into a transfinite sequence

* *Cf.* Mac Lane [15]. Curiously enough, these examples involve a use of the modular law of lattice theory!

of denumerable steps, each of which "preserves p-independence" in the sense discussed in §9.

The separability of these field towers is essential to get polynomials with distinct roots, in order to apply an implicit function theorem.* This is used in the proof of the structure theorem for p-adic fields (*cf.* Hasse-Schmidt [**10**]). These p-adic fields are fields topologically complete with respect to a suitable norm (or "absolute value"), obtained by extending the norm for the p-adic numbers of Hensel.† These p-adic fields are not themselves modular fields, but they determine a congruence relation $a \equiv b \pmod{p}$ from which modular fields can be obtained by the standard arithmetic device.

11. Troublesome examples. The extent of our ignorance of general modular fields can be forcibly illustrated by various startling examples. The field $P(t^{p^{-\infty}})$ used to illustrate §10 was still manageable, for though it had no separating transcendence basis, it at least was itself perfect. But can there be an imperfect field K which has no separating transcendence basis over some perfect sub-field P? There is indeed such a K, for which P may even be chosen as the maximum perfect sub-field. Over a finite field P choose a countable set of indeterminates $t_1, t_2, \cdots$, and then introduce additional algebraic elements in accord with the inseparable relations

$$(27) \qquad y_1^p = t_1 + t_2 t_3^p, \qquad y_2^p = t_2 + t_3 t_4^p, \qquad y_3^p = t_3 + t_4 t_5^p, \cdots.$$

Our example is the field $K = P(t_1, t_2, \cdots; y_1, y_2, \cdots)$. Since the y's are pth roots, the t's clearly cannot form a separating transcendence basis. One might try to invert the equations (27) to define everything in terms of the basis $t_1, t_2, y_1, y_2, y_3, \cdots$, but that still leaves the pth roots such as $t_3^p = (y_1^p - t_1)/t_2$. It can be shown that no method of picking a transcendence basis for K over P will yield a basis which is separating, and this example is but a taste of the trouble possible (*cf.* [**15**], [**16**]).

12. p-Algebras. The relevance of the study of inseparable extensions to other algebraic questions is clearly illustrated by the p-algebras, which are defined‡ as linear algebras over a field F of characteristic p which have as degree some power of the characteristic. The theory of these algebras, which culminates in the theorem that every such algebra is "similar" to a cyclic algebra, depends essentially on the construction of inseparable fields contained in the algebra (in technical parlance, every p-algebra has a purely inseparable splitting field). To illustrate this, choose as the base field the field $P(t)$ of all rational functions of t with coefficients in a perfect field P. Introduce a pth root u, with $u^p = t$, and a

* The so-called Hensel-Rychlik theorem; *cf.* Albert [**1**] or Mac Lane-Nilson [**19**, §11].

† See the description in C. C. MacDuffee [**14**].

‡ *Cf.* Albert [**2**, ch. 7]; and also Jacobson [**11**], Teichmüller [**25**].

quantity v with $v^p = v + t$. The set of all sums

$$w = \sum_{i,j} a_{ij} u^i v^j, \qquad (i = 0, \cdots, p-1; j = 0, \cdots, p-1; a_{ij} \text{ in } F),$$

then forms a linear algebra of degree p over F, if one uses the multiplication table

$$u^p = t, \qquad v^p = v + t, \qquad vu = u(v+1).$$

The essential point for the theory is that this algebra contains both the inseparable extension $F(u)$ and the cyclic separable extension $F(v)$ of the base field F.

There are many further ways in which modular fields can arise in other algebraic investigations. We mention here only the use of fields of characteristic 2 in discussing Boolean algebras (Stone [**24**]), the theory of matrices over a modular field (Albert [**5**]), the definition of modular fields by special polynomials (Carlitz [**7**]), and the quasi-algebraic closure of finite fields (Chevalley [**8**]).

13. Summary. Modular fields include finite fields, Galois extensions of fields, algebraic function fields, and fields for algebraic manifolds, as well as for more bizarre types. The study of such fields is suggested by their origin in arithmetic questions about congruences, p-adic numbers, and ideal theory. On the other hand, an independent survey of their structure is indicated by the program of abstract algebra: first the development of the abstract concept ("field") in order to cover the variegated known examples, then the derivation of general theorems touching this concept, and lastly a classification of the types of systems which fall under the concept. We have seen that the straightforward generalization of the known properties of number fields is but one phase of our structure theory. There is also the investigation of characteristic new phenomena, of inseparability, of p-independence and the like, which distinguish the modular fields from the non-modular. The presence of curious examples of fields, which must at present still be given individual treatment, indicates that the present situation abounds in new questions, and that abstract algebra can very well give rise to concrete conundrums.

Bibliography

Albert, A. A., [**1**] Modern Higher Algebra, Chicago, 1937. [**2**] Structure of Algebras, American Mathematical Society Colloquium Publications, vol. XXIV, New York, 1939. [**3**] p-Algebras over a field generated by one indeterminate, Bulletin of the American Mathematical Society, vol. 43, 1937, pp. 733–736. [**4**] Quadratic null forms over a function field, Annals of Mathematics, vol. 39, 1938, pp. 494–505. [**5**] Symmetric and alternate matrices in an arbitrary field, I, Transactions of the American Mathematical Society, vol. 43, 1938, pp. 386–436.

Becker, M. F., and Mac Lane, Saunders, [**6**] The minimum number of generators for inseparable algebraic extensions, Bulletin of the American Mathematical Society, vol. 46, 1940, pp. 182–186.

Carlitz, L., [**7**] A class of polynomials, Duke Mathematical Journal, vol. 43, 1938, pp. 167–182.

Chevalley, C., [**8**] Demonstration d'une hypothèse de M. Artin, Abhandlungen aus dem mathematischen Seminar der Universität Hamburg, vol. 11, 1936, pp. 73–75.

Hasse, H., [9] Theorie der Kongruenzzetafunktionen, Sitzungsbericht der Preussischen Akademie der Wissenschaften, Berlin, 1934, pp. 250–255.

Hasse, H., and Schmidt, F. K., [10] Die Struktur diskret bewerteter Körper, Journal für die reine und angewandte Mathematik, vol. 170, 1934, pp. 4–63.

Jacobson, N., [11] p-Algebras of exponent p, Bulletin of the American Mathematical Society, vol. 43, 1937, pp. 667–670. [12] Abstract derivation and Lie algebras, Transactions of the American Mathematical Society, vol. 42, 1937, pp. 206–224.

Krull, W., [13] Beiträge zur Arithmetik kommutativer Integritätsbereiche VII, Inseparable Grundkörpererweiterung, Bemerkungen zur Körpertheorie, Mathematische Zeitschrift, vol. 45, 1939, pp. 319–334.

MacDuffee, C. C., [14] The p-adic numbers of Hensel, this MONTHLY, vol. 45, 1938, pp. 500–508.

Mac Lane, Saunders, [15] Steinitz field towers for modular fields, Transactions of the American Mathematical Society, vol. 46, 1939, pp. 23–45. [16] Modular fields I, Separating transcendence bases, Duke Mathematical Journal, vol. 5, 1939, pp. 372–393. [17] Notes on Higher Algebra, (planographed) Ann Arbor, 1939. With Schmidt, F. K., [18], Ueber inseparable Körper, forthcoming in Mathematische Zeitschrift. With Nilson, E. N., [19] Algebraic functions, (planographed) Ann Arbor, 1940.

Schilling, O. F. G., [20] Foundations of an abstract theory of Abelian functions, American Journal of Mathematics, vol. 61, 1939, pp. 59–80. [21] Modern Aspects of the Theory of Algebraic Functions, Mimeographed, Chicago, 1938.

Schmidt, F. K., [22] Zur arithmetischen Theorie der algebraischen Funktionen II, Allgemeine Theorie der Weierstrasspunkte, Mathematische Zeitschrift, vol. 45, 1939, pp. 75–97.

Steinitz, E., [23] Algebraische Theorie der Körper, Journal für die reine und angewandte Mathematik, vol. 137, 1910, pp. 167–308; also edited by R. Baer and H. Hasse, Berlin, 1930.

Stone, M. H., [24] The theory of representations for Boolean algebras, Transactions of the American Mathematical Society, vol. 40, 1936, pp. 37–111.

Teichmüller, O., [25] p-Algebren, Deutsche Mathematik, vol. 1, 1936, pp. 362–388.

Weiss, Marie J., [26] Algebra for the undergraduate, this MONTHLY, vol. 46, 1939, pp. 635–642.

van der Waerden, B. L., [27] Moderne Algebra, vol. I, First edition, Berlin, 1930 (also second edition, 1938). [28] Zur algebraischen Geometrie XIV, Schnittpunktszahlen von algebraischen Mannigfaltigkeiten, Mathematische Annalen, vol. 115, 1938, pp. 619–644.

A COMBINATORIAL PROOF OF THE EXISTENCE OF GALOIS FIELDS*

R. C. MULLIN, University of Waterloo, Canada

In a first course in combinatorial mathematics one often desires to show the existence of finite fields of all orders p^n where p is a prime and n a positive integer. One may easily show that the ring of integers modulo p is a field (usually denoted $GF(p)$) but one must show that there exists an irreducible polynomial of degree n over $GF(p)$ in order to construct the nth degree extension $GF(p^n)$. The classical proof of this fact is algebraic in nature and requires a lengthy development of the structure of polynomials over $GF(p)$. The following proof, which I believe is new, is combinatorial in nature and somewhat shorter than

* From AMERICAN MATHEMATICAL MONTHLY, vol. 71 (1964), pp. 901–902.

the classical proof. For those familiar with Polya's theorems [1] the proof can be made somewhat shorter. By considering the product $(1+x+x^2+\cdots)^n$ or some other elementary method one notes that

$$(1) \qquad (1-x)^{-n} = \sum_{r=0}^{\infty} C(n, r)x^r,$$

where $C(n, r)$ is the number of combinations of n objects taken r at a time, repetitions allowed. Having made this observation we prove the

THEOREM. *The number of irreducible monic polynomials of degree n over any finite field K is strictly positive for any positive integer n.*

Let us assume that K contains s elements. Let f_n represent the number of irreducible monic polynomials of degree n over K. Clearly $f_n \leqq s^n$, the number of monic polynomials of degree n.

Since any polynomial over a field can be factored uniquely into a set of monic irreducible factors, and since there are s^n distinct monic polynomials of degree n,

$$(2) \qquad s^n = \sum_{(\alpha)} \prod_i C(f_i, \alpha_i),$$

where the sum is taken over all nonnegative integer solutions of $\alpha_1+2\alpha_2+\cdots+k\alpha_k=n$. Also

$$(3) \qquad \prod_{n=1}^{\infty} (1-x^n)^{-f_n} = \sum_{n=0}^{\infty}\left(\sum_{(\alpha)}\prod_i C(f_i,\alpha_i)\right)x^n = \sum_{n=0}^{\infty} s^n x^n = (1-sx)^{-1};$$

hence, taking logarithms

$$(4) \qquad \begin{aligned} \sum_{n=1}^{\infty} \frac{s^n x^n}{n} &= \ln(1-sx) = \sum_{n=1}^{\infty} f_n \ln(1-x^n) \\ &= \sum_{n=1}^{\infty}\sum_{m=1}^{\infty} f_n \frac{x^{mn}}{m} = \sum_{n=1}^{\infty}\sum_{d|n} df_d \frac{x^n}{n}. \end{aligned}$$

Equating coefficients yields

$$(5) \qquad s^n = \sum_{d|n} df_d \quad \text{or} \qquad (6) \qquad nf_n = s^n - \sum_{\substack{d|n \\ d<n}} df_d.$$

Since f_d enumerates polynomials, $f_d \geqq 0$, and hence by (6) $f_n \leqq s^n/n$. Therefore

$$(7) \qquad \begin{aligned} nf_n &\geqslant s^n - \sum_{k=1}^{n-1} kf_k \geqslant s^n - \sum_{k=1}^{n-1} s^k \\ &= s^n - \frac{s^n-1}{s-1} = \frac{1}{s-1} + s^n\left(1-\frac{1}{s-1}\right) \end{aligned}$$

which is positive for $s>1$.

Since $f_k \leqq s^k$ all infinite series and products are absolutely convergent for $|x| < |s^{-1}|$. This completes the proof.

Relation (5) may be used to verify that f_n is a polynomial $f_n(s)$ of degree n in s. The identity

$$\ln(1 - sx) = \sum_{n=1}^{\infty} f_n(s) \ln(1 - x^n) \tag{4(a)}$$

may be used to show an interesting property of the polynomials $f_n(s)$. Indeed

$$\ln(1 - x^m) = \sum_{l=1}^{m} \ln(1 - e^{2i\pi l/m}x) = \sum_{n=1}^{\infty} \ln(1 - x^n) \sum_{l=1}^{m} f_n(e^{2i\pi l/m}),$$

hence

$$\sum_{l=1}^{m} f_n(e^{2i\pi l/m}) = \delta_{n,m}, \tag{8}$$

where $\delta_{n,m}$ is the familiar Kronecker delta.

Reference

1. G. Polya, Kombinatorische Anzahlbestimmungen für Gruppen, Graphen, und chemische Verbindungen, Acta Math., 68 (1937) 145, 253.

THE NUMBER OF IRREDUCIBLE POLYNOMIALS OF DEGREE n OVER GF(p)*

G. J. SIMMONS, Sandia Laboratories and University of New Mexico

Many algebra texts [1, 2, for example] introduce the number theoretic Möbius inversion formula prior to discussing the cyclotomic extensions of a field, since it provides a simple means for computing the cyclotomic polynomials. It is apparently not well known, however, that the Möbius function can also be used to derive an expression for the number of irreducible monic polynomials of degree n over GF(p). Carmichael [3] gives a lengthy calculation of this quantity, which he calls N_n, by applying the inclusion-exclusion principle to the number of prime factors of the divisors of n. The following simple derivation of N_n can either be used in a course in number theory as an example at the time the Möbius inversion formula is introduced, or in a course in algebra as a supplement to the discussion of irreducible polynomials.

We include a brief derivation of the following theorem, which appears in most algebra texts [3, p. 258; 4, Problem 45.8, p. 369, etc.] and a statement of the Möbius inversion theorem to make this note self-contained.

THEOREM 1. *The polynomial* $x^{p^n} - x$ *is the product of all monic irreducible polynomials over* GF(p) *whose degree* d *divides* n.

* From AMERICAN MATHEMATICAL MONTHLY, vol. 77 (1970), pp. 743–745.

Proof. Let $p(x)$ be a monic irreducible polynomial of degree d over GF(p). Then $p(x)$ can be used to extend GF(p) to a field K containing p^d elements. If $p(x)$ divides $x^{p^n}-x$ then K may be embedded in the splitting field GF(p^n) of $x^{p^n}-x$. Then GF(p^n) is a vector space of dimension e over K. Viewing both fields as vector spaces over GF(p) and counting dimensions, $de=n$. Conversely, if $de=n$, then K is isomorphic to the fixed field of GF(p^n) under the automorphism $x \to x^{p^e}$ since each $\alpha \in K$ satisfies $\alpha^{p^d}=\alpha$. Thus $p(x)$ has a root α in GF(p^n). But α is also a root of $x^{p^n}-x$, hence $p(x)$ divides $x^{p^n}-x$.

Since each monic irreducible polynomial of degree d has d elements of GF(p^n) as roots, and no element of GF(p^n) is a root of more than one irreducible polynomial, counting shows that each irreducible polynomial occurs at most once in the factorization of $x^{p^n}-x$. ∎

If N_d is the number of irreducible monic polynomials of degree d over GF(p), then the sum of the weighted degrees is given by

$$p^n = \sum_{d|n} dN_d \tag{1}$$

where d ranges over all divisors of n.

The *Möbius function*, $\mu(d)$, of an integer d with the prime factorization

$$d = p_1^{\alpha_1} p_2^{\alpha_2} \cdots p_k^{\alpha_k}$$

is defined by

$$\begin{aligned} \mu(1) &= 1 \\ \mu(d) &= 0 \quad \text{if any } \alpha_i > 1 \\ \mu(d) &= (-1)^k \quad \text{if } \alpha_1 = \alpha_2 = \cdots = \alpha_k = 1. \end{aligned}$$

The Möbius inversion formula for an arithmetic function is given by the following theorem [1, 2, 4]:

THEOREM 2. *Let $f(n)$ and $g(n)$ be arithmetic functions satisfying $f(n)=\sum_{d/n} g(d)$. Then $g(n)=\sum_{d/n} \mu(d)f(n/d)$, where $\mu(d)$ is the Möbius function.* ∎

THEOREM 3. *The number of irreducible monic polynomials of degree n over* GF(p), N_n, *is*

$$N_n = \frac{1}{n} \sum_{d|n} \mu(d) p^{n/d}. \tag{2}$$

Proof. Equation (2) is the Möbius inversion of equation (1). ∎

For example, N_{12} for $p=2$ is given by:

$$N_{12} = \frac{1}{12}(2^{12} - 2^6 - 2^4 + 2^2) = 335.$$

The surprising rate at which N_n increases for even small values of p is seen in the following table for $p=5$ and 7.

n \ p	5	7
2	10	21
3	40	112
4	150	588
5	624	3,360
6	2,580	19,544
7	11,160	117,648
8	48,750	720,300
9	217,000	4,483,696

By considering the representation of N_n given by equation (2) to be a p-ary number, it is easily shown that $N_n>0$ for all positive n and p. This result provides another proof that there exists a monic irreducible polynomial of each degree n over GF(p) and, consequently, that there exists a finite field with p^n elements (obtained by extending GF(p) by the roots of the monic irreducible polynomial of degree n) for each prime p and integer n.

This work was supported by the US Atomic Energy Commission.

References

1. B. L. van der Waerden, Modern Algebra, Vol. 1, Ungar, New York, 1953, pp. 113–115.
2. Paul J. McCarthy, Algebraic Extensions of Fields, Blaisdell, Waltham, Mass., 1966, pp. 39–43 and p. 69.
3. R. D. Carmichael, Introduction to the Theory of Groups of Finite Order, Dover, New York, 1956 (Reprint), pp. 259–260.
4. J. B. Fraleigh, A First Course in Abstract Algebra, Addison-Wesley, Reading, Mass. 1968.

BIBLIOGRAPHIC ENTRY: FINITE FIELDS

1. Dennis Travis, On irreducible polynomials in Galois fields, AMERICAN MATHEMATICAL MONTHLY, 1963, 1089–1090.

Another method of computing the number of irreducible polynomials of degree n over GF(p^m).

(c)

REDUCIBILITY OF POLYNOMIALS

AN INDUCTIVE PROOF OF DESCARTES' RULE OF SIGNS*

A. A. ALBERT, University of Chicago

1. Introduction. The so-called proofs of Descartes' Rule generally given in college algebras are merely verifications of special cases. The only rigorous direct proof in the literature known to me is that of L. E. Dickson's *First Course in the Theory of Equations*. This proof uses a rather complicated notation and has always seemed to me to be difficult to follow.

I have recently discovered an alteration in the proof which enables the omission of any consideration of permanences. This results in a fundamental simplification of the notation. I present it here in full detail.

2. Sequence of real numbers. Let n be a positive integer and consider a sequence

$$a_0, a_1, \cdots, a_n \tag{1}$$

of $n+1$ real numbers a_i. We shall be interested in pairs of *consecutive non-zero* terms a_i, a_j of our sequence, and shall call a_i and a_j *consecutive* either if $j=i+1$ or if $j>i+1$ but every intervening a_k, with $i<k<j$, is zero.

A *variation* in sign of our sequence is a pair of consecutive non-zero terms of opposite sign. Let us count the number of variations and designate the integer so obtained by

$$V(a_0, a_1, \cdots, a_n). \tag{2}$$

It is the number of changes of sign if we take as our first sign that of the first non-zero a_i (reading from left to right) and then pass to the right through all the signs of its non-zero terms. What then if the first and last non-zero terms have the same sign? In this case an even number of changes of sign must have taken place, $V(a_0, a_1, \cdots, a_n)$ must be even. Similarly an odd number of changes of sign will result in a first and a last non-zero term which have opposite signs. We state a form of this result as

LEMMA 1. *Suppose that $a_0a_n \neq 0$. Then $V(a_0, a_1, \cdots, a_n)$ is even or odd according as a_0a_n is positive or negative.*

From this result we shall derive immediately

* From AMERICAN MATHEMATICAL MONTHLY, vol. 50 (1943), pp. 178–180.

LEMMA 2. *Let $r_0, \cdots, r_{n-1}$ be a sequence of positive real numbers and derive a second sequence $b_0, \cdots, b_n$ from* (1) *by the definitions*

$$b_0 = a_0, \qquad b_j = a_j + r_{j-1}b_{j-1} \qquad (j = 1, \cdots, n. \tag{3}$$

Then if a_0, a_n, and b_n are all not zero, and if $V(a_0, \cdots, a_n) = V(b_0, \cdots, b_n)$, the signs of a_n and b_n are the same.

For our hypothesis and Lemma 1 imply that a_0a_n and a_0b_n have the same signs. Hence so do a_n and b_n.

If we adjoin a term a_{n+1} to our sequence we will have $V(a_0, a_1, \cdots, a_n) = V(a_0, a_1, \cdots, a_{n+1})$ either if $a_{n+1}=0$ or if the sign of a_{n+1} is the same as that of a_n. It is also clear that in all cases

$$V(a_0, a_1, \cdots, a_n) \leqq V(a_0, a_1, \cdots, a_{n+1}) \leqq V(a_0, a_1, \cdots, a_n) + 1. \tag{4}$$

We shall use these remarks in an inductive proof of our principal.

LEMMA 3. *Define the sequence $b_0, b_1, \cdots, b_n$ as in* (3). *Then $V(b_0, b_1, \cdots, b_n) \leqq V(a_0, a_1, \cdots, a_n)$, and if a_0 and a_n are not zero but $b_n=0$ we have $V(b_0, b_1, \cdots, b_n) < V(a_0, a_1, \cdots, a_n)$.*

For if $n=1$ the number of variations in the sequence a_0, a_1 is zero or one according as a_0a_1 is not or is negative. Since $b_0b_1=a_0a_1+r_0a_0^2$ we see that if a_0a_1 is non-negative so is b_0b_1, the number of variations of sign is zero in both cases. Moreover if $a_0 \neq 0$ then b_0b_1 is positive, $b_1 \neq 0$. If a_0a_1 is negative then $V(a_0, a_1) = 1 \geqq V(b_0, b_1)$ and $V(a_0, a_1) > V(b_0, b_1) = 0$ if $b_1=0$.

Assume now that our result is true for all pairs of related sequences of our type with $n=m$, and consider two sequences $a_0, a_1, \cdots, a_{m+1}$, and $b_0, b_1, \cdots, b_{m+1}$. If $a_0=0$ then $b_0=0$, $b_1=a_1+r_0a_0=a_1$ the sequences $a_1, a_2, \cdots, a_{m+1}$ and $b_1, b_2, \cdots, b_{m+1}$ have precisely the same numbers of variations in sign as our original two sequences and are related sequences for $n=m$. Our conclusion then follows from the hypothesis of our induction. If some other $a_j=0$ then $b_j=a_j+r_{j-1}b_{j-1}=r_{j-1}b_{j-1}$ has the same sign as b_{j-1} and we may delete a_j from our first sequence and b_j from our second sequence without changing the number of variations of sign in either case. Moreover the new sequences are related in the prescribed fashion since $b_{j+1}=a_{j+1}+r_jb_j=a_{j+1}+(r_{j-1}r_j)b_{j-1}$. Hence our result follows again from the hypothesis of our induction.

There remains only the case where no one of the numbers $a_0, a_1, \cdots, a_{m+1}$ is zero. We note the trivial relations

$$V(b_0, b_1, \cdots, b_m) \leqq V(a_0, a_1, \cdots, a_m) \leqq V(a_0, a_1, \cdots, a_{m+1}). \tag{5}$$

If $b_{m+1}=0$ then the hypothesis $V(b_0, b_1, \cdots, b_m) < (a_0, a_1, \cdots, a_m)$ implies that $V(b_0, b_1, \cdots, b_{m+1}) < V(a_0, a_1, \cdots, a_{m+1})$. However if $V(b_0, b_1, \cdots, b_m) = V(a_0, a_1, \cdots, a_m)$ the hypothesis of our induction implies that $b_m \neq 0$ and has the same sign as a_m by Lemma 2. But then $b_{m+1}=a_{m+1}+r_mb_m=0$ only if a_{m+1} and b_m have opposite signs. Then a_{m+1} and a_m will have opposite signs, the sequence

$a_0, a_1, \cdots, a_{m+1}$ has one more variation in sign than the sequence $a_0, a_1, \cdots, a_m$ and one more than $b_0, b_1, \cdots, b_{m+1}$ as desired. Finally let $b_{m+1} \neq 0$. From (4) if $V(b_0, b_1, \cdots, b_m) < V(a_0, a_1, \cdots, a_m)$ then we have our desired inequality. The only possibility for it not to hold is indeed when $V(b_0, b_1, \cdots, b_m) = V(a_0, b_1, \cdots, a_m)$, when a_m and a_{m+1} have the same signs, and when b_m and b_{m+1} have opposite signs. But by Lemma 2 a_m and b_m have the same signs and if $b_m b_{m+1} = b_m^2 r_m + a_{m+1} b_m$ is negative so is $a_{m+1} b_m$. Then $a_{m+1} a_m$ is negative. This proves the lemma.

3. Polynomials with real coefficients. We shall assume, as is usual, the analytic result stating that if $f(x)$ is a polynomial with real coefficients and $a < b$ then there is an odd number or an even number of real roots in the interval $a < x < b$ according as $f(a) \cdot f(b)$ is negative or positive. We then have

LEMMA 4. *Let $a_0, \cdots, a_n$ be real numbers such that $a_0 a_n \neq 0$, and $f(x) = a_0 x^n + \cdots + a_n$. Then $f(x)$ has either an odd or an even number of positive roots according as $a_0 a_n$ is negative or positive.*

For the sign of $a_0 a_n$ is the same as that of $c_n = a_0^{-2}(a_0 a_n) = a_0^{-1} a_n$. But c_n is the constant term of $\phi(x) = a_0^{-1} f(x) = x^n + c_1 x^{n-1} + \cdots + c_n$. This polynomial has the same roots as $f(x)$ and $\phi(0) = c_n$, $\phi(h) > 0$ for $h > g$ where g may actually be taken to be any number greater than all the numbers $1 + |c_j|$. Then the positive roots of $f(x)$ lie in the interval $0 < x < h$ and, by the result assumed, there is an odd number of such roots if $c_n > 0$, an even number if $c_n < 0$.

We now arrive quickly at

DESCARTES' RULE OF SIGNS. *Let $f(x) = a_0 x^n + a_1 x^{n-1} + \cdots + a_n$ have real coefficients and t be the number of positive roots of $f(x) = 0$. Then the difference*

$$V(a_0, a_1, \cdots, a_n) - t,$$

is a non-negative even integer.

We may clearly assume that a_0 and a_n are not zero. If ρ is a positive real root of $f(x)$ we may write $f(x) = (x - \rho)\phi(x)$ where $\phi(x) = b_0 x^{n-1} + \cdots + b_{n-1}$, the b_i are computed as in (3) with the $r_j = \rho$, $b_n = 0$. By Lemma 3 we have $V(b_0, \cdots, b_{n-1}) \leqq V(a_0, a_1, \cdots, a_n) - 1$. After t such steps we obtain $f(x) = (x - \rho_1) \cdots (x - \rho_t)\psi(x)$, where $\psi(x) = c_0 x^{n-t} + c^{n-t-1} + \cdots + c_{n-t}$, $0 \leqq V(c_0, c_1, \cdots, c_{n-t}) \leqq V(a_0, a_1, \cdots, a_n) - t$. Hence $V(a_0, a_1, \cdots, a_n) \geqq t$ as desired. The evenness of their difference follows from the fact that the criteria for evenness of t and $V(a_0, a_1, \cdots, a_n)$ in Lemmas 4 and 1 are the same.

AN APPLICATION OF DETERMINANTS*

HELEN SKALA, University of Massachusetts, Boston

A simple and elegant application of the theory of determinants for the beginning student is the following criterion of Sylvester, a well-known theorem of algebraic lore: let K be a field and $f(x)=a_mx^m+\cdots+a_1x+a_0$, $g(x)=b_nx^n+\cdots+b_1x+b_0$, where $a_m\neq 0\neq b_n$, be two polynomials in $K[x]$; then $f(x)$ and $g(x)$ have a nonconstant factor in $K[x]$ if and only if the determinant of the following $(m+n)\times(m+n)$ matrix A is zero:

$$A=\begin{bmatrix} a_m & a_{m-1} & \cdots & a_1 & a_0 & 0 & 0 & \cdots & 0 \\ 0 & a_m & \cdots & a_2 & a_1 & a_0 & 0 & \cdots & 0 \\ \vdots & \vdots & \cdots & \vdots & \vdots & \vdots & \vdots & & \vdots \\ 0 & 0 & \cdots & & & & & \cdots & a_0 \\ b_n & b_{n-1} & \cdots & & b_0 & 0 & & \cdots & 0 \\ 0 & b_n & \cdots & & b_1 & b_0 & & \cdots & 0 \\ \vdots & \vdots & \cdots & \vdots & \vdots & \vdots & & & \vdots \\ 0 & 0 & \cdots & & & & & \cdots & b_0 \end{bmatrix}.$$

We present a simple proof of this theorem which requires only knowledge of the fact that the determinant of the product of two matrices is the product of the determinants; no use of the theory of linear equations is needed. Set

$$B=\begin{bmatrix} x^{n+m-1} & 0 & 0 & \cdot & \cdot & \cdot & 0 \\ x^{n+m-2} & 1 & 0 & \cdot & \cdot & \cdot & 0 \\ x^{n+m-3} & 0 & 1 & \cdot & \cdot & \cdot & 0 \\ \vdots & \vdots & \vdots & & & \vdots & \vdots \\ x & 0 & 0 & \cdot & \cdot & 1 & 0 \\ 1 & 0 & 0 & \cdot & \cdot & \cdot & 1 \end{bmatrix}.$$

Then $|B|=x^{n+m-1}$ and

$$|AB| = |A|\,x^{n+m-1} = \begin{vmatrix} x^{n-1}f(x) & a_{m-1} & a_{m-2} & \cdots & 0 \\ x^{n-2}f(x) & a_m & a_{m-1} & \cdots & 0 \\ \vdots & \vdots & \vdots & & \vdots \\ f(x) & \cdot & \cdot & \cdots & a_0 \\ x^{m-1}g(x) & b_{n-1} & b_{n-2} & \cdots & 0 \\ \vdots & \vdots & \vdots & & \vdots \\ g(x) & \cdot & \cdot & \cdots & b_0 \end{vmatrix} = f(x)h(x)+g(x)k(x),$$

where $h(x)$ and $k(x)$ are polynomials in $K[x]$ of degree at most $n-1$ and $m-1$, respectively, calculated by expanding $|AB|$ by the first column.

If $f(x)$ and $g(x)$ have a nonconstant factor $r(x)$, then

$$|A|\,x^{n+m-1} = f(x)h(x)+g(x)k(x) = r(x)q(x),$$

* From AMERICAN MATHEMATICAL MONTHLY, vol. 78 (1971), pp. 889–890.

for some polynomial $q(x)$. If $q(x)=0$, then clearly $|A|=0$. If $q(x)\neq 0$, then $r(x)$ is a multiple of some power of x. But since $r(x)$ is a factor of both $f(x)$ and $g(x)$, both a_0 and b_0 must be zero, whence the last column of A consists of zeros and again $|A|=0$.

Conversely, suppose $|A|=0$. Then $f(x)h(x)=-g(x)k(x)$. Factoring both sides of this equality into irreducible factors over K we must obtain the same factors, and hence all factors of $f(x)$ must divide either $g(x)$ or $k(x)$. But since $k(x)$ is of at most degree $m-1$, not all factors of $f(x)$ can divide $k(x)$, hence $f(x)$ and $g(x)$ have a common factor.

BIBLIOGRAPHIC ENTRIES: REDUCIBILITY OF POLYNOMIALS

Except for the entry labeled MATHEMATICS MAGAZINE, the references below are to the AMERICAN MATHEMATICAL MONTHLY.

1. A. C. Burnham, On some cases when the quintic is solvable by elementary methods, 1898, 99–101.

(Self-explanatory)

2. H. L. Dorwart, Irreducibility of polynomials, 1935, 369–381.

Gives a survey of known irreducibility criteria.

3. M. S. Klamkin and D. J. Newman, On the number of distinct zeroes of polynomials, 1959, 494–96.

Proves that if $P(x)$ is of degree n, $Q(x)$ of degree m, $n>m\geq 0$, then there are $n-m+1$ distinct x's for which $P(x)=0$ or $P(x)=Q(x)$. Uses this fact to obtain results on polynomial diophantine equations.

4. P. V. Krishnaiah, A simple proof of Descartes' rule of signs, MATHEMATICS MAGAZINE, 1963, 190.

5. D. B. Lloyd, Reducibility of polynomials of odd degree, 1968, 1081–1084.

Proves that if $f(x)$ is a polynomial of odd degree with rational coefficients and roots $\theta_1,\ldots,\theta_n$, and if $S_N(x)=\prod_{i,j}(x-(\theta_i+\theta_j))$ has a rational root, then $f(x)$ is reducible over the rationals.

6. D. B. Lloyd, Reducibility of polynomials of odd degree (II), 1969, 919–921.

Proves the multiplicative analogue of the previous theorem.

(d)

VALUATION THEORY

THE p-ADIC NUMBERS OF HENSEL*†

C. C. MacDUFFEE, University of Wisconsin

1. Introduction. One cannot blame a respectable mathematician for looking twice at the equation

$$-1 = 4 + 4\cdot 5 + 4\cdot 5^2 + 4\cdot 5^3 + \cdots.$$

However, if we add 1 to both sides of this equation, we have

$$\begin{aligned} 0 &= 5 + 4\cdot 5 + 4\cdot 5^2 + 4\cdot 5^3 + \cdots \\ &= 0 + 5\cdot 5 + 4\cdot 5^2 + 4\cdot 5^3 + \cdots \\ &= 0 + \quad 0 + 5\cdot 5^2 + 4\cdot 5^3 + \cdots \\ &= 0 + \quad 0 + \quad 0 + 5\cdot 5^3 + \cdots \\ &= 0 + \quad 0 + \quad 0 + \quad 0 + \cdots \end{aligned}$$

with 0's as far out as we care to carry it.

It may also seem a trifle strange to write

$$2/3 = 4 + 1\cdot 5 + 3\cdot 5^2 + 1\cdot 5^3 + 3\cdot 5^4 + \cdots,$$

where the coefficients beyond the first are alternately 1 and 3. Yet multiplying by 3 gives

$$\begin{aligned} 2 &= 12 + 3\cdot 5 + 9\cdot 5^2 + 3\cdot 5^3 + 9\cdot 5^4 + \cdots \\ &= \ 2 + \quad 0 + \quad 0 + \quad 0 + \quad 0 + \cdots. \end{aligned}$$

Furthermore

$$\sqrt{7} = 1 + 1\cdot 3 + 1\cdot 3^2 + 0\cdot 3^3 + 2\cdot 3^4 + \cdots.$$

For if we square this series, retaining only terms whose exponents are $\leqq 4$, we have

$$\begin{aligned} 7 &= 1 + 2\cdot 3 + 3\cdot 3^2 + 2\cdot 3^3 + 5\cdot 3^4 + \cdots \\ &= 1 + 2\cdot 3 + \quad 0 + \quad 0 + \quad 0 + \cdots. \end{aligned}$$

2. Justification of the p-adic numbers. No one will deny that the above examples put a heavy strain on our earlier conceptions of the terms *equality* and

* From American Mathematical Monthly, vol. 45 (1938), pp. 500–508.

† Presented for the Slaught Memorial Volume of the Monthly.

convergence. It is obvious that the statement

$$-1 = 4 + 4\cdot 5 + 4\cdot 5^2 + 4\cdot 5^3 + \cdots$$

is absurd if ordinary convergence is intended. The whole point to Hensel's theory is that this is not ordinary convergence, but a new type of convergence which, from the point of view of abstract algebra, is equally worthy of the name.

A relation of equality for a mathematical system Σ is defined as follows. Let a, b, and c be elements of Σ. Then

1. Either $a=b$ or $a\neq b$ (Determinative property).
2. $a=a$ (Reflexive).
3. If $a=b$, then $b=a$ (Symmetric).
4. If $a=b$ and $b=c$, then $a=c$ (Transitive).

These four properties of equality are all that are needed in mathematics, and consequently constitute an abstract formulation of the concept.

When Hensel* introduced the p-adic numbers, his treatment was somewhat informal, but he had a perfectly sound feeling for what he was doing. The present vogue is to introduce the p-adic numbers by a method due to Kürschák,† similar to the well known development of the real numbers by Cauchy sequences.‡

Let $a, b, \cdots$ be rational numbers. A function ϕ is called a *valuation* if

1. $\phi(a)$ is a positive number or 0,
2. $\phi(a) > 0$ for $a\neq 0$, $\phi(0)=0$,
3. $\phi(ab)=\phi(a)\cdot\phi(b)$,
4. $\phi(a+b)\leqq\phi(a)+\phi(b)$.

From (3) with $b=1$, we have $\phi(1)=1$. From (3) with $a=b=-1$, we have $\phi(-1)=1$. Then, with $a=-1$, we have $\phi(-b)=\phi(b)$.

Clearly ordinary absolute value, $\phi(a)=|a|$, is a valuation. Furthermore, the four properties listed above constitute an abstract formulation of the concept of absolute value in the sense that only these properties are needed for the development of the real numbers from the rational numbers by the method of regular sequences.

We recall that the ordinary integers or whole numbers $0, \pm 1, \pm 2, \pm 3, \cdots$ are called the *rational integers*, to distinguish them from algebraic integers such as $-\frac{1}{2}-\frac{1}{2}\sqrt{-3}$ which are not rational. A *rational prime* such as $\pm 2, \pm 3, \pm 5, \pm 7 \cdots$ is a rational integer neither 0 nor ± 1 such that, if it is resolved into a product of two rational integral factors, one of the factors must be 1 or -1. Two integers are *relatively prime*, or *prime to each other*, if their only common divisors are ± 1.

Let p be a fixed rational prime. Every rational number $a\neq 0$ is uniquely expressible in the form

$$a = (r/s)p^n, \qquad s > 0,$$

* K. Hensel, Theorie der algebraischen Zahlen, Teubner, 1908.

† J. Kürschák, Journal fur die reine und angewandte Mathematik, vol. 142, 1913, pp. 211–253.

‡ See B. L. van der Waerden, Moderne Algebra, 2nd ed. I, Springer 1937, p. 221.

where r and s are rational integers prime to each other and to p, and n is a rational integer. We define

$$\phi(a) = p^{-n}, \qquad a \neq 0, \qquad \phi(0) = 0.$$

The function $\phi(a)$ is a valuation for the rational field.

Properties (1) and (2) are evident. If

$$a = (r_1/s_1)p^m, \qquad b = (r_2/s_2)p^n,$$

where r_1, s_1, r_2, and s_2 are prime to p, then

$$ab = (r_1r_2/s_1s_2)p^{m+n},$$

where r_1r_2 and s_1s_2 are prime to p. Hence

$$\phi(ab) = p^{-m-n} = \phi(a)\cdot\phi(b).$$

Without loss of generality assume that $m \leqq n$. Then

$$a + b = \frac{r_1s_2 + r_2s_1p^{n-m}}{s_1s_2}\, p^m,$$

where s_1s_2 is prime to p, so that

$$\phi(a + b) \leqq p^{-m} = \phi(a),$$
$$\phi(a + b) \leqq \phi(a) + \phi(b).$$

Let p be a fixed prime, and let ϕ be defined relative to p. A sequence

$$\{a_i\} = (a_1, a_2, a_3, \cdots, a_i, \cdots)$$

of rational numbers is called *regular* if for every positive rational number ϵ there is a positive integer n_ϵ such that

$$\phi(a_i - a_j) < \epsilon \qquad i, j > n_\epsilon.$$

Denote by Ω_p the set of all regular sequences $\{a_i\}$. Two such sequences are defined to be *equal* if, for every ϵ, there is an n_ϵ such that

$$\phi(a_i - b_i) < \epsilon \qquad i > n_\epsilon.$$

Equality as defined above is determinative, reflexive, symmetric, and transitive.

The determinative property is evident. The reflexive property follows from the definition of regularity. Symmetry follows from the fact that $\phi(-a) = \phi(a)$. Transitivity follows from the "triangle property" $\phi(a+b) \leqq \phi(a) + \phi(b)$.

The theory now proceeds as in the usual treatment of real numbers as regular sequences. We define

$$\{a_i\} + \{b_i\} = \{a_i + b_i\}, \qquad \{a_i\}\cdot\{b_i\} = \{a_ib_i\}.$$

The sum and product of regular sequences are regular. The set Ω_p of all regular sequences, with equality, addition, and multiplication as we have defined them, is a field of characteristic zero—that is, it is a field which contains a subfield isomorphic with the rational field. Indeed, two fields Ω_p and Ω_q where p and q are distinct primes are non-isomorphic so that we obtain infinitely many essentially different fields, each, of course, different from the real field. But like the real field every Ω_p is perfect—that is, incapable of further extension by means of regular sequences based on a valuation which extends the valuation by which Ω_p was defined.

It is one of the standard procedures in analysis to show that every real number can be represented as an infinite decimal—that is, a series of the type

$$a_{-\nu}\left(\frac{1}{10}\right)^{-\nu} + \cdots + a_0 + a_1\left(\frac{1}{10}\right) + a_2\left(\frac{1}{10}\right)^2 + \cdots$$

or the negative of such a series, with $0 \leqq a_i < 10$. This process can be carried over intact to the p-adic fields. It may be proved that every regular p-adic sequence is equal in the p-adic sense to a sequence

$$p^{-\nu}\{d_i\}, \qquad \{d_i\} = (a_0,\ a_0 + a_1 p,\ a_0 + a_1 p + a_2 p^2, \cdots), \qquad 0 \leqq a_i < p.$$

That is, every p-adic number may be represented by a power series

$$a_{-\nu}p^{-\nu} + \cdots + a_0 + a_1 p + a_2 p^2 + \cdots, \qquad 0 \leqq a_i < p.$$

Just as the infinite decimal is automatically convergent in the Cauchy sense, so is the above series automatically convergent in the p-adic sense.

It is to be emphasized that for every rational prime p there is a field Ω_p quite comparable with the field of real numbers, but not isomorphic with it, nor with any other Ω_p. The field Ω_2 is the field of all *diadic* numbers, Ω_3 of all *triadic* numbers, Ω_5 of all *pentadic* numbers, etc. Once the field has been selected, all calculations remain in this field. We cannot add or multiply a triadic number and a pentadic number, for instance.

3. Solution of equations. Now that we understand the meaning of a statement such as

$$\sqrt{7} = 1 + 1\cdot 3 + 1\cdot 3^2 + 0\cdot 3^3 + 2\cdot 3^4 + \cdots,$$

it remains to show how any desired number of terms of the expansion can be derived.

First we note that every p-adic number

$$a_{-\nu}p^{-\nu} + \cdots + a_0 + a_1 p + a_2 p^2 + \cdots, \qquad 0 \leqq a_i < p,$$

is the sum of a rational number

$$a_{-\nu}p^{-\nu} + \cdots + a_{-1}p^{-1}$$

and a number

$$a_0 + a_1 p + a_2 p^2 + \cdots$$

having no negative exponents, which is called an *integral p-adic number*, or a *p-adic integer*.

Two p-adic integers

$$\alpha = a_0 + a_1 p + a_2 p^2 + \cdots, \qquad 0 \leqq a_i < p,$$
$$\beta = b_0 + b_1 p + b_2 p^2 + \cdots, \qquad 0 \leqq b_i < p,$$

are equal in the p-adic sense, according to the definition of §2, if for every $\epsilon > 0$ there is an n_ϵ such that for $i > n_\epsilon$,

$$\phi((a_0 - b_0) + (a_1 - b_1)p + (a_2 - b_2)p^2 + \cdots + (a_i - b_i)p^i) < \epsilon.$$

Suppose that $a_0 = b_0$, $a_1 = b_1$, $\cdots$, and that k is the first integer for which $a_k \neq b_k$, so that $a_k - b_k$ is prime to p. Then

$$\phi((a_k - b_k)p^k + (a_{k+1} - b_{k+1})p^{k+1} + \cdots + (a_i - b_i)p^i) = 1/p^k,$$

so that if we take $\epsilon < 1/p^k$, the condition that $\alpha = \beta$ is not met. Thus *if* $\alpha = \beta$, *corresponding coefficients are equal.* The converse is evident.

Now if

$$\alpha = a_0 + a_1 p + a_2 p^2 + \cdots, \qquad 0 \leqq a_i < p,$$

then clearly

$$\alpha \equiv a_0 \pmod{p},$$
$$\alpha \equiv a_0 + a_1 p \pmod{p^2},$$
$$\alpha \equiv a_0 + a_1 p + a_2 p^2 \pmod{p^3},$$
$$\cdots\cdots\cdots\cdots\cdots\cdots\cdots$$
$$\alpha \equiv a_0 + a_1 p + \cdots + a_{i-1} p^{i-1} \pmod{p^i}.$$

Thus the coefficients $a_0, a_1, a_2, \cdots$, in the expansion of α can be successively determined from the residues of α modulo p^i, $i = 1, 2, \cdots$. In other words, $\alpha = \beta$ *if and only if*

$$\alpha \equiv \beta \pmod{p^i}$$

for every positive integer i.

Let $f(x)$ be a polynomial with rational integral coefficients, and let p be a fixed rational prime. We wish to find out if $f(x) = 0$ has a solution α in Ω_p, and to determine an arbitrary number of its coefficients.

First, suppose that $f(x) = 0$ has an integral p-adic solution, *e.g.*,

$$\alpha = a_0 + a_1 p + a_2 p^2 + a_3 p^3 + \cdots, \qquad 0 \leqq a_i < p.$$

Denote

$$\alpha_{n-1} = a_0 + a_1 p + a_2 p^2 + \cdots + a_{n-1} p^{n-1}.$$

Thus α is a solution of $f(x) = 0$ in Ω_p if and only if

$$f(\alpha) \equiv 0 \pmod{p^i} \qquad (i = 1, 2, 3, \cdots);$$

that is to say, if and only if each of the infinitely many congruences

$$f(\alpha_0) \equiv 0 \pmod{p},$$
$$f(\alpha_1) \equiv 0 \pmod{p^2},$$
$$f(\alpha_2) \equiv 0 \pmod{p^3},$$
$$\cdots\cdots\cdots$$
$$f(\alpha_{n-1}) \equiv 0 \pmod{p^n},$$
$$\cdots\cdots\cdots$$

holds.

The problem is now reduced to a familiar one in congruences. Whether $f(x) \equiv 0 \pmod{p}$ has a solution or not is usually best determined by trial. If there is a solution, there is a solution α_0, $0 \leqq \alpha_0 < p$.

Now there is a well known step-by-step process for finding a solution (mod p^{n+1}) when a solution (mod p^n) is known.* Suppose that

$$\alpha_{n-1} = a_0 + a_1 p + a_2 p^2 + \cdots + a_{n-1}p^{n-1}, \qquad f(\alpha_{n-1}) \equiv 0 \pmod{p^n}.$$

We wish to find a_n so that

$$\alpha_n = \alpha_{n-1} + a_n p^n, \qquad f(\alpha_n) \equiv 0 \pmod{p^{n+1}}.$$

By the binomial theorem,

$$\begin{aligned} f(\alpha_n) &= f(\alpha_{n-1}) + a_n f'(\alpha_{n-1})p^n + \cdots \\ &\equiv f(\alpha_{n-1}) + a_n f'(\alpha_{n-1})p^n \pmod{p^{n+1}}, \end{aligned}$$

where f' denotes the derivative. Since $f(\alpha_{n-1}) \equiv 0 \pmod{p^n}$, there exists an integer h_{n-1} such that

$$f(\alpha_{n-1}) \equiv h_{n-1}p^n \pmod{p^{n+1}}, \qquad 0 \leqq h_{n-1} < p.$$

Hence a_n can be determined from the congruence

$$a_n f'(\alpha_{n-1})p^n + h_{n-1}p^n \equiv 0 \pmod{p^{n+1}},$$

which is equivalent to the congruence

$$a_n f'(\alpha_{n-1}) + h_{n-1} \equiv 0 \pmod{p}. \tag{1}$$

Clearly a_n will exist unless $f'(\alpha_{n-1}) \equiv 0$, $h_{n-1} \not\equiv 0 \pmod{p}$. If it exists, it can be chosen in the interval $0 \leqq a_n < p$.

In order that

$$x^2 = 7$$

be solvable in triadic numbers, it is first necessary that 7 be a quadratic residue†

* L. E. Dickson, Introduction to the theory of numbers, University of Chicago Press, 1929, p. 16, ex. 4.

† Dickson, *l.c.*, p. 30.

modulo 3. This condition is met, for both 1 and 2 are solutions of $x^2 \equiv 7 \pmod 3$. Let us take the first solution. Then $\alpha_0 = a_0 = 1$.

$$f(x) = x^2 - 7, \qquad f(\alpha_0) = -6 \equiv 3 \pmod 9, \qquad h_0 \equiv 1 \pmod 3,$$
$$f'(x) = 2x, \qquad f'(\alpha_0) \equiv 2 \pmod 3.$$

Then (1) becomes

$$2a_1 + 1 \equiv 0 \pmod 3,$$

which has the solution $a_1 = 1$. Thus $\alpha_1 = 1 + 1\cdot 3 = 4$ is a solution of $x^2 \equiv 7 \pmod 9$.

The second step gives $\alpha_1 = 4$,

$$f(\alpha_1) = 9, \qquad h_1 = 1, \qquad f'(\alpha_1) = 8 \equiv 2 \pmod 3,$$
$$2a_2 + 1 \equiv 0 \pmod 3, \qquad a_2 = 1, \qquad \alpha_2 = 1 + 1\cdot 3 + 1\cdot 3^2.$$

Two more steps give

$$\alpha_4 = 1 + 1\cdot 3 + 1\cdot 3^2 + 0\cdot 3^3 + 2\cdot 3^4,$$

which was checked in §1.

The other value of a_0, namely 2, is the first term of another triadic solution,

$$\beta = 2 + 1\cdot 3 + 1\cdot 3^2 + 2\cdot 3^3 + 0\cdot 3^4 + \cdots.$$

There are no other triadic solutions of $x^2 = 7$.

The equation

$$x^2 + x + 1 = 0,$$

whose roots in the complex field are not real has no solution in pentadic numbers, since

$$x^2 + x + 1 = 0 \pmod 5$$

has no solution. However, it has two heptadic solutions,

$$\alpha = 2 + 4\cdot 7 + 6\cdot 7^2 + \cdots,$$
$$\beta = 4 + 2\cdot 7 + 0\cdot 7^2 + \cdots$$

with the usual relations $\alpha^2 = \beta$ and $\beta^2 = \alpha$. It is incorrect to think of these solutions as being complex numbers—they belong to the field Ω_7.

So far we have looked only for integral p-adic solutions of $f(x) = 0$. But if the leading coefficient of $f(x)$ is divisible by p while not all of the other coefficients are divisible by p, $f(x) = 0$ may have a p-adic solution which is not integral. But this situation involves no difficulty, for a simple transformation reduces this case to the preceding.

Consider the equation

$$9x^2 = 7, \qquad \Omega_3.$$

This has no integral triadic solution, since

$$9x^2 - 7 \equiv 0 \pmod 3)$$

has no solution. But a transformation $3x = y$ yields an equation $y^2 = 7$ which we have solved in Ω_3. Then the given equation has as solutions the fractional triadic numbers

$$\alpha = 3^{-1} + 1 + 1\cdot 3 + 0\cdot 3^2 + 2\cdot 3^3 + \cdots,$$
$$\beta = 2\cdot 3^{-1} + 1 + 1\cdot 3 + 2\cdot 3^2 + 0\cdot 3^3 + \cdots.$$

4. Rational numbers in Ω_p. The close analogy of the p-adic series with infinite decimals is well exemplified in the behavior of the rational numbers. A rational number r/s with r prime to s can be expressed as a finite decimal if and only if every prime factor of s is 2 or 5. A positive rational number r/s with r prime to s can be expressed by a finite p-adic series if and only if s is a power of p.

A decimal is finite or periodic if and only if it is equal to a rational number. Analogously

A p-adic series is finite or periodic if and only if it is equal to a rational number.

It will be sufficiently general to consider the series

$$\alpha = A + p^k B + p^{k+l} B + p^{k+2l} B + \cdots,$$

where

$$A = a_0 + a_1 p + \cdots + a_{k-1} p^{k-1}, \qquad 0 \leqq a_i < p,$$
$$B = b_0 + b_1 p + \cdots + b_{l-1} p^{l-1}, \qquad 0 \leqq b_i < p.$$

We shall call B the *period* of α. Then

$$\alpha - A = p^k B + p^l [p^k B + p^{k+l} B + \cdots] = p^k B + p^l [\alpha - A];$$

that is,

$$\alpha = A + \frac{p^k B}{1 - p^l},$$

which is clearly rational.

To prove the converse, first suppose that $\alpha = r/s$ is a negative proper rational fraction, r prime to s, s prime to p and positive. There exist positive integers l such that

$$p^l \equiv 1 \pmod{s}$$

by Euler's theorem. Let l be the smallest such integer—that is, l is the exponent to which p belongs* modulo s. Let

$$1 - p^l = ms, \qquad m < 0, \qquad mr > 0.$$

Then

$$\alpha = r/s = mr/(1 - p^l).$$

* Dickson, *l.c.*, p. 16.

Since α is proper, mr is expressible in the form

$$mr = B = b_0 + b_1 p + \cdots + b_{l-1}p^{l-1}, \qquad 0 \leqq b_i < p.$$

Then

$$\alpha = B + p^l B + p^{2l} B + \cdots$$

is periodic.

If α is positive, it can be written as the sum of a polynomial in p and a negative proper fraction. The development of $-\alpha$ can be obtained by subtracting the development of α from

$$0 = p\cdot 1 + (p-1)\cdot p + (p-1)\cdot p^2 + (p-1)\cdot p^3 + \cdots.$$

Neither of these operations will destroy the eventual periodicity of the series.

The methods of this paragraph are quicker and more effective for rational numbers than the more general method of §3.

5. Generalizations. The ideas which were disclosed in the development of the p-adic numbers have inspired much modern research. Hensel* himself extended the theory far beyond the simple p-adic fields: to g-adic rings where g is not a prime, to p-adic extensions of algebraic fields, and to functions over such fields and rings. The concepts of valuation and p-adic extension are of great importance in the modern theory of linear algebras,† and their ramifications are still being explored.

ON DIVIDING A SQUARE INTO TRIANGLES‡

PAUL MONSKY, Brandeis University and Kyoto University

Sometime ago in this MONTHLY, Fred Richman and John Thomas [1] asked the following puzzling question:

Can a square **S** *be divided into an odd number of nonoverlapping triangles* T_i, *all of the same area?*

In [2], the answer was shown to be no, provided $S = [0, 1] \times [0, 1]$ and the coordinates of the vertices of the T_i are rational numbers with odd denominators. In this note we shall show that the answer is always no. In fact we shall prove the following more general result.

Suppose that $S = [0, 1] \times [0, 1]$ *is divided into* m *nonoverlapping triangles* T_i; *let* $a_i =$ area T_i. *Then there is a polynomial* f *with integer co-efficients such that* $f(a_1, \cdots, a_m) = 1/2$.

There are two parts to the proof: one combinatorial, the other valuation theoretic. The combinatorial argument generalizes an argument made in [2]. By itself it may be made to prove the desired result when the vertices of the T_i all have rational coordinates. But to handle the case of arbitrary vertices, it

* K. Hensel, Zahlentheorie, Berlin 1913. Mathematische Zeitschrift, vol. 2, 1918, pp. 433–452.

† Deuring, Algebren, Ergebnisse der Mathematik, vol. 4, Springer, 1935, p. 99.

‡ From AMERICAN MATHEMATICAL MONTHLY, vol. 77 (1970), pp. 161–164.

becomes necessary to argue with "congruences mod 2 in the reals." This is where valuation theory comes in; we make use of absolute values on the reals extending the 2-adic absolute value of the rationals. In the course of the proof the theorem of the extension of valuations plays a remarkable and unexpected role.

We begin with the combinatorial argument. Let R be a region in the plane bounded by a simple closed polygon. Suppose R is divided into m nonoverlapping triangles T_i. By a *vertex* we shall mean a vertex of some T_i, by a *side* a side of some T_i or of R. Two vertices are called *adjacent* if they are in the same side and the line segment joining them contains no other vertices. A *basic segment* is a line segment joining two adjacent vertices. Note that the boundary of each T_i is a union of nonoverlapping basic segments; the same is true of the boundary of R. Suppose now that the vertices are divided into three disjoint sets, $\mathfrak{A}$, $\mathfrak{B}$, and $\mathfrak{C}$. We shall say that a side or a basic segment is of *type* $\mathfrak{A}\mathfrak{B}$ if it has one endpoint in $\mathfrak{A}$ and one in $\mathfrak{B}$.

Lemma. *Suppose that no side contains vertices of all three types and that R has an odd number of sides of type $\mathfrak{A}\mathfrak{B}$. Then some T_i has vertices of all three types.*

To prove the lemma note the following. A side of type $\mathfrak{A}\mathfrak{B}$ contains an odd number of basic segments of type $\mathfrak{A}\mathfrak{B}$, while a side not of type $\mathfrak{A}\mathfrak{B}$ contains an even number of basic segments of type $\mathfrak{A}\mathfrak{B}$. (Use the fact that a side contains vertices of at most two types to prove this.) Suppose that no T_i has vertices of all three types. Then each T_i has either 0 or 2 sides of type $\mathfrak{A}\mathfrak{B}$. Hence the boundary of T_i contains an even number of basic segments of type $\mathfrak{A}\mathfrak{B}$. Similarly, the boundary of R contains an odd number of basic segments of type $\mathfrak{A}\mathfrak{B}$. But this is impossible; in an obvious sense the boundary of R is congruent to the sum of the boundaries of the T_i modulo 2.

We now come to the valuation theoretic part of the proof and need to introduce some further terminology. Let K be a field. By an *ultranorm* (sometimes called a *non-Archimidean absolute value*) on K we mean a function $\| \;\|$ from K to the nonnegative real numbers satisfying:

$$\|xy\| = \|x\| \cdot \|y\| \tag{1}$$

$$\|x + y\| \leqq \max(\|x\|, \|y\|) \tag{2}$$

$$\|x\| = 0 \Leftrightarrow x = 0. \tag{3}$$

We can easily prove that $\|1\| = \|-1\| = 1$, and that equality holds in equation (2) unless $\|x\| = \|y\|$.

As an example let K be the field of rational numbers. Any $x \neq 0$ in K may be written as $2^t(r/s)$, where r and s are odd integers and t is an integer. Set $\|x\| = (1/2)^t$. In this way we get an ultranorm on the rationals in which $\|2\| < 1$. This ultranorm is known as the *2-adic absolute value*. The more general fact that we shall need is this: *There is an ultranorm on the field of real numbers* (*or more generally on any extension of the rational numbers*) *such that* $\|2\| < 1$. This follows from the theorem of the extension of valuations whose proof may be found in many places; for example see [3].

Granting the above facts we may argue as follows. Choose an ultranorm on the reals for which $\|2\|<1$. Divide the points of the plane into three sets in the following way:

(1) (x, y) is in $\mathcal{A}$ if $\|x\| < 1$ and $\|y\| < 1$,

(2) (x, y) is in $\mathcal{B}$ if $\|x\| \geqq 1$ and $\|x\| \geqq \|y\|$,

(3) (x, y) is in $\mathcal{C}$ if $\|y\| \geqq 1$ and $\|y\| > \|x\|$.

Suppose now that $P=(x, y)$ and $P'=(x', y')$ are points and that P' is a translate of P by a point of type $\mathcal{A}$; in other words, that both $\|x'-x\|<1$ and $\|y'-y\|<1$. *Then P and P' have the same type.* If P is of type $\mathcal{A}$ this is obvious. If P is of type $\mathcal{B}$, then $\|x'\|=\|x\|\geqq 1$, while $\|y'\|\leqq \max(1, \|y\|)\leqq\|x\|=\|x'\|$; so P' is of type $\mathcal{B}$ too. If P is of type $\mathcal{C}$ the argument is similar.

It is now easy to see that a line L cannot contain points of all three types. For by translating a point of type $\mathcal{A}$ on L to the origin we may assume that $(0, 0)$ is on L. Let (x, y) and (x', y') be points of L of types $\mathcal{B}$ and $\mathcal{C}$. Then $\|x\|\geqq\|y\|$, $\|y'\|>\|x'\|$, and $\|xy'\|>\|x'y\|$. This is absurd as $xy'=x'y$. Note also that if a triangle T has vertices of all three types then $\|\text{area } T\|>1$. For we may assume that the vertex of T of type $\mathcal{A}$ is $(0, 0)$. Let (x, y) and (x', y') be the vertices of types $\mathcal{B}$ and $\mathcal{C}$. Then area T, up to sign, is equal to $\frac{1}{2}(xy'-x'y)$. But $\|xy'\|>\|x'y\|$. So $\|\text{area } T\|=\|\frac{1}{2}\|\,\|xy'\|=\|\frac{1}{2}\|\cdot\|x\|\cdot\|y'\|>1$.

Suppose now that $S=[0,1]\times[0,1]$ is divided into m nonoverlapping triangles T_i each of area $1/m$. Obviously S has exactly one side of type $\mathcal{A}\mathcal{B}$; by the lemma, some T_i has vertices of all three types. By the paragraph above, $\|\text{area } T_i\|=\|1/m\|>1$. So m is even. (Note that if all vertices have rational coordinates, then we can argue directly with the 2-adic absolute value of the rationals, and avoid the theorem of the extension of valuations; this is essentially what was done in [2].)

Finally, we indicate the proof of the more general theorem mentioned at the beginning of this paper. Let A be the ring $Z[a_1,\cdots,a_m]$. If 2 generates the unit ideal in A, then $1=2f(a_1,\cdots,a_m)$ and we are done. If $2A\neq A$, then 2 is contained in a height 1 prime ideal P of A. If A' is the integral closure of A and P' a prime of A' lying over P, then the local ring of P' on A' is a discrete valuation ring. This ring gives rise to an ultranorm on the quotient field of A such that $\|a_i\|\leqq 1$, while $\|2\|<1$. Extend this ultranorm to the reals, and use it to subdivide the plane into points of three types as above. Then, as above, some T_i has vertices of all three types, and $\|a_i\|=\|\text{area } T_i\|>1$, a contradiction. (By using valuation rings instead of ultranorms we could simplify the proof a little.)

The above proof is not so wildly nonconstructive as it first appears. For the entire argument is carried out in the field generated by the coordinates of the vertices. So it is only necessary to extend our ultranorm from Q to this finitely generated field, not to the entire field of real numbers.

References

1. Fred Richman and John Thomas, Problem 5479, this MONTHLY, 74 (1967) 329.
2. John Thomas, A dissection problem, Math. Mag., 41 (1968) 187–190.
3. Serge Lang, Algebra, Addison-Wesley, Reading, Mass., 1965, p. 299.

BIBLIOGRAPHIC ENTRY: VALUATION THEORY

1. H. S. Thurston, The solution of p-adic equations, AMERICAN MATHEMATICAL MONTHLY, 1943, 142–148.

(e)

ALGEBRAIC NUMBER THEORY

A NUMBER FIELD WITHOUT A RELATIVE INTEGRAL BASIS*

ROBERT MACKENZIE AND JOHN SCHEUNEMAN, Indiana University

In this note we describe an example of a number field F and a quadratic extension K/F that does not have an integral basis (minimal basis) relative to F.

As usual, denote by $\mathbf{Q}$ the field of rational numbers and by $\mathbf{Z}$ the ring of integers. Let $F=\mathbf{Q}(\gamma)$, where $\gamma^2+14=0$. Let $\mathfrak{O}_F$ be the ring of integers of F. Then $\mathfrak{O}_F=\mathbf{Z}+\mathbf{Z}\gamma$, as is well known. Let $\mathfrak{p}=7\mathfrak{O}_F+\gamma\mathfrak{O}_F$. Then $\mathfrak{p}^2=7\mathfrak{O}_F$, so since $[F:\mathbf{Q}]=2$, $\mathfrak{p}$ is a prime ideal and 7 is ramified in F.

LEMMA. *$\mathfrak{p}$ is not a principal ideal.*

Proof. Suppose $\mathfrak{p}=(a+b\gamma)\mathfrak{O}_F$, with $a, b\in\mathbf{Z}$. Then there are $u, v, x, y\in\mathbf{Z}$ such that $7=(a+b\gamma)\cdot(u+v\gamma)$ and $\gamma=(a+b\gamma)(x+y\gamma)$. Hence

$$au-14bv=7 \qquad ax-14by=0$$
$$bu+av=0 \qquad bx+ay=1.$$

Eliminating u, we obtain $-(a^2+14b^2)v=7b$. This implies $b=0$, since otherwise $a^2+14b^2>|7b|$, which would contradict the equation above. It follows that $\mathfrak{p}=a\mathfrak{O}_F$ with $ay=1$. This says $\mathfrak{p}=\mathfrak{O}_F$, a contradiction.

Now let $K=F(\Gamma)$, where $\Gamma^2+7=0$. It is easily checked that $\Delta=\frac{1}{2}(1+\Gamma)$ is an integer in K.

PROPOSITION. *K does not have an integral basis over F.*

Proof. Suppose $\{\mathrm{A}, \mathrm{B}\}$ is an integral basis of K over F. Then there are $\alpha_1, \beta_1, \alpha_2, \beta_2\in\mathfrak{O}_F$ such that $1=\alpha_1\mathrm{A}+\beta_1\mathrm{B}$ and $\Delta=\alpha_2\mathrm{A}+\beta_2\mathrm{B}$. Let $\overline{\mathrm{A}}, \overline{\mathrm{B}}, \overline{\Delta}$ be the

* From AMERICAN MATHEMATICAL MONTHLY, vol. 78 (1971), pp. 882–883.

conjugates of A, B, Δ, respectively, over F. Then

$$\begin{pmatrix} \alpha_1 & \beta_1 \\ \alpha_2 & \beta_2 \end{pmatrix} \begin{pmatrix} \mathrm{A} & \bar{\mathrm{A}} \\ \mathrm{B} & \bar{\mathrm{B}} \end{pmatrix} = \begin{pmatrix} 1 & 1 \\ \Delta & \bar{\Delta} \end{pmatrix}.$$

Taking determinants of both sides and squaring gives $\theta^2\delta = -7$, where

$$\theta = \det \begin{pmatrix} \alpha_1 & \beta_1 \\ \alpha_2 & \beta_2 \end{pmatrix} \in \mathfrak{O}_F \quad \text{and} \quad \delta = \det \begin{pmatrix} \mathrm{A} & \bar{\mathrm{A}} \\ \mathrm{B} & \bar{\mathrm{B}} \end{pmatrix}^2 \in \mathfrak{O}_F.$$

Then $\mathfrak{p}^2 = 7\mathfrak{O}_F = (\theta\mathfrak{O}_F)^2(\delta\mathfrak{O}_F)$, which implies $\theta\mathfrak{O}_F = \mathfrak{O}_F$ and $\mathfrak{p}^2 = \delta\mathfrak{O}_F$ by the unique factorization of ideals in $\mathfrak{O}_F$ and the fact that $\mathfrak{p}$ is not principal.

Since θ is a unit, $\{1, \Delta\}$ must be an integral basis for K over F. However. γ/Γ is an integer in K, so $\gamma/\Gamma = \alpha + \beta\Delta$ with $\alpha, \beta \in \mathfrak{O}_F$. Taking conjugates over F, $-\gamma/\Gamma = \alpha + \beta\bar{\Delta}$, so $2\gamma/\Gamma = \beta(\Delta - \bar{\Delta}) = \beta\Gamma$, or $2\gamma = -7\beta$. Thus $(2\mathfrak{O}_F)(\gamma\mathfrak{O}_F) = \mathfrak{p}^2(\beta\mathfrak{O}_F)$. But the left side has order 1 at $\mathfrak{p}$ while the right side has order $\geqq 2$ at $\mathfrak{p}$, which is impossible.

ALGEBRAIC NUMBER FIELDS AND THE DIOPHANTINE EQUATION $m^n = n^m$*

ALVIN HAUSNER, The City College of New York

1. Introduction. In the 1960 Putnam competition, the problem was posed of determining all integral solutions of the Diophantine equation

$$m^n = n^m \qquad (m \neq n). \tag{1}$$

This question has a long history starting, apparently, with Euler who treated it in *Introductio in Analysin Infinitorum II*, page 294 (see [2], p. 687; [8], pp. 150–151). It is not difficult to show that the pair 2, 4 is the only solution of (1) in positive integers (we do not count 4, 2 as another solution because of the symmetry of the equation). Let us quickly demonstrate this fact (incidentally, not by Euler's procedure). Without limiting generality, suppose $m > n$ and write $m = n + r$, where r is a positive integer. Substituting in (1), we find that $(n+r)^n = n^{n+r}$. This means that $\{(n+r)/n\}^n = n^r$ or $\{1 + (r/n)\}^n = n^r < e^r$. Hence $n = 1$ or 2. If $n = 1$, then $m = 1$ and this case is excluded. The case $n = 2$ yields $m = 4$.

Suppose we seek all integral solutions of (1). Now, $m = 0$ would make $n = 0$ and this is excluded. If (1) holds with m negative, then clearly n must also be negative. A simple discussion, whose details we omit, brings us back to the case already treated, and we get the one additional solution $-2, -4$.

The question of suitably extending (1) to arbitrary algebraic number fields quite naturally presents itself. Let K denote an algebraic number field, *i.e.*, a finite algebraic extension of the field R of rational numbers ([6], p. 35). If

* From AMERICAN MATHEMATICAL MONTHLY, vol. 68 (1961), pp. 856–861.

$\alpha \in K$, then $N_{K/R}(\alpha)=N(\alpha)$ will denote the norm of α ([6], p. 72). $N(\alpha)$ is a rational number and, if α is an algebraic integer in K, then $N(\alpha)$ is a rational integer. We note that $N(\alpha)=0$ if and only if $\alpha=0$, and that $N(\alpha\beta)=N(\alpha)N(\beta)$ for any $\alpha, \beta \in K$ ([6], p. 72). Further, if a is a rational integer in K of degree n over R, then $N_{K/R}(a)=a^n$.

A reasonable extension of (1) to the field K is the following: Find the algebraic integers $\alpha, \beta \in K$ which satisfy

$$\alpha^{N(\beta)} = \beta^{N(\alpha)} \qquad (\alpha \neq \beta). \tag{2}$$

This question generalizes the earlier one because $N(m)=m$ for integers in the field of rational numbers. Solutions of (2) certainly exist. For example, in $R(\sqrt{2})$ we have $\alpha=2+\sqrt{2}$, $\beta=6+4\sqrt{2}$; again, in $R(\sqrt{(-2)})$ we have $\alpha=2$, $\beta=\sqrt{(-2)}$. Equation (2) is satisfied by $\alpha=4048+1530\sqrt{7}$, $\beta=45+17\sqrt{7}$, and $\alpha=126+16\sqrt{62}$, $\beta=8+\sqrt{62}$ in $R(\sqrt{7})$ and $R(\sqrt{62})$, respectively. If the degree $[K:R]$ is even, we have infinitely many solutions $\alpha=2k$, $\beta=-2k$, $k=1, 2, \cdots$. This can be proved by substituting in (2) and using the fact that $N(-1)=1$ in fields of even degree. We might term this infinite family of solutions "trivial." Later we will show that there exist infinitely many (real) quadratic fields $R(\sqrt{D})$ such that (2) has infinitely many nontrivial algebraic integer solutions in each of these fields. Several such examples were given above.

2. Solutions in algebraic integers. Suppose K is an arbitrary algebraic number field and that (2) holds for $\alpha, \beta \in K$. Taking the norm of both sides of (2) and using the multiplicativity of the norm, we find that $N(\alpha)^{N(\beta)}=N(\beta)^{N(\alpha)}$. By the Euler problem, three cases can arise:

(a) $N(\alpha)=2$, $N(\beta)=4$; (b) $N(\alpha)=-2$, $N(\beta)=-4$; (c) $N(\alpha)=N(\beta)$, $\alpha \neq \beta$. We shall study each of these possibilities in turn.

In Case (a), we find $N(\alpha)^4=N(\beta)^2$ and this implies $N(\alpha^2)=\pm N(\beta)$ or $N(\alpha^2/\beta)=\pm 1$. Hence $\beta=\eta\alpha^2$, where η is a unit in K. Substituting in (2) we find $\alpha^{N(\eta\alpha^2)}=(\eta\alpha^2)^{N(\alpha)}$ or $\alpha^{4N(\eta)}=\eta^2\alpha^4$; thus $\eta=\pm 1$. Conversely, if α is such that $N(\alpha)=2$ in K and β is defined as α^2, then (2) holds for the pair α, β; this can be seen by direct verification. The possible solution $\alpha, -\alpha^2$ remains if $N(\alpha)=2$. The reader will easily see that $\alpha, -\alpha^2$ is a solution of (2) if and only if $[K:R]$ is even, since the minus sign in $-\alpha^2$ requires that $N(-1)=1$ in K.

Case (b) is similar. If $[K:R]$ is odd so that $N(-1)=-1$, and if $N(\alpha)=-2$, $\beta=-\alpha^2$, then α, β is a solution of (2) as is seen by checking. The other possibility α, α^2 is never a solution-pair quite independently of whether $N(-1)$ is ± 1.

Before discussing Case (c), we complete our study of Cases (a) and (b). We leave to the reader the easy proofs of the following: If α, α^2 is a Case (a) solution of (2) and η is a unit in K with $N(\eta)=1$, then $\alpha\eta, \alpha^2\eta^2$ is also a solution. If $[K:R]$ is even and $\alpha, -\alpha^2$ is a Case (a) solution, then $\alpha\eta, -\alpha^2\eta^2$ is likewise a solution when $N(\eta)=1$. If $N(\alpha)=-2$, then $\alpha\eta, -\alpha^2\eta^2$ is a solution if $[K:R]$ is odd and $N(\eta)=1$ or if $[K:R]$ is even and $N(\eta)=-1$. The preceding statements provide a complete list of all Case (a) and (b) solutions of (2).

Our discussion to this point shows the necessity of dealing with the four Diophantine equations

$$N(x_1\omega_1 + \cdots + x_n\omega_n) = \pm 1, \pm 2, \tag{3}$$

where $\omega_1, \cdots, \omega_n$ constitute an integral basis of K over R ([6], p. 63) and rational integer solutions of (3) are sought. The first two of equations (3) would be used to determine the units of K. It is therefore of interest to know which of equations (3) are indeed solvable and, if so, to determine all solutions. The equation $N(\alpha)=2$ will be studied below for quadratic fields.

We now deal with Case (c) which is different. If α, β satisfy (2) in Case (c), then $\alpha^{N(\alpha)}=\beta^{N(\alpha)}$ with $\alpha\neq\beta$. This means that $\alpha=\beta\zeta$, where ζ is an $|N(\alpha)|$th root of unity. Every algebraic number field K contains the roots of unity ± 1 and if there are other roots of unity present in K (such roots would be necessarily nonreal), these can occur only in a finite number ([6], p. 133). If $\zeta=1$, then $\alpha=\beta\zeta=\beta$ and we exclude this case in (2). But if $\zeta=-1$, we have an rth root of unity for all even r. Thus, if both $N_{K/R}(\alpha)$ and $[K:R]$ are even, then $\alpha, -\alpha$ provides a solution-set of (2) as is easily seen. The "trivial" solutions $2k, -2k$, $k=1, 2, \cdots$, mentioned earlier over fields of even degree belong to this family of solutions. We see that these are included in an even more extensive family of "trivial" solutions when $[K:R]$ is even: $2\lambda, -2\lambda$, where λ is any integer in K. Suppose, however, that K happens to contain a nonreal $|N(\alpha)|$th root of unity ζ. Then the degree of ζ over R is $\phi(q)$, where $q\geqq 3$ and q divides $|N(\alpha)|$. Here ϕ is Euler's totient function and $\phi(x)$ is even if $x\geqq 3$. Thus $[K:R]$ is even because the degree of the field K is divisible by the degrees of the numbers it contains ([6], p. 51). In this case $\alpha, \alpha\zeta$ does actually provide a solution of (2), for $\alpha^{N(\alpha\zeta)} =\alpha^{N(\alpha)N(\zeta)}=\alpha^{N(\alpha)}=(\alpha\zeta)^{N(\alpha)}$ since $\zeta^{N(\alpha)}=\zeta^{|N(\alpha)|}=1$ and since $N(\zeta)=1$ for all roots of unity $\zeta\neq-1$. More solutions are present in this case. Suppose γ is divisible by α in the ring of integers of K. Then $\gamma=\alpha\delta$ and $\gamma, \gamma\zeta$ is a solution-set of (2): $(\alpha\delta)^{N(\alpha\delta\zeta)}=\alpha^{N(\alpha\delta)}\delta^{N(\alpha\delta)}=(\alpha\delta\zeta)^{N(\alpha\delta)}$ since $\zeta^{N(\alpha\delta)}=(\zeta^{N(\alpha)})^{N(\delta)}=1^{N(\delta)}=1$. The sets γ, $\gamma\zeta$ (α divides γ, ζ an $|N(\alpha)|$th root of unity) exhaust all Case (c) solutions.

In summary, for every $\alpha\in K$ with $N(\alpha)=2$ we have the solution-set α, α^2 of (2). If $[K:R]$ is even, then $\alpha, -\alpha^2$ is also a solution. If $N(\alpha)=-2$ and $[K:R]$ is odd, then $\alpha, -\alpha^2$ is a solution. If $N(\alpha)=-2$, then α, α^2 is never a solution irrespective of $[K:R]$. Every field K of even degree has the set $\alpha, -\alpha$ as a solution if $N(\alpha)$ is even. If $|N(\alpha)|\geqq 3$ and K contains a nonreal $|N(\alpha)|$th root of unity ζ, then the infinitely many pairs $\gamma, \gamma\zeta$ are solutions for any multiple γ of α. This can happen for only finitely many $\zeta\in K$ and only when $[K:R]$ is even. In fields of odd degree lacking integers of norm ± 2, (2) cannot be solved.

Let us close this section with a more thorough discussion of the fields $R(\sqrt{D})$ in Case (a) as promised. Suppose D is a positive square-free integer. We would like to show that there are infinitely many $R(\sqrt{D})$ each containing infinitely many solutions of Case (a) type and not merely of the $\alpha, -\alpha$ type. In view of our discussion above, we must show that there are infinitely many $\alpha\in R(\sqrt{D})$, for infinitely many D, with $N(\alpha)=2$. Integers in $R(\sqrt{D})$ are of the form $a+b\sqrt{D}$, where a, b are rational integers (we are excluding those integers $\frac{1}{2}(a+b\sqrt{D})$ with a, b both odd in the case $D\equiv 1 \pmod 4$). The norm of $a+b\sqrt{D}$ over $R(\sqrt{D})$ is $N(a+b\sqrt{D})=(a+b\sqrt{D})(a-b\sqrt{D})=a^2-Db^2$. The question is therefore whether there are infinitely many square-free $D>0$ such that

$$a^2 - Db^2 = 2 \tag{4}$$

has infinitely many solution-sets a, b for each D. That the answer is in the affirmative is shown by taking, for example, $D=2p$, where p is a prime such that $p\equiv 7 \pmod 8$ (there are infinitely many such p). With such D, (4) has a solution set a_0, b_0 ([4], p. 174; [1], p. 450). But from one solution-set we may generate infinitely many more sets by employing the units of $R(\sqrt{D})$. Write

$$a = a_0x + Db_0y, \qquad b = a_0y + b_0x. \tag{5}$$

If we choose x, y in (5) from among the infinitely many integral solutions of the Pell equation $x^2-Dy^2=1$, then a, b from (5) satisfy (4) (for a detailed presentation of the Pell equation, see [5], Ch. 8). If a, b is a solution of (4), then $\alpha=a+b\sqrt{D}$ has norm 2 in $R(\sqrt{D})$ and α, $\beta(=\pm\alpha^2)$ is a solution of (2) as we already saw. Thus, there are infinitely many quadratic fields each containing infinitely many solutions of (2) of the Case (a) type.

3. Solutions in ideals. Our extension of (1) to the form (2) yields results in great contrast to the rational integer case. Let us seek another version of "$m^n=n^m$", for algebraic number fields, which retains the finite-solution character of the rational field case. Suppose (2) is satisfied by a pair $\alpha, \beta\in K$. Then Cases (a), (b) and (c) of Section 2 show that $N(\alpha)$ and $N(\beta)$ have the same sign. If both are positive, then (2) implies $(\alpha)^{N((\beta))}=(\beta)^{N((\alpha))}$ where (α) and (β) denote the principal ideals, in the ring of integers of K, generated by α and β, respectively. $N((\alpha))$ denotes the norm of the ideal (α) and $N((\alpha))=N(\alpha)$ since $N(\alpha)>0$ ([6], p. 107). If, on the other hand, $N(\alpha)$, $N(\beta)<0$, then (2) implies $\alpha^{-N(\beta)}=\beta^{-N(\alpha)}$ and, once again, the ideal equation $(\alpha)^{N((\beta))}=(\beta)^{N((\alpha))}$ holds since $N((\alpha))=-N(\alpha)$ if $N(\alpha)<0$ ([6], p. 107). This suggests that we consider the problem: Find all integral ideals (not necessarily principal) A, B in K such that

$$A^{N(B)} = B^{N(A)} \qquad (A \neq B). \tag{6}$$

Here, $N(A)$ and $N(B)$ denote the norms of A and B ([6], p. 104). We need only consider the case where $N(A)\neq N(B)$ since $A^k=B^k$ implies $A=B$ by the unique factorization theorem for ideals into prime ideals ([6], p. 91, p. 95). We will discover that only *finitely* many pairs of ideals in K can satisfy (6) so that the ideal-theoretic formulation of "$m^n=n^m$" preserves the finite-solution result of the rational case.

Suppose A and B are (integral) ideals in K which solve (6). Since the norm of an ideal is multiplicative ([6], p. 108) we find, by taking norms of both sides of (6): $N(A)^{N(B)}=N(B)^{N(A)}$. Since $N(A)$, $N(B)$ are positive integers and

$N(A)\neq N(B)$ by hypothesis, we have $N(A)=2$ and $N(B)=4$. Substituting these values in (6), we get $A^4=B^2$. Now, since the norm of A is 2, A is a prime ideal in K ([6], p. 109) and by the unique factorization theorem for ideals we find that $B=A^2$. Conversely, if A is an ideal of norm 2 in K (and hence prime) and B is defined as A^2, then (6) is satisfied. Since there are only finitely many ideals in K of fixed norm ([6], p. 109) and, in particular, only finitely many of norm 2, we see that (6) has only finitely many solutions (perhaps none).

There is the question of the number of solutions of (6). The preceding discussion shows that the solvability of (6) depends on the existence of (prime) ideals of norm 2 in K and this in turn depends on how the principal ideal (2) factors into prime ideals in K. The reason is that every ideal P of norm 2 divides the ideal (2) since an ideal contains (divides) its norm ([6], p. 109). Thus P occurs in the factorization of (2) into prime ideals because of the unique factorization theorem for ideals. In other words, the only place to seek (prime) ideals of norm 2 in K is among the prime ideal factors of (2). Now, the manner in which any rational prime splits into prime ideals in an algebraic number field is known. However, simple and specific results on this question are available in two classic cases, namely, for the quadratic and cyclotomic fields. We must apply these results to the particular prime 2.

In a quadratic field $R(\sqrt{D})$, the decomposition of (2) depends on the Kronecker symbol $(d|2)$, where d is the discriminant of $R(\sqrt{D})$ ([3], p. 235; [4], pp. 156–158). We simply state the result here. If d is even, then $(d|2)=0$ and the factorization of (2) is $(2)=P^2$ with $N(P)=2$. Thus, there is one solution of (6). If d is odd, then $(d|2)=(-1)^{(d^2-1)/8}$ and (2) is itself prime of norm 4 or $(2)=P_1P_2$ with $N(P_1)=N(P_2)=2$ depending on whether $(d|2)$ is -1 or $+1$, respectively. Thus, (6) has two distinct solution-sets if $d=8s\pm1$ and none if $d=8s\pm3$. More explicitly, if $D\not\equiv1 \pmod 4$ (in particular, if D is even), then $d=4D$ ([6], p. 67) and (6) has one solution in $R(\sqrt{D})$. If, however, $D\equiv1 \pmod 4$, then $d=D$ ([6], p. 67) and there is one or there are no solutions of (6) in $R(\sqrt{D})$ depending on the value of $(d|2)=(D|2)$ as we already explained.

If K is a cyclotomic field, *i.e.*, one obtained by adjoining a primitive mth root of unity to the rationals ($m\geqq2$), then the following is true (see [3], pp. 198–199; [4], p. 185). The principal ideal (2) remains prime in every cyclotomic field except that $(2)=P^{2^{\nu-1}}$ in the field where $m=2^\nu$. In this latter case, $N(P)=2$. As far as (6) is concerned, we see that there are no solutions for any cyclotomic field except those where $m=2^\nu$ ($\nu\geqq1$). In this case there is precisely one ideal P of norm 2 and thus P, P^2 is the only solution of (6).

Let us look at further examples. In the fields $R(\sqrt[n]{2})$ and $R(\sqrt[n]{-2})$ there is only one solution of (6). For, in $R(\sqrt[n]{(\pm2)})$, we see that $(2)=(\sqrt[n]{(\pm2)})^n$ and the ideals $(\sqrt[n]{2})$ and $(\sqrt[n]{(-2)})$ are such that $N((\sqrt[n]{(\pm2)}))=2$. This is true because $N((\sqrt[n]{2}))=|N(\sqrt[n]{2})|=2$; similarly for $N((\sqrt[n]{-2}))$. For example, in $R(\sqrt[9]{2})$, $A=(\sqrt[9]{2})$ and $B=A^2$ is the only solution of (6).

If K is any cubic field over R and if (2) is nonprime, then the only possible prime factorizations of (2) are P_1^3, $P_1^2\,P_2$, $P_1P_2P_3$ with $N(P_i)=2$ $(1\leqq i\leqq 3)$ or $(2)=P_4P_5$ with $N(P_4)=2$, $N(P_5)=4$. This is true because $N((2))=2^3=8$ over cubic fields. Hence there are either no, one, two or three solutions of (6). We give three examples of the splitting of (2) in cubic fields ([7] pp. 292–293). In $R(\vartheta)$ where ϑ satisfies the irreducible equation $x^3+x+1=0$, the ideal (2) does not decompose. Equation (6) has no solutions. There are two solutions in $R(\vartheta)$ where $\vartheta^3+6\vartheta+8=0$ because $(2)=(2,\ 1+\vartheta+\frac{1}{2}\vartheta^2)^2(2,\ \frac{1}{2}\vartheta^2)$. In $R(\vartheta)$, where $\vartheta^3+8\vartheta-4=0$, we find $(2)=P^3$; as a matter of fact, $(2)=(\frac{1}{2}\vartheta^2)^3(132\vartheta^2+68\vartheta-1023)$, where $132\vartheta^2+68\vartheta-1023$ is a unit. Equation (6) has one solution in this latter case.

Having finished with the examples, we may now state the general result. It is known by a theorem of Dedekind that a rational prime p, in an arbitrary field K, is divisible by the square of a prime ideal if and only if p divides the discriminant d of K ([6], p. 101). Clearly, then, the *maximum* number of solution-sets in any K of degree n over R of (6) is n, corresponding to the case where (2) is completely unramified, *i.e.*, (2) can be factored into n distinct prime ideals each of norm 2. A necessary, but not sufficient, condition for this maximum number of solutions is that the discriminant of K be odd.

Summarizing, (6) has solutions if and only if the principal ideal (2) is composite in K. If n is the degree of K, then there are at most n distinct solutions. A necessary condition for the maximum number of solutions is that 2 does not divide the discriminant of K. Hence, if the discriminant of K is even, there are less than n different solution-sets, perhaps none. If K is the quadratic field $R(\sqrt{D})$ and (2) nonprime, then (6) has one or two sets of solutions depending on whether D is even or odd. There are no solutions of (6) in the cyclotomic fields with the exception that one solution is present in each of the fields $R(\zeta)$ where ζ is a primitive 2^νth root of unity. In all cases the number of solution sets, in any field K, is identical with the number of prime ideals P in K with $N(P)=2$.

In closing, we remark that if we take (2) and make the transition to the principal ideal equation, namely $(\alpha)^{N((\beta))}=(\beta)^{N((\alpha))}$, we do not get infinitely many different ideal solutions even though (2) may have infinitely many solutions in K. This is no contradiction since the resulting principal ideals (α) and (β) give rise to only finitely many different ideals in K because associated numbers α and $\alpha\eta$ (where η is a unit) generate the same principal ideals.

References

1. G. Chrystal, Algebra, vol. II (2nd ed.), London, 1906.
2. L. E. Dickson, History of the Theory of Numbers, vol. II (Diophantine Analysis), New York, 1952.
3. R. Fricke, Lehrbuch der Algebra, Bd. 3 (Algebraische Zahlen), Braunschweig, 1928.
4. L. Holzer, Zahlentheorie, Teil I, Leipzig, 1958.
5. W. J. LeVeque, Topics in Number Theory, vol. I, Reading, Mass., 1956.
6. H. Pollard, The Theory of Algebraic Numbers (Carus Monograph No. 9) New York, 1950.
7. J. Sommer, Introduction à la Théorie des Nombres Algébriques (translated from the German by A. Lévy), Paris, 1911.
8. G. Wertheim, Anfangsgründe der Zahlenlehre, Braunschweig, 1902.

ON THE PRIME DIVISORS OF POLYNOMIALS*

IRVING GERST, State University of New York at Stony Brook, and
JOHN BRILLHART, University of Arizona

1. Introduction. The purpose of this paper is to prove in a fairly elementary way certain results in the theory of higher congruences. These results, which can readily be considered a part of elementary number theory, have previously been established with the use of ideal theory in algebraic number fields. In limiting our means to the basic ideas of field theory and the theory of equations, we hope to make accessible to a fairly wide audience an interesting part of number theory, which is usually not found in beginning treatments of this subject.

Insofar as the theorems presented here are concerned, this paper is purely expository, with the possible exception of the first part of Theorem 11, which we have not seen elsewhere. In other aspects of the paper, however, such as the arrangement of the material, the proofs, and the various applications, we have given a development which we hope will be of interest in itself.

We shall be concerned with those primes p for which the congruence $f(x) \equiv 0 \pmod{p}$ is solvable, where $f(x)$ is a polynomial with rational integer coefficients which is not identically zero (mod p). Such primes will be called *prime divisors* of f, since they divide the value $f(x_0)$ for some integer x_0. To indicate this we write $p|f(x)$. In this case $f(x)$ also has at least one linear factor (mod p). The set of *all* prime divisors of f will be written $P(f)$.

To illustrate these ideas we recall the following well-known facts: any prime p, $p\nmid a$, will divide $f(x)=ax+b$. Also, for $f(x)=x^2-a$, $P(f)$ can be determined by using the Quadratic Reciprocity Law. The general problem, however, of characterizing the prime divisors of a polynomial of degree >2 is still unsolved, except in certain special cases. (See Hasse [10].)

In what follows we present some known results regarding $P(f)$ for various classes of polynomials f. These results are of two types: (1) Infinitely many primes of a certain kind are shown to exist in $P(f)$. For example, it will be shown (Theorem 9) that any nonconstant polynomial possesses infinitely many prime divisors of the form $kn+1$, where n is a given positive integer. (2) Information concerning the sets of primes dividing several polynomials is derived from the relationships that hold between the extensions of the rational field Q defined by these polynomials. For example, if f and g are irreducible over Q with roots α and β respectively, and if $Q(\alpha)=Q(\beta)$, then f and g have the same prime divisors, with at most a finite number of exceptions.

* From AMERICAN MATHEMATICAL MONTHLY, vol. 78 (1971), pp. 250–266.

Irving Gerst received his Ph.D. in 1947 at Columbia University under J. F. Ritt in complex variables. From 1945–1961, he was a staff member and then head of applied mathematics groups at Burroughs and RCA. Since 1961 he has been Professor and Chairman of the Department of Applied Analysis at SUNY at Stony Brook, and during 1968 he spent a sabbatical at the University of Arizona. His principal research has been in network theory.

John Brillhart was lecturer and instructor at the University of San Francisco from 1955 to 1965, and an NSF Science Faculty Fellow in 1965–66. He received his Ph.D. at Berkeley under D. H. Lehmer in 1967, and since has been Associate Professor at the University of Arizona. His main areas of research are algebra and number theory. *Editor.*

The usual approach to these results by means of ideal theory is based on the important theorem of Dedekind (cf. [5] or [13, Thm. 8.1, p. 63]) which relates the factorization of $f(x)$ (mod p) to the factorization of p into prime ideals in the algebraic number field defined by a root of $f(x)$. As the ideal theoretic approach is quite powerful and far-reaching in the development of the theory of this subject, we recommend that the reader who is interested in the prime divisors of polynomials also become acquainted with the subject from this point of view. For an introduction to ideal theory in algebraic number fields we suggest Pollard [17]. More recent treatments can be found in Borevich and Shaferevich [4] or Samuel [18]. In the remarks following some of the theorems of the paper, we shall indicate where the reader can find material relevant to the ideal theoretic proofs of these theorems. In referring to this material, which is scattered and rather fragmentary, one must note that the prime divisors of $f(x)$, except for at most a finite number, are exactly those primes which have a prime ideal of the first degree as a factor.

In conclusion, it is important to point out that the methods employed in this paper are constructive and lend themselves to numerical calculation, in that they use the polynomials themselves, their discriminants, and the polynomials relating their roots. This fact allows us to give practical tests to answer various questions concerning polynomials and their fields. For example, it is not difficult to devise a simple and computable sufficient condition for showing that two irreducible polynomials do *not* define the same extensions of Q. (Remark (b) after Theorem 3.)

2. Preliminaries. (A) In this paper we shall always use f, g, and h to denote nonconstant polynomials in one variable with integer coefficients; that is, $f(x) \in Z[x]$. The symbol p will denote a rational prime.

We also use $P_i(f)$, $i=1, 2, \cdots, n$, where $\deg f=n$, to denote the set of prime divisors of f for which the congruence $f(x)\equiv 0 \pmod p$ has *exactly* i distinct solutions. (In this definition a multiple root is counted only once.) Certainly $P(f)$ will then be the union of the sets $P_i(f)$.

A special role is played in this theory by the primes $p \in P(f)$ for which f factors completely into a product of linear factors (mod p). Among such primes are those p for which f will be said to *split completely.*

In defining this special kind of factorization into linear factors we recall first that f (which may be reducible) can be uniquely expressed as $a\prod f_i^{\alpha_i}$, where a, $\alpha_i \in Z$, $\alpha_i > 0$, and the $f_i \in Z[x]$ are distinct, primitive, irreducible polynomials with positive leading coefficients [3, pp. 74–76]. Let $g=\prod f_i$. We then say f *splits completely* for $p\in P(f)$ if g is congruent to a product of *distinct* linear factors (mod p), that is, if $p \in P_m(f)$, where $m = \deg g$. (Clearly the maximum number of *distinct* linear factors of f (mod p) is m.) In the important special case where f has no multiple roots, that is, where $n=\deg f=m$, the condition that f splits completely is then $p\in P_n(f)$.

In what follows we shall often be concerned with the comparison of the sets of primes which divide two or more polynomials. If, for instance, the polynomials f and g have the same prime divisors, except possibly for a finite number, we say they have *essentially* the same prime divisors, and write $P(f)=P(g)$. Thus an $=$ sign, when used in conjunction with "$P(\ )$" or with "$P_i(\ )$", must be understood in this sense. For example, the inequality $P(f)\neq\varnothing$ would mean that $P(f)$ is infinite. We also use $P(f)\subseteq P(g)$ to denote that the prime divisors of f are also divisors of g with the possible exception of a finite number of primes, and say in this case that *almost all* prime divisors of f are divisors of g. (We note that our use here of "$\subseteq$" corresponds to the use of "$\leqq$" by Hasse [9, v. 2, p. 141].)

The preceding considerations can be generalized to the case where the coefficients of f and g are integers in an algebraic number field, with p being replaced by a prime ideal in the field. In this elementary introduction, however, we shall restrict our attention solely to rational integral coefficients.

(B) Next recall the relationship between polynomials and fields (cf. [3], [21]). The fields we shall be dealing with are all finite extensions of the rational field Q (i.e., algebraic number fields). Such extensions will be designated by K, L, and M, and can always be defined (in many ways) by adjoining to Q a root of a polynomial irreducible over Q. (Irreducibility will always be with respect to Q, unless otherwise stated.) Such a root is called a *primitive element* of the extension. We further say a polynomial f *belongs to* K, written $f:K$, if f is irreducible, with a primitive element of K as a root. There exist infinitely many polynomials belonging to any field K.

If $f:K$ and f has the roots $\alpha_1=\alpha, \alpha_2, \cdots, \alpha_n$, where $K=Q(\alpha)$, then $Q(\alpha_i)$ are called the "conjugate fields" of K. If all the conjugate fields are the same, K is said to be *normal* (or Galois). In this case any polynomial belonging to K is also called *normal*. (A normal polynomial is always irreducible.) Evidently when f is normal, the conjugates of a root α can be expressed as polynomials in α with rational coefficients. Conversely, if the roots of an irreducible polynomial f are related to any one of them by such polynomials, then f belongs to the simple, normal extension of Q generated by this root.

Let $K=Q(\alpha)$, $f:K$, $\deg f=n$, and $f(\alpha)=0$. Our proofs will be based, in the main, upon the following two standard results in field theory: (i) *If a polynomial* $\phi(x)\in Q[x]$ *has* α *as a root, then* $\phi(x)$ *is a multiple of* $f(x)$ *in* $Q[x]$. (ii) *Every element* $\gamma\in K$ *has a unique representation of the form* $\gamma=\sum_{j=1}^{n} r_j\,\alpha^{j-1}$, $r_j\in Q$.

Additional results in field theory required for our proofs will be introduced later as the need for them arises.

(C) In the sequel we shall always assume a polynomial belonging to a field is *monic*, so that its roots will be algebraic integers. If a given polynomial $f(x)=a_0x^n+\cdots+a_n$, $a_0\neq 0$, is not monic, we can replace it in our considerations by the monic polynomial g, defined by $g(x)=a_0^{n-1}f(x/a_0)$. It is clear from this definition that $P(g)=P(f)$, and also that $P_i(g)=P_i(f)$ for $i=1, 2, \cdots, n$, since the only difference between the two sets of primes is at most the finite number of primes dividing a_0. It is also clear that if $f:K$, then $g:K$. These observations imply that all theorems proved for monic polynomials in this paper are also true for nonmonic polynomials.

We conclude this section by recalling several elementary matters in the arithmetic of rational numbers (mod p). A fraction a/b (mod p) is defined for $p \nmid b$ as the unique solution x (mod p) of the congruence $bx \equiv a$ (mod p). The usual laws for operating with congruences continue to hold for these fractions. Moreover, if $p \nmid b$ and $f(a/b) = A/B$, A, $B \in Z$, and $p \nmid B$, then $f(a/b) \equiv A/B$ (mod p), where a/b and A/B are interpreted as fractions (mod p). In particular, if $p | A$ then $f(a/b) \equiv 0$ (mod p) and $p \in P(f)$. This particular argument will appear in many of the proofs.

3. Basic theorems. In this section we give three theorems which will be fundamental in much that follows. The first theorem is quite well known.

THEOREM 1 (Schur [20]). *Every nonconstant polynomial f has an infinite number of prime divisors; that is, $P(f) \neq \varnothing$.*

Proof. We may suppose $f(0) = c$ is not zero, since otherwise every prime divides $f(x)$. Furthermore, there exists at least one prime divisor of $f(x)$, for a polynomial can take the values ± 1 at most a finite number of times. Assume $p_1, p_2, \cdots, p_n$ are the only prime divisors of $f(x)$ and let $a = p_1 p_2 \cdots p_n$. Then $f(acx) = cg(x)$, where $g(x) = 1 + c_1 x + c_2 x^2 + \cdots$, and $a | c_i$. A prime divisor p of $g(x)$ will also divide $f(x)$, and so must be one of the p_i. But then $p | c_i$, which implies $p | 1$. Hence, $g(x)$ has no prime divisors, i.e., $g(x) = \pm 1$ for all integral x, which is impossible. ∎

In the second theorem we show that the primes dividing a polynomial belonging to a field essentially contain those that divide a polynomial belonging to any extension of that field.

THEOREM 2. *If $K \subseteq L$, $f{:}K$, and $g{:}L$, then $P(f) \supseteq P(g)$.*

Proof. We must show that almost all the prime divisors of g are also divisors of f. Let $f(\alpha) = g(\beta) = 0$, $\alpha \in K$, $\beta \in L$. Then, since $K \subseteq L$, we can write $\alpha = \phi(\beta)$, $\phi(x) \in Q[x]$. But then $f(\phi(x))$ and $g(x)$ have β as a common root, so the irreducibility of $g(x)$ implies

$$f(\phi(x)) = g(x)g_1(x), \quad g_1(x) \in Q[x]. \tag{1}$$

Now suppose $p \in P(g)$ does not divide any of the denominators in the coefficients of ϕ (assumed to be in lowest terms). (Since $g(x)$ is monic by assumption (cf. Sec. 2), the same will be true for the coefficients of $g_1(x)$.) Then from (1) if $g(b) \equiv 0$ (mod p), $b \in Z$, $\phi(b)$ will be a root of $f(x) \equiv 0$ (mod p). Thus $p | f(x)$. ∎

REMARKS. (a) It can be shown using ideal theory (Schinzel [19]), that there exist K and L, K a *proper* subfield of L, for which nonetheless $P(f) = P(g)$.

(b) For a discussion using ideal theory relating to Theorem 2, see [12, Prop. 20, p. 19].

Theorem 2 may be of use in showing a field K defined by f is not contained in a field L defined by g, $\deg f \leq \deg g$. (Certainly $K \not\subseteq L$ if $\deg f \nmid \deg g$.) For, if $g(\beta)=0$, consider all root polynomials $\phi(x)$, where $\phi(\beta)$ may represent some root α of f. If a prime p can be found such that (i) $p \mid g(x)$, (ii) $p \notin P(f)$, and (iii) p does not divide any denominator in all of the $\phi(x)$, then $K \not\subseteq L$.

From a computational point of view it is clear how one would attempt to find a prime p satisfying (i) and (ii). (See example following Theorem 3.)

To verify whether such a p also satisfies (iii), we need to identify the primes which may possibly occur in the denominators of $\phi(x)$. This is accomplished in the following Lemma.

LEMMA 1. *Let $L=Q(\beta)$ where β is an algebraic integer of degree n over Q with defining polynomial g. Also, let the canonical field representation of any algebraic integer $\alpha \in L$ be $\alpha=\phi(\beta)=\sum_{j=1}^{n} r_j\beta^{j-1}$ where the $r_j \in Q$ are in lowest terms. If p divides the denominator of any r_j, then it must divide the discriminant $D(g)$ of g.*

Proof. In establishing this lemma, which will be used in numerical applications, we shall require somewhat more than the rudiments of field theory, in that we assume the basis theorem for the domain of integers in an algebraic number field (see [17, p. 65]):

There exist algebraic integers $\gamma_i \in L$, $i=1, 2, \cdots, n$, such that if $\alpha \in L$ is an algebraic integer, then α has a unique representation of the form

$$\alpha = \sum_{i=1}^{n} d_i\gamma_i, \qquad d_i \in Z. \tag{2}$$

Since the powers of β are also integers in L,

$$\beta^{j-1} = \sum_{i=1}^{n} c_{ji}\gamma_i, \quad c_{ji} \in Z, \qquad j = 1, 2, \cdots, n. \tag{3}$$

Denote the field conjugates of β and the γ_i respectively as $\beta=\beta_1, \beta_2, \cdots, \beta_n$ and $\gamma_i=\gamma_i^{(1)}, \gamma_i^{(2)}, \cdots, \gamma_i^{(n)}$, $i=1, 2, \cdots, n$.

Using (3) we can obtain the *matrix* relation

$$(\beta_i^{j-1}) = (c_{ij})(\gamma_i^{(j)}), \qquad i,j = 1, 2, \cdots, n.$$

Taking the determinant of each side and squaring we obtain on the left the square of the Vandermonde determinant, which is well known to be the discriminant D of g. Thus,

$$D(g) = J^2 \cdot \Delta, \tag{4}$$

where $J=\det(c_{ij})$ and $\Delta=[\det(\gamma_i^{(j)})]^2$. In this equation J and $D(g)\neq 0$ are rational integers, and since Δ is thus an algebraic integer equal to a rational number, it must also be a rational integer.

It is worth mentioning that $|J|$ and Δ, customarily called the "index" of β and the "discriminant" of L respectively, can be shown to be independent of the choice of the basis $\{\gamma_i\}$.

Next, using (3) in $\alpha = \sum_{j=1}^{n} r_j \beta^{j-1}$, we get

$$\alpha = \sum_{i=1}^{n} \left(\sum_{j=1}^{n} r_j c_{ji} \right) \gamma_i.$$

Comparing this equation with (2) and equating corresponding coefficients of γ_i yields the system

$$d_i = \sum_{j=1}^{n} r_j c_{ji}, \qquad i = 1, 2, \cdots, n.$$

Solving for r_j (which is in lowest terms), $r_j = S_j/J$, $j = 1, 2, \cdots, n$, where the rational integers S_j are the numerators in Cramer's rule. Thus, any prime p dividing the denominator of r_j must divide J, which by (4) gives $p \mid D(g)$. ∎

REMARKS. (a) We observe from the above proof that Lemma 1 could have been stated with the index $|J|$ replacing $D(g)$. However, since $D(g)$ is much easier to compute (mod p) than the index, it is sufficient in numerical applications to use the lemma as stated.

(b) In ideal theoretic terms the condition $p \nmid D(g)$ implies that p is "unramified" in L, that is, p factors into distinct prime ideals in L.

The final theorem of this section gives information about the sets of primes that divide two different polynomials belonging to the *same* field.

THEOREM 3. *If $f{:}K$ and $g{:}K$, then $P(f) = P(g)$ and $P_i(f) = P_i(g)$ for $i = 1, 2, \cdots, n$, where $\deg f = \deg g = n$.*

Proof. By Theorem 2, $P(f) \subseteq P(g)$ and $P(f) \supseteq P(g)$. Hence $P(f) = P(g)$.

To prove the second part of the theorem, we first construct a 1-1 mapping of the incongruent roots of $g(x) \equiv 0 \pmod{p}$ *into* the set of incongruent roots of $f(x) \equiv 0 \pmod{p}$ for almost all primes p dividing both f and g.

Suppose $f(\alpha) = g(\beta) = 0$, $\alpha, \beta \in K$. Then there exist polynomials $\phi(x)$, $\psi(x) \in Q[x]$ for which $\beta = \psi(\alpha)$ and $\alpha = \phi(\beta)$. With this ϕ, define the map: $b_i \to \phi(b_i)$, where the b_i are the incongruent roots of $g(x) \equiv 0 \pmod{p}$. As in the proof of Theorem 2, we can derive both identity (1), and, if p does not divide the denominators in ϕ, the conclusion that follows from (1), which shows $\phi(b_i)$ is also a root of $f(x) \equiv 0 \pmod{p}$.

Next we show that $\phi(b_i) \not\equiv \phi(b_j) \pmod{p}$, $i \neq j$. From the equation $\psi(\phi(\beta)) = \beta$ we obtain the identity

$$(5) \qquad \psi(\phi(x)) - x = g(x) g_2(x), \qquad g_2(x) \in Q[x].$$

Suppose that $\phi(b_i) \equiv \phi(b_j) \pmod{p}$. Then setting $x = b_i$ and b_j in (5), if p does not divide the denominators in ψ, we find $b_i \equiv \psi(\phi(b_i)) \equiv \psi(\phi(b_j)) \equiv b_j \pmod{p}$, which is a contradiction. If we now reverse the roles of f and g in the preceding argument, we obtain a 1-1 mapping between the two sets of incongruent roots. ∎

REMARKS. (a) Since Theorem 3 is implied by Dedekind's theorem, ideal theoretic proofs can be found in the references cited in the introduction.

(b) It is clear from Lemma 1 and the proof of Theorem 3, if $p\nmid D(f)D(g)$, that $p|f(x)$ if and only if $p|g(x)$. Thus, by way of application, to show two irreducible polynomials of the same degree do *not* determine the same field, we have only to find a prime p such that $p\nmid D(f)D(g)$, $p|f(x)$, and either $p\nmid g(x)$, or, if $p|g(x)$, then f and g do not have the same number of incongruent roots (mod p).

Example. Consider the two irreducible polynomials $f_1(x)=x^4+4x^3-4x^2-40x-56$ and $f_2(x)=x^4-8x^2-24x-20$, both with discriminant $D=-2^{12}\cdot 3^2\cdot 31$. (Irreducibility can be shown by using undetermined coefficients and simple parity arguments.) We note $13|f_1(2)=-2^3\cdot 13$, but $13\nmid f_2(x)$ for $x=0, 1, \cdots, 12$. Also $13\nmid D$. Hence, the polynomials belong to different fields.

4. Polynomials belonging to normal fields. We next investigate the behavior of a *normal* polynomial with respect to its prime divisors.

THEOREM 4. *A normal polynomial splits completely for almost all its prime divisors.*

Proof. Let $f(x)=x^n+a_1x^{n-1}+\cdots+a_n$ be normal. Since $P_n(f)\subseteq P(f)$, we have only to show $P(f)\subseteq P_n(f)$. From the normality of f its roots can be written as

$$\alpha_1=\alpha,\ \alpha_2=\phi_2(\alpha),\ \cdots,\ \alpha_n=\phi_n(\alpha),\qquad \phi_i(x)\in Q[x].$$

Using these ϕ_i a useful two-variable identity can be obtained by expanding the product

$$(6)\qquad (y-x)(y-\phi_2(x))\cdots(y-\phi_n(x))=y^n+r_1(x)y^{n-1}+\cdots+r_n(x).$$

If we set $x=\alpha$ in (6), the left side becomes $f(y)$, while the coefficients on the right reduce to a_i. Hence, $r_i(x)-a_i$ and $f(x)$ have α as a common root, from which the irreducibility of f implies $r_i(x)-a_i=f(x)g_i(x)$, $i=1, 2, \cdots, n$, $g_i(x)\in Q[x]$. Replacing the $r_i(x)$ in (6) with these results, we obtain the desired identity

$$(7)\ (y-x)(y-\phi_2(x))\cdots(y-\phi_n(x))=f(y)+f(x)g_1(x,y),\ g_1(x,y)\in Q[x,y].$$

Now suppose p does not divide any of the denominators in the ϕ_i (and hence also in g_1). Also, let $p\in P(f)$; say $p|f(a)$. Then, substituting $x=a$ in (7) and considering (7) as a congruence (mod p), we have

$$(8)\qquad (y-a)(y-\phi_2(a))\cdots(y-\phi_n(a))\equiv f(y)(\text{mod } p).$$

If $p\nmid D(f)$ as well, the roots on the left side of (8) are distinct, which implies $p\in P_n(f)$. ∎

REMARKS. (a) For an ideal theoretic proof of Theorem 4 see Lang [12, Cor. 2, p. 21] or Mann [13, p. 69].

(b) Note by virtue of Lemma 1 the single condition $p\nmid D(f)$ encompasses all the restrictions on p in the above proof, and consequently f will split completely for such $p \in P(f)$.

By way of application, to show that a given irreducible polynomial f is *not* normal, we need only find a prime $p \in P(f)$, $p \nmid D(f)$, for which f does not split completely.

Example. Let $f(x) = x^4 - 2x^3 - 2x^2 + 3x + 1$. Then $f(x)$ is irreducible, since it is irreducible (mod 2). Now $f(5) = 11 \cdot 31$ and $11 \nmid D(f) = 5^2 \cdot 29$. With $p = 11$, $f(x) \equiv (x-5)(x-7)(x^2 - x - 5)$ (mod 11), where the quadratic factor is irreducible (mod 11). Thus f is not normal.

(c) It is worth mentioning in the above proof that the functions ϕ_i, which relate the roots of f over the rationals, are the same functions reduced (mod p), which relate the roots of f in Z_p, the integers (mod p).

Example. The polynomial $f(x) = x^3 - 3x + 1$ is normal with roots α, $\phi_2(\alpha) = \alpha^2 - 2$, and $\phi_3(\alpha) = -\alpha^2 - \alpha + 2$, where α is any root of $f(x)$. Also, $D(f) = 3^4$. Now $f(8) = 3 \cdot 163$, so we can choose $p = 163$ since $163 \nmid D(f)$. The two remaining roots of $f(x)$ (mod 163) can then be obtained by putting $a = 8$ in ϕ_2 and ϕ_3, giving the complete factorization

$$x^3 - 3x + 1 \equiv (x - 8)(x - 62)(x - 93) \pmod{163}.$$

In practice, if it happens that $f(x)$ splits completely for each prime divisor tried, we have a good indication that f is normal, since the converse of Theorem 4 is true [9, vol. 2, p. 141].

(d) Theorem 4 is a special case of the following more general theorem, which shows how a normal polynomial f factors (mod p), where p does not necessarily divide $f(x)$.

THEOREM. *If f is a normal polynomial and $p \nmid D(f)$, then f factors (mod p) into irreducible factors of the same degree, where the degree depends on p.*

An elementary proof of this result can be given using (7). However, we omit the proof here, because of the emphasis of this paper on linear factors. We note, however, the following interesting result:

COROLLARY. *If f is a normal polynomial of prime degree and $p \nmid D(f)$, then f is either irreducible (mod p) or splits completely (mod p).*

We next consider the special case of Theorem 2 in which the overfield L is assumed to be normal. In this case we can obtain more precise information about the prime divisors of g.

THEOREM 5. *If $K \subseteq L$ with L normal, $f:K$, and $g:L$, then $P_n(f) \supseteq P(g)$, where $\deg f = n$.*

Proof. We must show that f splits completely for almost all the prime divisors of g. By hypothesis the roots of f can be written as $\alpha_1 = \phi_1(\beta)$, $\alpha_2 = \phi_2(\beta)$, $\cdots$,

$\alpha_n = \phi_n(\beta)$, where β is any root of g and $\phi_i(x) \in Q[x]$. Using the same arguments as in the proof of Theorem 4, we can establish the identity

$$\prod_{i=1}^{n} [y - \phi_i(x)] = f(y) + g(x)g_1(x, y), \qquad g_1(x, y) \in Q[x, y]. \tag{9}$$

As before, if $g(b) \equiv 0 \pmod{p}$, $b \in Z$, but p does not divide any denominator in the ϕ_i, then

$$f(y) \equiv \prod_{i=1}^{n} [y - \phi_i(b)] \pmod{p},$$

where if $p \nmid D(f)$, the roots on the right side are distinct (mod p). ∎

COROLLARY. *Every nonconstant polynomial has an infinite number of prime divisors for which it splits completely.*

Proof. Consider the normal extension $L = Q(\alpha_1, \cdots, \alpha_n)$, where α_i are the roots of the given polynomial f. Let $g:L$. Each irreducible factor f_i of f over Z (as defined in Section 2(A)) defines a simple extension of Q which is a subfield of L. From Theorem 5 applied to these subfields, each f_i will split completely for almost all prime divisors p of g. By Theorem 1, g possesses infinitely many prime divisors. Hence, omitting from $P(g)$ the finite number of p which divide $D(h)(\neq 0)$, where $h(x)$ is the product of all the f_i, the corollary follows. ∎

We now inquire whether the inclusion in Theorem 5 can be reversed. Although this cannot be done in general for *any* normal extension, the reversal is legitimate in the following special case.

THEOREM 6. *Let $K \subseteq L$ with L the smallest normal extension of K. If $f:K$ and $g:L$, then $P_n(f) = P(g)$, where $\deg f = n$.*

REMARK. The field L can be described, alternatively, as the splitting field of f.

Proof. By Theorem 5 we need only prove $P_n(f) \subseteq P(g)$. Suppose $\alpha_1, \cdots, \alpha_n$ are the roots of f and $p \in P_n(f)$. By hypothesis we then have

$$f(x) = \prod_{i=1}^{n} (x - \alpha_i) \equiv \prod_{i=1}^{n} (x - a_i) \pmod{p}, \tag{10}$$

with the $a_i \in Z$ distinct. From (10) we conclude

$$G_j(\alpha_i) \equiv G_j(a_i) \pmod{p}, \qquad j = 1, 2, \cdots, n, \tag{11}$$

where G_j denotes the jth elementary symmetric function.

Next consider the construction of L from K. It is well known for the *smallest* normal extension of K that there exists a primitive element $\beta \in L$ of the form $\beta = \sum_{i=1}^{n} c_i\alpha_i$, where the $c_i \in Z$. Without loss of generality we can suppose that β is already a root of g, since by Theorem 3, any two polynomials belonging to the same field have essentially the same prime divisors. Now form the two polynomials

$$h(x) = \prod_{s_j \in S_n} \left[x - s_j\left(\sum_i c_i\alpha_i \right) \right] = \sum_k d_k x^k, \quad \text{where} \quad d_{n!} = 1$$

and

$$h^*(x) = \prod_{s_j \in S_n} \left[x - s_j\left(\sum_i c_i a_i \right) \right] = \sum_k d_k^* x^k, \quad \text{where} \quad d_{n!}^* = 1.$$

In these two products the s_j range over all permutations of the symmetric group S_n on n letters, where for example, by $s_j(\sum_i c_i\alpha_i)$ we mean $\sum_i c_i\alpha_l$, $l = s_j(i)$.

The coefficients d_k, as symmetric functions of the α_i, are of course integers, and may be written as $d_k = F_k(G_1(\alpha_i), G_2(\alpha_i), \cdots)$, where F_k is a polynomial with integer coefficients. By construction the coefficients d_k^* of $h^*(x)$ are determined by exactly the same polynomials F_k, with the arguments $G_j(\alpha_i)$ replaced by $G_j(a_i)$. It then follows from (11) that $d_k^* \equiv d_k \pmod p$, and therefore

$$h(x) \equiv h^*(x) \pmod p. \tag{12}$$

Since $h^*(x)$ is a product of linear factors (mod p), the same is true for $h(x)$ (mod p). But $g(x)$ has the root β in common with $h(x)$. Hence, $g(x)$ divides $h(x)$, so $p \mid g(x)$. ∎

Remarks. (a) For material relating to an ideal theoretic handling of Theorem 6 see Mann [13, Cor. 13.5.1, p. 113].

(b) From the last sentence of the preceding proof (possibly renumbering the s_j) we get that

$$g(x) \equiv \prod_{j=1}^{m} \left[x - s_j\left(\sum_i c_i a_i \right) \right] \pmod p, \qquad m = \deg g.$$

In this product it can be shown that s_j ranges over those permutations of S_n for which $\beta_j = s_j(\sum_i c_i\alpha_i)$ are the m conjugates of β. (These s_j, of course, constitute the Galois group of f.) The roots of $g(x)$ (mod p) are then given in terms of the roots of $f(x)$ (mod p) in the same way the roots of $g(x)$ are given in terms of the roots of $f(x)$ over Q.

Example. Let $f(x) = x^3 - 2$ with roots $\alpha_1 = \theta$, $\alpha_2 = \omega\theta$, and $\alpha_3 = \omega^2\theta$, $\theta = 2^{1/3}$ and $\omega = [-1 + (-3)^{1/2}]/2$. The smallest normal extension of Q containing the α_i is of degree 6, since f is not normal. (Remark (b) following Theorem 4 with $p = 5$ and $x = 3$.) The expression $\zeta = \alpha_1 + a\alpha_2 + b\alpha_3 = \theta(1 + a\omega + b\omega^2)$, $a, b \in Z$ is now chosen so that ζ will take 6 different values under the 6 permutations of the α_i. For example, let $a = 2$ and $b = -1$. Then $\zeta = \theta(1 + 2\omega - \omega^2)$ is a root of the (resolvent) equation $g(x) = x^6 + 40x^3 + 1372 = 0$. Now $31 \in P_3(f)$ since $f(x) \equiv (x-4) \cdot (x-7)(x-20) \pmod{31}$. The 6 roots b_i of $g(x) \equiv 0 \pmod{31}$ can now be computed by carrying out the 6 permutations on $a_1 \equiv 4$, $a_2 \equiv 7$, and $a_3 \equiv 20 \pmod{31}$ in the expression $a_1 + 2a_2 - a_3$; that is, $(a_1, a_2, a_3) = (4, 7, 20)$ gives $b_1 \equiv -2$, (a_1, a_3, a_2) gives $b_2 \equiv 6 \pmod{31}$, etc. Finally,

$$g(x) \equiv (x-6)(x-12)(x-21)(x-26)(x-29)(x-30) \pmod{31}.$$

(c) In Theorem 6 no information about $P_i(f)$ is obtainable for $i=1, 2, \cdots, n-2$, since all but at most a finite number of primes in these sets lie outside of $P(g)$.

(d) If in Theorem 6 the field K is already normal, then $K=L$ and Theorem 6 reduces to Theorem 4.

5. Applications. We are now in a position to prove several striking results about the prime divisors of arbitrary polynomials. As far as we know, these theorems, due to Nagell [15], have been proved previously only with the use of ideal theory. (See also Fjellstedt [6].)

THEOREM 7. *For any two nonconstant polynomials f and g, there exist infinitely many primes which split both f and g completely.*

Proof. Adjoin all the roots of f and g to Q, forming the normal field L. Let $h:L$. Then each irreducible factor of f or g over Z defines a simple extension of Q, which is a subfield of L. As in the proof of the Corollary to Theorem 5, it now follows that f and g individually split completely for almost all of the prime divisors of h. But then there must exist infinitely many primes for which, simultaneously, f and g split completely. ∎

COROLLARY. *Theorem 7 holds for a finite number of nonconstant polynomials.*

REMARK. Recently Nagell [16], by an elementary method different from our own, has proved the weaker result that there exist infinitely many common prime divisors of any finite number of nonconstant polynomials.

We next quote the well-known theorem on the form of the prime divisors of the cyclotomic polynomial $Q_n(x)$, that is, the monic polynomial whose roots are the primitive nth roots of unity [14, Th. 94, p. 164].

THEOREM 8. *If $p\nmid n$, then $p\mid Q_n(x)$ if and only if $p\equiv 1 \pmod{n}$.*

As an application, if in the Corollary to Theorem 7 we consider the set of $m+1$ polynomials $f_1, \cdots, f_m, Q_n$ and use Theorem 8, we obtain Theorem 9.

THEOREM 9. *If $f_1, \cdots, f_m$, $m\geqq 1$, are any nonconstant polynomials, then for each fixed $n\geqq 1$ there exist infinitely many primes of the form $kn+1$ for which each of the f_i splits completely.*

6. Polynomials belonging to composite fields. In this section we consider a more complicated situation than previously, in that we study the prime divisors of polynomials belonging to composite fields.

Recall the following facts about such fields: If $K=Q(\alpha)$ and $L=Q(\beta)$ are two finite, simple extensions of Q, then the composite field $M=KL$ is defined as $Q(\alpha, \beta)$. The field M is also a finite, simple extension of Q, for which a primitive element γ exists in the form $\gamma=\alpha+a\beta$, where a is a suitably chosen rational integer. The defining polynomial $h(y)$ of γ will then be the irreducible factor of

$$v(y) = \prod_{i=1}^{n} \prod_{j=1}^{m} [y - (\alpha_i + a\beta_j)], \qquad v(y) \in Q[y], \tag{13}$$

which has γ as a root. In this definition the α_i and β_j are the conjugates of $\alpha_1 = \alpha$ and $\beta_1 = \beta$, and a is chosen so that $v(y)$ has no multiple roots.

In particular, it follows if K and L are normal that $M = KL$ will also be normal. To see this, observe that any conjugate γ_k of γ will be of the form $\gamma_k = \alpha_i + a\beta_j$ for some i and j. The normality of K and L then implies the particular roots α_i and β_j must lie in K and L respectively. But K and L are subfields of M, so α_i and β_j, and hence γ_k are rationally expressible in terms of γ. Hence, M is normal.

THEOREM 10. *Let K and L be normal and let $M = KL$. If $f:K$, $g:L$, and $h:M$, then $P(h) = P(f) \cap P(g)$.*

Proof. Since $K \subseteq M$ and $L \subseteq M$, it follows from Theorem 2 that $P(f) \supseteq P(h)$ and $P(g) \supseteq P(h)$. Thus, $P(f) \cap P(g) \supseteq P(h)$. It remains to show that $P(f) \cap P(g) \subseteq P(h)$. If $\alpha_1 = \alpha, \alpha_2, \cdots, \alpha_n$ and $\beta_1 = \beta, \beta_2, \cdots, \beta_m$ are the roots of f and g respectively, then the normality of f and g implies there exist polynomials $\phi_i(x), \psi_j(x) \in Q[x]$ such that $\alpha_i = \phi_i(\alpha)$, $i = 1, 2, \cdots, n$, and $\beta_j = \psi_j(\beta)$, $j = 1, 2, \cdots, m$, where we take $\phi_1(x) = \psi_1(x) = x$. As previously, assume $\gamma = \alpha + a\beta$ is already a root of h, a a suitably chosen integer.

We next establish an identity analogous to (9), involving both f and g. If in identity (7) y is replaced by $y - a\psi_j(z)$, z a new indeterminate, we obtain

$$\prod_{i=1}^{n} [y - a\psi_j(z) - \phi_i(x)] = f(y - a\psi_j(z)) + f(x)q_j(x, y, z),$$

$q_j(x, y, z) \in Q[x, y, z]$. If j ranges over the values $1, 2, \cdots, m$ and the resulting identities are multiplied together, then

$$\prod_{j=1}^{m} \prod_{i=1}^{n} [y - a\psi_j(z) - \phi_i(x)] = \prod_{j=1}^{m} f(y - a\psi_j(z)) + f(x)r(x, y, z), \tag{14}$$

$r(x, y, z) \in Q[x, y, z]$. Expanding the product on the right gives

$$\prod_{j=1}^{m} f(y - a\psi_j(z)) = y^{mn} + r_1(z)y^{mn-1} + \cdots + r_{mn}(z),$$

$r_i(z) \in Q[z]$. Since substituting $z = \beta$ in the left side of this expansion gives

$$\prod_{j=1}^{m} f(y - a\beta_j) = \prod_{j=1}^{m} \prod_{i=1}^{n} [(y - a\beta_j) - \alpha_i] = v(y)$$

by (13), we can proceed as we did in establishing (7) to obtain

$$\prod_{j=1}^{m} f(y - a\psi_j(z)) = v(y) + g(z)s(y, z), \qquad s(y, z) \in Q[y, z].$$

Putting this result in (14), we obtain the desired identity

$$\prod_{j=1}^{m}\prod_{i=1}^{n}[y-\phi_i(x)-a\psi_j(z)] = v(y)+f(x)r(x,y,z)+g(z)s(y,z). \tag{15}$$

Now suppose $p\in P(f)\cap P(g)$ and p does not divide any denominator in (15); that is, there exist b, $c\in Z$ for which $f(b)\equiv g(c)\equiv 0 \pmod p$. Then it is clear $v(y)$ splits into (not necessarily distinct) linear factors (mod p) when we put $x=b$ and $z=c$ in (15), and consider (15) (mod p). Since $h(y)$ is a factor of $v(y)$, it follows that $p\in P(h)$. ∎

REMARKS. (a) For an ideal theoretic proof see Hasse [9, vol. 1, p. 50].

(b) It can be readily established by examining the above proof that the only primes which need to be excluded as divisors of the denominators in (15) are those dividing $D(f)D(g)$.

Example. Let $f(x)=x^2+1$ and $g(x)=x^2-2$. The corresponding (normal) fields are $K=Q(i)$ and $L=Q(2^{1/2})$. If, say, we choose $\gamma=2^{1/2}+i$ as a primitive element for KL, then the defining polynomial for γ is $h(x)=x^4-2x^2+9$. We shall determine $P(h)$.

We have by the first and second supplements to the Quadratic Reciprocity Law that

$$P(f) = \{p \mid p = 2 \text{ or } p\equiv 1 \pmod 4\}$$

and

$$P(g) = \{p \mid p = 2 \text{ or } p\equiv \pm 1 \pmod 8\}.$$

As $D(f)D(g)=-32$, it follows from Remark (b) that every prime in $P(f)\cap P(g)$ with the possible exception of $p=2$ is in $P(h)$. By inspection, we see that also $2\in P(h)$. (Observe the prime 2, though it is a divisor of both $D(f)$ and $D(g)$, still occurs as a divisor of f, g, and h.) On the other hand, since $D(h)=2^{14}\cdot 3^2$, $p=3$ is the only possible prime divisor of $h(x)$ which is not also in $P(f)\cap P(g)$. Evidently $3\mid h(x)$ and so

$$P(h) = \{p \mid p = 2,\ 3 \text{ or } p\equiv 1 \pmod 8\}.$$

That this result is correct can also be seen from Theorems 3 and 8 and the observation that $Q(\gamma)=Q(\gamma')$ where $\gamma'=(1+i)/2^{1/2}$ is a root of the cyclotomic polynomial $Q_8(x)=x^4+1$. Here $D(Q_8)=2^8$, which implies that $P(h)$ and $P(Q_8)$ are the same except for the possible existence of prime divisors $p=2$ or $p=3$ which may be present in just one of these sets.

COROLLARY. *If $f:K$, $g:L$, and $h:M$ with deg $f=n$, deg $g=m$, and deg $h=k$, where $M=KL$, then $P_k(h)=P_n(f)\cap P_m(g)$.*

Proof. Let K_1, L_1, and M_1 be the *smallest* normal extensions of K, L, and M respectively, and let $f_1:K_1$, $g_1:L_1$, and $h_1:M_1$. Then by Theorem 6, $P_n(f)=P(f_1)$,

$P_m(g)=P(g_1)$, and $P_k(h)=P(h_1)$. Since it can be shown that $M_1=K_1L_1$, the Corollary follows from Theorem 10. ∎

We now present a generalization of Theorem 10, in which only one of the fields K and L is normal. We have presented Theorem 10 separately, however, since the proof explicitly exhibits the solutions of the congruences involved.

THEOREM 11. *Let $M=KL$, where K is normal, and let $f:K$, $g:L$, and $h:M$, where $\deg g=m$. Also, if $\alpha\in K$ is a root of f, let the degree of α relative to L be s. Then $\deg h=ms$, and for $1\leqq i\leqq ms$ we have $P_i(h)=\varnothing$, if $s\nmid i$, and $P_i(h)=P(f)\cap P_r(g)$, if $i=rs$, $1\leqq r\leqq m$. Also, $P(h)=P(f)\cap P(g)$.*

Proof. Let $\alpha_1=\alpha, \alpha_2, \cdots, \alpha_n$ and $\beta_1=\beta, \beta_2, \cdots, \beta_m$ be the roots of f and g respectively. Also, let $\alpha+a\beta$, $a\in Z$, be a root of h, and $L=Q(\beta)$. Consider the factorization

$$f(x) = \prod_{k=1}^{t} f_k(x;\beta), \tag{16}$$

where the f_k are monic and irreducible over L. If f_1 has α as a root, then by hypothesis $\deg f_1=s$. Thus, since $M=L(\alpha)$, it follows that $\deg h=ms$. The normality of f also implies $M=L(\alpha_i)$ for each i, so the degree of all the α_i relative to L is s. Hence, the degree of all the f_k is s, and thus, comparing the degrees of the two sides of (16), we have $n=st$.

A factorization like that in (16) holds if β is replaced by any of its conjugates, since the fields $Q(\beta_j)$ are isomorphic. Thus, substituting β_j for β and $x-a\beta_j$ for x in (16), we obtain

$$f(x-a\beta_j) = \prod_{k=1}^{t} f_k(x-a\beta_j;\beta_j), \qquad j=1,2,\cdots,m.$$

Multiplying these m equations together and using (13) gives

$$v(x) = \prod_{j=1}^{m} f(x-a\beta_j) = \prod_{k=1}^{t} h_k(x),$$

where $h_k(x)=\prod_{j=1}^{m} f_k(x-a\beta_j;\beta_j)$. Since this latter product is symmetric in the β_j, $h_k(x)\in Z[x]$, where $\deg h_k=ms$. Clearly $h_1(x)=h(x)$. Thus, all the $h_k(x)$ belong to M, for certainly $\alpha_i+a\beta$ will be a root of $h_k(x)$ for some i, so, by the normality of f, $Q(\alpha_i+a\beta)=Q(\alpha+a\beta)=M$. Consequently, the $h_k(x)$ are irreducible over Q for all k.

Next consider the respective roots α_i^*, $i=1,2,\cdots,n$, and β_j^*, $j=1,2,\cdots,m$ of f and g over Z_p. Forming the polynomial

$$v^*(x) = \prod_{i=1}^{n}\prod_{j=1}^{m}(x-\alpha_i^*-a\beta_j^*),$$

we have $v^*(x)\in Z_p[x]$, and then, by the argument used to establish (12), we find

$$v(x) \equiv v^*(x) \pmod{p}. \tag{17}$$

Also, if $p \nmid D(f)D(g)$ and $p \in P(f) \cap P_r(g)$ for some r, $1 \leqq r \leqq m$, then exactly r distinct β_j^* will lie in Z_p, and by Theorem 4, all the α_i^* will be distinct in Z_p. Hence, if $p \nmid D(v)$, v^* will have at least nr distinct linear factors (mod p). But v^* cannot have more than this number, for the presence of a linear factor implies for some i and j that $\alpha_i^* + a\beta_j^* \in Z_p$, and since the $\alpha_i^* \in Z_p$ and $p \nmid a$, then $\beta_j^* \in Z_p$. (Actually $p \nmid a$ follows directly from $p \nmid D(v)$.)

Now by (17), $p | v(x)$, and hence p divides at least one of the factors $h_k(x)$ of $v(x)$. Say $p \in P_i(h_k)$. Then by Theorem 3, almost all the primes dividing this factor will also divide every other factor of $v(x)$, and the number of incongruent solutions (mod p) in each of these will be the same. Hence, the number of distinct linear factors of $v(x)$ will be ti. Equating this number to the number of factors in $v^*(x)$, we obtain $ti = nr$, which implies $i = rs$ (using $n = st$). Thus,

$$P_i(h) \supseteq P(f) \cap P_r(g), \qquad i = rs, \quad r = 1, 2, \cdots, m. \tag{18}$$

We next prove the inclusion in (18) can be reversed and that $P_i(h) = \varnothing$ if $s \nmid i$. Suppose $p \in P_i(h)$ for some i, $1 \leqq i \leqq ms$. Since K and L are subfields of M, by Theorem 2, $P_i(h) \subseteq P(f) \cap P(g)$. Also, each $p \in P_i(h)$, with a finite number of exceptions, belongs to some $P_r(g)$, $1 \leqq r \leqq m$. If $p \in P_r(g)$, then since f splits completely (mod p), we again find that v^* has nr distinct linear factors, while v has ti linear factors. Thus, $r = i/s$. This shows r is uniquely determined by i, and when $s \nmid i$, at most a finite number of primes belong to $P_i(h)$. Also, when i is a multiple of s, then that multiple must be r, which implies $P_i(h) \subseteq P(f) \cap P_r(g)$.

Finally,

$$P(h) = \bigcup_{i=1}^{ms} P_i(h) = \bigcup_{r=1}^{m} [P(f) \cap P_r(g)] = P(f) \cap \left[\bigcup_{r=1}^{m} P_r(g) \right] = P(f) \cap P(g). \blacksquare$$

7. Concluding remarks. In the preceding sections we have obtained information about the prime divisors of polynomials from properties of their roots and associated fields. We now consider the converse problem: Do the prime divisors of polynomials determine their algebraic properties? This question was first investigated in 1880 by Kronecker [11], who laid the foundations for much of the research that has followed.

In our brief discussion of this question, we shall limit ourselves to considering true converses of Theorems 2 and 3, which represent quite well the results in this field. We shall present these converses without proof, however, since all the known proofs employ ideal theoretic and transcendental methods, which lie outside the scope of this paper. The reader who would like further information on these matters is referred to the rather full discussion in Hasse [9, vol. 2, pp. 138–146], where the basic references can be found.

We begin with the converse of Theorem 3: If two irreducible polynomials f and g of degree n are such that $P_i(f) = P_i(g)$, $i = 1, 2, \cdots, n$, then they belong to the same field. That this statement is *false* was shown by Gassmann [7] in 1926, when he proved the existence of two 180th degree polynomials (!) belonging to *nonconjugate* fields, which not only satisfy the required conditions, but also factor in exactly the same way (mod p) for almost all primes p. (See also Schinzel [19].) One of the authors [8, p. 138] has recently given a pair of polynomials $f(x)=x^8-3\cdot 2^4$, $g(x)=x^8-3^7$, which has the Gassmann property and which provides a simpler counterexample to the converse of Theorem 3.

We can obtain a *true* converse to Theorem 3, however, if we impose the further condition that f and g be normal (see Bauer [1]). (We note in this case that $P_n(f)=P(f)$ and $P_n(g)=P(g)$.)

If we modify the hypothesis of the converse of Theorem 2 in a similar way, we obtain the THEOREM: *If* $f:K$ *with* f *normal,* $g:L$, *and* $P(f)\supseteq P(g)$, *then* $K\subseteq L$. (See Bauer [2].)

As a particular application of this theorem in the case K is the nth cyclotomic field, we have the result: *A field L will contain K if and only if for* $g:L$, $P(g)$, *with a finite number of exceptions, contains only primes* $p\equiv 1 \pmod{n}$. (See Theorem 9.)

In closing, we would like to remark it would be of some interest if elementary proofs of the theorems of this section could be constructed.

References

1. M. Bauer, Über einen Satz von Kronecker, Arch. Math. und Physik, Bd. 6, 218–219.

2. ———, Zur Theorie der algebraischen Zahlkörper, Math. Ann., 77 (1916) 353–356.

3. G. Birkhoff and S. MacLane, A Survey of Modern Algebra, 3rd ed., Macmillan, New York, 1965, Ch. XIV.

4. Z. I. Borevich and I. R. Shafarevich, Number Theory, Academic Press, New York, 1966.

5. R. Dedekind, Über den Zusammenhang zwischen der Theorie der Ideale und der höheren Kongruenzen, Abh. Akad. Wiss. Göttingen, 23 (1878) 3–37. (Repr. in Vol. 1, pp. 202–232 of the Gesammelte Mathematische Werke, Chelsea, New York, 1969. The theorem in question is Thm. I, pp. 212–213.)

6. L. Fjellstedt, Bemerkungen über gleichzeitige Lösbarkeit von Kongruenzen, Ark. Mat., 3 (1955) 193–198.

7. F. Gassmann, Bemerkungen zur vorstehenden Arbeit von Hurwitz, Math. Z., 25 (1926) 665–675.

8. I. Gerst, On the theory of nth power residues and a conjecture of Kronecker, Acta. Arith., 17 (1970) 121–139.

9. H. Hasse, Bericht über neuere Untersuchungen und Probleme aus der Theorie der algebraischen Zahlkörper, 2nd ed., Physica, Würzburg, 1965.

10. ———, Über das Problem der Primzerlegung in Galoisschen Zahlkörpern, S.-B. Berlin. Math. Ges., (1951/52) 8–27.

11. L. Kronecker, Über die Irreducibilität von Gleichungen, Monatsb. Deutsch. Acad. Wiss. Berlin, (1880) 155–163.

12. S. Lang, Algebraic Numbers, Addison-Wesley, Reading, Mass., 1964.

13. H. B. Mann, Introduction to Algebraic Number Theory, The Ohio State University Press, Columbus, 1955.

14. T. Nagell, Introduction to Number Theory, 2nd ed., Chelsea, New York, 1964, pp. 164–167.

15. ———, Zahlentheoretische Notizen I, Ein Beitrag zur Theorie der höheren Kongruenzen, Videnskapsselskapets Skrifter I, Matem.-Naturv. Klasse No. 13, Kristiania, 1923, 3–6.

16. T. Nagell, Sur les diviseurs premiers des polynômes, Acta Arith., 15 (1969) 235–244.

17. H. Pollard, The Theory of Algebraic Numbers, MAA Carus Mathematical Monographs No. 9, Wiley, New York, 1950.

18. P. Samuel, Théorie Algébrique des Nombres, Hermann, Paris, 1967.

19. A. Schinzel, On a theorem of Bauer and some of its applications, Acta Arith., 11 (1966) 333–344.

20. I. Schur, Über die Existenz unendlich vieler Primzahlen in einiger speziellen arithmetischen Progressionen, S.-B. Berlin. Math. Ges., 11 (1912) 40–50.

21. B. van der Waerden, Modern Algebra, Ungar, New York, 1953, vol. 1, Ch. 4 and 5.

WHAT IS A RECIPROCITY LAW?*

B. F. WYMAN, Stanford University

1. Introduction. The Law of Quadratic Reciprocity has fascinated mathematicians for over 300 years, and its generalizations and analogues occupy a central place in number theory today. Fermat's glimmerings (1640) and Gauss's proof (1796) have been distilled to an amazing abstract edifice called **class field theory.**

As a graduate student I learned the great cohomological machine and studied **Artin's Reciprocity Law**, one form of which gives an isomorphism between two cohomology groups. A little later I read Shimura's paper [**19**], called "A non-solvable reciprocity law," and couldn't understand the title at all. Where were the cohomology groups? Why was Shimura's theorem a reciprocity law?

It was an embarrassing, but healthy ignorance, because it made me go back and figure out the number theory that lay behind all those cohomology groups. Such a reassessment is especially important nowadays, because it seems more and more certain that the *next* generalization of the Law of Quadratic Reciprocity will require new techniques, and nobody is quite sure which techniques will work.

In this paper I should like to discuss reciprocity laws from a rather general but very concrete point of view. Suppose $f(X)$ is a monic irreducible polynomial with integral coefficients, and suppose p is a prime number. Reducing the coefficients of $f(X)$ modulo p gives a polynomial $f_p(X)$ with coefficients in the field $\mathbf{F}_p$ of p elements. The polynomial $f_p(X)$ may factor (even though the original $f(X)$ was irreducible). If $f_p(X)$ factors over $\mathbf{F}_p$ into a product of distinct linear factors, we say that $f(X)$ **splits completely modulo** p, and we define $\mathrm{Spl}(f)$ to be the set of all primes such that $f(X)$ splits completely modulo p.

The general **reciprocity problem** we shall be considering is: *Given $f(X)$ as above, describe the factorization of $f_p(X)$ as a function of the prime p.* Sometimes we ask for less: *give a rule to determine which primes belong to* $\mathrm{Spl}(f)$. This vague

* From AMERICAN MATHEMATICAL MONTHLY, vol. 79 (1972), pp. 571–586.

Bostwick Wyman received his Ph.D. at Berkeley in 1966 under G. Hochschild and A. Ogg. He was an Instructor at Princeton for two years and has been an Assistant Professor at Stanford since then. He spent a year's leave at the University of Oslo. His main research interest is algebraic number theory. *Editor.*

question is hard to make precise until it is answered. What is a "rule"? What is an acceptable method for describing the factorization of $f_p(X)$? Anyway, a satisfactory answer to this unsatisfactory question will be called a **reciprocity law.**

Quadratic polynomials are easiest to handle, and Section 2 shows how the usual Law of Quadratic Reciprocity gives a reciprocity law. (If it did not, our language would be all wrong.) Section 3 treats cyclotomic polynomials, and Sections 4 and 5 take up general results. It turns out that the reciprocity problem has been solved satisfactorily for polynomials which have an abelian Galois group, but that very little is known about polynomials whose Galois group is not abelian.

For an arbitrary polynomial $f(X)$ and a specific prime p, it only takes a finite number of steps to decide whether p is in $\mathrm{Spl}(f)$. Sections 6 and 7 give a description of an efficient algorithm for doing this calculation and report on results obtained for a family of quintic polynomials. These results probably do *not* constitute a reciprocity law, and the last section tries to answer the main question, "What is a reciprocity law?"

Prerequisites. Section 2 assumes only knowledge of the Law of Quadratic Reciprocity. The later sections assume somewhat more: acquaintance with cyclotomic polynomials, Galois groups, and the division algorithm in polynomial rings. Parts of Sections 4 and 5 assume the rudiments of algebraic number theory, but they can be skipped.

Notation. We use $\boldsymbol{Z}$, $\boldsymbol{Q}$, and $\boldsymbol{C}$ for the integers, rational numbers, and complex numbers, respectively. If q is a prime or prime power, then $\mathbf{F}_q$ is the field with q elements. If R is a ring, then $R[X]$ is the ring of polynomials with coefficients in R; mostly we deal with $\boldsymbol{Z}[X]$ and $\mathbf{F}_p[X]$.

2. Quadratic Polynomials. Suppose that $f(X)$ is an irreducible quadratic polynomial with integral coefficients. If p is a prime number, let $f_p(X)$ be the corresponding polynomial in $\mathbf{F}_p[X]$ obtained by reducing the coefficients of $f(X)$ modulo p. The reduced polynomial $f_p(X)$ can factor in one of three ways:

(0) $f_p(X) = l(X)^2$, where $l(X)$ is linear.

(1) $f_p(X) = l_1(X) \cdot l_2(X)$, where $l_1(X)$ and $l_2(X)$ are two distinct linear polynomials. In this case we say that $f(X)$ **splits** modulo p.

(2) $f_p(X)$ is irreducible in $\mathbf{F}_p[X]$.

In this paper we shall stick to polynomials of the form $X^2 - q$, where q is prime. If $f(X) = X^2 - q$, then Case (0) occurs modulo p when $p = q$, and also when $p = 2$. (The prime 2 behaves strangely for quadratic polynomials.) To distinguish Cases (1) and (2) we need to know whether q is a quadratic residue modulo p. If q is a quadratic residue, and $q \equiv a^2 \pmod p$, we get $X^2 - q \equiv (X + a)\,(X - a) \pmod p$. This puts us in Case (1) if $p \neq 2$. If q is not a quadratic residue, we are in Case (2).

Using the Legendre symbol, and ignoring the prime 2 and the exceptional Case (0) (a widespread practice!), we summarize:

(1) $X^2 - q$ splits modulo p if $(q/p) = +1$.

(2) $X^2 - q$ is irreducible modulo p if $(q/p) = -1$.

Remember that we are trying to describe the set $\mathrm{Spl}(X^2 - q)$ of primes p such that $X^2 - q$ splits modulo p, and now we know that *p is in $\mathrm{Spl}(X^2 - q)$ if and only if $(q/p) = +1$.*

The reader should still be skeptical, because this translation of the problem does not do much for us. The symbol (q/p) is not easy to evaluate, and besides, if we change p we have to start all over again. Since there are infinitely many primes p, this naive approach requires an infinite amount of work to describe $\mathrm{Spl}(X^2 - q)$. Can we find a better description?

Since q is fixed and p varies, things would be better if we could use the symbol (p/q) instead of (q/p). For fixed q, the value of (p/q) depends only on the residue class of p modulo q. There are only q residue classes, and therefore only q symbols to evaluate. This suggests looking for a relationship between (p/q) and (q/p) in hopes of using (p/q) to describe $\mathrm{Spl}(X^2 - q)$. Now you can guess where we are; we have sneaked up behind the Law of Quadratic Reciprocity. Legendre's statement goes like this [**10**, p. 455 ff.]:

THEOREM 2–1 (Law of Quadratic Reciprocity): *Let p be an odd prime. Then*

1. $(1/p) = (-1)^P$, *where* $P = \frac{1}{2}(p-1)$.
2. $(2/p) = (-1)^R$, *where* $R = (p^2-1)/8$.
3. *If q is another odd prime, then* $(q/p) = (-1)^{P\cdot Q}(p/q)$, *where* $P = \frac{1}{2}(p-1)$ *and* $Q = \frac{1}{2}(q-1)$.

Gauss gave the first proof of this theorem [**6**, Article 131 ff.], and a modern proof can be found in almost any number theory text, for example, Niven and Zuckerman [**17**, p. 74].

This venerable law is really exactly what we need to compute $\mathrm{Spl}(X^2 - q)$. We start with a less fancy but quite useful form of the theorem.

THEOREM 2–2. *Let p and q be distinct odd primes.*

1. *If* $q \equiv 1 \pmod 4$, *then* $(q/p) = (p/q)$.
2. *If* $q \equiv 3 \pmod 4$, *then* $$(q/p) = \begin{cases} (p/q) & \text{if } p \equiv 1 \pmod 4 \\ -(p/q) & \text{if } p \equiv 3 \pmod 4. \end{cases}$$

The derivation of Theorem 2–2 from Theorem 2–1 is an easy exercise.

Now we are ready to give a prescription for computing (q/p) for fixed q and variable p: First, compute (b/q) for all integers b such that $1 \leqq b \leqq q-1$. Second, given p, find the b such that $1 \leqq b \leqq q-1$ and $b \equiv p \pmod q$. We have therefore

$(b/q) = (p/q)$. Third, use the tables in Theorem 2 to convert knowledge of (p/q) into knowledge of (q/p).

Example 1. $q = 17$. The squares modulo 17 are 1, 2, 4, 8, 9, 13, 15, and 16, so that we have $(b/17) = +1$ for b equal one of these, and $(b/17) = -1$ for $b = 3, 5, 6, 7, 10, 11, 12$, or 14. That is (second step), $(p/17) = +1$ if and only if $p \equiv 1, 2, 4, 8, 9, 13, 15$, or 16 (mod 17). Finally, (third step), $17 \equiv 1 \pmod 4$ so that $(17/p) = (p/17)$. If we return to the language of polynomials splitting modulo a prime, we can say that

$$p \in \mathrm{Spl}(X^2 - 17) \text{ if and only if}$$

$$p \equiv 1, 2, 4, 8, 9, 13, 15, \text{ or } 16 \pmod{17}.$$

That is, the set $\mathrm{Spl}(X^2 - 17)$ can be defined by "congruence conditions modulo 17."

Example 2. $q = 11$. By finding the quadratic residues modulo 11, we conclude that $(p/11) = +1$ if and only if $p \equiv 1, 3, 4, 5$, or 9 (mod 11). In this case $11 \equiv 3 \pmod 4$ so $(11/p) = \pm (p/11)$ with a sign that depends on the residue of p modulo 4. For example, $23 \equiv 1 \pmod{11}$, and $23 \equiv 3 \pmod 4$, so that $(11/23) = -(23/11) = -(1/11) = -1$. On the other hand, $89 \equiv 1 \pmod{11}$ but $89 \equiv 1 \pmod 4$ and $(11/89) = +(89/11) = +(1/11) = +1$. Using the Chinese Remainder Theorem, we see that the value of $(11/p)$ depends on the residue class of p modulo 44, and after some calculation we get:

$$p \in \mathrm{Spl}(X^2 - 11) \text{ if and only if}$$

$$p \equiv 1, 5, 7, 9, 19, 25, 35, 37, 39, \text{ or } 43 \pmod{44}.$$

In this case the set $\mathrm{Spl}(X^2 - 11)$ can be described by congruence conditions modulo 44.

The results of the last two examples are actually quite general.

THEOREM 2–3. *Suppose that q is an odd prime. Then the set* $\mathrm{Spl}(X^2 - q)$ *can be defined by congruence conditions modulo q if $q \equiv 1$* (mod 4) *and modulo $4q$ if $q \equiv 3$* (mod 4). *Furthermore,* $\mathrm{Spl}(X^2 - 2)$ *can be described by congruence conditions modulo* 8.

In this theorem the phrase "congruence conditions" is interpreted as in the examples. The first part follows from Theorem 2–2, and the second part from Theorem 2–1, part 2. Details are left as an exercise for the reader.

Theorem 2–3 shows that the Law of Quadratic Reciprocity gives a "reciprocity law" in the sense of Section 1. That is, it yields a nice description of sets $\mathrm{Spl}(f)$ for quadratic polynomials. In the next section we shall try to find such a reciprocity law for certain special polynomials (the *cyclotomic* ones) of higher degree.

3. Cyclotomic polynomials. Suppose ζ is a primitive nth root of unity; for instance, $\zeta = e^{2\pi i/n}$ is one choice. Then the minimal polynomial of ζ over $\boldsymbol{Q}$ is written

$\Phi_n(X)$ and is called the **n-th cyclotomic polynomial.** One knows that $\Phi_n(X)$ has coefficients in $\boldsymbol{Z}$ and has degree $\phi(n)$, where ϕ is the Euler phi-function. It can be computed conveniently from the formula

$$X^n - 1 = \prod_{d|n} \Phi_d(X),$$

where the product runs are all divisors of n, including 1 and n itself. For example, $\Phi_1(X) = X - 1$, and if p is a prime, then $X^p - 1 = (X-1) \cdot \Phi_p(X)$ and

$$\Phi_p(X) = X^{p-1} + X^{p-2} + \cdots + X + 1.$$

Proofs of these facts and more information about $\Phi_n(X)$ can be found in Lang [**14**, p. 206], van der Waerden [**20**, Sec. 53] and in many other algebra textbooks.

The goal of this section is a "reciprocity law" for these cyclotomic polynomials. We want a description of the set $\mathrm{Spl}(\Phi_n(X))$, and, just as in the quadratic case, the description will be given in terms of congruence conditions with respect to a modulus which depends on the polynomial. The theorem follows.

THEOREM (Cyclotomic Reciprocity Law). *The cyclotomic polynomial $\Phi_n(X)$ factors into distinct linear factors modulo p if and only if $p \equiv 1 \pmod{n}$.*

First we give a lemma about finite fields, and then use the lemma to prove the theorem. To avoid excessive notation we also use the symbol $\Phi_n(X)$ to denote the cyclotomic polynomial with coefficients reduced modulo a prime p.

LEMMA. *Suppose p is a prime number, and a is an element of $\mathbf{F}_p$ with $a^n = 1$. If $a^d \neq 1$ for all proper divisors d of n, then $X - a$ divides $\Phi_n(X)$ in $\mathbf{F}_p[X]$.*

Proof. The relation $X^n - 1 = \prod_{d|n}\Phi_d(X)$ holds in $\mathbf{F}_p$, so that $a^n - 1 = 0 = \prod_{d|n}\Phi_d(a)$. Since $\mathbf{F}_p$ is a field, it follows that $\Phi_m(a) = 0$ for some divisor m of n, and that $a^m - 1 = \prod_{d|n}\Phi_d(a) = 0$. This gives $a^m = 1$ which can only happen if $m = n$. Therefore, $\Phi_n(a) = 0$, and $X - a$ divides $\Phi_n(X)$.

Proof of theorem. Recall that the multiplicative group $\mathbf{F}_p^*$ of non-zero elements of $\mathbf{F}_p$ is cyclic of order $p - 1$. Therefore, $\mathbf{F}_p^*$ has a cyclic subgroup of order n if and only if n divides $p - 1$. Such a subgroup has $\phi(n)$ generators, so that $\mathbf{F}_p^*$ contains $\phi(n)$ distinct primitive nth roots of 1 (these generators!) if and only if it contains one, and this happens exactly when $p \equiv 1 \pmod{n}$.

Now assume $p \equiv 1 \pmod{n}$, so that $\mathbf{F}_p$ contains $\phi(n)$ distinct primitive roots of 1. These must be roots of $\Phi_n(X)$, by the lemma, so that $\Phi_n(X)$ splits into a product of distinct linear factors.

Conversely, assume that $\Phi_n(X)$ splits into linear factors modulo p. If these factors are distinct, then p cannot divide n (exercise: start from Lang [**14**, p. 206]), and it follows easily that $X^n - 1$ also has distinct roots modulo p. Let a be a root of $\Phi_n(X)$ in $\mathbf{F}_p$, so that $a^n = 1$. If d is the smallest divisor of n such that $a^d = 1$, then $\Phi_d(a) = 0$

by the lemma. If $d \neq n$, the basic relationship $X^n - 1 = \prod_{d|n} \Phi_d(X)$ shows that a is at least a double root of $X^n - 1$, a contradiction. Therefore a generates a cyclic subgroup of order n in $\mathbf{F}_p^*$, and $p \mid n-1$. This completes the proof of the "cyclotomic reciprocity law."

4. Abelian polynomials. In the first two sections we saw that if $f(X)$ is a quadratic or cyclotomic polynomial, then the set $\mathrm{Spl}(f)$ can be described by congruences with respect to a certain modulus. This gives a rather precise solution to the vague "reciprocity problem."

Unfortunately, such a nice description of $\mathrm{Spl}(f)$ is not always possible. We can, however, describe exactly the set of polynomials for which congruence conditions give the answer we need.

First we must recall some Galois theory. Associated to each polynomial of degree n is the **root field** $K_f = Q(\alpha_1, \cdots, \alpha_n)$, where $\alpha_1, \cdots, \alpha_n$ are the complex roots of $f(X)$. (We avoid the more common term, "splitting field," because of possible confusion with polynomials "splitting modulo p.") The field K_f is a finite Galois extension of Q, uniquely determined by $f(x)$. The Galois group of K_f/Q is often called the **Galois group of $f(X)$**, and $f(X)$ is called an **abelian polynomial** if its Galois group is abelian.

The next theorem shows the importance this notion has for the reciprocity problem.

ABELIAN POLYNOMIAL THEOREM. *The set* $\mathrm{Spl}(f)$ *can be described by congruences with respect to a modulus depending only on* $f(X)$ *if and only if* $f(X)$ *is an abelian polynomial.*

Why should Galois groups have anything to do with polynomials splitting modulo primes? What are "congruence conditions" exactly? Enough machinery is developed in the rest of this section to establish the importance of the Galois groups, and to give a precise form of the theorem. A complete proof is far beyond the scope of this paper. In fact, the proof of the theorem involves almost all of "class field theory over the rationals." Perhaps the best avenue for an ambitious reader is to work through a basic text in algebraic number theory, and then go on to the cohomological treatment in Cassels and Fröhlich [3], or the analytic approaches of Lang [15], Weil [21], or Goldstein [7].

At this point we must escalate the prerequisites: the reader should be familiar with integral dependence, Dedekind domains, and the factorization of prime ideals in Galois extensions, or else be willing to suspend his disbelief. It is safe to skip this discussion and go on to Section 5.

Let K be an algebraic extension of Q. The elements of K whose (monic) minimal polynomial has coefficients in Z make up the ring of **algebraic integers in K**, written $\mathcal{O}_K$. The ring $\mathcal{O}_K$ is a Dedekind domain if K/Q is finite.

If p is a prime in $\boldsymbol{Z}$, the ideal $p\mathcal{O}_K$ factors uniquely into a product of prime ideals:

$$p\mathcal{O}_K = \mathfrak{P}_1 \cdots \mathfrak{P}_r.$$

If $\mathfrak{P}$ is one of the factors of p, the residue class ring $\mathcal{O}_K/\mathfrak{P}$ is a finite field extension of $\boldsymbol{Z}/p\boldsymbol{Z}$. This **residue class field extension** is cyclic, with Galois group generated by the **Frobenius map** ϕ: $\phi(a) = a^p$ for all a in $\mathcal{O}_K/\mathfrak{P}$.

Except for a finite number of exceptions (called **ramified primes**) the $\mathfrak{P}_i$ appearing in $p\mathcal{O}_K$ are all distinct. If $K/\boldsymbol{Q}$ is Galois with group G, and p is not ramified, then for each $\mathfrak{P}_i$ there is a unique $\sigma \in G$ such that σ reduces to the Frobenius map modulo $\mathfrak{P}_i$. This automorphism is called the **Artin symbol** corresponding to $\mathfrak{P}$. We denote it by $\sigma_{\mathfrak{P}}$, so that the defining formula is

$$\sigma_{\mathfrak{P}}(a) \equiv a^p (\operatorname{mod} \mathfrak{P}) \text{ for all } a \in \mathcal{O}_K.$$

These Artin symbols $\sigma_{\mathfrak{P}}$ are not good enough for our purposes. We need to define an Artin symbol σ_p corresponding to a prime number p "downstairs." This is not possible in general, because different choices of the ideal $\mathfrak{P}$ may give different $\sigma_{\mathfrak{P}}$ in G. How are these various $\sigma_{\mathfrak{P}}$ related? If $\mathfrak{P}$ and $\mathfrak{Q}$ are two factors of $p\mathcal{O}_K$, then there is an automorphism τ in G such that $\tau(\mathfrak{P}) = \mathfrak{Q}$. It turns out that $\sigma_{\mathfrak{Q}} = \tau\sigma_{\mathfrak{P}}\tau^{-1}$. All the $\sigma_{\mathfrak{P}}$ corresponding to a single p are conjugate, and we call this *conjugacy class* the **Artin symbol corresponding to** $\boldsymbol{p}$. In the good case that G is abelian, we can identify a conjugacy class with its unique member, so that the Artin symbol for p is an element σ_p in G.

EXERCISE. If you are familiar with number theory in quadratic fields, try to work out the Artin symbols for them. Start with the field $\boldsymbol{Q}(\sqrt{q})$ where q is an odd prime, and identify the Galois group with $\{\pm 1\}$. Check that after this identification, the Artin symbol σ_p is exactly the Legendre symbol (q/p). (Were you wondering why σ_p is called a "symbol"?) What about more complicated quadratic fields? Finally, try to compute the Artin symbols σ_p for the cyclotomic field $\boldsymbol{Q}(\zeta_m)$. (Goldstein [7, p. 96 ff.] is one of many possible references.)

From here on, $\boldsymbol{K}/\boldsymbol{Q}$ is an abelian extension with group G. We denote by $\boldsymbol{Q}^*$ the multiplicative group of non-zero rational numbers, and we think of $\boldsymbol{Q}^*$ as the (multiplicative) free abelian group generated by the primes. For a fixed field $\boldsymbol{K}$, let $\Gamma \subseteq \boldsymbol{Q}^*$ be the free abelian subgroup generated by the unramified primes in $\boldsymbol{K}/\boldsymbol{Q}$. We extend the definition of the Artin symbol by setting $\sigma_{pq} = \sigma_p \cdot \sigma_q$, and $\sigma_a = \sigma_p^{-1}$ if $a = 1/p$. This procedure gives a *group homomorphism*, $\sigma: \Gamma \to G$, called the **Artin map**.

Can we find the kernel and image of this homomorphism? The image is easy to describe: *the Artin map σ is surjective.* We shall get some idea of the proof in the next section.

What about the kernel? The result here is more complicated and requires some more terminology. If a is an integer, the **ray group** Γ_a is defined as follows: a rational number $r \neq 0$ is in Γ_a if r can be written as c/d with c and d prime to a and $c \equiv d \pmod{a}$. Then *the kernel of the Artin map for K/Q contains the ray group Γ_a for some $a = p_1^{e_1} \cdots p_s^{e_s}$, where $p_1, \cdots, p_s$ are the ramified primes in K, and $e_i \geqq 1$.*

The two italicized statements above make up the **Artin Reciprocity Law**. Emil Artin conjectured it in 1923 [**1**, p. 98], and proved it in 1927 [**1**, p. 131]. (Artin worked over arbitrary number fields, not just over Q.) The theorem is central in all modern treatments of class field theory. It is proved in all the books recommended above, and in many others as well. We state it again for reference.

Artin Reciprocity Law: *Let K/Q be a finite abelian extension with Galois group G, and let Γ be the subgroup of Q^* generated by the primes unramified in K. Then the Artin symbol gives a surjective group homomorphism $\sigma: \Gamma \to G$ whose kernel contains the ray group Γ_a, where a is an appropriate product of the ramified primes.*

The Artin Reciprocity Law is a precise form of half the Abelian Polynomial Theorem: if $f(X)$ is an abelian polynomial, then $\mathrm{Spl}(f)$ can be described by congruence conditions. To see why, we start with a crucial lemma.

Lemma. *Suppose $f(X)$ is an abelian polynomial with root field K, Galois group G, and Artin map $\sigma: \Gamma \to G$. Then except perhaps for a finite number of exceptional primes, $f(X)$ splits modulo p if and only if σ_p is trivial.*

Proof. We can only give an outline here. If p is unramified and $p\mathcal{O}_K = \mathfrak{P}_1 \cdots \mathfrak{P}_s$, then the Chinese Remainder Theorem gives

$$\mathcal{O}_K/p\mathcal{O}_K \cong \oplus_{i=1}^{s} \mathcal{O}_K/\mathfrak{P}_i.$$

On the other hand, except for a finite number of p,

$$\mathcal{O}_K/p\mathcal{O}_K \cong \mathbf{F}_p[X]/(f_p(X)),$$

where $f_p(X)$ is the reduction of $f(X)$ modulo p. (This is a hard exercise; the exceptions all divide the discriminant of $f(X)$.) Therefore, except for finitely many p,

$$\mathbf{F}_p[X]/(f_p(X)) \cong \oplus_{i=1}^{s} \mathcal{O}_K/\mathfrak{P}_i.$$

When is the Artin symbol σ_p trivial? By definition σ_p induces the Frobenius map, $x \to x^p$, on each direct summand. The Frobenius map is trivial on $\mathcal{O}_K/\mathfrak{P}$ only if $\mathcal{O}_K/\mathfrak{P} \cong \mathbf{F}_p$, so that $\mathbf{F}_p[X]/(f_p(X)) \cong \mathbf{F}_p^n$ when σ_p is trivial, and this is only possible when $f_p(X)$ factors into linear factors. All the steps are reversible, so the converse holds too.

This lemma, combined with the Artin Reciprocity Law, guarantees that the set $\mathrm{Spl}(f)$ contains all primes p such that $p \equiv 1 \pmod{a}$, with at most finitely many exceptions. (Check this!)

We need to change $\mathrm{Spl}(f)$ slightly at this point. Add to $\mathrm{Spl}(f)$ any primes $p \equiv 1 \pmod{a}$ not already there, and throw away any divisors of a. Call the resulting set S; this is the set we can describe by explicit congruence conditions.

Let $Q^*(a)$ be the multiplicative subgroup generated by all primes p which do not divide a. (A fraction b/c in lowest terms is in $Q^*(a)$ if both b and c are prime to a.) Let S' be the subgroup of Q^* generated by S. The set S has been chosen so that $\Gamma_a \subseteq S' \subseteq Q^*$, and the importance of these inclusions comes out in the next lemma.

LEMMA. *$Q^*(a)/\Gamma_a \cong (Z/aZ)^*$, where $(Z/aZ)^*$ is the group of invertible elements in Z/aZ.*

Proof. Define $\theta: Q^*(a)/\Gamma_a \to (Z/aZ)^*$ by $\theta(b/c) = bc^{-1} \pmod{a}$. Check as an exercise that θ is a surjective homomorphism with kernel exactly Γ_a.

This lemma supplies us with congruence conditions. Starting with $\mathrm{Spl}(f)$, pass to S, and consider the set $\theta(S')$ of residue classes modulo a. A given prime p will lie in S if and only if its residue class modulo a lies in $\theta(S')$. Since S and $\mathrm{Spl}(f)$ differ in at most a finite number of primes, we shall be content with this result.

Next we attack the other half of the Abelian Polynomial Theorem: If $\mathrm{Spl}(f)$ can be defined by congruences, then $f(X)$ must be an abelian polynomial. We shall need a hard theorem which says (roughly) that the root field K_f of $f(X)$ is uniquely determined by the set $\mathrm{Spl}(f)$. We introduce some notation: If S and T are two sets of primes, then $S \subseteq^* T$ means that except for at most a finite number of exceptions every member of S is a member of T. The precise statement is then:

INCLUSION THEOREM. *Suppose $f(X)$ and $g(X)$ are polynomials with root fields K_f and K_g, respectively. Then $K_f \subseteq K_g$ if and only if $\mathrm{Spl}(g) \subseteq^* \mathrm{Spl}(f)$.*

Note the reversal! The similarity to Galois theory can be made very precise for abelian polynomials and is an important part of class field theory. The theorem itself holds for arbitrary $f(X)$ and $g(X)$.

It is not hard to prove that $K_f \subseteq K_g$ implies $\mathrm{Spl}(g) \subseteq^* \mathrm{Spl}(f)$. The converse requires analytic techniques, and is a corollary of the Tchebotarev Density Theorem discussed in the next section. See Cassels and Fröhlich [3, Exercise 6.1, p. 362] or Goldstein [7, Theorem 9-1-13, p. 164] for a proof.

Assume now that $\mathrm{Spl}(f)$ can be defined by congruences modulo an integer a. Actually we assume more: namely, that $\mathrm{Spl}(f)$ contains the ray group Γ_a. (Exercise: What's the difference between these assumptions?) According to Section 3, Γ_a is $\mathrm{Spl}(\Phi_a(X))$, and the root field of $\Phi_a(X)$ is the cyclotomic field $Q(\zeta_a)$, which is abelian over Q. Since $\Gamma_a \subseteq \mathrm{Spl}(f)$, the Inclusion Theorem gives $K_f \subseteq Q(\zeta_a)$, so that K_f must also be abelian over Q.

One corollary of this discussion deserves special mention.

KRONECKER'S THEOREM. *Every abelian extension of Q is contained in a cyclotomic extension.*

Proof. Exercise: Combine the Artin Reciprocity Law with the argument above. (There is an elementary proof in Gaal [**5**, p. 242].)

5. General polynomials. The Tchebotarev Density Theorem. If $f(X)$ is an irreducible polynomial in $\boldsymbol{Z}[X]$ which is not abelian, then very little can be said about the set $\mathrm{Spl}(f)$. The best general result is a statement about the relative "size" of $\mathrm{Spl}(f)$. First we describe a numerical measure of sets of primes called the **density**.

Let Π be the set of all prime numbers, and let $T \subseteq \Pi$ be any subset. For any real $x \geqq 1$, let

$$\delta(T,x) = \frac{\mathrm{card}\,\{p \in T \mid p < x\}}{\mathrm{card}\,\{p \in \Pi \mid p < x\}}.$$

DEFINITION. *If T is a set of primes such that* $\lim_{x\to\infty} \delta(x,T) = \delta(T)$, *then T has density $\delta(T)$.*

Note that the limit may not exist. In that case we say, naturally enough, that T "does not have a density." If T does have a density, then $0 \leqq \delta(T) \leqq 1$. Since Π is infinite, any finite set of primes has density 0, and it is easy to see that if S and T differ by a finite set of primes, then $\delta(S) = \delta(T)$. Clearly $\delta(\Pi) = 1$.

One can prove that a set of primes is infinite by showing that it has a non-zero density. The first theorem of this type was proved by Lejeune Dirichlet in 1837.

DIRICHLET'S THEOREM. *Suppose m is a positive integer and a is an integer relatively prime to m. Then the set of all primes congruent to a modulo m has a density equal to $1/\phi(m)$.*

In particular, the set of all primes congruent to a modulo m is infinite. Although this much can be proved directly for some a and m (see Hardy and Wright [**8**, p. 13]), no general proof avoids analysis and the notion of density. Proofs of the theorem can be found all over; one is in Davenport [**4**, pp. 1 and 28].

The density result we need for the reciprocity problem is the *Tchebotarev Density Theorem.* We give a weakened version first.

WEAK TCHEBOTAREV THEOREM. *Let $f(X)$ be an irreducible polynomial in $\boldsymbol{Z}[X]$ with root field $\boldsymbol{K}_f$, and suppose that $[\boldsymbol{K}_f : \boldsymbol{Q}] = n$. Then $\mathrm{Spl}(f)$ has a density equal to $1/n$.*

This theorem implies part of Dirichlet's Theorem. Take $f(X)$ to be the cyclotomic polynomial $\Phi_m(X)$ so that $[\boldsymbol{K}_f : \boldsymbol{Q}] = \phi(m)$, so that the theorem gives $\delta[\mathrm{Spl}(\Phi_m)] = 1/\phi(m)$. By Section 3, a prime p is in $\mathrm{Spl}(\Phi_m)$ if and only if $p \equiv 1 \pmod{m}$ and putting all this together gives Dirichlet's result for $a = 1$. The rest of Dirichlet's Theorem follows from the full Tchebotarev Theorem discussed below.

The interested reader should go back to Section 2 and examine quadratic polynomials from the point of view of density results. The following main result can be derived from either of the two theorems above: *Suppose a is not a perfect square. Then the set of primes p such that $(a/p) = +1$ has density* $\frac{1}{2}$. (What about those p with $(a/p) = -1$? What about primes dividing a?)

To explain the strong form of Tchebotarev's theorem, we need to use Artin symbols again. To read the rest of this section you need either the last part of Section 4 or faith. It is safe to skip to Section 6.

Let $f(X)$ in $\mathbf{Z}[X]$ have root field $\mathbf{K}_f$ and Galois group G. The group G is not necessarily abelian, and the Artin symbol corresponding to p is a conjugacy class C_p of elements of G. (There are a finite number of ramified p for which C_p cannot be defined. We ignore these.)

Tchebotarev proved his theorem in 1925 and his methods inspired Artin's proof of the Reciprocity Law.

TCHEBOTAREV DENSITY THEOREM. *Let $f(X) \in Z[X]$ be irreducible with Galois group G, and let C be a fixed conjugacy class of elements of G. Let S be the set of primes p whose Artin symbol C_p equals C. Then S has a density, and*

$$\delta(S) = \frac{\operatorname{card}(C)}{\operatorname{card}(G)}.$$

In particular, if $C = \{1\}$, then $S = \operatorname{Spl}(f)$ (by a lemma in Section 4) and $\delta(S) = 1/\operatorname{card}(G)$. We recover the weak theorem. If the group G is abelian, then each conjugacy class has one member and the corresponding sets of primes each have density $1/\operatorname{card}(G)$. This shows immediately that *the Artin map is surjective.* (Why?) Also explicit calculation of Artin symbols in cyclotomic fields gives a proof of Dirichlet's Theorem from Tchebotarev's Theorem.

6. An algorithm for the reciprocity problem. What have we learned so far about the reciprocity problem? Not much, in general, but we can claim to understand abelian polynomials completely. This knowledge at least gives a starting place for the study of polynomials with *solvable* Galois group. We do not discuss this here, but see Hasse [**9**, pp. 64–69] and Cassels and Fröhlich [**3**, Ex. 2.15, p. 354]. For polynomials with non-solvable groups, the only progress is the tantalizing example of Shimura mentioned in the introduction.

No satisfactory description of general sets $\operatorname{Spl}(f)$ has been given up to now, but for fixed $f(X)$ and a particular prime p, we can at least ask whether p lies in $\operatorname{Spl}(f)$. This involves factoring $f(X)$ modulo p, which is a finite process. The point of this section is to do the factoring *efficiently*. The method we use is essentially due to Berlekamp [**12**, Chapter 6]. Our formulation, designed to give only that information relevant to the reciprocity problem, is slightly different from Berlekamp's.

The prerequisites for the discussion are the Chinese Remainder Theorem for polynomial rings, and some knowledge of finite fields. (The material needed is covered in Berlekamp [**2**] and Lang [**14**], especially pages 63 and 182.)

Suppose given a polynomial $f(X)$ in $\mathbf{Z}[X]$ of degree n, with no repeated factors.

and let $f_p(X)$ be its reduction modulo p. Assume $f_p(X) = g_1(X) \cdots g_r(X)$ where $g_i(X)$ is irreducible of degree d_i. Our problem is to compute $d_1, \cdots, d_r$; we know $d_1 + \cdots + d_r = n$. For example, $p \in \mathrm{Spl}(f)$ if $r = n$ and each $d_i = 1$.

First we compute the discriminant $D(f)$ by the classical formula (e.g., Lang [**14**, p. 139]). If p divides $D(f)$, then $f_p(X)$ has a repeated factor. We declare such p "bad" and do not consider them further. If p does not divide $D(f)$, then the $g_i(X)$ are distinct irreducible polynomials and are therefore relatively prime. The Chinese Remainder Theorem gives:

(*) $$\mathbf{F}_p[X]/(f_p(X)) \cong \oplus_{i=1}^r \mathbf{F}_p[X]/(g_i(X)).$$

We write $\Lambda = \mathbf{F}_p[X]/(f_p(X))$ and $k_i = \mathbf{F}_p[X]/(g_i(X))$. Since $g_i(X)$ is irreducible of degree d, then $k_i = \mathbf{F}_q$, the unique finite field with $q = p^{d_i}$ elements. Since $[k_i : \mathbf{F}_p] = d_i$, we can recover all we need by computing the dimensions of the summands on the righthand side.

Here we have a case in which two isomorphic structures cannot be identified: the ring Λ is given very concretely as an n-dimensional $\mathbf{F}_p$ space, with basis $1, x^2, \cdots, x^{n-1}$, where x is the residue class of X modulo $f(X)$. Addition is vector space addition, and multiplication is carried out modulo $f(X)$. Our problem is to extract the direct sum decomposition, or at least compute the d_i, from *this* description of Λ.

As preparation, consider a finite extension k of $\mathbf{F}_p$, with $[k : \mathbf{F}_p] = d$, say. The mapping $\phi(z) = z^p : k \to k$ is a field isomorphism called the Frobenius map, and $\phi(z) = z$ if and only if $z \in \mathbf{F}_p$. Moreover, $\phi^i(z) = z$ for $1 \leqq i \leqq d$ if and only if $z \in \mathbf{F}_q \subseteq k$, where $q = p^i$. Thus, d can be computed as the smallest integer such that $\phi^d =$ identity on k.

The Frobenius map $z \to z^p$ on Λ, which we also denote by ϕ, is a ring isomorphism useful in studying the structure of Λ. For example, if $\Lambda \cong \mathbf{F}_p \oplus \cdots \oplus \mathbf{F}_p$ (n summands), then $\phi =$ identity. More generally, the smallest d such that $\phi^d =$ identity on Λ (the order of ϕ) equals the least common multiple of the d_i. Since x generates Λ as a ring, the order is the smallest d such that $\phi^d(x) = x$, so it is easy to compute. We shall see in the next section that the order can give a lot of information in special cases. In general, however, we need a refinement.

Suppose γ denotes the isomorphism in the Chinese Remainder Theorem:

$$\gamma : \Lambda \cong k_1 \oplus \cdots \oplus k_r.$$

Then it is easy to see that

$$\gamma(\ker(\phi - I)) = \mathbf{F}_p \oplus \cdots \oplus \mathbf{F}_p, \ r \text{ summands},$$

where $I : \Lambda \to \Lambda$ is the identity map, and $\ker(\phi - I)$ is the kernel of the linear transformation $(\phi - I) : \Lambda \to \Lambda$.

Similarly,

$$\gamma(\ker(\phi^2 - I)) = \boldsymbol{l}_1 \oplus \cdots \oplus \boldsymbol{l}_r,$$

where $\boldsymbol{l}_i = \mathbf{F}_{p^2}$ if $\mathbf{F}_{p^2} \subseteq \boldsymbol{k}_i$, and $\boldsymbol{l}_i = \mathbf{F}_p = \boldsymbol{k}_i$, otherwise.

Therefore, $\ker(\phi^2 - I)$ has $\mathbf{F}_p$-dimension equal to $2r-$(the number of summands with $d_i = 1$).

DEFINITION. *For each integer i, let $\nu_i =$ nullity $(\phi^i - I) = \dim(\ker(\phi^i - I))$, where "dim" denotes vector space dimension over the prime field $\mathbf{F}_p$.*

For each integer j, let $\mu_j =$ the number of factors in the decomposition (*) which have dimension exactly equal to j.

In this notation $\nu_1 = r$, the total number of factors, and $\nu_2 = 2r - \mu_1$. The reader should verify that

$$\begin{aligned} \nu_3 &= \mu_1 + 2\mu_2 + 3(r - \mu_1 - \mu_2) \\ &= 3r - 2\mu_1 - \mu_2. \end{aligned}$$

Generally, it is not hard to see that

$$(\#) \qquad \nu_k = kr - (k-1)\mu_1 - (k-2)\mu_2 - \cdots - \mu_{k-1}.$$

This relationship is very important. Knowing the μ_i is the same as knowing $d_1, d_2, \cdots, d_r$, so they give the factorization of $f_p(X)$. On the other hand, we shall see below that the ν_i are relatively easy to compute. The reader should use equation (#) to verify the following inversion formula:

$$(\#\#) \qquad \mu_k = 2\nu_k - \nu_{k-1} - \nu_{k+1}.$$

We summarize these facts in the theorem.

THEOREM. *Suppose, given $\Lambda = \mathbf{F}_p[X]/(f_p(X)) = \boldsymbol{k}_1 \oplus \cdots \oplus \boldsymbol{k}_r$, and let $d_i = [\boldsymbol{k}_i : \mathbf{F}_p]$. Let ϕ be the Frobenius automorphism of Λ, and let $\nu_i =$ nullity $(\phi^i - I)$. Then $r = \nu_1$, and there are exactly $\mu_j = 2\nu_j - \nu_{j-1} - \nu_{j+1}$ summands with $d_i = j$, $j = 1, \cdots, d$. Here d is the smallest integer such that $\phi^d = I$.*

This theorem forms the basis of an efficient algorithm. First compute the matrix $[\phi]$ with respect to the basis $\{1, x, \cdots, x^{n-1}\}$ of Λ. (Berlekamp calls this the **Q-matrix.**) Then compute successively $\nu_i =$ nullity $([\phi]^i - I)$. Finally, compute the μ_i from the theorem. If $\mu_1 = n$, then p belongs to $\mathrm{Spl}(f)$, and in more complicated situations the μ_i give information about the Artin symbol belonging to p.

Of course, we must examine this proposed algorithm. How hard is it? How long does it take? Can it produce significant results and lead to a better theoretical understanding of the problem?

First of all I have to admit that it is completely unreasonable to do the algorithm by hand. I worked on $f(X) = X^5 - X - 1$ with $p = 11$ for an hour and could not make it come out. It is much easier to factor by trial and error when p is small, but large primes are impossible.

Fortunately it is not too difficult to write a FORTRAN program which will do calculations in the ring Λ. Since Λ is an n-dimensional vector space over $\mathbf{F}_p$ with a nice basis $\{1, x, x^2, \cdots, x^{n-1}\}$, its elements can be represented as a $1 \times n$ FORTRAN array. The program written for the next section uses FORTRAN's integer arithmetic and works modulo a variable prime p.

The algorithm is very efficient in that *the number of operations required to factor* $f(X)$ *modulo* p *is proportional to* $\log p$. In fact, the only part of the algorithm that depends essentially on p is computing x^p in the ring Λ. Abstractly speaking, how many steps does it take to compute x^p? Certainly less than $2 \cdot \log_2 p$, since x^p can be computed by successively squaring together some multiplications by x. (Are you skeptical? If $p = 23 = 10111$ (binary), the steps are $x, x^2, x^4, x^5, x^{10}, x^{11}, x^{22}, x^{23}$, which requires $7 < 2 \cdot \log_2(23)$ steps.) The fascinating subject of number theory algorithms and the time needed to do them is discussed in Lehmer [**16**]. Knuth [**13**, p. 388 ff.] goes into more detail and discusses algorithms very similar to this one.

7. Numerical results. With the help of R. W. Latzer I have written a FORTRAN program to carry out the algorithm for the polynomials $X^5 - X - a$, where a is an integer. This is the "Bring-Jerrard Quintic" which has the non-solvable Galois group $\mathfrak{S}(5)$ for general a, and in particular for $a = 1$, and $a = 2$. The program factored $X^5 - X - 1$ for all p up to 23,099 in about two minutes, at which time the program overflowed the FORTRAN integer capacity. (I have learned that Professor J. D. Brillhart, using other methods, has factored many members of a more general family of quintics up to $p = 1000$.)

If $f(X)$ is any irreducible quintic polynomial, then $f_p(X)$ can factor in one of eight ways:

Type 0:	$p \mid D(f)$	
Type 1:	Five linear factors	1/120
Type 23:	(Quadratic) (Quadratic) (Linear)	15/120
Type 24:	(Quadratic) (Three Linear)	10/120
Type 3:	(Cubic) (Linear) (Linear)	20/120
Type 4:	(Quartic) (Linear)	30/120
Type 5:	(Quintic)	24/120
Type 6:	(Quadratic) (Cubic)	20/120

The factors are irreducible and distinct, when displayed. The type is the order d

of the Frobenius map when p does not divide $D(f)$ except that Type 23 means (order $= 2$, nullity $\nu_1 = 3$) and Type 24 means (order $= 2$, nullity $\nu_1 = 4$). Thus, no nullities have to be computed, except when the order $= 2$. The fractions give the density of primes of each type, according to the Tchebotarev Density Theorem.

Finally, we give some examples of actual numerical results.

1. $f(X) = X^5 - X - 1$.

(a) $D(f) = 19 \cdot 151$, so 19 and 151 are bad.

(b) The primes of Type 1 (those in $\mathrm{Spl}(f)$) which are less than 23099 are 1973, 3769, 5101, 7727, 8161, 9631, 11093, 14629, 16903, 17737, 17921, 18097, 19477, 20759, 21727, and 22717. There are 16 primes in this list, giving a ratio of 16/2350 $\approx .0068$, as compared with a density of $1/120 \approx .00833$.

(c) The primes less than 500 are classified as follows:

Type 0. 19, 151.
Type 1. None.
Type 23. 67, 71, 239, 251, 313, 421, 433, and 491.
Type 24. 163, 193, 227, 307, 467, 487, and 499.
Type 3. 17, 41, 43, 47, 53, 107, 113, 179, 181, 191, 229, 281, 293, 311, 317, 347, 349, 373, 409, 457, and 463.
Type 4. 23, 29, 31, 61, 97, 101, 127, 131, 157, 173, 223, 241, 263, 269, 331, 359, 389, 439, 443, and 479.
Type 5. 3, 5, 11, 13, 79, 89, 109, 137, 139, 211, 257, 337, 379, 397, 431, 449, and 461.
Type 6. 2, 7, 37, 59, 73, 83, 103, 149, 167, 197, 199, 233, 271, 277, 283, 353, 367, 383, 401, and 419.

2. $f(X) = X^5 - X - 2$.

(a) $D(f) = 2^4 \cdot 3109$, so 2 and 3109 are bad.

(b) The primes of type 1 less than 23099 are 229, 271, 1637, 2647, 2857, 3673, 6323, 7103, 8123, 8999, 11161, 12197, 14341, 14503, 14929, 17183, 18679, 19457 and 20563. There are 19 primes in this list, giving 19/2350 $\approx .00809$.

3. It is also possible to fix p and let the coefficient a in $X^5 - X - a$ vary modulo p. So far I have done this for all p up to 239. For example, if $p = 31$, we get:

Type 0. $a \equiv 11, 20 \pmod{31}$.
Type 1. None.
Type 23. $a \equiv 2, 3, 28, 29 \pmod{31}$.
Type 24. $a \equiv 0, 15, 16 \pmod{31}$.
Type 3. $a \equiv 7, 24 \pmod{31}$.
Type 4. $a \equiv 1, 5, 8, 9, 14, 17, 22, 23, 26, 30 \pmod{31}$.
Type 5. $a \equiv 6, 10, 12, 13, 18, 19, 21, 25 \pmod{31}$.
Type 6. $a \equiv 4, 27 \pmod{31}$.

8. What is a reciprocity law? A general reciprocity law should provide a description of the set $\mathrm{Spl}(f)$ associated with a polynomial $f(X)$. The algorithm discussed in this paper is such a description, but few number theorists would consider it a reciprocity law. More is wanted, but the exact requirements are still vague and undefined.

A good general reciprocity law should specialize to the Artin Reciprocity Law in the case of abelian polynomials. A very good reciprocity law should include a one-to-one correspondence between certain sets of prime numbers and field extensions, giving more substance to the Inclusion Theorem in Section 5. Such a correspondence should generalize the known abelian theorems of class field theory. Y. Ihara [**11**] is beginning to make some progress toward this goal in the function field case.

Even if a good correspondence cannot be set up, any reciprocity law must be set in a general framework, and should unify various kinds of number theoretic phenomena. The examples in Shimura [**19**] are related to the theory of elliptic curves, but they are very special, and it is not clear how to use them as a foundation for a general reciprocity law. (The specialist should look at Ihara's discussion of this question.)

I would like to mention briefly another direction of research which may lead to reciprocity laws. The Artin Reciprocity Law can be interpreted as a theorem about certain classes of analytic functions: see Artin's original paper [**1**, p. 97] or the section "Abelian *L*-functions are Hecke *L*-functions" in Goldstein [**7**, p. 182]. There seem to be important non-abelian analogues to this viewpoint which involve group representations and automorphic forms, and the interested reader should look at the introduction to Jacquet-Langlands' book [**12**] or Shalika's paper [**18**].

Finally, I have to confess that I still do not know what a reciprocity law is, or what one should be. The reciprocity problem, like so many other number theory problems, can be stated in a fairly simple and concrete way. However, the simply stated problems are often the hardest, and a complete solution seems to be far out of reach. In fact, we probably will not know what we are looking for until we have found it.

This research was supported in part by a grant from the National Science Foundation under grant GP 29696.

References

1. E. Artin, Collected Papers, Addison-Wesley, Reading, Mass., 1965.
2. E. R. Berlekamp, Algebraic Coding Theory, McGraw-Hill, New York, 1968.
3. J. W. S. Cassels and A. Fröhlich, Algebraic Number Theory, Thompson, Washington, 1967.
4. H. Davenport, Multiplicative Number Theory, Markham, Chicago, 1971.
5. L. Gaal, Classical Galois Theory with Examples, Markham, Chicago, 1971.
6. C. F. Gauss, Disquisitiones Arithmeticae, transl. A. A. Clarke, Yale, New Haven, 1966.
7. L. Goldstein, Analytic Number Theory, Prentice-Hall, Englewood Cliffs, N. J., 1971.
8. G. Hardy and E. Wright, An Introduction to the Theory of Numbers, 4th ed. Oxford, 1960.
9. H. Hasse, Bericht über neuere Untersuchungen und Problemen aus der Theorie der algebraischen Zahlkörper, Teil I, Ia, II, 2nd edit., Physica-Verlag, Würzburg, 1965.
10. A. M. Legendre, Recherches d'Analyse Indéterminée, Hist. Acad., Paris, 1785.
11. Y. Ihara, Non-abelian class fields over function fields in special cases, to appear in Proc. of the Intern. Congress of Math., Nice, 1970.

12. H. Jacquet and R. P. Langlands, Automorphic Forms on Gl(2), Springer-Verlag Lecture Notes in Mathematics, No. 114, Berlin, 1970.

13. D. Knuth, The Art of Computer Programming, Volume 2: Seminumerical Algorithms, Addison-Wesley, Reading, Mass., 1969.

14. S. Lang, Algebra, Addison-Wesley, Reading, Mass., 1965.

15. ———, Algebraic Number Theory, Addison-Wesley, Reading, Mass., 1971.

16. D. H. Lehmer, Computer Technology Applied to the Theory of Numbers, Studies in Number Theory, MAA, Prentice-Hall, Englewood Cliffs, N.J., 1969.

17. I. Niven and H. S. Zuckerman, An Introduction to the Theory of Numbers, 2nd ed., Wiley, New York, 1966.

18. J. Shalika, Some Conjectures in Class Field Theory, in AMS, Proc. of Symposia in Pure Math., Volume XX: Stony Brook Number Theory Institute, Providence, 1971.

19. G. Shimura, A non-solvable reciprocity law, J. Reine Angew. Math., 221 (1966) 209–220.

20. B. van der Waerden, Modern Algebra, Vol. I, Revised English Edition, Ungar, New York, 1953.

21. A. Weil, Basic Number Theory, Springer, New York, 1967.

CORRECTION TO "WHAT IS A RECIPROCITY LAW?"

This MONTHLY, 79:571–586 (June–July, 1972)

B. F. WYMAN, Ohio State University

I wish to thank Karl A. Beres and Lawrence J. Dickson for pointing out that the theorem of Section 6 (p. 287) is incorrect. Professor Dickson has derived correct formulas for the v_m and μ_j (defined on p. 287):

$$v_m = \sum_{j=1}^{n} (j,m)\mu_j, \quad n = \deg(f). \tag{1}$$

$$\mu_j = \sum_{r=1}^{[n/j]} \Sigma_{m|jr}(\mu(r)\mu(m)/\phi(jr))v_{jr/m}, \tag{2}$$

where ϕ and μ are the Euler and Möbius functions.

Since the theorem referred to was not used in the preparation of the computer program mentioned in Section 7, the numerical results reported there are unaffected.

DENSITY QUESTIONS IN ALGEBRAIC NUMBER THEORY*

L. J. GOLDSTEIN, University of Maryland

Very often, the number theorist bases conjectures on empirical investigations. Even before the invention of the electronic computer, number theorists spent much time doing calculations, the results of which suggested possibly true statements. After the empirical stage of his investigation is completed, the number theorist then tries to supply proofs for his conjectures. It is here where the number theorist applies a formidable armada of high-powered machinery, ranging from analytic function theory to algebraic geometry. It is most surprising that even the most innocently conceived conjecture may lead into a vast jungle of very difficult and technical mathematics. But such is the nature of number theory. In this lecture, I should like to discuss a set of conjectures which typify the process of number-theoretic creation as we have described it: These conjectures originate out of empirical investigation and those few that we are able to prove seem to lead us far afield for their proofs.

1. Gauss' conjecture. Let us denote by $\mathbf{Z}$ the rational integers, p an odd prime, a an arbitrary integer, and $\mathbf{Z}_p^\times$ the group of nonzero residue classes mod p. Since $\mathbf{Z}_p^\times$ is the multiplicative group of a finite field, a well-known result asserts that $\mathbf{Z}_p^\times$ is cyclic of order $p-1$. We say that a is a *primitive root* modulo p if $(a, p)=1$ and if its residue class $\bar{a}$ in $\mathbf{Z}_p^\times$ is a generator of $\mathbf{Z}_p^\times$.

LEMMA 1.1. *The number a is a primitive root modulo p if and only if $(a,p)=1$ and $a^\nu \not\equiv 1 \pmod p$ for $\nu=1,2,\cdots,p-2$.*

Note that by Fermat's Little Theorem, if $(a, p)=1$, then

$$a^{p-1} \equiv 1 \pmod p.$$

From now on, let us fix a, and let us define

$$\mathfrak{A}(a) = \{p \mid p \text{ is prime and } a \text{ is a primitive root modulo } p\}.$$

It may be that $\mathfrak{A}(a)$ is empty. For example, if a is a perfect square, say x^2, with $(p, a)=1$, then

$$x^{p-1} \equiv 1 \pmod p$$

by Fermat's Little Theorem, so that

$$a^{(p-1)/2} \equiv 1 \pmod p.$$

Therefore, if $p \nmid x$, p odd, then $p \notin \mathfrak{A}(a)$. However, if $p \mid x$, then it is certainly true

* From AMERICAN MATHEMATICAL MONTHLY, vol. 78 (1971), pp. 342–351.

Larry Goldstein received his Princeton PhD under G. Shimura in 1967. He was a Gibbs lecturer at Yale for two years before his present associate professorship at Maryland. His main research is in analytic and algebraic number theory, and his book, *Analytic Number Theory,* is scheduled to appear. *Editor.*

that $p \notin \mathfrak{A}(a)$. Therefore, we have shown that $\mathfrak{A}(a)=\varnothing$ if a is a perfect square. Moreover, since $(-1)^2=+1$, we see that $p\notin \mathfrak{A}(-1)$ if $p-1>2$. Therefore, since -1 is a primitive root modulo 3, we have proved that $\mathfrak{A}(-1)=\{3\}$.

By means of laborious calculations, Gauss investigated the case $a=10$ and arrived at the following conjecture, which is stated in Article 316 of his *Disquisitiones Arithmeticae*:

CONJECTURE A: $\mathfrak{A}(10)$ *is infinite.*

In the next sections, we shall present some heuristic evidence for this conjecture, as well as some more general conjectures which seem to be true.

2. Artin's conjecture. In a conversation with Hasse in 1927, Artin made the following conjecture:

CONJECTURE B: *Suppose that a is not -1 and not a perfect square. Then $\mathfrak{A}(a)$ is infinite.*

This conjecture was not just a wild guess, but followed from a very compelling probabilistic argument which Artin advanced. In order to trace Artin's line of thought, we must first define a few notions.

Let $\mathcal{S}$ be a set of primes (finite or infinite), and let x be a positive real number. Let $\pi(x)$ denote the number of primes $\leqq x$, and let $\pi(x, \mathcal{S})$ denote the number of primes in $\mathcal{S}$ which are $\leqq x$. We say that $\mathcal{S}$ has a *(natural) density* if

$$\lim_{x\to\infty} \pi(x, \mathcal{S})/\pi(x)$$

exists. The value of the limit is called the *density of* $\mathcal{S}$ and is denoted $d(\mathcal{S})$. We clearly have

$$0 \leqq d(\mathcal{S}) \leqq 1.$$

Moreover, if $d(\mathcal{S})>0$, then $\mathcal{S}$ is infinite since $\pi(x)\to\infty$ as $x\to\infty$. In a few moments we shall reformulate Conjecture B in the form of a statement about densities. But first we must state some preliminary information about algebraic number theory.

Let K be an algebraic number field, that is, a finite, algebraic extension of Q. Let $\mathfrak{O}$ be the ring of integers of K, that is, the integral closure of Z in K. If p is an ordinary prime, then $p\mathfrak{O}$ is an ideal of $\mathfrak{O}$, but is usually no longer a prime ideal. However, $p\mathfrak{O}$ can be written as a product of powers of prime ideals of $\mathfrak{O}$ (since $\mathfrak{O}$ is a Dedekind domain):

$$p\mathfrak{O} = \mathfrak{P}_1^{e_1} \cdots \mathfrak{P}_g^{e_g}.$$

It is a general fact from algebraic number theory that $g\leqq \deg(K/Q)$. We say that p *splits completely* in K if $g=\deg(K/Q)$. Here is a basic theorem which one meets in the analytical portion of algebraic number theory.

THEOREM 2.1 (Dirichlet). *Let $n=\deg(K/Q)$ and let $\mathcal{S}$ denote the set of all* *primes which split completely in K. Then $\mathcal{S}$ has a density and*

$$d(\mathcal{S}) = 1/n.$$

Let q be a prime and let L_q denote the splitting field over Q of the polynomial $X^q - a$. We get L_q from Q in two steps. First we adjoin to Q a primitive qth root of unity ζ_q. Then we adjoin to $Q(\zeta_q)$ any qth root of a, say the real value of $a^{1/q}$. Then,

$$L_q = Q(\zeta_q, a^{1/q}). \tag{1}$$

$L_q/Q(\zeta_q)$ is a Galois extension of degree either 1 or q with cyclic Galois group. (The extension is a so-called Kummer extension.) Also, $Q(\zeta_q)/Q$ is a Galois extension of degree $q-1$ with cyclic Galois group. Thus, L_q/Q is a Galois extension with solvable Galois group and

$$\deg(L_q/Q) = q - 1 \quad \text{or} \quad q(q-1), \tag{2}$$

depending on the value of a.

From the tool box of the algebraic number theorist, we quote the following result:

THEOREM 2.2. *p splits completely in $L_q \Leftrightarrow p \equiv 1 \pmod q$ and*

$$a^{(p-1)/q} \equiv 1 \pmod q.$$

Combined with Lemma 1.1, this yields the following:

THEOREM 2.3. *a is a primitive root modulo p if and only if for each prime q, the prime p does not split completely in L_q.*

For K an algebraic number field, let $\mathrm{Spl}(K)$ denote the set of all primes which split completely in K; let $\mathcal{P}$ denote the set of all primes. Then, by Theorem 2.3, we can assert the following:

COROLLARY 2.4. $\mathcal{A}(a) = \bigcap_q (\mathcal{P} - \mathrm{Spl}(L_q))$.

Now for Artin's probabilistic argument: By Dirichlet's theorem, $\mathrm{Spl}(L_q)$ has a density and $d(\mathrm{Spl}(L_q)) = 1/\deg(L_q/Q)$. Therefore, $\mathcal{P} - \mathrm{Spl}(L_q)$ has a density and

$$d(\mathcal{P} - \mathrm{Spl}(L_q)) = 1 - 1/\deg(L_q/Q).$$

Therefore, from Corollary 2.4, we might guess that $\mathcal{A}(a)$ has a density and that

$$d(\mathcal{A}(a)) = \prod_q (1 - 1/n(q)), \qquad n(q) = \deg(L_q/Q). \tag{*}$$

Let us see how (*) fits in with Conjecture B. First of all, it is easy to check that if $a \neq -1$ then $n(q) = q(q-1)$ for all but a finite number of q. Therefore, the product converges for $a \neq -1$. For $a = -1$, the product diverges to 0. Thus, if $a \neq -1$, the product can converge to 0 if and only if one of the factors $= 0$, and this in turn if and only if $n(q) = 1$ for some q. But it is trivial to check that $n(q) \geqq q - 1 > 1$ if $q > 2$. Therefore, the product $= 0$ if and only if $n(2) = 1$. But

$L_2 = \mathbf{Q}(a^{1/2})$, so that $n(2)=1$ if and only if a is a perfect square. Therefore, we conclude that if $a \neq -1$ and $a \neq b^2$, then the product is positive, so that $d(\mathcal{A}(a)) > 0$, which implies Conjecture B.

Thus, the heuristic arguments of Artin seemed to fit the facts such as they were known at the time. However, experimental calculations by D. H. Lehmer cast a serious doubt as to whether the true value of the density of $\mathcal{A}(a)$ was given by (*). In the face of this disagreement between conjecture and evidence, it was necessary to reexamine the reasoning which led to (*). Let us consider the probabilistic event "a randomly chosen prime belongs to $\mathcal{P} - \mathrm{Spl}(L_q)$." Dirichlet's Theorem may be interpreted as saying that the probability of this event is $1/n(q)$. We then get the probability that a randomly chosen prime belongs to the intersection of all $\mathcal{P} - \mathrm{Spl}(L_q)$ by multiplying the corresponding probabilities. This is valid, as every student of probability knows, only when the events are pairwise independent. Therefore, what probably goes wrong is that something analogous to probabilistic independence is violated. Of course, all of our analogies with probability theory are only of heuristic value. But they seem to lead somewhere in this case! For, upon close inspection, we see that the fields L_q are not "independent" of one another, that is, it is not true that $L_q \cap L_{q'} = \mathbf{Q}$ for $q \neq q'$. Therefore, if we wish to make a statement like (*), it is necessary to somehow take into account this dependence.

By Corollary 2.4,

$$\mathcal{A}(a) = \mathcal{P} - \bigcup_q \mathrm{Spl}(L_q).$$

Note, however, that the primes which split completely in two fields L_{q_1} and L_{q_2} are subtracted twice on the right hand side of (3). In an attempt to count each prime in $\mathcal{A}(a)$ once and only once, let us add back in those primes which were removed twice to get

$$\mathcal{A}(a) = \mathcal{P} - \bigcup_q \mathrm{Spl}(L_q) + \bigcup_{\substack{q_1, q_2 \\ q_1 \neq q_2}} \mathrm{Spl}(L_{q_1}) \cap \mathrm{Spl}(L_{q_2}).$$

In adding the last term, however, we have counted twice the primes which split completely in three fields $L_{q_1}, L_{q_2}, L_{q_3}$. Therefore, let us correct this by writing

$$\mathcal{A}(a) = \mathcal{P} - \bigcup_q \mathrm{Spl}(L_q) + \bigcup_{\substack{q_1, q_2 \\ q_1 \neq q_2}} \mathrm{Spl}(L_{q_1}) \cap \mathrm{Spl}(L_{q_2})$$
$$- \bigcup_{\substack{q_1, q_2, q_3 \\ q_i \text{ distinct}}} \mathrm{Spl}(L_{q_1}) \cap \mathrm{Spl}(L_{q_2}) \cap \mathrm{Spl}(L_{q_3}).$$

But now the primes which split completely in four fields have been subtracted twice, so we must add them back in, and the process continues. Eventually, we arrive at a formula for $\mathcal{A}(a)$ in which each prime is counted exactly once. If $q_1, q_2, \cdots, q_r$ are distinct primes, $k = q_1 \cdots q_r$, let us define L_k to be the composite

$$L_k = L_{q_1} \cdot \cdots \cdot L_{q_r}.$$

Then

$$\mathrm{Spl}(L_{q_1}) \cap \cdots \cap \mathrm{Spl}(L_{q_r}) = \mathrm{Spl}(L_k).$$

Therefore, we may write our formula for $\mathcal{A}(a)$ in the form

$$\begin{aligned}\mathcal{A}(a) = \mathcal{P} - \bigcup_{q} \mathrm{Spl}(L_q) + \bigcup_{\substack{q_1, q_2 \\ q_i \text{ distinct}}} \mathrm{Spl}(L_{q_1 q_2}) \\ - \bigcup_{\substack{q_1, q_2, q_3 \\ q_i \text{ distinct}}} \mathrm{Spl}(L_{q_1 q_2 q_3}) + \cdots.\end{aligned} \tag{3}$$

We have defined L_k for each positive, square-free integer. Let $n(k) = \deg(L_k/Q)$. Then by Dirichlet's Theorem and (3), we can conjecture that

$$d(\mathcal{A}(a)) = 1 - \sum_{q} n(q)^{-1} + \sum_{\substack{q_1, q_2 \\ q_i \text{ distinct}}} n(q_1 q_2)^{-1} - \cdots. \tag{4}$$

By rewriting the right hand side of (4), we derive the following conjecture:

CONJECTURE C: *$\mathcal{A}(a)$ has a natural density, and*

$$d(\mathcal{A}(a)) = \sum_{k} \mu(k)/n(k), \qquad n(1) = 1,$$

where $\mu(k)$ denotes the Möbius function and the sum runs over all positive square-free integers k (including 1).

It is Conjecture C that agrees with the experimental evidence. Note, however, that from the form of the sum in Conjecture C, it is no longer evident that $d(\mathcal{A}(a)) > 0$ if $a \neq -1$ and $a \neq b^2$. Also, it must be checked that the series converges. Both points are answered by the following theorem.

THEOREM 2.5 (Hooley [3]). *Let k be a positive square-free integer, let h denote the largest positive integer such that a is an h-th power, and let*

$$k_1 = k/(h, k),$$

$$a_1 = \text{the square-free part of } a,$$

$$\epsilon(k) = \begin{cases} 2 & \text{if } k \text{ is divisible by } 2a_1 \text{ and } a_1 \equiv 1 \pmod 4 \\ 1 & \text{otherwise.} \end{cases}$$

Then $n(k) = k_1\phi(k)/\epsilon(k)$, where $\phi(k)$ denotes Euler's function.

As an immediate consequence of Hooley's theorem, we deduce two corollaries.

COROLLARY 2.6. *The sum $\sum_k \mu(k)/n(k)$ converges absolutely.*

COROLLARY 2.7. *If k and a are relatively prime, then*

$$n(k) = k\phi(k).$$

Using Hooley's theorem, we can write the sum of Conjecture C as a product, so that we may revise Conjecture C as follows:

CONJECTURE D: *$\mathfrak{A}(a)$ has a natural density and*

$$d(\mathfrak{A}(a)) = \begin{cases} C(k), & a_1 \not\equiv 1 \pmod 4) \\ C(k)\cdot\left[1 - \mu(|a_1|) \prod_{\substack{q|h \\ q|a_1}} (q-2)^{-1} \prod_{\substack{q\nmid h \\ q|a_1}} (q^2-q-1)^{-1}\right], & a_1 \equiv 1 \pmod 4, \end{cases}$$

where

$$C(k) = \prod_{q|h} (1-(q-1)^{-1}) \prod_{q\nmid h} (1-\phi(q^2)^{-1}).$$

This is our final form of Artin's conjecture. Implicit in the statement of Conjecture D is the statement that if $a \neq -1$ and a is not a perfect square, then $d(\mathfrak{A}(a)) > 0$. For then $|a_1| \neq 1$. Since $C(k) > 0$, we see that (mod Conjecture D)

$$d(\mathfrak{A}(a)) = 0 \Leftrightarrow a_1 \equiv 1 \pmod 4) \quad \text{and} \quad \mu(|a_1|) = 1$$

and

$$\prod_{\substack{q|h \\ q|a_1}} (q-2)^{-1} \prod_{\substack{q\nmid h \\ q|a_1}} (q^2-q-1)^{-1} = 1.$$

The last of the three conditions on the right can be satisfied only when $|a_1| = 1$, 2, or 3. But of these three possibilities only $|a_1| = 1$ is consistent with the remaining two conditions. Therefore Conjecture D implies

$$(5) \qquad \begin{aligned} d(\mathfrak{A}(a)) = 0 &\Leftrightarrow |a_1| = 1 \\ &\Leftrightarrow a = -1 \quad \text{or} \quad a = b^2. \end{aligned}$$

3. Bilharz's Theorem. Let k be a finite field with q elements, $k[t]$ the ring of polynomials over k in an indeterminate t, and $K = k(t)$ the field of rational functions in t with coefficients in K. The field K is the simplest example of an algebraic function in one variable. The arithmetic properties of such fields parallel the arithmetic of $\boldsymbol{Q}$, with $k[t]$ playing the role of the rational integers. In many ways, the arithmetic of K is even simpler than that of $\boldsymbol{Z}$, so that often number theorists use function fields as a testing ground for conjectures about the rational integers. This testing process consists of reformulating a problem about $\boldsymbol{Q}$ or $\boldsymbol{Z}$ into an analogous problem about K or $k[t]$, respectively, and then solving the analogous problem.

In 1935, Bilharz [1], a student of Hasse, formulated and proved the analogue of Artin's conjecture. The role of the rational primes is played by the monic, irreducible polynomials $P \in k[t]$. If P is such a polynomial, then the *norm of P*, denoted NP, is defined by

$$NP = q^r, \qquad r = \deg(P).$$

The quotient ring

$$K_P = k[t]/Pk[t], \qquad P \text{ monic, irreducible,}$$

is a finite field with NP elements. The multiplicative group $K_P^\times$ of K_P is cyclic. Suppose that $A \in K$ is not divisible by P. We say that A is a *primitive root modulo* P if A mod $Pk[t]$ generates $K_P^\times$. Given $A \in K$, we can define

$$\mathcal{A}(A) = \{P \mid A \text{ is a primitive root modulo } P\}.$$

It is easy to check that if A is an r-*th* power for some r dividing $q-1$, then $\mathcal{A}(A)=\varnothing$. In analogy with the situation in $\boldsymbol{Q}$, we can formulate a conjecture.

CONJECTURE A′: *If A is not an r-th power for any prime r dividing $q-1$, then $\mathcal{A}(A)$ is infinite.*

Conjecture A' was proved in the cited work of Bilharz. The most interesting feature of Bilharz's paper is that he proves Conjecture A′ only by assuming a deep result, at the time conjectured but not proved, known as the "Riemann hypothesis for function fields over finite fields." The conjecture was settled by André Weil in 1941 [4], so that the gap in Bilharz's argument was filled.

Let $\mathcal{S}_0$ be the set of all monic, irreducible polynomials in $k[t]$, and let $x \geqq 0$. For $\mathcal{S} \subseteq \mathcal{S}_0$, define

$$\pi_K(x, \mathcal{S}) = \sum_{\substack{P \in \mathcal{S} \\ NP \leqq x}} 1, \qquad \pi_K(x) = \pi_K(x, \mathcal{S}_0).$$

We say that $\mathcal{S}$ has a *natural density* if

$$\lim_{x\to\infty} \frac{\pi_K(x, \mathcal{S})}{\pi_K(x)}$$

exists. We can formulate the analogues of the density conjectures in the function field case. However, the situation here is very much different from the preceding case. The set $\mathcal{A}(A)$ usually does not have a natural density. However, it is possible to define a new concept of density (Dirichlet density) with respect to which the analogues of the density conjectures are true. The proofs of these results are contained in Bilharz's paper.

4. Hooley's Theorem. Let $\boldsymbol{L}$ be an algebraic number field. If $\mathfrak{A}$ is an ideal of the ring of integers $\mathfrak{O}_L$, the *norm of* $\mathfrak{A}$ denoted $N\mathfrak{A}$, is the number of elements in the (finite) ring $\mathfrak{O}_L/\mathfrak{A}$. The *Dedekind zeta function* of $\boldsymbol{L}$ is defined by

$$\zeta_L(s) = \sum_{\mathfrak{A}} N\mathfrak{A}^{-s},$$

where $\mathfrak{A}$ runs over all ideals of $\mathfrak{O}_K$ and s is a complex variable. The series on the right converges absolutely for $\mathrm{Re}(s) > 1$. Moreover, for s in this half-plane,

$$\zeta_L(s) = \prod_{\mathfrak{p}} (1 - N\mathfrak{p}^{-s})^{-1}, \tag{6}$$

where $\mathfrak{p}$ runs over all prime ideals of $\mathfrak{O}_L$. The product of (6) converges absolutely for $\text{Re}(s)>1$. Therefore,

$$\zeta_L(s) \neq 0 \qquad (\text{Re}(s) > 1). \tag{7}$$

It is possible to show that $\zeta_L(s)$ can be analytically continued to a meromorphic function on the whole s-plane. The continued function (also denoted $\zeta_L(s)$) has only one pole, a simple pole at $s=1$ with residue 1. Moreover, $\zeta_L(s)$ satisfies a functional equation connecting its behavior at s with its behavior at $1-s$. One consequence of this functional equation is that the zeros of $\zeta_L(s)$ in the half-plane $\text{Re}(s)<0$ are known. These zeros are called *trivial zeros*. By (7), all nontrivial zeros of $\zeta_L(s)$ lie in the strip

$$0 \leqq \text{Re}(s) \leqq 1.$$

There is strong evidence in favor of the following conjecture.

CONJECTURE (Riemann Hypothesis): *All nontrivial zeros of $\zeta_L(s)$ lie on the line* $\text{Re}(s)=1/2$.

The special case $L=Q$ of this celebrated conjecture was first stated by Riemann in 1860. Although the Riemann hypothesis has received the attention of many of the greatest mathematicians of the last 100 years, it remains unproved, and is one of the most significant unsolved problems of contemporary mathematics.

There is a link between the Riemann hypothesis and Conjecture C (the most general form of Artin's conjecture)—namely, Hooley [3], has proved the analogue of Bilharz's theorem:

THEOREM 4.1. *Assume that the Riemann hypothesis is true for each of the fields L_k. Then Conjecture* C *is true.*

5. Analogues of Artin's conjecture. It is possible to generalize the heuristic argument which gave rise to Conjecture C: Suppose that $\mathcal{S}$ is a set of rational primes, and suppose that for each $q\in\mathcal{S}$ there is given a number field L_q. Let $\mathcal{A}=\mathcal{A}(\mathcal{S}, \{L_q\})$ denote the set of rational primes which do not split completely in each L_q for $q\in\mathcal{S}$. Let us make a conjecture about the natural density of $\mathcal{A}$.

For $k=q_1 \cdots q_r$, $q_i\in\mathcal{S}$, set

$$L_k = L_{q_1} \cdots L_{q_r},$$
$$n(k) = \deg(L_k/Q).$$

Define $L_1=Q$, so that $n(1)=1$. Using the same arguments as in Paragraph 2, we can formulate another conjecture.

CONJECTURE E: *Suppose that*

$$\sum_k n(k)^{-1}$$

converges, where the sum runs over all k for which $n(k)$ is defined. Then $\mathcal{A}$ has a *natural density*

$$d(\mathfrak{A}) = \sum_k \mu(k)n(k)^{-1}.$$

Conjecture E clearly contains Conjecture C as a special case, namely for $\mathcal{S} = \{\text{all rational primes}\}$, $L_q = Q(\zeta_q, a^{1/q})$ $(q \in \mathcal{S})$. There are only two special cases for which Conjecture E has been verified. When $\mathcal{S}$ is finite, Conjecture E can be easily checked using Dirichlet's theorem. When $\mathcal{S}$ is infinite, however, Conjecture E is very difficult. The only case known is now given.

THEOREM 5.1 (Goldstein [2]). *Suppose that*

$$L_q \supseteq Q(\zeta_{q^2})$$

holds for all but a finite number of $q \in \mathcal{S}$. *Then Conjecture* E *is true. In particular, Conjecture* E *is true if* $\mathcal{S} = \{$*all rational primes*$\}$ *and*

$$L_q = Q(\zeta_{q^2}, a^{1/q}) \qquad (a \in Z, q \in \mathcal{S}).$$

Theorem 5.1 is tantalizingly close to Artin's conjecture. One might hope that the methods used to prove Theorem 5.1 could be appropriately generalized to prove Artin's conjecture. However, it appears that Conjecture E is of a much higher order of difficulty and any hopes in that direction are overly optimistic.

6. Conclusion. In this talk I have tried to indicate how a number theorist comes by his conjectures. In some sense, the combination of intuition, deduction, and heuristic arguments by means of which we have arrived at our conjectures, is a typical way in which many mathematicians work. There is much that we have been forced to omit. For example, it is possible to formulate Conjecture E as a conjecture about Haar measure on a certain compact topological group. In this formulation Conjecture E can be thought of as a generalization of Dirichlet's theorem to infinite-dimensional extensions L of Q. For an exposition of this theory, the reader is referred to [2]. If I have said little about methods of proof, it is because there are only a few theorems now proved in the subject. I hope that this talk will generate enough interest to remedy this appalling situation.

This paper is the text of an invited address delivered by the author under the title "On a Conjecture of Artin" at the Northeastern Sectional Meeting on June 20, 1969.
Research supported by NSF Grant GP-13872.

References

1. H. Bilharz, Primdivisoren mit vorgegebener Primitivwürzel, Math. Ann., 114 (1937) 476–492.
2. L. Goldstein, Analogues of Artin's conjecture, Trans. Amer. Math. Soc., (to appear).
3. C. Hooley, On Artin's conjecture, J. Reine Angew. Math., 225 (1967) 209–220.
4. A. Weil, Sur les Courbes Algébriques et les Variétés qui s'en Déduisent, Hermann, Paris, 1948.

BIBLIOGRAPHIC ENTRIES: ALGEBRAIC NUMBER THEORY

Except for the entry labeled MATHEMATICS MAGAZINE, the references below are to the AMERICAN MATHEMATICAL MONTHLY.

1. Donald L. Goldsmith, An old theorem in a new setting, 1963, 313–315.

Generalizes Euclid's proof to show that there are an infinite number of primes in the ring of integers of a quadratic number field.

2. G. Bachman, The decomposition of a rational prime ideal in cyclotomic fields, 1966, 494–497.

Uses valuation theory to determine the decomposition of a rational prime in the ring of integers of a cyclotomic field.

3. E. T. Bell, Gauss and the early development of algebraic integers, MATHEMATICS MAGAZINE, 1944, 188–204, and 219–233.

4

LINEAR ALGEBRA

(a)

EQUATIONS, DETERMINANTS, AND RELATED TOPICS

ON THE SOLUTION OF A SYSTEM OF LINEAR EQUATIONS*

G. A. MILLER, University of Illinois

A general system of m linear equations in n unknowns may be denoted as follows:

$$T = \begin{cases} a_{11}x_1 + a_{12}x_2 + \cdots + a_{1n}x_n + k_1 = 0, \\ a_{21}x_2 + a_{22}x_3 + \cdots + a_{2n}x_n + k_2 = 0, \\ \vdots \qquad\quad \vdots \qquad\quad \vdots \qquad\quad \vdots \\ a_{m1}x_1 + a_{m2}x_2 + \cdots + a_{mn}x_n + k_m = 0. \end{cases}$$

In the study of the solutions of this system the following two matrices are of especial importance:

$$A = \begin{vmatrix} a_{11}a_{12}\cdots a_{1n} \\ a_{21}a_{22}\cdots a_{2n} \\ \vdots \quad \vdots \quad \vdots \\ a_{m1}a_{m2}\cdots a_{mn} \end{vmatrix}, \qquad B = \begin{vmatrix} a_{11}a_{12}\cdots a_{1n}k_1 \\ a_{21}a_{22}\cdots a_{2n}k_2 \\ \vdots \quad \vdots \quad \vdots \\ a_{m1}a_{m2}\cdots a_{mn}k_m \end{vmatrix}.$$

Capelli exhibited the pedagogic advantage in employing, in the study of the system T, the concept of rank of these matrices, which are known, respectively, as the matrix and the augmented matrix of the system. The rank of a matrix is, according to Frobenius, the order of the largest non-vanishing determinant formed by elements of the matrix in order. The theorem proved by Capelli may be stated as follows: The necessary and sufficient condition that the system T is solvable is

* From AMERICAN MATHEMATICAL MONTHLY, vol. 17 (1910), pp. 137–139.

that the rank of the matrix of T is equal to the rank of its augmented matrix.* By solvable is meant that a set of n *finite* values may be assigned to the unknowns $x_1, x_2, \ldots, x_n$, so that each of the equations in T is satisfied.

When the system T is solvable it is also said to be consistent or compatible. If the system T is consistent and of rank r we may assign arbitrary values to at least one set of $n-r$ unknowns in this system so that after this is done it is possible to solve the resulting system. To each such $n-r$ arbitrary values there will correspond a single set of values for the remaining r unknowns, provided the $n-r$ unknowns to which arbitrary values were assigned were so selected that the rank of the matrix of the system is not diminished by omitting the coefficients of these $n-r$ unknowns from the matrix. As this interesting theory is so clearly presented in Bôcher's *Introduction to Higher Algebra* it does not appear necessary to enter into greater details here.

The main object of the present note is to consider the question: when can a given unknown in a consistent system of equations have only one value. To make this question perfectly clear we may consider the following system of three equations in three unknowns:

$$2x-y+2z=8,$$
$$4x-2y-z=-4,$$
$$6x-3y+z=-4.$$

It is evident that the rank of the matrix of this system is equal to the rank of the augmented matrix, each being 2. For every arbitrary value of x there is one and only one pair of values for y and z such that each of these three equations is satisfied. For instance, when $x=0$, y and z must have the value 0, 4 respectively; and when $x=1$ the values of y and z are 2, 4 respectively. Similarly, there is one and only one pair of values of x and z for every arbitrary value of y. On the contrary, z must always have the same value, namely, 4. We have therefore a system here in which the unknown z can have only one value although the system has an infinite number of solutions.

We are now in position to understand the following theorem:

The necessary and sufficient condition that a given unknown in a consistent system can have only one value is that the rank of the matrix of the system is diminished by omitting the coefficients of this unknown from this matrix.

That this condition is necessary follows directly from the general theory mentioned above; for, if the rank of this matrix were not diminished by omitting the given coefficients we could assign an arbitrary value to this unknown and solve the resulting system. Hence it remains only to prove that the given condition is also sufficient.

Suppose that the system T is consistent and that the rank of its matrix A is r,

---

* Capelli, *Rivista di Matematica*, Vol. 2 (1892), page 54. Capelli used the term characteristic instead of rank. The former of these terms is frequently employed in France and Italy. Cf. Pincherle, *Lezioni di algebra complementare*, 1909, page 92.

but that the rank of the matrix A' obtained by omitting the coefficients of x_a from A is less than r. The rank of A' must therefore be $r-1$ as the co-factor of at least one of these coefficients cannot vanish in a non-vanishing determinant of order r contained in A. As the matrix A' is of rank $r-1$ each of its rows can be expressed as a linear function of $r-1$ of these rows.* Hence we may replace the system T by a new system T' having the following form:

$$
\begin{array}{l}
a_{1a}x_a + l_1 + k_1 = 0, \\
a_{2a}x_a + l_2 + k_2 = 0, \\
\cdots\cdots\cdots\cdots\cdots\cdots \\
a_{r-1a}x_a + l_{r-1} + k_{r-1} = 0, \\
a_{ra}x_a + \phi_0(l_1, l_2, \ldots, l_{r-1}) + k_r = 0, \\
\cdots\cdots\cdots\cdots\cdots\cdots \\
a_{ma}x_a + \phi_{m-r}(l_1, l_2, \ldots, l_{r-1}) + k_m = 0,
\end{array}
$$

where $\phi_0, \ldots, \phi_{m-2}$ are the linear functions of $l_1, l_2, \ldots, l_{r-1}$.

The system T' is consistent since T is consistent and it must be of rank r since T has this rank. Hence it results that one and only one set of values for $x_a, l_1, \ldots, l_{r-1}$ will satisfy the first r equations. That is, x_a can have only one value in system T'. It can therefore have only one value in system T and the theorem in question has been established. From this theorem it results that each unknown in system T has either only one value or it has an infinite number of values whenever this system is consistent. In particular, a system of linear equations has always an infinite number of distinct solutions whenever it has more than one solution.

From the above it is evident that the language as regards solvability of a system of linear equations becomes much more concise by means of the concept of rank. Although this concept is comparatively new in mathematics it is of such fundamental importance that it should occupy a more prominent place in the courses in advanced algebra. It should be remembered that a thing can only appear simple when we can see clearly through it. In particular, the theory of linear equations appears simple only after an exhaustive study by means of such a powerful instrument as the concept of rank.

* Cf. Bôcher, *Introduction to Higher Algebra*, page 36.

ON THE MATRIX EQUATION $BX=C$*†

H. T. BURGESS, University of Wisconsin

Section 1. To Find the Matrix X. The problem is to calculate the elements of the matrix X to satisfy the matrix equation $BX=C$:

$$\begin{Vmatrix} b_{11} & b_{12} & \cdots & b_{1n} \\ b_{21} & b_{22} & \cdots & b_{2n} \\ \cdots & \cdots & \cdots & \cdots \\ b_{n1} & b_{n2} & \cdots & b_{nn} \end{Vmatrix} \begin{Vmatrix} x_{11} & x_{12} & \cdots & x_{1n} \\ x_{21} & x_{22} & \cdots & x_{2n} \\ \cdots & \cdots & \cdots & \cdots \\ x_{n1} & x_{n2} & \cdots & x_{nn} \end{Vmatrix} = \begin{Vmatrix} c_{11} & c_{12} & \cdots & c_{1n} \\ c_{21} & c_{22} & \cdots & c_{2n} \\ \cdots & \cdots & \cdots & \cdots \\ c_{n1} & c_{n2} & \cdots & c_{nn} \end{Vmatrix}.$$

If we compute the matrix product BX, we get

$$BX = \begin{Vmatrix} \Sigma b_{1\epsilon}x_{\epsilon 1} & \Sigma b_{1\epsilon}x_{\epsilon 2} & \cdots & \Sigma b_{1\epsilon}x_{\epsilon n} \\ \Sigma b_{2\epsilon}x_{\epsilon 1} & \Sigma b_{2\epsilon}x_{\epsilon 2} & \cdots & \Sigma b_{2\epsilon}x_{\epsilon n} \\ \cdots & \cdots & \cdots & \cdots \\ \Sigma b_{n\epsilon}x_{\epsilon 1} & \Sigma b_{n\epsilon}x_{\epsilon 2} & \cdots & \Sigma b_{n\epsilon}x_{\epsilon n} \end{Vmatrix},$$

where the summation runs for $\epsilon=1,2,\ldots,n$.

The conditions to be fulfilled are obtained by setting the elements of the product BX equal to the corresponding elements of C. Taking these by columns we get the following n-sets of simultaneous linear equations:

$$\begin{array}{l} b_{11}x_{1\kappa}+b_{12}x_{2\kappa}+\cdots+b_{1n}x_{n\kappa}=c_{1\kappa}, \\ b_{21}x_{1\kappa}+b_{22}x_{2\kappa}+\cdots+b_{2n}x_{nk}=c_{2\kappa}, \\ \cdots\cdots\cdots\cdots\cdots\cdots \\ b_{n1}x_{1\kappa}+b_{n2}x_{2\kappa}+\cdots+b_{nn}x_{n\kappa}=c_{n\kappa}, \end{array} \qquad \kappa=1,2,\ldots,n. \qquad \text{I.}$$

The matrix B on the n-sets of unknowns is the same for each of the n-sets of equations, but the column of c's is different for each set. These n-sets may all be solved at once by the following simple device: Write out the two matrices B and C in juxtaposition as one matrix in the form

$$\begin{Vmatrix} b_{11} & b_{12} & b_{13} & \cdots & b_{1n} & c_{11} & c_{12} & c_{13} & \cdots & c_{1n} \\ b_{21} & b_{22} & b_{23} & \cdots & b_{2n} & c_{21} & c_{22} & c_{23} & \cdots & c_{2n} \\ b_{31} & b_{32} & b_{33} & \cdots & b_{3n} & c_{31} & c_{32} & c_{33} & \cdots & c_{3n} \\ \cdots & \cdots & \cdots & \cdots & \cdots & \cdots & \cdots & \cdots & \cdots & \cdots \\ b_{n1} & b_{n2} & b_{n3} & \cdots & b_{nn} & c_{n1} & c_{n2} & c_{n3} & \cdots & c_{nn} \end{Vmatrix}.$$

The following operations may be performed on this matrix which give an equivalent matrix in the sense that the n-sets of equations written down from the resulting matrix will have the same solutions as the systems I.

---

* From AMERICAN MATHEMATICAL MONTHLY, vol. 23 (1916), pp. 152–155.

† For the elementary properties of matrices the reader may conveniently consult Bôcher's *Introduction to Higher Algebra*, using the index to find the appropriate sections.

(1) Any two rows may be interchanged,

(2) Any row may be multiplied by a constant not zero,

(3) Any row may be multipled by a constant and added to any other row.

By (1) the element b_{11} can be different from zero, by (2) it can then be made unity. Next by (3) with $-b_{21}, -b_{31}, \ldots, -b_{n1}$, as multipliers of the first row, all the remaining elements of the first column can be replaced by zeros.

If the matrix B is non-singular, this process can be continued with the successive columns of the matrix until the matrix B is reduced to the unit matrix I and the matrix C is simultaneously reduced to X. For when B is reduced to I, the n-sets of equations have the form

$$\begin{array}{lll} x_{1\kappa} & = \bar{c}_{1\kappa}, & \\ \quad x_{2\kappa} & = \bar{c}_{2\kappa}, & \\ \qquad x_{3\kappa} & = \bar{c}_{3\kappa}, & \kappa = 1,2,\ldots,n. \qquad \text{II.} \\ \cdots\cdots\cdots & & \\ \qquad\quad x_{n\kappa} & = \bar{c}_{n\kappa}, & \end{array}$$

Illustration: To determine the matrix X to satisfy the equation

$$\begin{Vmatrix} 1 & 5 & 1 \\ 3 & 4 & 2 \\ 2 & 1 & 3 \end{Vmatrix} \begin{Vmatrix} x_{11} & x_{12} & x_{13} \\ x_{21} & x_{22} & x_{23} \\ x_{31} & x_{32} & x_{33} \end{Vmatrix} = \begin{Vmatrix} 18 & 11 & 3 \\ 20 & 11 & 7 \\ 10 & 4 & 8 \end{Vmatrix}$$

we write

$$\begin{Vmatrix} 1 & 5 & 1 & 18 & 11 & 3 \\ 3 & 4 & 2 & 20 & 11 & 7 \\ 2 & 1 & 3 & 10 & 4 & 8 \end{Vmatrix}.$$

Reducing the first column by (3), we get

$$\begin{Vmatrix} 1 & 5 & 1 & 18 & 11 & 3 \\ 0 & -11 & -1 & -34 & -22 & -2 \\ 0 & -9 & 1 & -26 & -18 & 2 \end{Vmatrix}.$$

Reducing the third column by (3), we get

$$\begin{Vmatrix} 1 & 14 & 0 & 44 & 29 & 1 \\ 0 & -20 & 0 & -60 & -40 & 0 \\ 0 & -9 & 1 & -26 & -18 & 2 \end{Vmatrix}.$$

Dividing the second row by -20 by use of (2), and reducing the second column by (3), we get

$$\begin{Vmatrix} 1 & 0 & 0 & 2 & 1 & 1 \\ 0 & 1 & 0 & 3 & 2 & 0 \\ 0 & 0 & 1 & 1 & 0 & 2 \end{Vmatrix}. \qquad \text{Hence } X = \begin{Vmatrix} 2 & 1 & 1 \\ 3 & 2 & 0 \\ 1 & 0 & 2 \end{Vmatrix}.$$

The method applies to $XB = C$ by use of the conjugates, for $B'X' = C'$.

Section 2. To find the inverse of a matrix. A very useful application occurs when C is replaced by the unit matrix I, for in this case X becomes the inverse of B. The amount of work required to calculate the inverse of a matrix by this method is practically the same as that required to compute one of its elements by the ordinary method.

Illustration: To compute the inverse of the matrix

$$B=\begin{Vmatrix} 1 & -1 & -5 \\ -1 & 3 & -10 \\ -1 & 0 & 12 \end{Vmatrix} \quad \text{we write} \begin{Vmatrix} 1 & -1 & -5 & 1 & 0 & 0 \\ -1 & 3 & -10 & 0 & 1 & 0 \\ -1 & 0 & 12 & 0 & 0 & 1 \end{Vmatrix}.$$

Reducing column one by (3):

$$\begin{Vmatrix} 1 & -1 & -5 & 1 & 0 & 0 \\ 0 & 2 & -15 & 1 & 1 & 0 \\ 0 & -1 & 7 & 1 & 0 & 1 \end{Vmatrix}.$$

Interchanging rows two and three by (1), changing signs in row two by (2), and reducing column two by (3):

$$\begin{Vmatrix} 1 & 0 & -12 & 0 & 0 & -1 \\ 0 & 1 & -7 & -1 & 0 & -1 \\ 0 & 0 & -1 & 3 & 1 & 2 \end{Vmatrix}.$$

Changing the signs in the last row by (2) and reducing column three by (3):

$$\begin{Vmatrix} 1 & 0 & 0 & -36 & -12 & -25 \\ 0 & 1 & 0 & -22 & -7 & -15 \\ 0 & 0 & 1 & -3 & -1 & -2 \end{Vmatrix}.$$

Hence,

$$B^{-1}=\begin{Vmatrix} -36 & -12 & -25 \\ -22 & -7 & -15 \\ -3 & -1 & -2 \end{Vmatrix}.$$

A RECURRING THEOREM ON DETERMINANTS*†

OLGA TAUSSKY, National Bureau of Standards

1. Introduction. This note concerns a theorem (Theorem I) on determinants [0–21, 25–27] of which proofs are being published again and again; on the other hand, the theorem is not as well known as it deserves to be. The theorem has arisen in many varied connections as is indicated by the titles of the papers quoted. Although it can be proved in a very simple manner, some of the proofs that have been given are very complicated. The theorem deals with determinants of matrices with a "dominant" main diagonal. Such matrices are particularly useful.

In what follows the theorem and several generalizations are discussed. A rather important application to estimating characteristic roots of general matrices with complex elements is mentioned. By applying these estimates to the matrices with a "dominant" main diagonal more general results are obtained.

2. Complex matrices. It will be convenient to denote by A_i the sum of the moduli of the non-diagonal terms of the ith row of a matrix $\mathbf{A}=(a_{ij})$.

THEOREM I. *If (a_{ik}) is an $n\times n$ matrix with complex elements such that*

$$|a_{ii}| > A_i, \qquad i = 1, \cdots, n \tag{1}$$

then $|a_{ik}|\neq 0$.

Proof. Assume that $|a_{ik}|=0$. The system of equations

$$\begin{aligned} a_{11}x_1 + \cdots + a_{1n}x_n &= 0 \\ &\vdots \\ a_{n1}x_1 + \cdots + a_{nn}x_n &= 0 \end{aligned} \tag{2}$$

then has a non-trivial solution $x_1, \cdots, x_n$. Let r be one of the indices for which $|x_i|$, $(i=1, \cdots, n)$, is maximum. Consider the rth equation in System (2). It implies

$$|a_{rr}||x_r| \leq \sum_{k=1,k\neq r}^{n} |a_{rk}||x_k| \leqq |A_r| x_r|$$

which is in contradiction with (1). Hence $|a_{ik}|\neq 0$.

THEOREM II. *Let (a_{ik}) be an $n\times n$ matrix with complex elements such that*

$$|a_{ii}| \geqq A_i, \qquad i = 1, \cdots, n \tag{3}$$

with equality in at most $n-1$ cases. Assume further that the matrix cannot be trans-

* From AMERICAN MATHEMATICAL MONTHLY, vol. 56 (1949), pp. 672–676.

† In the x, t, plane.

formed to a matrix of the form

$$\begin{pmatrix} P & U \\ 0 & Q \end{pmatrix}$$

by the same permutation of the rows and columns, where P and Q are square matrices and 0 consists of zeros. It follows that the determinant $|a_{ik}| \neq 0$.

Proof. The proof is similar to that for Theorem I. Assume, for example, that the first relation in (3) is not an equality. From this it follows that $|x_r| > |x_k|$ for at least one value of k. Hence the rth equation of (2) is in contradiction with (3), provided not all $a_{ri} = 0$ for which $|x_r| > |x_i|$. If this, however, is the case, then the rth row contains $n-s$ zeros where s is the number of suffixes j for which $|x_j| = |x_r|$. All the s corresponding rows contain $n-s$ zeros in the same places. It follows that the matrix is of the form which was excluded.

3. Real matrices. If all the relations in (3) become equalities the theorem ceases to hold, as is shown by any real matrix (a_{ik}) with $a_{ik} \leqq 0$ for $i \neq k$ and $\sum_{k=1}^{n} a_{ik} = 0$, $i = 1, \cdots, n$.

In this connection the following result was established: [6, 10, 14]

THEOREM III. *Let (a_{ik}) be a real $n \times n$ matrix such that $a_{ii} \geqq 0$ and $a_{ik} \leqq 0$ for $i \neq k$. Assume in addition that*

$$a_{ii} \geqq A_i, \qquad i = 1, \cdots, n \tag{4}$$

and that the matrix is not of the type excluded in Theorem II. The determinant then vanishes if and only if $\sum_{k=1}^{n} a_{ik} = 0$, $i = 1, \cdots, n$.

Proof. The determinant obviously vanishes if $\sum_{k=1}^{n} a_{ik} = 0$. Assume, conversely, that the determinant vanishes. Consider the System (2). From the arguments used for the proof of Theorem I and II it follows that $|x_1| = |x_2| = \cdots = |x_n|$. Any equation of the system (2) then implies $\sum_{k=1}^{n} a_{ik} = 0$.

THEOREM IV. *If (a_{ik}) is a real $n \times n$ matrix such that*

$$a_{ii} > A_i, \qquad i = 1, \cdots, n \tag{5}$$

then $|a_{ik}| > 0$.

Proof. This theorem can be proved by induction [12]. A different proof is obtained by using the fact that Theorem IV is obviously true if $a_{ik} = 0$ for $i \neq k$. Using this and Theorem I a proof can be obtained by continuity arguments.

4. Generalizations.

THEOREM V. *If (a_{ik}) is an $n \times n$ matrix such that*

$$|a_{ii}|\,|a_{kk}| > A_i A_k, \qquad i, k = 1, \cdots, n;\ i \neq k, \tag{6}$$

then $|a_{ik}| \neq 0$. [14, 23]

Proof. Note that the relations (6) imply $|a_{ii}| > A_i$ for all i but one. If these inequalities are satisfied for all i, the relations (1) hold. In this case the theorem is known. Assume, for example, that

$$|a_{11}| < A_1; \quad |a_{ii}| > A_i, \qquad i = 2, \cdots, n;$$
$$|a_{ii}||a_{kk}| > A_i A_k, \quad i, k = 1, \cdots, n; i \neq k.$$

Without loss of generality it may be assumed that $a_{11} = 1$ so that the above relations can be replaced by

$$1 < A_1; \quad |a_{ii}| > A_i, \qquad i = 2, \cdots, n;$$
$$|a_{ii}| > A_1 A_i, \qquad i = 2, \cdots, n;$$
$$|a_{ii}||a_{kk}| > A_i A_k, \quad i, k = 2, \cdots, n; i \neq k.$$

Multiply the first column of the matrix (a_{ik}) by A_1. It will be sufficient to prove that this new matrix is non-singular. Denote its elements by a'_{ik} the numbers corresponding to A_i by A'_i. The following inequalities hold:

$$|a'_{11}| = A'_1 = A_1; \qquad |a'_{ii}| > A'_i, \qquad i = 2, \cdots, n.$$

Since the matrix (a'_{ik}) satisfies (3), it follows that $|a'_{ik}| \neq 0$. Hence also $|a_{ik}| \neq 0$.

THEOREM VI. *Theorems I and II are best possible insofar as the inequalities involved cannot be replaced by weaker ones.*

Proof. Suppose that

$$|a_{11}| + \epsilon > A_1; \qquad |a_{ii}| > A_i, \qquad i = 2, \cdots, n,$$

where $\epsilon > 0$, but is arbitrarily small. The result follows because the matrix

$$\begin{pmatrix} \epsilon/2 & \epsilon \\ 1/2 & 1 \end{pmatrix}$$

for which these relations are satisfied is clearly singular.

5. Application. If Theorem II is applied to the characteristic determinant of any $n \times n$ matrix (a_{ik}) with complex coefficients it follows that the characteristic roots must lie inside the circles with centres a_{ii} and radii A_i [**9, 10, 14, 18, 22, 24, 25, 28–30**]. A boundary point can only be a characteristic root if it is also on the boundary of the $n-1$ other circles.

Similarly, the application of Theorem VI shows that the roots lie inside or on the boundary of a set of $n(n-1)/2$ Cassini ovals.

Now apply the circles in particular to a real matrix (a_{ik}) of the type considered in Theorems III and IV. These circles may pass through the origin, but otherwise lie entirely to the right of the imaginary axis. This gives

THEOREM VII. *All the non-zero characteristic roots of matrices with real elements which satisfy* (4) *or* (5) *have positive real parts.*

If none of the non-diagonal elements is positive it has been shown that the root with the smallest real part is real [10].

References*

0. T. Muir, History of the Theory of Determinants.

1. L. Levy, Sur la possibilité de l'équilibre électrique, Comptes Rendus (Paris), 93, 2 (1881), 706–708.

2. J. Desplanques, Théorème d'algébre, J. Math. Spéc. (3) 1(1887), 12–13.

3. H. Minkowski, Zur Theorie der Einheiten in den algebraischen Zahlkörpern, Göttinger Nachr. (1900), 90–93. (Ges. Abh. 1, 316–317). Diophantische Approximationen (Leipzig, 1907), 143–144.

4. A. Besikovitch, J. d. Physik-math. Gesellschaft d. Stastsuniversität v. Permj, 1(1918).

5. J. Hadamard, Leçons sur la propagation des ondes, Paris, (1903), 13–14.

6. A. A. Markoff, Extension des théorèmes limités du calcul des probabilités à la somme de valeurs liéis en chaîne, Mémoires Acad. Petersbourg (8), 22, 9 (1908), 25–26.

7. H. von Koch, Über das Nichtverschwinden einer Determinante, Jahresbericht d. D. M. V. 22 (1913) 285–291.

8. R. Tambs Lyche, Un théorème sur les déterminants. Det. Kong. Vid. Selskab., Forh. I. Nr. 41 (1928), 119–120.

9. S. Gersgorin, Über die Abgrenzung der Eigenwerte einer Matrix, Izv. Akad. Nauk, S. S. S. R. 7 (1931), 749–754.

10. H. Rohrbach, Bemerkungen zu einem Determinantensatz von Minkowski, Jahresbericht d. D.M.V. 40 (1931), 49–53.

11. E. Artin, Über Einheiten relativ galoisscher Zahlkörper, J. f. Math. 167 (1932) ,153.

12. P. Furtwängler, Über einen Determinantensatz, Sitzungsber. Akad. Wiss. IIa, 145 (1936), 527–528.

13. H. T. Davis, The Theory of Linear Operators, Bloomington, Ind. (1936).

14. A. Ostrowski, Sur la détermination des bornes inférieures pour une classe des determinants, Bull. Sci. Math. 61 (1937), 1–14.

Über die Determinanten mit überwiegender Hauptdiagonale, Comm. Math. Helv. 10 (1937), 69–96.

15. A. Robson, Math. Gazette 26 (1942), 191.

16. H. G. Forder, Math. Gazette 28 (1944), 63–64.

17. O. Taussky, Math. Gazette 29 (1945), 15.

18. A. Brauer, Limits for the characteristic roots of a matrix, Duke Math. J. 13 (1946), 387–395.

19. C. Massonnet, Sur une condition suffisante pour qu'un déterminant soit positif, Bull. Soc. Roy. Sci. Liège, 14 (1945), 313–317.

20. M. Parodi, Sur l'existence des réseaux électriques, Comptes Rendus (Paris), 223 (1946) 23–25.

21. R. P. Boas, this MONTHLY, 55 (1948), 99.

22. H. Wittmeyer, Einfluss der Änderung einer Matrix auf die Lösung des zugehörigen. Gleichungssystems, Zeitschr. f. ang. Math. und Mech. 16 (1936), 287–300.

23. A. Brauer, Limits for the characteristic roots of a matrix II, Duke Math. J. 14 (1947), 21–26.

24. O. Taussky (Todd), A method for obtaining bounds for characteristic roots of matrices with applications to flutter calculations, Aeronautical Research Council of Great Britain, Report 10.508 (1947).

* I owe some of the references to G. B. Price who encouraged me to write this note.

25. T. Kojima, On a theorem of Hadamard and its application, Tohoku Math. Journal 5 (1914), 54–60.
26. D. G. Bourgin, Positive determinants, this MONTHLY, vol. 46 (1939), 225–226.
27. M. Müller, Ein Kriterium für das Nichtverschwinden von Determinanten, Math. Zeitschrift 51 (1948), 291–293.
28. A. Brauer, Limits for characteristic roots of a matrix III, Duke Math. J. 15 (1948), 871–877.
29. O. Taussky, Bounds for characteristic roots of matrices, Duke Math. J. 15 (1948), 1043–44.
30. M. Parodi, Remarque sur la stabilité, Comptes Rendus (Paris) 228 (1949), 51–52.
Application d'un théorème de M. Hadamard à l'étude de la stabilité des systèmes, Comptes Rendus (Paris) 228 (1949), 807–808.
Complément à un travail sur la stabilité Comptes Rendus (Paris) 228 (1949), 1198–1200.
Sur la détermination d'une limite supérieure de la partie réelle des racines de l'équation aux fréquences propres d'un réseau électrique, Comptes Rendus (Paris) 228 (1949), 1400–02.

ON AN ELEMENTARY DERIVATION OF CRAMER'S RULE*

D. E. WHITFORD AND M. S. KLAMKIN, Polytechnic Institute of Brooklyn

The purpose of this note is to point out an elementary derivation of Cramer's rule which should be easily understood by freshman students.

Consider the simultaneous set of equations

$$\begin{aligned} a_1x + b_1y + c_1z &= d_1 \\ a_2x + b_2y + c_2z &= d_2 \\ a_3x + b_3y + c_3z &= d_3, \end{aligned} \tag{1}$$

Now

$$x\begin{vmatrix} a_1 & b_1 & c_1 \\ a_2 & b_2 & c_2 \\ a_3 & b_3 & c_3 \end{vmatrix} = \begin{vmatrix} a_1x & b_1 & c_1 \\ a_2x & b_2 & c_2 \\ a_3x & b_3 & c_3 \end{vmatrix} = \begin{vmatrix} a_1x + b_1y + c_1z & b_1 & c_1 \\ a_2x + b_2y + c_2z & b_2 & c_2 \\ a_3x + b_3y + c_3z & b_3 & c_3 \end{vmatrix}$$

by the elementary transformations of a determinant. Hence if x is to satisfy equations (1) it is necessary that

$$x\begin{vmatrix} a_1 & b_1 & c_1 \\ a_2 & b_2 & c_2 \\ a_3 & b_3 & c_3 \end{vmatrix} = \begin{vmatrix} d_1 & b_1 & c_1 \\ d_2 & b_2 & c_2 \\ d_3 & b_3 & c_3 \end{vmatrix}$$

or

$$x = \begin{vmatrix} d_1 & b_1 & c_1 \\ d_2 & b_2 & c_2 \\ d_3 & b_3 & c_3 \end{vmatrix} \div \Delta, \quad \text{provided} \quad \Delta \neq 0,$$

* From AMERICAN MATHEMATICAL MONTHLY, vol. 60 (1953), pp. 186–187.

where Δ is the determinant of the coefficient matrix of the system (1). Similarly

$$y = \begin{vmatrix} a_1 & d_1 & c_1 \\ a_2 & d_2 & c_2 \\ a_3 & d_3 & c_3 \end{vmatrix} \div \Delta, \qquad z = \begin{vmatrix} a_1 & b_1 & d_1 \\ a_2 & b_2 & d_2 \\ a_3 & b_3 & d_3 \end{vmatrix} \div \Delta.$$

That these conditions are sufficient, when $\Delta \neq 0$, can be established by substituting back into (1), which gives

$$a_r \begin{vmatrix} d_1 & b_1 & c_1 \\ d_2 & b_2 & c_2 \\ d_3 & b_3 & c_3 \end{vmatrix} + b_r \begin{vmatrix} a_1 & d_1 & c_1 \\ a_2 & d_2 & c_2 \\ a_3 & d_3 & c_3 \end{vmatrix} + c_r \begin{vmatrix} a_1 & b_1 & d_1 \\ a_2 & b_2 & d_2 \\ a_3 & b_3 & d_3 \end{vmatrix} = d_r \begin{vmatrix} a_1 & b_1 & c_1 \\ a_2 & b_2 & c_2 \\ a_3 & b_3 & c_3 \end{vmatrix}.$$

That this is true follows from

$$\begin{vmatrix} a_r & b_r & c_r & d_r \\ a_1 & b_1 & c_1 & d_1 \\ a_2 & b_2 & c_2 & d_2 \\ a_3 & b_3 & c_3 & d_3 \end{vmatrix} = 0$$

since the top row is equivalent to one of the other rows.

The method can be extended immediately to n linear equations in n unknowns.

A SHORT PROOF OF CRAMER'S RULE*

STEPHEN M. ROBINSON, United States Military Academy, West Point†

Many texts on linear algebra (e.g., [1] through [7]) prove Cramer's rule by using the relationship $A^{-1} = \text{adj } A/\det A$ and comparing cofactor expansions. The following proof may provide more insight into what is actually happening when Cramer's rule is used.

Let $Ax = b$, with A $n \times n$ and nonsingular. Let the columns of A be $a_1, \cdots, a_n$ and those of the identity be $e_1, \cdots, e_n$. Define X_k by

$$X_k = [e_1, \cdots, e_{k-1}, x, e_{k+1}, \cdots, e_n].$$

Then

$$x_k = \det X_k = \det A^{-1}AX_k = \det AX_k/\det A$$
$$= \det [a_1, \cdots, a_{k-1}, b, a_{k+1}, \cdots, a_n]/\det A,$$

which is Cramer's rule.

* From MATHEMATICS MAGAZINE, vol. 43 (1970), pp. 94–95.

† Current address: Mathematics Research Center, University of Wisconsin, Madison.

This proof makes it easier to see what we are doing when we use Cramer's rule. We want to evaluate det X_k in order to find x_k. But X_k contains the unobservable vector **x**. We therefore take the determinant of the image of X_k under the transformation represented by A, and then, to compensate for the transformation, divide the result by det A.

I am indebted to Dr. T. H. M. Crampton for helpful comments on this subject.

The opinions expressed herein are those of the author, and do not necessarily reflect the position of the Department of the Army or the U. S. Government.

References

1. H. G. Campbell, An Introduction to Matrices, Vectors, and Linear Programming, Appleton-Century-Crofts, New York, 1965.
2. C. W. Curtis, Linear Algebra, Allyn and Bacon, Boston, 1963.
3. G. Hadley, Linear Algebra, Addison-Wesley, Reading, 1961.
4. K. Hoffman and R. Kunze, Linear Algebra, Prentice-Hall, Englewood Cliffs, 1961.
5. M. Marcus and H. Minc, Introduction to Linear Algebra, Macmillan, New York, 1965.
6. P. C. Shields, Linear Algebra, Addison-Wesley, Reading, 1964.
7. R. R. Stoll, Linear Algebra and Matrix Theory, McGraw-Hill, New York, 1952.

A NOTE ON THE SYLVESTER-FRANKE THEOREM*

HARLEY FLANDERS, University of California, Berkeley

L. Tornheim [1] has recently given a proof of the Sylvester-Franke theorem based on elementary transformations. We shall present here another proof which is based on the Grassmann exterior algebra [2].

Let L be a linear space of dimension n over a field of scalars k. For each non-negative integer p, there is associated to L in an intrinsic manner a new vector space, denoted by $\Lambda^p L$, whose elements are called p-vectors. This space may be characterized as follows. There is a function f which carries ordered p-tuples of vectors of L into $\Lambda^p L$, *i.e.*,

$$(x_1, \cdots, x_p) \to f(x_1, \cdots, x_p) \in \Lambda^p L \qquad \text{for } x_i \in L,$$

such that:

(a) f is linear in each variable. In short, f is multilinear.

(b) f is alternating. This means that if π is any permutation of the indices $1, 2, \cdots, p$, then

$$f(x_{\pi(1)}, \cdots, x_{\pi(p)}) = (\text{sgn } \pi) f(x_1, \cdots, x_p)$$

for all elements $x_1, \cdots, x_p$ of L. Here sgn π denotes the sign of the permutation π. It is 1 if π is even and -1 if π is odd.

(c) The collection of all images $f(x_1, \cdots, x_p)$ of p-tuples of L generates the space $\Lambda^p L$.

* From AMERICAN MATHEMATICAL MONTHLY, vol. 60 (1953), pp. 543–545.

(d) If $x_1, \cdots, x_n$ is a basis of L, then the $C_{n,p}$ elements $f(x_{i_1}, \cdots, x_{i_p})$, $(1 \leqq i_1 < i_2 < \cdots < i_p \leqq n)$ are linearly independent in $\Lambda^p L$.

The following are consequences of (a)–(d):

(e) The $C_{n,p}$ elements of (d) form a basis of $\Lambda^p L$.

(f) If $x_i = x_j$ for some $i \neq j$, then $f(x_1, \cdots, x_n) = 0$.

We now use the following notation. The *exterior product* of $x_1, \cdots, x_p$ is

$$x_1 \wedge \cdots \wedge x_p = f(x_1, \cdots, x_p).$$

Let M be a second linear space over k and A a linear transformation on L into M. This induces a linear transformation, denoted by $\Lambda^p A$, on $\Lambda^p L$ into $\Lambda^p M$ which is determined as follows: If $x_1, \cdots, x_p$ are in L, then

$$(\Lambda^p A)(x_1 \wedge \cdots \wedge x_p) = Ax_1 \wedge \cdots \wedge Ax_p. \tag{1}$$

In terms of matrices, this exterior power $\Lambda^p A$ of A is precisely the pth compound of A.

To see this, we must first introduce some more notation. Let $N = \{1, 2, \cdots, n\}$. We shall denote by H and K subsets of N of exactly p elements and we shall always write $H = \{h_1, h_2, \cdots, h_p\}$ where $h_1 < h_2 < \cdots < h_p$. Thus there are $C_{n,p}$ possible sets H.

Now let $x_1, \cdots, x_n$ be a basis of L and $y_1, \cdots, y_m$ a basis of M. We must here assume both $0 \leqq p \leqq n$ and $0 \leqq p \leqq m$. The equations

$$Ax_i = \sum a_{ij} y_j \tag{2}$$

yield the matrix $\|a_{ij}\|$ of A with respect to the given bases x_i, y_j. The $C_{n,p}$ p-vectors $x_H = x_{h_1} \wedge \cdots \wedge x_{h_p}$ form a basis of $\Lambda^p L$ and the $C_{m,p}$ p-vectors $y_K = y_{k_1} \wedge \cdots \wedge y_{k_p}$ form a basis of $\Lambda^p M$. We have

$$\begin{aligned} (\Lambda^p A) x_H &= Ax_{h_1} \wedge \cdots \wedge Ax_{h_p} \\ &= \sum a_{h_1,k_1} \cdots a_{h_p,k_p} y_{k_1} \wedge \cdots \wedge y_{k_p} = \sum a_{H,K} y_K, \\ &\qquad\qquad \text{where } a_{H,K} = | a_{h_i,k_j} |, \; i, j = 1, \cdots, p. \end{aligned} \tag{3}$$

This shows that the matrix of $\Lambda^p A$ with respect to the bases x_H and y_K of $\Lambda^p L$ and $\Lambda^p M$ is indeed the pth compound of the matrix of A.

It is useful to note that if $A = 0$, the zero-transformation on L into M, then $\Lambda^p A = 0$, and if $A = I$, the identity transformation on L onto itself, then $\Lambda^p A = I$, the identity on $\Lambda^p L$ onto itself.

LEMMA. $(\Lambda^p A)(\Lambda^p B) = \Lambda^p(AB)$.

Proof. We have

$$\begin{aligned} (\Lambda^p A)(\Lambda^p B)(x_1 \wedge \cdots \wedge x_p) &= (\Lambda^p A)(Bx_1 \wedge \cdots \wedge Bx_p) = ABx_1 \wedge \cdots \wedge ABx_p \\ &= \Lambda^p(AB)(x_1 \wedge \cdots \wedge x_p). \end{aligned}$$

COROLLARY. *If P is a non-singular linear transformation on L onto itself, then $\Lambda^p P$ is non-singular and $(\Lambda^p P)^{-1} = \Lambda^p(P^{-1})$. If in addition A is a linear trans-*

formation on L into itself, then $\Lambda^p(PAP^{-1})$ *is similar to* Λ^pA. *Consequently* $|\Lambda^p(PAP^{-1})| = |\Lambda^pA|$.

Proof. We have $PP^{-1}=I$. By the Lemma, $(\Lambda^pP)(\Lambda^pP^{-1})=\Lambda^pI=I$, hence $(\Lambda^pP)^{-1}=\Lambda^p(P^{-1})$. Likewise, $\Lambda^p(PAP^{-1})=(\Lambda^pP)(\Lambda^pA)(\Lambda^pP)^{-1}$.

It will be recalled that if A is a linear transformation on L into M, then there is induced a linear transformation ${}^tA$, called the transpose of A, on M^* to L^*, where L^* is the dual of L, or space of linear functionals on L to k. When computed with respect to dual bases, the matrix of ${}^tA$ is the usual transpose of the matrix of A. For completeness we mention the following result which is proved in [2, p. 103–4]:

$$ {}^t(\Lambda^pA) = \Lambda^p({}^tA). \tag{4} $$

This together with the lemma and corollary above yield all the corollaries of [1].

We can now state and prove the Sylvester-Franke theorem.

THEOREM. *If A is a linear transformation on L into itself, then*

$$ |\Lambda^pA| = |A|^{C_{n-1,p-1}}. \tag{5} $$

Case 1. The matrix of A is diagonal. Then we have a basis $x_1, \cdots, x_n$ of L such that $Ax_i=a_ix_i$ for scalars a_i. Thus $(\Lambda^pA)x_H=a_Hx_H$ where $a_H=a_{i_1}\cdots a_{i_h}$, and so

$$ |\Lambda^pA| = \prod_H a_H = (a_1\cdots a_n)^{C_{n-1,p-1}} = |A|^{C_{n-1,p-1}}. $$

The exponent represents the number of combinations H out of N which contain a specific a_i.

Case 2. The general case. If our field k were the field of complex numbers, we could finish the proof by a continuity argument, using the fact that the matrices with distinct characteristic roots are dense amongst all matrices. But each of these is similar to a diagonal matrix, so the result follows from Case 1 and the Corollary above. An algebraic proof that always works is analogous to this continuity proof. The desired result (4) is an algebraic identity amongst the coefficients of A. Thus it suffices to prove (4) in case $A=\|x_{ij}\|$ is a matrix with independent variables as coefficients. But such a matrix has distinct characteristic roots in an algebraic extension of the field $k(x_{ij})$, hence $PAP^{-1}=B$, a diagonal matrix, and our conclusion follows.

References

1. L. Tornheim, The Sylvester-Franke theorem, this MONTHLY, vol. 59, 1952, pp. 389–91.
2. N. Bourbaki, Algèbre Multilinéaire, Paris, 1948.

ON THE GENERALIZED INVERSE OF AN ARBITRARY LINEAR TRANSFORMATION*

D. W. ROBINSON, Brigham Young University

In 1920 E. H. Moore [9] (see also [10]) introduced the concept of a generalized inverse of an arbitrary (rectangular and/or singular) matrix. The notion was later studied independently from different points of view by J. von Neumann [11], K. O. Friedrichs [5], A. Bjerhammer [2, 3], and R. Baer [1]. In 1955 a thorough investigation of the idea was made by R. Penrose [12], and in the following year R. Rado [15] showed that the basic definitions of Moore and Penrose were equivalent. Also, several applications of the concept have been made (see, for example, [6, 13, 14]), and related results have been considered, (see, for example, [4, 8, 16]).

The purpose of this present note is to provide a treatment of the generalized inverse of an arbitrary linear transformation suitable for use in the classroom.

Let $\mathcal{V}$ and $\mathcal{W}$ be finite dimensional (left) vector spaces over a division ring. Let T be a linear transformation of $\mathcal{V}$ into $\mathcal{W}$. Finally, suppose that S is a linear transformation of $\mathcal{W}$ into $\mathcal{V}$ such that

$$TST = T. \tag{1}$$

We begin by considering some immediate consequences and applications of (1).

First, if T is a one-to-one mapping of $\mathcal{V}$ onto $\mathcal{W}$, then it is evident that $S = T^{-1}$, the unique inverse of T.

Second, if β is a vector of $\mathcal{W}$, then the equation

$$xT = \beta$$

has a solution if and only if $\beta ST = \beta$, and in this case the general solution of the equation is the collection of vectors in $\mathcal{V}$ of the form

$$\beta S + \alpha(I - TS),$$

where α is an arbitrary vector of $\mathcal{V}$ and I is the identity transformation on $\mathcal{V}$. For, if $\beta ST = \beta$, then

$$(\beta S + \alpha(I - TS))T = \beta ST + \alpha(T - TST) = \beta,$$

and if $\alpha T = \beta$, then

$$\beta ST = \alpha TST = \alpha T = \beta,$$

$$\alpha = \alpha(TS) + \alpha(I - TS) = \beta S + \alpha(I - TS).$$

Thus, one particular application of (1) is to provide an explicit solution, say, of a consistent algebraic system of n linear equations in m unknowns.

Another consequence of (1) is that the null-space $\mathfrak{N}$ and the range $\mathfrak{R}$ of T are given by

$$\mathfrak{N} = \mathcal{V}(I - TS), \qquad \mathfrak{R} = \mathcal{W}(ST). \tag{2}$$

* From American Mathematical Monthly, vol. 69 (1962), pp. 412–416.

The first equation follows from the preceding paragraph with $\beta=0$. The second is a consequence of $\alpha T=\alpha(TST)=(\alpha T)(ST)$ and $\beta(ST)=(\beta S)T$, where α and β are arbitrary vectors of $\mathcal{V}$ and $\mathcal{W}$ respectively.

Finally, suppose that

$$\mathfrak{M} = \mathcal{V}(TS), \qquad \mathcal{S} = \mathcal{W}(I - ST). \tag{3}$$

The space $\mathfrak{M}$ is isomorphic to the range $\mathfrak{R}$ under the restriction of T to $\mathfrak{M}$ into $\mathfrak{R}$. For $\alpha T=\alpha(TST)=(\alpha(TS))T$ and, if $(\alpha(TS))T=\alpha T=0$, then $\alpha(TS)=(\alpha T)S=0S=0$. Furthermore, since $(TS)^2=(TST)S=TS$ and $(ST)^2=S(TST)=ST$, both TS and ST are idempotent. Thus, (see for example [7], sec. 41) TS is the projection on $\mathfrak{M}$ along $\mathfrak{N}$ and ST is the projection on $\mathfrak{R}$ along $\mathcal{S}$, and the given vector spaces have the direct sum decompositions

$$\mathcal{V} = \mathfrak{M} \oplus \mathfrak{N}, \qquad \mathcal{W} = \mathfrak{R} \oplus \mathcal{S}. \tag{4}$$

Conversely, given any complements $\mathfrak{M}$ and $\mathcal{S}$ of the null-space $\mathfrak{N}$ and the range $\mathfrak{R}$ of T, respectively, it is shown below that a linear transformation S exists such that (1), (2), and (3) are satisfied. However, in general such an S is not unique. For example, if $T=0$, then any linear transformation of $\mathcal{W}$ into $\mathcal{V}$ satisfies these three equations. But, as is now demonstrated, there is only one such S that also satisfies the equation

$$STS = S, \tag{5}$$

which is symmetrically related to equation (1).

THEOREM. *Let $\mathcal{V}$ and $\mathcal{W}$ be finite dimensional vector spaces over a division ring. Let T be a linear transformation of $\mathcal{V}$ into $\mathcal{W}$ with null-space $\mathfrak{N}$ and range $\mathfrak{R}$. Finally, let $\mathfrak{M}$ and $\mathcal{S}$ satisfy* (4). *Then there is one and only one linear transformation S of $\mathcal{W}$ into $\mathcal{V}$ such that* (1), (2), (3), *and* (5) *are satisfied.*

We demonstrate this theorem by proving the following equivalent result.

COROLLARY 1. *Let the notation and conditions be as in the theorem above. Furthermore, let E be the projection on $\mathfrak{M}$ along $\mathfrak{N}$ and F be the projection on $\mathfrak{R}$ along $\mathcal{S}$. Then there is one and only one linear transformation S of $\mathcal{W}$ into $\mathcal{V}$ such that*

$$TS = E, \qquad ST = F, \qquad TST = T, \qquad STS = S. \tag{6}$$

Proof. First, if S_1 as well as S satisfies (6), then

$$S_1 = S_1TS_1 = S_1E = S_1TS = FS = STS = S.$$

Second, let β be any vector of $\mathcal{W}$. It is now shown that there is one and only one μ in $\mathcal{V}$ such that $\mu E=\mu$ and $\beta F=\mu T$. For, by definition of $\mathfrak{R}$, there exists a vector α in $\mathcal{V}$ such that $\alpha T=\beta F$. Let $\mu=\alpha E$. Then $\mu E=\alpha E^2=\alpha E=\mu$, and since $\alpha(I-E)$ is in $\mathfrak{N}$, $\mu T=\alpha ET=\alpha T=\beta F$. Suppose further that $\mu_1 E=\mu_1$, and $\beta F=\mu_1 T$. Since $\mu-\mu_1$ is in both $\mathfrak{M}$ and $\mathfrak{N}$, it follows that $\mu_1=\mu$.

Thus, we define

$$S\colon \beta S = \mu,$$

where $\mu E=\mu$ and $\beta F=\mu T$. The mapping S is clearly a linear transformation of $\mathcal{W}$ into $\mathcal{V}$. Furthermore, since $(\alpha E)E=\alpha E$ and $(\alpha T)F=\alpha T=(\alpha E)T$, by the preceding paragraph, $(\alpha T)S=\alpha E$. That is, for any α of $\mathcal{V}$, $\alpha(TS)=\alpha E$. Thus, $TS=E$ and $TST=ET=T$. Finally, since $\beta(ST)=(\beta S)T=\mu T=\beta F$ and $\beta(STS)=(\beta S)(TS)=(\beta S)E=\mu E=\mu=\beta S$, it follows that $ST=F$ and $STS=S$.

The unique linear transformation S of this theorem is called the *generalized inverse* of T relative to the direct sum decomposition (4). It is evident that $\mathcal{S}$ is the null-space and $\mathcal{M}$ is the range of S, and that T is the generalized inverse of S relative to (4). In other words, T and S are symmetrically related in the sense that TS is the projection on the range of S along the null-space of T and ST is the projection on the range of T along the null-space of S. (Compare [1], Lemma 1, p. 178; here $\mathcal{V}=\mathcal{W}$.)

If $\mathcal{V}$ and $\mathcal{W}$ are real (complex) inner product spaces, then it is natural to choose the complements of $\mathcal{N}$ and $\mathcal{R}$ to be their respective orthogonal complements. The projections E and F are then perpendicular projections, and Corollary 1 may be restated as follows. (See also [12], [15].)

COROLLARY 2. *Let $\mathcal{V}$ and $\mathcal{W}$ be finite dimensional real (complex) inner product spaces. Let T be a linear transformation of $\mathcal{V}$ into $\mathcal{W}$. Then there exists one and only one linear transformation S of $\mathcal{W}$ into $\mathcal{V}$ such that*

$$(TS)' = TS, \qquad (ST)' = ST, \qquad TST = T, \qquad STS = S, \tag{7}$$

where the prime denotes the (complex conjugate) transpose operation.

Proof. If $\mathcal{M}$ and $\mathcal{S}$ are chosen to be the orthogonal complements of the null-space and the range of T, respectively, then since every perpendicular projection is equal to its (complex conjugate) transpose, (7) follows immediately from (6).

Conversely, if S is such that (7) is satisfied, then $I-TS$ and ST are the perpendicular projections on the null-space and the range of T along their respective orthogonal complements. Thus, by the theorem above, S is unique.

Adopting the notation of other writers on this subject, the unique S of (7) is denoted by $T\dagger$.

We conclude this note with an illustration. Let $\mathcal{V}$ and $\mathcal{W}$ be real inner product spaces with respective orthonormal bases $(\alpha_1, \alpha_2, \alpha_3)$ and $(\beta_1, \beta_2, \beta_3, \beta_4)$. Let

$$\begin{aligned} \alpha_1 &\to \beta_1 + \beta_2, \\ \alpha_2 &\to \qquad\qquad\; \beta_3 - \beta_4, \\ \alpha_3 &\to \beta_1 + \beta_2 + \beta_3 - \beta_4, \end{aligned}$$

define a linear transformation T of $\mathcal{V}$ into $\mathcal{W}$. Since $\nu=(1/\sqrt{3})(\alpha_1+\alpha_2-\alpha_3)$ and $\rho_1=(1/\sqrt{2})(\beta_1+\beta_2)$, $\rho_2=(1/\sqrt{2})(\beta_3-\beta_4)$ are orthonormal bases of the null-space and the range of T, respectively, it follows that $\alpha E = \alpha - (\alpha, \nu)\nu$, $\beta F = (\beta, \rho_1)\rho_1 + (\beta, \rho_2)\rho_2$, and the projections E and F are described by

$$\alpha_1 \to \tfrac{1}{3}(2\alpha_1 - \alpha_2 + \alpha_3), \qquad \beta_1, \beta_2 \to \tfrac{1}{2}(\beta_1 + \beta_2),$$

$$\alpha_2 \to \tfrac{1}{3}(-\alpha_1 + 2\alpha_2 + \alpha_3), \qquad \beta_3 \to \tfrac{1}{2}(\beta_3 - \beta_4),$$

$$\alpha_3 \to \tfrac{1}{3}(\alpha_1 + \alpha_2 + 2\alpha_3), \qquad \beta_4 \to -\tfrac{1}{2}(\beta_3 - \beta_4).$$

Now, if P is the linear transformation of $\mathcal{W}$ into $\mathcal{V}$ such that β_1, $\beta_2 \to \frac{1}{2}\alpha_1$, $\beta_3 \to \frac{1}{2}\alpha_2$, and $\beta_4 \to -\frac{1}{2}\alpha_2$, then $PT = F$. Hence $T\dagger = T\dagger T T\dagger = F T\dagger = P T T\dagger = PE$ is given by

$$\beta_1 \to \tfrac{1}{6}(2\alpha_1 - \alpha_2 + \alpha_3)$$

$$\beta_2 \to \tfrac{1}{6}(2\alpha_1 - \alpha_2 + \alpha_3),$$

$$\beta_3 \to \tfrac{1}{6}(-\alpha_1 + 2\alpha_2 + \alpha_3),$$

$$\beta_4 \to \tfrac{1}{6}(\alpha_1 - 2\alpha_2 - \alpha_3).$$

The method used here to find $T\dagger$ is an adaptation of the familiar method for calculating the inverse of a nonsingular matrix by means of elementary row operations. The reader may find it of interest to rework this example in the language of matrices, using the results to illustrate further this material.

References

1. R. Baer, Linear Algebra and Projective Geometry, New York, 1952.
2. A. Bjerhammar, Rectangular reciprocal matrices with special reference to geodetic calculations, Bull. Géodésique (1951) 188–220.
3. ———, A Generalized Matrix Algebra, National Research Council of Canada, Division of Applied Physics, Ottawa, 1957.
4. M. P. Drazin, Pseudo-inverses in associative rings and semi-groups, this MONTHLY 65 (1958) 506–514.
5. K. O. Friedrichs, Functional Analysis and Applications, mimeographed lecture notes, New York University, 1949–50, pp. 144–147 and 151–153.
6. T. N. E. Greville, Some applications of the pseudoinverse of a matrix, SIAM Review 2 (1960) 15–22.
7. P. R. Halmos, Finite Dimensional Vector Spaces, Van Nostrand, Princeton, 1958.
8. M. R. Hestenes, Inversion of matrices by biorthogonalization and related results, J. Soc. Indust. Appl. Math., 6 (1958) 51–90.
9. E. H. Moore, On the reciprocal of the general algebraic matrix, Abstract, Bull. Amer. Math. Soc., 26 (1920) 394–395.
10. ———, General Analysis, Mem. Amer. Philos. Soc., 1 (1935) 197–209.
11. J. v. Neumann, On regular rings, Proc. Acad. Sci. U.S.A., 22 (1936) 707–713.
12. R. Penrose, A generalized inverse for matrices, Proc. Camb. Philos. Soc., 51 (1955) 406–413.
13. ———, On best approximate solutions of linear matrix equations, Proc. Camb. Philos. Soc., 52 (1956) 17–19.
14. L. D. Pyle, The generalized inverse in linear programming, Dissertation (1960), Purdue University.
15. R. Rado, Note on generalized inverses of matrices, Proc. Camb. Philos. Soc., 52 (1956) 600–601.
16. R. D. Sheffield, A general theory for linear systems, this MONTHLY, 65 (1958) 109–111.

SCALAR VALUED MAPPINGS OF SQUARE MATRICES*

S. CATER, University of Oregon

In this paper F is a field, n is a positive integer and $L_n(F)$ denotes the algebra of all n by n matrices with entries in F. Let ϕ be a mapping of $L_n(F)$ into F; we shall study conditions which require ϕ to be the determinant on $L_n(F)$.

In [3] Hensel characterized the determinant on $L_n(F)$ as a polynomial of degree n in n^2 variables (that is, $\det(x_{ij})$ is a polynomial in the variables x_{ij} of degree n) which is multiplicative on $L_n(F)$ under the usual definition of the product of two matrices. In [6] Stephanos showed that if F is the complex field, and if ϕ is a multiplicative mapping of $L_n(F)$ into F which is analytic in each entry (when the other n^2-1 entries are fixed), then ϕ is a power of the determinant. In [1] Cater showed that it suffices in Stephanos' Theorem that ϕ is analytic in one of the diagonal entries. In [4] Hosszu showed that if F is the complex field and if ϕ is a multiplicative mapping of $L_n(F)$ into F, then there exists a multiplicative mapping p of F into F, for which $\phi(A)=p[\det A]$ for all A in $L_n(F)$. In [1] Cater let F be the complex field, endowed $L_n(F)$ with the usual topology, and showed that if ϕ is a continuous multiplicative mapping of $L_n(F)$ into F and if $\phi(aI)\geqq 0$ for all scalars $a\geqq 0$, then ϕ is uniquely determined by the scalar $\phi[\exp(1+i)I]$; and in particular if $\phi[\exp(1+i)I]=\exp(n+ni)$ then ϕ is the determinant on $L_n(F)$. Most of these results are established by examining ϕ on elementary matrices. We introduce the following

DEFINITION. *The mapping ϕ is said to be submultiplicative if $\phi(ABC)=\phi(CBA)$ for all matrices A, B, C, in $L_n(F)$.*

Plainly if ϕ is multiplicative on $L_n(F)$, then ϕ is submultiplicative because $\phi(ABC)=\phi(A)\phi(B)\phi(C)=\phi(C)\phi(B)\phi(A)=\phi(CBA)$ for all A, B, C in $L_n(F)$. If p is any mapping of F into F then the mapping $A\rightarrow p(\det A)$ is submultiplicative on $L_n(F)$ because

$$\begin{aligned} p[\det(ABC)] &= p[(\det A)(\det B)(\det C)] \\ &= p[(\det C)(\det B)(\det A)] = p[\det(CBA)]. \end{aligned}$$

In the present paper we will show that any submultiplicative mapping ϕ can be represented in this manner; there must exist a mapping p of F into F for which $\phi(A)=p(\det A)$ for all A in $L_n(F)$. We will prove further that if ϕ is a submultiplicative mapping which is given by a polynomial in the matrix entries, then ϕ is given by a polynomial in the determinant; this generalizes the work of Hensel. Finally we will characterize all the continuous multiplicative mappings of $L_n(F)$ when F is the real field.

THEOREM 1. *Let F be a field not of characteristic 2 and let ϕ be a submultiplicative mapping of $L_n(F)$ into F. Then there exists a unique mapping p of F into F*

* From AMERICAN MATHEMATICAL MONTHLY, vol. 70 (1963), pp. 163-169.

such that $\phi(A) = p(\det A)$, *all* A *in* $L_n(F)$. *Furthermore* ϕ *is multiplicative on* $L_n(F)$ *iff* p *is multiplicative on* F.

Before proving this Theorem we define elementary matrices of Types I, II and III as Jacobson does in [5], pp. 19–20;

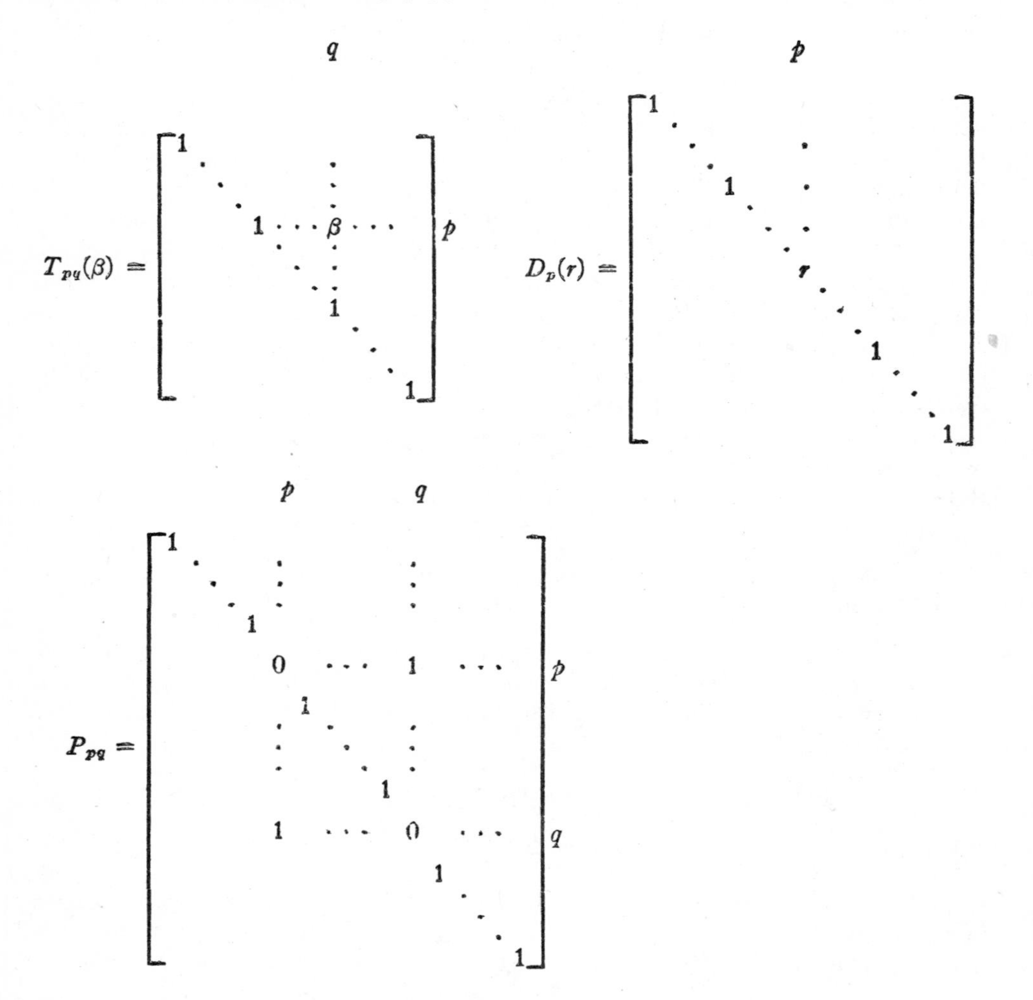

where β and r are in F and all the entries not indicated are 0. Any nonsingular matrix A in $L_n(F)$ is the product of elementary matrices in $L_n(F)$. Furthermore $\det[T_{pq}(\beta)] = 1$, $\det[D_p(r)] = r$ and $\det[P_{pq}] = -1$.

LEMMA 2. *Under the hypothesis of Theorem* 1

$$\phi[AD_p(r)] = \phi[AD_1(r)], \quad \phi[AP_{pq}] = \phi[AD_1(-1)], \quad \phi[AT_{pq}(\beta)] = \phi(A),$$

and in general $\phi(AE)=\phi[AD_1(\det E)]$ *for any elementary matrix* E *and for any matrix* A *in* $L_n(F)$.

Proof. Observe that

$$\phi(BC)=\phi(BIC)=\phi(CIB)=\phi(CB),$$
$$\phi(BCD)=\phi(DCB)=\phi(CBD)=\phi(BDC),$$

all B, C, D in $L_n(F)$. We first give a proof for $n=2$. For any A in $L_2(F)$ we have

$$\phi[AD_2(r)]=\phi\left[A\begin{pmatrix}1&0\\0&r\end{pmatrix}\begin{pmatrix}0&1\\1&0\end{pmatrix}\begin{pmatrix}0&1\\1&0\end{pmatrix}\right]=\phi\left[A\begin{pmatrix}0&1\\1&0\end{pmatrix}\begin{pmatrix}1&0\\0&r\end{pmatrix}\begin{pmatrix}0&1\\1&0\end{pmatrix}\right]$$
$$=\phi\left[A\begin{pmatrix}r&0\\0&1\end{pmatrix}\right]=\phi[AD_1(r)]$$

and

$$\phi[AP_{12}]=\phi\left[A\begin{pmatrix}0&1\\1&0\end{pmatrix}\begin{pmatrix}-\frac{1}{2}&\frac{1}{2}\\\frac{1}{2}&\frac{1}{2}\end{pmatrix}\begin{pmatrix}-1&1\\1&1\end{pmatrix}\right]=\phi\left[A\begin{pmatrix}-1&1\\1&1\end{pmatrix}\begin{pmatrix}0&1\\1&0\end{pmatrix}\begin{pmatrix}-\frac{1}{2}&\frac{1}{2}\\\frac{1}{2}&\frac{1}{2}\end{pmatrix}\right]$$
$$=\phi\left[A\begin{pmatrix}-1&0\\0&1\end{pmatrix}\right]=\phi[AD_1(-1)].$$

If β is a nonzero element in F, then

$$\phi(A)=\phi\left[A\begin{pmatrix}\beta^{-1}&0\\0&1\end{pmatrix}\begin{pmatrix}\beta&0\\0&1\end{pmatrix}\begin{pmatrix}1&0\\-1&1\end{pmatrix}\begin{pmatrix}1&0\\1&1\end{pmatrix}\right]$$
$$=\phi\left[A\begin{pmatrix}\beta^{-1}&0\\0&1\end{pmatrix}\begin{pmatrix}1&0\\1&1\end{pmatrix}\begin{pmatrix}\beta&0\\0&1\end{pmatrix}\begin{pmatrix}1&0\\-1&1\end{pmatrix}\right]=\phi\left[A\begin{pmatrix}1&0\\\beta-1&1\end{pmatrix}\right].$$

And

$$\phi(A)=\phi\left[A\begin{pmatrix}1&0\\\beta&1\end{pmatrix}\right]\qquad\text{if } \beta\neq-1.$$

But

$$\phi(A)=\phi\left[A\begin{pmatrix}1&0\\-1&1\end{pmatrix}\begin{pmatrix}1&0\\1&1\end{pmatrix}\right]=\phi\left[A\begin{pmatrix}1&0\\-1&1\end{pmatrix}\right],$$

and consequently $\phi(A)=\phi[AT_{21}(\beta)]$ for all β in F.

Now suppose that $n>2$. It follows that $\phi[AT_{pq}(\beta)]=\phi(A)$ from the above argument where p replaces 2 and q replaces 1; the ppth, qqth, pqth and qpth entries in the matrices in this argument are given above and we set any other ijth entry equal to δ_{ij} in these matrices (excluding A) where δ denotes the Kronecker delta. Likewise $\phi[AD_p(r)]=\phi[AD_1(r)]$ and $\phi[AP_{pq}]=\phi[AD_1(-1)]$ and the proof is complete.

LEMMA 3. *Under the hypothesis of Theorem* 1 $\phi(A)=\phi[D_1(\det A)]$ *for each nonsingular* A *in* $L_n(F)$.

Proof. Let A be a nonsingular matrix in $L_n(F)$ and put $A=A_1A_2\cdots A_m$ where each A_i is an elementary matrix. Then

$$\begin{aligned}\phi(A) &= \phi[A_1\cdots A_m] = \phi[A_1\cdots A_{m-1}D_1(\det A_m)]\\ &= \phi[D_1(\det A_m)A_1\cdots A_{m-1}]\end{aligned}$$

by Lemma 2; $\phi(A)=\phi[D_1(\det A_{m-1})D_1(\det A_m)A_1\cdots A_{m-2}]$ and repeated applications of this principle show that

$$\begin{aligned}\phi(A) &= \phi[D_1(\det A_1)\cdots D_1(\det A_m)] = \phi[D_1(\det A_1\cdots A_m)]\\ &= \phi[D_1(\det A)].\end{aligned}$$

LEMMA 4. *Under the hypothesis of Theorem* 1 $\phi(A)=\phi[D_1(\det A)]$ *for all* A *in* $L_n(F)$.

Proof. It suffices to show that $\phi(A)=\phi[D_1(0)]$ for singular matrices A in $L_n(F)$. Let V be an n-dimensional vector space over F and select a linear basis of V so that each matrix in $L_n(F)$ can be identified (in the usual manner) with a linear operator on V. Let A be a singular matrix in $L_n(F)$. Then there is an $(n-1)$-dimensional subspace U of V which contains the subspace VA and there is a singular matrix T mapping each vector in U into itself; hence $A=AT$. For each index j, $1\leqq j\leqq n$, there is a nonsingular S_j in $L_n(F)$ for which $S_jTS_j^{-1}=D_j(0)$. Then

$$\begin{aligned}\phi(A) &= \phi[AT] = \phi[ATS_1^{-1}S_1] = \phi[AS_1TS_1^{-1}] = \phi[S_1TS_1^{-1}A]\\ &= \phi[S_1TS_1^{-1}ATS_2^{-1}S_2] = \phi[S_1TS_1^{-1}AS_2TS_2^{-1}]\\ &= \phi[(S_2TS_2^{-1})(S_1TS_1^{-1})A]\end{aligned}$$

and repeated applications of this principle show that

$$\phi(A) = \phi[(S_nTS_n^{-1})\cdots(S_1TS_1^{-1})A] = \phi(N),$$

where N denotes the zero matrix. Since $D_1(0)$ is singular we have $\phi(A)=\phi(N)=\phi[D_1(0)]$.

Proof of Theorem 1. We have shown that $\phi(A)=\phi[D_1(\det A)]$ for all A in $L_n(F)$. Define the mapping p of F into F as follows; for each a in F put $p(a)=\phi[D_1(a)]$. Plainly $\phi(A)=p(\det A)$ for each A in $L_n(F)$. Indeed p is uniquely determined by ϕ; for if p' were another such mapping of F into F, then for each matrix A in $L_n(F)$ for which $a=\det A$ we have

$$p'(a) = p'(\det A) = \phi(A) = p(\det A) = p(a).$$

Now suppose p is multiplicative on F; then ϕ must be multiplicative on $L_n(F)$ because

$$\phi(AB) = p(\det AB) = p[(\det A)(\det B)] = p(\det A)p(\det B) = \phi(A)\phi(B).$$

On the other hand if ϕ is multiplicative on $L_n(F)$ then p is multiplicative on F because

$$p(ab) = p[(\det A)(\det B)] = p(\det AB) = \phi(AB)$$
$$= \phi(A)\phi(B) = p(\det A)p(\det B) = p(a)p(b)$$

for any matrices A, B in $L_n(F)$ for which $a=\det A$, $b=\det B$. This concludes the proof of Theorem 1.

Note that if tr is the trace mapping on $L_n(F)$, then $\mathrm{tr}(AB)=\mathrm{tr}(BA)$ for all A, B in $L_n(F)$. However for $n=2$ we have

$$\mathrm{tr}\left[\begin{pmatrix}-1 & 0\\ 1 & 0\end{pmatrix}\begin{pmatrix}0 & 1\\ 1 & 0\end{pmatrix}\begin{pmatrix}1 & 0\\ 0 & 0\end{pmatrix}\right] = \mathrm{tr}\begin{pmatrix}0 & 0\\ 0 & 0\end{pmatrix} = 0,$$

$$\mathrm{tr}\left[\begin{pmatrix}1 & 0\\ 0 & 0\end{pmatrix}\begin{pmatrix}0 & 1\\ 1 & 0\end{pmatrix}\begin{pmatrix}-1 & 0\\ 1 & 0\end{pmatrix}\right] = \mathrm{tr}\begin{pmatrix}1 & 0\\ 0 & 0\end{pmatrix} = 1,$$

and tr is not submultiplicative on $L_2(F)$ (in fact, tr is not submultiplicative on $L_n(F)$ for $n\geqq 2$). There is no mapping p of F into F for which $\mathrm{tr}(A)=p(\det A)$ on $L_n(F)$. Consequently in Theorem 1 the hypothesis that ϕ is submultiplicative may not be replaced by $\phi(AB)=\phi(BA)$ for all A, B in $L_n(F)$.

COROLLARY 5. *If in Theorem 1 all the elements in F have n-th roots in F and if ψ is another submultiplicative mapping of $L_n(F)$ into F for which $\psi(aI)=\phi(aI)$ for every scalar a in F, then $\psi=\phi$ on $L_n(F)$; in particular if $\psi(aI)=a^n$ for every a in F, then ψ is the determinant on $L_n(F)$.*

Proof. For any a in F there is an element b in F for which $b^n=a$ and by Lemma 4,

$$\phi(bI) = \phi[D_1(b^n)] = \phi[D_1(a)]$$

and likewise $\psi(bI)=\psi[D_1(a)]$. But $\psi(bI)=\phi(bI)$ and $\psi[D_1(a)]=\phi[D_1(a)]$. Again by Lemma 4, $\phi(A)=\phi[D_1(\det A)]=\psi[D_1(\det A)]=\psi(A)$ for all A in $L_n(F)$.

COROLLARY 6. *If in Theorem 1, F is a topological field (the set $L_n(F)$ can be regarded as the cartesian product of n^2 copies of F) and $L_n(F)$ is endowed with the cartesian product topology, then ϕ is continuous on $L_n(F)$ iff p is continuous on F.*

Proof. Suppose p is continuous on F. Then $\phi(A)=p(\det A)$ is continuous on $L_n(F)$ because $A\to\det A$ is a continuous mapping of $L_n(F)$ into F. Suppose ϕ is continuous on $L_n(F)$. Then $p(a)=\phi[D_1(a)]$ is a continuous mapping of F into F because $a\to D_1(a)$ is a continuous mapping of F into $L_n(F)$.

THEOREM 7. *Let F, $L_n(F)$ and ϕ be as in Theorem 1. Then the following statements are equivalent:*

(1) *There is a polynomial $Q(x_{ij})$ in n^2 variables with coefficients in F for which $Q(a_{ij})=\phi(a_{ij})$ for all (a_{ij}) in $L_n(F)$.*

(2) *There is a polynomial* $q(x_i)$ *in* n *variables with coefficients in* F *for which* $q(a_{ii}) = \phi(a_{ij})$ *for all diagonal* (a_{ij}) *in* $L_n(F)$.

(3) *There is a polynomial* $s(x)$ *in one variable with coefficients in* F *for which* $s[\det(a_{ij})] = \phi(a_{ij})$ *for all* (a_{ij}) *in* $L_n(F)$.

Proof. The implications (1)⇒(2) is trival. Assume (2). For all A in $L_n(F)$ we have $\phi(A) = \phi[D_1(\det A)]$ by Lemma 4. But

$$\phi[D_1(\det A)] = q(\det A, 1, \cdots, 1)$$

and it suffices to define the polynomial s in (3) to be $s(x) = q(x, 1, \cdots, 1)$. This proves that (2)⇒(3). Now assume (3); there is a polynomial s in one variable for which $\phi(a_{ij}) = s[\det(a_{ij})]$. But $\det(x_{ij})$ is a polynomial in the n^2 variables x_{ij}. It suffices to define the polynomial Q in (1) to be $Q(x_{ij}) = s[\det(x_{ij})]$. Hence (3)⇒(1) and the proof is complete.

COROLLARY 8. *If* F *is an infinite field in Theorem* 7, *if* (1), (2), (3) *hold and if* ϕ *is nonconstant, then*

(i) $n \leqq$ *degree* q *and* $n \leqq$ *degree* Q.

(ii) ϕ *is multiplicative iff* $\phi(a_{ij}) = [\det(a_{ij})]^m$ *for some positive integer* m *and all* (a_{ij}) *in* $L_n(F)$.

Proof. Because F is infinite there is no nonzero polynomial with coefficients in F of which every element in F is a root. Consequently each polynomial Q, q and s given in Theorem 7 is unique. Now $\det(x_{ij})$ is a polynomial in the n^2 variables x_{ij} of degree n. By the proof of (3)⇒(1) in Theorem 7 it follows that $s[\det(x_{ij})] = Q(x_{ij})$ and $n \leqq$ degree Q. By a similar argument $q(a_{ii}) = s[\det(a_{ij})]$ for all diagonal (a_{ij}) in $L_n(F)$ and $n \leqq$ degree q (the details are left to the reader). This proves (i).

To prove (ii) observe that if $\phi(a_{ij}) = [\det(a_{ij})]^m$ then ϕ is obviously multiplicative on $L_n(F)$. Now assume ϕ is multiplicative on $L_n(F)$. Let s be the polynomial given in (3); it suffices only to show that $s(x)$ is of the form x^m for some $m > 0$.

Let m be the least positive integer for which x^m has a nonzero coefficient in $s(x)$. We have $s[(\det A)(\det A)] = [s(\det A)]^2$ for all A in $L_n(F)$, $s(t^2) = [s(t)]^2$ for all t in F and $[s(x)]^2 - s(x^2) = 0$. Consequently the constant term in $s(x)$ is 0; for otherwise the coefficient of x^m in $[s(x)]^2 - s(x^2)$ would be nonzero. And finally $s(x) = x^m$; for if m' were the least integer exceeding m for which $x^{m'}$ has a nonzero coefficient in $s(x)$, then the coefficient of $x^{m'+m}$ in $[s(x)]^2 - s(x^2)$ would be nonzero. This concludes the proof.

Let F be the real field and let n be an even integer. Then the negative elements in F have no nth roots in F. We would not expect Corollary 5 to be valid for this choice of F and n; indeed if ϕ is either det or $|\det|$ on $L_n(F)$ we have $\phi(aI) = a^n$ for all a in F. This leads us to

THEOREM 9. *If* F *is the real field, if* n *is a positive even integer and if* ϕ *is a* *multiplicative mapping of $L_n(F)$ into F such that $\phi(aI)=a^n$ for all a in F, then either $\phi(A)=\det A$ for all A in $L_n(F)$ or $\phi(A)=|\det A|$ for all A in $L_n(F)$.*

Proof. By Theorem 1 there is a multiplicative mapping p of F into F for which $p(\det A)=\phi(A)$ for all A in $L_n(F)$. In particular $p(a^n)=p(\det aI)=\phi(aI)=a^n$ for all scalars $a\geqq 0$, and $p(a)=a$ for all $a\geqq 0$. On the other hand $p(-1)p(-1)=p(1)=1$ and $p(-1)=\pm 1$. Now plainly if $p(-1)=-1$, then $p(a)=a$ for all scalars a and $\phi=\det$ on $L_n(F)$; if $p(-1)=1$, then $p(a)=|a|$ for all scalars a and $\phi=|\det|$ on $L_n(F)$. This concludes the proof.

THEOREM 10. *If F is the real field, if $L_n(F)$ has the product topology, if n is any positive integer, and if ϕ is a nonconstant, continuous, multiplicative mapping of $L_n(F)$ into F, then there is a positive real number r for which either $\phi(A)=|\det A|^r$, all A in $L_n(F)$ or $\phi(A)=s(A)|\det A|^r$, all A in $L_n(F)$ where $s(A)=1$ if $\det A\geqq 0$ and $s(A)=-1$ if $\det A<0$; in particular if $\phi(2I)=2^n$ then either $\phi=\det$ or $\phi=|\det|$ on $L_n(F)$.*

Proof. There is a nonconstant, continuous, multiplicative mapping p of F into F for which $p(\det A)=\phi(A)$ for all A in $L_n(F)$. Let c be a scalar for which $p(c)\neq 1$. Then $p(0)=p(0)p(c)$ and it follows that $p(0)=0$. On the other hand there is no nonzero scalar c for which $p(c)=0$; for if there were we would have $p(x)=p(c)p(c^{-1}x)=0$ for all scalars x. Furthermore p maps the positive real axis into itself because $p(a^{1/2})p(a^{1/2})=p(a)$ for all scalars $a>0$. Because p is continuous and multiplicative we conclude that there is a real number r for which $p(c)=c^r$ for all $c>0$. To prove this observe that there is a number r for which $p(2)=2^r$; then $p(2^i)=2^{ir}$, $p(2^{i/j})=2^{ir/j}=(2^{i/j})^r$ for integers i and j with $j\neq 0$, and $p(c)=c^r$ for all $c>0$ follows from the continuity of p. Because $p(0)=0$ it follows that $r>0$. Furthermore $p(-1)p(-1)=p(1)=1$ and $p(-1)=\pm 1$.

Now $\phi(A)=p(\det A)=p[s(A)|\det A|]=p[s(A)]|\det A|^r$ and the conclusion is evident.

References

1. S. Cater, On multiplicative mappings of rings of operators, Proc. Amer. Math. Soc., 13 (1962) 55–58.

2. J. Gaspar, Eine neues Definition für Determinanten, Publ. Math. Debrecen, 3 (1954) 257–260.

3. K. Hensel, Über dem Zusammenhang zwischen der Systemen und ihren Determinanten, J. Reine Angew. Math., 159 (1928) 246.

4. M. Hosszu, A remark on scalar valued multiplicative functions of matrices, Publ. Math. Debrecen, 6 (1959) 288–289.

5. N. Jacobson, Lectures in Abstract Algebra, Vol. II, Van Nostrand, Princeton, N. J., 1953.

6. C. Stephanos, Sur une Propriété Caractéristique des Déterminants, Ann. Mat. Pura Appl., 21 (1913) 233–236.

A PROOF OF THE EQUALITY OF COLUMN AND ROW RANK OF A MATRIX*

HANS LIEBECK, University of Keele, Staffordshire, England

Let A be an $m \times n$ complex matrix, A^* its conjugate transpose, and let x and y denote $n \times 1$ matrices (column vectors).

LEMMA 1. *$y^*y=0$ if and only if $y=0$.*

LEMMA 2. *$Ax=0$ if and only if $A^*Ax=0$.*

Proof. If $A^*Ax=0$, then $y^*y=0$, where $y=Ax$. Hence, by Lemma 1, $y=0$. The converse is obvious.

Now let $R(A)$ denote the range of A, i.e. the vector space $\{Ax; \text{all } x\}$. Note that column rank (c.r.) $A = \dim R(A)$. We write r.r. A for the row rank of A.

LEMMA 3. $\dim R(A) = \dim R(A^*A)$.

Proof. By Lemma 2, $Ax_1, \cdots, Ax_k$ are linearly independent if and only if $A^*Ax_1, \cdots, A^*Ax_k$ are linearly independent.

THEOREM. c.r. A = c.r. A^* = r.r. A.

Proof. c.r. A = c.r. $A^*A = \dim\{A^*(Ax); \text{all } x\} \leqq \dim\{A^*y; \text{all } y\}$ = c.r. A^*. Thus also c.r. $A^* \leqq$ c.r. A^{**} = c.r. A, and so c.r. A = c.r. A^* = r.r. $\bar{A}$ = r.r. A.

BIBLIOGRAPHIC ENTRIES: EQUATIONS, DETERMINANTS, AND RELATED TOPICS

Except for the entry labeled TWO-YEAR COLLEGE MATHEMATICS JOURNAL, the references below are to the AMERICAN MATHEMATICAL MONTHLY.

1. John Williamson, The expansion of determinants of composite order, 1933, 65–69.

Presents a method for reducing a determinant of order mn to determinants of order $m(n-1)$.

2. G. B. Price, Some identities in the theory of determinants, 1947, 75–90.

Includes an interesting historical account.

3. Leonard Tornheim, The Sylvester-Franke theorem, 1952, 389–391.

G. B. Price in the preceding article asked for an elementary proof of this theorem. Compare the proof by exterior algebra in Flanders' article reprinted above (pp. 314–316).

4. Donald Greenspan, Methods of matrix inversion, 1955, 303–318.

5. Irving Reiner, Completion of primitive matrices, 1966, 380–381.

Proves that an r by n matrix with entries in a Dedekind domain R can be extended to an n by n unimodular matrix if and only if the r by r minors generate the unit ideal in R. See also Schenkman's paper (pp. 16–17) and Reiner's earlier article (pp. 109–110) reprinted above.

---

* From AMERICAN MATHEMATICAL MONTHLY, vol. 73 (1966), p. 1114.

6. N. S. Mendelsohn, Some elementary properties of ill conditioned matrices and linear equations, 1956, 285–295.

Discusses a process for replacing systems of equations with equivalent systems in order to reduce round-off errors.

7. S. B. Townes, The volume of a tetrahedron as a determinant, 1956, 574–575.
8. R. D. Sheffield, A general theory for linear systems, 1958, 109–111.
9. Elmar Zemgalis, On one-sided inverses of matrices, TWO-YEAR COLLEGE MATHEMATICS JOURNAL, Spring 1971, 45–48.

Discusses the existence of one-sided inverses of (non-square) matrices and how to find them.

10. C. G. Cullen and K. J. Gale, A functional definition of the determinant, 1965, 403–406.

Establishes properties of the determinant from its multiplicativeness and its value at transvections. Related to Cater's paper (pp. 321–327) reprinted above.

11. Marvin Marcus and Henryk Minc, Permanents, 1965, 577–591.
12. M. Machover, Matrices which take a given vector into a given vector, 1967, 851–852.

Given vectors x and y, shows how to find all matrices L with $Lx=y$.

(b)

CHARACTERISTIC POLYNOMIALS AND EIGEN VALUES

ON CHARACTERISTIC ROOTS OF MATRIX PRODUCTS*

W. M. SCOTT, University of Alabama

1. Introduction. It is well known that two square matrices C_1 and C_2 are such that C_1C_2 and C_2C_1 have the same determinant and that the trace of C_1C_2 is equal to the trace of C_2C_1. H. S. Thurston† proved that C_1C_2 and C_2C_1 have the same characteristic roots. His proof depended upon a continuity concept. MacDuffee‡ gives a proof and lists references to the theorem.

The proof given here is a purely matrix one.

2. We prove the following:

THEOREM 1. *If C_1 and C_2 are two square matrices, then C_1C_2 and C_2C_1 have the same characteristic roots.*

* From AMERICAN MATHEMATICAL MONTHLY, vol. 48 (1941), pp. 201–203.

† H. S. Thurston, On the characteristic equations of products of square matrices, this MONTHLY, vol. 38, 1931, pp. 322–324. See also, J. H. M. Wedderburn, Lectures on Matrices, American Mathematical Society Publications, 1934, p. 25.

‡ C. C. MacDuffee, Theory of matrices, Ergebnisse der Mathematik, Springer, Berlin, 1933.

Proof. It is well known that similar matrices have the same characteristic roots.

We have the two following cases to consider:

(i). If either C_1 or C_2 is non-singular, say $|C_1| \neq 0$, then

$$C_1C_2 \sim C_2C_1$$

for

$$C_1^{-1}C_1C_2C_1 = C_2C_1.$$

(ii). If both C_1 and C_2 are singular, let $C_1C_2 = X$. We know that there exist two non-singular matrices, M and N, such that

$$MC_1N = \begin{pmatrix} E_r & O_2 \\ O_3 & O_4 \end{pmatrix},$$

where E_r is a unit matrix of rank r, the rank of C_1, and the other three, O_2, O_3, O_4, are three zero-matrices of r rows and $n-r$ columns, $n-r$ rows and r columns, and $n-r$ rows and $n-r$ columns, respectively.

We also know that since $|M| \neq 0$ and $C_1C_2 = X$,

$$MC_1C_2M^{-1} \sim X$$

or

$$MC_1 NN^{-1}C_2M^{-1} \sim X.$$

Now we split $N^{-1}C_2M^{-1}$ into sub-matrices as follows:

$$N^{-1}C_2M^{-1} = \begin{pmatrix} C_{11} & C_{12} \\ C_{21} & C_{22} \end{pmatrix},$$

where C_{11} is r by r, C_{12} is r by $n-r$, C_{21} is $n-r$ by r, and C_{22} is $n-r$ by $n-r$.

We now have that

$$MC_1 NN^{-1}C_2M^{-1} = \begin{pmatrix} E_r & O_2 \\ O_3 & O_4 \end{pmatrix} \begin{pmatrix} C_{11} & C_{12} \\ C_{21} & C_{22} \end{pmatrix} = \begin{pmatrix} C_{11} & C_{12} \\ O & O \end{pmatrix} = P \sim X.$$

But the characteristic equation of P depends only upon C_{11}, that is, the characteristic roots of P are entirely determined from the sub-matrix C_{11}.

Similarly, we put $C_2C_1 = X'$, and write

$$N^{-1}C_2C_1N \sim X'$$

or

$$N^{-1}C_2M^{-1}MC_1N \sim X'.$$

But we have already written

$$N^{-1}C_2M^{-1} = \begin{pmatrix} C_{11} & C_{12} \\ C_{21} & C_{22} \end{pmatrix}$$

and

$$MC_1N = \begin{pmatrix} E_r & O_2 \\ O_3 & O_4 \end{pmatrix},$$

so that we have

$$N^{-1}C_2M^{-1}MC_1N = \begin{pmatrix} C_{11} & C_{12} \\ C_{21} & C_{22} \end{pmatrix} \begin{pmatrix} E_r & O_2 \\ O_3 & O_4 \end{pmatrix} = \begin{pmatrix} C_{11} & O \\ C_{21} & O \end{pmatrix} = Q \sim X'.$$

But this matrix, Q, depends only upon the sub-matrix C_{11} for its characteristic roots. That is, X' and X have the same characteristic roots, or, that is, C_2C_1 and C_1C_2 have the same characteristic roots.

As was shown by Thurston this can immediately be extended to any cyclic permutation of n matrices and we have the following:

THEOREM 2. *Any cyclic permutation of square matrices $C_1, C_2, C_3, \cdots, C_n$ will have the same characteristic roots, when the product is taken, as the matrix formed from the product $C_1 \cdot C_2 \cdot C_3 \cdots C_n$.*

THE MATRICES AB AND BA*

W. V. PARKER, Alabama Polytechnic Institute

There has been considerable interest recently in the pair of matrices AB and BA. It is well known that, for square matrices A and B, the matrices AB and BA have the same characteristic equation [1]. The proof given here may be of interest because it is short and elementary.

If A is non-singular, we have

$$A^{I}(AB)A = BA$$

and hence the matrices have the same characteristic equation since they are similar. If both A and B are singular Roth [2] has pointed out that AB and BA need not be similar. Flanders [3] has recently shown that AB and BA have the same elementary divisors except for those corresponding to the characteristic root zero even when A and B are not square matrices.

Now write

$$F(x, y) = |Ix - (A - Iy)B| = x^n + f_1(y)x^{n-1} + \cdots + f_n(y) \tag{1}$$

and

$$G(x, y) = |Ix - B(A - Iy)| = x^n + g_1(y)x^{n-1} + \cdots + g_n(y). \tag{2}$$

Since $F(x, y)$ and $G(x, y)$ are polynomials of degree n in x and y, $f_i(y)$ and $g_i(y)$ are polynomials of degrees not greater than n. Also from the statement above

* From AMERICAN MATHEMATICAL MONTHLY, vol. 60 (1953), p. 316.

$f_i(y) = g_i(y)$ whenever $|A - Iy| \neq 0$. Since these polynomials are equal for an infinite number of values of y it follows that $f_i(y) \equiv g_i(y)$ and $F(x, y) \equiv G(x, y)$. Setting $y = 0$ we have

$$|Ix - AB| \equiv |Ix - BA|.$$

The argument here is based on a field with an infinite number of elements. The referee has suggested that this is not necessary for the proof to be valid. It would suffice to have a field with more than $2n$ elements.

References

1. H. S. Thurston, On the characteristic equation of products of square matrices, this MONTHLY, vol. 38, 1931, pp. 322–324.
2. W. E. Roth, A theorem on matrices, this MONTHLY, vol. 44, 1937, p. 95.
3. Harley Flanders, Elementary divisors of AB and BA, Proceedings of the Amer. Math. Soc. vol. 2, 1951, pp. 871–874.

A REMARK ON CHARACTERISTIC POLYNOMIALS*

JOSEF SCHMID, University of Fribourg, Switzerland

The purpose of this note is to give a simple proof of the following

PROPOSITION. *If A and B are $n \times m$ and $m \times n$ matrices respectively, over a unitary, commutative ring R, I_n and I_m the unit-matrices of orders n and m, then for the characteristic polynomials of AB and BA holds the relation*

$$t^m \det(tI_n - AB) = t^n \det(tI_m - BA).$$

COROLLARY. *The characteristic coefficients of AB and BA are equal, as far as possible. In particular, the two extreme cases are $trAB = trBA$ and a (generalized) Lagrange-identity.*

COROLLARY. *If $AB = I_n$ and $BA = I_m$, then $n = m$.*

Proof. In this case the proposition yields $t^m(t-1)^n = t^n(t-1)^m$ and this implies $n = m$ (cf. [1, 2]).

Proof of the proposition. Consider the two quadratic matrices C, D of order $n+m$ over $R[t]$:

$$C = \begin{pmatrix} tI_n & A \\ B & I_m \end{pmatrix}, \qquad D = \begin{pmatrix} I_n & 0 \\ -B & tI_m \end{pmatrix}.$$

Multiplication gives

$$CD = \begin{pmatrix} tI_n - AB & tA \\ 0 & tI_m \end{pmatrix}, \qquad DC = \begin{pmatrix} tI_n & A \\ 0 & tI_m - BA \end{pmatrix},$$

and the result now follows from $\det CD = \det DC$.

* From AMERICAN MATHEMATICAL MONTHLY, vol. 77 (1970), pp. 998–999.

References

1. H. Flanders, Elementary divisors of AB and BA, Proc. Amer. Math. Soc., 2(1951) 871–874.
2. W. G. Leavitt, Modules over commutative rings, this MONTHLY, 71 (1964) 1112–1113.
3. C. C. MacDuffee, The theory of matrices, Ergebnisse der Math., 2(1933) 23.
4. W. M. Scott, On the characteristic roots of matrix products, this MONTHLY, 48(1941) 201–203.
5. H. S. Thurston, On the characteristic equation of products of square matrices, this MONTHLY, 38(1931) 322–324.
6. H. W. Turnbull and A. C. Aitken, An Introduction to the Theory of Canonical Matrices, Blackie, London, 1932, p. 181.

A NOTE ON SCALAR FUNCTIONS OF MATRICES*

RICHARD BELLMAN, Stanford University

It is well known that the scalar functions of the matrix A which occur as the coefficients of λ in the characteristic equation of A,

$$(1)\qquad |A-\lambda I| = 0 = \lambda^n - f_1(A)\lambda^{n-1} + f_2(A)\lambda^{n-2} - \cdots + (-1)^n f_n(A),$$

possess the important commutative property

$$(2)\qquad f_k(AB) = f_k(BA).$$

The purpose of this note is to demonstrate a converse result:

THEOREM. *If $\phi(A)$ is a polynomial in the elements a_{ij}, $i, j=1, 2, \cdots, n$, of A and has the property that $\phi(AB)=\phi(BA)$ for all square matrices A and B of order n, then $\phi(A)$ is a polynomial in $f_1(A), f_2(A), \cdots, f_n(A)$.*

Proof. Let us consider first the case where A has simple characteristic roots. There exists, then, a non-singular matrix T which reduces A to diagonal form,

$$(3)\qquad T^{-1}AT = L = \begin{pmatrix} \lambda_1 & 0 & \cdots & 0 \\ 0 & \lambda_2 & \cdots & 0 \\ \cdot & \cdot & \cdot & \cdot \\ 0 & 0 & \cdots & \lambda_n \end{pmatrix}$$

where $\lambda_1, \lambda_2, \cdots, \lambda_n$ are the characteristic roots of A. Thus,

$$(4)\qquad \phi(A) = \phi(T^{-1}LT) = \phi(LTT^{-1}) = \phi(L) = \psi(\lambda_1, \lambda_2, \cdots, \lambda_n),$$

where ψ is a polynomial in the λ_i. By choosing T suitably, we may change arbitrarily the order of appearance of the λ_i. From this it follows that ψ is a symmetric function of the λ_i, and hence a polynomial in the elementary symmetric functions of the λ_i, which are precisely the $f_k(A)$.

Since any square matrix may be arbitrarily closely approximated by matrices possessing simple characteristic roots, it follows that the representation obtained above is valid for general square matrices.

* From AMERICAN MATHEMATICAL MONTHLY, vol. 59 (1952), p. 391.

A NOTE ON MATRIX POWER SERIES*

I. M. SHEFFER†, Pennsylvania State College

Let $A : \|\alpha_{ij}\|$ be a square matrix of order k, and let

$$f(z) = \sum_{n=0}^{\infty} f_n z^n \tag{1}$$

be a power series with non-zero radius r. We consider the condition that the *matrix power series*

$$f(A) = \sum_{n=0}^{\infty} f_n A^n \tag{2}$$

be convergent. This problem has been studied by Hensel‡ and others, but a simpler treatment than theirs seems possible.

The characteristic equation for A is

$$\Delta(\lambda) \equiv \begin{vmatrix} \alpha_{11} - \lambda & \alpha_{12} \cdots \alpha_{1k} \\ \vdots & \vdots \\ \alpha_{k1} & \alpha_{k2} \cdots \alpha_{kk} - \lambda \end{vmatrix} = 0, \tag{3}$$

or, on expanding,

$$\Delta(\lambda) \equiv \delta_0 + \delta_1 \lambda + \cdots + \delta_k \lambda^k = 0, \tag{4}$$

where the δ_i are readily determined. It is well-known that a matrix satisfies its characteristic equation

$$\Delta(A) \equiv \delta_0 I + \delta_1 A + \cdots + \delta_k A^k = 0, \tag{5}$$

where I is the identity matrix. On multiplying through by A^n we obtain the relations

$$\delta_0 A^n + \delta_1 A^{n+1} + \cdots + \delta_k A^{n+k} = 0, \qquad n = 0, 1, \ldots. \tag{6}$$

Now set

$$A^n = \|\alpha_{ij}^{(n)}\|, \qquad n = 0, 1, \ldots, \tag{7}$$

so that

$$\alpha_{ij}^{(0)} = \begin{cases} 1, i = j \\ 0, i \neq j \end{cases}; \qquad \alpha_{ij}^{(1)} = \alpha_{ij}. \tag{8}$$

Then (6) can be re-written in terms of the $\alpha_{ij}^{(n)}$:

$$\delta_0 \alpha_{ij}^{(n)} + \delta_1 \alpha_{ij}^{(n+1)} + \cdots + \delta_k \alpha_{ij}^{(n+k)} = 0, \qquad n = 0, 1, \ldots. \tag{6'}$$

* From AMERICAN MATHEMATICAL MONTHLY, vol. 37 (1930), pp. 228–231.

† National Research Fellow.

‡ K. Hensel, *Über Potenzreihen von Matrizen*, Journal für die Reine und Angewandte Mathematik, vol. 155 (1926), pp. 107–110.

That is, for each (i,j), $\alpha_{ij}^{(n)}$ considered as a function of n satisfies the difference equation of order k with constant coefficients:

$$\delta_k u_{n+k} + \delta_{k-1} u_{n+k-1} + \cdots + \delta_1 u_{n+1} + \delta_0 u_n = 0, \qquad n = 0, 1, \ldots. \tag{9}$$

Let the characteristic equation (4) have the zeros $\lambda_1, \lambda_2, \ldots, \lambda_q$, of order $s_1, s_2, \ldots, s_q$, so that $s_1 + \cdots + s_q = k$. Then, as is well-known in the theory of such difference equations, the general solution of (9) is given by

$$u_n = \sum_{p=1}^{q} \left(k_{p0}\lambda_p^n + k_{p1} n \lambda_p^{n-1} + \cdots + k_{p,s_p-1} n(n-1)\cdots(n-s_p+2)\lambda_p^{n-s_p+1}\right), \tag{10}$$

where the k_{pr}'s are arbitrary constants.

Let us now return to the series (2). It converges* if for each (i,j) the series $\Sigma_0^\infty f_n \alpha_{ij}^{(n)}$ converges, and then (definition)

$$f(A) = \left\| \sum_0^\infty f_n \alpha_{ij}^{(n)} \right\| \qquad (i,j = 1, \ldots, k). \tag{11}$$

On appealing to (10) and to the fact that $\alpha_{ij}^{(n)}$ satisfies (9), we see:

THEOREM 1. *Series* (2) *converges, and to the sum-matrix* (11), *if the following series converge.*†

$$\Sigma f_n \lambda_p^n, \Sigma f_n n \lambda_p^{n-1}, \ldots, \Sigma f_n n(n-1)\cdots(n-s_p+2)\lambda_p^{n-s_p+1}, \qquad p = 1, 2, \ldots, q.$$

COROLLARY. *Series* (2) *converges if*

(i) $|\lambda_p| \leqslant r$, $p = 1, \ldots, q$;

(ii) *for every* λ_p *such that* $|\lambda_p| = r$, *the power series*‡ *for* $f(z), f'(z), \ldots, f_p^{(s-1)}(z)$ *converge at* $z = \lambda_p$.

The converse of this theorem is not strictly true, as may be seen by the following example: Choose $A = I$ (the identity), and let $f(z) = \Sigma_0^\infty f_n z^n$ be any analytic function satisfying the conditions: (i) the radius of convergence is unity; (ii) $\Sigma_0^\infty f_n$ converges; (iii) $\Sigma_0^\infty n f_n$ diverges. Now $f(A) = f(I) = I(\Sigma_0^\infty f_n)$, so that $\Sigma_0^\infty f_n I^n$ converges. But I has $\lambda = 1$ as a k-fold root of its characteristic equation, so that were the converse to hold universally, the series for $f'(z), \ldots, f^{(k-1)}(z)$ should all converge for $z = 1$. But this is not the case.

Let M be a matrix of order k. It generates a linear transformation in vector-space of k dimensions: $\Sigma_{j=1}^k m_{ij} x_j = y_i$ or $M(\mathbf{x}) = \mathbf{y}$ carries vector $\mathbf{x}$ into vector $\mathbf{y}$. A direction is *invariant* under M if there is a scalar λ such that for every vector $\mathbf{x}$ in this direction, $M(\mathbf{x}) = \lambda \mathbf{x}$.

---

* We take this as the *definition* of the convergence of (2).

† It may be noted that the convergence of the last of these series implies the convergence of all that precede it, as Hensel (loc. cit.) points out, so that it suffices merely to demand that $\Sigma f_n n^{s_p-1} \lambda_p^n$ converges, $p = 1, 2, \ldots, q$.

‡ See footnote †.

LEMMA. *If matrix A of order k has k linearly independent invariant directions, there exists a matrix Φ with non-vanishing determinant such that*

$$A=\Phi^{-1}A^{*}\Phi, \tag{12}$$

where

$$A^{*}=\begin{Vmatrix} \lambda_1 & & 0 \\ & \ddots & \\ 0 & & \lambda_k \end{Vmatrix}, \tag{13}$$

$\lambda_1,\ldots,\lambda_k$ being the zeros* of the characteristic equation (3).

Proof. Setting $\Phi=\|\phi_{ij}\|$ and expanding the equation $\Phi A=A^{*}\Phi$, we obtain a system of k equations in the k unknowns $\phi_{i1},\phi_{i2},\ldots,\phi_{ik}$ (i fixed$=1,2,\ldots,k$), the determinant being $\Delta(\lambda_i)=0$. The hypothesis on the invariant directions assures that there are k linearly independent sets $(\phi_{i1},\ldots,\phi_{ik})$, $i=1,2,\ldots,k$, so that Φ has a non-vanishing determinant.

DEFINITION. A is *regular* if it has k linearly independent invariant directions.

COROLLARY 1. *If A is regular then α_{ij} is a linear function of the k quantities $\lambda_1,\lambda_2,\ldots,\lambda_k$.*

COROLLARY 2. *If A is regular, then*

$$P(A)=\Phi^{-1}\begin{Vmatrix} P(\lambda_1) & & 0 \\ & \ddots & \\ 0 & & P(\lambda_k) \end{Vmatrix}\Phi \tag{14}$$

for every polynomial $P(z)$.

From the definition of convergence it follows that $f(A)=\Sigma_0^{\infty}f_nA^n=\lim_{s=\infty}\Sigma_{n=0}^{s} f_nA^n$; whence from corollaries 1 and 2,

$$f(A)=\lim_{s=\infty}\Phi^{-1}\begin{Vmatrix} \sum_{n=0}^{s} f_n\lambda_1^n & & 0 \\ & \ddots & \\ 0 & & \sum_{n=0}^{s} f_n\lambda_k^n \end{Vmatrix}\Phi$$

$$=\Phi^{-1}\lim_{s=\infty}\begin{Vmatrix} \sum_0^{s} f_n\lambda_1^n & & 0 \\ & \ddots & \\ 0 & & \sum_0^{s} f_n\lambda_k^n \end{Vmatrix}\Phi=\Phi^{-1}\begin{Vmatrix} f(\lambda_1) & 0 \\ 0 & f(\lambda_k) \end{Vmatrix}\Phi. \tag{15}$$

* They need not be distinct.

Hence we have the sharper result:

THEOREM 2. *If A is regular then a necessary and sufficient condition that $\Sigma_0^\infty f_n A^n$ converge is that the series for $f(z)$ converge at $z=\lambda_1,\dots,\lambda_k$.*

BIBLIOGRAPHIC ENTRIES: CHARACTERISTIC POLYNOMIALS AND EIGENVALUES

The references below are to the AMERICAN MATHEMATICAL MONTHLY.

1. E. T. Browne, Limits to the characteristic roots of a matrix, 1939, 252–265.
2. A. B. Farnell, Limits for the field of values of a matrix, 1945, 448–493.

This and the preceding article establish bounds on the eigenvalues and other quantities related to a matrix.

3. Olga Taussky, Matrices with trace zero, 1962, 40–42.

Gives short proofs of a number of results involving the eigenvalues and the field of values of a matrix.

4. Ali R. Amir-Moéz and Alfred Horn, Singular values of a matrix, 1958, 742–748.

Derives relations between the eigenvalues of a complex matrix and those of the matrices occurring in its polar decomposition.

5. D. W. Robinson, A matrix application of Newton's identities, 1961, 367–369.

Presents another approach to the problem considered in the papers of Scott, Parker, and Schmid reprinted above (pp. 329–333). Additional results on traces.

6. W. Gilbert Strang, Eigenvalues of Jordan Products, 1962, 37–40.

Finds bounds for the eigenvalues of $AB+BA$ in terms of those for A and B.

7. W. N. Everitt, Two theorems in matrix theory, 1962, 856–859.

Derives inequalities involving the characteristic roots of related Hermitian matrices.

(c)

INNER PRODUCTS AND QUADRATIC FORMS

A NOTE ON NORMAL MATRICES*

L. MIRSKY, University of Sheffield, England

A matrix is said to be *normal* if it commutes with its transposed conjugate. The product of two normal matrices is not necessarily normal but the following result, proved some years ago by Wiegmann [1], is available.

* From AMERICAN MATHEMATICAL MONTHLY, vol. 63 (1956), p. 479.

THEOREM. *If A, B, and AB are normal matrices, then BA is also normal.*

The aim of the present note is to give a very simple proof of Wiegmann's theorem. We denote by X^* the transposed conjugate of the matrix X, and write

$$f(X) = \text{trace}\,(X^*X), \qquad g(X) = |\lambda_1|^2 + \cdots + |\lambda_n|^2,$$

where $\lambda_1, \cdots, \lambda_n$ are the characteristic roots of X. Our proof is based upon the following well-known criterion of Schur [2]:

$$X \text{ is normal if and only if } f(X) = g(X). \tag{1}$$

Since A, B are normal matrices and since, moreover,

$$\text{trace}\,(XY) = \text{trace}\,(YX),$$

we have

$$\begin{aligned} f(AB) &= \text{trace}\,(B^*A^*AB) = \text{trace}\,(BB^*A^*A) \\ &= \text{trace}\,(B^*BAA^*) = \text{trace}\,(A^*B^*BA), \end{aligned}$$

so that

$$f(AB) = f(BA). \tag{2}$$

Now, AB and BA have the same characteristic roots. Therefore

$$g(AB) = g(BA). \tag{3}$$

Again, since AB is normal, we have by (1)

$$f(AB) = g(AB). \tag{4}$$

From (2), (3), and (4) we at once infer that $f(BA) = g(BA)$; and the required conclusion now follows by (1).

References

1. N. A. Wiegmann, Normal products of matrices, Duke Math. J., vol. 15, 1948, pp. 633–638.
2. I. Schur, Über die charakteristischen Wurzeln einer linearen Substitution mit einer Anwendung auf die Theorie der Integralgleichungen, Math. Annalen, vol. 66, 1909, pp. 488–510.

TOPOLOGICAL ASPECTS OF SYLVESTER'S THEOREM ON THE INERTIA OF HERMITIAN MATRICES*

HANS SCHNEIDER, University of Wisconsin, Madison

1. A set of $n \times n$ matrices with complex elements has a natural topology associated with it. One may therefore look for a topological interpretation of some results in the theory of matrices. We shall show that Sylvester's classical theorem on the inertia (signature) of Hermitian matrices concerns the connected components of the space of all Hermitian matrices of fixed rank r.

Most of the arguments used in the proof of our theorem are elementary and familiar. Yet our result does not appear in the literature. The reason may well be that matrix theorists tend to use "continuity properties" as they arise, without formalizing them, while topologists do not usually study equivalence relations on matrices. This note is offered as an illustration that even on a fairly elementary level, something is gained by looking for inter-connections between different mathematical fields.

2. Let n be a positive integer and let $\omega = (\pi, \nu, \delta)$ be an ordered triple of non-negative integers with $\pi + \nu + \delta = n$. Let E_ω be the diagonal matrix $E_\omega = \operatorname{diag}(1, \cdots, 1, -1, \cdots, -1, 0, \cdots, 0)$ with π diagonal elements 1, ν diagonal elements -1 and δ elements 0. Sylvester's theorem ([7] p. 100, [3] p. 83) on the inertia of Hermitian matrices asserts that for each $n \times n$ Hermitian matrix H there exists just one matrix E_ω for which there exists a nonsingular matrix X such that $X^*HX = E_\omega$.

But can we pick out the triple ω that occurs in Sylvester's theorem, without use of that theorem and directly in terms of the matrix H? One possibility, which we mention merely because of its intrinsic interest, is to proceed geometrically. Let V be the space of all positive n-tuples and associate with H the quadratic form Δ: $(x, x) = x^*Hx$. If it should happen that for some $y \in V$, $(y, y) > 0$ then also $(x, x) > 0$ for $x = \alpha y$ if $\alpha \neq 0$. Thus we have found a subspace W of V of which $(x, x) > 0$ for all $x \neq 0$. In other words, Δ is positive definite on W. Now suppose that π is the dimension of a subspace W of largest dimension on which Δ is positive and similarly suppose ν is the dimension of a subspace W' of largest dimension on which Δ is negative definite (i.e., $(x, x) < 0$, if $0 \neq x \in W'$). If $\delta = n - \pi - \nu$, then it may be proved that $\omega = (\pi, \nu, \delta)$ is the ω of Sylvester's theorem, (see [1] pp. 148–150).

3. We shall use an entirely different approach. Since H is Hermitian all its eigenvalues are real. We shall define π, ν, δ in terms of the eigenvalues of H.

DEFINITION 1. *Let $\pi(H) = \pi$ be the number of positive eigenvalues of H, $\nu(H) = \nu$ the number of negative eigenvalues of H, and $\delta(H) = \delta$, the number of zero eigenvalues of H. Then the ordered triple $\omega = (\pi, \nu, \delta)$ will be called the inertia of H. We shall write $\omega = \operatorname{In} H$.*

---

* From AMERICAN MATHEMATICAL MONTHLY, vol. 73 (1966), pp. 817–821.

DEFINITION 2. *Two Hermitian matrices H, K are inertially equivalent if* In $H =$ In K. *We shall write* $H \overset{i}{\sim} K$.

The next definition is standard.

DEFINITION 3. (e.g. [6] p. 99, p. 84). *Two Hermitian matrices H and K are conjunctive (conjunctively equivalent) if there exists a nonsingular X such that $X^*HX = K$. We shall write $H \overset{c}{\sim} K$.*

It is evident that $\overset{i}{\sim}$ and $\overset{c}{\sim}$ are equivalence relations on any set of Hermitian matrices.

4. Our next two equivalence relations are of a different kind, since they may be defined on any topological space.

If E is a topological space, the space is called connected if the empty set and E are the only subsets of E which are both open and closed (see [5] p. 117). A subset U of E is connected if and only if it is a connected space in the topology induced on U by E. Thus U is connected if and only if, for any set $F \subseteq E$ which is both open and closed, either $U \subseteq F$ or $U \subseteq E \backslash F$, the complement of F in E. This motivates

DEFINITION 4. *Let E be a topological space. We call $x, y \in E$ connectable in E if for every open and closed set U in E both $x, y \in U$ or both $x, y \notin U$. We shall write $x \overset{u}{\sim} y$.*

An *arc* in a topological space E is a continuous image of the interval (0, 1) on the real line in the space E ([5] p. 139).

DEFINITION 5. *Let E be a topological space. We call $x, y \in E$ arc connectable in E if there exists an arc in E joining x, y, i.e., if there exists a continuous function f from the unit interval (0, 1) on the real line into E with $f(0) = x$ and $f(1) = y$. We shall write $x \overset{a}{\sim} y$.*

The following lemma is a restatement of a well-known result (see [5] p. 141).

LEMMA 1. *If E is a topological space, then $x \overset{a}{\sim} y$ implies that $x \overset{u}{\sim} y$.*

Proof. Let U be any open and closed set containing x and let f be a continuous function of (0, 1) into E with $f(0) = x$ and $f(1) = y$. Then $f^{-1}(U)$ is an open and closed subset of (0, 1) which contains 0, and the only such set is (0, 1) itself. Thus $y = f(1) \in U$ and so $x \overset{u}{\sim} y$.

5. Let S be any set of $n \times n$ matrices. We can norm S in many ways. For example we can put $\|A\| = \max_{i,j} |a_{ij}|$ for $A \in S$. To turn S into a topological space we choose as the open sets arbitrary unions of finite intersections of all cubes $N(A, \epsilon) = \{B \in S: \max_{i,j} |b_{ij} - a_{ij}| = \|B - A\| < \epsilon\}$, with $A \in S$ and $\epsilon > 0$. Thus a subset T of S is open if and only if for $A \in T$ we can find on $\epsilon = 0$ such that $N(A, \epsilon) \subseteq T$. Observe that S need not be a linear space, nor will our topological space S necessarily be complete.

In this section, we shall consider the space N of all nonsingular complex matrices normed as above.

LEMMA 2. *For all $A, B \in N$, A is arc connectable to B in N.*

Proof. It is enough to prove that for all $A \in N$, $A \overset{a}{\sim} I$, the identity matrix. Choose σ so that $e^{i\sigma}A$ has no negative eigenvalue, and set $f(t) = e^{i\sigma t}A$, $0 \leqq t \leqq 1$. If A belongs to N, so does $e^{i\sigma t}A$ and clearly f is continuous. Thus $A \overset{a}{\sim} C$. Next set $g(t) = (1-t)C + tI$, $0 \leqq t \leqq 1$. Evidently g is again continuous, and since the eigenvalues of $g(t)$ are of the form $\gamma(t) = (1-t)\gamma + t$ where γ is an eigenvalue of C and here γ is not negative, it follows that $\gamma(t) \neq 0$ and so $g(t)$ is nonsingular. Thus $C \overset{a}{\sim} I$. Hence $A \overset{a}{\sim} I$, and this completes the proof.

6. We now require a lemma of a different type. Usually it is expressed by asserting that the eigenvalues of a matrix are continuous functions of the elements of the matrix. We shall state the result precisely:

LEMMA 3. *Let A be a matrix with distinct eigenvalues $\alpha_1, \cdots, \alpha_s$ of multiplicities $m_1, \cdots, m_s$ respectively. Let $\epsilon > 0$ and let $\Gamma(\alpha_i, \epsilon)$ be the circle with center α_i and radius ϵ. Then there is a positive σ, such that every matrix B, for which $\|B-A\| < \sigma$, has exactly m_i eigenvalues in the circle $\Gamma(\alpha_i, \epsilon)$.*

Proof. Let $p(t) = \lambda^n + p_{n-1}\lambda^{n-1} + \cdots + p_0$, and $q(t) = \lambda^{n-1} + q_{n-1}\lambda^{n-1} + \cdots + q_0$. We use a theorem on the zeros of a polynomial (see [4], p. 3). If the zeros of $p(\lambda)$ are α_i with multiplicity m_i, $i = 1, \cdots, s$, $\alpha_i \neq \alpha_j$, if $i \neq j$, and if $(q_j - p_j) < \eta$, $j = 0, 1, \cdots, n-1$, where η is sufficiently small, then m_i zeros of $q(t)$ lie in the circle $\Gamma(\alpha_i, \epsilon)$. Now the eigenvalues of A and B are simply the zeros of the characteristic polynomials $\det(\lambda I - A)$ whose coefficients are sums of products of elements of A, and similarly for B. Since addition and multiplication of complex numbers is continuous, we deduce that for sufficiently small $\sigma > 0$, $\|B-A\| < \sigma$, implies that $|q_i - p_i| < \eta$, $j = 0, \cdots, n-1$, and the result follows.

Our proof of Lemma 3 is not really much of a proof, since it refers the result for the spectra of a matrices back to the corresponding theorem for the zeros of polynomials ("continuity of zeros of polynomials"). This latter result is deeper than any other theorem we have used in this note, and we shall not attempt to prove it here. We may note that in the application of Lemma 3, the matrices A and B are both Hermitian. For normal, and therefore for Hermitian matrices, a more precise result is given in [2]:

LEMMA 4. *If A and B are normal matrices with eigenvalues $\alpha_1, \cdots, \alpha_n$ and $\beta_1, \cdots, \beta_n$ respectively, then there exists a suitable numbering of the eigenvalues such that*

$$\sum_i |\alpha_i - \beta_i|^2 \leqq \sum_{i,j} |a_{ij} - b_{ij}|^2.$$

This result rests on the famous theorem of Birkhoff that the permutation matrices are the vertices of the convex polyhedron of doubly stochastic matrices. For a proof of Birkhoff's theorem see [3] p. 97, or [2], and for a proof of

Lemma 4 see [2]. Of course, Lemma 4 implies Lemma 3 since $\sum_{i,j} |a_{ij}-b_{ij}|^2 \leqq n^2\|A-B\|^2$.

7. From now on our space will be the space H_r^n of all $n\times n$ Hermitian matrices of *fixed rank r*. We shall first examine a trivial situation. The space of all Hermitian 1×1 matrices is just the real line R and hence H_1^1 is the real line with the origin removed. The connectivity properties of R and H_1^1 are quite different. R is connected and H_1^1 is not. Similarly the connectivity properties of H_r^n will be quite different from these of the space of all $n\times n$ Hermitian matrices. The reason for focusing on H_r^n is that this space yields an interesting theorem.

Notation. Let $H\in H_r^n$. Then the set of all $K\in H_r^n$ such that $H\overset{i}{\sim}K$ will be denoted by $\mathbf{I}(H)$, and the eigenvalue class $\mathbf{I}(H)$ will be called an inertial component of H_r^n. Similarly we define $\mathbf{C}(H)$, $\mathbf{U}(H)$, $\mathbf{A}(H)$ to be the equivalence classes of H for $\overset{c}{\sim}$, $\overset{u}{\sim}$, $\overset{a}{\sim}$, respectively, and we call $\mathbf{C}(H)$ a conjunctive component, $\mathbf{U}(H)$ a connected component and $\mathbf{A}(H)$ an arc component of H_r^n.

THEOREM. *Let H_r^n be the topological space of all $n\times n$ Hermitian matrices of rank r. Then the four equivalence relations $\overset{i}{\sim}$, $\overset{c}{\sim}$, $\overset{a}{\sim}$, $\overset{u}{\sim}$, coincide on H_r^n. Equivalently, for any $H\in H_r^n$, $\mathbf{I}(H)=\mathbf{C}(H)=\mathbf{A}(H)=\mathbf{U}(H)$.*

Proof. We shall prove that $H\overset{i}{\sim}K$ implies $H\overset{c}{\sim}K$, $H\overset{c}{\sim}K$ implies $H\overset{a}{\sim}K$, $H\overset{a}{\sim}K$ implies $H\overset{u}{\sim}K$, and $H\overset{u}{\sim}K$ implies $H\overset{i}{\sim}K$.

(a) $H\overset{i}{\sim}K$ implies $H\overset{c}{\sim}K$: Suppose $\operatorname{In} H=\operatorname{In} K=\omega=(\pi,\ \nu,\ \delta)$ say. It is enough to prove that $H\overset{c}{\sim}E_\omega$, where E_ω is defined in Section 1. (Note that $\pi+\nu=r$ and that $E_\omega\in H_r^n$.) Since H is Hermitian there exists a unitary Y such that $Y^*HY=\operatorname{diag}(\alpha_1,\cdots,\alpha_n)$. By definition of H, and since the α_i are the eigenvalues of H, it follows that π of the α_i are positive, ν are negative and the remaining δ are zero. Replacing Y, if necessary, by the unitary matrix YP, where P is a permutation matrix, we may suppose that $Y^*HY=\operatorname{diag}(\alpha_1,\cdots,\alpha_n)$ where $\alpha_i>0$, $i=1,\cdots,\pi$, $\alpha_{i+1}<0$, $i=\pi+1,\cdots,\pi+\nu$, and $\alpha_i=0$, $i=\pi+\nu+1,\cdots,n$.

Thus if $D=\operatorname{diag}(\sqrt{\alpha_1},\cdots,\sqrt{\alpha_\pi},\sqrt{-\alpha_{\pi+1}},\cdots,\sqrt{-\alpha_{\pi+\nu}},1,\cdots,1)$ and $X=YD^{-1}$, then $X^*HX=E_\omega$.

(b) $H\overset{c}{\sim}K$ implies $H\overset{a}{\sim}K$: Let $X^*HX=K$, where X is nonsingular. Since by Lemma 2 any two nonsingular matrices are connected in the space N of nonsingular complex matrices, there is a continuous function $t\rightarrow X(t)$ of (0, 1) into N such that $X(0)=I$ and $X(1)=X$. Further, transposition and matrix multiplication are continuous operations. Hence the function $f(t)=X^*(t)HX(t)$ is continuous. But rank $X^*(t)HX(t)=r$, since $X(t)$ is nonsingular, whence $X(t)HX(t)\in H_r^n$. Further $f(0)=H$ and $f(1)=K$, and so $H\overset{a}{\sim}K$.

(c) $H\overset{c}{\sim}K$ implies $H\overset{u}{\sim}K$. This is just Lemma 1.

(d) $H\overset{u}{\sim}K$ implies $H\overset{i}{\sim}K$. We shall prove the equivalent result that $\mathbf{I}(H)$ contains $\mathbf{U}(H)$, for $H\in H_r^n$: Let $K\in\mathbf{I}(H)$, and let $\operatorname{In} K=\omega=(\pi,\ \nu,\ \delta)$, and suppose the eigenvalues α_i, $i=1,\cdots,\pi$, of K are positive, the eigenvalues α_i, $i=\pi+1,\cdots,\pi+\nu$ are negative and the eigenvalues α_i, $i=\pi+\nu+1,\cdots,n$ are zero.

Let $0<\epsilon<\min\{|\alpha_i|:\alpha_i\neq 0\}$. By Lemma 3 there exists a neighborhood $N(K,\sigma)$ of K in H_r^n (thus each $L\in N(K,\sigma)$ is, by assumption, Hermitian of rank r) so that the spectrum of each $L\in N(K,\sigma)$ is contained in the union of the n circles $\Gamma(\alpha_i,\epsilon)$. It follows that each L in this neighborhood of K has at least as many positive (negative) eigenvalues as K has positive (negative) eigenvalues, i.e. $\pi(L)\geqq\pi$, and $\nu(L)\geqq\nu$. But as $L\in H_r^n$, $\pi(L)+\nu(L)=r=\pi+\nu$, whence $\pi(L)=\pi$ and $\nu(L)=\nu$. Thus In $L=$ In K. It follows for each $K\in \mathrm{I}(H)$, there exists an $N(K,\sigma)\subseteq \mathrm{I}(H)$, and so $\mathrm{I}(H)$ is open. Now $H_r^n\backslash \mathrm{I}(H)=\bigcup\{\mathrm{I}(M): M\in H_r^n, M\notin \mathrm{I}(H)\}$ and a union of open sets is open. Hence $\mathrm{I}(H)$ is also closed. By the definition of $\mathrm{U}(H)$, we see that $\mathrm{U}(H)$ is contained in every open and closed set containing H, whence $\mathrm{U}(H)\subseteq \mathrm{I}(H)$. This completes the proof of the theorem.

COROLLARY. *The topological space H_r^n has precisely $r+1$ distinct inertial components (or conjunctive components, or connected components, or arc components).*

Proof. Obviously, each $\omega=(\pi,\nu,\delta)$ with $\pi+\nu=r$ corresponds to one inertial component of H_r^n, and there are just $r+1$ such ω. By the theorem, each inertial component is a conjunctive component (and a connected component and an arc component).

8. It should be noted that in the proof of our theorem (a) is just a standard proof that there exists an ω such that $H\overset{c}{\sim}E_\omega$. The direct proof that this ω is unique is simple (see [7] pp. 92, 100), but we do not need this proof. For distinct ω, the corresponding E_ω obviously lie in distinct inertial components and the uniqueness now follows from the equality of $\overset{c}{\sim}$ and $\overset{i}{\sim}$ on H_r^n.

The concept of inertia may be extended to matrices that are not Hermitian. For some results in this direction see [6].

I wish to thank Marvin Marcus, Judith Molinar, and Robert Thompson for comments which have helped to improve this paper.

References

1. E. Artin, Geometric Algebra, Interscience, New York, 1957.

2. A. J. Hoffman and H. Wielandt, The variation of the spectrum of a normal matrix, Duke M. J., 20 (1953) 37–40.

3. M. Marcus and H. Minc, A survey of matrix theory and matrix inequalities, Allyn and Bacon, Boston, 1964.

4. M. Marden, The geometry of the zeros of a polynomial in a complex variable, Math. Surveys III, American Math. Soc., 1949.

5. B. Mendelson, Introduction to Topology, Allyn and Bacon, Boston, 1962.

6. A. Ostrowski and H. Schneider, Some theorems on the inertia of matrices, J. Math. Anal ysis Appl., 4(1962) 72–84.

7. S. Perlis, Theory of matrices, Addison-Wesley, Reading, Mass., 1952.

THE DEFINITION OF A QUADRATIC FORM*

ANDREW M. GLEASON, Harvard University

Let V be a vector space over a field F. We exclude, once and for all, the possibility that F have characteristic 2, but otherwise F may be arbitrary. The vector space V may have finite or infinite dimension.

A quadratic form on V is classically defined as a function Q from V to F which can be expressed in the form

$$Q(v) = \sum_{\alpha,\beta} \lambda_{\alpha,\beta} f_\alpha(v) f_\beta(v),$$

where $\{f_\alpha\}$ is the family of linear functionals dual to some basis $\{x_\alpha | \alpha \in A\}$ of V and $\{\lambda_{\alpha,\beta}\}$ is any family of constants indexed on $A \times A$. For each v this sum will be effectively finite. A more intrinsic definition describes a quadratic form as the composition of some bilinear form from $V \times V$ to F with the diagonal injection of V into $V \times V$; that is,

$$Q(v) = B(v, v),$$

where B is bilinear.

Each bilinear form thus defines a quadratic form, but the correspondence is not one-to-one since B and the symmetric bilinear form S defined by

$$S(v, w) = \tfrac{1}{2}(B(v, w) + B(w, v))$$

define the same quadratic form. On the other hand, there is a one-to-one correspondence between symmetric bilinear forms and quadratic forms since we can recover S from Q by the formula

$$4S(v, w) = Q(v + w) - Q(v - w). \tag{1}$$

The study of quadratic forms is thereby reduced to the study of symmetric bilinear forms.

There is something inappropriate about defining a quadratic form which is a function of one variable, in terms of a bilinear form which involves two variables. This raises the question of what requirements can be imposed on a function from V to F to define the set of all quadratic forms.

The best known identity satisfied by quadratic forms is the parallelogram law

$$Q(v + w) + Q(v - w) = 2Q(v) + 2Q(w). \tag{2}$$

For the convenience of the reader, we reprove the theorem of von Neumann and Jordan [3] concerning the parallelogram law.

0.1 THEOREM. *Let Q be a function from V to F which satisfies* (2). *The function S defined by* (1) *is symmetric and biadditive and* $Q(v) = S(v, v)$.

* From AMERICAN MATHEMATICAL MONTHLY, vol. 73 (1966), pp. 1049–1056.

Proof. Putting $v=w=0$ in (2), we see that $Q(0)=0$. Then putting $v=0$, we find that $Q(-w)=Q(w)$. From this it follows that S is symmetric. Finally, putting $v=w$, we get $Q(2v)=4Q(v)$ and derive $S(v, v)=Q(v)$.

Since S is symmetric, to prove biadditivity it will be sufficient to prove it is additive in its first argument. Let x, y, and z be any three vectors in V.

$$\begin{aligned} 8S(x,z)+8S(y,z) &= 2Q(x+z)+2Q(y+z)-2Q(x-z)-2Q(y-z) \\ &= Q(x+y+2z)+Q(x-y)-Q(x+y-2z)-Q(x-y) \\ &= 4S(x+y,\, 2z), \end{aligned}$$

whence

$$S(x, z)+S(y, z) = \tfrac{1}{2}S(x+y,\, 2z). \tag{3}$$

Consider the special case $y=0$. Using $S(0, z)=S(z, 0)=0$ (direct from (1)), we obtain the identity $S(x, z)=\frac{1}{2}S(x, 2z)$. Hence (3) becomes

$$S(x, z)+S(y, z) = S(x+y,\, z).$$

0.2 COROLLARY. *Let V be a vector space over F of dimension at least two. If Q is a function from V to F such that the restriction of Q to each two dimensional subspace of V is a quadratic form, then Q is a quadratic form.*

Proof. Let v and w be any two vectors in V. Since the restriction of Q to the linear subspace spanned by v and w is a quadratic form the parallelogram identity (2) must hold. Hence $Q(v)=S(v, v)$ where S is a symmetric biadditive function from $V\times V$ to F. To check that S is homogeneous, i.e., that $S(\lambda v, w)=\lambda S(v, w)$, for a fixed v and w requires only reference to the subspace spanned by v and w. Since Q is a quadratic form on this subspace, S is homogeneous.

The corollary does not help us to find a suitable definition of a quadratic form. As is well known, additivity implies homogeneity with respect to scalars in the prime subfield. However, the parallelogram identity cannot possibly imply the homogeneity of S when F is not a prime field. In that case there exists an additive map θ of F into itself which is not homogeneous over F; then if Q is a nontrivial quadratic form, $\theta \circ Q$ satisfies (2) but the corresponding biadditive form $\theta \circ S$ is not homogeneous. Obviously some identity involving the scalar multiplication is necessary to define a quadratic form.

While teaching an elementary course in which the only field involved was the real numbers, I conjectured that the parallelogram law together with the identity

$$Q(\lambda x) = \lambda^2 Q(x) \tag{4}$$

would suffice. This seemed plausible since almost any regularity assumption in conjunction with additivity suffices to guarantee linearity for functions from the reals to the reals. However, the conjecture is false. For fields of characteristic 0, the identities (2) and (4) guarantee the homogeneity of S with respect to algebraic scalars, but not with respect to transcendental scalars!

In this paper we shall prove this and discuss some other identities satisfied by quadratic forms.

1. Quasi-quadratic forms. We shall denote by $\mathcal{Q}$ the set of all quadratic forms defined on V, and by $\mathcal{Q}_0$ the set of all functions from V to F which satisfy the identities (2) and (4). Evidently $\mathcal{Q}_0$ is a linear subspace of the space of all functions from V to F and $\mathcal{Q}$ is a linear subspace of $\mathcal{Q}_0$. We shall refer to a member of $\mathcal{Q}_0$ as a *quasi-quadratic* form.

We associate with each quasi-quadratic form Q a symmetric biadditive form S using the definition (1) and a function E from $F\times V\times V$ to F defined by

$$E(\lambda, v, w) = S(\lambda v, w) - \lambda S(v, w). \tag{5}$$

It is obvious that E is additive in each of its three arguments. Furthermore, it is clear that E is identically zero if and only if S is actually linear in its first argument. Since S is symmetric, this amounts to the assertion that E is zero if and only if $Q\in\mathcal{Q}$, i.e., Q is a genuine quadratic form.

Let $\mathcal{E}$ be the set of all functions E from $F\times V\times V$ to F which are

(a) linear and skew-symmetric in their second and third arguments, and

(b) additive derivations in their first argument, i.e.,

$$E(\lambda + \mu, v, w) = E(\lambda, v, w) + E(\mu, v, w) \quad \text{and}$$

$$E(\lambda\mu, v, w) = \lambda E(\mu, v, w) + \mu E(\lambda, v, w).$$

The set of such functions is obviously a linear space over F.

1.1 THEOREM. *The map Θ: $Q\to E$ defined by* (1) *and* (5) *is a linear surjection from $\mathcal{Q}_0$ to $\mathcal{E}$ with kernel $\mathcal{Q}$.*

Proof. Clearly Θ is a linear map from $\mathcal{Q}_0$ to the set of all functions from $F\times V\times V$ to F. We have already seen that its kernel is $\mathcal{Q}$. The essential step is to show that E, as defined in (1) and (5), is actually in $\mathcal{E}$.

It follows directly from (1) and (2) that $Q(x)+2S(x, y)+Q(y)=Q(x+y)$. Replace y by λx, expand using (4), and cancel terms to get

$$S(\lambda x, x) = \lambda S(x, x).$$

Polarize the latter: Replace x by $x+y$, expand by biadditivity and cancel terms.

$$S(\lambda x, y) + S(\lambda y, x) = \lambda(S(x, y) + S(y, x)). \tag{6}$$

Transposing terms, we obtain $E(\lambda, x, y)=-E(\lambda, y, x)$. Thus E is skew-symmetric in its second and third arguments.

Replace x by λx in (6). Since $S(\lambda x, \lambda y)=\lambda^2 S(x, y)$ by (1) and (4), we get

$$S(\lambda^2 x, y) + \lambda^2 S(x, y) = 2\lambda S(\lambda x, y).$$

Polarize again, replacing λ by $\lambda+\mu$, etc. Recall that S is homogeneous with respect to the scalar 2.

$$S(\lambda\mu x, y) + \lambda\mu S(x, y) = \lambda S(\mu x, y) + \mu S(\lambda x, y). \tag{7}$$

Transposing terms, we find $E(\lambda, \mu x, y)=\mu E(\lambda, x, y)$, that is, E is homogeneous in its second argument. By skew symmetry it is also homogeneous in its third argument. This shows that E satisfies condition (a) in the definition of $\mathcal{E}$. Subtracting $2\lambda\mu S(x, y)$ from both sides of (7), we find that E also satisfies condition (b). Hence $E\in\mathcal{E}$.

Finally we must show that Θ is surjective. Suppose $E_0\in\mathcal{E}$. We will construct a quasi-quadratic form Q with $\Theta Q=E_0$.

Choose a basis $\{x_\alpha | \alpha\in A\}$ of V and let $\{f_\alpha\}$ be the dual family of linear functionals on V. Then we have $v=\sum_\alpha f_\alpha(v)x_\alpha$ for any $v\in V$, the sum being effectively finite.

Define a function B from $V\times V$ to F by

$$B(v, w) = \sum_\alpha E_0(f_\alpha(v), x_\alpha, w)$$

and put $Q(v)=2B(v, v)$. Since B is additive in its first argument and linear in its second, it follows immediately that Q satisfies the parallelogram identity (2). Moreover

$$\begin{aligned} Q(\lambda v) &= 2B(\lambda v, \lambda v) = 2\lambda B(\lambda v, v) \\ &= 2\lambda \sum_\alpha E_0(\lambda f_\alpha(v), x_\alpha, v) \\ &= 2\lambda \sum_\alpha \lambda E_0(f_\alpha(v), x_\alpha, v) + 2\lambda \sum_\alpha f_\alpha(v) E_0(\lambda, x_\alpha, v) \end{aligned}$$

since E_0 is a derivation in its first argument. Since E_0 is linear in its second argument and skew symmetric in its second and third, the second sum is

$$\sum_\alpha E_0(\lambda, f_\alpha(v)x_\alpha, v) = E_0(\lambda, v, v) = 0.$$

Thus $Q(\lambda v)=\lambda^2 Q(v)$ and Q is a quasi-quadratic form.

The symmetric biadditive form S associated with Q is given by

$$S(v, w) = B(v, w) + B(w, v)$$

and $\Theta Q=E$, where $E(\lambda, v, w)=B(\lambda v, w)-\lambda B(v, w)+B(w, \lambda v)-\lambda B(w, v)$. Since B is linear in its second argument, the last two terms cancel. Hence,

$$\begin{aligned} E(\lambda, v, w) &= \sum_\alpha [E_0(\lambda f_\alpha(v), x_\alpha, w) - \lambda E_0(f_\alpha(v), x_\alpha, w)] \\ &= \sum_\alpha f_\alpha(v) E_0(\lambda, x_\alpha, w) = E_0(\lambda, v, w) \end{aligned}$$

because E_0 is a derivation in its first argument and linear in its second. This shows that $\Theta Q=E_0$ and concludes the proof.

Let $\mathfrak{D}$ be the set of all additive derivations of F, that is the set of all functions D from F to F which are additive and satisfy

$$D(\lambda\mu) = \lambda D(\mu) + \mu D(\lambda).$$

Let $\mathfrak{I}$ be the space of skew-symmetric bilinear forms on V. The map $\langle D, T\rangle \rightarrow E$, where $E(\lambda, v, w) = D(\lambda)T(v, w)$, is a bilinear map from $\mathfrak{D}\times\mathfrak{I}$ to $\mathcal{E}$ which induces an injective linear map from $\mathfrak{D}\otimes\mathfrak{I}$ to $\mathcal{E}$. (See [1] or [2] for the definition of the tensor product $\mathfrak{D}\otimes\mathfrak{I}$.) When V and hence $\mathfrak{I}$ is finite dimensional, it is easy to see that this map is also surjective. Hence we have the following corollary.

1.2 COROLLARY. *If V has finite dimension, $\mathcal{Q}_0/\mathcal{Q}$ is naturally isomorphic to* $\mathfrak{D}\times\mathfrak{I}$.

Let Q be a quasi-quadratic form and let S be the corresponding biadditive form. S will be homogeneous with respect to the scalar μ if and only if $E(\mu, v, w) = 0$ for all v and w. Since $\lambda \rightarrow E(\lambda, v, w)$ is a derivation of F for each v and w, S will certainly be homogeneous with respect to those scalars μ for which every derivation vanishes. Conversely, suppose that D is a derivation of F which does not vanish at μ and assume that V has dimension at least two. Define $E_0(\lambda, v, w) = D(\lambda)T(v, w)$, where T is some nontrivial skew-symmetric bilinear form on V. Any of the quasi-quadratic forms Q for which $\Theta Q = E_0$ will be associated with a symmetric biadditive form S which is not homogeneous with respect to the scalar μ.

Suppose that F has characteristic 0. Every derivation of F vanishes at every algebraic number, hence the form S associated with a quasi-quadratic form will be homogeneous with respect to algebraic numbers. On the other hand, if μ is transcendental, there is a derivation of F which does not vanish at μ (see [4], particularly Corollaries 1′ and 2′, pp. 124–5). Hence there are quasi-quadratic forms for which the associated biadditive form is not homogeneous with respect to μ.

Suppose that F has characteristic p. The set of all pth powers is a subfield F' of F and every derivation must vanish on all of F'. Moreover, if μ is in F but not F', there is a derivation of F which does not vanish at μ (see [4], particularly the remark on p. 126). Hence the biadditive form associated with a quasi-quadratic form is always homogeneous with respect to scalars in F' but need not be with respect to scalars not in F'.

2. Other identities for quadratic forms. Since the identities (2) and (4) fail to characterize quadratic forms, what other identity should we use? We can provide directly for the linearity of the biadditive function S by requiring the identity

$$Q(x + \lambda y) - Q(x - \lambda y) = \lambda(Q(x + y) - Q(x - y)), \tag{8}$$

in addition to the parallelogram law (2), since (8) becomes immediately $S(x, \lambda y) = \lambda S(x, y)$.

It turns out that (8) alone almost guarantees that Q is a quadratic form without requiring the parallelogram law. Both constant and linear functions satisfy the identity (8). We shall refer to a function as quadratic if it can be represented as the sum of a constant, a linear functional, and a quadratic form.

We shall see that (8) characterizes the quadratic functions except when F has three elements. (Note that every function from V to F satisfies (8) if F has only three elements.)

We begin with a simple extension of Corollary 0.2.

2.1 LEMMA. *Let V be a vector space over a field F of dimension at least two. Let Q be a function from V to F such that the restriction of Q to each two dimensional subspace of V is a quadratic function. Then Q is a quadratic function.*

Proof. Define

$$Q_1(x) = \tfrac{1}{2}(Q(x) + Q(-x) - 2Q(0)),$$
$$Q_2(x) = \tfrac{1}{2}(Q(x) - Q(-x)), \qquad \text{and}$$
$$Q_3(x) = Q(0).$$

A brief calculation shows that the restriction of Q_1 to each two dimensional subspace of V is a quadratic form. Hence, by corollary 0.2, Q_1 is a quadratic form. The restriction of Q_2 to each two dimensional subspace of V is a linear functional. Hence Q_2 is a linear functional. Since $Q = Q_1 + Q_2 + Q_3$, Q is a quadratic function.

2.2 THEOREM. *Let V be a vector space over a field F. Assume that F contains more than three elements and has characteristic different from 2. Every function Q from V to F satisfying the identity* (8) *is a quadratic function.*

Proof. In view of the lemma it is sufficient to prove this theorem when the dimension of V is zero, one, or two. When V has dimension zero, it is trivial.

Let Φ be the set of all functions ϕ from F to $\mathfrak{F}$ which satisfy the identity (8). If $\phi \in \Phi$ and $\lambda \in F$, then

$$\begin{aligned}\phi(2\lambda) &= \phi(0) + (\phi(\lambda + \lambda) - \phi(\lambda - \lambda)) \\ &= \phi(0) + \lambda(\phi(\lambda + 1) - \phi(\lambda - 1)) \\ &= \phi(0) + \lambda(\phi(1 + \lambda) - \phi(1 - \lambda)) + \lambda(\phi(0 + (1 - \lambda)) - \phi(0 - (1 - \lambda))) \\ &= \phi(0) + \lambda^2(\phi(2) - \phi(0)) + \lambda(1 - \lambda)(\phi(1) - \phi(-1)).\end{aligned}$$

This shows explicitly that Φ contains only quadratic functions. This proves the theorem for one dimensional vector spaces. It also shows that any member of Φ which vanishes at as many as three points is everywhere zero.

Now suppose that V is two dimensional. Let $\mathfrak{Q}_1$ be the set of all functions from V to F which satisfy (8). We know that $\mathfrak{Q}_1$ contains the six dimensional space of quadratic functions on V. It will therefore be sufficient to prove that $\mathfrak{Q}_1$ has dimension at most six over F.

Let L be a line in V, say $L = \{x + \lambda y \mid \lambda \in F\}$. Suppose that $Q \in \mathfrak{Q}_1$; then $\phi(\lambda) = Q(x + \lambda y)$ defines a function $\phi \in \Phi$. Hence, if Q vanishes at three points of L it vanishes at all points of L.

Let L_1, L_2, and L_3 be three lines of V in general position. Let x_1, x_2, and x_3

be their three points of intersection and let y_1, y_2, and y_3 be three other points on the lines L_1, L_2, and L_3, respectively. Suppose that $Q \in \mathcal{Q}_1$ and Q vanishes at each of the six points x_1, x_2, x_3, y_1, y_2, and y_3. We shall prove that Q vanishes on all of V. This will prove that $\mathcal{Q}_1$ has dimension at most six and thus finish the proof.

Since Q vanishes at three points of each of the lines L_1, L_2, and L_3, it vanishes at all points of $L_1 \cup L_2 \cup L_3$. Let z be any other point of V and assume F has at least seven elements. Among the at-least-eight lines through z, there must be one which does not contain x_1, x_2, or x_3 and is not parallel to L_1, L_2, or L_3. This line M meets L_1, L_2, and L_3 in three distinct points. Thus Q vanishes at three points of M. Therefore Q vanishes at z. We have proved that Q vanishes on all of V. The fields with two, three, and four elements have been excluded from the theorem. If F has five elements, it is readily checked that there are many lines which meet L_1, L_2, and L_3 in distinct points; then, repeating the argument, we see that Q vanishes on all of V.

The calculation involved in the proof of the lemma shows that we can distinguish the quadratic forms among the quadratic functions by requiring that

$$Q(-x) = Q(x) \tag{9}$$

and

$$Q(0) = 0, \tag{10}$$

both of which are implied by (4). Similarly, the linear functionals are singled out by the identity

$$Q(-x) = -Q(x). \tag{11}$$

With the aid of (9) we can write (8) in the form

$$Q(\lambda y + x) - Q(\lambda y - x) = \lambda(Q(y + x) - Q(y - x)). \tag{12}$$

This latter identity by itself implies (9) by taking $\lambda = 0$; hence (12) is equivalent to (8) and (9) together.

2.3 COROLLARY. *Let V be a vector space over a field F. Assume that F contains more than three elements and has characteristic different from* 2. *Then every function from V to F satisfying* (10) *and* (12) *is a quadratic form.*

The field of three elements must still be excluded because for this field (12) is equivalent to (9) which is insufficient to characterize quadratic forms.

Combining (8) and (11), we obtain the identity

$$Q(\lambda y + x) + Q(\lambda y - x) = \lambda(Q(y + x) + Q(y - x)). \tag{13}$$

Again setting $\lambda = 0$, we see that (13) is equivalent to (8) and (11). Therefore (13) characterizes linear functionals except over fields of characteristic two and the field of three elements.

References

1. N. Bourbaki, Eléments de Mathématique VII, Algèbre Chapitre 3, Algèbre Multilinéaire, Hermann et Cie., Paris, 1948.
2. C. Chevalley, Fundamental Concepts of Algebra, Academic Press, New York, 1956.
3. P. Jordan and J. von Neumann, On inner products in linear metric space, Annals of Math., 36 (1935) 719–723.
4. O. Zariski and P. Samuel, Commutative Algebra, Vol. 1, Van Nostrand, Princeton, 1958.

A SIMPLIFIED PROOF OF A SUFFICIENT CONDITION FOR A POSITIVE DEFINITE QUADRATIC FORM*

S. M. SAMUELS, Purdue University

We present an inductive proof of the statement that positivity of the leading principal minors of the matrix of a quadratic form implies positive definiteness of the form. The chief interest of the proof is its simplicity. It compares favorably with the proof given by C. J. Seelye [1], which, in turn was offered in part as a simplification of the proof given in most textbooks.

The proof is an immediate consequence of the following simple

LEMMA. *Let*

$$\mathbf{A} = \begin{bmatrix} \mathbf{A}_{n-1} & \mathbf{a}_n \\ \mathbf{a}_n' & a_{nn} \end{bmatrix}$$

be an n-th order symmetric matrix, with $\mathbf{A}_{n-1}$ *the (symmetric) submatrix consisting of the first* $n-1$ *rows and columns of* $\mathbf{A}$. *Then, if* $\mathbf{A}_{n-1}$ *is nonsingular, there is a nonsingular matrix P such that*

$$\mathbf{P}'\mathbf{A}\mathbf{P} = \begin{bmatrix} \mathbf{A}_{n-1} & \mathbf{0} \\ \mathbf{0}' & b_{nn} \end{bmatrix}.$$

If, in addition, $\det \mathbf{A}_{n-1}$ *and* $\det \mathbf{A}$ *are positive, then* b_{nn} *is positive.*

Proof. Since $\mathbf{A}_{n-1}$ is nonsingular, $\mathbf{a}_n$ has a unique representation as a linear combination of the columns of $\mathbf{A}_{n-1}$. Let $\alpha_1, \cdots, \alpha_{n-1}$ be the coefficients of the corresponding columns of $\mathbf{A}_{n-1}$ in this representation. Let $\mathbf{P}'$ be the matrix obtained from the identity matrix by replacing its last row by

$$[-\alpha_1, \cdots, -\alpha_{n-1}, 1].$$

Then $\mathbf{P}$ is nonsingular and, by the symmetry of $\mathbf{A}$, is the desired matrix.

Since

$$\begin{aligned} \det \mathbf{P}'\mathbf{A}\mathbf{P} &= (\det \mathbf{P})^2 \det \mathbf{A} \\ &= b_{nn} \det \mathbf{A}_{n-1}, \end{aligned}$$

the second part of the lemma follows immediately.

We now prove the desired theorem, assuming, without loss of generality, that the matrix of the quadratic form is symmetric.

* From AMERICAN MATHEMATICAL MONTHLY, vol. 73 (1966), pp. 297–298.

THEOREM. *If the leading principal minors of the n-th order symmetric matrix* $\mathbf{A}$ *are all positive, then the quadratic form* $\mathbf{x}'\mathbf{A}\mathbf{x}$ *is positive definite.*

Proof. The theorem is obviously true for $n=1$. Assume it is true for $n-1$ and let $\mathbf{A}$ be an nth order symmetric matrix with positive leading principal minors. Then $\mathbf{A}$ satisfies the hypotheses of the lemma. Now positive definiteness of $\mathbf{x}'\mathbf{A}\mathbf{x}$ is equivalent to positive definiteness of

$$\mathbf{x}'\mathbf{P}'\mathbf{A}\mathbf{P}\mathbf{x} = \mathbf{x}_{n-1}'\mathbf{A}_{n-1}\mathbf{x}_{n-1} + x_n^2 b_{nn}$$

where $\mathbf{x}_{n-1}$ denotes the first $n-1$ variables. The form is positive definite since the first term is positive by the inductive hypothesis (every leading principal minor of $\mathbf{A}_{n-1}$ is a leading principal minor of $\mathbf{A}$) and the second term is positive by the lemma.

Reference

1. C. J. Seelye, Conditions for a positive-definite quadratic form established by induction, this MONTHLY, 65 (1958) 355–356.

THE SCARCITY OF CROSS PRODUCTS ON EUCLIDEAN SPACES*

BERTRAM WALSH, University of California, Los Angeles

When first introduced to the dot product on $\mathbf{R}^n$ (n arbitrary) and the cross product on $\mathbf{R}^3$, students are not unlikely to inquire about the definition of a cross product for $\mathbf{R}^n$, $n \neq 3$. This note points out that by proving the elementary propositions and theorem below, students may answer this question themselves. Namely, they may show that if reasonable demands are made of "cross products" (that they satisfy the axioms for cross products in Apostol's recent Calculus text [2]), then only on $\mathbf{R}^1$, $\mathbf{R}^3$ and $\mathbf{R}^7$ can cross products exist—the one on $\mathbf{R}^1$ being trivial and the ones on $\mathbf{R}^7$ somewhat pathological. We indicate the elementary proofs, for whose rediscovery students will need knowledge of the fact that orthonormal (o.n.) sets in $\mathbf{R}^n$ have cardinalities $\leq n$, with equality if and only if the o.n. set is a basis; orthogonal projection is required a couple of times, but projection is onto a subspace for which an orthonormal basis has already been constructed, so that the projection can be written out explicitly, componentwise.

The axioms [2, p. 275] are

(5.28) $\qquad \mathbf{a} \times \mathbf{b} = -(\mathbf{b} \times \mathbf{a})$

(5.29) $\qquad \mathbf{a} \times (\mathbf{b} + \mathbf{c}) = (\mathbf{a} \times \mathbf{b}) + (\mathbf{a} \times \mathbf{c})$

(5.30) $\qquad \lambda(\mathbf{a} \times \mathbf{b}) = (\lambda\mathbf{a}) \times \mathbf{b}$

(5.31) $\qquad \mathbf{a} \cdot (\mathbf{a} \times \mathbf{b}) = 0$

(5.32) $\qquad |\mathbf{a} \times \mathbf{b}| = [|\mathbf{a}|^2 |\mathbf{b}|^2 - (\mathbf{a} \cdot \mathbf{b})^2]^{1/2}$

* From AMERICAN MATHEMATICAL MONTHLY, vol. 74 (1967), pp. 188–194.

for all scalars λ and all vectors $\mathbf{a}, \mathbf{b}, \mathbf{c}$. One can easily verify that in the presence of the other four axioms, (5.32) is equivalent to the statement that the cross product of two orthogonal unit vectors is also a unit vector, so that in spite of its complicated appearance the axiom is natural.

Our principal tools are the identities

$$(1) \qquad (\mathbf{a} \times \mathbf{b}) \cdot \mathbf{c} = -[\mathbf{b} \cdot (\mathbf{a} \times \mathbf{c})]$$

and

$$(2) \qquad \mathbf{a} \times (\mathbf{b} \times \mathbf{c}) - \mathbf{c} \times (\mathbf{a} \times \mathbf{b}) = 2(\mathbf{a} \cdot \mathbf{c})\mathbf{b} - (\mathbf{b} \cdot \mathbf{c})\mathbf{a} - (\mathbf{a} \cdot \mathbf{b})\mathbf{c};$$

the first of these, which expresses the fact that the linear transformation $\mathbf{b} \to (\mathbf{a} \times \mathbf{b})$ is skew-symmetric, is easily derived from (5.31) by setting $\mathbf{a}+\mathbf{c}$ where $\mathbf{a}$ occurs, using linearity and symmetry, and canceling what one already knows about. The second is derived by two applications of the "polarization" trick: writing (5.32) as

$$(\mathbf{a} \times \mathbf{b}) \cdot (\mathbf{a} \times \mathbf{b}) = (\mathbf{a} \cdot \mathbf{a})(\mathbf{b} \cdot \mathbf{b}) - (\mathbf{a} \cdot \mathbf{b})^2$$

and setting $\mathbf{b}+\mathbf{d}$ where $\mathbf{b}$ is, using bilinearity and symmetry to get an expansion of both sides and canceling what one already knows about, one gets

$$(\mathbf{a} \times \mathbf{b}) \cdot (\mathbf{a} \times \mathbf{d}) = (\mathbf{a} \cdot \mathbf{a})(\mathbf{b} \cdot \mathbf{d}) - (\mathbf{a} \cdot \mathbf{b})(\mathbf{a} \cdot \mathbf{d}).$$

This can be written, using (1) and the linearity of the dot product, as

$$-[\mathbf{a} \times (\mathbf{a} \times \mathbf{b})] \cdot \mathbf{d} = [(\mathbf{a} \cdot \mathbf{a})\mathbf{b} - (\mathbf{a} \cdot \mathbf{b})\mathbf{a}] \cdot \mathbf{d}$$

and since $\mathbf{d} \in \mathbf{R}^n$ is arbitrary, this gives

$$(3) \qquad \mathbf{a} \times (\mathbf{a} \times \mathbf{b}) = (\mathbf{a} \cdot \mathbf{b})\mathbf{a} - (\mathbf{a} \cdot \mathbf{a})\mathbf{b},$$

a useful special case of (2). Now setting $\mathbf{a}+\mathbf{c}$ wherever $\mathbf{a}$ is in (3), expanding by linearity, canceling what one already knows about and using the skew-symmetry of the cross product (i.e. (5.28)) leads directly to (2). A useful special case, incidentally, is

(4) For mutually orthogonal $\mathbf{a}, \mathbf{b}, \mathbf{c}$, $\quad \mathbf{a} \times (\mathbf{b} \times \mathbf{c}) = \mathbf{c} \times (\mathbf{a} \times \mathbf{b})$

which one sees by observing that the right side of (2) is zero whenever $\mathbf{a}$, $\mathbf{b}$ and $\mathbf{c}$ are mutually orthogonal. The identities (1), (2), (3) and (4) hold for any cross product satisfying the axioms, on any $\mathbf{R}^n$.

Given such a cross product on $\mathbf{R}^n$, we make two definitions:

(I) A vector subspace A of $\mathbf{R}^n$ is *closed under* $\times$ if whenever $\mathbf{a}_1 \in A$ and $\mathbf{a}_2 \in A$, then also $\mathbf{a}_1 \times \mathbf{a}_2 \in A$.

(II) A subspace B of $\mathbf{R}^n$ is *stable under* $A\times$, where $A \subseteq \mathbf{R}^n$, if whenever $\mathbf{a} \in A$ and $\mathbf{b} \in B$, then $\mathbf{a} \times \mathbf{b} \in B$.

So A is closed under $\times$ iff stable under $A\times$. We now have

PROPOSITION 1. *Let* $A \subseteq \mathbf{R}^n$, *B be a subspace of* $\mathbf{R}^n$. *If B is stable under* $A\times$, *then so is its orthogonal complement* $B^{\perp} = \{\mathbf{c} \mid \mathbf{c} \in \mathbf{R}^n, \mathbf{b} \cdot \mathbf{c} = 0$ *for all* $\mathbf{b} \in B\}$.

Proof. If $\mathbf{c} \in B^{\perp}$, then for any $\mathbf{a} \in A$ and $\mathbf{b} \in B$

$$\mathbf{b} \cdot (\mathbf{a} \times \mathbf{c}) = -[(\mathbf{a} \times \mathbf{b}) \cdot \mathbf{c}] = 0 \quad \text{since } \mathbf{a} \times \mathbf{b} \in B.$$

Thus for any $\mathbf{b} \in B$, $\mathbf{b} \perp (\mathbf{a} \times \mathbf{c})$, and so $\mathbf{a} \times \mathbf{c} \in B^{\perp}$. ∎

PROPOSITION 2. *Let A be a subspace of $\mathbf{R}^n$ which is closed under $\times$ and possesses an orthonormal basis $\{\mathbf{f}_1, \cdots, \mathbf{f}_k\}$. Let $\mathbf{b} \in A^{\perp}$. Then the vectors $\{\mathbf{b}, \mathbf{f}_1 \times \mathbf{b}, \cdots, \mathbf{f}_k \times \mathbf{b}\}$ lie in $A^{\perp}$ and are mutually orthogonal and of the same length as $\mathbf{b}$. In particular if $\mathbf{b}$ is a unit vector they form an orthonormal set of $k+1$ elements in $A^{\perp}$.*

Proof. That $\{\mathbf{b}, \mathbf{f}_1 \times \mathbf{b}, \cdots\} \subseteq A^{\perp}$ follows from Proposition 1. The orthogonality relations and lengths follow from $\mathbf{b} \cdot (\mathbf{f}_i \times \mathbf{b}) = 0$ and

$$\begin{aligned}(\mathbf{f}_i \times \mathbf{b}) \cdot (\mathbf{f}_j \times \mathbf{b}) &= (\mathbf{b} \times \mathbf{f}_i) \cdot (\mathbf{b} \times \mathbf{f}_j) \\ &= -[\mathbf{b} \times (\mathbf{b} \times \mathbf{f}_i) \cdot \mathbf{f}_j] \quad \text{by (1)} \\ &= -[(\mathbf{b} \cdot \mathbf{f}_i)\mathbf{b} - (\mathbf{b} \cdot \mathbf{b})\mathbf{f}_i] \cdot \mathbf{f}_j = (\mathbf{f}_i \cdot \mathbf{f}_j)(\mathbf{b} \cdot \mathbf{b}). \blacksquare\end{aligned}$$

It is easy to verify that the only cross product on $\mathbf{R}^1$ is identically zero, and that there is no cross product on $\mathbf{R}^2$: it couldn't be zero, and there aren't enough dimensions for it to be nonzero. Suppose we have a cross product on $\mathbf{R}^n$, then, $n \geqq 3$. If $\mathbf{e}_1$ and $\mathbf{e}_2$ are an orthonormal pair of elements of $\mathbf{R}^n$, then $\mathbf{e}_1 \times \mathbf{e}_2$ is a unit (by (5.32)) vector normal to both $\mathbf{e}_1$ and $\mathbf{e}_2$; set $\mathbf{e}_3 = \mathbf{e}_1 \times \mathbf{e}_2$. Let H be the linear subspace of $\mathbf{R}^n$ spanned by $\{\mathbf{e}_1, \mathbf{e}_2, \mathbf{e}_3\}$. Regardless of what n may be,

PROPOSITION 3. *H is closed under $\times$.*

Proof. Bilinearity of $\times$ insures that it suffices to check this on basis vectors. By identity (3)

$$\begin{aligned}\mathbf{e}_2 \times \mathbf{e}_3 = \mathbf{e}_2 \times (\mathbf{e}_1 \times \mathbf{e}_2) &= -[\mathbf{e}_2 \times (\mathbf{e}_2 \times \mathbf{e}_1)] \\ &= -[(\mathbf{e}_2 \cdot \mathbf{e}_1)\mathbf{e}_2 - (\mathbf{e}_2 \cdot \mathbf{e}_2)\mathbf{e}_1] = \mathbf{e}_1\end{aligned}$$

and similarly $\mathbf{e}_3 \times \mathbf{e}_1 = \mathbf{e}_2$. ∎

Thus the basis $\{\mathbf{e}_1, \mathbf{e}_2, \mathbf{e}_3\}$ has the same multiplication table as the basis $\{\mathbf{i}, \mathbf{j}, \mathbf{k}\}$ for $\mathbf{R}^3$ with the usual cross product.

One can resolve any vector $\mathbf{a} \in \mathbf{R}^n$ into

$$\mathbf{a} = \sum_{i=1}^{3} (\mathbf{a} \cdot \mathbf{e}_i)\mathbf{e}_i + \left[\mathbf{a} - \sum_{i=1}^{3} (\mathbf{a} \cdot \mathbf{e}_i)\mathbf{e}_i\right]$$

with the first component in H and the second in $H^{\perp}$. Thus either $H = \mathbf{R}^n$, in which case $n = 3$, or else it is possible to find a unit vector $\mathbf{m} \in H^{\perp}$. Suppose the latter. Then by Proposition 2, the set $\{\mathbf{m}, \mathbf{e}_1 \times \mathbf{m}, \mathbf{e}_2 \times \mathbf{m}, \mathbf{e}_3 \times \mathbf{m}\}$ is an orthonormal set in $H^{\perp}$, and $\{\mathbf{e}_1, \mathbf{e}_2, \mathbf{e}_3, \mathbf{m}, \mathbf{e}_1 \times \mathbf{m}, \mathbf{e}_2 \times \mathbf{m}, \mathbf{e}_3 \times \mathbf{m}\}$ is an orthonormal set in $\mathbf{R}^n$, so $n \geqq 7$. Let C denote the subspace of $\mathbf{R}^n$ spanned by this last set.

PROPOSITION 4. *C is closed under $\times$.*

Proof. One need only check the five types of products which can occur:

(i) $\mathbf{e}_i \times \mathbf{e}_j$. Belongs to $H \subseteq C$, by Proposition 3.

(ii) $\mathbf{e}_i \times \mathbf{m}$. Belongs to C by definition.

(iii)
$$\begin{aligned}\mathbf{e}_i \times (\mathbf{e}_j \times \mathbf{m}) &= 2(\mathbf{e}_i\cdot\mathbf{m})\mathbf{e}_j - (\mathbf{e}_j\cdot\mathbf{m})\mathbf{e}_i - (\mathbf{e}_i\cdot\mathbf{e}_j)\mathbf{m} + \mathbf{m} \times (\mathbf{e}_i \times \mathbf{e}_j)\\ &= -(\mathbf{e}_i \times \mathbf{e}_j) \times \mathbf{m} - (\mathbf{e}_i\cdot\mathbf{e}_j)\mathbf{m} \in C.\end{aligned}$$

(iv)
$$\mathbf{m} \times (\mathbf{e}_i \times \mathbf{m}) = -\mathbf{m} \times (\mathbf{m} \times \mathbf{e}_i) = (\mathbf{m}\cdot\mathbf{m})\mathbf{e}_i - (\mathbf{m}\cdot\mathbf{e}_i)\mathbf{m} = \mathbf{e}_i \in C.$$

(v)
$$\begin{aligned}(\mathbf{e}_i \times \mathbf{m}) \times (\mathbf{e}_j \times \mathbf{m}) &= (\mathbf{m} \times \mathbf{e}_i) \times (\mathbf{m} \times \mathbf{e}_j)\\ &= \mathbf{e}_j \times [(\mathbf{m} \times \mathbf{e}_i) \times \mathbf{m}] \quad \text{by (4)}\\ &= \mathbf{e}_j \times [\mathbf{m} \times (\mathbf{e}_i \times \mathbf{m})]\\ &= \mathbf{e}_j \times \mathbf{e}_i \quad \text{by (iv) above.} \ \blacksquare\end{aligned}$$

One can resolve any vector $\mathbf{a} \in \mathbf{R}^n$ into

$$\begin{aligned}\mathbf{a} = &\left[\sum_{i=1}^{3} (\mathbf{a}\cdot\mathbf{e}_i)\mathbf{e}_i + (\mathbf{a}\cdot\mathbf{m})\mathbf{m} + \sum_{i=1}^{3} (\mathbf{a}\cdot(\mathbf{e}_i \times \mathbf{m}))(\mathbf{e}_i \times \mathbf{m})\right]\\ &+ \left[\mathbf{a} - \sum_{i=1}^{3} (\mathbf{a}\cdot\mathbf{e}_i)\mathbf{e}_i - (\mathbf{a}\cdot\mathbf{m})\mathbf{m} - \sum_{i=1}^{3} (\mathbf{a}\cdot(\mathbf{e}_i \times \mathbf{m}))(\mathbf{e}_i \times \mathbf{m})\right]\end{aligned}$$

with the first component in C and the second in $C^\perp$. Thus either $C = \mathbf{R}^n$, in which case $n = 7$, or it is possible to find a unit vector $\mathbf{n} \in C^\perp$.

THEOREM. *If there exists a cross product on $\mathbf{R}^n$ which satisfies Apostol's axioms, then $n = 1$, 3 or 7. Conversely, there exist cross products on these three spaces.*

Proof. We have seen that for $n \leqq 3$, cross products exist precisely when $n = 1$ or 3, the cross product on $\mathbf{R}^1$ being identically zero and the two cross products on $\mathbf{R}^3$ being the classical ones, and that if $\mathbf{R}^n$ has a cross product for $n > 3$, then $n \geqq 7$. If $n > 7$, then one can find a unit vector $\mathbf{n} \in C^\perp$, where C is constructed as above. We shall show, however, that the existence of such a vector would lead to a contradiction. Indeed, should such an $\mathbf{n}$ exist, then $\mathbf{m} \times \mathbf{n}$ is also a unit vector in $C^\perp$; set $\mathbf{p} = \mathbf{m} \times \mathbf{n}$. Just as in Proposition 3, we have $\mathbf{n} \times \mathbf{p} = \mathbf{m}$ and $\mathbf{p} \times \mathbf{m} = \mathbf{n}$. Let us compute some products: we find that, for $i \neq j$,

$$\begin{aligned}&(\mathbf{e}_i \times \mathbf{m}) \times (\mathbf{e}_j \times \mathbf{n})\\ &\quad = (\mathbf{e}_i \times \mathbf{m}) \times (\mathbf{e}_j \times (\mathbf{p} \times \mathbf{m})) \quad \text{by (4)}\\ &\quad = (\mathbf{e}_i \times \mathbf{m}) \times (\mathbf{m} \times (\mathbf{e}_j \times \mathbf{p})) \quad \text{by (4)}\\ &\quad = (\mathbf{e}_j \times \mathbf{p}) \times ((\mathbf{e}_i \times \mathbf{m}) \times \mathbf{m})\\ &\quad = (\mathbf{e}_j \times \mathbf{p}) \times (\mathbf{m} \times (\mathbf{m} \times \mathbf{e}_i)) = (\mathbf{e}_j \times \mathbf{p}) \times [(\mathbf{m}\cdot\mathbf{e}_i)\mathbf{m} - (\mathbf{m}\cdot\mathbf{m})\mathbf{e}_i]\\ &\quad = -(\mathbf{e}_j \times \mathbf{p}) \times \mathbf{e}_i = \mathbf{e}_i \times (\mathbf{e}_j \times \mathbf{p}) = \mathbf{p} \times (\mathbf{e}_i \times \mathbf{e}_j).\end{aligned}$$

Similarly

$$\begin{aligned}
(\mathbf{e}_j \times \mathbf{n}) \times (\mathbf{e}_i \times \mathbf{m}) &= \\
(\mathbf{e}_j \times \mathbf{n}) \times (\mathbf{e}_i \times (\mathbf{n} \times \mathbf{p})) &= \\
-(\mathbf{e}_j \times \mathbf{n}) \times (\mathbf{e}_i \times (\mathbf{p} \times \mathbf{n})) &= \qquad \text{by (4)} \\
-(\mathbf{e}_j \times \mathbf{n}) \times (\mathbf{n} \times (\mathbf{e}_i \times \mathbf{p})) &= \qquad \text{by (4)} \\
-(\mathbf{e}_i \times \mathbf{p}) \times ((\mathbf{e}_j \times \mathbf{n}) \times \mathbf{n}) &= \quad \cdots \text{ by (3)} \\
&= -(\mathbf{e}_i \times \mathbf{p}) \times (-\mathbf{e}_j) \\
&= \mathbf{p} \times (\mathbf{e}_i \times \mathbf{e}_j).
\end{aligned}$$

(The computations indicated by the ellipsis are straightforward.) But putting these together gives $(\mathbf{e}_i\times\mathbf{m})\times(\mathbf{e}_j\times\mathbf{n}) = (\mathbf{e}_j\times\mathbf{n})\times(\mathbf{e}_i\times\mathbf{m})$; this contradicts the skew-symmetry of $\times$ unless this particular product is zero. However, $\mathbf{p}\times(\mathbf{e}_i\times\mathbf{e}_j)$ is the product of two perpendicular unit vectors and thus has unit norm. This contradiction shows that the case $n > 7$ cannot arise.

It remains to exhibit a cross product on $\mathbf{R}^7$ which satisfies the axioms. Such a product can be constructed from the familiar one on $\mathbf{R}^3$ as follows: $\mathbf{R}^7$ corresponds 1-1 with the set of all triples $(\mathbf{a}, \lambda, \mathbf{b})$ where $\mathbf{a}$ and $\mathbf{b}$ are in $\mathbf{R}^3$ and $\lambda \in \mathbf{R}$ under the correspondence

$$a_1\mathbf{i}_1 + \cdots + a_7\mathbf{i}_7 \to (a_1\mathbf{i} + a_2\mathbf{j} + a_3\mathbf{k},\ a_4,\ a_5\mathbf{i} + a_6\mathbf{j} + a_7\mathbf{k}).$$

It is easy to see that this correspondence turns addition and scalar multiplication in $\mathbf{R}^7$ into component-by-component addition and scalar multiplication of triples; defining the dot product of triples by

$$(\mathbf{a}_1, \lambda_1, \mathbf{b}_1)\cdot(\mathbf{a}_2, \lambda_2, \mathbf{b}_2) = (\mathbf{a}_1\cdot\mathbf{a}_2) + \lambda_1\lambda_2 + (\mathbf{b}_1\cdot\mathbf{b}_2)$$

we see easily that the dot product of two triples equals the dot product of the two elements of $\mathbf{R}^7$ to which they correspond. We actually define the cross product on these triples: if one recalls what Propositions 3 and 4 told one the multiplication table of C should be, and if one thinks of $(\mathbf{a}, \lambda, \mathbf{b})$ as representing a vector whose component in H is $\mathbf{a}$, whose component along $\mathbf{m}$ is λ, and whose component in the space spanned by $\{\mathbf{e}_1\times\mathbf{m}, \mathbf{e}_2\times\mathbf{m}, \mathbf{e}_3\times\mathbf{m}\}$ is $\mathbf{b}\times\mathbf{m}$, then one is led to the definition

$$\begin{aligned}
(\mathbf{a}_1, \lambda_1, \mathbf{b}_1) \times (\mathbf{a}_2, \lambda_2, \mathbf{b}_2) = ([&\lambda_1\mathbf{b}_2 - \lambda_2\mathbf{b}_1 + (\mathbf{a}_1 \times \mathbf{a}_2) \\
&- (\mathbf{b}_1 \times \mathbf{b}_2)],\ [-(\mathbf{a}_1\cdot\mathbf{b}_2) + (\mathbf{b}_1\cdot\mathbf{a}_2)],\ [\lambda_2\mathbf{a}_1 \\
&- \lambda_1\mathbf{a}_2 - (\mathbf{a}_1 \times \mathbf{b}_2) - (\mathbf{b}_1 \times \mathbf{a}_2)]).
\end{aligned}$$

(Note that the cross products inside the brackets in the triple are products in $\mathbf{R}^3$ and thus well defined.) Knowing that the cross product in $\mathbf{R}^3$ satisfies the axioms, one easily verifies that this cross product on $\mathbf{R}^7$ (via identification of $\mathbf{R}^7$ with the triples) satisfies the axioms also. ▮

Remark: The pathological behavior of the cross product on $\mathbf{R}^7$ begins with its lack of anything near uniqueness: in $\mathbf{R}^3$ there are only two choices for the cross product of a given orthonormal pair of vectors, while in $\mathbf{R}^7$ our construction can be modified to yield cross products which choose any of the unit vectors in the five-dimensional space orthogonal to the given pair. The cross products on $\mathbf{R}^7$ also fail to satisfy the Lie identity

$$\mathbf{a} \times (\mathbf{b} \times \mathbf{c}) + \mathbf{c} \times (\mathbf{a} \times \mathbf{b}) + \mathbf{b} \times (\mathbf{c} \times \mathbf{a}) = 0;$$

indeed

$$\mathbf{m} \times (\mathbf{e}_1 \times \mathbf{e}_2) + \mathbf{e}_2 \times (\mathbf{m} \times \mathbf{e}_1) + (\mathbf{e}_1 \times (\mathbf{e}_2 \times \mathbf{m})) = -3(\mathbf{e}_3 \times \mathbf{m}).$$

Any number of other standard identities also fail in $\mathbf{R}^7$.

Postscript. The discussion above has concentrated on the geometric problem of defining a cross product on a real inner product space, and has attacked this problem via reasoning processes close to those of three-dimensional vector geometry. The theorem it produces is actually equivalent to a classical theorem of Hurwitz which states that a bilinear "multiplication" operation (denoted by juxtaposition) with the property $|\mathbf{xy}| = |\mathbf{x}|\,|\mathbf{y}|$ can be introduced on a finite-dimensional real inner product space if and only if the space has dimension 1, 2, 4 or 8, and that the operation is then "essentially" the multiplication of the reals, complexes, Hamilton quaternions, or Cayley numbers, and actually is one of those multiplications if an identity is present. Indeed, assuming Hurwitz's theorem and being given an $\mathbf{R}^n$ with a cross product as before, write $\mathbf{R}^{n+1}$ as (isometric to) $\mathbf{Re} \oplus \mathbf{R}^n$ where $\mathbf{e} \in \mathbf{R}^{n+1}$, and define a multiplication on $\mathbf{R}^{n+1}$ by

$$(\lambda\mathbf{e} + \mathbf{a})(\mu\mathbf{e} + \mathbf{b}) = (\lambda\mu - \mathbf{a}\cdot\mathbf{b})\mathbf{e} + (\mu\mathbf{a} + \lambda\mathbf{b} + \mathbf{a} \times \mathbf{b});$$

then it is easy to verify, using the cross product axioms, that this bilinear multiplication has the property $|\lambda\mathbf{e}+\mathbf{a}|\,|\mu\mathbf{e}+\mathbf{b}| = |(\lambda\mathbf{e}+\mathbf{a})(\mu\mathbf{e}+\mathbf{b})|$ and hence $n+1=2$, 4 or 8, $n=1$, 3, or 7. On the other hand, Hurwitz's theorem follows quite readily from ours: given a real inner product space A with a multiplication satisfying $|\mathbf{xy}| = |\mathbf{x}|\,|\mathbf{y}|$ and a left-and-right identity $\mathbf{e}$, one finds that $(\mathbf{x}\cdot\mathbf{x})(\mathbf{y}\cdot\mathbf{y}) = (\mathbf{xy}\cdot\mathbf{xy})$ leads, after the usual couple of polarizations, to

$$2(\mathbf{z}\cdot\mathbf{x})(\mathbf{y}\cdot\mathbf{w}) = (\mathbf{zy}\cdot\mathbf{xw}) + (\mathbf{zw}\cdot\mathbf{xy}), \tag{5}$$

of which we shall need only the cases

$$(\mathbf{e}\cdot\mathbf{x})(\mathbf{y}\cdot\mathbf{y}) = (\mathbf{y}\cdot\mathbf{xy}), \qquad (\mathbf{x}\cdot\mathbf{x})(\mathbf{e}\cdot\mathbf{w}) = (\mathbf{x}\cdot\mathbf{xw}) \tag{6, 6'}$$

and

$$2(\mathbf{z}\cdot\mathbf{e})(\mathbf{y}\cdot\mathbf{e}) = (\mathbf{zy}\cdot\mathbf{e}) + (\mathbf{z}\cdot\mathbf{y}). \tag{7}$$

Now let V denote the subspace $\mathbf{Re}^{\perp}$ of A, and define an operation $\times$ on V by $\mathbf{a}\times\mathbf{b} = \mathbf{ab} + (\mathbf{a}\cdot\mathbf{b})\mathbf{e}$.

Since $(\mathbf{a}\times\mathbf{b})\cdot\mathbf{e}=(\mathbf{ab}\cdot\mathbf{e})+(\mathbf{a}\cdot\mathbf{b})=2(\mathbf{a}\cdot\mathbf{e})(\mathbf{b}\cdot\mathbf{e})=0$ by (7), $\times$ maps V bilinearly into itself, i.e. satisfies (5.29) and (5.30), and the version of (5.30) with λ applied to $\mathbf{b}$. It follows easily from (6) and (6′) that $(\mathbf{a}\times\mathbf{b})\cdot\mathbf{b}=0=\mathbf{a}\cdot(\mathbf{a}\times\mathbf{b})$, so (5.31) holds; moreover, it is not difficult to show from these two versions of (5.31) that (5.28) also holds. Finally, (5.32) follows quickly from (7). Thus by our theorem (which, it is easy to check, does not rest on finite-dimensionality hypotheses) V is 7, 3, 1 or possibly 0-dimensional, A is 8, 4, 2 or 1-dimensional, and quite easily seen to be isomorphic to the Cayley numbers, quaternions, complexes or reals. An elegant algebraic treatment of Hurwitz's theorem is to be found in [3].

Hurwitz's classical theorem suggests generalizations in various directions. We have noted in passing that (in the presence of an identity) finite dimensionality is a useless hypothesis; Wright [4] has shown that a nonassociative real normed division algebra with $|xy|=|x|\,|y|$ must actually have an inner-product norm, thus be one of our four standard algebras. A "nonassociative Gelfand-Mazur theorem," with merely $|xy|\leqq|x|\,|y|$, seems still to be lacking. In the situation of the classical Hurwitz theorem, the unit sphere of $\mathbf{R}^n$ ($n=1, 2, 4$ or 8) comes equipped with a continuous product operation, with identity; a long series of profound topological results, culminating in [1], has shown that these values of n are the only ones for which such a continuous product on the unit sphere can exist.

The author thanks the referee for his suggestion that material be appended which would explicitly point out the relation of the preceding material to the Hurwitz theorem and to the recent topological results.

References

1. J. F. Adams, On the non-existence of elements of Hopf invariant one, Ann. Math., 72 (1960) 20–104.

2. T. M. Apostol, Calculus, vol. I, Blaisdell, New York-London, 1961.

3. C. W. Curtis, The four and eight square problem and division algebras, MAA Studies in Math., vol. 2, Studies in Modern Algebra.

4. F. B. Wright, Absolute valued algebras, Proc. Nat. Acad. Sci. U. S. A., 39 (1953) 330–332.

REFLECTIONS HAVE REVERSED VECTORS*

A. M. ADELBERG, Grinnell College

1. Introduction. In this note we prove the following elementary theorem, which gives some geometric insight into the notion of a reflection of a metric vector space:

THEOREM A. *Every reflection has a reversed vector.*

We also show that the preceding theorem is almost immediately equivalent to the following one:

THEOREM B. *Every rotation of a space of odd dimension and every reflection of a space of even dimension has a fixed vector.*

In spite of the elementary nature of these results, we have not been able to locate Theorem A in the literature except in [1] where both theorems are given, but under restrictive hypotheses, namely for real, anisotropic vector spaces. The proof given there rests on properties of the reals, and does not generalize. Theorem B can be found in the literature (see [2], page 131 or [3], Proposition 187.1), but the proofs make essential use of the relatively deep Cartan-Dieudonné Theorem, so it is of some interest to have a direct proof. It will be seen that the judicious use of determinants may substantially simplify many of the arguments on metric vector spaces that occur in the existing literature.

Finally we shall use Theorem B to provide a proof for the special case of the Cartan-Dieudonné Theorem for anisotropic spaces that is very likely briefer than anything currently available.

2. Preliminaries. We recall the main definitions and the basic results that are needed as follows:

Let k be a field of characteristic $\neq 2$, which will be the base field for all vector spaces. If V is a finite-dimensional vector space, V is called a **metric vector space** (mvs) if there is given a symmetric bilinear form $\beta: V \times V \to k$, called the inner product; if X and Y are elements of V, $\beta(X, Y)$ is usually denoted by XY, and in particular $\beta(X, X)$ by X^2.

We say that X and Y are **orthogonal** if $XY=0$, and call X a **null-vector** if $X^2=0$. If $S \subset V$, the **orthogonal complement of** S is $S^* = \{X \in V \mid XY=0 \text{ for all } Y \in S\}$. Clearly S^* is a vector subspace of V. In particular, $V^* = \text{Rad } V$ is a subspace, called the **radical** of V. We say that V is **nonsingular** if $\text{Rad } V = \{0\}$. It is possible to show that if V is non-singular and S is a subspace, then $\dim S + \dim S^* = \dim V$, hence $S^{**} = S$. V is called **anisotropic** if 0 is the only null-vector. An anisotropic mvs is clearly non-singular, but the converse is not true. An important example of a non-singular mvs which is not anisotropic is the **hyperbolic plane** $V = k^2$, with inner product $(x, y)(x', y') = xx' - yy'$. The null-vectors for this mvs are the vectors (x, y), where $y = \pm x$, so there are 2 null-lines.

* From AMERICAN MATHEMATICAL MONTHLY, vol. 79 (1972), pp. 59–62.

If V is a mvs and $\sigma: V \to V$ is a linear isomorphism, σ is called an **isometry** if $(\sigma X)(\sigma Y) = XY$ for all X and Y in V. If $E_1, \cdots, E_n$ is a basis of V, the $n \times n$ matrix $M = (E_i E_j)$ is called the **matrix of the product** with respect to the basis. It is easy to show that V is non-singular if and only if M is, and that σ is an isometry if and only if $(\sigma E_i)(\sigma E_j) = E_i E_j$ for $1 \leqq i, j \leqq n$. It follows immediately that if A is the matrix of σ with respect to the given basis, then σ is an isometry if and only if $M = A^t M A$.

Hence if V is non-singular and σ is an isometry, then with the same notations, $\det M = \det A^t \det M \det A = (\det A)^2 \det M$, so $\det \sigma = \det A = \pm 1$. If $\det \sigma = 1$, then σ is called a **rotation** (proper isometry), while if $\det \sigma = -1$, then σ is called a **reflection** (improper isometry). Clearly the set of all rotations is a subgroup of index 2 of the group of all isometries of V.

If $f: V \to V$, a non-zero vector X in V is **fixed** if $f(X) = X$ and **reversed** if $f(X) = -X$. If V is a mvs and U and W are subspaces, we write $V = U \perp W$ when $V = U \oplus W$ and $XY = 0$ for all X in U and Y in W, and call V the **orthogonal sum** of U and W. Clearly if σ is an isometry of U and τ is an isometry of W, there is a unique isometry of V extending σ and τ, denoted by $\sigma \perp \tau$. It is not hard to show that if H is a non-singular hyperplane of a non-singular mvs V, then H^* is 1-dimensional, $V = H \perp H^*$, and there is a unique isometry of V called the **symmetry with respect to H** which fixes the vectors in H and reverses the vectors in H^*, namely $I_H \perp (-I_{H^*})$. Symmetries are clearly reflections (since H^* is a line) and are involutions.

3. Proofs of the theorems. To prove Theorem A, let σ be a reflection of a non-singular mvs V. Then, the notations being as above, to show that σ has a reversed vector we must show that $A + I$ is a singular matrix. But, using basic properties of determinants,

$$\begin{aligned} \det M \det (A + I) &= -\det A^t \det M \det (A + I) = -\det (A^t M A + A^t M) \\ &= -\det (M + A^t M) = -\det (I + A^t) M \\ &= -\det (I + A^t) \det M \\ &= -\det (I + A)^t \det M = -\det (I + A) \det M. \end{aligned}$$

Therefore $2 \det M \det (I + A) = 0$, hence $\det (I + A) = 0$, so $A + I$ is singular.

To demonstrate the equivalence of Theorems A and B, observe that if σ is an isometry of an n-dimensional space, then $-\sigma$ is an isometry, which is a reflection if and only if σ is a rotation and n is odd or σ is a reflection and n is even, i.e., if and only if σ satisfies the hypothesis of Theorem B. Since a vector is fixed for σ if and only if it is reversed for $-\sigma$, the two theorems are clearly equivalent.

4. Remarks. Theorem B can be used to give a quick proof of the special case of the Cartan-Dieudonné Theorem (see [2], p. 129) for anisotropic spaces. The Cartan-Dieudonné Theorem, which is important in determining the structure of the group of isometries, says that any isometry of a non-singular n-dimensional mvs is the product of at most n symmetries. If we assume that the mvs V is **anisotropic** then the following inductive proof is legitimate:

For $n=1$, the Cartan-Dieudonné Theorem is obvious since the only isometries of a non-singular line are $\pm I$.

If σ satisfies the hypothesis of Theorem B, then σ has a fixed vector X, which is a fortiori not a null-vector. Letting $V_1=\langle X\rangle^*$, V_1 is an anisotropic hyperplane of V, and σ induces an isometry σ_1 of V_1. By the induction hypothesis, σ_1 is the product of at most $n-1$ symmetries of V_1. If τ_1 is an isometry of V_1, then $\tau=I_{\langle X\rangle}\perp\tau_1$ is the only isometry of V which fixes X and extends τ_1. If τ_1 is the symmetry with respect to the non-singular hyperplane H_1 of V_1, then τ is the symmetry with respect to $\langle X\rangle\perp H_1$. Hence $\sigma=I_{\langle X\rangle}\perp\sigma_1$ is the product of at most $n-1$ symmetries of V in this case.

If σ does not satisfy the hypothesis of Theorem B, and τ is any symmetry of V, then $\tau\sigma$ does satisfy the hypothesis, hence by the preceding case, $\tau\sigma$ is the product of at most $n-1$ symmetries of V and $\sigma=\tau(\tau\sigma)$ is the product of at most n symmetries (with one chosen arbitrarily).

Attempting to modify the preceding argument for the general case of a non-singular mvs V of dimension n we run into the problem that a fixed vector X may be null, and consequently $\langle X\rangle^*$ may be singular. Thus for the induction step, we would probably need a Cartan-Dieudonné type theorem for singular spaces.

It is possible to adapt the argument to the general case for $n\leqq 3$ as follows: For $n=1$, non-singular coincides with anisotropic. For $n=2$, it is a nice exercise to show that I is the only isometry of a non-singular plane with a fixed null-vector (the hyperbolic plane is essentially the only example), hence a reflection is a symmetry, and a rotation is the product of 2 symmetries, one being arbitrary. For $n=3$, we show that any isometry with a fixed null-vector is the product of at most 2 symmetries as follows: If σ has a fixed null-vector X, then σ induces an isometry of the singular plane $\langle X\rangle^*$ which leaves X fixed. Hence if $Y\in\langle X\rangle^*-\langle X\rangle$, then Y is non-null (otherwise $\langle X\rangle^*$ would be a null-plane, while in fact $\langle X\rangle=\text{Rad }\langle X\rangle^*$), and $\sigma(Y)=aX\pm Y$ for some $a\in k$. Let τ_1 be the symmetry of V with respect to the non-singular plane $\langle Y\rangle^*$, so that τ_1 fixes X and reverses Y, and τ_2 be the symmetry of V with respect to the non-singular plane $\langle (a/2)X-Y\rangle^*$, so that τ_2 fixes X and reverses $(a/2)X-Y$. We assert that if $\sigma(Y)=aX-Y$, then $\sigma=\tau_2$, while if $\sigma(Y)=aX+Y$, then $\sigma=\tau_1\tau_2$. To see this, let $\sigma'=\tau_2\sigma^{-1}$ in the first case and $\sigma'=\tau_1\tau_2\sigma^{-1}$ in the second. Then σ' is an isometry of V leaving X and Y fixed. σ' induces an isometry of the non-singular plane $\langle Y\rangle^*$ leaving the null-vector X fixed, hence according to the discussion for $n=2$, σ' restricted to $\langle Y\rangle^*$ is the identity. Thus $\sigma'=I$, which completes the proof for $n\leqq 3$.

It is unlikely that this type of elementary argument will suffice to prove the general case of the Cartan-Dieudonné Theorem for $n\geqq 4$ or even for $n=4$, since it is known that there are isometries of a non-singular 4-dimensional mvs with a fixed null-vector, which cannot be represented as the product of fewer than 4 symmetries.

References

1. W. H. Greub, Linear Algebra, 3-rd ed., Springer, New York, 1967, p. 222.
2. E. Artin, Geometric Algebra, Interscience, New York, 1957.
3. E. Snapper and R. Troyer, Affine and Metric Geometry, Notes 1967.

BIBLIOGRAPHIC ENTRIES: INNER PRODUCTS AND QUADRATIC FORMS

The references below are to the AMERICAN MATHEMATICAL MONTHLY.

1. H. B. Mann, Quadratic forms with linear constraints, 1943, 430–433.

Proves necessary and sufficient conditions for a quadratic form to be positive (or negative) for all values satisfying given linear constraints.

2. C. G. Cullen, A note on normal matrices, 1965, 643–644.

Proves that if A and B are normal matrices such that one has no two eigenvalues of the same absolute value, then AB is normal if and only if $AB=BA$. The proofs use the spectral theorem.

3. Alvin Hausner, Uniqueness of the polar decomposition, 1967, 303–304.
4. V. G. Sigillito, An application of the Schwarz inequality, 1968, 656–658.

Uses the Schwarz inequality to show that a norm-preserving linear transformation is unitary.

5. C. R. Johnson, Positive definite matrices, 1970, 259–264.

This article was the winning entry in a contest for undergraduates.

6. P. R. Halmos, Finite-dimensional Hilbert spaces, 1970, 457–464.

Discusses three subjects in finite-dimensional linear algebra motivated by the study of operator theory in infinite-dimensional Hilbert space: (1) characterizing pairs of subspaces algebraically (2) characterizing linear transformation geometrically (3) lattices of invariant subspaces.

For articles on related topics, see also items (1), (2), (4) and (7) in the bibliography of Section B above (p. 337).

(d)

CANONICAL FORMS

A CONDITION FOR DIAGONABILITY OF MATRICES*

H. K. FARAHAT and L. MIRSKY, University of Sheffield, England

A square matrix is said to be *diagonable* if it is similar to a diagonal matrix. Various necessary and sufficient conditions for the diagonability of matrices are to be found in the literature. (See, for example, **1.**) In this note we establish yet another such condition, which does not appear to have been stated previously.

By a *group matrix* we shall understand any matrix which is an element in

* From AMERICAN MATHEMATICAL MONTHLY, vol. 63 (1956), pp. 410–412.

some multiplicative group of matrices. Every non-singular matrix is, of course, a group matrix; but groups of singular matrices also exist and were first investigated by Ranum (2). We shall prove the following

THEOREM. *A matrix A is diagonable if and only if $A-\omega I$ is a group matrix for every (complex) value of ω.*

All matrices considered below are assumed to be of type $n\times n$. We denote by I the unit matrix, by $R(A)$ the rank of A, and by diag $(\omega_1, \cdots, \omega_n)$ the diagonal matrix with diagonal elements $\omega_1, \cdots, \omega_n$. If $\mathfrak{B}$ is a vector space, then $d(\mathfrak{B})$ denotes its dimension.

A few preliminary results will be required.

LEMMA 1. *All matrices in a multiplicative matrix group have the same rank.*

Let A be an arbitrary element and U the unit element of a matrix group Γ. Denoting by B the inverse element of A, we have $U=AB$, $A=AU$. Therefore

$$R(U) = R(AB) \leqq R(A) = R(AU) \leqq R(U),$$

so that $R(A)=R(U)$ for all $A\in\Gamma$.

LEMMA 2. *If $R(AB)=R(B)$, then $R(ABC)=R(BC)$.*

Denote by $\mathfrak{N}_T$ the space of all vectors x such that $Tx=0$. Then

$$d(\mathfrak{N}_T) = n - R(T). \tag{1}$$

Hence, by hypothesis, $d(\mathfrak{N}_{AB})=d(\mathfrak{N}_B)$. But, trivially, $\mathfrak{N}_{AB}\supset\mathfrak{N}_B$, and so

$$\mathfrak{N}_{AB} = \mathfrak{N}_B. \tag{2}$$

Next, let $x\in\mathfrak{N}_{ABC}$, so that $ABCx=0$. Then $Cx\in\mathfrak{N}_{AB}$ and therefore, by (2), $Cx\in\mathfrak{N}_B$, *i.e.*, $BCx=0$. Thus $x\in\mathfrak{N}_{BC}$, and we have shown that $\mathfrak{N}_{ABC}\subset\mathfrak{N}_{BC}$. But, trivially, $\mathfrak{N}_{ABC}\supset\mathfrak{N}_{BC}$; and therefore $\mathfrak{N}_{ABC}=\mathfrak{N}_{BC}$. The assertion now follows by (1).

LEMMA 3. *If $\mu(x)$ is the minimum polynomial of A and $h(x)$ is a polynomial such that $h(A)=O$, then $\mu(x)$ divides $h(x)$.*

LEMMA 4. *If all zeros of the minimum polynomial of A are simple, then A is diagonable.**

Both these results are well-known.

Proof of the theorem. (i) Suppose that A is diagonable, and write $S^{-1}AS=\text{diag}\,(\omega_1, \cdots, \omega_n)$. Then

$$S^{-1}(A - \omega I)S = \text{diag}\,(\omega_1 - \omega, \cdots, \omega_n - \omega). \tag{3}$$

Now the matrix on the right-hand side of (3) is an element in the matrix group Γ consisting of the matrices

$$\text{diag}\,(\epsilon_1 x_1, \ldots, \epsilon_n x_n),$$

* The converse statement is also true, but not necessary for our purpose.

where $x_1, \cdots, x_n$ take all non-zero (complex) values and ϵ_r is defined as 0 or 1 according as $\omega = \omega_r$ or $\omega \neq \omega_r$. Hence $A - \omega I$ is an element of the group $S\Gamma S^{-1}$, and so is a group matrix.

(ii) Suppose, next, that $A - \omega I$ is a group matrix for every value of ω. Then, by Lemma 1,

$$R(A - \omega I) = R\{(A - \omega I)^2\} \qquad \text{(all } \omega). \tag{4}$$

Now let ω be any zero of the minimum polynomial $\mu(x)$ of A, so that $\mu(x) = (x - \omega)^k \phi(x)$, where $k \geqq 1$ and $\phi(\omega) \neq 0$. There exist polynomials $f(x)$ and $g(x)$ such that

$$(x - \omega)^k f(x) + \phi(x) g(x) = 1. \tag{5}$$

This implies that

$$(A - \omega I)^k f(A) + \phi(A) g(A) = I. \tag{6}$$

Writing $E = \phi(A) g(A)$, multiplying both sides of (6) by E, and using the fact that $\mu(A) = O$, we obtain $E^2 = E$. Putting $Z = (A - \omega I)E$, we therefore have

$$Z^m = (A - \omega I)^m E, \tag{7}$$

so that

$$Z^k = O. \tag{8}$$

In view of (4), (7), and Lemma 2, we at once obtain $R(Z) = R(Z^2)$. Hence, by Lemma 2 and (8),

$$R(Z) = R(Z^2) = \cdots = R(Z^k) = 0,$$

so that $Z = O$, *i.e.*,

$$(A - \omega I)\phi(A) g(A) = O.$$

It follows by Lemma 3 that $(x - \omega)^k \phi(x)$ divides $(x - \omega)\phi(x) g(x)$. But, by (5), $g(\omega) \neq 0$ and therefore $k = 1$. All zeros of $\mu(x)$ are therefore simple; and by Lemma 4, A is diagonable.

References

1. M. P. Drazin, On diagonable and normal matrices, Quart. J. Math. (Oxford) (2), vol. 2, 1951, pp. 189–198.

2. A. Ranum, The group-membership of singular matrices, Amer. J. Math., vol. 31, 1909, pp. 18–41.

ON THE REDUCTION OF A MATRIX TO DIAGONAL FORM*

MARVIN EPSTEIN and HARLEY FLANDERS, University of California, Berkeley

Given an $n\times n$ matrix $A=(a_{ij})$ with elements in a field F, one wishes to know when it is possible to find a matrix P so that the matrix $D=PAP^{-1}$ is in diagonal form. Of course even when this is possible it will be necessary, in general, to take the elements of P and therefore of D in some extension field of F, specifically in the algebraic closure of F. Nevertheless it is desirable to have a rational criterion for the diagonalizability of A, *i.e.*, a criterion which can be verified entirely within the given field F irrespective of where the elements of P and D lie. In this paper we give such a criterion.

We shall restrict attention to the case of fields of characteristic zero, and we denote by I the identity $n\times n$ matrix.

By the degree of a matrix A we shall mean the degree of its minimal polynomial. Thus if A has degree s then $I, A, \cdots, A^{s-1}$ are linearly independent while $I, A, \cdots, A^{s-1}, A^s$ are linearly dependent over F.

THEOREM. *Let A be an $n\times n$ matrix of degree s with elements in a field F of characteristic zero. Let S denote the trace and Δ the determinant*

$$\Delta = \det\left[S(A^{i+j})\right], \qquad (0\leqq i,j\leqq s-1).$$

Then A can be reduced to diagonal form (in some extension field of F) if and only if $\Delta\neq 0$.

Proof. If A can be reduced to diagonal form then

$$PAP^{-1} = B = \begin{pmatrix} \lambda_1 I_1 & & 0 \\ & \ddots & \\ 0 & & \lambda_r I_r \end{pmatrix}$$

where $\lambda_1, \cdots, \lambda_r$ are the distinct characteristic roots of A and I_k is the n_k-rowed identity matrix $(n_1+\cdots+n_r=n)$. Evidently $r=s$ and for $u=1, 2, 3, \cdots, S(A^u)=S(B^u)$, hence

$$S(A^{i+j}) = n_1\lambda_1^{i+j} + \cdots + n_r\lambda_r^{i+j}, \qquad (0\leqq i,j\leqq r-1).$$

This implies, by the product rule for determinants

$$\Delta = \det\left[S(A^{i+j})\right] = \begin{vmatrix} n_1 & n_2 & \cdots & n_r \\ n_1\lambda_1 & n_2\lambda_2 & \cdots & n_r\lambda_r \\ \cdots & \cdots & \cdots & \cdots \\ n_1\lambda_1^{r-1} & n_2\lambda_2^{r-1} & \cdots & n_r\lambda_r^{r-1} \end{vmatrix} \begin{vmatrix} 1 & \lambda_1 & \cdots & \lambda_1^{r-1} \\ 1 & \lambda_2 & \cdots & \lambda_2^{r-1} \\ \cdots & \cdots & \cdots & \cdots \\ 1 & \lambda_r & \cdots & \lambda_r^{r-1} \end{vmatrix}$$

$$= n_1\cdots n_r V^2,$$

* From AMERICAN MATHEMATICAL MONTHLY, vol. 62 (1955), pp. 168–171.

where V is the (non-vanishing) Vandermonde determinant of $\lambda_1, \cdots, \lambda_r$.

Conversely, suppose A cannot be put in diagonal form. We see from the Jordan canonical form that if $\lambda_1, \cdots, \lambda_r$ are the distinct characteristic roots of A and since s is the degree of A, $s \geqq r$. As before let λ_k have multiplicity n_k so for some k, $n_k > 1$ (or else A is diagonalizable). It is then easily seen (again from the Jordan canonical form) that $(A-\lambda_1 I)(A-\lambda_2 I) \cdots (A-\lambda_r I) \neq 0$, so that $s > r$. We again have

$$S(A^{i+j}) = n_1\lambda_1^{i+j} + \cdots + n_r\lambda_r^{i+j}, \qquad (0 \leqq i, j \leqq s-1),$$

but since Δ is the determinant of an $s \times s$ matrix we have

$$\Delta = \det\,[S(A^{i+j})] = \begin{vmatrix} n_1 & \cdots & n_r & 0 \cdots 0 \\ n_1\lambda_1 & \cdots & n_r\lambda_r & 0 \cdots 0 \\ \cdot & \cdot & \cdot & \cdot \\ n_1\lambda_1^{s-1} & \cdots & n_r\lambda_r^{s-1} & 0 \cdots 0 \end{vmatrix} \begin{vmatrix} 1 & \lambda_1 \cdots \lambda_1^{s-1} \\ \cdot & \cdot \\ 1 & \lambda_r \cdots \lambda_r^{s-1} \\ 0 & 0 \cdots 0 \\ \cdot & \cdot \\ 0 & 0 \cdots 0 \end{vmatrix} = 0.$$

Example. Let

$$A = \begin{pmatrix} 1 & 2 \\ 3 & 4 \end{pmatrix}.$$

Clearly the degree of A is 2. We have

$$A^2 = \begin{pmatrix} 7 & 10 \\ 15 & 22 \end{pmatrix} \quad \text{so} \quad \Delta = \begin{vmatrix} S(I) & S(A) \\ S(A) & S(A^2) \end{vmatrix} = \begin{vmatrix} 2 & 5 \\ 5 & 29 \end{vmatrix} = 33 \neq 0;$$

hence A is reducible to diagonal form. We can choose

$$P = \begin{pmatrix} -3+\sqrt{33} & 4 \\ -3-\sqrt{33} & 4 \end{pmatrix} \quad \text{and then} \quad D = \begin{pmatrix} \frac{1}{2}(5+\sqrt{33}) & 0 \\ 0 & \frac{1}{2}(5-\sqrt{33}) \end{pmatrix}.$$

The above theorem follows also from certain well-known results in the theory of linear algebras. To consider the problem from this viewpoint, we fix the matrix A and the field F as well as an algebraic closure $\bar{F}$ of F.

We say that A is reducible over F if, by a multiplication of the type QAQ^{-1}, where Q is an $n \times n$ matrix with elements in F, A can be transformed to a matrix of the form

$$\begin{pmatrix} M_1 & M_2 \\ 0 & M_3 \end{pmatrix}$$

where, for some m, $1 \leqq m \leqq n-1$, M_1 is an $m \times m$ matrix, M_3 is an $(n-m) \times (n-m)$ matrix, and M_2 is an $m \times (n-m)$ matrix. Otherwise we say that A is irreducible. We say that A is completely reducible if, by a multiplication of the above type, A can be transformed to a matrix of the type

$$\begin{bmatrix} N_1 & 0 & \cdots & 0 \\ 0 & N_2 & \cdots & 0 \\ \cdot & \cdot & \cdot & \cdot \\ 0 & 0 & \cdots & N_t \end{bmatrix}$$

where each N_i is an irreducible $n_i \times n_i$ matrix ($n_1 + \cdots + n_t = n$). In particular, (for $t=1$) an irreducible matrix is completely reducible.

It can be shown that if A is irreducible over F then its minimal polynomial is irreducible over F; more generally, A is completely reducible over F if and only if its minimal polynomial is the product of distinct factors each irreducible over F, or equivalently, if and only if its minimal polynomial is the product of distinct linear factors over $\bar{F}$. On the other hand, over $\bar{F}$ the latter condition on the minimal polynomial is necessary and sufficient for the diagonalizability of A. Since the minimal polynomial of A is an invariant, depending only on the field containing the elements of A, it follows that A is diagonalizable over $\bar{F}$ if and only if A is completely reducible over F.

A set of $n \times n$ matrices with elements in F is called an algebra over F if it is a vector space over F (necessarily finite-dimensional) and is closed under matrix multiplication. The algebra over F generated by A is the smallest algebra of $n \times n$ matrices over F which contains A and I. This algebra consists of linear combinations over F of powers of A (including $I = A^0$) and a basis for the vector space underlying the algebra is $I, A, \cdots, A^{r-1}$ when A is of degree r.

A set $\mathfrak{M}$ of $n \times n$ matrices over F may be regarded as a set of linear endomorphisms of an n-dimensional vector space V over F with a fixed basis. A subspace W of V is said to be invariant under $\mathfrak{M}$ if each endomorphism in $\mathfrak{M}$ maps W into itself. An invariant subspace is said to be simple under $\mathfrak{M}$ if it contains no other invariant subspace except (0).

An algebra $\mathfrak{A}$ over F of $n \times n$ matrices containing I is said to be semi-simple if V is the direct sum of simple subspaces invariant under $\mathfrak{A}$. If $\mathfrak{A}$ is generated by A then it is known that $\mathfrak{A}$ is semi-simple if and only if A is completely reducible over F.

A theorem in the theory of algebras states that an algebra $\mathfrak{A}$ of matrices over a field F of characteristic zero is semi-simple if and only if the function $S(XY)$ (where $X, Y \in \mathfrak{A}$ and S denotes the trace function) is non-degenerate, *i.e.*, if $X \in \mathfrak{A}$ is such that $S(XY) = 0$ for every $Y \in \mathfrak{A}$, then $X = 0$.

Combining these results we see that if F is of characteristic zero and $\mathfrak{A}$ is the algebra over F generated by A, then the following are equivalent:

(1) A is diagonalizable over $\bar{F}$;

(2) the minimal polynomial of A is the product of distinct factors each irreducible over F;
(3) A is completely reducible over F;
(4) $\mathfrak{A}$ is semi-simple;
(5) $S(XY)=0$ for fixed $X\in\mathfrak{A}$ and all $Y\in\mathfrak{A}$ implies $X=0$.

Now let $x_0, \cdots, x_{r-1}$ be indeterminates and let $X=\sum_{i=0}^{r-1}x_iA^i$. Then

$$S(XA^j) = S\left(\sum_{i=0}^{r-1} x_iA^{i+j}\right) = \sum_{i=0}^{r-1} S(A^{i+j})x_i, \quad (0 \leqq j \leqq r-1).$$

It is clear that the function $S(XY)$ will be degenerate on $\mathfrak{A}$ if and only if there exist elements $\alpha_0, \cdots, \alpha_{r-1}\in F$, not all zero, such that

$$\sum_{i=0}^{r-1} S(A^{i+j})\alpha_i=0, \qquad (0 \leqq j \leqq r-1).$$

Thus the non-degeneracy of $S(XY)$ is equivalent to the requirement that there exists only the trivial solution for a set of r homogeneous linear equations over F in r unknowns. This requirement is satisfied if and only if the determinant of coefficients, det $[S(A^{i+j})]$, is non-vanishing.

ON THE MINIMIZATION OF MATRIX NORMS*

L. MIRSKY, University of Sheffield

A well-known inequality, due to Schur [1], states that *if A is a complex $n\times n$ matrix with euclidean norm $\|A\|$ and characteristic roots $\omega_1, \cdots, \omega_n$, then*

$$\|A\|^2 \geqq \sum_{i=1}^{n} |\omega_i|^2. \tag{1}$$

Equality occurs in (1) *if and only if A is normal.* Commenting on this result, Wedderburn [2] observed: "Since replacing a matrix by a similar one corresponds to a change of coordinates when the matrix is regarded as a linear transformation, it follows from Schur's work that, when the elementary divisors of A are simple, $\|A\|$ has its minimum value when A is represented as a diagonal matrix, that is, in its normal form; and it seems probable that the normal form also gives the minimum value even if the elementary divisors are not simple." Wedderburn thus conjectured that, for a fixed matrix A, $\|S^{-1}AS\|$ is minimal when $S^{-1}AS$ is the classical canonical form of A. This would imply, in particular, that, for a nondiagonable matrix A and any nonsingular matrix S,

$$\|S^{-1}AS\|^2 \geqq \sum_{i=1}^{n} |\omega_i|^2 + 1.$$

We shall demonstrate that Wedderburn's conjecture is incorrect by establishing the following result.

* From AMERICAN MATHEMATICAL MONTHLY, vol. 65 (1958), pp. 106–107.

THEOREM. *Let A be a complex $n \times n$ matrix with characteristic roots $\omega_1, \ldots, \omega_n$. Then*

$$\inf \|S^{-1}AS\|^2 = \sum_{i=1}^{n} |\omega_i|^2, \tag{2}$$

where the lower bound is taken with respect to all nonsingular matrices S. Furthermore, the lower bound is attained if and only if A is diagonable.

Denote by T_1 a matrix such that $T_1^{-1}AT_1$ is triangular, say

$$T_1^{-1}AT_1 = \begin{bmatrix} \omega_1 & b_{12} & \cdots & b_{1n} \\ 0 & \omega_2 & \cdots & b_{2n} \\ \cdots & \cdots & \cdots & \cdots \\ 0 & 0 & \cdots & \omega_n \end{bmatrix}$$

Let δ be an arbitrary positive number $\leqq 1$, and write $T_2 = \operatorname{diag}(1, \delta, \cdots, \delta^{n-1})$, $S_0 = T_1T_2$. Then

$$\|S_0^{-1}AS_0\|^2 = \sum_{i=1}^{n} |\omega_i|^2 + \sum_{1 \leq i < k \leq n} |b_{ik}|^2 \delta^{2(k-i)}$$

$$\leqq \sum_{i=1}^{n} |\omega_i|^2 + \frac{1}{2} n(n-1) b^2 \delta^2,$$

where $b = \max_{i,k} |b_{ik}|$. The relation (2) now follows since, by (1),

$$\|S^{-1}AS\|^2 \geqq \sum_{i=1}^{n} |\omega_i|^2$$

for every nonsingular matrix S. Furthermore, if A is diagonable, then the lower bound of $\|S^{-1}AS\|$ is clearly attained. If, on the other hand, the bound is attained so that, for some S,

$$\|S^{-1}AS\|^2 = \sum_{i=1}^{n} |\omega_i|^2,$$

then $S^{-1}AS$ is normal and A is diagonable.

References

1. I. Schur, Über die charakteristischen Wurzeln einer linearen Substitution mit einer Anwendung auf die Theorie der Integralgleichungen, Math. Ann., vol. 66, 1909, pp. 488–510.

2. J. H. M. Wedderburn, The absolute value of the product of two matrices, Bull. Amer. Math. Soc., vol. 31, 1925, pp. 304–308.

BIBLIOGRAPHIC ENTRIES: CANONICAL FORMS

The references below are to the AMERICAN MATHEMATICAL MONTHLY.

1. E. T. Browne, On the reduction of a matrix to a canonical form, 1940, 437–450.

Presents a unified treatment of three canonical forms.

2. B. E. Mitchell, Unitary multiples of a matrix, 1954, 610–623.

Discusses canonical forms for one-sided multiples of a matrix.

3. M. P. Epstein, On a class of determinants associated with a matrix, 1956, 160–162.

Extends some results in the article of Epstein and Flanders reprinted above, (pp. 365–368).

4. D. W. Robinson, A note on diagonable matrix polynomials, 1960, 173–174.

Discusses when a polynomial in a given matrix is diagonable.

5. S. Cater, An elementary development of the Jordan canonical form, 1962, 391–393.

This development draws more from the analyst's than the algebraist's bag of tricks.

6. D. E. Daykin, The rational canonical form of a function of a matrix, 1963, 1082–1085.

Discusses the relationship between the rational canonical forms for A and $f(A)$, where f is any polynomial.

7. P. A. Fillmore, On similarity and the diagonal of a matrix, 1969, 167–169.

Generalizes the result that a matrix with trace zero is similar to one with zero main diagonal.

8. G. P. Barker, Topological properties of the row echelon form, 1973, 787–789.

Inspired by Schneider's article reprinted above (pp. 339–343).

(e)

LINEAR ALGEBRA OVER FINITE FIELDS

REGULAR POLYGONS OVER $GF[3^2]$*

D. W. CROWE, University College, Ibadan, Nigeria

Introduction. Shephard [6] has introduced the notion of regular complex polygon, and has enumerated all such polygons. An example of a regular quaternion polygon has also been given by Crowe [4]. It is, in fact, apparent that any field with a nontrivial involutory automorphism permits an analogous definition of "unitary" and thus of "regular polygon." The purpose of the present note is to illustrate this by finding the regular polygons in the plane over the field with 9 elements, $GF[3^2]$.

* From AMERICAN MATHEMATICAL MONTHLY, vol. 68 (1961), pp. 762–765.

Definitions and properties of $GF[p^{2n}]$. There is an involutory automorphism, $x \to x^{p^n}$, in any field $GF[p^{2n}]$ (p prime, $n=1, 2, \cdots$). This defines a *conjugate*, $\bar{x}=x^{p^n}$, in this field. The two-dimensional vector space over this field, with the following unitary structure, will be designated $UG(2, p^{2n})$. Let (x, y) be a point of the space, and let $(x', y')=(x, y)A$, where

$$A = \begin{pmatrix} a & b \\ c & d \end{pmatrix}.$$

The matrix A is said to be *unitary* if $x\bar{x}+y\bar{y}=x'\bar{x}'+y'\bar{y}'$ for all (x, y). Necessary and sufficient conditions for A to be unitary are $a\bar{a}+b\bar{b}=c\bar{c}+d\bar{d}=1$ and $a\bar{c}+b\bar{d}=0$. In fact:

Every 2×2 *unitary matrix of determinant* Δ *is of the form*

$$\begin{pmatrix} a & b \\ -\bar{b}\Delta & \bar{a}\Delta \end{pmatrix},$$

where $a\bar{a}+b\bar{b}=1$.

Proof. Let

$$\begin{pmatrix} a & b \\ c & d \end{pmatrix}$$

be unitary. Then $a\bar{a}+b\bar{b}=1$, $\bar{a}c+\bar{b}d=0$, and $\Delta=ad-bc$. Thus $a\bar{a}d+b\bar{b}d=d$, and substituting for ad and $\bar{b}d$ from the other two equations yields $\bar{a}(\Delta+bc)-\bar{a}cb=d$, that is, $d=\bar{a}\Delta$. Similarly, $c=-\bar{b}\Delta$.

A (*unitary*) *reflection* in $UG(2, p^{2n})$ is a unitary matrix exactly one of whose eigenvalues is 1. A *regular* (*unitary*) *polygon in* $UG(2, p^{2n})$ is a configuration of points and lines ("vertices" and "edges") which is transformed into itself by two unitary reflections, one, R, which cyclically permutes the vertices on an edge, and another, S, which cyclically permutes the edges at one of these vertices. In practice we usually choose this vertex to be $(1, 0)$, so that S has the form

$$\begin{pmatrix} 1 & 0 \\ 0 & \Delta \end{pmatrix}.$$

Regular polygons in $UG(2, 3^2)$. We represent the nonzero elements of $GF[3^2]$ as powers, $\gamma, \gamma^2, \cdots, \gamma^8=1$, of a root, γ, of the irreducible polynomial x^2+x+2 (mod 3). (See, for example, [1], ch. IX.) Thus $\gamma^2+\gamma+2=0$, so that $\gamma^2=2\gamma+1$, $\gamma^3=2\gamma^2+\gamma=(\gamma+2)+\gamma=2\gamma+2$, etc. We note that the only reflections in $UG(2, 3^2)$ are of period two or four. For if

$$P = \begin{pmatrix} a & b \\ -\bar{b}\Delta & \bar{a}\Delta \end{pmatrix}$$

is of period eight then $\Delta=\gamma^r$, r odd. But $(-\bar{b}\Delta)(\overline{-\bar{b}\Delta})+(\bar{a}\Delta)(\overline{\bar{a}\Delta})=\Delta\bar{\Delta}=\gamma^r\bar{\gamma^r}$ $=\gamma^{4r}=-1\neq 1$, so that P is not unitary. The pairs, R, S, of generating reflections for regular polygons in $UG(2, 3^2)$ are thus of three types: (*i*) both reflections of period four, (*ii*) both reflections of period two, and (iii) one reflection each of periods two and four. We treat the three cases separately.

Case (*i*). Let

$$R=\begin{pmatrix} a & b \\ -\bar{b}\Delta & \bar{a}\Delta \end{pmatrix}, \qquad S=\begin{pmatrix} 1 & 0 \\ 0 & \gamma^2 \end{pmatrix}.$$

Since R (or its inverse) has eigenvalues $1, \gamma^2$ we have $\Delta=\gamma^2$, and $a+\bar{a}\Delta=a+a^3\gamma^2$ $=1+\gamma^2=\gamma^3$. The solutions to the latter equation are $a=\gamma^2$, 1, γ^7. The first two solutions yield "degenerate polygons" having only one edge and one vertex respectively. Corresponding to $a=\gamma^7$ there are four matrices, namely

$$R=\begin{pmatrix} \gamma^7 & \gamma \\ \gamma & \gamma^7 \end{pmatrix}, \quad SRS^{-1}, \quad S^2RS^2, \quad \text{and} \quad S^{-1}RS.$$

Thus the only nondegenerate polygon is generated by

$$R=\begin{pmatrix} \gamma^7 & \gamma \\ \gamma & \gamma^7 \end{pmatrix} \quad \text{and} \quad S=\begin{pmatrix} 1 & 0 \\ 0 & \gamma^2 \end{pmatrix}$$

Its 24 vertices are the images of (1, 0) under R and S. Using Shephard's notation, ${}_nx$, to designate the n numbers obtained by multiplying x by the nth roots of unity, we can write them as $({}_41, 0)$, $(0, {}_41)$, and $({}_4\gamma, {}_4\gamma)$. They are exactly the 24 points of the unit circle, $x\bar{x}+y\bar{y}=1$, in UG $(2, 3^2)$. These vertices lie by fours on the 24 edges, $x={}_4\gamma$, $y={}_4\gamma$, ${}_4x+{}_4y=1$. Any one of the 96 figures consisting of an edge and a vertex on it can be transformed into any other by an element of the group $\{R, S\}$ generated by R and S. Thus $\{R, S\}$ has order at least 96. But R and S satisfy $R^4=I$ and $RSR=SRS$, so that $\{R, S\}$ has order at most 96 ([3], p. 79). It is thus seen that this polygon is an isomorphic copy of the regular complex polygon $4\{3\}4$, or 4(96)4, in Shephard's notation. In fact, its group is the group of all unitary transformations in $UG(2, 3^2)$. As such it is discussed in detail by Edge in [5].

Case (ii). If R and S are both of period two we have $\Delta=-1$, and

$$R=\begin{pmatrix} a & b \\ \bar{b} & -\bar{a} \end{pmatrix}$$

Thus $a-\bar{a}=1-1=0$, which has solutions $a=\pm 1$, 0. The nonzero solutions yield degenerate polygons, as in Case (i). Corresponding to $a=0$ we get

$$R=\begin{pmatrix} 0 & 1 \\ 1 & 0 \end{pmatrix} \quad \text{and} \quad SRS^{-1}, \quad \text{where} \quad S=\begin{pmatrix} 1 & 0 \\ 0 & -1 \end{pmatrix}.$$

Thus the only nondegenerate polygon is generated by this R and S, and has the four vertices $(\pm 1, 0)$, $(0, \pm 1)$. It is an isomorphic copy of the square, $2\{4\}2$. The group $\{R, S\}$ is the dihedral group of order eight, defined by $R^2 = S^2 = I$, $(RS)^2 = (SR)^2$.

Case (iii). If R and S have different periods there are two possibilities (aside from degenerate polygons analogous to those in the preceding cases). For one of these we take

$$R = \begin{pmatrix} 0 & 1 \\ 1 & 0 \end{pmatrix} \quad \text{and} \quad S = \begin{pmatrix} 1 & 0 \\ 0 & \gamma^2 \end{pmatrix}.$$

Its eight vertices, $({}_4 1, 0)$ and $(0, {}_4 1)$, lie in pairs on the 16 edges, ${}_4x + {}_4y = 1$, and there are four of these edges at each vertex. Its group, $\{R, S\}$, is of order 32, and has the abstract definition $R^2 = S^4 = I$, $(RS)^2 = (SR)^2$. This polygon is also an isomorphic copy of a regular complex polygon, $2\{4\}4$, or $2(32)4$ in Shephard's notation. The other polygon is the dual, $4\{4\}2$, of this. It is most efficiently obtained by interchanging the roles of R and S. Thus we put $R' = S$ and $S' = R$, and find the images of a point, say (γ, γ), left invariant by S'. The 16 vertices, $({}_4\gamma, {}_4\gamma)$, thus obtained lie by fours on the eight edges, $x = {}_4\gamma$, $y = {}_4\gamma$, and there are two of these edges at each vertex.

Another example. We conclude with another "representation" of a complex polygon. Coxeter ([2], pp. 107, 108) has represented the regular complex polygon $3\{3\}3$ in $EG(2, 3)$. But his generators R_1, R_2 cannot qualify as reflections, for their eigenvalues are all 1's. (Furthermore, they are not unitary.) In $EG(2, 7)$ the situation is improved somewhat, for the matrices

$$R = \begin{pmatrix} 4 & 1 \\ 1 & 6 \end{pmatrix} \quad \text{and} \quad S = \begin{pmatrix} 1 & 0 \\ 0 & 2 \end{pmatrix}$$

(entries mod 7) each have eigenvalues 1 and 2, that is, 1 and a cube root of 1, and could thus conceivably be called "reflections." Applying them to (1, 0) yields eight vertices, $(\pm 1, 0)$ and $\pm(3, {}_3 3)$, lying by threes on eight edges, $x + {}_3 y = \pm 1$ and $x = \pm 3$. The group $\{R, S\}$ is defined by $R^3 = I$, $RSR = SRS$, and has order 24. This is thus another representation of $3\{3\}3$, but this R and S are still not unitary.

The smallest field permitting a representation of the generators of $3\{3\}3$ as unitary reflections is $GF[5^2]$. In fact, if β is a root of the irreducible polynomial, $x^2 + 3x + 3$ (mod 5), then

$$R = \begin{pmatrix} \beta^{15} & \beta^2 \\ \beta^6 & \beta^{11} \end{pmatrix} \quad \text{and} \quad S = \begin{pmatrix} 1 & 0 \\ 0 & \beta^8 \end{pmatrix}$$

yield such a representation. The eight vertices are $\pm(1, 0)$ and $\pm(\beta^3, {}_3\beta^6)$, lying by threes on the eight edges $x = \pm\beta^3$ and $x + {}_3\beta^7 y = \pm 1$.

References

1. R. D. Carmichael, Introduction to the Theory of Groups of Finite Order, Boston, 1937.
2. H. S. M. Coxeter, Factor groups of the braid group, Proceedings of the Fourth Canadian Mathematical Congress, 1959, pp. 95–122.
3. H. S. M. Coxeter and W. O. J. Moser, Generators and Relations for Discrete Groups, Ergeb. Math. (N.F.), vol. 14, Berlin, 1957.
4. D. W. Crowe, A regular quaternion polygon, Canad. Math. Bull., vol. 2, 1959, pp. 77–79.
5. W. L. Edge, The simple group of order 6048, Proc. Cambridge Phil. Soc., vol. 56, 1960, pp. 189–204.
6. G. C. Shephard, Regular complex polytopes, Proc. London Math. Soc. (3), vol. 2, 1952, pp. 82–97.

MATRICES OVER A FINITE FIELD*

S. D. FISHER, University of Wisconsin, and M. N. ALEXANDER, Cornell University

This paper is concerned with matrices over the field of q elements, where, of course, $q=p^N$, p a prime and N a positive integer.

Theorems 1–5 are computations of the orders of certain subsets of these matrices while Theorem 6 describes some of the properties of the matrices themselves. Theorems 1 and 2 are well-known results but the proof of Theorem 2 appears to be new.

Some of the proofs make use of orbits and stabilizers. We give the basic definitions and theorem below; additional details may be found in Chevalley (2).

Let H be a group, S a set, and G_S the group of one-to-one functions from S onto S. H acts on S if there is a homomorphism, f, from H into G_S. If this is the case, the orbit of a fixed s in S is the set of all elements of S of the form $f(h)(s)$, for h in H. The stabilizer of a fixed s in S is the set of all h in H such that $f(h)(s)=s$. In the case when H is finite, we have the following relationship: the order of the orbit of s is the quotient of the order of H and the order of the stabilizer of s. The proof of this theorem is obtained by noting that the function which associates with each left coset, hH_s, of the stabilizer of s in H the element $f(h)(s)$, in the orbit of s, is both one-to-one and onto.

We denote by F_{nm} the vector space of $n\times m$ matrices and by G_n the group of $n\times n$ nonsingular matrices over this field.

THEOREM 1. *The order of* G_n *is* $\prod_{j=0}^{n-1}(q^n-q^j)$.

Proof. Given s $n\times 1$ vectors, there are q^s linear combinations of them and hence q^n-q^s $n\times 1$ vectors linearly independent of them. ∎

THEOREM 2. *The number of* $n\times m$ *matrices of rank* k *is*

$$\frac{\prod_{j=0}^{k-1}(q^n-q^j)\prod_{j=0}^{k-1}(q^m-q^j)}{\prod_{j=0}^{k-1}(q^k-q^j)}.$$

* From AMERICAN MATHEMATICAL MONTHLY, vol. 73 (1966), pp. 639–641.

Proof. A matrix A has rank k if and only if A is equivalent to the $n\times m$ matrix

$$H=\begin{bmatrix} I_k & 0 \\ 0 & 0 \end{bmatrix}$$

where I_k is the $k\times k$ identity.

Let $G_n\oplus G_m$, the direct sum of G_n and G_m, act on M, the set of $n\times m$ matrices of rank k, by $(g_n, g_m)A\to g_nAg_m^{-1}$ for $g_n\in G_n$, $g_m\in G_m$, and $A\in M$. The orbit of H under this action is all of M. Hence, to find the order of M we need only find the order of S, the stabilizer of H.

Suppose $U=(u_{ij})$ and $V=(v_{ij})$ and $(U, V)\in S$. Then $UH=HV$ and consequently we find

$$U=\begin{bmatrix} W & \mathrm{I} \\ 0 & \mathrm{II} \end{bmatrix} \quad \text{and} \quad V=\begin{bmatrix} W & 0 \\ \mathrm{III} & \mathrm{IV} \end{bmatrix}$$

where W is $k\times k$ and W, I, II, III, IV are arbitrary, provided only that U and V are nonsingular. By Theorem 1, W may be formed in $\prod_{j=0}^{k-1}(q^k-q^j)$ ways.

Once W is formed it follows that the remainder of U may be formed in $\prod_{j=k}^{n-1}(q^n-q^j)$ ways so that U is nonsingular. Similarly, the remainder of V may be formed in $\prod_{j=k}^{m-1}(q^m-q^j)$ ways making V nonsingular. Therefore, the order of S is

$$\prod_{j=0}^{k-1}(q^k-q^j)\prod_{j=k}^{n-1}(q^n-q^j)\prod_{j=k}^{m-1}(q^m-q^j)$$

and the conclusion follows from Theorem 1. ∎

The proof of the next theorem is immediate and is therefore omitted.

THEOREM 3. *Let v be a nonzero fixed element of F_{m1}. Let $S_0=\{T\in F_{nm}\mid Tv=0\}$. Then F_{nm}/S_0 is isomorphic to F_{n1} as groups; hence, if $w\in F_{n1}$, then the number of $T\in F_{nm}$ such that $Tv=w$ is $q^{n(m-1)}$.*

THEOREM 4. *Let v be a fixed nonzero element of F_{n1}. The number of $T\in G_n$ such that $Tv=v$ is $\prod_{k=1}^{n-1}(q^n-q^k)$; hence, each nonzero v in F_{n1} induces a decomposition of G_n into q^n-1 disjoint subsets, $S_w=\{T\mid Tv=w\}$, each of order*

$$\prod_{k=1}^{n-1}(q^n-q^k),$$

where w ranges over the nonzero elements of F_{n1}.

Proof. Let G_n act on the nonzero elements of F_{n1} by $T\to Tv$. Then the orbit of v is all of these elements and therefore the order of the stabilizer of v is the order of G_n divided by the order of F_{n1}. ∎

Let T be an $n\times m$ matrix. We say that the ith row of T has character k if $\sum_{j=1}^{m} t_{ij}=k$. If each row has character k, then T is a matrix of character k. Note that T has character k if and only if $T\theta=k\theta$, where θ is the column vector of all 1's.

THEOREM 5. (1) *The number of* $n \times m$ *matrices with* n_k *rows of character* k, *where* $\sum_{k=0}^{q-1} n_k = n$, *is* $[n!/n_0! \cdots n_{q-1}!]q^{n(m-1)}$.

(2) *The number of nonsingular matrices with* n_k *rows of character* k, *where* $\sum_{k=0}^{q-1} n_k = n$, *is* $[n!/n_0! \cdots n_{q-1}!]\prod_{j=1}^{n-1} (q^n - q^j)$.

Proof. There are $[n!/n_0! \cdots n_{q-1}!]$ elements of F_{n1} with exactly n_k k's. Conclusions (1) and (2) now follow from Theorems 3 and 4, respectively. ∎

THEOREM 6. (1) *If* T *is nonsingular and has character* k, *then* T^{-1} *has character* k^{-1} (*note that it is impossible that* $k=0$).

(2) *If* A *has character* nk *and* B *has character* mk, *then* $A+B$ *has character* $(n+m)k$.

Hence, the nonsingular matrices of character k^t, t *an integer, form a subgroup of* G_n *which by Theorem* 5 *has order*

$$\epsilon_k \prod_{j=1}^{n-1} (q^n - q^j),$$

where ϵ_k *is the number of distinct powers of* k (mod q); *the matrices of character* sk, s *an integer, form a subgroup of* F_{nm} *of order* $\eta_k q^{n(m-1)}$ *where* η_k *is the number of distinct multiples of* k (mod q).

Proof. The proof is immediate by noting the effect of multiplying θ by the prescribed matrices. ∎

This paper was written while the authors were employed for the summer at the Office of Research and Development, United States Patent Office.

References

1. G. Birkhoff and S. MacLane, A Survey of Modern Algebra, rev. ed., Macmillan, New York, 1953.

2. C. Chevalley, Fundamental Concepts of Algebra, Academic Press, New York, 1957.

3. L. E. Dickson, Linear Groups, Teubner, Leipzig, 1901.

BIBLIOGRAPHIC ENTRY: LINEAR ALGEBRA OVER FINITE FIELDS

1. J. H. Hodges, The matrix equation $AX=B$ in a finite field, AMERICAN MATHEMATICAL MONTHLY, 1956, 243–244.

Computes the number of solutions of this equation.

(f)

APPLICATIONS

LINEAR RECURRENCE RELATIONS*†

J. L. BRENNER, State College of Washington

1. Introduction. The relation

$$u_{n+1} = u_n + u_{n-1},$$

which is satisfied by the Fibonacci sequence $u_{-1}=0$, $u_0=1$, $u_1=1$, $u_2=2$, $u_3=3$, $u_4=5, \cdots$, can be generalized. For example, the sequence of Lucas [5] can be defined similarly from the relation

$$u_{n+1} = Pu_n - Qu_{n-1}, \tag{1}$$

where u_{-1}, u_0 have the same values as before, and $u_1, u_2, \cdots$ are calculated in succession from (1). Another linear recurrence relation is discussed in [6] in connection with a divisibility question. The most general linear recurrence relation (over integers) can be written in the form

$$u_{n+1} = \alpha_0 u_n + \alpha_1 u_{n-1} + \cdots + \alpha_{k-1} u_{n-k+1} + a, \tag{2}$$

where k is positive, a and α_i are integers ($i=0, \cdots, k-1$), and α_{k-1} is not 0. The integer k is called the degree of the relation, and is fixed henceforth. If (initial) integral values $u_{-1}, u_0, \cdots, u_{k-2}$ are prescribed, the values $u_{k-1}, u_k, \cdots$ can be computed in succession from (2). If α_{k-1} happens to be 1 or -1, the sequence u_k can be computed for negative values of k also, by solving (2) for u_{n-k+1}, as in [6]. Linear recurrence relations and the sequences associated with them are useful in solving certain problems concerning primality or divisibility; Lehmer [4] gave a test for the primality of Mersenne numbers 2^p-1 (p, a prime) which has been used to discover a prime with several hundred digits; this test was based on properties of the Lucas sequence. The relation $u_n | u_{nr}$ holds generally for Lucas sequences. An illuminating account of linear recurring sequences is given in [3].

Two interesting (known) properties of recurring sequences are studied in this article by the use of matrix methods. Firstly, the residues mod p of the u_n must be periodic; so that secondly, if u_{-1} is 0, then some u_n must be divisible by

* From AMERICAN MATHEMATICAL MONTHLY, vol. 61 (1954), pp. 171–173.

* The writing of this article was sponsored by the Office of Ordnance Research, U. S. Army, under contract DA-04-200-ORD-113. This is technical report 7 under that contract.

p, if p is a prime not dividing α_{k-1}. General theorems about matrices lead quickly to fairly precise results concerning the size of this period. On the other hand, it would be possible to establish theorems concerning the period of a matrix belonging to a certain class of matrices by starting with results concerning recurrence relations which have already been established in other ways.

2. The matrix formula. Let A stand for the $(k+1)\times(k+1)$ matrix

$$\begin{pmatrix} 1, & 0, & 0, & 0, & \cdots & 0, & a \\ 0, & 0, & 0, & 0, & \cdots, & 0, & \alpha_{k-1} \\ 0, & 1, & 0, & 0, & \cdots, & 0, & \alpha_{k-2} \\ 0, & 0, & 1, & 0, & \cdots, & 0, & \alpha_{k-3} \\ \cdot & \cdot & \cdot & \cdot & \cdot & \cdot & \cdot \\ \cdot & \cdot & \cdot & \cdot & \cdot & \cdot & \cdot \\ \cdot & \cdot & \cdot & \cdot & \cdot & \cdot & \cdot \\ 0, & 0, & 0, & 0, & \cdots, & 1, & \alpha_0 \end{pmatrix}.$$

In case k is 1, A is the matrix $\left(\begin{smallmatrix} 1, & a \\ 0, & \alpha_0 \end{smallmatrix}\right)$. Let B stand for the $(k+2)\times(k+2)$ matrix

$$\begin{pmatrix} 0, & 1, \; u_{-1}, \; \cdots, \; u_{k-2} \\ 0 & \\ 0 & A \\ \vdots & \\ 0 & \end{pmatrix}.$$

From (2), the formula

$$B^{n+1} = \begin{pmatrix} 0, & 1, \; u_{n-1}, \; \cdots, \; u_{n+k-2} \\ 0 & \\ 0 & A^{n+1} \\ \vdots & \\ 0 & \end{pmatrix} \qquad (n > -1) \tag{3}$$

is easily proved by induction.

3. Periodicity. Apparition of primes. Since the vector $(u_{n-1}, u_n, \cdots, u_{n+k-2})$ can take at most p^k values which are distinct modulo the prime p, the following theorem is true.

THEOREM 1. (Lagrange) *Modulo an arbitrary prime, the values u_n given by the linear recurrence relation* (2) *are periodic with period not exceeding p^k.*

Not all periods less than $1+p^k$ are actually possible. To show this, it is convenient to use some information from the theory of automorphisms of abelian groups.

Let G be an elementary abelian group of order p^{k+1}, that is, an abelian group generated by $k+1$ independent generators, each of order p. [G is isomorphic to the additive group of $(k+1)$-vectors $(\beta_0, \beta_1, \cdots, \beta_k)$, the components of the vectors being integers mod p.] An automorphism of G can be described by mapping the first generator $(1, 0, \cdots, 0)$ onto any of the $p^{k+1}-1$ elements which have order p; by mapping the second generator $(0, 1, \cdots, 0)$ onto any of the $p^{k+1}-p$ elements independent of this element; *etc.* Thus the order of the group of automorphisms of G is $(p^{k+1}-1)(p^{k+1}-p) \cdots (p^{k+1}-p^k)$, which is later denoted by $\phi(p, k)$.

COROLLARY. *The order of any automorphism of G is a divisor of* $\phi(p, k)$.

Actually, more than this is known [1].

THEOREM. (Burnside) *The group of automorphisms of G is isomorphic to the group of* $(k+1)\times(k+1)$ *nonsingular matrices* mod p [2, p. 242].

COROLLARY. *The order of any such matrix is a divisor of* $\phi(p, k)$.

Thus if A mod p is non-singular, its order is a divisor of this number. The condition for this is that α_{k-1} shall not be divisible by p. Here, the order of A mod p means the smallest positive exponent r such that the relation $A^r \equiv I$ (mod p) holds. When A mod p is non-singular, such an exponent must exist.

THEOREM. *Let r be the order of A* mod p. *Then the relation* $B^{r+1}\equiv B$ (mod p) *holds.*

This theorem follows from the relation $B^{r+1}=BB^r$ together with the displayed matrix equation (3).

COROLLARY. *Modulo an arbitrary prime p which does not divide* α_{k-1}, *the values* u_n *given by the linear recurrence relation* (2) *are periodic with period dividing* $(p^{k+1}-1)(p^{k+1}-p) \cdots (p^{k+1}-p^k)$; *indeed the period divides the order r of A* mod p.

COROLLARY. *If any one of* $u_{-1}, u_0, \cdots, u_{k-2}$ *is* 0, *then every prime p, not a divisor of* α_{k-1}, *has a finite rank of apparition; that is, an infinite number of terms* u_n *are divisible by the fixed prime p.*

The methods of this article can be generalized in various directions, which will be noted by the sophisticated reader.

References

1. Brenner, J. The linear homogeneous group. Annals of Math. (2) 39, 1938, pp. 472–493.
2. Burnside, W. The Theory of Groups of Finite Order. Cambridge, 1897.
3. Dickson, L. E. History of the Theory of Numbers, I, ch. xvii.
4. Lehmer, D. H. An extended theory of Lucas' functions. Annals of Math. (2) 31, 1930, pp. 419–448.
5. Lucas, E. Amer. J. Math. 1, 1878, pp. 184–220.
6. Mills, W. H. A system of quadratic diophantine equations. Pacific J. Math. 3, 1953, pp. 209–220.

ON AN APPLICATION OF THE VANDERMONDE DETERMINANT*

W. V. PARKER, Alabama Polytechnic Institute

In this MONTHLY for December 1953, Harley Flanders gave a solution of the following problem. Let $x_1, x_2, \cdots, x_n$ be n numbers such that $\sum x_i = \sum x_i^2 = \cdots = \sum x_i^n = 0$; to prove that $x_1 = x_2 = \cdots = x_n = 0$. He used the Vandermonde determinant and an induction process.

The following solution, which uses the Vandermonde determinant but avoids the induction process, may be of interest. Let the solution be such that n_i of the x's are equal to $a_i \neq 0$, $i = 1, 2, \cdots, k$, and $a_i \neq a_j$, $i \neq j$, and the remaining x's are zero. Then we have $\sum n_i a_i = \sum n_i a_i^2 = \cdots = \sum n_i a_i^n = 0$. The first k of these equations in $n_1, n_2, \cdots, n_k$ constitute a homogeneous system whose determinant is $a_1 \cdot a_2 \cdot \cdots \cdot a_k \cdot V(a_1, a_2, \cdots, a_k)$ where $V(a_1, a_2, \cdots, a_k)$ is a nonvanishing Vandermonde determinant. Hence, $n_i = 0$, $i = 1, 2, \cdots, k$, and $x_1 = x_2 = \cdots = x_n = 0$ is the only solution.

INTEGRATION BY MATRIX INVERSION†

WILLIAM SWARTZ, Montana State College

The integration of several functions using differential operators was considered by Osborn.‡ The integration of these and certain other functions by matrix inversion can furnish an application of several aspects of matrix theory of interest to the student of matrix algebra.

Let V be the vector space of differentiable functions. Let the n-tuple f be a basis spanning a subspace S of V which is closed under differentiation. Then differentiation comprises a linear transformation T of S into itself. If the matrix A represents T relative to f, then when A is nonsingular the elements of fA^{-1} yield antiderivatives of the elements of f.

To integrate $e^{ax} \sin bx$ and $e^{ax} \cos bx$ consider $f = (e^{ax} \sin bx, e^{ax} \cos bx)$. Then

$$fT = (ae^{ax} \sin bx + be^{ax} \cos bx, \ -be^{ax} \sin bx + ae^{ax} \cos bx)$$

and

$$A = \begin{pmatrix} a & -b \\ b & a \end{pmatrix}.$$

Furthermore

$$A^{-1} = \frac{1}{a^2 + b^2} \begin{pmatrix} a & b \\ -b & a \end{pmatrix}$$

* From AMERICAN MATHEMATICAL MONTHLY, vol. 61 (1954), p. 639.

† From AMERICAN MATHEMATICAL MONTHLY, vol. 65 (1958), pp. 282–283.

‡ Roger Osborn, Note on integration by operators, this MONTHLY, vol. 64, 1957, p. 431.

and then

$$fA^{-1} = \left(\frac{e^{ax}}{a^2+b^2}(a\sin bx - b\cos bx), \frac{e^{ax}}{a^2+b^2}(b\sin bx + a\cos bx)\right)$$

yields antiderivatives of the elements of f.

To derive the formula

$$\text{(1)} \qquad \int x^n e^x dx = e^x[x^n - nx^{n-1} + n(n-1)x^{n-2} - \cdots + (-1)^n n!]$$

for positive integers n consider $f = (e^x, xe^x, x^2e^x, \cdots, x^ne^x)$. Then

$$fT = (e^x, e^x + xe^x, \cdots, nx^{n-1}e^x + x^ne^x)$$

and there follows the interesting matrix

$$A = \begin{pmatrix} 1 & 1 & 0 & \cdot & \cdot & \cdot & 0 & 0 & 0 \\ 0 & 1 & 2 & \cdot & \cdot & \cdot & 0 & 0 & 0 \\ 0 & 0 & 1 & \cdot & \cdot & \cdot & 0 & 0 & 0 \\ \cdot & \cdot & \cdot & \cdot & \cdot & \cdot & \cdot & \cdot & \cdot \\ \cdot & \cdot & \cdot & \cdot & \cdot & \cdot & \cdot & \cdot & \cdot \\ \cdot & \cdot & \cdot & \cdot & \cdot & \cdot & \cdot & \cdot & \cdot \\ 0 & 0 & 0 & \cdot & \cdot & \cdot & 1 & n-1 & 0 \\ 0 & 0 & 0 & \cdot & \cdot & \cdot & 0 & 1 & n \\ 0 & 0 & 0 & \cdot & \cdot & \cdot & 0 & 0 & 1 \end{pmatrix}$$

Since only an antiderivative of the last element of f is required, one is only interested in the last column of A^{-1}. Due to the peculiar form of A the inverse is easily deduced. One surmises that the last column of A^{-1} is the transpose of the row

$$\text{(2)} \qquad \left((-1)^n n!, (-1)^{n-1}n!, (-1)^{n-2}\frac{n!}{2!}, \cdots, n(n-1), -n, 1\right).$$

That this supposition is correct may be verified by induction on n. In this connection it is useful to consider the $(n+2)$ rowed matrix corresponding to A as a partitioned matrix containing A as a principal submatrix. Finally one notes that multiplication of (2) by f yields the required formula (1).

VARIABLE MATRIX SUBSTITUTION IN ALGEBRAIC CRYPTOGRAPHY*

JACK LEVINE, North Carolina State College

1. Introduction. The use of algebraic methods in cryptography is well-known through two important papers by Hill [1], [2]. Briefly, the basic idea can be formulated in the following way. Consider the system of simultaneous congruences

$$y_i = \sum_{j=1}^{n} a_{ij}x_j \pmod{26}, \qquad i = 1, \cdots, n, \tag{1.1}$$

where the constants a_{ij} are chosen so that the determinant $|a_{ij}|$ is *prime* to 26. By means of (1.1) the set of n variables $(x_1, \cdots, x_n)$ is transformed to the set $(y_1, \cdots, y_n)$ and, conversely, the set $(y_1, \cdots, y_n)$ will be transformed to the unique set $(x_1, \cdots, x_n)$ by means of the inverse transformation which exists by the assumption on $|a_{ij}|$.

To each of the 26 letters of the alphabet we associate an integer from the set 0, 1, $\cdots$, 25, so that no two letters correspond to the same integer. For simplicity we illustrate with the correspondence (used throughout this paper)

	A	B	C	D	E	F	G	H	I	J	K	L	M	N	O	P	Q	R	S	T	U	V	W	X	Y	Z
(1.2)	1	2	3	4	5	6	7	8	9	10	11	12	13	14	15	16	17	18	19	20	21	22	23	24	25	0

Now to encipher a message, or plain text, by means of (1.1), first replace each letter of the text by means of its numerical equivalent, using for illustration, (1.2). Then divide the resulting sequence of numbers into groups containing n numbers each. Call these

$$p_{11}p_{12} \cdots p_{1n} \quad p_{21}p_{22} \cdots p_{2n} \quad \cdots \quad p_{i1}p_{i2} \cdots p_{in} \quad \cdots. \tag{1.3}$$

Each group of (1.3) is then used in (1.1) for $x_1 \cdots x_n$, and the transformed set $y_1 \cdots y_n$ calculated. Call the sequence of these sets

$$c_{11}c_{12} \cdots c_{1n} \quad c_{21}c_{22} \cdots c_{2n} \quad \cdots \quad c_{i1}c_{i2} \cdots c_{in} \quad \cdots. \tag{1.4}$$

Convert the numbers of (1.4) into their letter equivalents by (1.2). These letters will constitute the cipher text corresponding to the given plain text.

The decipherment is accomplished by means of the inverse to (1.1).

As a concrete example, select $n=3$, and (1.1) as

$$\begin{aligned} y_1 &\equiv x_1 + 2x_2 + 3x_3 \\ y_2 &\equiv 2x_1 + 5x_2 + 6x_3 \\ y_3 &\equiv x_1 + 2x_2 + 4x_3 \end{aligned} \tag{1.5}$$

To encipher the text CRYPTOGRAPHIC, divide into sets of three letters, add-

* From AMERICAN MATHEMATICAL MONTHLY, vol. 65 (1958), pp. 170–179.

ing, say, XX to complete the last set:

C	R	Y	P	T	O	G	R	A	P	H	I	C	X	X
3	18	25	16	20	15	7	18	1	16	8	9	3	24	24

The sequence (1.3) is here 3 18 25 16 20 15 $\cdots$. Substitute the first set 3 18 25 $(=p_{11}p_{12}p_{13})$ in (1.5) for $x_1x_2x_3$ to give

$$y_1 \equiv 3 + 36 + 75 = 114 \equiv 10 = \text{J},$$
$$y_2 \equiv 6 + 90 + 150 = 246 \equiv 12 = \text{L},$$
$$y_3 \equiv 3 + 36 + 100 = 139 \equiv 9 = \text{I},$$

Here the first cipher sequence $c_{11}c_{12}c_{13}$ of (1.4) is 10 12 9, which converted to letters by (1.2) gives JLI as shown.

The complete encipherment proceeds as above, and produces

JLI WNL TFU GVP SJQ

To decipher, obtain the inverse of (1.5),

$$\begin{aligned} x_1 &\equiv 8y_1 + 24y_2 + 23y_3 \\ x_2 &\equiv 24y_1 + y_2 \\ x_3 &\equiv 25y_1 + y_3 \end{aligned} \qquad (1.6)$$

(The congruences are of course taken mod 26, in which $25 = -1$, $24 = -2$, *etc.*) The reciprocal of a prime p, mod 26, is q, where $pq \equiv 1$ mod 26.

Now using JLI $= 10$ 12 9 as $y_1y_2y_3$ in (1.6) gives

$$x_1 = 80 + 288 + 207 = 575 \equiv 3 = \text{C}$$
$$x_2 \equiv 240 + 12 = 252 \equiv 18 = \text{R}$$
$$x_3 \equiv 250 + 9 = 259 = 25 = \text{Y}$$

or CRY, the first plain-text group. The rest of the plain text is found in like manner. (In actual practice we would use -1 for 25, -2 for 24, *etc.*, in (1.6).)

In Hill's papers the transformation (1.1) is generalized by the use of matrix coefficients, but the above is sufficient for our purpose.

(The author notes in passing that simultaneous equations were used by him for cryptographic purposes to a limited extent several years prior to the appearance of Hill's papers.)

2. Fixed substitution. The cryptographic method represented by (1.1) is known as a *fixed substitution system.* This means that any given plain-text group will always be replaced by the same cipher-text group. This is true because the coefficients a_{ij} remain fixed throughout the encipherment of a message.

From a cryptographic point of view there is a distinct advantage in using a *variable substitution* method, whereby the various appearances of a given plain-text group will be replaced by different cipher groups. It is our purpose to indicate some simple ways to accomplish this based on (1.1).

3. Variable substitution, first method. It is convenient to represent (1.1) as a matrix congruence

$$C \equiv AP \pmod{26}, \tag{3.1}$$

where matrices A, C, P are defined by

$$A = \begin{bmatrix} a_{11} & \cdots & a_{1n} \\ \cdot & & \cdot \\ a_{n1} & \cdots & a_{nn} \end{bmatrix}, \quad P = \begin{bmatrix} p_1 \\ \cdot \\ p_n \end{bmatrix}, \quad C = \begin{bmatrix} c_1 \\ \cdot \\ c_n \end{bmatrix} \tag{3.2}$$

and P, C are one-column matrices representing corresponding plain- and cipher-text groups.

Now in classical cryptography several variable substitution methods are well-known. These can be represented by the congruences

$$c_i \equiv p_i + k_i \pmod{26}, \qquad i = 1, \cdots, N, \tag{3.3}$$

where p_i is the numerical value of the ith plain-text letter according to some correspondence as (1.2), c_i is the numerical value of the corresponding cipher-text letter, N is number of letters in the message, and the sequence of numbers $k_1k_2 \cdots$ has one of the following properties:

(a) $k_1k_2 \cdots$ is a periodic sequence, say $k_1k_2 \cdots k_mk_1k_2 \cdots k_mk_1 \cdots$, where the numbers $k_1 \cdots k_m$ of the period are selected in any preassigned manner.

(b) The number k_i is chosen by the relation

$$k_i = c_{i-1} \tag{3.4}$$

so

$$c_i \equiv p_i + c_{i-1} \qquad (c_0 \text{ chosen in advance}). \tag{3.5}$$

(c) The number k_i is chosen by the relation

$$k_i = p_{i-1}, \tag{3.6}$$

so

$$c_i \equiv p_i + p_{i-1} \qquad (p_0 \text{ chosen in advance}). \tag{3.7}$$

Note that in each of the above three methods p_i is uniquely determined in the decipherment process. This is, of course, a prime requisite in any cryptographic system.

In matrix form (3.3) can be written as

$$C \equiv P + K \tag{3.8}$$

where the indicated matrices are each of one row and one column, since (3.3) represents encipherment one letter at a time.

Now to obtain a variable substitution analogous to (3.1) we generalize (3.8) to

$$C \equiv AP + BK \pmod{26}, \tag{3.9}$$

where $A = [a_{ij}]$, $B = [b_{ij}]$ are $n \times n$ matrices with fixed elements (and $|A|$ prime to 26). Matrices C and P are as given in (3.2), and K is a one-column matrix,

$$K = \begin{bmatrix} k_1 \\ \cdot \\ \cdot \\ k_n \end{bmatrix}.$$

Corresponding to the three cases (a), (b), (c) above for choosing the k_i, we have:

(a′) Define matrices

$$C_i = \begin{bmatrix} c_{i1} \\ \cdot \\ \cdot \\ c_{in} \end{bmatrix}, \quad P_i = \begin{bmatrix} p_{i1} \\ \cdot \\ \cdot \\ p_{in} \end{bmatrix}, \quad K_i = \begin{bmatrix} k_{i1} \\ \cdot \\ \cdot \\ k_{in} \end{bmatrix}, \tag{3.10}$$

using (1.3), (1.4), and

$$K_i = K_{i+m}, \qquad i = 1, 2, \cdots, \tag{3.11}$$

where $K_1, \cdots, K_m$ are chosen in any preassigned manner.

Then

$$C_i \equiv AP_i + BK_i, \qquad i = 1, 2, \cdots, \tag{3.12}$$

from (3.9) gives the substitution. Also,

$$P_i \equiv A^{-1}C_i - A^{-1}BK_i. \tag{3.13}$$

(b′) In this case we choose $K_i = C_{i-1}$, so

$$C_i \equiv AP_i + BC_{i-1} \qquad (C_0 \text{ chosen in advance}).$$

(c′) Choose $K_i = P_{i-1}$, so

$$C_i \equiv AP_i + BP_{i-1} \qquad (P_0 \text{ chosen in advance}). \tag{3.14}$$

To obtain involutory transformations (in which a transformation and its inverse are identical) we have from (3.12), (3.13),

$$A = A^{-1}, \qquad B = -A^{-1}B = -AB, \tag{3.15}$$

and (3.13) becomes $P_i = AC_i + BK_i$. A solution of (3.15) is

$$A^2 = I, \qquad B = A - I \qquad (I = \text{identity matrix}).$$

To obtain A such that $A^2 = I$, a formula in [2] may be used,

$$(3.16)\qquad a_{ij} \equiv \delta_{ij} - \tau\lambda_i\lambda_j, \qquad \sigma\tau \equiv 2 \pmod{26}, \qquad \sigma \equiv \sum_1^n \lambda_i^2 \pmod{26},$$

σ must be prime to 26.

We illustrate case (c′) using

$$A = \begin{bmatrix} 1 & 2 & 3 \\ 2 & 5 & 6 \\ 1 & 2 & 4 \end{bmatrix}, \qquad B = \begin{bmatrix} 4 & 1 & 1 \\ 2 & 0 & 3 \\ 1 & 2 & 0 \end{bmatrix}, \qquad P_0 = \begin{bmatrix} 1 \\ 2 \\ 3 \end{bmatrix},$$

$$(3.17)\qquad \begin{bmatrix} c_{i1} \\ c_{i2} \\ c_{i3} \end{bmatrix} = \begin{bmatrix} 1 & 2 & 3 \\ 2 & 5 & 6 \\ 1 & 2 & 4 \end{bmatrix} \begin{bmatrix} p_{i1} \\ p_{i2} \\ p_{i3} \end{bmatrix} + \begin{bmatrix} 4 & 1 & 1 \\ 2 & 0 & 3 \\ 1 & 2 & 0 \end{bmatrix} \begin{bmatrix} p_{i-1,1} \\ p_{i-1,2} \\ p_{i-1,3} \end{bmatrix}.$$

To encipher CRYPTOGRAPHIC, we have

$$\begin{bmatrix} c_{11} \\ c_{12} \\ c_{13} \end{bmatrix} = \begin{bmatrix} 1 & 2 & 3 \\ 2 & 5 & 6 \\ 1 & 2 & 4 \end{bmatrix} \begin{bmatrix} 3 \\ 18 \\ 25 \end{bmatrix} + \begin{bmatrix} 4 & 1 & 1 \\ 2 & 0 & 3 \\ 1 & 2 & 0 \end{bmatrix} \begin{bmatrix} 1 \\ 2 \\ 3 \end{bmatrix} = \begin{bmatrix} 19 \\ 23 \\ 14 \end{bmatrix} = \begin{bmatrix} \text{S} \\ \text{W} \\ \text{N} \end{bmatrix},$$

$$\begin{bmatrix} c_{21} \\ c_{22} \\ c_{23} \end{bmatrix} = \begin{bmatrix} 1 & 2 & 3 \\ 2 & 5 & 6 \\ 1 & 2 & 4 \end{bmatrix} \begin{bmatrix} 16 \\ 20 \\ 15 \end{bmatrix} + \begin{bmatrix} 4 & 1 & 1 \\ 2 & 0 & 3 \\ 1 & 2 & 0 \end{bmatrix} \begin{bmatrix} 3 \\ 18 \\ 25 \end{bmatrix} = \begin{bmatrix} 0 \\ 17 \\ 25 \end{bmatrix} = \begin{bmatrix} \text{Z} \\ \text{Q} \\ \text{Y} \end{bmatrix},$$

$$\begin{bmatrix} c_{31} \\ c_{32} \\ c_{33} \end{bmatrix} = \begin{bmatrix} 1 & 2 & 3 \\ 2 & 5 & 6 \\ 1 & 2 & 4 \end{bmatrix} \begin{bmatrix} 7 \\ 18 \\ 1 \end{bmatrix} + \begin{bmatrix} 4 & 1 & 1 \\ 2 & 0 & 3 \\ 1 & 2 & 0 \end{bmatrix} \begin{bmatrix} 16 \\ 20 \\ 15 \end{bmatrix} = \begin{bmatrix} 15 \\ 5 \\ 25 \end{bmatrix} = \begin{bmatrix} \text{O} \\ \text{E} \\ \text{Y} \end{bmatrix}, \textit{ etc.}$$

The complete encipherment is SWN ZQY OEY BMG VQW, using CXX as the last plain group.

The decipherment can be obtained from the inverse to (3.17),

$$P_i = \begin{bmatrix} 8 & 24 & 23 \\ 24 & 1 & 0 \\ 25 & 0 & 1 \end{bmatrix} C_i + \begin{bmatrix} 1 & 24 & 24 \\ 6 & 2 & 25 \\ 3 & 25 & 1 \end{bmatrix} P_{i-1}.$$

4. Variable substitution, second method. We return to the basic relation (3.1) and attempt to replace the matrix A of fixed elements by a matrix with variable elements. A general situation is obtained if the elements a_{ij} of A be considered as polynomial functions of a set of parameters $t, u, v, \cdots$ in such a way that the determinant $|A|$ of A is *independent of the parameters* and is a prime number mod 26. The inverse A^{-1} of A will then exist for all parameter values, and hence $P = A^{-1}C$ can always be found. We consider one of the simpler cases here.

Any triangular matrix

$$T = \begin{bmatrix} t_{11} & 0 & \cdots & 0 \\ t_{21} & t_{22} & \cdots & \cdot \\ \cdot & \cdot & \cdots & \cdot \\ t_{n1} & t_{n2} & \cdots & t_{nn} \end{bmatrix}$$

with $t_{ij}(t, u, v \cdots)$, $(i \neq j)$, such that t_{ii} is a prime mod 26 will have for determinant $|T| = t_{11}t_{22} \cdots t_{nn}$, a prime mod 26. If T be transformed by elementary transformations leaving $|T|$ unchanged we can obtain a general matrix A of the desired property.

For example, using

$$T = \begin{bmatrix} 1 & 0 & 0 & 0 \\ t & 3 & 0 & 0 \\ 2t+1 & 2 & 5 & 0 \\ 1 & t & t+1 & 7 \end{bmatrix}, \qquad |T| \equiv 1, \text{ mod } 26,$$

we can transform T to

$$A(t) \equiv \begin{bmatrix} 1 & 1 & t & 2t \\ t & t+3 & t & 2t^2 \\ 2t+1 & 2t+3 & 2t^2+t+5 & 2t^2+2t \\ 1 & t+1 & 2t+1 & 2t+7 \end{bmatrix}, \qquad |A(t)| \equiv 1, \text{ mod } 26.$$

Place $C = A(t)P$. For each $P = P_i$ give t a value k_i determined in some preassigned manner,

$$C_i = A(k_i)P_i, \qquad P_i = A^{-1}(k_i)C_i. \tag{4.1}$$

Any of the methods (a′), (b′), (c′) can be used, taking for example,

$$k_i = c_{i-1,1} \text{ or } k_i = p_{i-1,1}, \text{ or } k_i = p_{i-1,1} + p_{i-1,2}, \textit{ etc.},$$

in the latter two cases.

One disadvantage of this procedure to obtain $A(t)$ is that the elements of $A^{-1}(t)$ will in general be high degree polynomials in the parameters, thus causing computational difficulties.

One way to avoid this difficulty is to assume $A(t)$ is linear in t and impose the condition that $A^{-1}(t)$ is likewise. Thus, place

$$A(t) = G + tH, \tag{4.2}$$

with G, H constant element matrices, and $|G|$ a prime mod 26. It is easily shown $A^{-1}(t)$ will be linear in t if $H = XG$, $X^2 = 0$. Then

$$A^{-1}(t) = G^{-1} - tG^{-1}X. \tag{4.3}$$

To obtain a general matrix X satisfying $X^2=0$, define N by

$$(4.4) \qquad N = \begin{bmatrix} 0 & n_1 & & & & & & & & \\ 0 & 0 & & & & & & & & \\ & & 0 & n_2 & & & & & & \\ & & 0 & 0 & & & & & & \\ & & & & \ddots & & & & & \\ & & & & & 0 & n_q & & & \\ & & & & & 0 & 0 & & & \\ & & & & & & & 0 & & \\ & & & & & & & & \ddots & \\ & & & & & & & & & 0 \end{bmatrix}, \qquad q \leqq [n/2].$$

N consisting of all zeros except $n_1, n_2, \cdots, n_q$ placed immediately to the right of the main diagonal terms in alternate rows as shown. The n_i are arbitrary constants. It is evident that $N^2=0$. X is now defined by

$$(4.5) \qquad X = QNQ^{-1},$$

Q being an arbitrary constant-term matrix with an inverse. From (4.5), $X^2=0$.

From (4.2) we then define $A(t)$ by

$$(4.6) \qquad A(t) = G + tXG = G + tQNQ^{-1}G$$

$A^{-1}(t)$ being given by (4.3).

Example. Take

$$N = \begin{bmatrix} 0 & 1 & 0 \\ 0 & 0 & 0 \\ 0 & 0 & 0 \end{bmatrix}, \qquad Q = \begin{bmatrix} 17 & 6 & 12 \\ 6 & 13 & 24 \\ 12 & 24 & 23 \end{bmatrix},$$

$$X = \begin{bmatrix} 24 & 13 & 18 \\ 10 & 0 & 14 \\ 20 & 0 & 2 \end{bmatrix}, \qquad G = \begin{bmatrix} 25 & 2 & 16 \\ 2 & 25 & 10 \\ 16 & 10 & 3 \end{bmatrix} (=G^{-1}),$$

$$XG = \begin{bmatrix} 4 & 7 & 22 \\ 6 & 4 & 20 \\ 12 & 8 & 14 \end{bmatrix}, \qquad G^{-1}X = \begin{bmatrix} 4 & 13 & 16 \\ 4 & 0 & 16 \\ 24 & 0 & 18 \end{bmatrix},$$

$$A(t) = \begin{bmatrix} 25 & 2 & 16 \\ 2 & 25 & 10 \\ 16 & 10 & 3 \end{bmatrix} + t \begin{bmatrix} 4 & 7 & 22 \\ 6 & 4 & 20 \\ 12 & 8 & 14 \end{bmatrix},$$

$$A^{-1}(t) = \begin{bmatrix} 25 & 2 & 16 \\ 2 & 25 & 10 \\ 16 & 10 & 3 \end{bmatrix} - t \begin{bmatrix} 4 & 13 & 16 \\ 4 & 0 & 16 \\ 24 & 0 & 18 \end{bmatrix}.$$

Using these in (4.1) gives

$$C_i = \begin{bmatrix} 25 + 4k_i & 2 + 7k_i & 16 + 22k_i \\ 2 + 6k_i & 25 + 4k_i & 10 + 20k_i \\ 16 + 12k_i & 10 + 8k_i & 3 + 14k_i \end{bmatrix} P_i, \tag{4.7}$$

$$P_i = \begin{bmatrix} 25 + 22k_i & 2 + 13k_i & 16 + 10k_i \\ 2 + 22k_i & 25 & 10 + 10k_i \\ 16 + 2k_i & 10 & 3 + 8k_i \end{bmatrix} C_i. \tag{4.8}$$

Take $k_i = p_{i-1,1} + p_{i-1,2} + p_{i-1,3}$, $(k_1 = 1)$, and encipher CRYPTOGRAPHIC(XX),

$$C_1 = \begin{bmatrix} 3 & 9 & 12 \\ 8 & 3 & 4 \\ 2 & 18 & 17 \end{bmatrix} \begin{bmatrix} 3 \\ 18 \\ 25 \end{bmatrix} = \begin{bmatrix} 3 \\ 22 \\ 1 \end{bmatrix} = \begin{bmatrix} \text{C} \\ \text{V} \\ \text{A} \end{bmatrix}, \qquad k_2 = 3 + 18 + 25 \equiv 20,$$

$$C_2 = \begin{bmatrix} 1 & 12 & 14 \\ 18 & 1 & 20 \\ 22 & 14 & 23 \end{bmatrix} \begin{bmatrix} 16 \\ 20 \\ 15 \end{bmatrix} = \begin{bmatrix} 24 \\ 10 \\ 15 \end{bmatrix} = \begin{bmatrix} \text{X} \\ \text{J} \\ \text{O} \end{bmatrix}, \qquad k_3 = 16 + 20 + 15 \equiv 25,$$

$$C_3 = \begin{bmatrix} 21 & 21 & 20 \\ 22 & 21 & 16 \\ 4 & 2 & 15 \end{bmatrix} \begin{bmatrix} 7 \\ 18 \\ 1 \end{bmatrix} = \begin{bmatrix} 25 \\ 2 \\ 1 \end{bmatrix} = \begin{bmatrix} \text{Y} \\ \text{B} \\ \text{A} \end{bmatrix}, \qquad k_4 = 7 + 18 + 1 \equiv 0,$$

$$C_4 = \begin{bmatrix} 25 & 2 & 16 \\ 2 & 25 & 10 \\ 16 & 10 & 3 \end{bmatrix} \begin{bmatrix} 16 \\ 8 \\ 9 \end{bmatrix} = \begin{bmatrix} 14 \\ 10 \\ 25 \end{bmatrix} = \begin{bmatrix} \text{N} \\ \text{J} \\ \text{Y} \end{bmatrix}, \qquad k_5 = 16 + 8 + 9 \equiv 7,$$

$$C_5 = \begin{bmatrix} 1 & 25 & 14 \\ 18 & 1 & 20 \\ 22 & 14 & 23 \end{bmatrix} \begin{bmatrix} 3 \\ 24 \\ 24 \end{bmatrix} = \begin{bmatrix} 3 \\ 12 \\ 18 \end{bmatrix} \quad \begin{bmatrix} \text{C} \\ \text{L} \\ \text{R} \end{bmatrix}.$$

The cipher-text is thus CVA XJO YBA NJY CLR.
To decipher, use (4.8),

$$P_1 = \begin{bmatrix} 21 & 15 & 0 \\ 24 & 25 & 20 \\ 18 & 10 & 11 \end{bmatrix} \begin{bmatrix} 3 \\ 22 \\ 1 \end{bmatrix} = \begin{bmatrix} 3 \\ 18 \\ 25 \end{bmatrix} = \begin{bmatrix} \text{C} \\ \text{R} \\ \text{Y} \end{bmatrix}, \ \textit{etc.}$$

There is an advantage in using an involutory transformation (4.1), *i.e.*, one such that $A(t)=A^{-1}(t)$. To obtain this we require from (4.3), (4.6),

$$G = G^{-1}, \qquad XG = -GX. \tag{4.9}$$

Define matrix J by

$$\begin{bmatrix} 1 & j_1 & & & & & & & & \\ 0 & -1 & & & & & & & & \\ & & 1 & j_2 & & & & & & \\ & & 0 & -1 & & & & & & \\ & & & & \ddots & & & & & \\ & & & & & 1 & j_q & & & \\ & & & & & 0 & -1 & & & \\ & & & & & & & a_1 & & \\ & & & & & & & & \ddots & \\ & & & & & & & & & a_r \end{bmatrix} = J, \tag{4.10}$$

constants $j_1, \cdots, j_q$ arbitrary, and $a_1, \cdots, a_r$ all $=\pm 1$, $(r+2q=n)$.
Then by (4.4), (4.10),

$$J^2 = I, \qquad JN = -NJ. \tag{4.11}$$

Place

$$G = QJQ^{-1}, \qquad X = QNQ^{-1}, \tag{4.12}$$

giving

$$G^2 = I, \qquad XG = -GX, \qquad (X^2 = 0),$$

satisfying (4.9).

From (4.12), $XG=QNJQ^{-1}=-QNQ^{-1}=-X$, since direct calculation shows $NJ=-N$. Hence (4.6) gives

$$A(t) = G - tX = A^{-1}(t), \tag{4.13}$$

which can also be expressed as

$$A(t) = Q(J - tN)Q^{-1} = A^{-1}(t). \tag{4.14}$$

Example. Use N, Q, X of the previous example, and

$$J = \begin{bmatrix} 1 & 1 & 0 \\ 0 & -1 & 0 \\ 0 & 0 & -1 \end{bmatrix}.$$

By (4.12), (4.13)

$$G = \begin{bmatrix} 3 & 9 & 10 \\ 6 & 19 & 2 \\ 12 & 14 & 3 \end{bmatrix}, \qquad A(t) = \begin{bmatrix} 3 + 2t & 9 + 13t & 10 + 8t \\ 6 - 10t & 19 & 2 + 12t \\ 12 + 6t & 14 & 3 + 24t \end{bmatrix} = A^{-1}(t).$$

In case $n=2$ it can be verified that

$$A(t) = A^{-1}(t) = \begin{bmatrix} a + bt & c + dt \\ e + ft & -(a + bt) \end{bmatrix}$$

if $a = bcd' \pm 1$, $e = -b^2c(d')^2 \mp 2bd'$, $f = -b^2d'$, and b, c, d are arbitrary (d prime mod 26), and $dd' \equiv 1$ mod 26.

The modulus 26 used throughout this paper is not essential. Other moduli can be used with suitable modifications where necessary.

References

1. L. S. Hill, Cryptography in an algebraic alphabet, this MONTHLY, vol. 36, 1929, pp. 306–312.
2. ———, Concerning certain linear transformation apparatus of cryptography, this MONTHLY, vol. 38, 1931, pp. 135–154.

THE SEARCH FOR HADAMARD MATRICES*

SOLOMON W. GOLOMB AND LEONARD D. BAUMERT,
Jet Propulsion Laboratory, California Institute of Technology

Consider the square array in Figure 1.

$$\begin{bmatrix} + & + & + & + \\ + & + & - & - \\ + & - & + & - \\ + & - & - & + \end{bmatrix}.$$

FIG. 1

Any two rows of this array "agree" in half their places, and "disagree" in the other half. The same phenomenon occurs for the columns of the array as well. Any such array of + and − signs, with an equal number of agreements and disagreements between any pair of rows, is called an "Hadamard matrix," after the French mathematician Jacques Hadamard [1]. In matrix notation, these arrays satisfy the relation $AA^T = kI$, where k is the order of the matrices. The problem is to discover all the possible Hadamard matrices, or at least to describe the *sizes* for which such arrays exist.

In 1933, R. E. A. C. Paley [2] described a variety of methods for constructing Hadamard matrices. Beyond the 1×1 case, "+", and the 2×2 case, "$\begin{smallmatrix}+&+\\+&-\end{smallmatrix}$",

* From AMERICAN MATHEMATICAL MONTHLY, vol. 70 (1963), pp. 12–17.

Paley shows that all such arrays must have a number of rows (and of columns) which is a multiple of 4. Up to 200×200, he was able to describe examples of $4t\times4t$ Hadamard matrices in all but six cases, the exceptional values of $4t$ being 92, 116, 156, 172, 184, 188. Using black and white squares, instead of plus and minus signs, Hadamard matrices of the first eight sizes (1×1, 2×2, 4×4, 8×8, 12×12, 16×16, 20×20, 24×24) are shown in Figure 2.

The six exceptional cases of Paley have never been proved *not* to exist. It was merely that Paley was unable to construct them. In fact, in 1944, John Williamson [3] succeeded in constructing an Hadamard matrix of size 172×172. Moreover, no one has shown that Hadamard matrices *ever* fail to exist for any size $4t\times4t$, and since 1933 several attempts have been made to prove or disprove the existence of the 92×92 array, as an indication of whether or not to believe the conjecture that an Hadamard matrix of size $4t\times4t$ exists for all values of $t=1, 2, 3, 4, 5, \cdots$.

There is a simple method of getting bigger Hadamard matrices from smaller ones. For example, to get an 8×8 array from a 2×2 array and a 4×4 array, one substitutes the 2×2 array instead of each $+$ sign in the 4×4 array, and one substitutes the *negative* of the 2×2 array instead of each $-$ sign in the 4×4 array, as shown in Figure 3.
This method of combining two smaller matrices to get a bigger matrix is called the "tensor product" of the two matrices. Letting H_n stand for the Hadamard matrix of order n, the illustrations in Figure 2 satisfy $H_4=H_2 * H_2$, $H_8=H_4 * H_2$, $H_{16}=H_8 * H_2$, and $H_{24}=H_{12} * H_2$.

Starting with the 2×2 Hadamard matrix, repeated tensor products give the 4×4 matrix, the 8×8 matrix, the 16×16 matrix, the 32×32 matrix, the 64×64 matrix, the 128×128 matrix, etc. Starting with the 12×12 matrix (which comes from a special construction that works because $4t-1=11$ is a *prime*), we also get a 24×24 matrix, a 48×48 matrix, a 96×96 matrix, a 192×192 matrix, etc., using repeated tensor products with the 2×2 matrix; and also the 144×144 matrix, the 1728×1728 matrix, etc., taking the tensor product of the 12×12 matrix with itself.

A variety of special circumstances were used for the other examples in Paley's catalog of Hadamard matrices. Thus, the 20×20 example arises from $4t-1=19$ being a prime, while the 28×28 example comes from $4t-1=3^3$ being a perfect *power* of a prime.

Several years ago, at the Jet Propulsion Laboratory of Caltech, we became interested in the problem of the optimum codes for communicating through space. The rows of an Hadamard matrix form an ideal set of "code words" for this purpose, because of the high degree of mutual distinguishability (as many disagreements as agreements) between any two such rows. A row of $+$'s and $-$'s can be regarded as a "square wave" of pulses and spaces, or of $+90°$ and $-90°$ phase shifts, for purposes of radio communication. We resolved to determine once and for all, whether or not the 92×92 Hadamard matrix exists.

There are 8464 entries in a 92×92 matrix, and hence 2^{8464} ways to form a

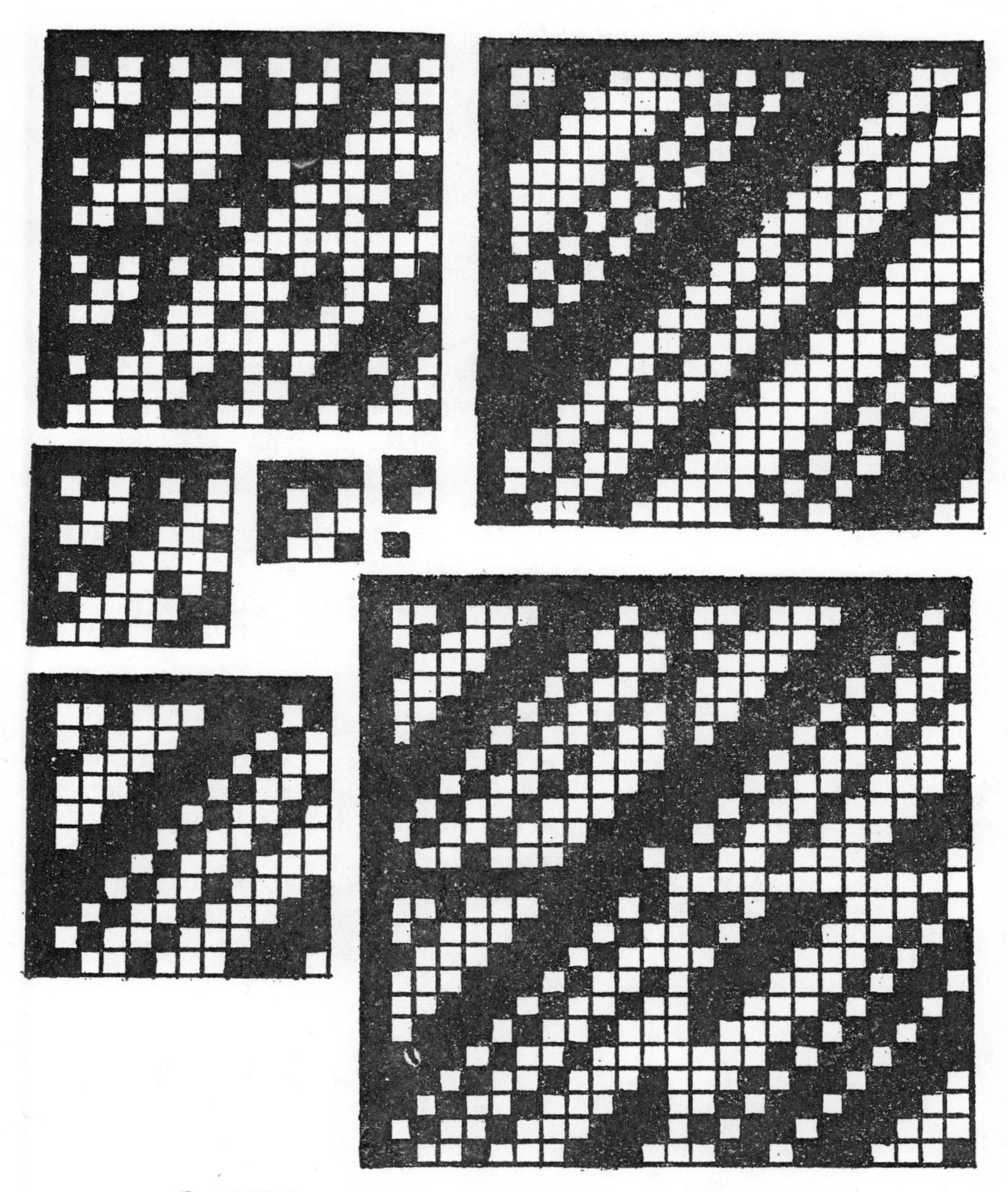

FIG. 2. Hadamard Matrices of orders 1, 2, 4, 8, 12, 16, 20, and 24.

$$\begin{bmatrix} + & + \\ + & - \end{bmatrix} * \begin{bmatrix} + & + & + & + \\ + & + & - & - \\ + & - & + & - \\ + & - & - & + \end{bmatrix} = \begin{bmatrix} \begin{bmatrix} + & + \\ + & - \end{bmatrix} & \begin{bmatrix} + & + \\ + & - \end{bmatrix} & \begin{bmatrix} + & + \\ + & - \end{bmatrix} & \begin{bmatrix} + & + \\ + & - \end{bmatrix} \\ \begin{bmatrix} + & + \\ + & - \end{bmatrix} & \begin{bmatrix} + & + \\ + & - \end{bmatrix} & \begin{bmatrix} - & - \\ - & + \end{bmatrix} & \begin{bmatrix} - & - \\ - & + \end{bmatrix} \\ \begin{bmatrix} + & + \\ + & - \end{bmatrix} & \begin{bmatrix} - & - \\ - & + \end{bmatrix} & \begin{bmatrix} + & + \\ + & - \end{bmatrix} & \begin{bmatrix} - & - \\ - & + \end{bmatrix} \\ \begin{bmatrix} + & + \\ + & - \end{bmatrix} & \begin{bmatrix} - & - \\ - & + \end{bmatrix} & \begin{bmatrix} - & - \\ - & + \end{bmatrix} & \begin{bmatrix} + & + \\ + & - \end{bmatrix} \end{bmatrix}$$

$$= \begin{bmatrix} + & + & + & + & + & + & + & + \\ + & - & + & - & + & - & + & - \\ + & + & + & + & - & - & - & - \\ + & - & + & - & - & + & - & + \\ + & + & - & - & + & + & - & - \\ + & - & - & + & + & - & - & + \\ + & + & - & - & - & - & + & + \\ + & - & - & + & - & + & + & - \end{bmatrix}$$

FIG. 3

92×92 matrix of $+$ and $-$ signs. Even if a computer could form and investigate these matrices at the rate of a million a second, the problem would run for too many eons. It was necessary to restrict the problem in some drastic fashion in order to succeed by computer search.

Professor Marshall Hall suggested that we run a program using the same method which enabled Williamson to find a 172×172 matrix. Williamson's idea for finding an Hadamard matrix of size $4t \times 4t$ was to look for a matrix of the form shown in Figure 4,

$$\begin{bmatrix} A & B & C & D \\ -B & A & -D & C \\ -C & D & A & -B \\ -D & -C & B & A \end{bmatrix}$$

FIG. 4

where each of A, B, C, and D is a $t \times t$ matrix. An example of such a structure is shown in Figure 5.

The other conditions which Williamson imposed were that each of A, B, C, and D be a cyclic symmetric matrix (each row being merely a cyclic permutation of the previous row); and that $A^2+B^2+C^2+D^2=4tI$, where A^2, B^2, C^2, D^2

$$\begin{bmatrix}
+ & + & + & + & - & - & + & - & - & + & - & - \\
+ & + & + & - & + & - & - & + & - & - & + & - \\
+ & + & + & - & - & + & - & - & + & - & - & + \\
- & + & + & + & + & + & - & + & + & + & - & - \\
+ & - & + & + & + & + & + & - & + & - & + & - \\
+ & + & - & + & + & + & + & + & - & - & - & + \\
- & + & + & + & - & - & + & + & + & - & + & + \\
+ & - & + & - & + & - & + & + & + & + & - & + \\
+ & + & - & - & - & + & + & + & + & + & + & - \\
- & + & + & - & + & + & + & - & - & + & + & + \\
+ & - & + & + & - & + & - & + & - & + & + & + \\
+ & + & - & + & + & - & - & - & + & + & + & +
\end{bmatrix}$$

FIG. 5

refer to the squares of the respective matrices as usually defined for matrices (*not* in the sense of tensor products), and I is the unit matrix. For purposes of these operations, $+$ is considered to be $+1$, and $-$ is considered to -1. For the case of the 12×12 matrix in Figure 5, Williamson's "four squares" relation takes the form:

$$A^2 + B^2 + C^2 + D^2$$

$$= \begin{bmatrix} + & + & + \\ + & + & + \\ + & + & + \end{bmatrix}^2 + \begin{bmatrix} + & - & - \\ - & + & - \\ - & - & + \end{bmatrix}^2 + \begin{bmatrix} + & - & - \\ - & + & - \\ - & - & + \end{bmatrix}^2 + \begin{bmatrix} + & - & - \\ - & + & - \\ - & - & + \end{bmatrix}^2$$

$$= \begin{bmatrix} 3 & 3 & 3 \\ 3 & 3 & 3 \\ 3 & 3 & 3 \end{bmatrix} + \begin{bmatrix} 3 & - & - \\ - & 3 & - \\ - & - & 3 \end{bmatrix} + \begin{bmatrix} 3 & - & - \\ - & 3 & - \\ - & - & 3 \end{bmatrix} + \begin{bmatrix} 3 & - & - \\ - & 3 & - \\ - & - & 3 \end{bmatrix}$$

$$= \begin{bmatrix} 12 & 0 & 0 \\ 0 & 12 & 0 \\ 0 & 0 & 12 \end{bmatrix} = 12 \begin{bmatrix} 1 & 0 & 0 \\ 0 & 1 & 0 \\ 0 & 0 & 1 \end{bmatrix} = 12I.$$

Williamson's construction [3] is intimately related to Lagrange's theorem that every positive integer is the sum of four squares.

The process of searching through all possible matrices of this special form was programmed by one of us (L. D. Baumert) for the I.B.M. 7090 computer at the Jet Propulsion Laboratory, and on the night of September 27, 1961, an example of the long-awaited 92×92 Hadamard matrix was discovered, in less than an hour of machine time. In fact, there turned out to be *one and only one* example of the Williamson type for size 92. This is shown in Figure 6. (There are many "trivial" rearrangements of A, B, C, D, and of row elements and column elements within A, B, C, D, which lead to superficially different examples. It is by excluding such manipulations that uniqueness is achieved.)

Taking the tensor product of the 92×92 Hadamard matrix with the 2×2 Hadamard matrix yields a 184×184 Hadamard matrix, thereby removing yet another entry from Paley's list of six doubtful cases, and leaving only the sizes 116, 156, and 188.

Elimination of all six cases from Paley's list will justify the "engineering conclusion" that Hadamard matrices "always exist" (that is, for size $4t\times4t$). However, the mathematical goal of *proving* this conclusion will still be far from attained. It does, however, seem quite likely that not merely Hadamard ma-

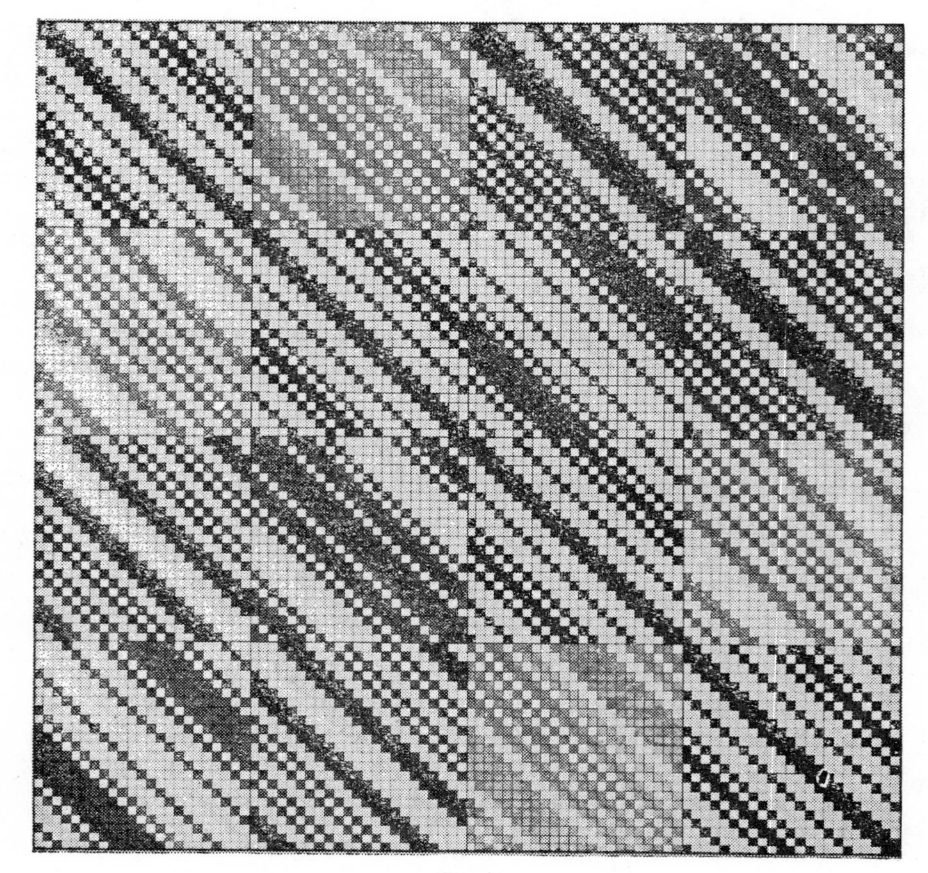

FIG. 6

trices, but Hadamard matrices of the Williamson type, "always exist," and this assertion, stating as it does some specific structural properties for Hadamard matrices, may be easier to prove than the *unrestricted* existence theorem. In any case, there is still much work to be done.

References

1. Jacques Hadamard, Résolution d'une question relative aux déterminants, Bull. Sci. Math. (2), Part 1, 17 (1893) 240–246.

2. R. E. A. C. Paley, On Orthogonal Matrices, J. Math. Phys., 12 (1933) 311–320.

3. John Williamson, Hadamard's Determinant Theorem and the Sum of Four Squares, Duke Math. J., 11 (1944) 65–81.

A POLYGON PROBLEM*

E. R. BERLEKAMP, E. N. GILBERT, AND F. W. SINDEN,
Bell Telephone Labs., Murray Hill, N. J.

Let P_0 be a closed polygon. Let P_1 be the polygon obtained by joining the midpoints of P_0's sides in order (see Figure 1). We write

$$P_1 = TP_0.$$

Let $P_2 = TP_1$, $P_3 = TP_2$, etc. We will call $P_1, P_2, P_3, \cdots$ the *descendants* of P_0.

We consider the question: for what polygons P_0 does there exist a *convex* descendant P_n?

It is easy to show that all descendants of convex polygons are convex. Hence, the existence of one convex descendant implies the existence of infinitely many.

For polygons with fewer than five sides the results are immediate:

2-gons: all descendants have coincident vertices.

3-gons: all descendants are similar to the parent P_0, which is necessarily convex.

4-gons: all descendants are parallelograms, hence convex, though the parent P_0 may be nonconvex.

(To see the last result observe that the sides of P_1 are parallel to the diagonals of P_0.)

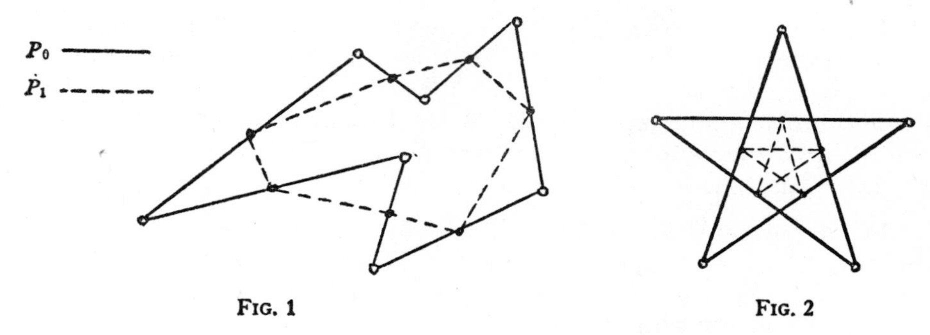

FIG. 1 FIG. 2

* From AMERICAN MATHEMATICAL MONTHLY, vol. 72 (1965), pp. 233–241.

Polygons with five or more sides are more complicated, but in many examples we found that repeated application of T led quickly to convex polygons. This did not seem surprising, for intuitively T is a smoothing operator. We began with the conjecture that every polygon ultimately has convex descendants.

Figure 2 shows this conjecture to be false. The regular pentagram begets only its own kind.

This example, though, may seem unfair since the pentagram intersects itself. Figure 3 shows a non-selfintersecting polygon without convex descendants.

This polygon, in fact, is less innocent than it looks. Not only are its decendants nonconvex but infinitely many of them are self-intersecting. (To see this, note that the relation

$$\frac{X_2 - X_1}{X_3 - X_2} = \frac{1 + \sqrt{5}}{2}$$

between vertex projections X_1, X_2, X_3 holds for all descendants as well as for the polygon itself. P_i is self-intersecting whenever P_{i-1} is not.)

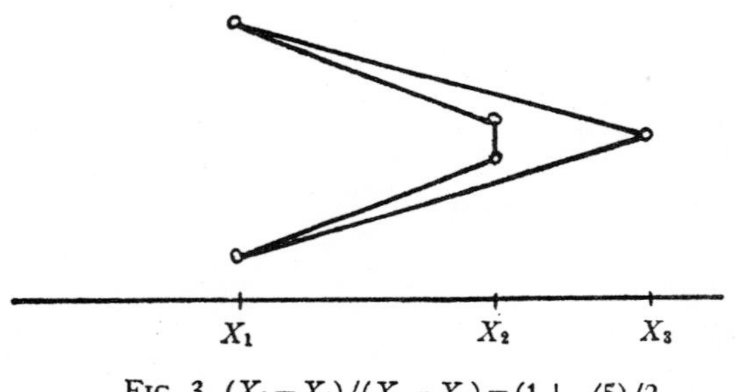

FIG. 3. $(X_2-X_1)/(X_3-X_2)=(1+\sqrt{5})/2$

In spite of these examples, the conjecture that all polygons have convex descendants is not entirely empty. It turns out that *almost* all polygons have convex descendants.

This problem originated with G. R. MacLane, to whom we are indebted for some helpful suggestions. We have recently heard of an independent solution by A. M. Gleason.

Planar polygons. To analyze planar polygons it is convenient to use complex numbers. Let the N-tuple $\mathbf{z}=(z_1, \cdots, z_N)$ represent the vertices of a closed polygon in order. We will refer to the polygon simply as $\mathbf{z}$. The first descendant of $\mathbf{z}$ is a polygon $\mathbf{w}$ with vertices

$$w_1 = \tfrac{1}{2}(z_1 + z_2), \cdots, w_N = \tfrac{1}{2}(z_N + z_1).$$

This may be written $\mathbf{w}=T\mathbf{z}$, where

$$T = \frac{1}{2}\begin{bmatrix} 1 & 1 & 0 & \cdots & 0 \\ 0 & 1 & 1 & \cdots & 0 \\ & & \cdots & & \\ 0 & & \cdots & 1 & 1 \\ 1 & & \cdots & 0 & 1 \end{bmatrix}$$

The eigenvalues and eigenvectors of T, i.e., the solutions of $T\mathbf{u}=\lambda\mathbf{u}$, are of particular interest, for if $\mathbf{u}$ is an eigenvector then the polygon represented by $\mathbf{u}$ is merely rotated and uniformly diminished by T. Specifically it is rotated about the origin through the angle arg λ and diminished by the factor $|\lambda|$.

The eigenvalues and eigenvectors can be written down explicitly. The eigenvalues are

$$\lambda_k = \tfrac{1}{2}(1 + e^{ik(2\pi/N)}), \qquad k = 0, 1, \cdots, N-1;$$

(see Figure 4) and the corresponding eigenvectors are

$$\mathbf{u}_k = (e^{i1k(2\pi/N)}, e^{i2k(2\pi/N)}, \cdots, e^{iNk(2\pi/N)}).$$

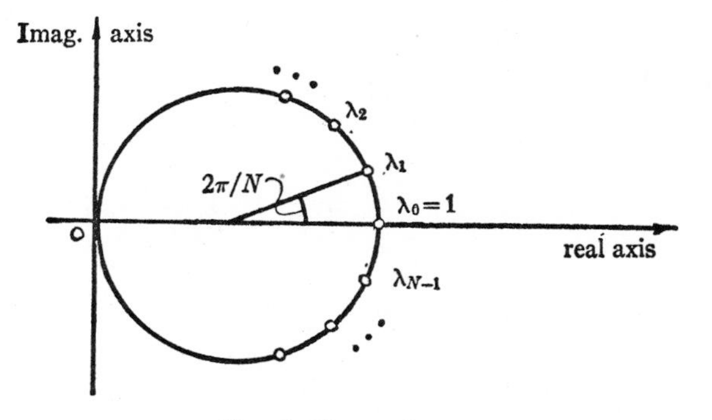

FIG. 4. Eigenvalues

The normalization $\mathbf{u}_k \cdot \mathbf{u}_k^* = N$ seems a little more convenient in the present context than the usual $\mathbf{u}_k \cdot \mathbf{u}_k^* = 1$. Note that the components of the eigenvectors are all roots of unity. Thus the polygons represented by the eigenvectors (eigenpolygons) are all inscribed in the unit circle. The kth one is obtained by joining in order the points on the unit circle with arguments

$$k1\phi,\ k2\phi,\ \cdots,\ kN\phi, \qquad \phi = 2\pi/N.$$

The eigenpolygons for $N=5, 6$ are shown in Figure 5. It is evident geometrically that these polygons are only rotated and diminished by the transformation T.

We will call any polygon that is similar to the kth eigen-N-gon a ***regular N-gon of order k***. Thus, we include among regular polygons self-intersecting, multiply-traversed and degenerate polygons.

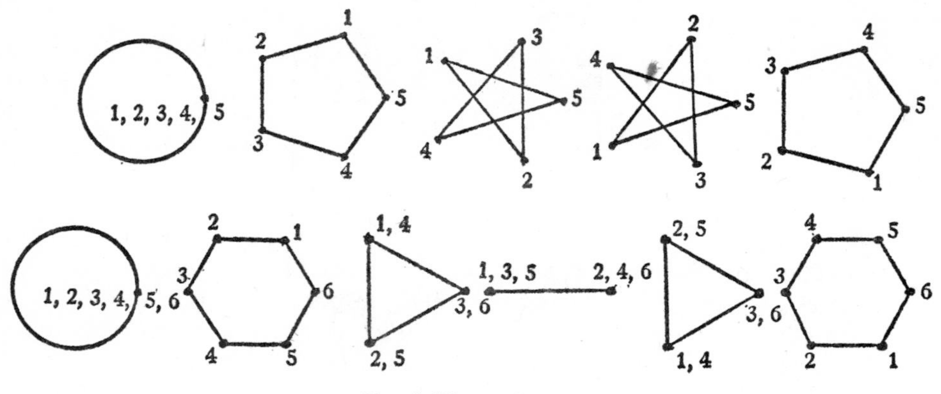

FIG. 5. Eigenpolygons

We observe that regular N-gons have the following properties: (1) The regular N-gons of order 0 are totally degenerate, i.e., all vertices coincide. (2) The regular N-gons of orders 1 and $N-1$ are the regular convex N-gons, counterclockwise and clockwise, respectively. We will regard all self-intersecting, multiply-traversed and degenerate polygons as nonconvex. (3) The regular N-gons of orders k and $N-k$ are the same except for the sense in which they are traversed. (4) The regular N-gon of order k is multiply-traversed if and only if k and N are not relatively prime.

The crucial fact is that an arbitrary N-gon can be written as a sum of regular N-gons. This is the meaning of the following eigenvector expansion, whose validity follows readily (see [1]) from the fact that the eigenvalues are distinct.

$$\mathbf{z} = \sum_{k=0}^{N-1} a_k \mathbf{u}_k \quad (a_k \text{ complex}) \tag{1}$$

or explicitly:

$$z_j = \sum_{k=0}^{N-1} a_k e^{ijk(2\pi/N)}.$$

The coefficients are given by the formula

$$a_k = \frac{1}{N} \sum_{j=1}^{N} z_j e^{-ijk(2\pi/N)},$$

This formula can be given a geometric interpretation as follows. Consider the transformation $\mathbf{w} = W\mathbf{z}$, where

$$w_j = z_j \cdot e^{ij(2\pi/N)}.$$

The effect of W is to rotate the jth vertex about the origin through the angle

$-j(2\pi/N)$. The vertex z_N makes a complete revolution, returning to its starting position, while the other vertices turn through a fraction j/N of a revolution proportional to their indices. This diminishes the angle between successive vertices by the constant amount $2\pi/N$. We will say that W *winds* polygon $\mathbf{z}$ about the origin. Winding carries the kth eigenpolygon into the $(k-1)$-st mod N. In particular it carries the convex eigenpolygon $\mathbf{u}_1$ into the degenerate eigenpolygon $\mathbf{u}_0=(1, 1, \cdots, 1)$ and it carries $\mathbf{u}_0$ into the convex eigenpolygon $\mathbf{u}_{N-1}$. Applied to an arbitrary polygon, winding simply permutes cyclically the coefficients of the component eigenpolygons:

$$a_1 \to a_0,\ a_2 \to a_1,\ \cdots,\ a_0 \to a_{N-1}.$$

The centroid of a polygon $\mathbf{z}$ is $\sum z_j/N$. All of the eigenpolygons except $\mathbf{u}_0$ have centroid O. Hence, the centroid of an arbitrary polygon is the centroid of its zeroth component, namely a_0. It follows that the centroid of the polygon obtained by winding $\mathbf{z}$ is a_1.

The coefficients a_k, then, can be given this geometric interpretation:

The coefficient a_k is the centroid of the polygon obtained by winding $\mathbf{z}$ k times.

We return now to the midpoint-joining transformation T. Consider a polygon $\mathbf{z}$ in decomposed form:

$$\mathbf{z} = \sum_{k=0}^{N-1} a_k\mathbf{u}_k.$$

Since $T\mathbf{u}_k=\lambda_k\mathbf{u}_k$, the first descendant of $\mathbf{z}$ is

$$\mathbf{z}_1 = T\mathbf{z} = \sum_{k=0}^{N-1} a_k\lambda_k\mathbf{u}_k$$

and the nth descendant is

$$\mathbf{z}_n = T^n\mathbf{z} = \sum_{k=0}^{N-1} a_k\lambda_k^n\mathbf{u}_k.$$

As n increases, the components with the largest eigenvalues in absolute value become relatively dominant. As can be seen from Figure 4, the eigenvalues are ordered as follows:

$$1 = \lambda_0 > |\lambda_1| = |\lambda_{N-1}| > |\lambda_2| = |\lambda_{N-2}| > \cdots$$

Since $\lambda_0=1$, the centroids (zeroth coefficients) of $\mathbf{z}_n$ remain fixed. The other components die out. Thus all vertices of $\mathbf{z}_n$ converge to the centroid. Relatively, though, the convex components $(k=1,\ N-1)$ die out least rapidly.

The normalized polygons

$$\mathbf{z}_n' = \frac{1}{|\lambda_1|^n}(\mathbf{z}_n - a_0\mathbf{u}_0)$$

have centroids O and convex components of constant magnitude. As n increases all nonconvex components go to zero.

The sum of the two convex components of $\mathbf{z}$ is the polygon with vertices

$$w_j = a_1 e^{ij(2\pi/N)} + a_{N-1} e^{i(N-1)j(2\pi/N)}.$$

This polygon is the affine image of a regular convex N-gon. Its vertices all lie on the ellipse

$$w(t) = a_1 e^{it} + a_{N-1} e^{-it}$$

at the points $w_j = w(t_j)$, $t_j = j(2\pi/N)$. The ellipse $w(t)$ circumscribes not only the sum $\mathbf{w}$ of the two convex components of $\mathbf{z}$, but also the sums $\mathbf{w}_n'$ of the two convex components of $\mathbf{z}$'s normalized descendants $\mathbf{z}_n'$. As n increases, $\mathbf{z}_n'$ approaches $\mathbf{w}_n'$. If the circumscribing ellipse $w(t)$ is nondegenerate, then for sufficiently large n $\mathbf{z}_n'$ is convex. If, on the other hand, $w(t)$ is degenerate, then all $\mathbf{z}_n'$ are self-intersecting or degenerate, hence by our definition nonconvex. A necessary and sufficient condition for $\mathbf{z}$ to have convex descendants, therefore, is that $w(t)$ be nondegenerate.

The semimajor axis of $w(t)$ has length $|a_1| + |a_{N-1}|$ and the semiminor axis has length $\big||a_1| - |a_{N-1}|\big|$. The ellipse degenerates if and only if $|a_1| = |a_{N-1}|$, i.e., if and only if the two convex components of $\mathbf{z}$ are of equal magnitude.

We can now state the answer to the question posed at the beginning of the paper as follows:

Represent the vertices of polygon P_0 by the complex numbers $z_1, \cdots, z_N$. Let $a_1, \cdots, a_N$ be the complex numbers given by:

$$a_k = \frac{1}{N} \sum_{j=1}^{N} z_j e^{-ijk(2\pi/N)}.$$

P_0 has a convex descendant P_n if and only if

$$|a_1| \neq |a_{N-1}|. \tag{2}$$

Stated geometrically the condition is this: P_0 has a convex descendant if and only if the centroids of the polygons P_+ and P_-, obtained by winding P_0 positively and negatively about the origin, lie at different distances from the origin.

Comments on planar polygons.

1. In many cases the descendants of a polygon display a curious alternation. The descendants of a rectangle, for example, are alternately rhombuses and rectangles. In the general case, the even descendants $P_2, P_4, P_6, \cdots$, if normalized in size, approach a fixed polygon P and the odd descendants approach another fixed polygon P'. P and P' are affine images of regular polygons and are inscribed in the same ellipse.

2. Without essentially complicating the problem the transformation T, which had the formula $w_j = \frac{1}{2}(z_j + z_{j+1})$ can be replaced by the more general trans-

formation T' with the formula

$$w'_j = A_0 z_j + A_1 z_{j+1} + \cdots + A_{N-1} z_{j+N-1}$$

where $A_0, A_1, \cdots$ are any constants and where the N subscripts on the z_{j+k} are to be computed modulo N. The eigenvectors are the same $\mathbf{u}_k$ as before but the eigenvalues are now

$$\lambda_k = \sum_r A_r e^{irk2\pi/N}.$$

If $|\lambda_1|$ or $|\lambda_{N-1}|$ exceeds all of $|\lambda_2|, \cdots, |\lambda_{N-2}|$ then, for almost all polygons P, all but a finite number of descendants of P are convex. Two examples of this kind are:

(i) $$w'_j = \alpha z_j + \beta z_{j+1} \qquad (\alpha > 0, \beta > 0, \alpha + \beta = 1)$$

and

(ii) $$w'_j = \tfrac{1}{3}(z_j + z_{j+1} + z_{j+2}).$$

The midpoints of the sides are replaced by points which subdivide the sides in ratio α/β in (i) and by the centroids of successive vertex triples in (ii).

3. There exist orphan polygons which have no ancestors. Any quadrilateral that is not a parallelogram is such a one. Orphan polygons are those even-sided polygons containing a nonzero component of order $N/2$. The vanishing eigenvalue $\lambda_{N/2} = 0$ wipes out this component in the first generation. Thus no polygon containing the $N/2$ component can be the descendant of another polygon.

Similar remarks apply to the more general transformation T' of the preceding paragraph. If T' has no zero eigenvalue (as in example (i) if $\alpha \neq \beta \neq \frac{1}{2}$) then all polygons have unique ancestors. In this case the family tree may be traced backwards as well as forwards. It is interesting to note that almost all convex polygons have remote ancestors which are not convex.

Nonplanar polygons. The nonplanar case, surprisingly, is hardly more complicated than the planar case. To make the generalization it is convenient to write the eigenpolygon decomposition in a new form. Combining the kth and $(N-k)$th terms in (1) we obtain a sum of $[N/2]+1$ terms each representing an affine image of a regular polygon; here $[N/2]$ means the integer part of $N/2$. For d-dimensional space this sum generalizes as follows. If X is a closed N-gon in d-space then X can be written

$$X = \sum_{k=0}^{[N/2]} W_k,$$

where again each component polygon W_k is an affine image of a regular polygon. *This differs from the planar case only in that the W_k need not all lie in the same plane.*

This may be derived explicitly as follows. Let

$$\mathbf{x}(t) = \sum_{k=0}^{[N/2]} \mathbf{u}_k \cos kt + \mathbf{v}_k \sin kt,$$

where the symbols $\mathbf{x}$, $\mathbf{u}$, $\mathbf{v}$ stand for vectors with d real components. It is always possible [2] to choose $\mathbf{u}_k$ and $\mathbf{v}_k$ so that $\mathbf{x}(t)$ assumes given values at N given points $t_1, \cdots, t_N$, in particular so that

$$\mathbf{x}\left(j\frac{2\pi}{N}\right) = \mathbf{x}_j,$$

where $\mathbf{x}_j$ are the vertices of the given polygon X. Then X can be written as

$$X = \sum_{k=0}^{[N/2]} W_k,$$

where W_k is the polygon with vertices

$$\mathbf{w}_{kj} = \mathbf{u}_k \cos kj\frac{2\pi}{N} + \mathbf{v}_k \sin kj\frac{2\pi}{N}\cdot$$

Polygon W_k lies in the plane determined by $\mathbf{u}_k$ and $\mathbf{v}_k$ and is inscribed in the eliipse

$$\mathbf{w}(t) = \mathbf{u}_k \cos t + \mathbf{v}_k \sin t.$$

W_k may be regarded as the sum of a clockwise and a counterclockwise regular N-gon of order k as was done in the first section. Alternatively W_k may be regarded as the projection onto a plane of a regular N-gon of order k. In general, a d-dimensional polygon may be regarded as the projection onto d-space of a $(d+1)$-dimensional polygon all of whose components are regular.

The midpoint-joining transformation T may be analyzed in terms of the decomposition in the same way as before. The only new element is the dimension of the descendants. (A polygon has the dimension of the smallest space it can be imbedded in.) The following example shows that T can reduce the dimension: Let X be a 3-dimensional polygon with vertices alternately in the planes $z=1$ and $z=-1$. The first descendant of X lies in the x, y-plane. This, however, is as far as the dimension reduction can go. Further application of T leaves the dimension unchanged. In general the situation is this:

All descendants $X_1, X_2, \cdots$ of a d-dimensional polygon X_0 have the same dimension. X_1 may be of dimension one less than X_0, but this can occur only if X_0 is an orphan, i.e., only if X_0 contains a nonzero component of order $N/2$.

Proof. Suppose X_n has dimension d and X_{n+1} has dimension $d'<d$. Let $\overline{X}_n$ and $\overline{X}_{n+1}$ be the projections of X_n and X_{n+1} onto a line orthogonal to the d'-space containing X_{n+1}. All vertices of $\overline{X}_{n+1}$ lie in a single point p. Since p must

be the midpoint of all sides of $\overline{X}_n$, the vertices of $\overline{X}_n$ must lie alternately in two points p_1, p_2 equidistant from p. $\overline{X}_n$, then, is a regular N-gon of order $N/2$. If X_n has the decomposition

$$X_n = \sum_{k=0}^{[N/2]} W_k$$

then $\overline{X}_n$ has the decomposition

$$\overline{X}_n = \sum_{k=0}^{[N/2]} \overline{W}_k.$$

But in the latter decomposition only $\overline{W}_{N/2}$ can be nonzero. If $\overline{W}_{N/2}$ is nonzero then so is $W_{N/2}$. In this case X_n is an orphan, hence $n=0$. This completes the proof.

In conclusion, we may say that almost all nonplanar polygons lack planar descendants. If the first descendant is nonplanar then so are all the rest.

On the other hand, all nonplanar polygons have descendants which differ arbitrarily little (relative to their size) from planar polygons. And *almost* all nonplanar polygons have descendants which differ arbitrarily little (relative to their size) from planar *convex* polygons.

The first author is now at the University of California, Berkeley.

References

1. F. R. Gantmacher, Theory of Matrices, vol. 1, Chelsea, New York, 1959, p. 72.
2. Ch. de la Vallée Poussin, Leçons sur l'approximation des fonctions d'une variable réelle, Gauthier-Villars, Paris, 1919, p. 93.

MATRICES IN THE MARKET PLACE*

F. D. PARKER, SUNY at Buffalo

Introduction. One often hears the story of the young American in a foreign bank who borrows ten dollars, exchanges them for francs, the francs for rupees, the rupees for pounds sterling, and the pounds sterling for dollars, all of which allows him to return the loan of ten dollars and pocket a profit. True or not, it would seem that such transactions are susceptible to mathematical treatment if an appropriate model is available. Such a model might be expected to find discrepancies in the foreign exchange system or any similar system. Since such systems are subjected to frequent, almost continuous, fluctuations (as well as incomplete information) it would seem that a shrewd manipulator could take advantage of such discrepancies to his financial gain.

As it happens, there is such a model, a model which has some nontrivial properties. The same model occurs in one aspect of group theory.

* From MATHEMATICS MAGAZINE, vol. 38 (1965), pp. 125–128.

Equitable matrices. We begin by considering a system of n sets. Each set contains elements to which a value can be assigned (each element of a set has the same value). Each element of set S_i can be exchanged for a_{ij} elements of set S_j. Now the system can be described by a square matrix of order n whose entry in row i, column j is a_{ij}. To keep an application in mind, the elements of a given set can be dollars, bushels of wheat, shares of stock, one day's labor, etc. For such a system to be equitable, i.e., that there is no financial advantage to be gained by bartering, there must be certain conditions on the matrix.

1. $a_{ij}>0$ for all i, j.
2. $a_{ii}=1$ for all i.
3. $a_{ij}a_{ji}=1$ for all i, j.
4. $a_{ij}a_{jk}=a_{ik}$ for all i, j, k.

The fourth condition is the important one which demands that a transaction involving, for example, exchanging dollars for francs, then francs for rupees, yields as many rupees as the direct exchange of dollars for rupees.

In fact, these four conditions can be replaced by the following definition:

An *equitable matrix* is a square matrix of order n with positive entries such that $a_{ij}a_{jk}=a_{ik}$ for all i, j, k.

Equitable matrices have some interesting properties, some of which are described by the following theorems.

THEOREM I. *If M is an equitable matrix of order n, then $M^2=nM$.*

Proof. The element of M^2 in row i column j is given by

$$\sum_{k=1}^{n} a_{ik}a_{kj} = \sum_{k=1}^{n} a_{ij} = na_{ij}.$$

THEOREM II. *If S is a square matrix of order n consisting entirely of unity elements, then S is an equitable matrix, and any equitable matrix (of order n) is similar to S.*

Proof. It is immediate that S is an equitable matrix. To show that M is similar to S, we can actually find the matrix P such that $P^{-1}MP=S$. Let $P=\text{diag }(1, a_{21}, a_{31}, \cdots a_{n1})$, then $P^{-1}=\text{diag }(1, a_{12}, a_{13}, \cdots, a_{1n})$. Then the element in the ij position of MP becomes $a_{ij}a_{j1}=a_{i1}$. Premultiplication by P^{-1} now provides in the ij position of $P^{-1}MP$ the element $a_{1i}a_{i1}=a_{11}=1$.

THEOREM III. *All equitable matrices of order n are similar, and their characteristic values are $(n, 0, 0, \cdots, 0)$.*

Proof. Since any equitable matrix is similar to S, they are all similar to each other. The characteristic values of S are easily shown to be $(n, 0, 0, \cdots, 0)$.

THEOREM IV. *Equitable matrices form a commutative group under Hadamard multiplication.* (The Hadamard product of $[a_{ij}]$ and $[b_{ij}]$ is $[a_{ij}b_{ij}]$.)

Proof. Consider two equitable matrices A and B. The entry in row i, column j of the Hadamard product is $a_{ij}b_{ij}$, and the entry in row j, column k is $a_{jk}b_{jk}$. But $a_{ij}b_{ij}a_{jk}b_{jk}=a_{ik}b_{ik}$, which is the entry in row i, column k. Hence equitable matrices are closed with respect to Hadamard multiplication. Associativity and commutativity are immediate, S serves as the identity and the transpose of a given equitable matrix is its inverse.

The group itself does not appear to be very interesting, being isomorphic to the direct product of $n-1$ groups, each group being the multiplicative group of positive real numbers. This is proved by realizing that only $n-1$ entries of an equitable matrix can be independent. To put it another way, as soon as the exchange rate is known for a unit of a given class in terms of a unit of each of the other classes (a row of the matrix), then the matrix is uniquely determined.

THEOREM V. *A matrix which diagonalizes an equitable matrix M is given by*

$$R=\begin{bmatrix} 1 & 0 & 0 & 0 & \cdots & 0 & -1 \\ a_{21} & a_{21} & 0 & 0 & \cdots & 0 & 0 \\ a_{31} & -a_{31} & a_{31} & 0 & \cdots & 0 & 0 \\ a_{41} & 0 & -a_{41} & a_{41} & \cdots & 0 & 0 \\ \vdots & & & & & \vdots & \vdots \\ a_{n1} & 0 & 0 & 0 & \cdots & -a_{n1} & a_{n1} \end{bmatrix}$$

Proof. Direct multiplication yields $MR=R\Lambda$, where $\Lambda=\text{diag}(n, 0, 0, \cdots, 0)$.

Inequitable matrices. What about inequitable matrices? How can they be discovered and exploited? Consider an ideal situation in which only one exchange rate is altered from an equitable situation. If, for example, the rate of exchange of a unit of commodity i for a unit of commodity j changes from a_{ij} to $a_{ij}+h$, then the matrix M^2-nM no longer is identically zero. The change occurs in row i, and in column j, and then M^2-nM is

$$\begin{bmatrix} & & ha_{1j} & & \\ & 0 & ha_{2j} & 0 & \\ & & \vdots & & \\ ha_{i1} & ha_{i2} \cdots & 2h & \cdots & ha_{in} \\ & & \vdots & & \\ & 0 & ha_{nj} & 0 & \end{bmatrix}$$

An individual who owns k units of any commodity p can exchange them for ka_{pi} units of commodity i, then $ka_{pi}(a_{ij}+h)$ units of commodity j, and finally $ka_{pi}(a_{ij}+h)(a_{jp})$ units of his original commodity, thus realizing a profit of kha_{ji} units of commodity p. If h is negative, the transactions are made in the reverse order, yielding a profit of $-kh/(a_{ij}+h)$ units of commodity p.

Of course, this is an extremely unsophisticated model, but it is conceivable that an individual with reliable information on a wide market, and with high-speed computing facilities for keeping an almost continuous record of the matrix M^2-nM might be able to take advantage of a market discrepancy before the usual forces could react to restore equilibrium. The model does show, moreover, that a profit can be made by an individual owning any commodity, and that only three transactions are necessary. Thus we reach the conclusion that some of the transactions carried out by the "young American" were superfluous.

So far we have disregarded the usual fees which are associated with such transactions (commissions, taxes, brokers' fees, etc.). Suppose that such fees are directly proportional to the amount of the transaction, and that one unit of commodity p no longer is worth a_{pi} units of commodity i, but rather ra_{pi} units of commodity $i (r<1)$. If r is too small, then it is possible that no transaction can be profitable; in practice, r has a damping effect on the trading activity of a market. Again considering the case in which only one exchange rate is altered from an equitable rate a_{ij} to $a_{ij}+h$, we find now that the conversion of one unit of commodity p to commodity i to commodity j to commodity p now produces

$$r^3 a_{pi}(a_{ij}+h)a_{jp}$$

units of commodity p. Unless h is large enough, there will be no profit. The condition for a profitable transaction is $hr^3>a_{ij}(1-r^3)$. If h is negative, the transactions are carried out in the opposite direction, and one unit of commodity p now becomes

$$r^3 a_{pj}\frac{1}{a_{ij}+h}a_{ip}$$

units of the same commodity, and the condition for a profitable transaction is $-h>(1-r^3)a_{ij}$.

An application to group theory. In abstract algebra we frequently encounter a finite set G with a binary operation. If for any two elements a and b of G there is a unique solution to the equation $ax=b$ and a unique solution to the equation $ya=b$, then G is a quasigroup. If at the same time there is a distinguished element e such that $ae=ea=a$ for all a in G, then G is a loop. If, in addition, the associative law $a(bc)=(ab)c$ holds for all a, b, c in G, then G is a group. Since groups have been extensively studied and possess much stronger theorems than do loops, it is often desirable to determine whether a given loop is a group. If the multiplication table of the loop is arranged so that the distinguished element e lies on the main diagonal (the normal form) and the table is considered as a matrix, then Zassenhaus [1] has shown that $a_{ij}a_{jk}=a_{ik}$ is a necessary condition that the operation is associative. This is precisely the requirement for an equitable matrix. Therefore, a necessary condition that a loop be a group is that $P^{-1}MP=S$, where M is the Cayley table in normal form, P^{-1} and P are the diagonal matrices described in Theorem II, S is composed of entries all of which are the distinguished element e, and the operations are carried out in the binary operation of the system.

References

1. H. Zassenhaus, The theory of groups, Chelsea, New York, 1949.
2. F. D. Parker, When is a loop a group?, Amer. Math Monthly, 72 (1965) No. 7.

BIBLIOGRAPHIC ENTRIES: APPLICATIONS

The references below are to the AMERICAN MATHEMATICAL MONTHLY.

1. C. C. MacDuffee, Some applications of matrices in the theory of equations, 1950, 154–161.

Discusses situations in which the companion matrix may be used in place of the theory of symmetric functions.

2. J. P. Ballantine, Inverse matrices in interpolation, 1952, 178–180.
3. F. D. Parker, Inverses of Vandermonde matrices, 1964, 410–411.

The above two articles deal with the use of Vandermonde matrices in finding a polynomial passing through a given finite set of points. Ballantine only considers low dimensions.

4. Mark Lotkin, The treatment of boundary problems by matrix methods, 1953, 11–19.

Discusses systems of differential equations.

5. W. G. Leavitt, Systems of linear differential equations, 1956, 335–337.

Uses elementary divisor theory.

6. C. Lanczos, Linear systems in self-adjoint form, 1958, 665–679.

Extends properties of symmetric matrices to arbitrary matrices using orthogonal transformations. The results are applicable to boundary value problems.

7. Mark Lotkin, The partial summation of series by matrix methods, 1957, 643–647.

Formulas involving binomial coefficients are proved by matrix methods.

8. R. A. Rosenbaum, An application of matrices to linear recursion relations, 1959, 792–793.

9. Alan G. Konheim, A generalized independence theorem and error correction codes, 1960, 228–231.

A generalization of the notion of independence of a set of vectors over a finite field, with an application to coding theory.

10. A. Nerode and H. Shank, An algebraic proof of Kirchhoff's network theorem, 1961, 244–247.

A proof of Ohm's and Kirchhoff's laws via inner products.

11. Fergus Gaines, On the arithmetic mean-geometric mean inequality, 1967, 305–306.

Derives this inequality from Schur's matrix inequality.

12. R. B. Kirchner, An explicit formula for e^{At}, 1967, 1200–1204.

Computes e^{At} in terms of A and its eigenvalues.

13. A. W. Roberts, The derivative as a linear transformation, 1969, 632–638.

Applies linear algebra to multivariable calculus. The theorems proven here reach conclusions about a function on an open set from hypotheses on the matrix of the derivative of that function (e.g., orthogonal, skew symmetric).

14. Norman Levinson, Coding theory: a counterexample to G. H. Hardy's conception of applied mathematics, 1970, 249–258.

An introduction to the algebraic theory of error correcting codes.

(g)

OTHER TOPICS

METHODS OF PROOF IN LINEAR ALGEBRA*

HARLEY FLANDERS, University of California, Berkeley

1. Introduction. In a problem in linear algebra, one has, generally speaking, a definite field of coefficients given in advance. Often the first question is to find an appropriate linear space so that the problem reduces to one concerning linear transformations on that space. This in turn may be handled in one of several ways. For example the older approach is to consider the space of n-tuples $(a_1, \cdots, a_n)$ as the linear space, in which case one is committed to a definite coordinate basis $(1, 0, 0, \cdots), (0, 1, 0, \cdots), \cdots$. The modern approach is to use invariant basis-free methods. Of course bases cannot be completely discarded. In many cases the essential finite dimensionality of the situation is exploited precisely by using a basis. In such problems the best technique is to select a basis which most fits the given data. At all times however one should be perfectly willing to change the basis if a more natural choice appears, to change the space itself if that can help matters, and even to change the field of coefficients for a more useful one.[(1)]†

In this article we shall emphasize the idea of abandoning what has already been selected for something better. We shall give proofs of several known theorems in linear algebra which will show the various possibilities. These proofs, while probably not the shortest or most elegant possible, at least seem to us to bring out the reasons for the validity of the theorems they demonstrate. Some of these theorems have perhaps not appeared heretofore to be as elementary as they really are.

In Section 2 we give the familiar characterization of automorphisms of a total matric algebra. In the proof one sees at the outset that one basis is as good as another, but after a basis is fixed one soon sees that there is another definite one to be reckoned with. In Section 3 we give a result on fields of matrices. It is pretty clear in this situation what the right linear space is, but the point of the proof is to consider the space over a new field of coefficients. The remaining sections are centered around the idea of superdiagonalization, which is a technique of selecting a basis naturally suited to a problem. This is introduced in Section 4, applied to a single transformation in Section 5, to a commutative set of transformations in Section 6, and to a nil associative algebra in Section 8; some basic concepts of linear associative algebra theory are given in Section 7. A combination of the results of Sections 6 and 8 is given in Section 9. One of the main tools in superdiagonalization is the fact that if M is a subspace of a space

* From AMERICAN MATHEMATICAL MONTHLY, vol. 63 (1956), pp. 1–15.

† See the notes at the conclusion of the paper.

L which reduces a given transformation, then that transformation induces transformations on both M and on the quotient or coset space L/M. Thus attention is shifted to other linear spaces. In order to study this further we introduce the idea of a Lie algebra in Section 10 and then prove the basic theorems of Engel and Lie in Sections 11–14. A final remark on characteristic roots is given in Section 15.

2. Automorphisms of matrices. We begin with the following theorem.[2]

THEOREM 1. *Let $\mathfrak{M}_n$ be the algebra of all $n \times n$ matrices over a field k and let σ be an automorphism of this algebra. Then σ is an inner automorphism, that is, there exists a non-singular matrix P such that $\sigma(A) = PAP^{-1}$ for all A.*

First of all, we turn this into a linear space problem by noting that $\mathfrak{M}_n$ is isomorphic to the algebra of all linear transformations on any n-dimensional linear space L over k. Let us restate the proposition this way.

THEOREM 1′. *Let $\mathfrak{A} = \text{Lin}(L)$ be the algebra of all linear transformations on an n-dimensional linear space L over a field k and let σ be an automorphism of $\mathfrak{A}$. Then there exists a non-singular linear transformation $\mathbf{P}$ of L such that $\sigma(\mathbf{A}) = \mathbf{PAP}^{-1}$ for all $\mathbf{A}$ in $\mathfrak{A}$.*

The proof consists of little more than a precise formulation of the possible isomorphisms between $\mathfrak{A} = \text{Lin}(L)$ and $\mathfrak{M}_n$ plus the observation that one is not, after all, forced to stick to any particular one of these isomorphisms.

We select *any* basis $\mathbf{v}_1, \cdots, \mathbf{v}_n$ of L. Each $\mathbf{A} \in \text{Lin}(L)$ yields a matrix $A = \|a_{ij}\|$ according to the equations

$$\mathbf{v}_i\mathbf{A} = \sum_{j=1}^{n} a_{ij}\mathbf{v}_j,$$

and this correspondence $\mathbf{A} \leftrightarrow A = \|a_{ij}\|$ is the kind of isomorphism we deal with between $\text{Lin}(L)$ and $\mathfrak{M}_n$. The basic matric units E_{ij} (with 0 everywhere except for 1 in i-th row and j-th column) correspond to transformations $\mathbf{E}_{ij}$ defined as follows:

$$\mathbf{v}_i\mathbf{E}_{jk} = \delta_{ij}\mathbf{v}_k$$

on the basis of L we have selected. They satisfy the important algebraic relations

$$\mathbf{E}_{ij}\mathbf{E}_{kl} = \delta_{jk}\mathbf{E}_{il}. \tag{1}$$

Here is the crux of the matter.

LEMMA. *Let $\mathfrak{A} = \text{Lin}(L)$ as before and let $\mathbf{F}_{ij}$ be n^2 elements of $\mathfrak{A}$, not all 0 and satisfying the relations*

$$\mathbf{F}_{ij}\mathbf{F}_{kl} = \delta_{jk}\mathbf{F}_{il}.$$

Then there exists a basis $\mathbf{w}_1, \cdots, \mathbf{w}_n$ of L such that

$$\mathbf{w}_i\mathbf{F}_{jk} = \delta_{ij}\mathbf{w}_k. \tag{2}$$

Proof. Some $\mathbf{F}_{ij} \neq 0$ and $\mathbf{F}_{ij} = \mathbf{F}_{i1}\mathbf{F}_{11}\mathbf{F}_{1j}$, hence $\mathbf{F}_{11} \neq 0$. We select $\mathbf{w} \in L$ so that $\mathbf{w}\mathbf{F}_{11} \neq 0$ and set

$$\mathbf{w}_1 = \mathbf{w}\mathbf{F}_{11},\ \mathbf{w}_2 = \mathbf{w}\mathbf{F}_{12},\ \cdots,\ \mathbf{w}_n = \mathbf{w}\mathbf{F}_{1n}.$$

These vectors $\mathbf{w}_1, \cdots, \mathbf{w}_n$ form a basis of L; all we need show is that they are linearly independent. But a dependence relation $\sum a_i \mathbf{w}_i = 0$ yields $\sum a_i \mathbf{w}\mathbf{F}_{1i} = 0$. We simply multiply on the right by $\mathbf{F}_{j1}$ to obtain $a_j \mathbf{w}\mathbf{F}_{11} = 0$, hence $a_j = 0$ for $j = 1, \cdots, n$. Having this, one readily verifies that the $\mathbf{w}_j$ satisfy the relations (2).

Now we can complete the proof of Theorem 1′. We start with any basis $\mathbf{v}_1, \cdots, \mathbf{v}_n$ of L and the corresponding basis $\mathbf{E}_{ij}$ of $\mathfrak{A}$ $(i, j = 1, \cdots, n)$. We set $\mathbf{F}_{ij} = \sigma(\mathbf{E}_{ij})$. Since σ is an algebra automorphism of $\mathfrak{A}$, the relations (1) satisfied by the $\mathbf{E}_{ij}$ imply the relations for the $\mathbf{F}_{ij}$ in the hypothesis of the lemma, hence there is a basis $\mathbf{w}_1, \cdots, \mathbf{w}_n$ of L such that $\mathbf{w}_i \mathbf{F}_{jk} = \delta_{ij}\mathbf{w}_k$.

Now we have two bases of L: $\mathbf{v}_1, \cdots, \mathbf{v}_n$; $\mathbf{w}_1, \cdots, \mathbf{w}_n$. We define an automorphism (non-singular linear transformation) $\mathbf{P}$ of L by sending one basis into the other: $\mathbf{w}_i\mathbf{P} = \mathbf{v}_i$; hence $\mathbf{v}_i\mathbf{P}^{-1} = \mathbf{w}_i$. We now have

$$\mathbf{w}_i\mathbf{P}\mathbf{E}_{jk}\mathbf{P}^{-1} = \mathbf{v}_i\mathbf{E}_{jk}\mathbf{P}^{-1} = \delta_{ij}\mathbf{v}_k\mathbf{P}^{-1} = \delta_{ij}\mathbf{w}_k = \mathbf{w}_i\mathbf{F}_{jk}.$$

Hence $\mathbf{P}\mathbf{E}_{jk}\mathbf{P}^{-1} = \mathbf{F}_{jk}$, for they agree on a basis of L. We have proved that $\mathbf{P}\mathbf{E}_{jk}\mathbf{P}^{-1} = \sigma(\mathbf{E}_{jk})$. Since the $\mathbf{E}_{jk}$ form a linear basis of $\mathfrak{A}$, we finally have $\mathbf{P}\mathbf{A}\mathbf{P}^{-1} = \sigma(\mathbf{A})$ for all $\mathbf{A} \in \text{Lin}\,(L)$.

We might mention that this result has some fairly deep consequences in associative algebra theory; for example, it follows rather easily that every automorphism of a central simple associative algebra is inner.[3]

3. Fields of matrices. In this section we consider a result which at first sight is rather surprising.[4]

THEOREM 2. *Let $\mathfrak{M}_n$ be the algebra of all $n \times n$ matrices over a field k and consider k as a subalgebra of $\mathfrak{M}_n$ by identifying elements a of k with the corresponding scalar multiples aI of the identity. Let k' be another field such that $k \leqq k' \leqq \mathfrak{M}_n$ and suppose $[k' : k] = m$ is the linear dimension of k' over k. Let $\mathfrak{C}$ be the set of all elements of $\mathfrak{M}_n$ which commute with every element of k'. Then m divides n so that $n = mq$ and $\mathfrak{C}$ is isomorphic to $\mathfrak{M}_q(k')$, the algebra of $q \times q$ matrices over k'.*[5]

The result looks somewhat far-fetched and complicated; indeed quite complicated proofs have been given. Nevertheless when looked at in the proper invariant light it is transparent.

Again we start with an n-dimensional linear space L over k and write $n = \dim_k (L)$. As before, $\mathfrak{A} = \text{Lin}\,(L)$ and now we have $k \leqq k' \leqq \mathfrak{A}$. Since k' is a field of linear transformations on L, we may consider L as a linear space over k' (and this is the crucial observation). As such it is obviously finite dimensional, say of dimension $q = \dim_{k'}(L)$. That $n = mq$ is a consequence of the plausible relation

$$\dim_k(L) = \dim_{k'}(L) \cdot \dim_k(k')$$

which is quickly proved by showing that a basis $\mathbf{v}_1, \cdots, \mathbf{v}_q$ of L over k' and a basis $\mathbf{A}_1, \cdots, \mathbf{A}_m$ of k' over k make up a basis $\{\mathbf{v}_i\mathbf{A}_j; i=1, \cdots, q, j=1, \cdots, m\}$ of qm elements of L over k. Finally, for a linear transformation on L over k to commute with all of the elements of k' is exactly the same as its being a linear transformation on L over k'. Hence the subalgebra $\mathfrak{C}$ of $\mathfrak{A} = \text{Lin}_k(L)$ of all linear transformations which commute elementwise with the transformations in k' may be identified with $\text{Lin}_{k'}(L)$ which, of course, is isomorphic (over k') to $\mathfrak{M}_q(k')$.

4. Superdiagonalization. There is a powerful technique for handling individual linear transformations or sets of transformations which is closely related to the idea of invariant subspace. Let L be an n-dimensional linear space over a field k and let $\mathbf{A}$ be a linear transformation on L into L. Suppose M is a subspace of L; we say that M *reduces* $\mathbf{A}$ if whenever $\mathbf{v} \in M$ then $\mathbf{vA} \in M$. By using the notation

$$M\mathbf{A} = \{\mathbf{vA} \mid \mathbf{v} \in M\},$$

the condition is simply $M\mathbf{A} \leqq M$. Similarly if $\mathfrak{S}$ is a set of linear transformations, we say that M reduces $\mathfrak{S}$ if $M\mathfrak{S} \subseteq M$, meaning $M\mathbf{A} \leqq M$ for each $\mathbf{A}$ in $\mathfrak{S}$.

Since the sets of transformations also include the sets consisting of a single transformation, we may restrict attention to sets. Suppose then that $\mathfrak{S}$ is a set of transformations on L and that we have found subspaces $M_1, M_2, M_3, \cdots, M_n$ of L, each of which reduces $\mathfrak{S}$ and such that

$$L = M_1 > M_2 > M_3 > \cdots > M_n > 0 = M_{n+1}$$

so that (since $\dim(L) = n$) $\dim(M_2) = n-1$, $\dim(M_3) = n-2, \cdots$. We fit a basis of L to this chain of subspaces by selecting vectors $\mathbf{v}_1, \cdots, \mathbf{v}_n$ where $\mathbf{v}_i \in M_i$ but $\mathbf{v}_i \notin M_{i+1}$. If $\mathbf{A} \in \mathfrak{S}$, then $\mathbf{v}_i\mathbf{A} \in M_i\mathbf{A} \leqq M_i$, hence $\mathbf{v}_i\mathbf{A}$ is a linear combination of $\mathbf{v}_i, \mathbf{v}_{i+1}, \cdots, \mathbf{v}_n$; that is,

$$\begin{cases} \mathbf{v}_1\mathbf{A} = a_{11}\mathbf{v}_1 + & & \cdots + a_{1n}\mathbf{v}_n \\ \mathbf{v}_2\mathbf{A} = & a_{22}\mathbf{v}_2 + & \cdots + a_{2n}\mathbf{v}_n \\ \mathbf{v}_3\mathbf{A} = & & a_{33}\mathbf{v}_3 + \cdots + a_{3n}\mathbf{v}_n \\ \cdots\cdots & \cdots\cdots & \cdots\cdots \\ \mathbf{v}_n\mathbf{A} = & & a_{nn}\mathbf{v}_n \end{cases}$$

so that the matrix of $\mathbf{A}$ with respect to the basis $(\mathbf{v}_i)$ is in superdiagonal (or triangular) form; and this is true for each $\mathbf{A}$ in $\mathfrak{S}$.

The remaining sections deal with some cases in which this superdiagonalization of a set of transformations is possible. An important tool in this theory is the following. If the subspace M reduces a set $\mathfrak{S}$ of transformations, then $\mathfrak{S}$ induces a set of linear transformations on the subspace M. Also, $\mathfrak{S}$ induces a set of transformations on the quotient space L/M, the space of cosets of L modulo

M. In fact, if $\mathbf{A} \in \mathfrak{S}$ and $\mathbf{v} \in L$, then the induced transformation, also denoted $\mathbf{A}$, on L/M has the effect

$$(\mathbf{v} + M)\mathbf{A} = (\mathbf{v}\mathbf{A}) + M$$

on the coset $\mathbf{v}+M$. One readily checks that this is well defined, *i.e.*, independent of the particular representative $\mathbf{v}$ of the coset, exactly because $M\mathbf{A} \leqq M$. Speaking in a general way, algebraic properties of the original set $\mathfrak{S}$ carry over to the induced sets both on M and on L/M.

5. Single linear transformation. We begin by showing that if the ground field k is sufficiently large then a single transformation $\mathbf{A}$ may be superdiagonalized. The field in fact must contain all of the characteristic roots of $\mathbf{A}$. We take care of this for *all* $\mathbf{A}$ by assuming that k is algebraically closed, *i.e.*, that every polynomial $t^n + a_1t^{n-1} + \cdots + a_n$ with coefficients in k has a root in k.

THEOREM 3. *Let* $\mathbf{A}$ *be linear on a space* L *of dimension* $n>1$ *over an algebraically closed field* k. *Then there is a subspace* M *reduced by* $\mathbf{A}$ *such that* $0<M<L$.

We simply set $M = k\mathbf{v}$, the space generated by a single characteristic vector $\mathbf{v}$. Since $\mathbf{v}\mathbf{A} = c\mathbf{v}$ we have $M\mathbf{A} \leqq M$.

COROLLARY. *The transformation* $\mathbf{A}$ *may be superdiagonalized.*

Proof. We proceed by induction on $n = \dim (L)$. Since $\mathbf{A}$ induces a single linear transformation on M and $\dim (M) < n$, we have a sequence

$$0 < M_1 < M_2 < \cdots < M_r = M$$

of subspaces reduced by $\mathbf{A}$, each of dimension one greater than the preceding; since $\mathbf{A}$ induces a single linear transformation on L/M and $\dim (L/M) = \dim (L) - \dim (M) < n$, we have another such sequence

$$M < M_{r+1} < \cdots < M_n = L.$$

Together they yield a sequence

$$0 < M_1 < M_2 < \cdots < M_n = L$$

which provides the superdiagonalization of $\mathbf{A}$.

We give the proof in this manner as a model to be followed in more complicated situations.

6. Commutative sets. Our next result also requires a restriction on k.

THEOREM 4. *Let* $\mathfrak{S}$ *be a commutative set of linear transformations on a space* L *of dimension* $n>1$ *over an algebraically closed field* k. *Then there is a subspace* M *reduced by* $\mathfrak{S}$ *such that* $0<M<L$.

Proof. If each $\mathbf{A}$ in $\mathfrak{S}$ is a scalar multiple of the identity $\mathbf{I}$, then any proper subspace M of L will reduce $\mathfrak{S}$. If this is not the case, we select an $\mathbf{A} \in \mathfrak{S}$ such that $\mathbf{A} \neq c\mathbf{I}$ for any c. Let c be a characteristic root of $\mathbf{A}$ and set

$$M = \{\mathbf{v} \mid \mathbf{v}\mathbf{A} = c\mathbf{v}\}.$$

Then $0 < M < L$. If $\mathbf{B}$ is an arbitrary transformation in $\mathfrak{S}$, then for each $\mathbf{v} \in M$ we have

$$(\mathbf{v}\mathbf{B})\mathbf{A} = (\mathbf{v}\mathbf{A})\mathbf{B} = (c\mathbf{v})\mathbf{B} = c(\mathbf{v}\mathbf{B}),$$

hence $\mathbf{v}\mathbf{B} \in M$. It follows that $M\mathbf{B} \leqq M$, *i.e.*, that M reduces $\mathbf{B}$ and consequently M reduces $\mathfrak{S}$.

As a corollary, *any commutative set may be superdiagonalized.*

Proof. Exactly as for the corollary to Theorem 3. It must only be checked that the sets induced by $\mathfrak{S}$ on M and on L/M are also commutative; this follows immediately from their definitions.(6)

7. Algebras and their representations.(4) An *associative algebra* is the structure consisting of a (finite dimensional) linear space $\mathfrak{A}$ over a field k which is also an associative ring and for which the identities $(ax)y = x(ay) = a(xy)$ hold, connecting the multiplication by a scalar a with the ring multiplication of algebra elements x, y. A *subalgebra* $\mathfrak{B}$ of $\mathfrak{A}$ is a subset of $\mathfrak{A}$ which is also an algebra under the operations of $\mathfrak{A}$. If $\mathfrak{B}$ and $\mathfrak{C}$ are two subalgebras, then $\mathfrak{B}\mathfrak{C}$ denotes the totality of finite sums $\sum y_i z_i$ where $y_i \in \mathfrak{B}$, $z_i \in \mathfrak{C}$. Clearly $\mathfrak{B}\mathfrak{C}$ is a linear subspace of $\mathfrak{A}$, but it is not necessarily a subalgebra itself. It is not even true in general that $\mathfrak{B}\mathfrak{C}$ includes $\mathfrak{B}$ or $\mathfrak{C}$. A subalgebra $\mathfrak{B}$ of $\mathfrak{A}$ is an *ideal* if $\mathfrak{A}\mathfrak{B} \leqq \mathfrak{B}$ and $\mathfrak{B}\mathfrak{A} \leqq \mathfrak{B}$. If $\mathfrak{B}$ and $\mathfrak{C}$ are ideals, then $\mathfrak{B}\mathfrak{C}$ is also an ideal. Also, in general $(\mathfrak{B}\mathfrak{C})\mathfrak{D} = \mathfrak{B}(\mathfrak{C}\mathfrak{D})$.

A *homomorphism* is an operation preserving mapping on an algebra $\mathfrak{A}_1$ into an algebra $\mathfrak{A}_2$. The *kernel*, consisting of all elements of $\mathfrak{A}_1$ mapping onto 0 is an ideal of $\mathfrak{A}_1$; indeed each ideal of an algebra is the kernel of some homomorphism. The image of $\mathfrak{A}_1$ is a subalgebra of $\mathfrak{A}_2$.

Especially interesting are the *matric algebras*, or *algebras of linear transformations*. These are the subalgebras of Lin (L), the algebra of all linear transformations on a finite-dimensional space L over k. Each algebra $\mathfrak{A}$ is homomorphic to an algebra of linear transformations on $L = \mathfrak{A}$ given by the *regular representation*: each $y \in \mathfrak{A}$ is associated to the linear transformation $\mathbf{R}_y$, called the *right multiplication* of y, and defined by

$$x\mathbf{R}_y = xy.$$

An element x of an algebra $\mathfrak{A}$ is *nilpotent* if some power of x vanishes: $x^n = 0$. The algebra $\mathfrak{A}$ is called a *nil algebra* if each of its elements is nilpotent. The algebra $\mathfrak{A}$ is called *nilpotent* if some power of the whole algebra vanishes: $\mathfrak{A}^n = 0$. Here $\mathfrak{A}^n = \mathfrak{A}\mathfrak{A}\mathfrak{A} \cdots$ to n factors. Thus $\mathfrak{A}^n$ is the set of all sums $\sum x_1 y_2 \cdots z_n$, and $\mathfrak{A}^n = 0$ is the same as $x_1 y_2 \cdots z_n = 0$ for every n elements $x_1, y_2, \cdots, z_n$ of $\mathfrak{A}$. Clearly a nilpotent algebra is nil.

8. Nil algebras. The following result presupposes no special properties of the ground field.

THEOREM 5. *Let $\mathfrak{A}$ be a nil algebra of linear transformations on a space L of* *dimension* $n>1$. *Then there is a subspace M of L*, $0<M<L$, *and M reduces* $\mathfrak{A}$.

Proof. If $\mathfrak{A}=0$, it is clear. Assuming $\mathfrak{A}\neq 0$, there is an element $\mathbf{v}\in L$ such that $M=\mathbf{v}\mathfrak{A}\neq 0$. M is a subspace of L simply because $\mathfrak{A}$ is a linear space. Suppose $M=L$. Then $\mathbf{v}\in M$, hence $\mathbf{v}=\mathbf{vA}$ for some $\mathbf{A}\in\mathfrak{A}$. Hence $\mathbf{v}=\mathbf{vA}=\mathbf{vA}^2=\mathbf{vA}^3=\cdots$. But $\mathbf{A}$ is nilpotent, hence for some r, $\mathbf{A}^r=0$. This implies $\mathbf{v}=0$ which contradicts $M\neq 0$. Finally $M\mathfrak{A}=(\mathbf{v}\mathfrak{A})\mathfrak{A}=\mathbf{v}(\mathfrak{A}\mathfrak{A})\leqq\mathbf{v}\mathfrak{A}=M$ since $\mathfrak{A}$ is a ring (closed under multiplication).

It quickly follows that there is a basis of L with respect to which $\mathfrak{A}$ is superdiagonalized. But each $\mathbf{A}\in\mathfrak{A}$ is nilpotent, hence the diagonal elements in the matrix $A=\|a_{ij}\|$ representing $\mathbf{A}$ must all vanish.

COROLLARY 1. *Each nil algebra* $\mathfrak{A}$ *of linear transformations on an n-dimensional linear space L is nilpotent, in fact*

$$\mathfrak{A}^n = 0.$$

For we have seen that there is a basis of L such that each element $\mathbf{A}$ of $\mathfrak{A}$ has the representation

$$A = \begin{bmatrix} 0 & a_{12} \cdots & a_{1,n-1} & a_{1n} \\ \cdot & \cdot & \cdot & \cdot \\ \cdot & \cdot & \cdot & \cdot \\ \cdot & \cdot & \cdot & \cdot \\ 0 & 0 & \cdots 0 & a_{n-1,n} \\ 0 & 0 & \cdots 0 & 0 \end{bmatrix}.$$

But the product of any n such matrices vanishes.

Surprisingly enough, we can now generalize this corollary to include *all* algebras, not just the algebras of linear transformations.

COROLLARY 2. *Let* $\mathfrak{A}$ *be a nil algebra of dimension n over k. Then* $\mathfrak{A}$ *is nilpotent, in fact* $\mathfrak{A}^{n+1}=0$.

Proof. The regular representation of $\mathfrak{A}$ yields a nil algebra $\mathfrak{A}_1$ of linear transformations on $L=\mathfrak{A}$. By Corollary 1, $\mathfrak{A}_1{}^n=0$. Now if $x_0, x_1, \cdots, x_n\in\mathfrak{A}$, then

$$x_0x_1\cdots x_n = x_0(x_1\cdots x_n) = x_0(\mathbf{R}_{x_1}\mathbf{R}_{x_2}\cdots\mathbf{R}_{x_n}) = x_0 0 = 0.$$

Thus $\mathfrak{A}^{n+1}=0$.

9. A generalization. In a certain sense, the following result includes both Theorems 4 and 5.[7]

THEOREM 6. *Let k be an algebraically closed field and let* $\mathfrak{A}$ *be an algebra of linear transformations on a linear space L over k of dimension* $n>1$. *Suppose there is a nil ideal* $\mathfrak{B}$ *of* $\mathfrak{A}$ *such that the quotient algebra* $\mathfrak{A}/\mathfrak{B}$ *is commutative. Then there exists a subspace M of L*, $0<M<L$, *and M reduces* $\mathfrak{A}$.

Proof. If $\mathfrak{B}=0$, then $\mathfrak{A}$ is commutative and the result follows from Theorem 4. If $\mathfrak{B}\neq 0$, there is a $\mathbf{v}$ in L such that $M=\mathbf{v}\mathfrak{B}\neq 0$. Exactly as in the proof of Theorem 5 we conclude that $0<M<L$. Finally $M\mathfrak{A}=(\mathbf{v}\mathfrak{B})\mathfrak{A}=\mathbf{v}(\mathfrak{B}\mathfrak{A})\leqq\mathbf{v}\mathfrak{B}$ since $\mathfrak{B}$ is an ideal.

The hypotheses on $\mathfrak{A}$—that $\mathfrak{A}$ has a nil ideal and that the quotient algebra modulo this ideal is commutative—go over to any homomorphic image of $\mathfrak{A}$. But both the restriction of $\mathfrak{A}$ to M and the algebra induced by $\mathfrak{A}$ on L/M are homomorphic images of $\mathfrak{A}$, hence we may apply induction to superdiagonalize $\mathfrak{A}$ in the usual manner.

10. Lie algebras. Theorems 4 and 5 allow very broad generalizations which must be expressed in the language of Lie algebras. A *Lie algebra* over a field k consists of a linear space $\mathfrak{A}$ over k in which is defined a multiplication $[x, y]$. Thus if $x, y \in \mathfrak{A}$, then $[x, y] \in \mathfrak{A}$. We assume $[x, y]$ is linear in each variable, that $[x, x]=0$, $[x, y]+[y, x]=0$, and the Jacobi identity

$$[x, [y, z]] + [y, [z, x]] + [z, [x, y]] = 0.$$

We particularly note that a Lie algebra is not usually an associative algebra. A *subalgebra* (or *Lie subalgebra*) of a Lie algebra $\mathfrak{A}$ is a subset $\mathfrak{B}$ of $\mathfrak{A}$ which is itself a Lie algebra with respect to the operations of $\mathfrak{A}$.

If $\mathfrak{A}$ is an associative algebra, we may turn it into a Lie algebra by defining the *Lie product*

$$[x, y] = xy - yx$$

in terms of the associative product. This applies in particular to the algebra Lin (L) of linear transformations on a linear space L. A Lie subalgebra of this particular algebra is called a *Lie algebra of linear transformations.* Via the usual matric representation, we obtain a *Lie algebra of matrices.* (Example: the $n \times n$ matrices of trace zero. Note that this is closed under the Lie product but not under the ordinary product.)

If $\mathfrak{B}$ and $\mathfrak{C}$ are Lie subalgebras of a Lie algebra $\mathfrak{A}$, then $[\mathfrak{B}, \mathfrak{C}]$ denotes the set of all sums $\sum x_i y_i$ where $x_i \in \mathfrak{B}$, $y_i \in \mathfrak{C}$. A subalgebra $\mathfrak{B}$ of $\mathfrak{A}$ is called an *ideal* if $[\mathfrak{A}, \mathfrak{B}] \leqq \mathfrak{B}$. (It is unnecessary to assume $[\mathfrak{B}, \mathfrak{A}] \leqq \mathfrak{B}$ as in the associative case because $[y, x] = -[x, y]$.) If $\mathfrak{B}$ and $\mathfrak{C}$ are ideals, then so is $[\mathfrak{B}, \mathfrak{C}]$. This follows from the Jacobi identity as we may indicate symbolically

$$[\mathfrak{A}, [\mathfrak{B}, \mathfrak{C}]] \leqq [\mathfrak{B}, [\mathfrak{C}, \mathfrak{A}]] + [\mathfrak{C}, [\mathfrak{A}, \mathfrak{B}]] \leqq [\mathfrak{B}, \mathfrak{C}] + [\mathfrak{C}, \mathfrak{B}] \leqq [\mathfrak{B}, \mathfrak{C}].$$

In particular, $\mathfrak{A}' = [\mathfrak{A}, \mathfrak{A}]$ is the *derived algebra* of $\mathfrak{A}$. If $\mathfrak{A}^{(r)}$ has been defined, then we define $\mathfrak{A}^{(r+1)} = [\mathfrak{A}^{(r)}, \mathfrak{A}^{(r)}]$. Evidently $\mathfrak{A}', \mathfrak{A}'', \cdots$ are all ideals of $\mathfrak{A}$ and

$$\mathfrak{A} \geqq \mathfrak{A}' \geqq \mathfrak{A}'' \geqq \cdots \geqq \mathfrak{A}^{(r)} \geqq \cdots.$$

The Lie algebra $\mathfrak{A}$ is called *solvable* if $\mathfrak{A}^{(r)}=0$ for some r.

11. Engel's theorem.[8] The theorem of Engel we now discuss is the generalization of Theorem 5 we have in mind.

THEOREM 7. *Let $\mathfrak{A}$ be a Lie algebra of linear transformations on a linear space L over k such that each* **A** *in $\mathfrak{A}$ is nilpotent. Then there exists a non-zero vector* **v** *in L such that* $\mathbf{v}\mathfrak{A}=0$.

Here is the difference between this result and that of Theorem 5. In Theorem 5 we assumed not only that each $\mathbf{A}$ in $\mathfrak{A}$ is nilpotent, but that $\mathfrak{A}$ is an associative algebra, hence closed under ordinary multiplication: if $\mathbf{A}$ and $\mathbf{B}\in\mathfrak{A}$, then $\mathbf{AB}\in\mathfrak{A}$. Now we are only assuming $\mathbf{AB}-\mathbf{BA}=[\mathbf{A}, \mathbf{B}]\in\mathfrak{A}$.

Proof. We proceed by induction on $m=\dim(\mathfrak{A})$. When $m=1$, the result is obvious since $\mathfrak{A}=k\mathbf{A}$ where $\mathbf{A}$ is nilpotent. We henceforth assume $m>1$ and that the result is true for all lower dimensions. Since $\mathfrak{A}$ contains a one-dimensional Lie subalgebra,[9] there exists a subalgebra $\mathfrak{B}$, $0<\mathfrak{B}<\mathfrak{A}$, of *maximal dimension.* By the induction hypothesis, there is a non-zero $\mathbf{w}$ in L such that $\mathbf{w}\mathfrak{B}=0$. We set $\mathfrak{B}_1=\{\mathbf{A}\in\mathfrak{A} \mid \mathbf{wA}=0\}$. Then $\mathfrak{B}_1$ is a Lie subalgebra of $\mathfrak{A}$ and $\mathfrak{B}\leqq\mathfrak{B}_1\leqq\mathfrak{A}$. Since $\mathfrak{B}$ has maximal dimension there are only two possibilities: $\mathfrak{B}_1=\mathfrak{A}$ or $\mathfrak{B}_1=\mathfrak{B}$. If $\mathfrak{B}_1=\mathfrak{A}$, then $\mathbf{w}\mathfrak{A}=\mathbf{w}\mathfrak{B}_1=0$ and $\mathbf{v}=\mathbf{w}$ is the sought-for vector. We may consequently suppose that the contrary case $\mathfrak{B}_1=\mathfrak{B}$ prevails.

Thus *if* $\mathbf{A}\in\mathfrak{A}$ *and* $\mathbf{wA}=0$, *then* $\mathbf{A}\in\mathfrak{B}$. We set $M=\mathbf{w}\mathfrak{A}$ so that M is a subspace of L (since $\mathfrak{A}$ is a linear space). Clearly $M>0$ since $\mathbf{wA}\neq 0$ for each $\mathbf{A}$ not in $\mathfrak{B}$. Also $\mathbf{w}\mathfrak{A}<L$, for if $\mathbf{w}\in\mathbf{w}\mathfrak{A}$, then $\mathbf{w}=\mathbf{wA}$ for some $\mathbf{A}\in\mathfrak{A}$; but each $\mathbf{A}$ in $\mathfrak{A}$ is by hypothesis nilpotent. Thus $0<M<L$.

We next prove that $M\mathfrak{B}\subseteq M$. For if $\mathbf{wA}\in M$ and $\mathbf{B}\in\mathfrak{B}$, then

$$(\mathbf{wA})\mathbf{B} = \mathbf{w}(\mathbf{AB}) = \mathbf{w}[\mathbf{A}, \mathbf{B}] + \mathbf{wBA} = \mathbf{w}[\mathbf{A}, \mathbf{B}] \in \mathbf{w}\mathfrak{A} = M.$$

Thus $\mathfrak{B}$ *induces a Lie algebra of linear transformations on* M, each element of which is nilpotent since that is so of $\mathfrak{B}$. We apply the induction hypothesis once again to find a non-zero vector $\mathbf{z}$ in M such that $\mathbf{z}\mathfrak{B}=0$. Thus $\mathbf{z}=\mathbf{wA}_0$ for some $\mathbf{A}_0\in\mathfrak{A}$.

We next prove that $[\mathbf{A}_0, \mathfrak{B}]\subseteq\mathfrak{B}$. For if $\mathbf{B}\in\mathfrak{B}$, then

$$\mathbf{w}[\mathbf{A}_0, \mathbf{B}] = \mathbf{wA}_0\mathbf{B} - \mathbf{wBA}_0 = \mathbf{zB} = 0,$$

hence $[\mathbf{A}_0, \mathbf{B}]\in\mathfrak{B}$. It follows from this and the identity $[\mathbf{A}_0, \mathbf{A}_0]=0$, that the set

$$\mathfrak{A}_1 = \mathfrak{B} + k\mathbf{A}_0$$

is a Lie subalgebra of $\mathfrak{A}$ and that $\mathfrak{B}$ is an ideal in $\mathfrak{A}_1$. By the maximality of $\mathfrak{B}$ we must have $\mathfrak{A}_1=\mathfrak{A}$,

$$\mathfrak{A} = \mathfrak{B} + k\mathbf{A}_0,$$

and $\mathfrak{B}$ is an ideal in $\mathfrak{A}$.

Now let $N=\{\mathbf{u}\in L \mid \mathbf{u}\mathfrak{B}=0\}$ so that $\mathbf{w}\in N$, hence $0<N\leqq L$. If $\mathbf{u}\in N$ and $\mathbf{B}\in\mathfrak{B}$, then

$$(\mathbf{uA}_0)\mathbf{B} = \mathbf{u}(\mathbf{A}_0\mathbf{B}) = \mathbf{u}[\mathbf{A}_0, \mathbf{B}] + \mathbf{uBA}_0 = 0,$$

hence $\mathbf{uA}_0\in N$, $N\mathbf{A}_0\leqq N$, *i.e.*, N reduces $\mathbf{A}_0$. Since $\mathbf{A}_0$ is nilpotent, there exists a non-zero vector $\mathbf{v}\in N$ such that $\mathbf{vA}_0=0$. But $\mathbf{v}\mathfrak{B}=0$ since $\mathbf{v}\in N$, hence finally $\mathbf{v}\mathfrak{A}=0$.[10]

The following corollaries are now easy consequences.

COROLLARY 1. *Any Lie algebra of nilpotent transformations may be superdiagonalized.*

COROLLARY 2. *If* $\mathfrak{A}$ *is a Lie algebra of nilpotent transformations on an n-dimensional space L, and if* $\mathbf{A}_1, \cdots, \mathbf{A}_n \in \mathfrak{A}$, *then*

$$\mathbf{A}_1\mathbf{A}_2 \cdots \mathbf{A}_n = 0.$$

12. A lemma of Jacobson. As preparation for the proof in Section **14** we need the following result.(11)

LEMMA. *Let L be an n-dimensional linear space over a field k of characteristic* 0. *Let* **A**, **B**, **C** *be linear transformations on L such that*

$$\mathbf{C} = [\mathbf{A}, \mathbf{B}] \quad \text{and} \quad [\mathbf{A}, \mathbf{C}] = 0.$$

Then **C** *is nilpotent.*

Characteristic zero means that k contains the field of rationals or, what is the same thing, that if

$$n \cdot a = a + a + \cdots (n \text{ summands}) = 0,$$

then $a=0$ (for $n=1, 2, 3, \cdots$).(12)

Proof. The proof is based on a trace argument. The trace function S is linear and satisfies $S(\mathbf{XY}) = S(\mathbf{YX})$. We shall show that $S(\mathbf{C}^r) = 0$ for $r = 1, 2, \cdots$. In fact

$$S(\mathbf{C}) = S([\mathbf{A}, \mathbf{B}]) = S(\mathbf{AB}) - S(\mathbf{BA}) = 0,$$

and

$$\mathbf{C}^r = \mathbf{C}^{r-1}[\mathbf{A}, \mathbf{B}] = \mathbf{C}^{r-1}\mathbf{AB} - \mathbf{C}^{r-1}\mathbf{BA}.$$

But $[\mathbf{A}, \mathbf{C}] = \mathbf{AC} - \mathbf{CA} = 0$, hence $\mathbf{AC} = \mathbf{CA}$ and this implies $\mathbf{AC}^{r-1} = \mathbf{C}^{r-1}\mathbf{A}$. We have

$$S(\mathbf{C}^r) = S(\mathbf{AC}^{r-1}\mathbf{B}) - S(\mathbf{C}^{r-1}\mathbf{BA}) = S(\mathbf{AC}^{r-1}\mathbf{B}) - S(\mathbf{AC}^{r-1}\mathbf{B}) = 0.$$

The lemma now follows from the known trace criterion (over characteristic zero) for nilpotence: **C** is nilpotent when the first n powers of **C** have trace zero. For completeness we include a discussion of this result.

13. The trace argument. The precise result is this.

THEOREM 8. *Let* **C** *be a linear transformation on an n-dimensional space L over a field k of characteristic zero and suppose that*

$$S(\mathbf{C}) = S(\mathbf{C}^2) = S(\mathbf{C}^3) = \cdots = S(\mathbf{C}^n) = 0.$$

Then **C** *is nilpotent.*

We may give an irrational proof based on Theorem 3. We first must enlarge the field k so that it includes the characteristic roots of **C**. Then we may take a basis of L so that the matrix $C=\|c_{ij}\|$ of **C** is superdiagonal, the diagonal elements being the characteristic roots of **C**. Each power C^r of C is also superdiagonal and we realize from this that if $c_1, \cdots, c_n$ are the diagonal elements of C, then $c_1^r, \cdots c_n^r$ are those of C^r, hence the trace conditions become

$$\begin{cases} c_1 + c_2 + \cdots + c_n = 0 \\ c_1^2 + c_2^2 + \cdots + c_n^2 = 0 \\ c_1^n + c_2^n + \cdots + c_n^n = 0. \end{cases}$$

We deduce either by the Vandermonde determinant or by the Newton identities (on symmetric functions) that $c_1=c_2= \cdots =c_n=0$, and hence that C is nilpotent.

More in keeping with the spirit of this paper is a rational proof of Theorem 8 based on the technique rather than the result of Theorem 3. We begin with the characteristic polynomial

$$f(t) = \det (tI - \mathbf{C}) = t^n + a_1t^{n-1} + \cdots + a_n$$

of **C**. By the Cayley-Hamilton theorem,

$$\mathbf{C}^n + a_1\mathbf{C}^{n-1} + \cdots + a_n\mathbf{I} = 0.$$

Taking traces:

$$na_n = 0.$$

Hence $a_n=0, f(0)=0$. It follows that 0 is a characteristic root of **C**, hence there is a non-zero vector $\mathbf{v}_n$ in L such that $\mathbf{v}_n\mathbf{C}=0$. We make this the last element in a basis $\mathbf{v}_1, \mathbf{v}_2, \cdots, \mathbf{v}_{n-1}, \mathbf{v}_n$ of L so that the corresponding matrix of **C** is of the form

$$C = \begin{pmatrix} C_1 & * \\ 0 & 0 \end{pmatrix}$$

where C_1 is an $(n-1)$-rowed square matrix and the asterisk denotes some $(n-1)$-element column. Clearly

$$C^r = \begin{pmatrix} C_1{}^r & * \\ 0 & 0 \end{pmatrix}$$

so that $S(C^r)=S(C_1^r)=0$ for $r=1, 2, \cdots, (n-1)$ (and also n), *i.e.*, the hypotheses are satisfied for C_1. An induction argument suffices to finish the proof.

14. The theorems of Lie. We return to the study of Lie algebras. We recall that a Lie algebra A is *solvable* if the derived sequence of ideals $\mathfrak{A} \geqq \mathfrak{A}' \geqq \mathfrak{A}'' \geqq \cdots$ terminates with 0.

LEMMA. *Each Lie subalgebra and each homomorphic image of a solvable Lie algebra $\mathfrak{A}$ is solvable.*

Proof. If $\mathfrak{B} \leqq \mathfrak{A}$, then $\mathfrak{B}' = [\mathfrak{B}, \mathfrak{B}] \leqq [\mathfrak{A}, \mathfrak{A}] = \mathfrak{A}'$, $\mathfrak{B}'' \leqq \mathfrak{A}''$, *etc.* Hence $\mathfrak{A}^{(r)} = 0$ implies $\mathfrak{B}^{(r)} = 0$. The solvability of a homomorphic image is proved similarly.

The basic theorem of Lie shows that solvability implies a strong property of the derived algebra.

THEOREM 9. *Let $\mathfrak{A}$ be a solvable Lie algebra of linear transformations on an n-dimensional space L. Then each element of $\mathfrak{A}'$ is nilpotent.*

Proof. The solvability of $\mathfrak{A}$ implies

$$\mathfrak{A} > \mathfrak{A}' > \mathfrak{A}'' > \cdots > \mathfrak{A}^{(r)} = 0,$$

for equality at any step would entail equality forever afterwards. We set $m = \dim(\mathfrak{A})$ and select a linear subspace $\mathfrak{B}$ of $\mathfrak{A}$ such that

$$\mathfrak{A}' \leqq \mathfrak{B} < \mathfrak{A}$$

and $\dim(\mathfrak{B}) = m - 1$. Then

$$[\mathfrak{B}, \mathfrak{B}] \subseteq [\mathfrak{A}, \mathfrak{A}] = \mathfrak{A}' \leqq \mathfrak{B},$$

hence $\mathfrak{B}$ is a subalgebra. Even more:

$$[\mathfrak{A}, \mathfrak{B}] \subseteq [\mathfrak{A}, \mathfrak{A}] = \mathfrak{A}' \leqq \mathfrak{B},$$

hence $\mathfrak{B}$ is an ideal in $\mathfrak{A}$. We can actually do a little better than this. For if $\mathbf{A}_0$ is any element of $\mathfrak{A}$ not in $\mathfrak{B}$, then $\mathfrak{A} = \mathfrak{B} + k\mathbf{A}_0$; since $[\mathbf{A}_0, \mathbf{A}_0] = 0$, we have

$$\mathfrak{A}' = [\mathfrak{A}, \mathfrak{A}] = [\mathfrak{B} + k\mathbf{A}_0, \mathfrak{B} + k\mathbf{A}_0] \subseteq [\mathfrak{B}, \mathfrak{B}] + [\mathfrak{B}, \mathbf{A}_0] + [\mathbf{A}_0, \mathfrak{B}] \subseteq [\mathfrak{A}, \mathfrak{B}],$$

consequently

$$\mathfrak{A}' = [\mathfrak{A}, \mathfrak{B}].$$

We proceed with the proof of the theorem by double induction on (m, n). If $m = 1$, then $\mathfrak{A} = k\mathbf{A}_0$, $\mathfrak{A}' = 0$. If $n = 1$, then $\dim(L) = n = 1$, $\mathfrak{A}$ is a commutative set, $\mathfrak{A}' = 0$. Thus we may assume the theorem whenever either m or n (or both) are decreased, but neither is increased.

In particular we may assume that each element of $\mathfrak{B}'$ is nilpotent.

Case 1. $\mathfrak{B}' = 0$ Then if $\mathbf{A} \in \mathfrak{A}$ and $\mathbf{B} \in \mathfrak{B}$ we have $\mathbf{C} = [\mathbf{A}, \mathbf{B}] \in \mathfrak{B}$, $[\mathbf{C}, \mathbf{B}] \in \mathfrak{B}'$, hence $[\mathbf{C}, \mathbf{B}] = 0$. By the Jacobson Lemma $\mathbf{C}$ is nilpotent. Each element of $\mathfrak{A}' = [\mathfrak{A}, \mathfrak{B}]$ is a sum of such elements $\mathbf{C}$. Any two such, $\mathbf{C}_1$, $\mathbf{C}_2$ commute since $[\mathbf{C}_1, \mathbf{C}_2] \in \mathfrak{B}' = 0$, hence each element of $\mathfrak{A}'$, $\sum \mathbf{C}_j$, is nilpotent.[13]

Case 2. $\mathfrak{B}' \neq 0$. By Theorem 7, if $M = \{\mathbf{v} \in L \mid \mathbf{v}\mathfrak{B}' = 0\}$, then $0 < M < L$. We shall prove that M reduces $\mathfrak{A}$. To do this we must prove that if $\mathbf{v} \in M$ and $\mathbf{A} \in \mathfrak{A}$, then $(\mathbf{v}\mathbf{A})\mathfrak{B}' = 0$. But each element of $\mathfrak{B}'$ is a sum of elements of the form $[\mathbf{B}_1, \mathbf{B}_2]$ where $\mathbf{B}_1, \mathbf{B}_2 \in \mathfrak{B}$. Now

$$(\mathbf{v}\mathbf{A})[\mathbf{B}_1, \mathbf{B}_2] = \mathbf{v}[\mathbf{A}, [\mathbf{B}_1, \mathbf{B}_2]] + \mathbf{v}[\mathbf{B}_1, \mathbf{B}_2]\mathbf{A}$$
$$= -\mathbf{v}[\mathbf{B}_1, [\mathbf{B}_2, \mathbf{A}]] - \mathbf{v}[\mathbf{B}_2, [\mathbf{A}, \mathbf{B}_1]].$$

Since $\mathfrak{B}$ is an ideal of $\mathfrak{A}$, $[\mathbf{B}_2, \mathbf{A}] \in \mathfrak{B}$, $[\mathbf{B}_1, [\mathbf{B}_2, \mathbf{A}]] \in \mathfrak{B}'$. Likewise $[\mathbf{B}_2, [\mathbf{A}, \mathbf{B}_1]] \in \mathfrak{B}'$. Hence

$$(\mathbf{v}\mathbf{A})[\mathbf{B}_1, \mathbf{B}_2] = 0.$$

Since $\dim(M) < \dim(L)$, the induction hypothesis implies that each element of $\mathfrak{A}'$ induces a nilpotent linear transformation on M. By the theorem of Engel, Theorem 7, we find a non-zero vector $\mathbf{w}$ in M such that $\mathbf{w}\mathfrak{A}' = 0$.

We now set $N = \{\mathbf{w} \in L \mid \mathbf{w}\mathfrak{A}' = 0\}$ so that $0 < N < L$. (If $N = L$, then $\mathfrak{A}' = 0$, $\mathfrak{B}' = 0$; but we are still in Case 2.) The same argument we used on M may now be used on N to prove that $\mathfrak{A}$ reduces N: $N\mathfrak{A} \subseteq N$. Consequently $\mathfrak{A}$ acts on the quotient space L/N which is of lower dimension than L. By induction once again we deduce that each element of $\mathfrak{A}'$ induces a nilpotent transformation on L/N. Thus if $\mathbf{C} \in \mathfrak{A}'$, then for some s, $L\mathbf{C}^s \leqq N$, hence $L\mathbf{C}^{s+1} \leqq N\mathbf{C} = 0$. Thus each element of $\mathfrak{A}'$ is nilpotent on L.[14]

The superdiagonalization theorem for solvable Lie algebras holds, as we shall now see. This generalizes Theorem 6.

THEOREM 10. *Let $\mathfrak{A}$ be a solvable Lie algebra of linear transformations on a linear space L of dimension $n > 1$ over an algebraically closed field k of characteristic zero. Then there is a subspace M of L such that $0 < M < L$ and M reduces $\mathfrak{A}$.*

Proof. We set

$$N = \{\mathbf{v} \in L \mid \mathbf{v}\mathfrak{A}' = 0\}.$$

By Theorems 9 and 7 we have $0 < N \leqq L$. If $N = L$, then $\mathfrak{A}' = 0$ which means that $\mathfrak{A}$ is a commutative set; we simply apply Theorem 4. If $0 < N < L$, then we take $M = N$ and must prove that N reduces $\mathfrak{A}$. But $\mathbf{v} \in N$, $\mathbf{A} \in \mathfrak{A}$, and $\mathbf{C} \in \mathfrak{A}'$ implies

$$(\mathbf{v}\mathbf{A})\mathbf{C} = \mathbf{v}[\mathbf{A}, \mathbf{C}] + \mathbf{v}\mathbf{C}\mathbf{A} = 0$$

since $[\mathbf{A}, \mathbf{C}] \in \mathfrak{A}'$.

COROLLARY. *Under the hypotheses of Theorem 10, there exists a basis of L with respect to which each transformation $\mathbf{A}$ of $\mathfrak{A}$ has a superdiagonal matrix.*

By using the regular representation, these results may be applied to abstract Lie algebras. If $\mathfrak{A}$ is a Lie algebra over k and $y \in \mathfrak{A}$, we set $L = \mathfrak{A}$ and define $\mathbf{R}_y$ on L by

$$x\mathbf{R}_y = [x, y].$$

We have

$$\begin{aligned} x\mathbf{R}_{[y,z]} &= [x, [y, z]] = -[z, [x, y]] - [y, [z, x]] \\ &= [[x, y], z] - [[x, z], y] \\ &= [x\mathbf{R}_y, z] - [x\mathbf{R}_z, y] \\ &= x\mathbf{R}_y\mathbf{R}_z - x\mathbf{R}_z\mathbf{R}_y \\ &= x(\mathbf{R}_y\mathbf{R}_z - \mathbf{R}_z\mathbf{R}_y) = x[\mathbf{R}_y, \mathbf{R}_z], \end{aligned}$$

hence

$$\mathbf{R}_{[y,z]} = \mathbf{R}_y, \mathbf{R}_z.$$

This means that the mapping $y \to \mathbf{R}_y$ is a homomorphism on the Lie algebra $\mathfrak{A}$ into a Lie algebra of linear transformations on L. (This may be interpreted as the content of the Jacobi identity.) Here is a sample of how this is used.

THEOREM 11. *Let $\mathfrak{A}$ be a solvable Lie algebra over a field of characteristic zero. Then there is an integer r such that whenever $x \in \mathfrak{A}$ and $y_1, y_2, \cdots, y_r \in \mathfrak{A}'$, then*

$$[\cdots [[x, y_1], y_2] \cdots, y_r] = 0.$$

This follows by applying the theorems of Lie and Engel to the regular representation of $\mathfrak{A}$.

15. Characteristic roots. Here we make a further remark on super-diagonal sets. If $\mathfrak{S}$ is a set of superdiagonal $n \times n$ matrices, each $A \in \mathfrak{S}$ has the form

$$A = \begin{pmatrix} a_{11} & \cdots & a_{1n} \\ & \ddots & \vdots \\ 0 & & a_{nn} \end{pmatrix}$$

and the i-th diagonal element may be denoted by $\lambda_i(A)$ to denote its dependence on A. Thus $\lambda_1(A), \cdots, \lambda_n(A)$ are the characteristic roots of A. It is clear that if $B \in \mathfrak{S}$, then

$$\begin{aligned} \lambda_i(aA + bB) &= a\lambda_i(A) + b\lambda_i(B), \\ \lambda_i(AB) &= \lambda_i(A)\lambda_i(B). \end{aligned}$$

Thus if $F(X_1, \cdots, X_r)$ is any (non-commutative) polynomial in r variables with scalar coefficients, if $A_1, \cdots, A_r \in \mathfrak{S}$, and if $B = F(A_1, \cdots, A_r)$, then

$$\lambda_i(C) = F(\lambda_i(A_1), \cdots, \lambda_i(A_r)).$$

Notes

Numbers in square brackets refer to the bibliography below.

1. The basic material on linear spaces is found in the very readable book of Halmos [3]. We differ from the notation there in writing $\mathbf{v}A$ for the effect of a linear transformation A on a vector $\mathbf{v}$, instead of the customary $A\mathbf{v}$. We use the notation $M \leqq L$ for "M is a subspace of L" and $M < L$ for "M is a proper subspace of L."

2. Proofs will also be found in [1, p. 90] and [6, p. 302].
3. See [2, especially Ch. IV].
4. The necessary elementary material on associative algebras is given in [2, Ch. I] and [7, Ch. XVI].
5. See [1, p. 244].
6. The superdiagonalization of commutative sets is due to Frobenius. There is an interesting application to complex euclidean space (also a corresponding one for real euclidean space, but we restrict interest to the complex case.) Suppose that L is an n-dimensional complex space with an inner product denoted $(\mathbf{v}, \mathbf{w})$. From a sequence $L = M_1 > M_2 > \cdots > M_n > 0 = M_{n+1}$ of subspaces we now extract an orthonormal basis $\mathbf{v}_1, \cdots, \mathbf{v}_n$ by selecting $\mathbf{v}_j$ in M_j of unit length and orthogonal to M_{j+1}. Suppose $\mathfrak{S}$ is a commutative set of normal transformations. Thus $\mathrm{A} \in S$ implies $\mathrm{AA}^* = \mathrm{A}^*\mathrm{A}$ where A^* is the adjoint of A defined by $(\mathbf{v}\mathrm{A}, \mathbf{w}) = (\mathbf{v}, \mathbf{w}\mathrm{A}^*)$. Then by Theorem 4 and the remark we have just made, there is an orthonormal basis $\mathbf{v}_1, \cdots, \mathbf{v}_n$ of L with respect to which the matrix $A = \|a_{jk}\|$ of each A in $\mathfrak{S}$ is superdiagonal. But A is a normal matrix since A is normal and the basis is orthonormal. This means $AA^* = A^*A$ where A^* is the transposed conjugate of A. It quickly follows, by computing the diagonal elements of AA^* and A^*A, that A must be in diagonal, or spectral, form. Thus *a commutative set of normal transformations may be simultaneously brought to spectral form.*
7. See [5].
8. See [4], [8].
9. This is quite general. If $\mathfrak{A}$ is any Lie algebra over k and x is any non-zero element of $\mathfrak{A}$, then $\mathfrak{B} = kx$ is a (one-dimensional) Lie subalgebra of $\mathfrak{A}$. We must prove that $[\mathfrak{B}, \mathfrak{B}] \subseteq \mathfrak{B}$. But $[x, x] = 0$, hence $[\mathfrak{B}, \mathfrak{B}] = 0$.
10. This is not an easy result to prove by any means. The closure of $\mathfrak{A}$ under $[\mathrm{A}, \mathrm{B}] = \mathrm{AB} - \mathrm{BA}$ together with the nilpotence of each element of $\mathfrak{A}$ may be used to systematically derive identities amongst the elements of $\mathfrak{A}$: this is the basis of the interesting proof in [8] which, however, considers only the special case in which $L = \mathfrak{A}$ and $\mathfrak{A}$ acts on L through the regular representation (called the *adjoint* representation in Lie algebra parlance). The proof we have given minimizes the use of identities by the device of selecting a maximal $\mathfrak{B}$ with the desired property.
11. See [4].
12. The lemma and proof are also valid if k is a field of characteristic p provided that $p > n$. The same applies to the applications in Section 14.
13. If C_1, C_2 are nilpotent, $\mathrm{C}_1^p = \mathrm{C}_2^q = 0$, and $\mathrm{C}_1\mathrm{C}_2 = \mathrm{C}_2\mathrm{C}_1$, then $(\mathrm{C}_1 + \mathrm{C}_2)^{p+q-1} = 0$ since the binomial expansion of the power leads to a sum of terms $\mathrm{C}_1^\alpha \mathrm{C}_2^\beta$ where $\alpha + \beta = p + q - 1$. Hence either $\alpha \geqq p$ or $\beta \geqq q$. The same reasoning applies to any sum of commutative nilpotent transformations.
14. This proof is closely modeled on that of [4]. By using double induction we avoid the enveloping algebra that Jacobson employs.

Bibliography

1. A. A. Albert, Modern Higher Algebra, Chicago, 1937.
2. A. A. Albert, Structure of Algebras, New York, 1939.
3. P. R. Halmos, Finite Dimensional Vector Spaces, Princeton, 1942.
4. N. Jacobson, Rational methods in the theory of Lie algebras, Ann. Math., vol. 36, 1935, pp. 875–81.
5. N. H. McCoy, On the characteristic roots of matric polynomials, Bull. Amer. Math. Soc., vol. 42, 1936, pp. 592–600.
6. F. D. Murnaghan, Theory of Group Representations, Baltimore, 1938.
7. Van der Waerden, Modern Algebra II, 2nd edition, New York, 1950.
8. M. Zorn, On a theorem of Engel, Bull. Amer. Math. Soc., vol. 42, 1936, pp. 401–4.

LINEAR OPERATIONS ON MATRICES*†

MARVIN MARCUS, University of British Columbia and National Bureau of Standards

I. Introduction. In this article we survey some of the problems that arise in studying linear operations on finite matrices. As we shall see, these problems come from several interesting sources, e.g. multilinear algebra, combinatorial analysis, commutativity. As might be expected, there are a large number of unanswered questions that appear immediately and we will describe some of these in the course of this article.

The usual elementary linear operations on matrices are:

(i) multiplying a row (column) by a fixed nonzero constant;
(ii) interchanging rows (columns);
(iii) adding a multiple of one row (column) to another;
(iv) taking the transpose.

The reason that these operations are studied is that they leave fixed some important property of the matrix and at the same time bring the matrix into manageable form. For example, we know that a square matrix with entries in a field may be brought to diagonal form with only 1 and 0 appearing on the main diagonal, without altering its rank. This gives us a simple algorithm for computing the rank, $\rho(A)$, of any matrix A.

Let T denote a fixed linear operation obtained as a composite of linear operations of the kind described above. Then the important property that T has insofar as ρ is concerned is embodied in the equation: $\rho(T(A))=\rho(A)$ for all matrices A. A composite of linear operations of types (i), (ii), (iii) can be accomplished by pre- and post-multiplication by fixed nonsingular matrices. By this is meant that there exist fixed U and V such that

$$T(A) = UAV \qquad \text{for all } A. \tag{1}$$

This statement is false for (iv). That is, there exist no fixed U and V such that $UAV=A'$ for all A. We know of course that A and A' have the same similarity invariants and hence are similar, but the transforming matrix S_A satisfying $A'=S_AAS_A^{-1}$ depends on A. To see that U and V must depend on A first set $A=I$ so that $V=U^{-1}$. Then $UA=A'U$ and by varying A over diagonal matrices we conclude that U is a diagonal matrix. By next letting $A=E_{ij}$ (E_{ij} is the matrix with 1 in position (i, j), 0 elsewhere) we conclude that $U=I$. But then $A'=UAU^{-1}=A$ would imply that every matrix is symmetric.

We next consider a somewhat more involved example of a linear operation on matrices that holds fixed an important scalar valued function of the matrix. Let R_4 be the ordinary four dimensional vector space of real 4-tuples. Let $f(x)$

* From AMERICAN MATHEMATICAL MONTHLY, vol. 69 (1962), pp. 837–847.

† Supported in part by the Office of Naval Research. Present address: University of California, Santa Barbara.

be a function on R_4 to the real numbers defined by

$$f(x) = x_1^2 - x_2^2 - x_3^2 - x_4^2.$$

If L is a linear mapping of R_4 into itself satisfying

$$f(L(x)) = f(x)$$

for all x, then L is called a *Lorentz transformation.* Let ϕ be a mapping of R_4 into the set H_2 of complex 2-square Hermitian matrices defined by

$$\phi(x) = \begin{bmatrix} x_1 + x_2 & x_3 + ix_4 \\ x_3 - ix_4 & x_1 - x_2 \end{bmatrix}.$$

It is simple to verify that ϕ is an isomorphism of R_4 onto H_2, regarded as a vector space over the real numbers with $E_{11}+E_{22}$, $E_{11}-E_{22}$, $E_{12}+E_{21}$ and $i(E_{12}-E_{21})$ as a basis. We check immediately that

$$d(\phi(x)) = f(x),$$

where d denotes the determinant. Hence, if we define $T=\phi L\phi^{-1}$, a mapping on H_2 to itself, we have

$$d(\phi L\phi^{-1}(H)) = f(L\phi^{-1}(H)) = f(\phi^{-1}(H)) = d\phi(\phi^{-1}(H)) = d(H),$$

for all $H \in H_2$. Thus if $f(Lx)=f(x)$ we have $d(T(H))=d(H)$ and conversely. Therefore, determining the structure of the mappings L is the same as determining the structure of the mappings T that send 2-square Hermitian matrices into such and preserve the determinant.

II. Invariance. The general problem can be stated as follows. Let $M_n(F)$ denote the vector space of n-square matrices over a field F. Let Ω be a subspace of $M_n(F)$ and let f be a function on Ω with values in F. Let T be a linear transformation of Ω into itself with the property

$$f(T(A)) = f(A) \qquad \text{for all } A \in \Omega. \tag{2}$$

Problem. Determine the structure of the set $\Omega(f)$ of all T satisfying (2).

Sometimes the problem is modified by assuming that Ω is just a subset of some subspace R of $M_n(F)$ and that T is defined on R. For example, we can ask for all T that map the unitary group Ω into itself [11]. Ω is not a subspace of $M_n(F)$ and we must assume that T is defined on all of $M_n(F)$.

A natural choice for $f(A)$ is some fixed coefficient in the characteristic polynomial of A. Let $E_r(A)$ denote the sum of all principal r-square subdeterminants of A; if F is the complex field then $E_r(A)$ is the rth elementary symmetric function of the characteristic roots of A; e.g. $E_1(A)=tr(A)$, $E_n(A)=d(A)$. The problem of finding all T such that $d(T(A))=d(A)$ was solved originally by Kantor [6] and Frobenius [3]. These results were extended and generalized by Schur [26], Hua [4], and Dieudonné [2], and very recent results in this direction may be found in [16]. On the other hand not much can be said about T if we assume only that $E_1(T(A))=E_1(A)$. Next suppose $f(A)=E_2(A)$. Let T_{ij} be the linear map that interchanges the entries in positions (i, j) and (j, i) and leaves all other entries fixed. Clearly $E_2(T_{ij}(A))=E_2(A)$ for all A but T_{ij} cannot be represented (unless $n=2$) in the form

$$T(A) = UAV \quad \text{or} \quad T(A) = UA'V. \tag{3}$$

There are several other types of linear operations that hold E_2 invariant and are also not representable in the form (3):

(a) multiply the (i,j) element by λ and the (j,i) element by λ^{-1}, T_{ij}^{λ};

(b) interchange the (i, j), (j, i) pair with the (k, l), (l, k) pair of entries, T_{ijkl}.

We can easily prove that $\Omega(E_r)$ is a multiplicative group for $r\geqq 2$ and it is an open question whether or not $\Omega(E_2)$ is generated by the transformations T_{ij}, T_{ij}^{λ}, T_{ijkl} together with operations of type (3) with $U=V^{-1}$ (i.e. the *similarity transformations*). Clearly $\Omega(E_2)$ is closed under multiplication so it suffices in showing that $\Omega(E_2)$ is a group to prove simply that $E_r(T(A))=E_r(A)$ implies that T is nonsingular. Suppose then that $T(A)=0$. We have

$$\begin{aligned} E_r(A + X) &= E_r(T(A + X)) = E_r(T(A) + T(X)) \\ &= E_r(T(X)) \\ &= E_r(X), \end{aligned}$$

for all X. For $r\geqq 2$ one can make judicious choices of X to show that $A=0$. For $n>r\geqq 4$, $\Omega(E_r)$ is entirely generated by the similarity transformations and the transpose operation. That is, $T\in\Omega(E_r)$ if and only if T has the form (3) with $U=V^{-1}$ [12]. For $r=n$, $\Omega(E_n)$ consists of transformations of the type (3) with $d(UV)=1$. The structure of the group $\Omega(E_3)$ is an open question.

In 1925 I. Schur proved that if $r\geq 2$ and the rth order subdeterminants of $T(A)$ are fixed nonsingular linear homogeneous functions of the rth order subdeterminants of A then T must have the form (3). We may restate this result in terms of compound matrices. Recall that if A is an n-square matrix then $C_r(A)$, the rth compound of A, is the $\binom{n}{r}$-square matrix whose entries are the rth order subdeterminants of A arranged in doubly lexicographic order. Then Schur's result can be stated as follows: If there exists a nonsingular mapping S of the space $M_{\binom{n}{r}}(F)$ into itself such that $C_r(T(A))=SC_r(A)$ for all $A\in M_n(F)$ then T has the form (3). Here F is assumed to be the field of complex numbers.

An important fact that underlies many of the above results is the following: *If T maps $M_n(F)$ into itself and satisfies*

$$\rho(T(A)) = \rho(A) \tag{4}$$

for all $A\in M_n(F)$ then T must have the form (3). Hua [4] proved this for a somewhat more general class of transformations in which, however, it is assumed initially that T^{-1} exists and also has the property (4). Jacob [5] proved the same type of result when the conditions on F are weakened. In [9] the hypothesis is only that T preserves rank 1. That is, (4) is replaced by the assumption that *if* $\rho(A)=1$ *then* $\rho(T(A))=1$. Of course, if we know *a priori* that T is nonsingular and maps the set of rank 1 matrices *onto* itself then this constitutes no improvement. For then, if $\rho(A)=r$ then

$$A = \sum_{i=1}^{r} A_i, \qquad \rho(A_i) = 1 \quad \text{and} \quad \rho(T(A)) \leqq \sum_{i=1}^{r} \rho(T(A_i)) \leqq r.$$

The same argument can be applied to T^{-1} to conclude (4) for any A. The point of these remarks is that we can determine the structure of $\Omega(f)$ in many cases by showing that *if* $T\in\Omega(f)$ *then* T *preserves rank* 1. To illustrate this idea suppose that f is the determinant function. It is not difficult to show that *if* $A\neq 0$ *then* deg $d(xA+B)\leqq 1$ *for all* $B\in M_n$ *if and only if* $\rho(A)=1$. Thus if $T\in\Omega(d)$ and $\rho(A)=1$, it follows

$$d(xA + B) = d(xT(A) + T(B))$$

must have degree 1 for all $B\in M_n(F)$. But since $\Omega(d)$ is a multiplicative group, $T(B)$ ranges over $M_n(F)$ as B does. Hence the above remark applies to allow us to conclude that $\rho(T(A))=1$. We assume here that the underlying field has characteristic zero.

The theorem that determines the rank 1 preservers leads naturally into a structure problem for linear maps on spaces of tensors.

III. Mappings on tensor product spaces. Let U be an n-dimensional vector space over a field F. The space $U^{(2)}=U\otimes U$ is defined to be the dual space of the space of all bilinear functionals on U to the field F. We call $U^{(2)}$ the *tensor product* of U with itself. If y_1 and y_2 are in U we define $\beta=y_1\otimes y_2$ to be the element of $U^{(2)}$ satisfying $\beta(\phi)=\phi(y_1, y_2)$ for any bilinear functional ϕ. It is easy to show that if $x_1, \cdots, x_n$ is a basis of U then $x_i\otimes x_j$, $i, j=1, \cdots, n$ is a basis of $U^{(2)}$. If $w\in U^{(2)}$ we define $\omega(w)$ to be the least number of terms of the form $y_1\otimes y_2$ used in any representation of w as a sum. We call $\omega(w)$ the *length* of w. Next define an isomorphism θ between $M_n(F)$ and $U^{(2)}$ as follows:

$$\theta(E_{ij}) = x_i \otimes x_j$$

and extend linearly to all of $M_n(F)$. It is then easy to show that θ has the important property:

$$\rho(A) = r \quad \text{if and only if} \quad \omega(\theta(A)) = r.$$

If T is a linear transformation on $M_n(F)$ to itself then we may define S on $U^{(2)}$ to itself by

$$S = \theta T\theta^{-1}.$$

Then T will preserve rank 1 if and only if S preserves length 1. We can then proceed to show that *if S satisfies $\omega(S(\beta))=\omega(\beta)$ for all $\beta\in U^{(2)}$ then S must have the form*

$$S(x \otimes y) = Ux \otimes Vy \quad \text{or} \quad S(x \otimes y) = Uy \otimes Vx$$

for fixed matrices U and V in $M_n(F)$. It is an open problem to determine the structure of those T that preserve rank k for some fixed k, $1<k<n$. A question that does not have a simple matrix analogue involves higher order tensor spaces $U^{(k)}$. Let $U^{(k)}$ be the dual space of the space of all multilinear functionals of k vectors in U. Again, if $x_1, \cdots, x_k$ are in U we may define $x_1\otimes \cdots \otimes x_k\in U^{(k)}$ by $x_1\otimes \cdots \otimes x_k(\phi)=\phi(x_1, \cdots, x_k)$ for all multilinear functionals ϕ. The elements $x_{i_1}\otimes \cdots \otimes x_{i_k}$, $1\leqq i_t\leqq n$, $t=1, \cdots, k$ form a basis for $U^{(k)}$ and we may define the length of an element of $U^{(k)}$ in precisely the same way as we define it for $k=2$. Tensors of length 1 are sometimes called *pure vectors* in $U^{(k)}$ and it would be interesting to find all linear maps S on $U^{(k)}$ that map pure vectors into pure vectors. In a recent doctoral dissertation, R. Westwick considered a similar problem for mappings on Grassmann product spaces [**29**]. The question treated there is the following: Let $G_k(U)$ denote the dual space of the space of all alternating multilinear functionals on sets of k vectors in U, and for $x_j\in U$, $j=1, \cdots, k$ let $x_1\wedge \cdots \wedge x_k\in G_k(U)$ be defined by

$$x_1 \wedge \cdots \wedge x_k(\phi) = \phi(x_1, \cdots, x_k)$$

for all alternating multilinear functionals ϕ. The problem then is to determine all linear S mapping $G_k(U)$ into itself and having the property that any nonzero vector of the form $x_1\wedge \cdots \wedge x_k$ (a *pure vector of grade k*) is mapped into such a vector. In this case the map S turns out to be the kth compound of a matrix in $M_n(F)$ except when $n=2k$. This result is close to one of Chow [**1**] in which both S and S^{-1} are assumed to exist and to map the entire set of nonzero pure vectors of grade k *onto* itself. In [**8**] this result is used to determine $\Omega(f)$ when $f(A)=E_{2r}(A)$ and Ω is the space of skew-symmetric matrices over the real numbers. This is related to some work of Morita [**19**] who showed that if Ω is the space of skew hermitian matrices and $f(A)$ is the Hilbert norm of A (i.e. the nonnegative square root of the largest characteristic root of A^*A) then to within scalar multiples every transformation in $\Omega(f)$ is of the form (3) with U and V unitary except for $n=4$. Quite recently H. Minc and the present author extended this to symmetric functions of the eigenvalues of A^*A.

IV. The permanent function. A considerably less tractable matrix function to deal with is the permanent of A, defined by

$$p(A) = \sum_{\sigma} \prod_{i=1}^{n} a_{i\sigma(i)},$$

where the summation extends over all $n!$ permutations of $1, \cdots, n$. Recent interest in the permanent stems from its application to certain combinatorial problems [**24**] and an unresolved conjecture of van der Waerden [**18**]. We will give a short description of the v, k, λ problem and indicate where the permanent makes its appearance there. Let $0<\lambda<k<v$ and suppose we have v items $x_1, \cdots, x_v$ to be distributed into v sets $T_1, \cdots, T_v$ in such a way that $T_i \cap T_j$ has exactly λ of the x's in it for $i \neq j$ and each T_i has exactly k of the x's in it. We construct the *incidence matrix* A of this configuration as follows:

$$a_{ij} = 0 \quad \text{if } x_i \notin T_j,$$
$$a_{ij} = 1 \quad \text{if } x_i \in T_j.$$

Clearly the configuration is completely determined by the matrix A, and it can be proved [23] that

(a) A is normal and satisfies $AA' = A'A = (k-\lambda)I + \lambda J$ where J is the v-square matrix of 1's,
(b) $\lambda = k(k-1)/(v-1)$,
(c) $|d(A)| = k(k-\lambda)^{(v-1)/2}$.

Suppose that $T_1, \cdots, T_v$ and $R_1, \cdots, R_v$ are the sets used in two v, k, λ configurations of the items $x_1, \cdots, x_v$. We say that the two configurations are *isomorphic* if there is a permutation σ of $1, \cdots, n$ and a 1-1 correspondence θ between the sets T_j and R_j such that $\theta(T_i) = R_i$ and $x_i \in T_j$ if and only if $x_{\sigma(i)} \in R_j$.

A *system of distinct representatives* for a v, k, λ configuration is just a particular ordering of the items $x_{\mu(1)}, \cdots, x_{\mu(v)}$ such that

$$x_{\mu(i)} \in T_i, \qquad i = 1, \cdots, v.$$

That is, a system of distinct representatives corresponds uniquely to a term with value 1 in the permanent expansion of the incidence matrix A. Hence $p(A)$ is precisely the number of systems of distinct representatives for the v, k, λ configuration with incidence matrix A. In (c) above we noted that $|d(A)|$ is a function of v, k, λ only and does not depend on which one of the set of nonisomorphic v, k, λ configurations is represented by A. The same question may be asked about the number of distinct representatives, $p(A)$. Very recently Nikolai [20], using an identity of Ryser, described some interesting machine computations in which nonisomorphic $(v, k, \lambda) = (15, 7, 3)$ configurations with different permanents are exhibited. (There are no nonisomorphic v, k, λ configurations for $v < 15$ and in the case of finite projective planes with $(v, k, \lambda) = (n^2+n+1, n+1, 1)$ the first case of nonisomorphism arises for $n=9$, $v=91$.)

It is thus of interest to know in general what kinds of linear operations may be performed on a class Ω of matrices in such a way that the permanent of each $A \in \Omega$ is held fixed. In other words the problem is to determine the structure of $\Omega(p)$. Very recently it was proved [17] that except for $n=2$ the only linear T that satisfy

$$p(T(A)) = p(A)$$

for all $A \in M_n(F)$ are of the type (3) where U and V both have the special form

$$(5) \qquad \sum_{i=1}^{n} a_i E_{i\sigma(i)}, \qquad \prod_{i=1}^{n} a_i = 1,$$

where σ is a permutation of $1, \cdots, n$. (The field F is required to have enough elements.) The same result can be proved, with the further consequence that each $a_i = 1$ in (5), when it is assumed that T maps the polyhedron of doubly stochastic matrices into itself and preserves the permanent [15]. It might be conjectured that there exists some uniform linear operation that converts the permanent of A into the determinant:

$$(6) \qquad p(T(A)) = d(A).$$

In this way we would be able to use the extensive results available concerning determinant preservers. An old result of G. Pòlya [22] shows that no affixing of $\pm$ signs to the elements of A can (except for $n=2$) convert the permanent into the determinant uniformly; and very recently (again except for $n=2$) it was proved that *no* linear T can satisfy (6) for all $A \in M_n(F)$ [14]. In these theorems the problem was in every case reduced to a consideration of rank 1 preservers.

V. Commutativity. We indicate briefly how the idea of a linear mapping on a space of matrices can be very fruitful in studying commutativity. A comprehensive survey and bibliography of the subject of commutativity in finite matrices is contained in the excellent account of Taussky [27].

Suppose that A is a fixed n-square matrix and we consider a corresponding linear mapping T_A on $M_n(F)$ defined by

$$(7) \qquad T_A(X) = AX - XA.$$

It can be checked [7] that with respect to the basis E_{ij}, $1 \leqq i, j \leqq n$, ordered lexicographically, the matrix representation of T_A is $A \otimes I - I \otimes A'$. Here $C \otimes D$ is the Kronecker product of C and D; i.e. the n^2-square matrix which has $c_{ij}D$ as its (i, j) block when it is partitioned into n^2 nonoverlapping n-square submatrices. To see how this can be used we assume for example that A has the same minimum and characteristic polynomial. Then A is similar over F to the companion matrix C of its characteristic polynomial and it is trival to check that A may be replaced by C in examining the null space of T_A. Thus we may take

$$C = A = \begin{bmatrix} 0 & 1 & & 0 \\ \cdot & \cdot & & \cdot \\ \cdot & & \cdot & \cdot \\ \cdot & & & 0 \\ 0 & & 0 & 1 \\ a_0 & a_1 \cdots & a_{n-2} & a_{n-1} \end{bmatrix}$$

and hence

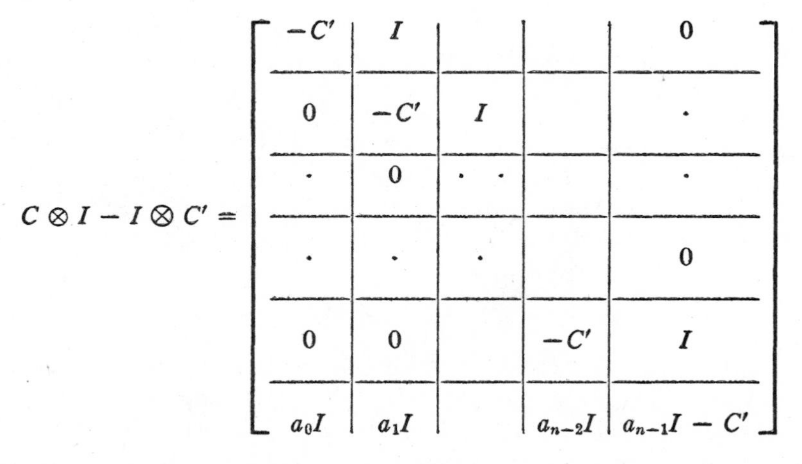

Clearly we may bring $C\otimes I-I\otimes C'$ by elementary row operations to the form

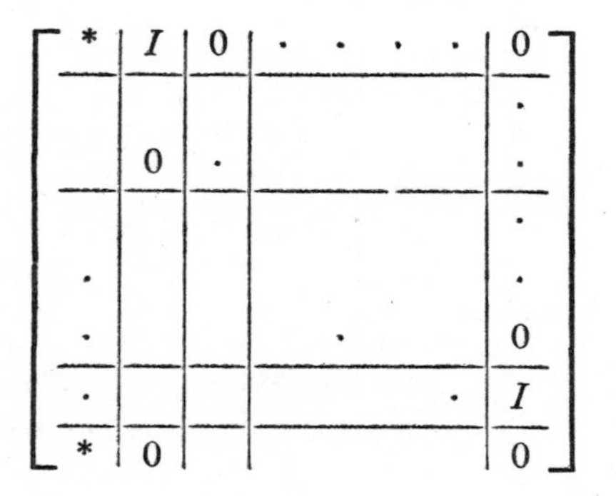

a matrix of rank at least $n(n-1)$. Hence the nullity of T_A is at most $n^2-n(n-1)=n$. On the other hand the matrices A^s, $s=0, \cdots, n-1$ are in the null space of T_A (they commute with A) and they are linearly independent (the degree of the minimum polynomial is n). Thus the null space of T_A is exactly n-dimensional and must therefore be generated by $I, A, \cdots, A^{n-1}$. In other words, the only matrices that commute with A are polynomials in A. Of course, this well-known result may be proved in other ways, but this technique has been applied successfully to the much more complicated study of higher order commutators [10], [25].

We indicate one more example of the approach. Suppose that A is a given linear transformation on a finite dimensional vector space U to itself, where U is equipped with a fixed nonsingular symmetric bilinear functional $\{u, v\}$. A problem considered in a recent paper by Osborn [21] is the following: Find the structure of the set Ω of all bilinear functionals α with respect to which A is self-adjoint:

$$\alpha(Au, v) = \alpha(u, Av), \qquad \text{all } u, v. \tag{8}$$

It is elementary that there exists a unique linear transformation X on U such that

$$\alpha(u, v) = \{Xu, v\}, \qquad \text{all } u, v.$$

Thus the equation (8) takes the form

$$\alpha(Au, v) = \{XAu, v\} = \alpha(u, Av) = \{Xu, Av\} = \{A^*Xu, v\}, \tag{9}$$

where A^* is the unique linear transformation satisfying

$$\{A^*u, v\} = \{u, Av\} \qquad \text{all } u, v.$$

By selecting a basis for U, and letting A_1 and X_1 denote the matrix representations of A and X in this basis respectively, we conclude from (9) and the nonsingularity of $\{u, v\}$ that

$$X_1A_1 - A_1'X_1 = 0.$$

Thus Ω is known as soon as we know the null space of the mapping

$$T_{A_1}(X_1) = X_1A_1 - A_1'X_1. \tag{10}$$

In interesting papers Osborn [21] and Taussky and Zassenhaus [28] examined the null space of T_{A_1} from considerably different points of view, and still further results along these lines are found in [13].

VI. Symmetry operators. Returning to the space of tensors $U^{(k)}$ introduced in III we define a *permutation operator* on $U^{(k)}$ by the equation

$$\sigma(x_1 \otimes \cdots \otimes x_k) = x_{\sigma^{-1}(1)} \otimes \cdots \otimes x_{\sigma^{-1}(k)}. \tag{11}$$

The meaning in (11) is this: σ is a permutation in the symmetric group of degree k, S_k. We can then regard σ as a linear transformation on $U^{(k)}$ by defining its action on pure vectors (which span $U^{(k)}$) by the equation (11) and extending linearly. If λ_σ is a scalar for each $\sigma \in S_k$ then we can define a *symmetry operator* T_λ on $U^{(k)}$ by

$$T_\lambda = \sum_{\sigma \in S_k} \lambda_\sigma \sigma. \tag{12}$$

Associated with T_λ is a scalar valued function on $M_k(F)$:

$$d_\lambda(A) = \sum_{\sigma \in S_k} \lambda_\sigma \prod_{i=1}^{k} a_{i\sigma(i)}, \tag{13}$$

$A = (a_{ij}) \in M_k(F)$. For example, if $\lambda_\sigma = \epsilon(\sigma) = \pm 1$ according as σ is even or odd, then T_λ is the skew-symmetry operator and the corresponding d_λ is the determinant. Again, if $\lambda_\sigma = 1$ for all $\sigma \in S_k$ then d_λ specializes to the permanent function. Another example of interest in certain combinatorial problems is obtained by setting $\lambda_\sigma = 1$ for σ any power of the full cycle $(1, \cdots, k)$, $\lambda_\sigma = 0$ otherwise. In general, the invariance problem appears as follows: *for a given choice of the* λ_σ *investigate the structure of the set of linear transformations on* $M_k(F)$ *into itself which hold fixed the corresponding* d_λ. This problem, of course, subsumes both the permanent and determinant preservation problems. There are various choices for the λ_σ and research is currently under way on those symmetrizers T_λ that arise from the Young diagrams ([30] p. 120). We close by describing the bridge that allows us to move freely between the symmetrizer T_λ and its associated matrix function d_λ. Let $e_1, \ldots, e_k$ be a linearly independent set in U and suppose $h_1, \ldots, h_k$ is a dual set of linear functionals, $h_i(e_j) = \delta_{ij}$. Suppose $A \in M_k(F)$ and set

$$x_i = \sum_{j=1}^{k} a_{ij} e_j.$$

If ϕ is the multilinear functional whose value is defined by $\phi(y_1, \cdots, y_k) = h_1(y_1) \cdots h_k(y_k)$, we compute that

$$\begin{aligned} T_\lambda(x_1 \otimes \cdots \otimes x_k)(\phi) &= \sum_{\sigma \in S_k} \lambda_\sigma x_{\sigma^{-1}(1)} \otimes \cdots \otimes x_{\sigma^{-1}(k)}(\phi) \\ &= \sum_{\sigma \in S_k} \lambda_\sigma \phi(x_{\sigma^{-1}(1)}, \cdots, x_{\sigma^{-1}(k)}) \\ &= \sum_{\sigma \in S_k} \lambda_\sigma \prod_{t=1}^{k} h_t(x_{\sigma^{-1}(t)}) = \sum_{\sigma \in S_k} \lambda_\sigma \prod_{t=1}^{k} a_{\sigma^{-1}(t),t} \\ &= \sum_{\sigma \in S_k} \lambda_\sigma \prod_{t=1}^{k} a_{t,\sigma(t)} = d_\lambda(A). \end{aligned}$$

References

1. W. L. Chow, On the geometry of algebraic homogeneous spaces, Ann. of Math., 50 (1949) 32–67.

2. J. Dieudonné, Sur une Généralisation du Groupe Orthogonal à Quatre Variables, Arch. Math., 1 (1948) 282–287.

3. G. Frobenius, Über die Darstellung der Endlichen Gruppen durch Lineare Substitutionen, S.-B. Deutsch. Akad. Wiss. Berlin, pp. 994–1015.

4. L. K. Hua, Geometries of matrices, I. Generalizations of von Staudt's theorem, Trans. Amer. Math. Soc., 57 (1945) 441–481.

5. H. G. Jacob, Coherence invariant mappings on Kronecker products, Amer. J. Math., 77 (1955) 177–189.

6. S. Kantor, Theorie der Äquivalenz von Linearen ∞ Scharen Bilinearer Formen, S.-B. der Münchener Akademie (1897), 367–381.

7. C. C. MacDuffee, Theory of Matrices, Chelsea, New York, 1946, p. 90.

8. M. Marcus and R. Westwick, Linear maps on skew symmetric matrices: The invariance of elementary symmetric functions, Pacific J. Math., 10 (1960) 917–924.

9. M. Marcus and B. N. Moyls, Transformations on tensor product spaces, Pacific J. Math., 9 (1959) 1215–1221.

10. M. Marcus and N. A. Khan, On matrix commutators, Canad. J. Math., 12 (1960) 269–277.

11. M. Marcus, All linear operators leaving the unitary group invariant, Duke Math. J., 26 (1958) 155–164.

12. M. Marcus and R. Purves, Linear transformations on algebras of matrices: The invariance of the elementary symmetric functions, Canad. J. Math., 11 (1959) 383–396.

13. M. Marcus and N. A. Khan, On a commutator result of Taussky and Zassenhaus, Pacific J. Math., 10 (1960) 1337–1346.

14. M. Marcus and H. Minc, On the relation between the permanent and the determinant, Illinois J. Math., 5 (1961) 376–381.

15. M. Marcus, H. Minc and B. N. Moyls, Permanent preservers on doubly stochastic matrices, Canad. J. Math., 14 (1962) 190–194.

16. M. Marcus and F. May, On a theorem of I. Schur concerning matrix transformations, Arch. Math., 11 (1960) 401–404.

17. M. Marcus and F. May, The permanent function, Canad. J. Math., 14 (1962) 177–189.

18. M. Marcus and M. Newman, Permanents of doubly stochastic matrices, Proc. of Symposia in Appl. Math. (Amer. Math. Soc.) 10 (1960) 169–174.

19. K. Morita, Schwarz's lemma in a homogeneous space of higher dimensions, Japan J. Math., 19 (1944) 45–56.

20. P. J. Nikolai, Permanents of incidence matrices, Math. Comp., 14 (1960) 262–266.

21. H. A. Osborn, A class of bilinear forms, Trans. Amer. Math. Soc., 90 (1959) 485–498.

22. G. Pòlya, Aufgabe 424, Arch. Math. u. Phys., 20, p. 271.

23. H. J. Ryser, Geometries and incidence matrices, Slaught Paper no. 4, Math. Assoc. of Amer., 1955.

24. H. J. Ryser, Compound and induced matrices in combinatorial analysis, Proc. of Symposia in Appl. Math., Amer. Math. Soc., 10 (1960) 149–167.

25. M. F. Smiley, Matrix commutators, Canad. J. Math., 13 (1961) 353–355.

26. I. Schur, Einige Bemerkungen zur Determinanten Theorie, S.-B. der Preuss. Akad. Wiss. Berlin, 25 (1925) 454–463.

27. O. Taussky, Commutativity in finite matrices, this MONTHLY, 64 (1957) 229–235.

28. O. Taussky and H. Zassenhaus, On the similarity transformation between a matrix and its transpose, Pacific J. Math., 9 (1959) 893–896.

29. R. Westwick, Linear transformations on Grassmann product spaces, Doctoral dissertation, Univ. of British Columbia, 1959.

30. H. Weyl, The classical groups, Princeton Univ. Press, Princeton, N. J. 1946.

COMMUTATIVITY IN FINITE MATRICES*†

OLGA TAUSSKY, National Bureau of Standards

1. Introduction. The real and complex numbers have the properties

(1) $$ab = ba \text{ for all } a \text{ and } b$$

and

(2) $$\text{for every } a \neq 0 \text{ there is an inverse } a^{-1},$$

* From AMERICAN MATHEMATICAL MONTHLY, vol. 64 (1957), pp. 229–235.

† Invited address at the meeting of the Maryland-District of Columbia-Virginia Section of the Mathematical Association of America, May 5, 1956.

The preparation of this paper was supported (in part) by the Office of Naval Research.

which implies

$$(2') \qquad ab = 0 \text{ only if } a = 0 \text{ or } b = 0.$$

A classical theorem of Frobenius* states that there are no other hypercomplex systems which have these properties. If we want to consider more general systems, we must give up at least one of these properties. For example, if we do not insist on the commutative law (1), the quaternions of Hamilton become acceptable, but no others. The set of $n \times n$ matrices $A, B, \cdots$, with complex elements can be regarded as a hypercomplex system with n^2 base elements and in this system, in general, $AB \neq BA$. In general, also, there is no inverse A^{-1} even if $A \neq 0$ and, furthermore, AB can be zero without $A = 0$ or $B = 0$. The two axioms may even be violated simultaneously, so that when $AB = 0$ we may still have $BA \neq 0$. A simple example is

$$A = \begin{pmatrix} 0 & 1 \\ 0 & 0 \end{pmatrix}, \qquad B = \begin{pmatrix} 1 & 0 \\ 0 & 0 \end{pmatrix}.$$

As soon as $AB \neq BA$ for a pair A, B, various other things go wrong too. Of these, we mention three:

(a) If $A_n = (a_{ik}^{(n)})$ and we define $\lim_{n\to\infty} A_n$ to be $(\lim_{n\to\infty} a_{ik}^{(n)})$, we may define the exponential of a matrix X by the formal power series $e^X = \sum_{i=0}^{\infty} X^i/i!$. Then, in general, $e^A e^B \neq e^{A+B}$, but when $AB = BA$, we always have $e^A e^B = e^{A+B}$.

Example: If

$$A = \begin{pmatrix} 0 & 1 \\ 0 & 0 \end{pmatrix}, \qquad B = \begin{pmatrix} 0 & 0 \\ 1 & 0 \end{pmatrix},$$

then

$$A + B = \begin{pmatrix} 0 & 1 \\ 1 & 0 \end{pmatrix}, \qquad A^2 = 0, \qquad B^2 = 0, \qquad (A + B)^2 = I,$$

$$e^A = \begin{pmatrix} 1 & 1 \\ 0 & 1 \end{pmatrix}, \qquad e^B = \begin{pmatrix} 1 & 0 \\ 1 & 1 \end{pmatrix}, \qquad e^A e^B = \begin{pmatrix} 2 & 1 \\ 1 & 1 \end{pmatrix},$$

but

$$e^{A+B} = \begin{pmatrix} \cosh 1 & \sinh 1 \\ \sinh 1 & \cosh 1 \end{pmatrix}.$$

(b) Let $\lim_{n\to\infty} A^n = X$, and $\lim_{n\to\infty} B^n = Y$. Then $\lim_{n\to\infty} (AB)^n$ does not exist, in general, when $AB \neq BA$, while for $AB = BA$ we have $(AB)^n = A^n B^n$ for all $n \geq 1$, so that $\lim_{n\to\infty} (AB)^n = \lim_{n\to\infty} A^n \lim_{n\to\infty} B^n = XY$.

* References listed by sections will be found at the end of the paper.

Examples: If

$$A = \begin{pmatrix} 0 & 1 \\ 0 & 0 \end{pmatrix}, \qquad B = \begin{pmatrix} 0 & 0 \\ 1 & 0 \end{pmatrix},$$

then

$$AB = \begin{pmatrix} 1 & 0 \\ 0 & 0 \end{pmatrix},$$

so that $X=0$, $Y=0$, but

$$\lim_{n\to\infty} (AB)^n = \begin{pmatrix} 1 & 0 \\ 0 & 0 \end{pmatrix} \neq XY.$$

If

$$A = \begin{pmatrix} 1/2 & 2 \\ 0 & 0 \end{pmatrix}, \qquad B = \begin{pmatrix} 1/2 & 0 \\ 2 & 0 \end{pmatrix},$$

then

$$AB = \begin{pmatrix} 4 + 1/4 & 0 \\ 0 & 0 \end{pmatrix},$$

and $X=0$, $Y=0$, but $\lim_{n\to\infty} (AB)^n$ does not exist.

(c) Let the eigenvalues of A be $\alpha_1, \cdots, \alpha_n$ and those of B, $\beta_1, \cdots, \beta_n$. When $AB \neq BA$, the eigenvalues $\gamma_1, \cdots, \gamma_n$ of AB are, in general, not $\alpha_1\beta_1, \cdots, \alpha_n\beta_n$ for any pairing. When $AB=BA$, however, the eigenvalues of any polynomial $p(A, B)$ are $p(\alpha_i, \beta_i)$, by another classical theorem of Frobenius.

Example: If

$$A = \begin{pmatrix} 0 & 1 \\ 0 & 0 \end{pmatrix}, \qquad B = \begin{pmatrix} 0 & 0 \\ 1 & 0 \end{pmatrix},$$

then

$$AB = \begin{pmatrix} 1 & 0 \\ 0 & 0 \end{pmatrix},$$

and $\alpha_1=\alpha_2=0$, $\beta_1=\beta_2=0$, while $\gamma_1=1$, $\gamma_2=0$.

The main object of this paper is a study of the replacement of $AB=BA$ in (c) by a weaker rule.

2. Commutators. Two theorems of McCoy and Drazin that are stated in Section 3 involve the notion of the commutator of two matrices A and B. We call $AB-BA$ the commutator of A and B and denote it by (A, B). Since $(A, B)=0$ when $AB=BA$, the commutator may be said to "measure" how much

AB differs from BA. In the theory of groups, if $AB \neq BA$, then $A^{-1}B^{-1}AB$ is called the commutator of A and B. Since, for matrices, A^{-1} and B^{-1} do not always exist, but addition and multiplication are defined, we take instead, $AB-BA$. We can also study commutators of higher order, for example, $(A, (A, B)) = A(AB-BA)-(AB-BA)A$ or $(B, (A, B))$.

The following remarks, about a special case of commutators, while not immediately relevant to our topic, indicate an interesting area of study.

Consider the commutators in which the last factor is A^*, the others all A, where A^* is the transposed and conjugate complex matrix of A. We write $(A, A^*) = AA^* - A^*A = C_2$, $A(A, A^*) - (A, A^*)A = C_3$, *etc.* Matrices for which $C_2 = (A, A^*) = 0$, the so-called normal matrices, play an important role as a natural generalization of hermitian matrices.

Suppose we have a matrix for which $C_2 \neq 0$, we might ask whether or not $C_3 = 0$. It turns out that if $C_2 \neq 0$ then $C_3 \neq 0$. Further if, $C_2 \neq 0$ then $(C_2, C_3) \neq 0$. Let us next consider C_4. We find that $C_4 = 0$ is possible even if $C_2 \neq 0$, $C_3 \neq 0$.

Example: If

$$A = \begin{pmatrix} 1 & 1 \\ 0 & 1 \end{pmatrix},$$

then

$$A^* = \begin{pmatrix} 1 & 0 \\ 1 & 1 \end{pmatrix},$$

and we have

$$C_2 = \begin{pmatrix} 1 & 0 \\ 0 & -1 \end{pmatrix}, \quad C_3 = \begin{pmatrix} 0 & -2 \\ 0 & 0 \end{pmatrix}, \quad C_4 = 0, \quad (C_2, C_3) = \begin{pmatrix} 0 & -4 \\ 0 & 0 \end{pmatrix}.$$

However, if in the case $n=2$, we have $C_2 \neq 0$, $C_3 \neq 0$, $C_4 \neq 0$, then $C_n \neq 0$ for any n. For $n=3$ the behavior is different as is shown by the following examples:

$$A = \begin{bmatrix} 0 & 1 & 0 \\ 0 & 0 & 2 \\ 0 & 0 & 0 \end{bmatrix}, \quad C_4 \neq 0, \quad C_5 = 0; \quad A = \begin{bmatrix} 0 & 1 & 1 \\ 0 & 0 & 1 \\ 0 & 0 & 0 \end{bmatrix}, \quad C_5 \neq 0, \quad C_6 = 0.$$

3. Two theorems of McCoy and Drazin. McCoy proved that if $(A, B) \neq 0$, but $(A, (A, B)) = (B, (A, B)) = 0$, then, although $AB \neq BA$, all eigenvalues of the product AB are of the form $\alpha_i\beta_i$ for a certain pairing and, more generally, all polynomials $p(A, B)$ have eigenvalues $p(\alpha_i, \beta_i)$ for the same pairing.

Later, M. P. Drazin generalized McCoy's theorem by showing that, if all commutators of A and B of a fixed order k (*i.e.*, involving k brackets) vanish, then all polynomials $p(A, B)$ have eigenvalues $p(\alpha_i, \beta_i)$.

Apparently there are none or at most very few theorems known which say: "such and such a statement is true if and only if $AB = BA$"; it is always "if

$AB=BA$ then . . . ". Hence we are tempted to replace $AB=BA$ in various ways by weaker hypotheses. One way is to assume that instead of $AB-BA=0$ some of the higher commutators vanish. Another one is suggested by McCoy and Drazin's theorems: It is known that $AB=BA$ implies that all $p(A, B)$ have as eigenvalues $p(\alpha_i, \beta_i)$, but the converse is not true. Consider, for example,

$$A = \begin{pmatrix} 1 & 2 \\ 0 & 2 \end{pmatrix}, \qquad B = \begin{pmatrix} 1 & 1 \\ 0 & 1 \end{pmatrix}.$$

Here $AB \neq BA$, but it can be shown (*cf.* Condition 4.1 below) that all $p(A, B)$ have as eigenvalues $p(\alpha_i, \beta_i)$. Further, no commutator of the form $(A, (A, (A, \cdots (A, (A, B)) \cdots)))$, however long, can vanish.

4. Property *P*. McCoy asked and answered the question: which are the pairs of $n \times n$ matrices A, B, (with complex elements) for which all polynomials $p(A, B)$ have as eigenvalues $p(\alpha_i, \beta_i)$? Each of the following two conditions is necessary and sufficient.

Condition 4.1. *There exists a matrix S, such that $S^{-1}AS$, $S^{-1}BS$ are both left triangular, or both right triangular.*

Condition 4.2. *All matrices $f(A, B)(AB-BA)$ are nilpotent, where f is an arbitrary polynomial in A and B, and nilpotency of a matrix X means that $X^r=0$ for a certain integer r.*

A special case of McCoy's theorem was pointed out recently by H. Schneider, namely, if $AB=0$ and $BA\neq 0$. Then $AB-BA=-BA$ and $(-BA)^2=0$, further $A(BA)=0$, $(B(BA))^2=0$. Hence, by Condition 4.2, such a pair qualifies, and from Condition 4.1 it follows that a matrix S exists such that $S^{-1}AS$, $S^{-1}BS$ are both triangular.

Another example of pairs of matrices with property P was used by A. Brauer. Let A be any $n\times n$ matrix and v one of its eigenvectors. Let B be an $n\times n$ matrix of rank 1 whose columns are all multiples of v. Then $n-1$ of the eigenvalues of $A+B$ coincide with $n-1$ eigenvalues of A. One reason for this is that B has $n-1$ eigenvalues zero. Further it can be shown that A and B have property P. For, let SAS^{-1} be triangular with the eigenvalue corresponding to v in the upper left corner. Then v goes over into Sv which is the vector $(1, 0, \cdots, 0)$. Hence SBS^{-1} is triangular too, with $n-1$ zeros in the main diagonal.

5. Property *L*. We obtain a larger class of matrices if we do not assume that all polynomials $p(A,B)$ have as eigenvalues $p(\alpha_i, \beta_i)$, but only that a subset has this property. We may assume, for example, that all $\lambda A+\mu B$ have as eigenvalues $\lambda\alpha_i+\mu\beta_i$, where λ, μ are arbitrary, complex numbers. Such a pair A, B is said to have property L. We now obtain a larger class of matrices if $n\geq 3$ (for $n=2$ the classes coincide). For example,

$$A = \begin{pmatrix} 1 & 0 & 0 \\ 0 & 2 & 0 \\ 0 & 0 & 3 \end{pmatrix}, \qquad B = \begin{pmatrix} -2 & 1 & 2 \\ -1 & -2 & -1 \\ 1 & 1 & 1 \end{pmatrix}$$

have eigenvalues, 1, 2, 3, and -2, -2, 1. However, AB does not have the eigenvalues 1 or -2. On the other hand, it can be shown that the pair A, B has property L. Another example is

$$A = \begin{pmatrix} 0 & 1 & 0 \\ 0 & 0 & -1 \\ 0 & 0 & 0 \end{pmatrix}, \qquad B = \begin{pmatrix} 0 & 0 & 0 \\ 1 & 0 & 0 \\ 0 & 1 & 0 \end{pmatrix}.$$

In this case all matrices $\lambda A + \mu B$ have all eigenvalues 0, but

$$AB = \begin{pmatrix} 1 & 0 & 0 \\ 0 & -1 & 0 \\ 0 & 0 & 0 \end{pmatrix}$$

has some eigenvalues $\neq 0$.

If, however, A and B are both normal, then it is true that $AB = BA$ if we assume only that $\lambda A + \mu B$ have all eigenvalues $\lambda\alpha_i + \mu\beta_i$. This was proved by Wiegmann and Wielandt. Wielandt proved it from the following more general theorem:

Let A, B be normal and let $\gamma_i(z)$ be the eigenvalues of $A + zB$ and assume that $\sum_{i=1}^n |\gamma_i(z)|^2 \geq \sum_{i=1}^n |\alpha_i + z\beta_i|^2$ for at least three values of z which are the vertices of a triangle in the z-plane which contains O in the interior. It follows that $AB = BA$.

We, of course, assume much more, namely, that $\gamma_i(z) = \alpha_i + z\beta_i$ for all values of z.

6. Matrices in $A + zB$ with multiple eigenvalues. In the previous section we treated the matrices for which all $\lambda A + \mu B$ have eigenvalues $\lambda\alpha_i + \mu\beta_i$, and again, we can point out a larger class.

This can be done in a way that is easily described for $n = 2$. Here a pencil of matrices $A + zB$ has either all matrices with a double eigenvalue, or exactly two, or exactly one. There is no other possibility. If the pair A, B has property L, then it is very easy to see that either all matrices have a double eigenvalue, or exactly one does. However, the converse of this is true too. For $n > 2$ either all matrices in the pencil have a multiple eigenvalue or at most $n(n-1)$ do. If, however, the pair A, B has property L, then either all matrices in the pencil $A + zB$ have a multiple eigenvalue or at most $n(n-1)/2$ do. The converse of this is not true, as is seen by the pair

$$A = \begin{pmatrix} 0 & 1 & 0 \\ 0 & 0 & 1 \\ 0 & 0 & 0 \end{pmatrix}, \qquad B = \begin{pmatrix} 1 & 1 & 2 \\ 1 & 1 & 2 \\ -1 & -1 & -2 \end{pmatrix}.$$

Here both A and B have a triple eigenvalue 0, but $A + B$ is nonsingular, so that property L does not hold. On the other hand, the matrices A and B are the only ones in the whole pencil $\lambda A + \mu B$ which have a multiple eigenvalue. However, a partial converse is true even for $n > 2$.

7. Diagonable pencils. In a discussion of multiple eigenvalues we must consider eigenvectors too. We know, for example, that if all matrices $A+zB$ have the full number of eigenvectors (*i.e.*, if for all finite values of z and $z=\infty$, the matrix $A+zB$ is similar to a diagonal matrix), then $AB=BA$. If, in the whole pencil, even a single matrix is not similar to a diagonal matrix then AB need not be equal to BA. This is shown by the example,

$$A=\begin{pmatrix}0 & 0\\ 0 & 1\end{pmatrix},\qquad B=\begin{pmatrix}1 & 1\\ 0 & 1\end{pmatrix}.$$

On the other hand, if $AB=BA$, then the pencil $A+zB$ contains either only diagonable matrices, or none, or exactly one. For $n=2$ the case of "none" never happens. For $n=3$ an example of a pencil with no diagonable matrices apart from 0 is given by

$$A=\begin{bmatrix}0 & 1 & 0\\ 0 & 0 & 1\\ 0 & 0 & 0\end{bmatrix},\qquad B=\begin{bmatrix}0 & 0 & 1\\ 0 & 0 & 0\\ 0 & 0 & 0\end{bmatrix}.$$

A pencil with only one diagonable matrix is given by

$$A=\begin{pmatrix}0 & 1\\ 0 & 0\end{pmatrix},\qquad B=\begin{pmatrix}\beta & \beta_1\\ 0 & \beta\end{pmatrix}.$$

References to 1

1. G. N. Cebotarev, On the solution of the matrix equation $e^B\cdot e^C=e^{B+C}$, Doklady Akad. Nauk SSSR (N.S.), vol. 96, 1954, pp. 1109–1112.

2. M. Fréchet, Les solutions non commutables de l'équation matricielle $e^x\cdot e^y=e^{x+y}$, Rend. Circ. Mat. Palermo (2), vol. 1, 1952, pp. 11–27.

3. M. Fréchet, Rectification, Rend. Circ. Mat. Palermo (2), vol. 2, 1953, pp. 71–72.

4. G. Frobenius, Über vertauschbare Matrizen, S.-B. Berlin. Math. Ges., 1896, pp. 601–14.

5. H. Geiringer, On the solution of systems of linear equations by certain iteration methods, Reissner Anniversary Volume, Contributions to Applied Mechanics, Edwards, Ann Arbor, 1948, pp. 365–393, in particular, p. 379.

6. C. W. Huff, On pairs of matrices (of order two) A, B, satisfying the condition $e^Ae^B\neq e^{A+B}\neq e^Be^A$, Rend. Circ. Mat. Palermo (2) vol. 2, 1953–54, pp. 326–330.

7. A. G. Kakar, Non-commuting solutions of the matrix equation $\exp(X+Y)=\exp X \exp Y$, Rend. Circ. Mat. Palermo (2), vol. 2, 1953–54, pp. 331–345.

8. K. Morinaga and T. Nono, On the non-commutative solution of the exponential equation $e^xe^y=e^{x+y}$, J. Sci. Hiroshima Univ. Ser. A, vol. 17, 1954, pp. 345–358.

9. O. Taussky and J. Todd, Infinite powers of matrices, J. London Math. Soc., vol. 17, 1942, pp. 146–151.

References to 2

1. T. Kato and O. Taussky, Commutators of A and A^*, J. Washington Acad. Sci., vol. 46, 1956, pp. 38–40.

References to 3

1. N. H. McCoy, On quasi-commutative matrices, Trans. Amer. Math. Soc., vol. 36, 1934, pp. 327–40.

2. M. P. Drazin, Some generalizations of matrix commutativity, Proc. London Math. Soc., (3), vol. 1, 1951, pp. 222–231.

3. W. E. Roth, On k-commutative matrices, Trans. Amer. Math. Soc., vol. 39, 1936, pp. 483–95.

References to 4

1. A. Brauer, Limits for the characteristic roots of a matrix IV, Duke Math. J., vol. 19, 1952, pp. 75–91.

2. M. P. Drazin, J. W. Dungey, K. W. Gruenberg, Some theorems on commutative matrices, J. London Math. Soc., vol. 26, 1951, pp. 221–228.

3. L. S. Goddard, Note on a matrix theorem of A. Brauer and its extension, Canad. J. Math., vol. 7, 1955, pp. 527–530.

4. L. S. Goddard and H. Schneider, Pairs of matrices with a non-zero commutator, Proc. Cambridge Philos. Soc., vol. 51, 1955, pp. 551–553.

5. J. K. Goldhaber, The homomorphic mapping of certain matrix algebras onto rings of diagonal matrices, Canad. J. Math. vol. 4, 1952, pp. 31–42.

6. J. K. Goldhaber and G. Whaples, On some matrix theorems of Frobenius and McCoy, Canad. J. Math., vol. 5, 1953, pp. 332–335.

7. N. H. McCoy, On the characteristic roots of matrix polynomials, Bull. Amer. Math. Soc., vol. 42, 1936, pp. 592–600.

8. H. Perfect, Methods of constructing certain stochastic matrices, Duke Math J., vol. 20, 1953, pp. 395–404 (see also review in Math. Rev.); II Duke Math. J. vol. 22, 1955, pp. 305–311.

9. H. Schneider, A pair of matrices with property P, this MONTHLY, vol. 62, 1955, pp. 247–249.

References to 5–7

1. T. S. Motzkin and O. Taussky, Pairs of matrices with property L, Trans. Amer. Math. Soc., vol. 73, 1952, pp. 108–114.

2. T. S. Motzkin and O. Taussky, Pairs of matrices with property L, Proc. Nat. Acad. Sci. U.S.A., vol. 39, 1953, pp. 961–963.

3. T. S. Motzkin and O. Taussky, Pairs of matrices with property L (II), Trans. Amer. Math. Soc., vol. 80, 1955, pp. 387–401.

4. H. Wielandt, Pairs of normal matrices with property L, J. Res. Nat. Bur. Standards, vol. 51, 1953, pp. 89–90.

5. N. A. Wiegmann, A note on pairs of normal matrices with property L, Proc. Amer. Math. Soc., vol. 4, 1953, pp. 35–36.

6. I. Kaplansky, Completely continuous normal operators with property L, Pacific J. Math., vol. 3, 1953, pp. 721–4.

BIBLIOGRAPHIC ENTRIES: OTHER TOPICS

The references below are to the AMERICAN MATHEMATICAL MONTHLY.

1. H. E. Goheen, On a lemma of Stieljes on matrices, 1949, 328–329.

Gives a sufficient condition for the inverse of a matrix to have all its entries positive.

2. I. N. Herstein, A note on primitive matrices, 1954, 18–20.

Gives conditions for a power of a matrix to have all entries positive when the original matrix has all entries non-negative.

3. Marvin Marcus, Some properties and applications of doubly stochastic matrices, 1960, 215–221.

Treats various results on eigenvalues, singular values and quadratic forms formulated in terms of a single extreme value problem. Also discusses conjectures on permanents and on the partial ordering of doubly stochastic matrices.

4. Richard Sinkhorn, Two results concerning doubly stochastic matrices, 1968, 632–634.

Gives necessary and sufficient conditions for a doubly stochastic matrix to be idempotent or reducible.

5. C. J. Eliezer, A note on group commutators of 2×2 matrices, 1968, 1090–1091.

If A and B are non-singular 2×2 matrices both of which commute with $ABA^{-1}B^{-1}$, then A and B either commute or anti-commute.

5

HISTORY†

GREEK METHODS OF SOLVING QUADRATIC EQUATIONS*

WALTER C. EELLS, University of Chicago

The following Bibliography includes the chief references consulted in the preparation of this paper.

CANTOR, M.—*Vorlesungen über Geschichte der Mathematik.* Band 1. Zweite Auflage. Leipzig, 1894.

NESSELMANN, G. H. P.—*Die Algebra der Greichen.* Berlin, 1842.

GOW, J.—*A Short History of Greek Mathematics.* Cambridge, 1884.

ALLMAN, G. J.—*Greek Geometry from Thales to Euclid.* Dublin, 1889.

TANNERY, P.—*Diophanti Alexandrini; Opera omnia cum Graecis Commentariis.* 2 Vols. Lipsiae, 1893. (Complete Greek text of Diophantus with Latin translation).

WERTHEIM, G.—*Die Arithmetik und die Schrift über Polygonalzahlen des Diophantus von Alexandria.* Leipzig, 1890. (Translation in German of the Greek text of Diophantus).

HANKEL, H.—*Zur Geschichte der Mathematik in Alterthum und Mittelalter.* Leipzig, 1874.

ZEUTHEN, H. G.—*Histoire des Mathématiques.* Paris, 1902.

HEIBERG, I. L.—*Euclidis Elementa.* Lipsiae, 1883. 2 Vols. (Greek text with Latin translation).

HEIBERG, I. L.—*Archimedis, Opera Omnia.* Vol. I. Lipsiae, 1880. (Greek text with Latin translation).

HEATH, T. L.—*The Thirteen Books of Euclid's Elements.* Cambridge, 1908. 3 Vols. (Translation in English of the Greek text of Euclid).

HEATH, T. L.—*The Works of Archimedes.* Cambridge, 1897.

HEATH, T. L.—*Diophantus of Alexandria, a Study in the History of Greek Algebra.* Cambridge, 1885.

HEATH, T. L.—*Apollonius, Conic Sections.* Cambridge, 1896.

LA GRANGE, J. L.—*Oeuvres.* Paris, 1877. *Leçons Elémentaires sur les Mathématiques.* Vol. 7, p. 180.

---

* From AMERICAN MATHEMATICAL MONTHLY, vol. 18 (1911), pp. 3–14.

† **Editor's Note**: The articles which are reprinted or listed in the bibliography in this chapter concern the history of algebra in general, rather than the history of a specific area of algebra. There are a number of other historical articles in this volume which appear in the chapter appropriate to their subject matter; for example, articles on the history of group theory appear in the chapter on group theory.

BALL, W. W. R.—*A Short History of Mathematics*. London, 1893.

CAJORI, F.—*History of Mathematics*. New York, 1894.

FINK, K.—*History of Mathematics*. (Beman and Smith translation) Chicago, 1900.

File of the *Bibliotheca Mathematica*, G. Enestrom, Leipzig.

Principal Greek mathematicians referred to in this paper with the approximate dates at which they flourished—PYTHAGORAS, 530 B. C.; EUCLID, 290 B. C.; ARCHIMEDES, 250 B. C.; APOLLONIUS, 230 B. C.; HIPPARCHUS, 150 B. C.; HERON (HERO), 120 B. C.; DIOPHANTUS, 275 A. D.; THEON, 380 A. D.; PROCLUS, 450 A. D.

Introduction.

Before entering upon a discussion of the methods employed by the ancient Greeks for solving Quadratic Equations, a brief summary should be made of the mathematical knowledge which they possessed in historic times. Various forms of reckoning, including finger-reckoning, pebble-reckoning, and some use of the sand-board ('*αβα'ξ*) especially characteristic of the Attic Greeks, had by the time of Pythagoras given place to a well defined system of notation and computation by means of symbols. These consisted of the letters of the regular Greek alphabet, with three additions from an older alphabet, which were used to make up a decimal system of notation.* Although operations with these symbols were cumbrous and complicated, they were possible and the ordinary operations of addition, subtraction, multiplication, and division were quite fully developed.† They had made some progress in the theory of numbers which was denoted by the term '*αριθμητικη'* as contrasted with *λογιστικη'* the art of calculation. Their knowledge of Algebra, as an abstract science, was almost nothing, until after the time of Christ, although they had a slight conception of it, perhaps, from the Egyptians who were familiar with simple equations long before the time of Pythagoras.‡ But the Greeks had made unusual progress in Geometry, and of the problems of Algebra which can easily be given a geometric interpretation and solved by methods essentially geometrical, they had a very considerable knowledge which we shall investigate a little more in detail. Accordingly we shall take up the Greek methods of solving quadratic equations under two distinct heads, geometrical or constructive methods, and methods purely algebraic.

Geometric methods.

Pure Quadratics. 1. *Square Root.* Although not formulated by the Greeks in this way we may properly look upon their work in finding square roots of numbers as the solution of the pure quadratic $x^2 = a$, and accordingly as the first work which they did in the solution of quadratic equations.

---

* See Cantor, pp. 110–119; Cajori, p. 64; Heath, *Archimedes*, p. lxix; Nesselmann, p. 78, et seq.

† Numerous problems given in Gow; Heath, *Archimedes*; Tannery; Hankel.

‡ See Cantor, pp. 37, 38, for numerous examples of equations from the Ahmes papyrus.

The Pythagoreans are credited with the first knowledge of the fact that in a square

$$\text{diagonal} : \text{side} :: \sqrt{2} : 1,$$

and of the method by which it was established, but it was kept a profound secret from their contemporaries.* Although there was this geometrical equivalent for $\sqrt{2}$, they also recognized its irrationality but had no numerical expression for it, because they did not recognize irrationals as members of their number system; when they occurred they were rejected in the same way that imaginaries were in later centuries.† Lines were the indispensable symbols for irrationals because they avoided the necessity for numerical expression. But in the further development of Geometry, especially about the time of Euclid, it became necessary to obtain approximate values for some square roots, and the results and methods by which they were obtained are worth consideration.

Archimedes, perhaps the greatest mathematician of antiquity, is the first to give us any number of them. But unfortunately he gives only results and no suggestion of the method by which he reached them, nor is there any example of the actual complete calculation of a root extant by anyone before the time of Christ. Archimedes in his *Circuli Dimensio*‡ in dealing with the problem of finding an approximate value for the ratio of circumference to diameter says that $\sqrt{3}$ lies between $\frac{1351}{780}$ and $\frac{265}{153}$, that is, between 1.7320513 and 1.7320327. Now $\sqrt{3}$ correct to seven decimal places is 1.7320508, so that his approximation is remarkably accurate, his upper limit being exactly equal to six places. Many ingenious theories have been proposed to explain how Archimedes could secure this result, among them a method by continued fractions, one by an approximation in the form of a series of fractions, Theon's method of sexagesimal fractions based on Euclid (illustrated later, see p. 7), and others.§ But we have no positive knowledge as to how he secured this result. If it were for $\sqrt{3}$ alone we might accept the theory that his only method consisted in guessing at the result and gradually making successive guesses more nearly accurate. But this seems highly improbable in view of the following results which occur in the same connection in Archimedes.‖ (I have added the correct results to two places in the third column.)

$3013\ 3/4$ is greater than $\sqrt{9082321}$	3013.69
$1838\ 9/11$ is greater than $\sqrt{3380929}$	1838.73

* It is interesting to note the story that the Pythagorean who first divulged this knowledge of the irrationals perished in a shipwreck as an evidence of the displeasure of the gods. A similar story is told of the discloser of the knowledge of the dodecahedron. See Allman, pp. 25, 47.

† See Cantor, p. 170.

‡ Prop. 3. Heiberg, *Archimedes*, p. 262 et seq. Heath, p. 93.

§ Heath refers to Gunther, *Die quadratischen Irrationalitäten der Alten und deren Entwickelungsmethoden*, Leipzig, 1882, as an exhaustive paper discussing in detail all the hypotheses offered up to 1882 for finding $\sqrt{3}$, including those of Zeuthen, Tannery, DeLagny, Heilermann, and Rodet.

‖ Prop. 3. Heiberg, *Archimedes*, p. 262 et seq. These values occur in finding the value of π by the successive steps necessary in inscribing in a circle a regular polygon of 96 sides.

1009 1/6 is greater than	$\sqrt{1018405}$	1009.16
2017 1/4 is greater than	$\sqrt{4069284\ 1/36}$	2017.247
591 1/8 is less than	$\sqrt{349450}$	591.14
1172 1/8 is less than	$\sqrt{1373943\ 33/64}$	1172.15
2339 1/4 is less than	$\sqrt{5472132\ 1/16}$	2339.26

These are very close approximations and those are especially interesting where the number whose square root is sought is itself fractional. It is quite clear from these examples that Archimedes must have had a definite method to secure such accurate results with these large and varied numbers.* It is worth noting in passing that the value of π which he secured by means of these numbers was between 3 1/7 and 3 10/71, that is, between 3.1429 and 3.1408.†

A century and a quarter later Heron used as a formula for computing square root, $\sqrt{a^2 \pm b}$ is approximately $a \pm \frac{b}{2a}$.‡ He gives as roots determined thus:

$$\sqrt{50} \text{ is } 7+1/14, \textit{ i.e., } 7.071.$$
$$\sqrt{63} \text{ is } 8-1/16, \textit{ i.e., } 7.937.$$
$$\sqrt{75} \text{ is } 8+11/16, \textit{ i.e., } 8.687.$$

The first two are correct to three places, but the third where the "b" is not unity is correct only to the first place, the true value being 8.660. Heron's method is probably based on Euclid, being an adaptation of the process given in the next paragraph. He also used for greater accuracy $a \pm \frac{b}{2a} > \sqrt{a^2 \pm b} > a \pm \frac{b}{2a \pm 1}$, by which $\sqrt{75}$ would be between 8.687 and 8.647. From the fact that Archimedes' results all appear as "greater than" or "less than" the real value it would seem that he might have used some similar formula; but of course Heron's is not accurate enough for the results given. Heron also gives 7/5 for $\sqrt{2}$ (correct value 1.414+), and 26/15 for $\sqrt{3}$, *i.e.*, 1.733 (correct value 1.732+). In his *Stereometrica* is the first known attempt to express the square root of a negative number. Without method or comment $\sqrt{81-144}$ is stated to be 8 less than 1/16!

The last method of finding square root we shall give is found in the works of Theon,§ very late, in the fourth century after Christ. It makes use of the sexagesimal system of angles of the Babylonians but is not essentially different from our present method. Although based on Euclid, Theon's language would indicate that

* Heath says "There is no doubt that in obtaining the integral portion of the square root" he used the method of Theon, illustrated in the next paragraph. Gow: "It is clear that Archimedes did not use Theon's method!" And there are various shades of opinion between these extremes. See previous references to Gunther. Also Cantor, p. 301 et seq. Heath, *Archimedes*, p. lxxx, gives in detail Hultsch's theory of "tentative assumptions" ingeniously worked out for many of these results, with several circumstances leading to its probability. See also Nesselmann, p. 108 et seq.

† Heiberg, *Archimedes*, Vol. I, p. 270.

‡ Fink, p. 70.

§ In his commentary on Ptolemy's *Almagest* (A. D. 125).

the method itself is comparatively new. As showing the antiquity of our method in its essence and illustrating the procedure of at least the later Greeks we give the following example paraphrased from Theon.†

"I ought to mention how we extract the approximate root of a quadratic which has only one irrational root. We learn the process from Euclid, II, 4, where it is stated 'if a straight line be divided at any point, the square of the whole line is equal to the square of both the segments together with twice the rectangle contained by the segments.' So with a number like 144, which has a rational root, as the line $\alpha\beta$ (see Fig. I) we take a lesser square say 100, of which the root is 10, as $\alpha\gamma$. We multiply 10 by 2 because there are two rectangles, and divide 44 by 20. The remainder 4 is the square of $\beta\gamma$ which must be 2. Let us try the number 4500 (see Fig. II), of which the root is 67° 4′ 55″. Take a square $\alpha\beta\gamma\delta$ containing 4500 degrees (μοῖραι). The nearest square number is 4489, of which the side (root) is 67°. Take $\alpha\eta=67°$ and $\alpha\epsilon\xi\eta$ the square of $\alpha\eta$. The remaining gnomen $\beta\xi\delta$ contains 11° or 660′. Now divide 660′ by 2 $\alpha\eta$, *i.e.*, by 134. The quotient is 4′. Take $\epsilon\theta, \eta\kappa=4'$ and complete the rectangles $\theta\xi$, $\xi\kappa$. Both these rectangles contain 536′ (268′ each). There remains 124′ = 7440″. From this we must subtract the square $\xi\lambda$ containing 16″. The remaining gnomen $\beta\lambda\delta$ contains 7424″. Divide this by $2\alpha\kappa = 134°\ 8'$. The quotient is 55″. The remainder is 46″ 40‴, which is the square of $\lambda\gamma$, of which the side is 55″ nearly enough".

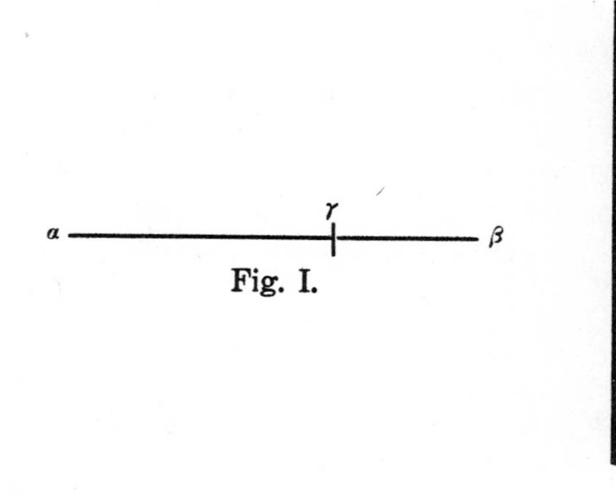

Fig. I.

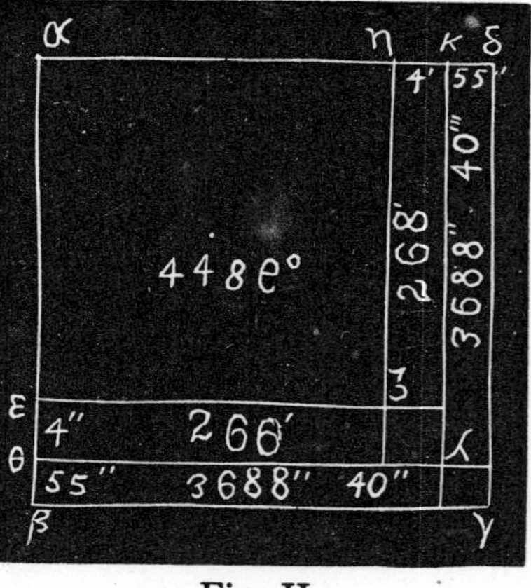

Fig. II.

Theon also gives another example with a figure for $\sqrt{2°\ 28'}$ which he finds to be 1° 34′ 15″. This is correct to .8′. The procedure is the same as above. Theon concludes: "When we seek a square root we first take the root of the nearest square number. We then double this and divide with it the remainder reduced to minutes, and subtract the square of the quotient. Then we reduce the remainder to seconds, and divide by twice the degrees and minutes (of the whole quotient). We thus

† Gow, p. 55. See also Cantor, pp. 460–461; Heath, *Archimedes*, p. lxxvi.

obtain nearly the root of the quadratic." Except for the sexagesimal notation this does not sound very different from the rule in our elementary texts today.

By this process Ptolemy gives $\sqrt{3}=\frac{103}{60}+\frac{55}{60^2}+\frac{23}{60^3}$, which in our notation is 1.7320509 and so correct to 6 places.

2. *Constructive methods.* If the pure quadratic is considered in the form "$x^2=ab$" it means in geometrical language to find a square equivalent to a given rectangle, and this was solved by Euclid and perhaps his predecessors, in the two propositions, "to construct a square equal to a given rectilinear figure"* and "to two given straight lines to find a mean proportional"† being the same methods given in our elementary geometries today.

If the quadratic is in the form "$x^2=pa^2$" it becomes geometrically the problem of the multiplication of the square, which is solvable by the Pythagorean theorem.‡ In its special form "$x^2=2a^2$" it becomes the problem of the duplication of the square, which we have already noted was first known by the Pythagoreans. Perhaps its successful solution suggested the more famous problem of antiquity, the duplication of the cube.

Affected Quadratics. By geometrical methods the Greeks were able to solve any equation of the type $x^2+px+q=0$ where a real solution was possible, although they did not consider it thus generally. Rather they considered quadratics under three forms,

$$\text{(a)} \quad x^2+px=q$$

$$\text{(b)} \quad x^2+q=px$$

$$\text{(c)} \quad x^2=px+q$$

due to the fact that the treatment of them was geometric. Their solution consisted principally in applying theorems on areas, one of the most powerful methods on which Greek geometry relied, or on proportion, in which they were also well versed.

The procedure may be shown by a typical example, of the first type,

$$x^2+ax=b.\text{§}$$

Expressed geometrically, this would be "To the segment $AB=a$ (see Fig. III) apply the rectangle DH, of known area, b, in such a way that CH shall be a square." The figure shows that for $CK=a/2$

$$FH=x^2+2(a/2)x+(a/2)^2=b^2+(a/2)^2.$$

But by the Pythagorean theorem

$$b^2+(a/2)^2=c^2,$$

* Euclid, II, 14. Heiberg, *Euclid*, Vol. I. p. 160.
† Euclid, VI, 13. Heiberg, *Euclid*, Vol. II, p. 110.
‡ Euclid, I, 47. Heiberg, *Euclid*, Vol. I, p. 110.
§ Fink, p. 79. Zeuthen, p. 36 et seq., gives several more.

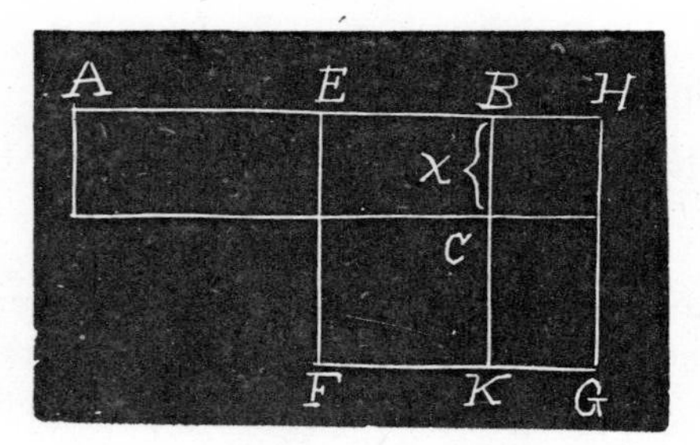

Fig. III.

whence

$$EH = c = a/2 + x \qquad \text{and} \qquad x = c - a/2 = BC.$$

In the same way Euclid solves all problems of this form and says that b must be greater than $a/2$ in $\sqrt{b^2-(a/2)^2}$ in order to give a solution. Thus imaginary roots are excluded.

The equation just solved is simply another way of expressing $x(x+p)=q$, which stated in Euclidean language would be "To produce a given line p to length $p+x$ so that the rectangle between the whole line so produced and the part produced, *i.e.*, $x(x+p)$ shall be equal to a given figure q." Similarly, $x^2=a(a-x)$ of the third type is Euclid's proposition "To cut a given straight line so that the rectangle contained by the whole and one of the segments is equal to the square on the remaining segment."* And Euclid finds $x=\sqrt{a^2+(a/2)^2}-a/2$.† It is certain that this particular problem, and probably others similar, were solved much earlier than Euclid, even by the Pythagoreans.‡

Their general method of solution by areas may be stated as follows. The problem was to apply to a given line a rectangle or more generally a parallelogram so that it would either contain a given area, or be greater or less than the given area by a constant. For these three conditions there arose probably even among the Pythagoreans the names παραβολη′ (parallel to, application, equal), 'υπερβολη′ (excess), and ″ελλειψις (falling short).§ After the time of Archimedes, however, Apollonius took these names for the conic sections because they were appropriate to the distinctive properties of the parabola, hyperbola, and ellipse as he defined them.

The solution of affected quadratics by proportionality of lines was more general, but a later development probably not extensively used much before Euclid, while we have seen that the one by areas antedated him two centuries or more. In

---

* Euclid II, 11. Heiberg, *Euclid*, Vol. 1, p. 152.1, p. 403.

† Cantor, pp. 249–250.

‡ Heath, *Euclid*, Vol. I, note to II, 5, p. 384, and II, 1.

§ For antiquity of these terms see Proclus' note to Euclid I, 44, in Allman, p. 24, and Heath, *Euclid*, Vol. I, p. 343, "These things (*i.e.*, method of areas) are ancient and are the Discoveries of the Muse of the Pythagoreans."

his sixth book Euclid gives by proportion the equivalent of the solution of the general equation $ax^2 \pm \frac{b}{c} x = S$ subject to the condition for real roots.*

Apollonius did the same thing by means of the conics, especially with the aid of the equation $y^2 = px \pm \frac{p}{a} x^2$, the equation which he uses to express the fundamental property of a central conic.**

We conclude accordingly that the Greeks before the beginning of the Christian era were able to solve any equation of the second degree, having two essentially different coefficients, *geometrically* for real positive roots.

Algebraic methods.

Beginnings. A great step forward in mathematical reasoning was made when the Greeks were enabled to divorce their Algebra from their Geometry and reason abstractly, without necessary connection with the concrete concepts of geometry. Only then did they commence to develop a real science of algebra as a distinct subject. As a separate science among the Greeks it seems to have had its beginnings at least as early as the second century before the Christian era, but it did not reach a comparatively consistent development until two or three centuries later under Diophantus, by far the most important name in the consideration of Greek Algebra.

Arabian authorities state that Hipparchus wrote a treatise on the solution of quadratic equations, but no traces of it exist today, and we have no way of knowing whether it was any advance over the geometrical methods we have been considering.†

Heron a little later was the first to adopt the algebraic method, demonstrated the first ten or twelve propositions of Euclid, Book II, by means of lines only, without reference to areas, and dealt with them analytically as representations of pure numbers. He seems to have solved the affected quadratic equation $ax^2 + bx = c$ by completing the square, but the evidence is not conclusive.‡

Diophantus. There is much dispute as to the date of Diophantus but probably it is safe to say he flourished in the last half of the third century of our era.§

La Grange speaks of him as "l'inventeur de l'Algèbre"|| while Tannery is unwilling to credit him with being anything more than a compiler. Nesselmann takes the intermediate view that the greater part of his propositions and ingenious methods are his own.# It seems fair to say that he did for Algebra what Euclid did

* VI, 27, 28, 29, 30. Heiberg, *Euclid*, Vol. II, pp. 158–172. Discussed in their algebraic bearing in Heath, *Euclid*, Vol. 2, p. 260 et seq. Cantor, p. 252.

** Heath, *Apollonius*, p. cx.

† Cantor, p. 346.

‡ See Cantor, p. 376 et seq.

§ Fully Discussed in Heath, *Diophantus*, pp. 3–17.

|| La Grange, Vol. 7, p. 219.

Nesselmann, p. 477. For full discussion of question of Diophantus' originality see Heath, *Diophantus*, Chap. VII, Cantor, p. 427 et seq.

for Geometry, organizing previous knowledge and adding much due to his own genius. His principal treatise is on "Arithmetic" ('Αριθμητικῶν βιβλι'α ιλ̄) in six books of which a portion is lost; it is the first treatise on Algebra extant.* He was the first to use symbols for operations and for an unknown quantity, although he does not claim this symbolism is original. The unknown quantity was called 'ο'αριθμο'ς and its symbol was ς' or ς$^{o'}$, plural ςς or ςς^{οι}. The square of the unknown was δυναμις and its symbol δ^{υ}. Symbols were used for as high as the sixth power of the unknown.† One of his equations looks like this

κ^{υ}	β	δ^{υ}	ᾱ	'ι'ση	ςς^{ois}	'^{o}ο	⋔	μ$^{o'}$	ι̅β̅
x^3	2	x^2	1	equals	x	4	minus	units	12

i. e., $2x^3+x^2=4x-12$. The part of Diophantus' work which deals with the solution of determinate quadratics is lost and we have little information as to its contents. Most of his extant work deals with the solution of indeterminate equations, including quadratics. His work shows deficiency in generalization but unusual ingenuity in the manipulation of processes, especially necessary since he never uses more than one symbol for the unknown, and other unknowns must be expressed in terms of it.

A single solution is sufficient for any set of conditions. Even when they lead to a quadratic he gives but one root, showing his ignorance of their true nature.‡ He refuses negative or imaginary roots as "impossible" or "absurd," but distinguishes between rational and irrational roots seeking only the former. His fundamental basis of classification of equations is not according to degree but *number of terms* when reduced to simplest form, *i.e.*, $ax^n=c$ is a "simple" equation whatever the power of x, while $ax^2+bx=c$ is a "mixed" equation with two terms. In what follows we take that part of his work, for the most part, which deals with what we know as quadratic equations, although, as pointed out, this distinction is not always made by him.

Pure Quadratics. In Definition 11 of the First Book§ he gives his method for solving problems leading to pure quadratics. "If a problem leads to an equation containing the same powers of the unknown (ε'ι'δη ταῦτα) on both sides but not with the same coefficients (μη' 'ομοπλη'θη) you must deduct like from like until only equal terms remain. But when on one side or both some terms are negative, ('εν 'ελλε'ιψεσι) you must add the negative terms to both sides till all the terms are positive ('ενυπα'ρχοντα) and then deduct as before stated. If he comes to an

* Diophantus announces 13 books. The seven existing mss. are in six books except one, which is in seven but has the same material. Probably we have the bulk of the original 13 books, with some sections lost (notably those on solution of determinate quadratics) and the remainder rearranged by later editors. See Nesselmann, p. 265 et seq; Gow, p. 101–102; Heath, *Diophantus*, Chap. II.

† For symbolism see Nesselmann, pp. 294–296; Heath, *Diophantus*, Chap. IV; Cantor, p. 439.

‡ Hankel (p. 162) thinks this is a relic of the old geometric practice. Gow suggests that it was because Algebra was the invention of practical men who needed only the one solution.

§ Tannery, Vol. I, p. 14.

equation such as $x^2 = ax$ he merely divides by the factor $x=0$ which he does not regard as a solution, *e.g.*,

$$\text{“}20x = 10x^2 \qquad \text{whence } x = 2.\text{”}*$$

Affected Quadratics. In Definition 11, Diophantus promises to give his method for solving “mixed” or affected quadratics. He continues: “We must contrive always, if possible, to reduce our equations so that they may contain one single term equated to one other. But afterward we will explain to you also how, when two terms are left equal to a third, such a question is solved.” That is, “reduce to the form $x^2 = a$ if possible. Later we will give a method for the complete quadratic $x^2 + ax = b$.” But this method, if ever written, is in the part which is lost. Without the method of solution he states numerous results, *e.g.*, “$84x^2 + 7x = 7$, whence x is found equal to $\frac{1}{4}$”**; “$630x^2 + 73x = 6$, whence the root is rational,”† and many similar ones. When a root is irrational he sometimes gives an approximation to it.‡ It is unfortunate that he nowhere gives an example of the complete solution of a single equation, but the form of his solution of equations of this type

$$ax^2 + bx + c = 0$$

is usually

$$x = \frac{(1/2)b \pm \sqrt{(1/4)b^2 - ac}}{a}.$$

This is exactly the form it would have, had he completed the square after first multiplying the equation through by the coefficient of the x^2 term. The consensus of opinion seems to be that this is the way in which he arrived at his results.§ From the variety of equations which arise in Diophantus' work we must believe that he had a general method of solution for all *determinate* quadratics, sufficient to determine one real root if such existed.

No sufficient idea can be given of the general methods he used in resolving *indeterminate* equations into solvable forms||, but two examples will be given to show his ingenuity and some of his methods of attack.

A common method of attack is that of *tentative assumptions*, in which a value is shrewdly assumed and then altered as occasion arises until it fits the conditions of the problem. It frequently involves the use of his single symbol for the unknown

---

* I. 31; Tannery, Vol. I, p. 68, and many others in first book.

**VI, 6. Tannery, p. 404.

† VI, 8. Tannery, p. 408.

‡ *E.g.* V, 30. Tannery, p. 384.

§ See Nesselmann, p. 319; Cantor, p. 443 et seq; Heath, *Diophantus*, pp. 90–92.

|| Hankel (pp. 164–165) says it is difficult after solving 100 of Diophantus' problems to have any idea how to attack the 101st. Nesselmann gives an unsatisfactory classification and discussion of methods of solution (pp. 315–328) and says himself that a complete discussion would mean copying Diophantus' entire work. Heath disagrees with both and devotes Chap. V. of his book on Diophantus to a fairly satisfactory classification and discussion.

quantity in different senses. For example, in book four* the problem is "to find three numbers such that their sum is a square and the square of any number plus the following one is a square." He assumes the three numbers first as $(x-1)$ $(4x)$ $(8x+1)$ where $(x-1)^2+4x$ and $(4x)^2+(8x+1)$ are both square numbers. Two of the conditions are thus satisfied, but the third is that the sum of the numbers, $13x$, must also be a square number. He says: "Take $13x=x^2$ with some square coefficient e.g., $13x=169x^2$. Then $x=13x^2$." A new use of "x" is thus introduced and $13x^2$ is substituted for the original x, the numbers now being, $13x^2+1$, $52x^2$ and $104x^2-1$. A fourth condition remains, viz., that

$$(104x^2+1)^2+(13x^2-1)$$

shall be a square. Diophantus then takes this expression equal to

$$x^2(104x^2+1)^2$$

and finds $x=\frac{111}{104}$, and substitutes this value, finally getting his three numbers

$$\frac{170989}{10816}, \frac{640692}{10816}, \frac{1270568}{10816}.$$

Of course this has all been expressed in modern notation. His last number, for example, is expressed

$$\frac{\alpha}{\rho\kappa\xi} : \frac{\omega\iota\varsigma}{\phi\xi\eta}.$$

Another frequent artifice which he uses is a method of *limits*. If he wishes to find a square number, between 10 and 11 for example, he multiplies these by successive squares until a square number lies between the product: thus between 40 and 44 or 90 and 99 no (rational) square lies, but between 160 and 176 is 169; hence $x^2=\frac{169}{16}$ will lie between the proposed limits. This is made use of in the problem: "To divide 1 into two parts such that if 3 be added to the first part and 5 to the other, the product of the two sums shall be a square."† If one part be $x-3$ the other is $4-x$. Then $x(9-x)$ must be a square. Suppose it equals $4x^2$, then $x=\frac{9}{5}$. But this will not suit the original assumption since x must be greater than 3 and less than 4. Now 5 is $4+1$. Hence what is wanted is to find a number y^2+1, such that $\dfrac{9}{y^2+1}$ is greater than 3 and less than 4. For such a purpose y^2 must be less than 2 and greater than $1\frac{1}{4}$. "I resolve these expressions into square fractions" says Diophantus, selecting $\frac{128}{64}$ and $\frac{80}{64}$ between which lies the square $\frac{100}{64}$ or $\frac{25}{16}$. He then puts $x(9-x)=\dfrac{25x^2}{16}$ instead of $4x^2$, whence $x=\frac{144}{41}$, and the two numbers are $\frac{21}{41}$ and $\frac{20}{41}$.

In general we may say that the type of indeterminate quadratic equations which he considers fully are limited to the case where one or two functions of the

* IV, 17. Tannery, pp. 222–225. Gow, p. 117.

† IV, 31; Tannery, pp. 278–283; Gow, p. 119. For this method see also Nesselmann, p. 318; Heath, *Diophantus*, pp. 115–120.

unknown in the form $Ax^2 + Bx + C$ must be a rational square, and are only fully treated in cases where the "C" vanishes. Otherwise his solution depends upon the particular equation involved, and consists of ingenious transformations or assumptions of the unknown.* The most characteristic feature of his work is the extraordinary artifices which he employs.

Such then are the principal methods known to the ancient Greeks for solving quadratic equations. In the classic period they had a comparatively complete knowledge of methods of solution from the purely geometric standpoint, practically none algebraically. Even in later times their knowledge of the algebraic solution was faulty and cramped, showing no true conception of the real nature of the quadratic, only familiarity with ingenious methods of manipulating particular equations. Considering the exceptionally high attainments of the Greeks in geometry as well as other branches of learning and culture, it seems a little strange that they did not make greater advances in algebra. This was reserved for the Hindoos some centuries later. Nevertheless it is instructive to trace out some of their attainments in this particular line in order to note their limitations as well as their achievements.

DIOPHANTUS OF ALEXANDRIA†

J. D. SWIFT, University of California at Los Angeles and Institute for Advanced Study

1. Introduction. The name of Diophantus of Alexandria is immortalized in the designation of indeterminate equations and the theory of approximation. As is perhaps more often the rule than the exception in such cases, the attribution of the name may readily be questioned. Diophantus certainly did not invent indeterminate equations. Pythagoras was credited with the solution $(2n+1, 2n^2+2n, 2n^2+2n+1)$ of the equation $x^2+y^2=z^2$; the famous Cattle Problem of Archimedes is far more difficult than anything in Diophantus, and a large number of other ancient indeterminate problems are known. Further, Diophantus did not even consider the most common type of problem called by his name, the linear equation or system of equations to be solved in integers.

Nevertheless, on at least three grounds the place of Diophantus in the development of mathematics is secure. On all the available data he was the first to introduce systematic algebraic procedures to the solution of non-linear indeterminate equations and the first to introduce extensive and consistent algebraic notation representing a tremendous improvement over the purely verbal styles of his predecessors (and many successors). Finally, the rediscovery of the book through Byzantine sources greatly aided the renaissance of mathematics in western Europe and stimulated many mathematicians, of whom the greatest was Fermat. (Much of Fermat's work is known from notes written in his copy of Diophantus.) [1]

---

* Heath, *Diophantus*, p. 115.

† From AMERICAN MATHEMATICAL MONTHLY, vol. 63 (1956), pp. 163–170.

Of Diophantus as an individual we have essentially no information. A famous problem in the Greek Anthology indicates that he died at the age of 84, but in what year or even in which century we have no definite knowledge. He quotes Hypsicles and is quoted by Theon, the father of Hypatia. Now Hypsicles, in the introduction to his book, the so-called Book XIV of Euclid, places himself within a generation or so of Apollonius of Perga whose time is definitely established by the rulers to whom he dedicates his works. Thus we may put Hypsicles in the early or middle part of the second century B.C. with reasonable accuracy [17]. Theon, on the other hand, definitely saw the eclipse of 364 A.D. [10]. Within this gap of five hundred years, historians are at liberty to place Diophantus wherever he best fits their theories of historical development [10, 14]. The majority follow [2] and, on the basis of a dubious reference by the Byzantine Psellus (C.1050), assign him to the third century A.D.

2. The Arithmetic. The surviving work of Diophantus consists of six books (sometimes divided into seven) of the *Arithmetic* and a fragment of a work on polygonal numbers. The introduction to the *Arithmetic* promises thirteen books. The position and content of the missing six or seven books is a matter of conjecture. (The reader is reminded that a "book" is a single scroll and represents the material contained in twenty to fifty pages of ordinary type.)

These books may be summarized as follows: Book I—Determinate systems of equations involving linear or quadratic methods. Books II–V—Equations and systems of equations, the majority of which are quadratic indeterminate although Books IV and V contain a selection of cubic equations, determinate and indeterminate. Book VI—Equations involving right triangles. All books consist of individual problems and their solutions in positive rationals. In the ordering of the problems some consideration has been given to relative difficulty and interrelation of material, but the over-all impression is of a disconnected assortment.

3. Notation. The numerical notation used by Diophantus is, of course, the standard Hellenistic notation which uses the letters of the Greek alphabet with three archaic letters added to give 27 different symbols; [6, 7] the first nine stand for units, the second for tens and the last for hundreds. Thus, for any further notation, either non-alphabetic symbols or monogrammatic characters were required.

There is a single symbol for the unknown quantity. This may be a monogram for *αριθμος*. The symbol for "minus" is apparently a monogram for the root of *λειψις* [5]. Addition is indicated by juxtaposition. The powers of the unknown are designated by easily recognizable monograms, the square by Δ^{υ} for *δυναμις*, the cube by K^{υ} for *κυβος*. Higher powers are formed from these by addition, *i.e.*, the fifth power is considered as square-cube. To avoid ambiguity it is necessary to have a special symbol for the zero-order terms also, a monogram for *μοναδος*, and to write all the negative terms together. Thus, if we adopt an equivalent set of conventions in English, retaining Arabic numerals and the letter x for the indeterminate, the expression: $6x^4+23x^3-2x^2+x-5$ would appear as $SS^q6C^u23X1MS^q2U^n5$.

Fractions were represented either in the inverse position to the present day or by inserting the word for "divided by" between the numerical expressions on the same line. Reciprocals of integers and negative powers of the unknown are designated by a special symbol placed after the number or power.

The most important limitation of this notation is the restriction to one unknown. Since practically all the problems require the determination of several quantities, a considerable part of Diophantus' work lies in the reduction to a single quantity. Further, no general solution in expressed parameters is possible. Even if a general method is indicated, it must be restricted in its presentation to a specific numerical case.

A particular problem will illustrate the situation. In problem 1, Book IV, it is desired, in modern terms, to solve the system: $x^3+y^3=a$, $x+y=b$. Essentially the method is to let $x=z+b/2$, $y=b/2-z$. Substitution in the first equation now yields a binomial quadratic. Let us look at this problem in a translation as bald as possible:

To partition a given number into two cubes of which the sum of the sides is given: Let the number to be partitioned be 370 and the sum of the sides U^n10. Let the side of the first cube be $x1U^n5$, the latter term of which is half the sum of the sides. Therefore, subtracting, the side of the other cube is U^n5Mx. Then the sum of the cubes will be S^q30U^n250. This is equal to U^n370 as is given and x becomes U^n2. As to the original numbers, the first side will be U^n7 and the second, U^n3. The first cube, 343; the second, 27.

4. Diophantine algebra. With this problem in mind let us turn to some aspects of Greek and Babylonian mathematics. A number of tablets [15, 16], both old Babylonian (1800–1600 B.C.) and Seleucid (300 B.C. and later), exist which teach the solution of equations which can be reduced to the forms $x+y=a$; $xy=b$ [10, 12, 13]. Again Euclid's Elements II, 5, 6 can best be viewed as giving solutions to these problems [12]. In modern notation, the procedure in both cases is to write $x=a/2+z$, $y=\pm(a/2-z)$; $xy=b=\pm(a^2/4-z^2)$; $z=\sqrt{a^2/4\pm b}$. Now Diophantus in I (27, 30) considers the same equations, solves them the same way and applies the basic idea repeatedly as in the quoted problem. Other examples can be followed in a similar way, *e.g.*, $x^2+y^2=a$, $xy=b$. (See [13] for a complete discussion of quadratic equations in antiquity.)

Let us now compare the treatment in the three cases: The tablets consist of lists of problems of varying complexity each framed in specific numbers and quantities. The problems are not "practical" nor in any sense rigorously geometric; men are added to days, lengths to areas, areas are multiplied, *etc.* It is clear that the basic thought is purely algebraic. The problems are so set that the solutions are positive integers or terminating sexagesimal fractions such that the roots can be obtained from tables of squares, but from other tablets we learn of approximations to non-terminating rationals like 1/7 or irrationals like $\sqrt{2}$.

The Euclidean problems are cast in the form of propositions about line segments, squares and rectangles. Their generalizations in II, 28, 29, concern parallelograms. The propositions are general and the result is deduced rigorously from the postulational basis. The results are line segments which may well be incommensurable with the original segments; *i.e.*, "irrational" answers are acceptable.

In Diophantus the problems are formulated in terms of abstract numbers but a "number" is always positive rational. The solutions are worked out in terms of particular numerical examples. This procedure may be considered analogous to carrying out a geometrical construction in terms of particular line segments and, indeed, Diophantus probably intended that his problems should be read in this manner. There is, however, no pretense at postulational development. No general propositions are stated even where the solution implies them. Restrictions on the choice of initial values are not always given; in the case of I, 27, we are informed that $a^2/4-b$ must be a square but in the problem in the previous section no restriction is mentioned. The most reasonable conclusion is that he did not know the form of the restriction or did not know how to express numbers that satisfied the restriction. The authors of [5] and [4] disagree but on the naive ground that since he did come up with workable numbers 370 and 10, he must have had some way of generating them. The answer is obvious; he generated them from the answer.

Like the Babylonians, Diophantus had no qualms about adding areas and lengths (See VI, 19 in 5) although, to be precise, he says that he adds "the number in the area" to "the number in the length." His algebraic technique is tremendously advanced beyond anything we possess of the Babylonians. The complicated cubic and higher degree equations and the indefinite equations are not even suggested in Babylonian algebra. The latter had examples of binary cubics and a few other higher degree equations soluble by tables; they also knew general forms for Pythagorean numbers and obtained solutions of $x^2-2y^2=1$ but this is as far as our present evidence takes them. Even in the quadratic case there may be a difference [13]. When a quadratic is to be solved, Diophantus makes some effort to choose the variable so that a binomial equation results, but if this is not practicable, the general quadratic formula (positive sign before the radical) is used without further comment. The question is still at issue whether the Babylonians ever solve a quadratic without bringing it into some normal form involving a known sum or difference and product.

It is useless even to try to guess what proportion of the advanced problems and methods are Diophantus' own. Most modern historians postulate a continuous underlying tradition of oriental algebraic methods in Greek mathematics rather than a sudden invasion in the Roman period. If this be so, texts and problem lists would certainly have existed. It is probable that the *Arithmetic* was in good part a compilation of such a quality that the predecessors were no longer held in repute. There are traces of the Diophantine notation elsewhere; Heron (60 A.D.) used the same minus sign for example, but no evidence exists that the semi-algebraic notation or the general methods it permitted were used before the publication of the *Arithmetic.*

To sum up, the basic algebraic approach in Diophantus is Babylonian. The generality and abstraction is Greek. The work may be viewed as an episode in the decline of Greek mathematics [12] or as the finest flowering of Babylonian algebra [10].

5. Indeterminate problems. In giving translations of several illustrative problems, I have avoided the usual practice of direct translation. Instead, I adhere carefully to the method of the original while replacing the particular numbers used by parameters. The rationale may thus be conveyed with less verbal explanation than if the presentation were given in its original special form. At the same time, the full power of the method is apparent.

II. 9: If $n=a^2+b^2$, find other representations of n as the sum of two squares: Modify a to $x+a$. The corresponding modification of b may be written $(rx-b)$. Here x is the unknown, a, b and r were assigned specific values.

$$n = a^2 + b^2 = (x + a)^2 + (rx - b)^2$$
$$= (r^2 + 1)x^2 + (2a - 2br)x + a^2 + b^2.$$

Thus $x=(2br-2a)/(r^2+1)$ where r may be any rational such that the required quantities are positive.

Note the clever choice of the unknown; fixing x and solving for r would leave a condition on x still to be met; b^2-x^2-2ax would have to be made a square. Again, if a is increased by a fixed amount and the unknown is taken as the corresponding decrease in b, the result does not come out at once. The choice of $rx-b$ instead of $b-rx$ was dictated solely by the numerical values selected which happened to make $b-rx$ negative. (But see V, 24, below.) Euler wrote $a^2+b^2=(a+rx)^2+(b-qy)^2$ which results in a more symmetric solution but this concept is foreign to Diophantus' notation and the solution above is quite general.

Here would be a perfect opportunity to state a proposition instead of a problem. A proof of the theorem: "Any number which is the sum of two squares can be represented as such in an infinite number of ways," is contained in the solution above.

III. 6: Find three numbers whose sum is a square and such that the sum of any two is a square:

$$x + y + z = t^2$$
$$x + y = u^2$$
$$y + z = v^2$$
$$x + z = w^2.$$

Here Diophantus assigns a definite value to w, or, in modern notation, lets it play the role of the parameter. He then chooses an unknown, r, restricting it as follows: Let $t=r+1$, $u=r$, $v=r-1$. Then $z=2r+1$, $y=r^2-4r$, $x=4r$ and $w^2=6r+1$. Thus $r=(w^2-1)/6$ where w is an arbitrary rational exceeding 5 (so that y is positive). So $x=(2w^2-2)/3$; $y=(w^2-1)(w^2-25)/36$; $z=(w^2+2)/3$.

This problem was chosen to illustrate two points. First Diophantus is not interested in generality except as an incidental by-product. A considerable increase in generality can be obtained merely by replacing $r\pm 1$ by $r\pm s$ in the solution and this possibility could easily have been indicated by the addition of a single phrase. Second, the choice of wording of the problems is often peculiar from a modern viewpoint. This problem is clearly equivalent to the single equation: $u^2+v^2+w^2=2t^2$.

Incidentally, using methods available to Diophantus but probably exceeding his control of notation, a much more general solution of this equation is available than the system $(u, v, w, t)=((w^2-1)/6, (w^2-7)/6, w, (w^2+t)/6)$ given above. The equation being homogeneous, it will be more convenient to solve in integers. Let $w=rs$, $u=s^2-p$, $v=s^2-q$, then

$$s^4+(r^2/2-p-q)s^2+(p^2+q^2)/2=t^2.$$

The left hand side is a perfect square if $p^2+q^2=2k^2$, $r^2/2-p-q=2k$. The first of these is the problem of finding three squares in arithmetic progression. It does not occur specifically in the *Arithmetic*, probably because it is too simple in the rational case, reducing essentially to $a=b=1$ in the problem above. It will be more convenient to take a solution derived from the solution given to the Pythagorean equation by Euclid. If $X^2+Y^2=Z^2$, $(X+Y)^2+(X-Y)^2=2Z^2$. Thus $p=-m^2+2mn+n^2$, $q=m^2+2mn-n^2$, $k=m^2+n^2$; $r^2=4(m+n)^2$ and $r=2(m+n)$. Thus $(u, v, w, t)=(s^2+m^2-2mn-n^2, s^2-m^2-2mn+n^2, 2(m+n)s, s^2+m^2+n^2)$. The previous solution is obtained by setting $m=2$, $n=1$ and dividing by 6.

V. 24: Find a solution of $x^4+y^4+z^4=t^2$.

If $t^2=(x^2-m)^2$, $x^2=(m^2-y^4-z^4)/2m$. Thus an integer m must be found so that $(m^2-y^4-z^4)/2m$ is a square. Let $m=y^2+z^2$ so $x^2=y^2z^2/(y^2+z^2)$. Thus y^2+z^2 must be a square, say $(y+r)^2$. Then $y=(z^2-r^2)/2r$. Thus

$$(x, y, z, t)=\frac{z^3-r^2z}{z^2+r^2},\quad \frac{z^2-r^2}{2r},\quad z,\quad \frac{z^8+14z^4r^4+r^8}{4r^2(z^2+r^2)^2}.$$

This example has been chosen for three reasons. First, it is of great historical interest. To this problem Fermat appended a note: "Why does Diophantus not ask for the sum of *two* biquadrates to be a square? This is, indeed, impossible . . . ." Later, Euler conjectured that it was also impossible to find three fourth powers whose sum was a fourth power; *i.e.*, to replace t^2 by t^4. This question remains unsolved.

Second, the problem indicates what happens when the notation is insufficient. First, the chosen unknown is x; m, y, z are assigned specific values to indicate that they play the role of parameters. But the problem cannot be completed, so the author turns to a sub-problem in which y is the unknown.

Finally, the problem contains a curious case of indifference to sign. The quantity x^2-m is, in fact, the negative root of $t^2=(x^2-m)^2$. Since only the square is used, no harm is done but we must remember that, to Diophantus, the quantity x^2-m, which he used, did not exist. The reader may find it interesting to see why x^2+m was not used by trying it. Why $m-x^2$ which is positive and produces the same result was not preferred is a matter for conjecture.

VI. 19: Find a right triangle such that its area added to one of its legs is a square while the perimeter is a cube.

First form the triangle $(2x+1,\ 2x^2+2x,\ 2x^2+2x+1)$. The perimeter is $4x^2+6x+2=(4x+2)(x+1)$. Since it is difficult to make a quadratic a cube, consider in turn the triangle $(2x+1)/(x+1)$, $2x$, $2x+1/(x+1)$ obtained by dividing through by $x+1$. The perimeter is $4x+2$ and the area is $(2x^2+x)/(x+1)$. Adding $(2x+1)/(x+1)$ to the latter we have $2x+1$. Thus $4x+2$ is required to be a cube and $2x+1$ a square. The obvious value for $2x+1$ is 4. Thus $x=3/2$ and the triangle is 8/5, 3, 17/5.

It is not clear whether or not Diophantus implies the more general solution $2x+1=4r^6$, $x=(4r^6-1)/2$; probably not.

This problem is illustrative of the rather peculiar problems considered throughout Book VI and of the complete freedom from geometrical considerations. To Euclid such phrases as "the sum of one side and the area" would have been shocking nonsense.

6. An approximation problem. In V. 9 it is required as a sub-problem to find two squares, both exceeding 6, whose sum is 13. Since we have, in the first example of the preceding section, a general method of partitioning a number into two squares when one such partition is given, it is merely necessary to set the two values equal, solve for the parameter and approximate this solution in rationals. If this is done with $a=2$ and $b=3$, we find $r=5+\sqrt{26}$. Approximating by $r=10$, $13=(258/101)^2+(257/101)^2$ and it is readily seen that the conditions are met.

Of course, Diophantus could not do this since the parameters were not expressed. He first finds a number slightly greater than $\sqrt{13/2}$. $13/2=26/4$; if $\sqrt{26}<5+1/x$, $x^2<10x+1$; let $x=10$, then $\sqrt{13/2}\sim 51/20$. Now $51/20=3-9/20=2+11/20$. Thus we wish to find a number near 1/20 such that $(3-9y)^2+(2+11y)^2=13$. Then $y=5/101$ and the squares are precisely those obtained above.

The problem is typical of the approximate methods used. To approximate the nth root of a rational, first write it in the form p/q^n by multiplying numerator and denominator by the necessary integer to make the denominator a perfect nth power. Then multiply p by the nth powers of successive integers until pa^n is sufficiently close to a perfect nth power, say b^n. The approximation is then b/aq. To improve an approximation a_1 to $\sqrt{a}$, set $(a_1+1/x)^2=a$ and approximate x.

7. Transmission of Diophantus. When the Arabs overran the Southeastern Mediterranean in the 7th century, they came into possession of manuscripts of works which had been published in sufficiently large editions to survive the wars attendant on the breakup of the Roman Empire and the lack of interest in learning of the early Christians. Among these was the *Arithmetic* or at least a portion of it. Translations and commentaries were published in Arabic. These have all been lost; their only trace is in bibliographers' references. When the Arabs formulated their own algebra, they apparently appealed directly to the basic Oriental tradition previously cited. The beginnings of an algebraic notation and the abstract numbers are nowhere to be seen. With the sole exception of the problems mentioned in section 3 as common to the whole ancient world (Diophantus—I, 27–30) not one problem from the *Arithmetic* is found in the algebra of Al-Khwarizmi or, as far as is known, in any other basic Oriental text [13]. Probably the Arabs found Diophantus too impractical for their utilitarian mathematics and the Hindus, if they ever saw the *Arithmetic*, were interested in other problems such as the theory of linear indeterminate problems.

In the other reservoir of learning, Byzantium, the manuscripts of Diophantus lay almost unnoticed for eight centuries. We do not know when the missing books were lost but the part which we now possess escaped the sack of Constantinople by the Crusaders in 1204 and later in the same century M. Planudes and G. Pachymeres wrote commentaries on the first part of the *Arithmetic*. At some time, probably in the course of the emigration of the Byzantine scholars during the Turkish conquests, copies were brought to Italy and Regiomontanus saw one there between 1461 and 1464.

The first translation to Latin was made by W. Holzmann who wrote under the Greek version of his name, Xylander. This translation was published in 1575. Meanwhile, Bombelli, in 1572, distributed all the problems in the first four books among problems of his own in a text on algebra. Bachet, borrowing liberally from Bombelli and Holzmann made another translation in 1621 and a second edition was published in 1670 including Fermat's marginal notes. In the next two centuries various translations were made into modern languages which were based primarily on the editions just mentioned by Holzmann and Bachet. Finally in 1890, P. Tannery prepared a definitive edition of the Greek text with a translation into simple mathematical Latin using modern numerical and algebraic notation. From this work the three excellent translations listed in the bibliography have been prepared. The references to the last two paragraphs are the commentaries in [2] and [6], particularly the latter which has been followed rather closely.

Bibliography

A. Greek Text with Latin Translation:

[1] C. G. Bachet: Diophanti Alexandrini Arithmeticorum libri sex, *etc.* Paris, second edition, 1670.

[2] P. Tannery: Diophanti Alexandrini opera omnia cum Graecis commentariis, Teubner, vol. 1, 1893, vol. ii, 1895.

B. Modern translations based on [2]:

[3] A. Czwalina: Arithmetik des Diophantos aus Alexandria, Göttingen, Vandenhoek, 1952.

[4] P. ver Eeke: Diophante d' Alexandrie. Les six livres arithmetiques et le livre des nombres polygones. Bruges, Descée, 1926.

[5] T. L. Heath: Diophantus of Alexandria, Cambridge, 1910.
C. Histories and Compilations of Ancient Mathematics:
[6] T. L. Heath: History of Greek Mathematics (Two volumes), Oxford, 1921.
[7] ——— A Manual of Greek Mathematics, Oxford, 1931.
[8] G. Loria: Le Scienze esatte nell' antica Grecia, 2nd edition, Milan, Hoepli, 1914.
[9] G. H. F. Nesslemann: Die Algebra der Griechen, Reimer, Berlin, 1842.
[10] O. Neugebauer: The Exact Sciences in Antiquity, Princeton, 1952.
[11] I. Thomas: Selections Illustrating the History of Greek Mathematics, Harvard and Cambridge (Loeb Classics) 1939.
[12] B. L. van der Waerden: Science Awakening, P. Noordhoff, Groningen, 1954 (English translation by A. Dresden).
D. Miscellaneous References:
[13] S. Gandz: The Origin and Development of the Quadratic Equations in Babylonian, Greek and Early Arabic Algebra, Osiris, vol. 3 (1938) pp. 405–557.
[14] J. Klein: Die griechische Logistik und die Entstehung der Algebra, Quellen and Studien zur Ges. der Math. Abt. B, vol. 3, 1934–6, pp. 18–105; 122–235.
[15] O. Neugebauer: Mathematische Keilschrift-Texte, 3 Vols., Springer, 1935–7 = Quellen und Studien zur Ges. der Math. Abt. A, vol. 2.
[16] ——— and A. Sachs: Mathematical Cuneiform Texts, American Oriental Society, New Haven, 1945 = Am. Oriental Series, vol. 29.
[17] Paulys Real Encyclopädie der Classischen Altertums Wissenschaft, Stuttgart, 1893.

CURRENT TRENDS IN ALGEBRA*

GARRETT BIRKHOFF, Harvard University

1. Introduction. Symbolic algebra is much older than many mathematicians suppose; it can be traced back at least to Diophantus of Alexandria (ca. 250 A.D.) and Brahmagupta (ca. 598–665 A.D.). For this early work, see Cajori [7, Arts. 101–5] and Ball [1, pp. 154–6]. Even so-called "modern" algebra is over a century old!

When you realize this, you should not find it too hard to believe that the availability of high-speed computers is giving rise to new trends in algebra. My ultimate aim is

* From AMERICAN MATHEMATICAL MONTHLY, vol. 80 (1973), pp. 760–782.

Garrett Birkhoff is the Putnam Professor of Pure and Applied Mathematics at Harvard, where he did his undergraduate and graduate work, was a Junior Fellow in the Society of Fellows, and has served on the faculty since. He has been a Visiting Lecturer at the University of Washington, University of Cincinnati, and the National University of Mexico, and held a Guggenheim Fellowship. He has served as President of SIAM, Vice-President of the AMS, the MAA, and the American Academy of Arts and Sciences, and Chairman of the CBMS. He is a member of the American Philosophical Society and the National Academy of Sciences, and has received honorary degrees from the National University of Mexico, the University of Lille, France, and the Case Institute of Technology.

His extensive publications in modern algebra, fluid mechanics, numerical analysis, and nuclear reactor theory include the books *Hydrodynamics* (Princeton University Press, 1950); *Lattice Theory* (American Mathematical Society Colloquium Publications, 1940, Third Edition 1967); *Survey of Modern Algebra* (with S. Mac Lane, Macmillan, 1941, 1953, 1965); *Jets, Wakes, and Cavities* (with E. H. Zarantonello, Academic Press, 1957); *Ordinary Differential Equations* (Ginn, 1962); *Algebra* (with S. MacLane, Macmillan, 1967); *Modern Applied Algebra* (with T. C. Bartee, McGraw-Hill, 1970). *Editor.*

to sketch for you, in §§8–10, what I think these trends are. But I wish to lead up to this theme by a brief résumé of the development of algebra as we know it today, over the past several centuries.

2. Classical algebra. The name *modern algebra* was originally intended (in 1930) to signify a contrast with *classical algebra*, which was generally understood to mean the *theory of equations*. This may be defined as the art of solving *numerical problems* by *manipulating symbols*, and seems to have originated with Al-Khwarizmi and other Islamic mathematicians during the period 800–1000 A.D. As we know, its most essential idea consists in replacing each verbal statement about *numerical* quantities by a symbolic *equation*, whose terms can be rearranged and combined by well-established general laws to give a sequence of equivalent but, hopefully, simpler equations. The original equation can be considered as "solved" when the unknown quantity has been isolated on one side of the equality symbol =, on the other side of which is some expression involving only known quantities.

Though the word "root" (of an equation) can be traced back to the Sanskrit,[1,2] and the word "power" (of a number) appears in al-Khwarizmi's *Algebra* ("al-jabr"), development of classical algebra in its present form was very gradual. The "al-jabr" of the Arabs did not become widely used in Western Europe, and the symbols + and − did not achieve their present significance, until nearly 1500. A major advance followed shortly thereafter, the solution of cubic and quartic equations by radicals being already contained in Cardan's *Ars Magna* (1545).

For the next two centuries, progress in algebra was mainly[3] in connection with its applications: to (analytic) geometry, which gave a vivid meaning to negative numbers, and to the calculus through the use of infinite series. Until after 1750, the significance of imaginary roots and complex numbers remained quite obscure, and even discussions of simultaneous linear equations and determinants were unsystematic and fragmentary.

But from 1750 to 1830, thanks especially to the work of Euler, Lagrange, and Gauss, classical algebra developed rapidly into approximately its present form. Thus the exponential function e^z became defined for all complex z as a power series, and as a result $a^z = e^{z \ln a}$ became well-defined for all positive a and complex z. The "Lagrange resolvent" was also invented by Euler [**13**, p. 27].

Above all, the Fundamental Theorem of Algebra was recognized as such, clearly stated, and proved. Euler considered its *real* forms, whose equivalence is easily shown. Two of these are:

(a) Every real polynomial of degree $n > 2$ has proper factors.

(b) Every real polynomial can be (uniquely) factored into real linear and quadratic factors.

Condition (a) follows for $n < 5$ from Cardan, and for $n = 5$ because every real

1. All notes are collected together at the end of the paper.

polynomial of odd degree has real roots. Euler satisfied himself that it was true for all n, but his proof is obscure.

Conditions (a) and (b) are easily shown to be equivalent also to the usual statement of the Fundamental Theorem of Algebra:

(c) Every complex polynomial can be factored into linear factors.

Gauss gave many relatively rigorous proofs of (c) from about 1800 on, and made it clear that all polynomial equations had solutions in terms of complex numbers $x + yi$, $i = \sqrt{-1}$, while the geometrical interpretation of complex numbers as points in the (x, y)-plane gave them a more than symbolic meaning.

Gauss also developed systematic iterative as well as elimination techniques for solving systems of simultaneous linear equations, and the laws of determinants also became generally known—all by 1825 or so.

A few years later, Galois and Abel showed that it was impossible to solve a general equation of the fifth degree by radicals[4], and after this mathematicians gradually began to turn their attention from the theory of equations to *non*-numerical applications of symbolic algebra (e.g., to groups, vectors and matrices).

MODERN ALGEBRA TO WORLD WAR I

3. "Modern" algebra to 1860. As a result, although real and complex algebra dominated the textbook literature for a full century after 1830, "modern" algebra had already achieved some notable successes by 1860.

Actually, already by 1770, Lagrange was interested in the "symmetric group" of all permutations of n letters and its subgroups, whose relevance to the solution by radicals of the general polynomial equation

$$x^n + a_1x^{n-1} + \cdots + a_n = (x - x_1)(x - x_2)\cdots(x - x_n) = 0, \tag{1}$$

he clearly recognized. A by-product of this interest was the Lagrange Theorem, that the order of any subgroup of a group G divides the order of G.

Ruffini, Galois, and Cauchy made further contributions to the development of group theory before 1845 [**13**, pp. 45–53]; Galois also made (in 1830) a fundamental contribution to the theory of fields, by constructing a *finite field* of each prime-power order p^r. (For formal definitions of groups and fields, see §3.)

Somewhat earlier Legendre and Gauss (1801) had initiated the study of *commutative rings*, by constructing the ring Z_n of the integers "modulo" n (i.e., in which integral multiples of n are set equal to zero) and the ring $Z[i]$ of all "Gaussian integers" $m + ni$, where $m, n \in Z$ are ordinary integers[5] and again $i = \sqrt{-1}$. Moreover Gauss, had proved that factorization into primes was *unique* in $Z[i]$.

By 1850, *noncommutative rings* were also being studied. Thus Hamilton introduced his *quaternions* in 1843; since they contain the complex numbers as a special case, they may be called *hypercomplex numbers*. And in the first edition of his book *Ausdehnungslehre* (1844) H. Grassmann discussed both *vector algebra* (a fairly natural generalization of Descartes' symbolic method for treating geometry) and, somewhat vaguely, hypercomplex numbers in general. These concepts were made much more precise (and their connections with n-dimensional geometry clarified) by Cayley; by Hamilton in the Preface to his book on *Quaternions* (1859); and by Grassmann in the second edition of his book (1878). Moreover, Cayley showed in 1858 [**13**, p. 84] that the theory of determinants of Vandermonde and Laplace was only one aspect of a much more powerful *matrix algebra.* Matrix algebra is much like ordinary algebra, except that for general matrices A and B, $AB \neq BA$: the multiplication of matrices, like that of quaternions, is *non*-commutative. Indispensable for all pure and applied mathematicians today, matrices were first introduced formally by Cayley in 1858, and gradually revolutionized linear algebra.[6]

Shortly before, two other novel areas of modern algebra had been opened up. In 1854, Boole had published his *Introduction to the Laws of Thought*, in which he showed that a substantial part of Aristotelian logic was described by an analog of ordinary algebra now called "Boolean algebra." This novel "algebra of logic" satisfied not only most of the laws of ordinary algebra, but also the curious identities

$$a^2 = a + a = a \text{ (which today would be written } a \wedge a = a, \vee a = a),$$

$$(a + b)a = a, \text{ and } (a + b)(a + c) = a + bc.$$

4. The axiomatic approach. We have just seen that many of the major branches of so-called "modern algebra" (rechristened "the new math" by the popular press in the post-Sputnik era) were already known to mathematicians by 1860. However, the axiomatic approach to the foundations of algebra did not come until later. Lagrange derived the Lagrange theorem for groups and Galois constructed Galois fields without ever thinking of groups or fields as defined by postulates at all; their assumptions were entirely intuitive! Even the names "commutative" and "distributive" for the corresponding laws of manipulation were not introduced (by Servois) until 1814,[7] nor the term "associative" (by Hamilton) until 1835.

The emancipation of algebra from exclusive concern with the real and complex fields owes much to the philosophical speculations about algebra of Peacock, Woodhouse,[8] Hamilton, de Morgan, Boole, and Cayley, but E. T. Bell's claim [**3**, pp. 180–1] that it was Peacock who: "first perceived common algebra as an abstract hypothetico-deductive science of the Euclidean pattern" goes too far. Though Peacock anticipated Hankel in announcing the "principle of permanence of equivalent forms," his "Symbolical Algebra" is mainly concerned with geometrical applications, and does not even mention axioms or postulates. In these qualities it resembles H. Grassmann's *Ausdehnungslehre* (1844).[9]

The role of axioms emerges much more clearly from the *Formenlehre* of R. Grassmann (1872); the *Operationskreis der Logikkalkul* of E. Schröder (1877); the axiomatic treatments of groups, fields, modules, and ideals by Cayley (1878), Frobenius and Stickelberger (1879), Dedekind,[10] Weber (1882, 1893), and E. H. Moore; and the independent contemporary work of Benjamin Peirce and his son, C. S. Peirce, at Harvard (1870–1881).[11]

Influenced by these writings, Peano[12] initiated in 1888 his axiomatic approach to arithmetic, about which I shall say much more later. A decade later, in his *Grundlagen der Geometrie* [9], Hilbert tried to improve on Euclid. He succeeded from the standpoint of rigor, but not from that of pedagogy! Perhaps his most fundamental contribution to axiomatics was his clear formulation of the notions of independence, consistency and completeness for axiom systems.

In 1902, E. H. Moore showed that Hilbert's own axioms were not independent, and during the next ten years E. V. Huntington, L. E. Dickson, and O. Veblen made other painstaking analyses of the independence of postulate systems for groups, fields in general, the real and complex fields in particular, the algebra of logic, and the foundations of geometry. One can get an excellent picture of this work by reading the papers by Moore and Huntington;[13] for a more colorful if less reliable survey, see [2, Ch. 3].

Partly as a result of such papers, the postulational approach to algebra finally became standard. Mathematicians found that amazingly few and simple postulates, many fewer than those of Euclidean geometry,[14] could provide a sufficient basis for very extensive algebraic theories. For example, all of group theory can be derived from general principles of logic and the following postulates, due to E. V. Huntington (1906).

Definition. A *group* G is a set of elements (to be denoted by small Latin letters), any two of which, say x and y, have a *product* xy which satisfies the following conditions:

G1. Multiplication is *associative*: $x(yz) = (xy)z$ for all $x, y, z \in G$.

G2. For any two elements $a, b \in G$, there exist $x, y \in G$ such that $xa = b$ and $ay = b$.

(We have used Peano's notation $x \in G$ above; it signifies that "the element x is a member of (belongs to) the set G.")

Ingenious arguments can be used to deduce from these postulates various other simple conditions, for example, that: (i) any group G contains a unique "idempotent" element e satisfying $ee = e$, (ii) this element satisfies $ex = xe = x$ for all $x \in G$ (acts as an "identity" for G), (iii) the elements x and y in G2 are uniquely determined by a and b, and so on.

Similarly, the entire theory of fields can be deduced from the following set of postulates, also due to Huntington.

Definition. A *field* is a set F of elements, any two of which have a *sum* $x + y$ and a *product* xy which satisfy the following conditions:

F1. Addition and multiplication are *commutative*:

$$x + y = y + x \text{ and } xy = yx \text{ for all } x, y \in F.$$

F2. Addition and multiplication are *associative*:

$$x + (y + z) = (x + y) + z \text{ and } x(yz) = (xy)z, \text{ all } x, y, z \in F.$$

F3. Multiplication is *distributive* on sums:

$$x(y + z) = xy + xz \text{ for all } x, y, z \in F.$$

F4. For any $a, b \in F$, there exists some $x \in F$ such that $a + x = b$.

F5. If $a + a \neq a$, then there exists some $y \in F$ such that $ay = b$. (Actually, Huntington weakened F5 by adding the condition $b + b \neq b$ to its hypothesis.) (Of course, the hypothesis $a + a \neq a$ is just an indirect way of assuming that $a \neq 0$, necessary here because Huntington wanted to avoid assuming the existence of a "zero" 0 in F.)

The postulational approach to algebra, combined with an awareness of the relevance of all kinds of algebraic systems, stimulated an interest in enumerating *all possible algebraic systems* satisfying specified conditions: all finite fields (Galois had found them all), all groups of given order n, and so on. In this enumeration, one must of course identify all groups (or fields) which are *isomorphic*, that is, whose elements are related by a *bijection which preserves group multiplication* (in fields, which preserves addition *and* multiplication). Such a bijection is called an *isomorphism*.

5. Morphisms and subalgebras. More generally, it is helpful to know when two algebraic systems A and B are related by a (homo)*morphism*, or mapping $\theta: A \rightarrow B$ which preserves all their defining operations. Finally, it is helpful to recognize the *subalgebras* of A, i.e., the subsets S of A which satisfy all postulates; under these circumstances, A is conversely called an *extension* of S. (Thus the complex field is an extension of the real field.) To test for being a subalgebra, it is usually sufficient to test for *closure* with respect to suitable operations. In a group, for example, a subgroup must contain: (i) the identity, (ii) with any x also x^{-1}; and (iii) with x and y also xy. In fields, one must require closure under addition, subtraction, multiplication, and division.

The preceding concepts apply to all of the usual kinds of algebraic systems; I shall come back to them in my next lecture.

6. Some deeper developments: 1860–1914. During the same decades that its foundations were being clarified by the postulational method, the scope and depth of algebra grew enormously. I can only indicate very sketchily a few especially remarkable results here.

First, Galois theory became clarified as follows; I shall stick to extensions of the rational field Q to fix ideas, but the results generalize to extensions of any field. Let $F = Q[x_1, \cdots, x_n]$ be the field generated by the roots of a polynomial

$$p(x) = (x - x_1)(x - x_2) \cdots (x - x_n) = x^n + a_1 x^{n-1} + \cdots + a_n = 0$$

with coefficients $a_k \in Q$. Define the *Galois group* $F(f: Q)$ of F (and of $p(x)$) over Q to be the group of all automorphisms α of F such that $\alpha(x) = x$ for all $x \in Q$. Then the theorem of Galois states that the equation $p(x) = 0$ is solvable by *radicals* in terms of the coefficients (i.e., over Q) if and only if the Galois group $G(F: Q)$ is "solvable" in the following sense.

DEFINITION. Define a *composition series* of a finite group G to be a chain of subgroups of G,

$$1 < S_1 < S_2 < \cdots < S_r = G,$$

each of which is a maximal normal subgroup of the one following. Form the associated quotient-groups S_k/S_{k-1}. Then G is called *solvable* when these quotient-groups are all *Abelian* (it is equivalent that they all be of prime order).

Second, pure group theory acquired depth. Among the many remarkable theorems about finite groups proved in the half-century 1860–1914, I shall mention only a few. First, it was shown that the set of S_k/S_{k-1} is the same (up to isomorphism and rearrangement) for all composition series (Jordan-Hölder Theorem). Again, it was shown that any group of prime-power order p^n is solvable. Finally, it was shown that if p^n divides the order of a group G, then G has a subgroup of order p^n (Sylow theorem).

Third, in the area of algebraic number theory, Dedekind developed ideal theory, and applied it to generalize the pioneer result of Gauss on the unique factorization of Gaussian integers, to a sweeping unique factorization theorem for any algebraic number field (i.e., any subfield of the complex field C having finite linear dimension over the rational subfield Q). Namely, he showed that factorization into prime ideals is unique.[15]

Dedekind's deep interest in ideal theory and in unique factorization into primes also led him to consider the operations of greatest common divisor (g.c.d.) and least common multiple (l.c.m.) from a postulational standpoint. Recognizing their analogy with "and" and "or" in Boolean algebra, he was led to develop and apply the elementary theory of *lattices* ("*Dualgruppen*"), *modular lattices*, *distributive lattices*, and *vector lattices* in two pioneer papers (1897, 1901), thus founding a major new branch of algebra which contained Boolean algebra as a special case.

7. Linear associative algebras. In 1870, at about the same time that Dedekind was developing ideal theory into a powerful tool, Benjamin Peirce of Harvard made a pioneer study of the systems of "hypercomplex numbers" vaguely adumbrated by Grassmann, Hamilton and Cayley. Peirce began by defining a "linear algebra" over a field F as a set A whose elements are arbitrary linear combinations

$$\mathbf{a} = (a_1, \cdots, a_r) = a_1 \mathbf{i}_1 + \cdots + a_r \mathbf{i}_r \tag{2}$$

of r basis elements $\mathbf{i}_l$, multiplied by some rule of the form

$$\mathbf{a} \cdot \mathbf{b} = (\Sigma a_l b_m)\mathbf{i}_l \cdot \mathbf{i}_m = \Sigma a_l b_m \gamma_{lmn} \mathbf{i}_n; \tag{2'}$$

the constants γ_{lmn} can be any scalars (elements of F). He called a linear algebra *associative* when the multiplication defined by (2′) is associative.

A very notable linear associative algebra is provided by Hamilton's quaternions, which have four basic elements $\mathbf{1}, \mathbf{i}, \mathbf{j}, \mathbf{k}$ and hence 64 constants (mostly zero) defined by the rules

$$\mathbf{1} \cdot \mathbf{a} = \mathbf{a} \cdot \mathbf{1} = \mathbf{a} \text{ for all } \mathbf{a}, \qquad \mathbf{i}^2 = \mathbf{j}^2 = \mathbf{k}^2 = -\mathbf{1}, \tag{3}$$

$$\mathbf{i} \cdot \mathbf{j} = -\mathbf{j} \cdot \mathbf{i} = \mathbf{k}, \quad \mathbf{j} \cdot \mathbf{k} = -\mathbf{k} \cdot \mathbf{j} = \mathbf{i}, \quad \mathbf{k} \cdot \mathbf{i} = -\mathbf{i} \cdot \mathbf{k} = \mathbf{j}. \tag{3'}$$

The identities of (3′) are clearly those for vector products. The quaternion algebra $R[\mathbf{i}, \mathbf{j}, \mathbf{k}]$ over the real field is also a *division algebra*: any nonzero quaternion $\mathbf{a} = a_0 + a_1\mathbf{i} + a_2\mathbf{j} + a_3\mathbf{k} \neq \mathbf{0}$ has an inverse, given by

$$\mathbf{a}^{-1} = (a_0 - a_1\mathbf{i} - a_2\mathbf{j} - a_3\mathbf{k})/(a_0^2 + a_1^2 + a_2^2 + a_3^2). \tag{3''}$$

Peirce[16] showed that the complex numbers and the quaternions formed the *only* hypercomplex division algebras over the real field.

Even more important is the *full matrix algebra* $M_n(F)$ of all n^2-matrices $A = \| a_{lm} \| = \Sigma a_{lm} e_{lm}$. The basis elements e_{lm} of $M_n(F)$ are multiplied by the rules that

$$e_{lm} e_{l'm'} = \begin{cases} e_{lm'} & \text{if } m = l' \\ 0 & \text{otherwise.} \end{cases} \tag{4}$$

Hence, the constants are given (in a slightly changed notation) by

$$\gamma_{lm,l'm',l''m''} = \begin{cases} 1 & \text{if } l' = m, l'' = l, m'' = m' \\ 0 & \text{otherwise.} \end{cases} \tag{4'}$$

From 1870 on, many mathematicians tried to classify linear associative algebras over the real and complex field, using the Fundamental Theorem of Algebra as a tool where convenient. Papers by Frobenius (1878, 1903), Molien (1893), and Cartan (1898) were especially noteworthy.[17]

In a remarkable paper published in 1907, Wedderburn showed that most of the structure theorems of Cartan and Frobenius could be proved for linear associative algebras over an arbitrary field! In particular, he proved the following basic results, whose precise meaning will be explained below. For further details, see [3, Ch. 11]. Wedderburn himself stated that "Most of the results contained in the present paper have already been given, chiefly by Cartan and Frobenius, for algebras over the rational field."

(i) Any linear associative algebra is the direct sum (in the vector space sense) of a "semisimple" subalgebra and a unique "nilpotent" invariant subalgebra;

(ii) The semi-simple summand in (i) is the direct sum of "simple" linear associative algebras, in a unique way;

(iii) Each *simple* summand in (ii) is, for some n, the "full matrix algebra" $M_n(D)$ of all $n \times n$ matrices $A = \| a_{ij} \|$ with entries a_{ij} in a suitable "division algebra" D over F, the field of scalars of the original linear associative algebra.

To explain (i), we recall that a linear algebra is called "nilpotent" when, for some finite integer n, all products $a_1 a_2 \cdots a_n$ of n factors vanish. A "subalgebra" of an algebra is a subset closed under addition and multiplication (as well as linear combination over the field of scalars); such a subalgebra K is called "invariant" when $k \in K$ implies $ak \in K$, and $ka \in K$ for any element a, even if not in K; this is the condition that K be an *ideal* in the sense of ring theory.

Not all linear algebras are associative. The most important family of non-associative algebras is the family of *Lie algebras*. In these, the associative law is replaced by the following three identities:

$$[aa] = 0, \ [ab] + [ba] = 0, \ [[ab]c] + [[bc]a] + [[ca]b) = 0,$$

true for all a, b, c. In the 1870's, Lie had shown that real and complex Lie algebras provided the key to the understanding of *continuous groups* based on a finite number of parameters. It was therefore most remarkable that Killing (1888–1890) and Élie Cartan (1894), were able to prove that Lie algebras satisfied structure theorems somewhat analogous to those for associative algebras stated above—and to determine all "simple" Lie algebras over C. This work of Killing and Cartan on the structure of Lie algebras came *before* the analogous work of Molien and Cartan on linear associative algebras[17a].

One can, perhaps, summarize the preceding developments in the statement that more was known about "modern algebra" by research algebraists in 1914 than most Ph.D's know today. However, algebra was still regarded as subordinate to classical analysis, and the complex field reigned supreme. Thus, of the two advanced texts on algebra (as distinguished from number theory) most widely used in 1900, Weber's began with a chapter on algebraic functions and Serret's with one on continued fractions!

8. Symbolic logic to Gödel. In retrospect, it seems not too surprising that the dramatic successes of 19th century algebraists and logicians should have encouraged some imaginative mathematicians to develop a *symbolic logic* which would reduce all theorem-proving to mechanical symbol manipulations according to prescribed rules or "axioms." Actually, this idea goes back at least to Leibniz, who envisioned around 1700 symbolic methods capable of "increasing the power of reason far more than any optical instrument has ever aided the power of vision." To his fertile mind, the powerful symbolic algebra of the differential and integral calculus (much of which he invented) must have seemed a direct confirmation of the potentialities of symbolic methods.

The symbolic approach was developed tremendously by Peano from 1889 on. His main contributions to it may be found in his *Formulario Matematico* (5th ed.,

1908), whose preface states that: "All progress in mathematics is in response to the introduction of symbols (ideographic signs). ... Among two symbolic systems, the one with fewer symbols is, in general, preferable. But the fundamental use of [symbolic methods] is to facilitate calculation." The preface continues with a review of the origin of various symbols, including $+$, $\times$, D (derivative), $\int$, and those for vector and Boolean algebra. It then proposes for general adoption the symbols $\in$ (for membership) and $\exists$ (there exists). Peano claims that with these, and a handful of other symbols and symbolic conventions, all mathematics can be presented in symbolic form.[18]

Actually, Peano was not the first mathematician to conceive of a purely symbolic mathematics. In 1634, Hérigone had written in the Preface of his *Cursus Mathematicus*: "I have invented a new method of making demonstrations, brief and intelligible, without the use of any language," and his symbolic style was adhered to by Wallis (1656) and Barrow (1655, 1660).[19]

Peano then substantiates his claim by 386 pages of text containing symbolic synopses of: (1) Mathematical Logic, (2) Arithmetic, (3) Algebra, (4) Vectors ("Geometry"), (5) Limits, (6) Derivatives, (7) Integrals. Most successful are Parts (1) and (2); the latter contains Peano's celebrated construction[20] of the nonnegative integers by his "successor function": $1 = 0+, 2 = 1+, 3 = 2+, \cdots$, and his derivation of the laws of arithmetic from it is superb. In 70 additional pages, Peano extends his purely symbolic treatment to plane curves, differential equations, and various other topics.

However, Peano's *Formulaire* must be viewed as primarily a thought-provoking *tour de force*, in spite of its wealth of ideas and insights. Nowhere does he list the rules of symbol-manipulation for passing from one formula to the next; he fails to provide a system of axioms for logic. His proofs, like Euclid's in geometry, can only be verified by attributing *meanings* to words.

This major gap was filled by Whitehead and Russell in their three volume masterpiece *Principia Mathematica* [**18**]. Here they specified carefully the symbol-manipulations ("rules of inference") which can be used infallibly in passing from hypotheses to conclusions in symbolic logic (mathematical reasoning).

Using their specified rules of inference as "axioms" for symbolic logic, Whitehead and Russell showed that one can paraphrase symbolically at least the construction of the real field R from the positive integers, as well as much of set theory and arithmetic. These major achievements were presented as empirical evidence supporting the thesis that *all mathematical theorem-proving can be reduced to mechanical symbol-manipulations* (i.e., to pure symbolic logic).[21]

Nobody disputes the claims of Whitehead and Russell, that their rules of inference for "Peanese" (the symbolic language of Peano), are (i) infallible subject to restrictions stated in English in their text, and (ii) sufficient for much of elementary mathematics. However, the actual mathematical coverage of *Principia* (in nearly

2000 pages!) is far less than Peano's, and it cannot be said that their symbolic methods used "increase the power of reason;" I think they *decrease* it, probably for psychological reasons.[22]

9. Hilbert and Gödel, 1918–31. Because of its capability of replacing special axioms for the different branches of mathematics by theorems (cf. *Principia Mathematica*, Preface, first paragraph), Hilbert said in 1918 that "Russell's Axiomatization of Logic is the crowning achievement of axiomatics."[23] And Hilbert spoke with authority, as the man who had rigorized the axioms of Euclid in his famous *Grundlagen der Geometrie* only 20 years before. I quote from the introduction to this book:

"Geometry—like number theory—requires for its deductive (*folgerichtige*) construction only a few basic theorems (*Grundsätze*). These theorems are called *axioms*,[24] and their connected development has had numerous treatments since Euclid.... The following book is a new attempt to develop the simplest possible complete axiom system for geometry ... so as to clarify the significance of the different groups of axioms and the consequences of the individual axioms."

In much the same spirit of *axiomatic analysis*, Hilbert and his collaborators, especially his co-authors W. Ackermann and P. Bernays,[25] made after 1918 major efforts to prove *deductively* (by metamathematical arguments) the adequacy of the axioms of Whitehead and Russell (the evidence of *Principia Mathematica* was empirical). They focused attention on two main questions:

(i) Are these axioms *contradiction-free*, i.e., using them, is it impossible to prove both p and its contradiction $\sim p$?

(ii) Can one test the truth or falsity of any given proposition (e.g., of arithmetic) in a finite number of steps?

Hilbert may have been attracted to these questions partly because he had established an analog of the first in his *Grundlagen der Geometrie*, by using Cartesian geometry as a model for Euclidean plane geometry, and of the second for polynomial ideal bases by general transfinite arguments using the "ascending chain condition."

Question (i) was given a positive answer by Ackermann (who had earlier proved the redundancy of one of the Whitehead-Russell axioms) and von Neumann in 1927, *under suitable restrictions*. These restrictions, which are quite technical,[26] seemed quite harmless at first sight, and led to a feeling of optimism about Hilbert's program in the years 1927–1930.

Question (ii), the *Entscheidungsproblem* or Decidability problem, was however given a negative answer, even for arithmetic propositions, by Gödel in 1931. By an ingenious use of metamathematical reasoning, ultimately based on Cantor's diagonal construction, he inferred from this undecidability the *incompleteness* of the Whitehead-Russell-Hilbert system in the following sense. Assuming as true the additional *consistency axiom*, that "false formulas are unprovable," one can prove a number-theoretic formula which would not be provable without it. It is a corollary that one cannot prove that Hilbert's axioms are contradiction-free, so that in particular Question (i) is undecidable.

Thus Gödel's paper shattered Hilbert's high hopes. To quote Hermann Weyl[27]: "Gödel enumerated the symbols, formulas, and sequences of formulas in Hilbert's formalism in a certain way, and thus transformed the assertion of consistency into an arithmetic proposition. He could show that this proposition can neither be proved nor disproved within the formalism. This can mean only two things: either the reasoning by which a proof of consistency is given must contain some argument that has no formal counterpart within the system, i.e., we have not succedeed in completely formalizing the procedure of mathematical induction; or hope for a strictly 'finitistic' proof of consistency must be given up altogether. When Gentzen (1936) finally succeeded in proving the consistency of arithmetic he trespassed those limits indeed by claiming as evident a type of reasoning that penetrates into Cantor's 'second class of ordinal numbers'."

Gödel's result ended abruptly a half-century of optimism about symbolic logic, at least as formalized by Peano, Whitehead, and Russell. It showed that their formalizations were incapable of resolving the paradoxes and ambiguities of Cantor's theory of infinite sets.[28]

THE REIGN OF MODERN ALGEBRA, 1930–1970.

10. The rise of "modern" algebra. Just before Gödel shattered the high hopes of symbolic logicians for formalizing all mathematics in terms of "Peanese", van der Waerden's *Moderne Algebra* (1930–31) precipitated a new revolution. The goal of this brilliantly written book is clearly stated in its preface.

"The 'abstract', 'formal' or 'axiomatic' direction, which has given to algebra renewed momentum,[29] has above all led to a series of new concepts in *group* theory, *field* theory, *valuation* theory, and the theory of *hypercomplex numbers*, to insight into new connections and to far-reaching results. The main aim of this book is to introduce the reader into this new world of concepts."

As I have indicated, both the axiomatic approach and much of the content of "modern" algebra dates back to before 1914. However, even in 1929, its concepts and methods were still considered to have marginal interest as compared with those of analysis in most universities, including Harvard. By exhibiting their mathematical and philosophical unity, and by showing their power as developed by Emmy Noether and her other students (most notably E. Artin, R. Brauer, and H. Hasse), van der Waerden made "modern algebra" suddenly seem central in mathematics. It is not too much to say that the freshness and enthusiasm of his exposition electrified the mathematical world—especially mathematicians under 30 like myself.

In particular, it made classical *real and complex algebra* seem passé, or at least a part of analysis and not of "algebra" in the true sense. This view is exemplified in

Moderne Algebra, where the real and complex fields are not even *defined* until after Galois theory has been presented, and the existence and uniqueness of a smallest algebraically closed extension of *any* field (Steinitz, 1910) are proved purely algebraically (by transfinite induction). What a contrast with the texts of Weber, Serret, and Perron!

11. Lattice theory. This new attitude was a major stimulus in the rebirth of lattice theory, which had lain dormant since the pioneer papers of Dedekind. In 1933, I wrote that lattice theory provided "a point of vantage from which to attack combinatorial problems in ··· abstract algebra."[30] And by 1938, enough progress had been made in applying it to logic, algebra, geometry, probability, measure and integration theory, and functional analysis to cause the American Mathematical Society to hold a symposium on the then very fresh subject.[31]

12. College algebra. The displacement of classical algebra by modern algebra took time. Thus it was not until after World War II that modern algebra became popular at the college level in our country—a popularity due partly to the *Survey of Modern Algebra* which Mac Lane and I had published in 1941. Actually, our approach seems quite conservative today! Thus, unlike van der Waerden, we presented the essentials of the theory of equations before defining groups, and the theory of real and complex matrices (including the Principal Axis Theorem for symmetric and Hermitian matrices) with geometric applications before Galois theory. We also included Boolean algebra, thinking it essential for students to understand the algebra of sets and logic; I shall return to this later.

13. Bourbaki's influence. Abstract mathematics, as reformulated by N. Bourbaki[32] in his *Éléments de Mathématique*, was popularized in French universities not long after. This many-volume treatise, mostly written in the decade 1945–55, attempts to develop all of (pure) mathematics systematically from the notions of *set* and *function*: it presents the content of mathematics as concerned with abstractly conceived relational *structures over sets* and mappings (especially *morphisms*) between them; cf. Book 1, Ch. 4.

Algebraic structures are treated in this spirit in Book 2, as defined by sets of *elements* with (internal or external) finitary *operations*. The reader is then led authoritatively and surely through a carefully polished and systematic sequence of definitions, examples, and theorems about groups, rings, fields, and most of the other kinds of systems I have mentioned. Other branches of mathematics are treated in much the same style in later books. The net effect is to make mathematics appear as a *polished monolith, built purely deductively from the notions of set and function.*

14. The flowering of abstract algebra. The enthusiasm generated by van der Waerden's book, reinforced in the ways that I have described, has given rise to an unprecedented flowering of all aspects of abstract algebra over the past 40 years. In particular, the theories of *groups, rings* and *fields* (to which the bulk of *Moderne Algebra* was devoted) have achieved new levels of depth and sophistication, of which perhaps the most dramatic example is the result that *every finite group of odd order is solvable.* This result, proved by Thompson and Feit in over 200 pages of very technical reasoning, had long been conjectured—but to prove it would have seemed hopeless to most mathematicians in 1930.

The last 40 years have also seen the theories of Lie, Jordan, and multilinear algebras mature to a point that makes what was known in 1930 seem amateurish if not naive. The same is true of lattice theory, semigroup and quasigroup theory, category theory, and homological and combinatorial algebra, all of which were either unknown or nearly so in 1930. Finally, algebraic geometry has become rigorized as a new branch of axiomatic algebra, based securely on deep results about commutative rings and their ideals and valuations.[33]

15. Wider repercussions. The tidal wave generated by enthusiasm about abstract algebra had wider repercussions. Thus to young men in 1930, like myself, van der Waerden's book made *classical analysis* stemming from the calculus ("*analyse infinitésimale*"), which had dominated mathematics for over two centuries, suddenly seem old and tired. Indeed, the abstract approach adopted by van der Waerden for algebra soon became fashionable in functional analysis and topology. The idea that all mathematics could be viewed as topological algebra gained a strong impetus from the solution of Hilbert's Fifth Problem, which showed that the hypothesis of differentiability could be replaced by mere continuity in the theory of Lie groups: any locally Euclidean continuous group is isomorphic to an analytic Lie group [**22**, p. 184]. Even research on partial differential equations, the traditional stronghold of the applied mathematician, has increasingly centered around the quest for new abstract concepts permitting one to prove extremely general existence and uniqueness theorems.

Partly because of such shifts in emphasis, by 1960 most younger mathematicians had come to believe that all mathematics should be developed axiomatically from the notions of set and function, and this approach had come to seem no longer modern but classical! By 1959, van der Waerden had changed his title from "*Moderne Algebra*" to "*Algebra.*" And in the 1960's, Mac Lane and I wrote another "Algebra" which went further in the direction of abstraction, by organizing much of pure algebra around the central concepts of morphism, category, and "universality." The "universal" approach to algebra, which I had initiated in the 1930's and 1940's stressing the role of lattices, was developed much further in two important books by Cohn and Grätzer. In a parallel development, Lawvere (1965) proposed "The category of categories as a foundation for mathematics," beginning with the statement[34]:

> In the mathematical development of recent decades one sees clearly the rise of the conviction that the relevant properties of mathematical objects are those which can be stated in terms of their abstract structure rather than in terms of the elements which the objects were thought to be made of. The question thus naturally arises whether one can give a foundation for mathematics

which expresses wholeheartedly this conviction concerning what mathematics is about, and in particular, in which classes and membership in classes do not play any role. Here by "foundation" we mean a single system of first-order axioms in which all usual mathematical objects can be defined and all their usual properties proved. A foundation of the sort we have in mind would seemingly be much more natural and readily-usable than the classical one when developing such subjects as algebraic topology, functional analysis, model theory of general algebraic systems, etc.

16. The "new mathematics" of 1960. In the post-Sputnik era of the early and middle 1960's, enthusiasm went even further. Especially in the United States, a vogue developed for exposing school children to formal concepts of set, function and axiom often only half-appreciated by their teachers! Its proponents encouraged the spread of the myth that these constituted a "New Mathematics," unknown fifty years earlier. One ostensible aim of this vogue was to indoctrinate young people so that they could fill a supposed shortage of mathematical teachers and research workers. This seemed highly desirable at a time when our postwar "baby bulge" and prosperity was quadrupling of the demand for college teachers of mathematics, while an unquestioning faith in the value of basic science was increasing the support for research in pure mathematics at a rate of 10–15 per cent annually. But as of 1972, it all seems strangely out-of-date!

To summarize, algebra developed harmoniously during the years 1930–60, with its main stream flowing smoothly, swiftly, and finally triumphantly in the channels I have described. Some measure of its triumph may be found in the fact that, whereas three of the first four Fields medals were awarded in Analysis (in 1936 and 1950), three of the four awarded in 1970 were in Algebra.

However, in the last 5–10 years, powerful new currents have become apparent. Some of these have arisen as countercurrents to extremism; thus René Thom has recently written a thought-provoking article entitled *'Modern' Mathematics: An Educational and Philosophical Error*,[35] in which he urges that geometry should replace algebra because "any question in algebra is either trivial or impossible to solve. By contrast, the classic problems of geometry present a wide variety of challenges."

However, I do not wish to dwell on the exaggerations of a decade which most of us recall with nostalgia. Extreme abstraction in research circles, attempts to inculcate premature sophistication in children, and uncritical expansionism in basic physical science have provoked reactions which by now threaten to go too far in the opposite direction.

Instead, I wish to describe four *positive* current trends in algebra which, in my opinion, hold great promise for the future.

FOUR COMPUTER-INFLUENCED CURRENT TRENDS

17. The new numerical algebra. Already in the 1940's a new revolution was brewing, whose ultimate implications for mathematics are unpredictable. Namely, the construction of efficient *high-speed digital computers* was making it feasible to solve mathematical problems whose effective solution would have previously been prohibitively costly and time-consuming. To many mathematicians, including myself, it had become evident by 1950 that the resulting *revolution in applied mathematics* would open up challenging new areas for basic research. In particular, since digital computers can only represent real numbers to a *finite* number of significant digits, and can only represent values of real functions at a *finite* number of points (approximate "nodal values" at "mesh-points"), their use in solving differential equations (e.g., from physics or engineering) requires a very careful *numerical analysis* of *roundoff* and *truncation* errors.[36]

Thus, to actually *solve* a system of differential equations (to a desired approximation), one usually first replaces it by an approximating system of *algebraic equations* (obtained perhaps by finite difference or finite element methods), whose unknowns typically represent nodal values at mesh-points, which is then solved (also approximately) on a digital computer. I shall say nothing about this first step of *discretization* here, because the theorems in numerical analysis and approximation theory required to justify it belong to classical analysis and not to algebra. Suffice it to say that it often leads to very large matrices and associated systems of simultaneous linear equations, which may involve 50,000 or more unknowns! The main problem is to solve these efficiently.

These matrices typically have many special properties, which must be exploited to achieve efficiency. They are usually very *sparse* (have mostly zero entries), and often symmetric, or symmetrizable by permutations or linear transformations. Their diagonal elements may be "dominant" (i.e., at least as great as the sum of the absolute values of the other entries), and they may have positive diagonal and negative off-diagonal entries. Matrices having all of the above properties are essentially what are called *Stieltjes matrices*; they arise naturally in *network flow problems.*

One usually wants to either: (i) *solve* the linear system (written symbolically $Ax = b$), or (ii) determine *eigenvalues* of A (the former are of course the roots of $|A - \lambda I| = 0$). As regards (i), most mathematicians imagined in 1940 that large linear systems should be solved (if at all!) by *Gaussian elimination*, and that the rest was sheer drudgery. A few eminent analysts (including Gauss, Jacobi, and von Mises) had appreciated the value of *iterative* methods (also used by Gauss) and had studied their rates of convergence, but these methods were (and still are!) totally ignored in textbooks on "linear algebra." Similar remarks apply to eigenvalue problems, where the experience of most mathematicians was limited to 3×3 (if not to 2×2) matrices $A = \| a_{ij} \|$, whose eigenvalues they might have found using textbook formulas to solve the cubic characteristic equation

$$\lambda^3 - (a_{11} + a_{22} + a_{33})\lambda^2 + \beta\lambda - A = 0,$$

where

$$\beta = a_{22}a_{33} + a_{33}a_{11} + a_{11}a_{22} - a_{23}a_{32} - a_{31}a_{13} - a_{12}a_{21}.$$

In practice, such textbook methods are extremely inaccurate and inefficient for most large matrices[37], and they were replaced in the 1950's by new algorithms, whose invention and analysis created a major new area of "classical" algebra: the *new numerical algebra.* Excellent surveys of what is now known about this area are contained in authoritative books by Varga [**17**], Wilkinson [**19**], and Young [**20**]; every forward-looking young algebraist should at least be cognizant of their contents!

18. Sparse matrices. The past five years have also seen substantial improvements (over Gauss) in *elimination* techniques for solving large systems with sparse coefficient-matrices. In particular, these have drawn on graph theory for ideas; see [**15**] for a cross-section of current work.

There are many other interesting new areas of research in (real and complex) numerical algebra. I shall just mention three of the most important; references to activity in them may be found in many review journals:

(a) Finding the roots of polynomial equations of degrees up to 100.
(b) "Unconstrained" minimization of functions of many variables.
(c) Linear programming and other techniques for finding minima of functions subjected to "constraints" by equations and inequalities.

Actually these "new" areas also originated in the 1940's, if not earlier. Thus by 1947, linear programming was defined, and the "simplex method" of solving its problems invented by George Dantzig; see p. 20 of G. Hadley's *Linear Programming* (Addison-Wesley, 1962). Moreover, its fundamental techniques were made accessible at the college freshman level by Kemeny, Snell, and Thompson 10 years later, in their popular *Introduction to Finite Mathematics* (Prentice-Hall, 1957).

19. Integer arithmetic. In programming languages for computers, a basic distinction is made between *exact* "integer arithmetic" and *approximate* "real arithmetic." I have omitted the problems of "integer programming" and of solving Diophantine equations on computer in the above discussion, because they involve integer and not real and complex numerical algebra. Nevertheless, activity in these fields represents another strong trend in contemporary numerical algebra.

20. Theory of automata. Although many mathematicians think of high-speed computers as simply "number-crunchers" or supersliderules whose primary mathematical role is to carry out elaborate numerical computations, and although "arithmetic units" may be the most highly organized special pieces of computer "hardware," computers are actually much more versatile. Large general purpose computers are designed to be *universal* instruments, capable of expediting all kinds of "mental" tasks. Much as the Industrial Revolution was made possible by machines which could perform all kinds of "physical" tasks more cheaply and efficiently than human beings, the Computer Revolution is aimed at doing the same with mental tasks. This prospect makes the study of computers especially fascinating. From a mathematical standpoint, partly because general purpose computers are *digital* assemblies of a finite set of components, their study is based on a new, *purely algebraic* concept which I shall now define axiomatically.

DEFINITION. A *finite state machine* (or "automaton") M consists of a collectiom A of "input symbols," a collection S of "states," and a collection Z of "output symbols," related by two operations $\nu: A \times S \to S$ and $\zeta: S \times Z \to Z$. The operation ν assigns to each "input symbol" $a \in A$ and "prior state" $s \in S$ a "new state" $\nu(a, s) \in S$; the operation ζ assigns to a and s a "printout" $\zeta(a, s) \in Z$. More concretely, such a finite state machine M can be thought of as evolving from a specified *starting state* s_0, *recursively* by $s_k = \nu(s_{k-1}, a_k)$, and as *printing out* $z_k = \zeta(s_{k-1}, a_k)$ for $k = 1, \cdots, n$ in succession. In this way, it converts strings of input symbols or *programs* $a_1, a_2, \cdots, a_n$ into printouts $z_1, z_2, \cdots, z_n$.

Abstractly, a finite state machine is clearly just a new kind of algebraic system $M = [A, S, Z; \nu, \zeta]$. If one simplifies M by ignoring Z and ζ (this is called a "forgetful functor" in category theory), the simplified M just describes the *action of a free semigroup* (the set A^* of all possible input "programs") *on a set* (the set S of states). The resulting theory of state machines without output fits nicely into axiomatic (or "modern") algebra and, as has recently been shown,[39] so-called "universal algebra" can be applied to it.

21. Turing machines. Quite similar to finite state machines, but a little more complicated, are the "Turing machines" invented by the logician Turing in 1936, before high-speed general purpose digital computers existed. Turing proved that they could indeed carry out most processes of mathematical "thought." Thus they are capable of printing out the binary or decimal expansion of any "definable" (alias "computable") real number, such as e, π, or the kth zero of the Bessel function $J_n(x)$, and they can "deduce all the provable formulas of the *restricted* Hilbert functional calculus," giving all true theorems and no false ones.

Some two decades after Turing showed that his machines could, *in principle*, carry out the kind of mechanical theorem-proving dreamed of by Leibniz, Whitehead and Russell, and Hilbert, Hao Wang did this *in practice*. Namely, he wrote a special program which produced "proofs" in minutes for all the 350 theorems in the predicate calculus with equality that were actually stated in Whitehead and Russell's *Principia Mathematica*![40]

22. Computational complexity; optimization. A third and very strong trend in algebra, and indeed in mathematics generally, is a concern with *computational complexity* and with *optimization*. In all *applications* of algebra, of course, the efficiency of symbol-manipulation is a prime consideration, but for many years it was taboo to discuss it in research journals devoted to pure mathematics.

This snobbish taboo against discussing efficiency obscured some very important basic facts. Thus, in the area of mathematical logic, the scholarly books by Whitehead and Russell and the Hilbert school did *not* seriously try to improve the efficiency of formal deductive schemes, whereas Leibniz and Peano were really trying to (and did; especially Leibniz!) develop symbolic techniques for making mathematical reasoning more efficient and, therefore, more powerful. This difference becomes painfully obvious if one compares the number of symbols required by Whitehead and Russell to derive the basic formal properties of sets and relations, with the number of words needed by mathematicians to get equally far. So far, it is only in the area of the propositional calculus of logic itself, and by using a powerful computer, that mechanical theorem-proving has been realized on a substantial scale (by Hao Wang, see §21).

Having finally recognized the importance of efficiency, mathematical logicians have begun to analyze the "computational complexity" of applying general definitions to particular cases. Their analysis has already borne fruit in the development of shorter procedures for multiplying numbers and matrices.

Concern with computational complexity in algebra has as its ultimate goal, of course, the *optimization* of symbolic methods. In turn, the question of optimization has already suggested a number of basic problems whose solution should be a continuing challenge, stimulating coming generations of *pure* algebraists. Two of these are, respectively: (i) the "shortest form" problem of Boolean algebra, and (ii) the "most efficient coding" problem of information theory.

Other fascinating optimization problems, concerning which surprising discoveries have recently been made, are: (iii) what is the least number of operations on digits required to multiply two n-digit integers? (iv) what is the smallest number of arithmetic operations required to multiply together two $n \times n$ matrices? (v) how can one solve n simultaneous linear equations in n unknowns with the fewest additions, subtractions, multiplications and divisions? I regret that I do not have time to discuss these problems here, and must refer you to the literature ([**21**] and [**11**, vol. 2, pp. 258–78]).

23. Combinatorial algebra. A fourth current trend in algebra is towards emphasis on *combinatorial* ideas,[41] and especially on those involving *graphs* or *networks*. This trend is surely due to an intuitive recognition of the fact that digital computers and the deductive procedures of mathematics have a structure whose analysis requires combinatorial methods. As Hermann Weyl wrote in 1949: "The network of nerves joining the brain with the sense organs is a subject that by its very nature invites combinatorial investigation. Modern computing machines translate our insight into the combinatorial structure of mathematics into practice by mechanical and electronic devices."[42]

From burgeoning elementary courses in "Discrete Mathematics" which are intended to *precede* courses in axiomatic algebra,[43] probability and statistics, to the ambitious 7-volume treatise [**11**] on *The Art of Computer Programming* being written by Donald Knuth, the new emphasis is the same: permutations, combinations, partitions, generating functions, trees, sorting, searching, recurrences, and difference equations, block designs, and so on. Even a casual reading of the books I have cited makes it very clear that the 200 year reign of the Calculus and Analysis has ended—and that they will continue to be displaced in our colleges by courses in Algebra in the broadest sense of discrete mathematics and the *science* (no longer just art!) of symbol manipulation.

In a sense, this trend continues the revolution begun by van der Waerden, but there has been a major change. No longer do axioms and deductive systems, patterned after Euclid's *Elements*, seem so fundamental. Neither do groups or rings, with their subgroups, subrings and morphisms. Their place is taken by various *relational structures* (including partial orderings and "complexes" in the sense of combinatorial topology) which are far less amenable to the general algebraic techniques which played such a central role in the "modern algebra" of 1930–1960.

Instead, the kinds of algebraic structures (as contrasted with "relational" structures) which are most relevant to digital computers and combinatorics are loops, monoids and lattices (or groupoids, semigroups and semilattices), which were largely ignored by most algebraists in 1930–1960. Loosely speaking, much as *groups* are related to *symmetries*, so *loops* are related to *designs* (or "patterns"), *monoids* to *actions* (e.g., of input instructions on the states of an automaton), and *lattices* to *structure*.

In particular, Rota[44] and his associates have shown that lattice theory provides a point of vantage from which to attack combinatorial problems in general, and not just those of algebra as I had stated in 1933 (see §7). Going even further, N. S. Mendelsohn has very recently applied concepts of universal algebra to generate combinatorial designs and vice-versa [**23**, pp. 123–32].

One naturally wonders where all these new trends will lead to. I am myself sure of only one thing: that they will *not* make the classical "modern algebra" expounded in van der Waerden obsolete, any more than this made real and complex algebra or the calculus obsolete. As Knuth emphasizes ([**11**, vol. 1, p. 1]; see also [**1**]) the word *algorithm* (or "algorism") which is so central to the mathematics of computation is just a corruption of the name Al-Khwarizmi, the originator of the word "algebra."

Indeed, the four current trends in algebra which I have been describing were merely *stimulated* by the consideration of digital computers, in much the same way that the calculus and analysis were stimulated by thinking about geometry, mechanics and mathematical physics. They are simply opening up new areas of mathematics for future generations to study, with an ever increasing variety and richness of interrelations and applications, in which old and new ideas will mingle and be reshaped. Within a few decades, new concepts and trends may well emerge from this mingling and reshaping. Certainly, this kind of continuing evolution is the only thing that can keep algebra perennially a fresh and exciting subject!

Notes

[1] See footnote on page 464.

[2] For this and other facts, I am indebted to Professor David Pingree of Brown University; Thomas Hawkins, Gian-Carlo Rota, Gerald Sachs, and John Tate made other very helpful comments.

[3] A notable exception is provided by the binomial theorem, discovered by Pascal in 1653. For readable accounts of the facts summarized in this section, see Rouse Ball [1] and E. T. Bell [3].

[4] Their expositions were very obscure; see G. Birkhoff, *Isis* 3 (1973), 260–7. That of Galois was clarified by Betti in 1852.

[5] We here follow the usual custom of letting Z (for the German "*Zahl*" meaning integer) stand for the set $\{0, \pm 1, \pm 2, \cdots\}$. Gauss attributed the consideration of integers mod n ("modular numbers") to Legendre.

[6] For penetrating historical surveys of linear and non-commutative algebra, see N. Bourbaki, [6, pp. 78–91 and 120–28]. For a readable summary of Cayley's contribution, see pp. 102–15 of E. T. Bell [2].

[7] Gergonne's Annales 5 (1814–15), p. 93; for Hamilton's ideas, see his *Mathematical Papers*, vol. III, Cambridge Univ. Press, 1967. Leibniz and Cramer had very fragmentary ideas about determinants; see [1, p. 375] and D. J. Struik, *A Source Book in Mathematics*, Harvard University Press, 1969, p. 180.

[8] R. Woodhouse, Phil. Trans. 91 (1801), 89–119; G. Peacock, Reps. Brit. Assn. Adv. Sci. 3 (1834), 185–32 and *Algebra*, 2 vols., 1845; A. de Morgan, Trans. Camb. Phil. Soc., 7 (1839) 173–87 and 287–300; G. Boole, Cambridge and Dublin Math. J., 3 (1848) 183–98.

[9] F. Klein, *Entwicklung der Mathematik im* 19*ten Jahrhundert*, vol. 1, p. 175, characterized this as "almost unreadable."

[10] In his supplements to Dirichlet's *Vorlesungen über Zahlentheorie*, 1863, 1871.

[11] Benjamin Peirce, *Linear Associative Algebra*, Boston, 1870; see also Amer. J. Math., 3 (1880) 15–57, and 4 (1881) 97–229 (reprinted from Proc. Am. Acad. Boston, 1875).

[12] See his *Collected Papers*. vol. 2, Cremonese, Rome, 1958, p. 134. In the Amer. Math. Society *Semicentennial Addresses*, vol. 2, p. 15, Bell attributed the postulational approach to Peano! Peano was also the first to *number* his theorems.

[13] Volumes 3–6 of the Transactions of the (then young) American Mathematical Society (1902–5) contain a dozen articles on postulate systems by the men named above.

[14] Eulcid's *Elements*, which included "axioms" for magnitudes (algebra) as well as "postulates" for geometry, were written in Alexandria, Egypt, around 300 B. C.; see Ball [1].

[15] For a historical discussion of ideal theory and Dedekind's work on algebraic numbers, see [3, Ch. 10].

[16] Op. cit. supra, pp. 216–29. The same result was proved independently by Frobenius, op. cit. infra.

[17] G. Frobenius, Crelle, 84 (1878) 1–63, and Berlin Sitzb. (1903) 504–37 and 634–5. Wedderburn's *Lectures on Matrices*, Amer. Math. Soc., 1934, contains a complete bibliography to 1933.

[17a] See Thomas Hawkins, Archive for History of Exact Sciences, 8 (1972) 243–87.

[18] A related symbolic style of writing was used by E. H. Moore in his *Introduction to a Form of General Analysis*, New Haven Colloquium, Yale Univ. Press, 1910.

[19] See F. Cajori, "Past struggles between symbolists and rhetoricians. . .", Proc. Int. Math. Congress Toronto (1924), vol. 2, pp. 937–41.

[20] First published in 1889 (*Arithmetices principia nova methodo exposita*).

[21] The fact that this was so had been airily asserted a decade earlier by Russell in his witty *Principles of Mathematics*, of which *Principia Mathematica* was originally intended to be comprised in a second volume!

[22] See G. Birkhoff, "Mathematics and Psychology," SIAM Review, 11 (1969) 429–69.

[23] *Werke*, vol. 3, p. 153; Math. Annalen 78, 405–15.

[24] Hilbert is here slurring over Euclid's distinction between "axioms" (for magnitudes in general) and "postulates" (for geometrical entities).

[25] Of the books **[10]** and *Grundlagen der Mathematik* (2 vols., 1939), respectively.

[26] See S. C. Kleene, *Introduction to Metamathematics*, Van Nostrand, 1932, pp. 204–5.

[27] This MONTHLY, 53 (1946) 1–18. Gödel's original paper was published in the Monats. Math. Phys., 38 (1931) 173–98.

[28] Careful historical reviews of the question touched on here may be found in N. Bourbaki, **[6,** Ch. 1**]**, and (by P. Bernays) in Hilbert's *Werke*, vol. 3, pp.. 196–217; this volume also contains Hilbert's papers on logic.

[29] In German, "der die Algebra ihren erneuten Aufschwung verdankt."

[30] Proc. Camb. Phil. Soc., 29 (1933) 441.

[31] Bull. Amer. Math. Soc., 44 (1938) 793–827.

[32] A pen-name assumed in 1937 by a group of then young French mathematicians, who wished to overthrow the domination of French mathematics by classical analysts. See this MONTHLY, 57 (1950) 221–32 for authentic statement of Bourbaki's opinions, including the view that the axiomatic method is "a *standardization* of mathematical technique," and that the principal mathematical structures are those of a group, of order, and of a topological space.

[33] For example, anyone doing serious research on algebraic "geometry" today is expected to consider the two-volume treatise on *Commutative Rings* by O. Zariski and P. Samuel as standard *preliminary* material, but not to know Newton's classification of real cubic curves!

[34] F. William Lawvere, "The category of categories as a foundation for mathematics," *Proc. Conf. Categorical Algebra*, La Jolla, 1965 (S. Eilenberg *et al*, eds.), Springer, 1966.

[35] *American Scientist*, Nov. – Dec., 1971.

[36] Mathematicians habituated to exclusively deductive reasoning should realize that, in practice, error analysis relies very heavily on empirical evidence as well as on theoretical principles.

[37] Though not as nearly inefficient as Cramer's Rule, which is still often the only prescription given to students!

[38] See Marvin Minsky, *Computation: Finite and Infinite Machines*, Prentice-Hall, 1967.

[39] G. Birkhoff and J. D. Lipson, "Heterogeneous Algebras," J. Comb. Analysis, 2 (1969).

[40] H. Wang, IBM J. Res. Develop., 4 (1960) 2–22. For the general question of the computer as a "brain," see the reference of note 22.

[41] Wallis, Tchirnhaus, and Leibniz all recognized before 1700 that combinatorics belonged to algebra. See **[13**, p. 14**]** and **[21**, p. 2**]**.

[42] E. F. Beckenbach (editor), *Applied Combinatorial Mathematics*, Wiley, 1964, p. 537.

[43] As currently recommended by the CUPM Panel on the Impact of Computing on Mathematics Courses. On an intermediate level, see C. L. Liu, *Introduction to Combinatorial Mathematics*, McGraw-Hill, 1968; on a more advanced level, see M. Hall, *Combinatorial Theory*, Ginn, 1967.

[44] "On the foundations of combinatorial theory," J. für Wahrsch., 2 (1966) 340–68; *Combinatorial geometries* (preliminary edition), M.I.T. Press, 1970; and refs. given there.

References

1. W. W. Rouse Ball, A Short History of Mathematics, 3rd ed., Macmillan, New York, 1901.

2. Eric T. Bell, Mathematics: Queen and Servant of Sciences, McGraw-Hill, New York, 1951.

3. ———, The Development of Mathematics, McGraw-Hill, New York, 1940.

4. Garrett Birkhoff and Thomas C. Bartee, Modern Applied Algebra, McGraw-Hill, New York, 1970.

5. Garrett Birkhoff and Saunders Mac Lane, A Survey of Modern Algebra, Macmillan, New York, 1941.

6. Nicolas Bourbaki, Éléments d'Histoire des Mathématiques, Hermann, Paris, 1960.

7. Florian Cajori, A History of Mathematical Notations, 2 vols., Open Court, Chicago, 1928–9.

8. George Grätzer, Universal Algebra, Van Nostrand, Princeton, N. J., 1968.

9. David Hilbert, Grundlagen der Geometrie, 1899; 2nd. ed., 1901. Authorized translation by E. J. Townsend, Open Court, Chicago, 1902, 1910.

10. David Hilbert and W. Ackermann, Grundzüge der theoretische Logik, 4th ed., 1949.

11. Donald Knuth, Algorithms, 7 projected volumes, Addison-Wesley, Reading, Mass., 1969.

12. S. Mac Lane and G. Birkhoff, Algebra, Macmillan, New York, 1967.

13. Uta Merzbach, "... Development of Modern Algebraic Structures from Leibniz to Dedekind," Ph. D. Thesis, Harvard, 1965.

14. Giuseppe Peano, Formulario Matematico, 4th ed., Torino, 1908.

15. Donald Rose and Ralph Willoughby (eds.), Sparse Matrices and their Applications, Plenum Press, New York, 1971.

16. B. L. van der Waerden, Moderne Algebra, 2 vols., Springer, New York, 1930–31.

17. Richard S. Varga, Matrix Iterative Analysis, Prentice-Hall, Englewood Cliffs, N.J., 1962.

18. Alfred N. Whitehead and Bertrand Russell, Principia Mathematica, 3 vols., Cambridge Univ. Press, 1911.

19. James Wilkinson, The Algebraic Eigenvalue Problem, Clarendon Press, Oxford, 1966.

20. David M. Young, Iterative Solution of Large Linear Systems, Academic Press, New York, 1971.

21. Garrett Birkhoff and Marshall Hall (eds.), Computers in Algebra and Number Theory, SIAM-AMS Proceedings, vol. IV, Amer. Math. Society, 1971.

22. Deane Montgomery and Leo Zippin, Topological Transformation Groups, Wiley-Interscience, New York, 1955.

23. W. Tutte (ed.), Recent Progress in Combinatorics, Academic Press, New York, 1969.

CORRECTIONS TO "CURRENT TRENDS IN ALGEBRA"

(This MONTHLY, 80 (1973) 760–782)

GARRETT BIRKHOFF, Harvard University

1. p. 466, last line. Change "Cayley (1878)" to "Cayley (1858)."

2. p. 468, lines –10, –9. Delete: "I shall... next lecture."

3. p. 474, Section 10, lines 6–7. Change: "*group... field... valuation*" to "*field... ideal... group*."

4. p. 474, line –6. Change "students" to "younger colleagues." For a list of Emmy Noether's students, see Auguste Dick, "Emmy Noether," *Elem. Math., Beiheft* No. 13, Birkhäuser, Basel, 1970.

BIBLIOGRAPHIC ENTRIES: HISTORY

Except for the entry labeled MATHEMATICS MAGAZINE, the references below are to the AMERICAN MATHEMATICAL MONTHLY

1. Louis C. Karpinski, Algebraical developments among the Egyptians and Babylonians, 1917, 257–265.

A short, interesting survey with some sample problems and solutions taken from ancient sources.

2. Frederick V. Waugh and Margaret W. Maxfield, Side-and-diagonal numbers, MATHEMATICS MAGAZINE, 1967, 74–83.

A history of rational approximations to irrational square roots, starting with the Greeks.

3. L. C. Karpinski, The algebra of Abu Kamil, 1914, 37–48.

Contains sample problems and solutions from the work of an important and evidently little known medieval algebraist. Rather technical.

4. H. Gray Funkhouser, A short account of the history of symmetric functions of roots of equations, 1930, 357–365.

Discusses elementary symmetric functions and the relation between the roots and coefficients of a polynomial, from the middle ages to the early nineteenth century.

5. W. H. Bussev, The origin of mathematical induction, 1917, 199–207.

Contains some interesting selections from Pascal's work on binomial coefficients. However, induction is older than this article indicates, since it was known to the Greeks in the form of "the method of descent."

6. Solomon Lefschetz, The early development of algebraic geometry, 1969, 451–460.

The point of view taken toward the foundations of algebraic geometry is old-fashioned and possibly unfamiliar to many readers. Nevertheless, informative and authoritative.

7. J. Dieudonné, The historical development of algebraic geometry, 1972, 827–866.

An outstanding article, full of important material.

8. Clark H. Kimberling, Emmy Noether, 1972, 136–149.

More emphasis on biography and on anecdote than on mathematics. Contains extensive quotes from articles by H. Weyl and Alexandroff.

9. Saunders MacLane, Some recent advances in algebra, 1939, 3–19.

What was recent in 1939. The article pays much attention to the structure theory of associative algebras.

10. G. A. Miller, Several historic problems which have not yet been solved, 1907, 6–8.

Of the five problems listed, only one (the existence of transcendental numbers) has seen much progress since this article was written.

6

ADDITIONAL TOPICS

SUMS OF SQUARES*

OLGA TAUSSKY, California Institute of Technology

1. Introduction. Sums of squares is a major concept in mathematics going back to ancient days and yet of great current interest. It is a subject which links many different branches of mathematics and produces results which have a certain similarity but whose complete connection is still not understood. In recent years logicians have been much interested in the subject too. There are applications to and from logic to sums of squares. Statistics from its beginnings has been involved in sums of squares.

This article describes some of these ideas, but is by no means comprehensive. In particular the chapter on number theory is very incomplete. The presentation is a spotlight treatment, sometimes putting very deep results next to easier ones, although the latter may have a particular appeal and even importance. Proofs are included only when they are very brief. Some new ideas are incorporated.

This account is devoted to algebra and number theory on the whole, apart from describing facts which link up with analysis and topology and ought not to be separated. But sums of squares also occur in the very definition of the Hilbert space and all its consequences, e.g. in Parseval's theorem, in the definition of L^2-convergence, in normed algebras and such like. The composition of infinite quadratic forms will not be discussed. Not even the theory of finite euclidean and unitary space will be included, nor facts concerning orthonormal vectors, nor the theory of norms of finite matrices. Hence orthogonal and unitary matrices are not treated, nor automorphs of quadratic forms.

---

* From AMERICAN MATHEMATICAL MONTHLY, vol. 77 (1970), pp. 805–830.

The author received her training under P. Furtwängler at the University of Vienna, held assistantships at Göttingen and Vienna, and a fellowship at Girton College, Cambridge, and was a lecturer at London University. She was Scientific Officer with the British Aircraft Production Ministry and served with the Department of Scientific and Industrial Research, London. After many years with the National Bureau of Standards, she transferred to Cal Tech where she is Research Associate (equivalent rank of Full Professor). Leaves of absence were spent at NYU, Bryn Mawr College, Vienna (as Fulbright Professor), and UCLA.

Olga Taussky's prolific research centers on algebraic number theory, quaternions, and matrix theory. She was co-editor of Hilbert's *Gesammelte Abhandlungen* and edited several volumes of the Nat. Bur. Stand. Applied Math. Series. *Editor.*

2. Pythagorean triangles and Fermat's last theorem. The first sum of squares we meet in our life is in Pythagoras' theorem

$$a^2 + b^2 = c^2, \tag{1}$$

where a, b are the sides of a right angled triangle with hypotenuse c. Later we meet in elementary trigonometry

$$\cos^2 \alpha + \sin^2 \alpha = 1 \tag{2}$$

and much later we meet

$$\cos^2 z + \sin^2 z \equiv 1 \tag{3}$$

for all complex values z. The Pythagorean triangles too turn up at an early stage in our education. They are right angled triangles whose sides have integral ratios like

$$3, 4, 5;\ 5, 12, 13; \cdots.$$

They are already mentioned in an old-Babylonian text discovered by Neugebauer and Sachs. It is known that there are infinitely many such triangles and that they are obtained from a parametric formula

$$\lambda(m^2 - n^2), \quad \lambda 2mn, \quad \lambda(m^2 + n^2) \tag{4}$$

with λ, m, n any integers. Sometimes the same triangle can be obtained several times by this formula, e.g. the triangle 6, 8, 10 is given by $\lambda = 1$, $m = 3$, $n = 1$, and by $\lambda = 2$, $m = 2$, $n = 1$. That every expression (4) leads to a Pythagorean triangle is clear. The converse will now be proved. We start with some general remarks.

At least one of a, b is even. For, if $a = 2n+1$, $b = 2m+1$, n, m integers, then $c^2 = 4N+2$, N an integer. This is impossible, for the square of an odd number is always of the form $4r+1$, r an integer, and that of an even number is divisible by 4. We shall assume that b is even. If a and b have a common factor $d \neq 2$, then $d \mid c$. Hence a/d, b/d also define a Pythagorean triangle. If, however, $d = 2$ and $b/2$ is odd then we cannot remove it, unless $a/2$ is even, in which case we interchange the role of a and b. If, however, both $a/2$ and $b/2$ are odd then $(c/2)^2$ would again be of the form $(4M+2)$, M an integer, which is impossible.

The following elementary proof for the converse uses the fact that the expressions (4) suggest the formulas for $\cos 2\alpha$, $\sin 2\alpha$. Let

$$a^2 + b^2 = c^2.$$

Define the angle $\alpha (0 < 2\alpha < \pi/2)$ by

$$\frac{a}{c} = \cos 2\alpha, \quad \frac{b}{c} = \sin 2\alpha.$$

Since $\cos 2\alpha = \cos^2\alpha - \sin^2\alpha$ and $\cos^2\alpha + \sin^2\alpha = 1$ we have

$$\cos^2 \alpha = \frac{1}{2}\left(1 + \frac{a}{c}\right) = r_1, \text{ a rational,}$$

$$\sin^2 \alpha = \frac{1}{2}\left(1 - \frac{a}{c}\right) = r_2, \text{ a rational.}$$

Since $\sin 2\alpha = 2 \sin \alpha \cos \alpha$ we have

$$\sin \alpha \cos \alpha = \frac{1}{2}\,\frac{b}{c} = r, \text{ a rational.}$$

Hence

$$r = \sqrt{r_1 r_2}.$$

Hence

$$\sin \alpha = \frac{r}{r_1} \cos \alpha = \frac{n}{m} \cos \alpha,$$

where n, m are integers which we may suppose without a common factor. Put

$$\frac{\cos^2 \alpha}{m^2} = \lambda_0 \quad \text{(a rational) so that } \sin^2 \alpha = \lambda_0 n^2.$$

This gives

$$a = \lambda(m^2 - n^2), \qquad b = \lambda \cdot 2mn,$$

where $\lambda = \lambda_0 c$. We claim that λ is necessarily integral. For, if λ were fractional with p, a prime, dividing the denominator, then $p \mid (m^2 - n^2)$, $p \mid 2mn$. If $p \neq 2$, then $p \mid m$ or $p \mid n$. Since $p \mid (m+n)$ or $p \mid (m-n)$ it follows that $p \mid m$ and $p \mid n$ which is a contradiction. If $p = 2$ and $p \nmid m$, $p \nmid n$ then b is not even as was assumed.

In contrast to the various elementary proofs of (4) a proof using Galois theory will now be given. It is based on Hilbert's Theorem 90 which concerns algebraic extension fields with a cyclic Galois group. This theorem is obtained nowadays as a special case of a theorem in Galois cohomology (see e.g. Jacobson, Algebra III.)

Let F be a cyclic extension of a field K of relative degree l. Let S be a generator of the Galois group of F over K. For any $\alpha \in F$ we write α^S for the automorphism defined by S. We then have

$$\operatorname{norm}_{F/K}(\alpha) = \operatorname{norm}_{F/K}(\alpha^S).$$

Hence by the multiplicativity of the norm we have

$$\operatorname{norm}_{F/K}(\alpha/\alpha^S) = 1.$$

Hilbert's theorem states that, conversely, any element $\beta \in F$ with norm $\beta = 1$, is of the form $\beta = \alpha/\alpha^S$ for a suitable $\alpha \in F$.

Apply the theorem to the situation where F is the extension obtained from the rational number field Q by adjoining $\sqrt{-1}$, i.e., the set of elements $m+in$, $m, n \in Q$. This field is cyclic with respect to Q and of degree 2. Further

$$(m + in)^S = m - in \quad \text{and} \quad \text{norm}(m + in) = m^2 + n^2.$$

Let then $a^2+b^2=c^2$ hold with a, b and c in Z, the ring of integers. This implies

$$\text{norm}\left(\frac{a}{c} + i\frac{b}{c}\right) = 1.$$

By Hilbert's theorem

$$\frac{a}{c} + i\frac{b}{c} = \frac{m + in}{m - in} = \frac{(m + in)^2}{m^2 + n^2}$$

for some $m, n \in Z$. Comparing the real and imaginary parts (4) emerges.

We add three further comments to the study of Pythagorean triangles:

(1) The expressions (4) can be given another interpretation: It is clear that Pythagorean triangles have much in common with complex numbers. The product of two complex numbers m_1+in_1, m_2+in_2 is $m_1m_2-n_1n_2+i(m_1n_2+m_2n_1)$. This associates two bilinear forms $m_1m_2-n_1n_2$, $m_1n_2+m_2n_1$ with the field of complex numbers. The corresponding quadratic forms

$$m^2 - n^2, 2mn$$

are exactly the expressions (4).

Similarly, one can associate a set of n bilinear (respectively quadratic) forms with any basis of an algebra. This idea is being investigated separately, particularly for algebraic number fields.

(2) The Pythagorean triangles form a group under a certain composition law: more precisely to every triangle a, b, c consider the whole set λa, λb, λc, with $\lambda=1, 2, \cdots$, as an element of the group under consideration. Further identify all the four pairs $\pm a$, $\pm b$. We may therefore assume a, b as coprime positive integers. Exactly one of these two integers is then even, because $a^2+b^2=c^2$ is a square, hence cannot be $\equiv 2(4)$. We will assume that b is an even number.

Let a_i, b_i, $i=1, 2$, be a pair which generate a Pythagorean triangle. Then it follows at once that

$$A = a_1a_2 + b_1b_2, \qquad B = a_1b_2 - a_2b_1$$

again generate a Pythagorean triangle. The set of triangles generated by A, B is the "product" of a_1, b_1 and a_2, b_2. If $a_1=a_2$, $b_1=b_2$, we obtain

$$A = a_1^2 + b_1^2, \qquad B = 0$$

which is equivalent with $A=1$, $B=0$. We consider this as the unit element in our group. For, this element when composed with a, b gives a, $-b$ which is equivalent with a, b. The associative law too follows for our composition if we again allow the above identification.

(3) The two quadratic forms $f=x^2-y^2$, $g=2xy$ associated with Pythagorean triangles, have a special property. Let

$$f_i = x_i^2 - y_i^2, \quad g_i = 2x_iy_i, \qquad i = 1, 2.$$

Then the pair of forms

$$f_1f_2 - g_1g_2, \qquad f_1g_2 + f_2g_1$$

are again the same forms f, g, but applied to the indeterminates:

$$x_1x_2 - y_1y_2, \qquad x_1y_2 + x_2y_1.$$

For $x_1=x_2$, $y_1=-y_2$ this gives the well-known relation $(x^2-y^2)^2+4x^2y^2 = (x^2+y^2)^2$.

The existence of these triangles made it desirable to know whether the equation

$$x^n + y^n = z^n \tag{5}$$

can be solved in integers x, y, z for $n>2$, apart from trivial cases. The not yet established statement that this is impossible is referred to as Fermat's last theorem. A similar question was raised concerning the relation (3) namely, do there exist two entire functions $f(z)$, $g(z)$, neither of them a constant, such that for some $n>2$

$$(f(z))^n + (g(z))^n \equiv 1. \tag{6}$$

A very brief proof was given by Iyer showing that such a pair does not exist. The equation (6) above is identical with

$$\prod(f(z) + \zeta g(z)) \equiv 1$$

when the product is taken over all solutions ζ of the equation $x^n+1=0$. Since none of the n factors can vanish the meromorphic function $f(z)/g(z)$ cannot assume any of the n values of ζ which would be in contradiction with the 'big' Picard theorem.

3. Sums of squares in number theory. This is an enormous subject which can only be touched briefly here. Again a very old theorem comes to our mind immediately: every positive integer is a sum of four squares of integers. Then we have the characterizations of integers which are sums of three squares and the well-known fact that an integer is a sum of two squares if and only if its square-free part is the product of primes $\equiv 1(4)$.

The quadratic form $x_1^2+x_2^2+x_3^2+x_4^2$ is not the only form $\sum a_ix_i^2$, a_i positive integers, which represents all positive integers. Forms of this nature are called universal and have been studied, e.g. by L. E. Dickson, Kloosterman, Linnik,

Pall, and Ramanujan. Forms which represent all but one positive integer were examined by Halmos. Heilbronn showed that there exist four continuous functions $f_i(x)$ such that every rational x is represented by $\sum f_i(x)^2$. To represent positive integers by more than four, in particular by many, squares has been studied too and the function giving the number of representations of the fixed integer n as the sum of exactly s squares has been of much interest for a long time. Recently Bateman pointed out that the function $f_s(n)=(2s)^{-1}r_s(n)$ is multiplicative precisely for $s=1, 2, 4, 8$, where $r_s(n)$ is the number of representations of n as a sum of s integral squares.

The problem of representing algebraic integers as sums of squares in the same field has been much studied, but will not be discussed here.

Next we look at *integral* symmetric matrices, a subject not fully investigated so far.

(The fact that a *real* symmetric matrix has real characteristic roots links their study with sums of squares, but this aspect will be discussed in Chapter 6.)

Even the subject of rational symmetric matrices is not fully explored. For 2×2 matrices the characteristic roots of such a matrix must lie in a field $Q(\sqrt{m})$, m a sum of two squares. This can be checked easily from the formula for the zeros of the (quadratic) characteristic polynomial. However, it can also be obtained as a special case of the following fact: Let F be an extension of degree n of the rationals and let it have a symmetric $n\times n$ Q-representation. Then this is equivalent with the existence of n elements $\alpha_1, \cdots, \alpha_n$ in F such that $\sum_i \alpha_i^{(r)}\alpha_i^{(s)}=0$, $r, s=1, \cdots, n$, $r\neq s$, where the upper suffices denote the conjugate elements with respect to Q. First we show that this is necessary: Let A be the Q-matrix which represents the primitive element α. Since A is symmetric its characteristic vectors are orthogonal. Since A is a Q-matrix these vectors can be chosen in $Q(\alpha^{(i)})$, $i=1, \cdots, n$, and as the conjugates of the vector corresponding to the characteristic root α. Hence the components of this vector are of the form of the above $\alpha_1, \cdots, \alpha_n$. Next we show sufficiency: the matrix

$$(\alpha_i^{(j)})\begin{bmatrix} \alpha & & & \\ & \alpha^{2} & & \\ & & \ddots & \\ & & & \alpha^{(n)} \end{bmatrix}(\alpha_i^{(j)})'$$

is rational and symmetric; here and later the prime indicates the transpose. Hence we have obtained a rational symmetric representation of $Q(\alpha)$.

From the orthogonality it also follows that F is totally real.

The case $n=2$ gives an alternative proof for a fact mentioned above, for the two elements α_1, α_2 are expressible as

$$\alpha_1 = a + b\sqrt{m}, \qquad \alpha_2 = c + d\sqrt{m},$$

where a, b, c, d, m are in Q and m is not a square in Q. Then

$$a^2 - mb^2 + c^2 - md^2 = 0.$$

Hence

$$a^2 + c^2 = m(b^2 + d^2).$$

Symmetric $n \times n$ matrices over the integers with a given characteristic polynomial of degree n, with integer coefficients and 1 as coefficient of x^n can sometimes best be studied by looking first for general matrices, i.e. not necessarily symmetric ones. Such matrices fall into classes: two matrices being considered equivalent if they belong to the same integral unimodular similarity class. For irreducible polynomials this leads to an integral $n \times n$ representation of the ring generated by a zero of the polynomial and hence to a rational $n \times n$ representation for the algebraic number field generated by it. Let A be a suitable matrix. Then A', the transpose, is a matrix-zero of the same polynomial. Under special circumstances it can belong to the class of A. We then have

$$A' = S^{-1}AS$$

when S is integral and unimodular. If in addition S is p.d. (positive definite) and even of the form $S = TT'$ with T integral (this follows from p.d. for $n \leqq 7$, (see Chapter 4)), then

$$T^{-1}AT = (T^{-1}AT)'$$

Hence the class of T contains a symmetric matrix and conversely. Faddeev, Shapiro, Bender studied symmetric matrices over algebraic number fields with given characteristic polynomials. However, not all polynomials which can turn up for symmetric matrices have been characterized so far.

This whole chapter belongs partly to the theory of positive definite matrices. They are treated in the next chapter which also includes further number theoretic results. Also the similarity between a matrix and its transpose and the connection with symmetric matrices will turn up there again for the case of the real number field.

4. Positive definite (p.d.) matrices. These are real symmetric matrices with positive characteristic roots. Again a very old fact is our starting point: a positive definite quadratic form with real coefficients is a sum of squares of linear forms. Positive definite hermitian forms are expressible as $\sum l_i(x)\bar{l}_i(\bar{x})$, where $l_i(x)$ is a linear form and $\bar{l}_i(x)$ is the form with the conjugate coefficients.

For the matrix itself this means in the real case that it can be factorized as AA' where A' is the transpose of A and in the complex case as BB^* where $B^* = \bar{B}'$ is the transposed complex conjugate.

One of the most important uses of p.d. matrices is to generalize facts in many different branches of mathematics, by replacing the identity matrix by a given p.d. matrix. We begin with two specific examples.

(1) *The orthogonal matrices.* They leave $\sum x_i^2$ unchanged. This was generalized in the study of the 'automorphs' of p.d. quadratic forms.

(2) *The field of values of a complex $n \times n$ matrix A.* It is the set of numbers in the complex plane given by x^*Ax/x^*x, where $x \neq 0$ is an arbitrary complex n-vector. Givens introduced the generalized field of values x^*AHx/x^*Hx when H is a p.d. form. This was recently even extended to operators.

In differential geometry Riemannian geometry extends Euclidean geometry. In the subject of partial differential equations the theory of elliptic equations extends that of the Laplace equation. There are many examples in the calculus of variations. There are examples in number theory on all levels.

We begin with the discussion of products of real symmetric matrices. It can be shown that every real matrix is the product of two symmetric matrices—more generally a matrix with elements in an arbitrary field F can be expressed as the product of two symmetric matrices in the same field. In the case of the reals: if one of the two symmetric factors is positive definite then the product is similar to a real diagonal. If both factors are p.d. then the product is similar to a p.d. diagonal. Connected with this is the following fact: while every matrix with elements in a field F is similar to its transpose via a symmetric matrix over F, the latter can be chosen in the form AA' (and therefore p.d. for the reals) if the original matrix is similar to a symmetric matrix, (i.e. has real characteristic roots and is diagonalizable in the case of the reals). Real matrices with positive determinant can always be expressed as products of p.d. matrices, in fact only four factors are needed, unless the matrix is a negative scalar in which case five factors are required, in general. This was shown by Ballantine who also characterized products of three p.d. factors. Products of two p.d. matrices had been characterized by Taussky.

Positive definite matrices can be employed to determine the signs of the real parts of the characteristic roots of a general matrix. This is an important practical problem. By a theorem of D. C. Lewis, Jr., one can find for any matrix A with simple elementary divisors a p.d. hermitian G such that the roots λ of $\det(GA+A^*G-2\lambda G)=0$ are the real parts of the characteristic roots of A. Also there is the matrix version of Lyapunov's stability criterion:

A matrix A is stable if and only if a p.d. G exists such that $AG+GA^*$ is negative definite.

This was generalized to give statements concerning the signs of the real parts of the characteristic roots.

The fact that every real symmetric matrix can be transformed to diagonal form by an orthogonal matrix can be generalized by saying that a pair of symmetric matrices, one of which is p.d., can be transformed to diagonal form simultaneously. This again is generalized by saying that a pair of symmetric matrices S_1, S_2 which generate a pencil $\lambda S_1+\mu S_2$ which contains a p.d. matrix can be simultaneously diagonalized. Such pencils have been studied recently and were linked up with the convex cone formed by the p.d. $n \times n$ matrices in the $n(n+1)/2$ dimensional space.

The cone of p.d. matrices H is invariant under the transformation

$$AHA'$$

when A is any non-singular matrix of the same dimension. Thus they form a 'positivity domain' like the positive vectors which are invariant under transformation by a positive matrix (in this case even a non-negative ($\neq 0$) matrix). The transformation defined by A on the linear space of symmetric matrices can be regarded as a 'positive operator' and the finite version of the Krein-Rutman theorem concerning such operators can be applied to it. From this it follows that the matrix which corresponds to the operator has a positive dominant characteristic root. The matrices H which are 'characteristic vectors' are of interest too.

We now turn to consideration of number theory and we mention some facts concerning the factorization XX' and the problem of representing integers by p.d. forms.

Positive definite matrices play a big role in number theory. In particular, unimodular integral matrices A have been studied by Hermite and Minkowski who showed that for $n \leqq 7$ every such matrix is of the form BB', with B an integral $n \times n$ matrix. For $n=8$ this is not any longer true as an example by Korkine and Zolotareff shows. Mordell showed that there are two classes in this case if we count two matrices A, B as belonging to the same class if $B=SAS'$ where S is an integral and unimodular matrix. The number of classes is finite in all cases.

A special case of p.d. unimodular matrices arises from the set of group matrices for the finite group G (a ring isomorphic with the integral group ring of G). These matrices are of the form $(a_{rs^{-1}})$, where r, s range over the elements of G in a fixed order. The unimodular ones correspond to the units in the group ring. A factorization of the above type is possible only if B is a group matrix for the same group, times an integral matrix P with $PP'=I$. For $n \leqq 7$ such a factorization is possible always, for $n=8$ there are again two classes. For $n=9$, however, there is only one class. These results were obtained by Taussky, M. Newman, R. C. Thompson, M. Kneser, some still unpublished.

The results concerning the factorization of positive definite unimodular matrices have been extended to hermitian matrices and to matrices over quadratic number fields. For instance, any matrix

$$A = \begin{pmatrix} a & \alpha \\ \bar{\alpha} & b \end{pmatrix}$$

with a, b positive integers, $\alpha = x+iy$ a Gaussian integer and $ab-\alpha\bar{\alpha}=1$ can be factorized into BB^* with B a square matrix of Gaussian integers. This result leads to an alternative proof for Lagrange's theorem on expressing a positive integer a as a sum of four squares. The following fact is needed: Given a number a, integers $b>0$, x, y can be found such that

$$x^2 + y^2 = -1 + ab.$$

Starting with the positive integer a the matrix A is now determined. The factorization $A = BB^*$ then expresses a as a sum of four squares. (This was pointed out by M. Newman during a discussion with the author.)

Even if the p.d. integral matrix is not of the form XX', X an integral square matrix, it may still be expressible in this form with X a rectangular matrix with more columns than rows. For the corresponding quadratic form this means that it is still expressible as a sum of squares of integral linear forms, but with more than n terms. However, even this is not always possible. This was studied by Mordell, Erdös, Chao Ko, Pall and Taussky.

Alternatively, the problem of expressing a matrix A in the form XBX' has been studied in full generality by Siegel, where now A and B do not need to have the same dimension. Hence this includes the representation of numbers by p.d. quadratic forms.

The extension of the problems to algebraic number fields led Siegel to a conjecture which was established quite recently. It concerns 'classes' and 'genera' of forms. It is known that over the rationals the forms

$$\sum_1^4 x_i^2, \qquad \sum_1^8 x_i^2$$

are in a genus of one class. Over totally real fields the form $\sum_1^4 x_i^2$ lies in a genus of one class only for $Q(\sqrt{2})$, $Q(\sqrt{5})$. This is what Siegel had conjectured. After some initial progress by Dzewas it was established by Barner.

5. Formally real fields. The congruence

$$x^2 + y^2 \equiv -1(p), \qquad p \text{ any prime},$$

used in Chapter 4, can be used as a link with this theory since the integers modulo a prime p form a field. A formally real field is characterized by the fact that -1 cannot be expressed as a sum of squares in that field. The fact that the above congruence can be solved for any p not only shows that the residues mod p do not form a formally real field (a fact obvious from the finite characteristic), but it gives the actual expression for -1.

If the field is not formally real and if its characteristic $\neq 2$ then every element in the field is a sum of squares.

If -1 is a sum of squares then it is of interest to study the minimum number of terms in the representation of -1 for various fields F. It is easy to see that 3 cannot occur as a minimum. For, assume that 3 is the minimum and let

$$-1 = x_1^2 + x_2^2 + x_3^2, \qquad x_i \in F, \quad x_i \neq 0.$$

This implies

$$0 = x_0^2 + x_1^2 + x_2^2 + x_3^2, \qquad x_i \in F.$$

Hence

$$0 = (x_0^2 + x_1^2)^2 + (x_0^2 + x_1^2)(x_2^2 + x_3^2)$$
$$= (x_0^2 + x_1^2)^2 + (x_0x_2 - x_1x_3)^2 + (x_0x_3 + x_1x_2)^2,$$

so that transferring $(x_0^2+x_1^2)^2$ to one side and dividing across by it (using the fact that $x_0^2+x_1^2=0$ by assumption) we get a representation of -1 as a sum of two squares. This contradicts our assertion. This proof depends on the multiplication of complex numbers and their norms. A similar proof using the multiplication of quaternions and the multiplicativity of their norms shows that 5, 6, 7 cannot occur as a minimum. The same idea, using Cayley numbers, shows that the numbers 9, $\cdots$, 15 cannot occur as a minimum. The question concerning the possible minima had been raised by van der Waerden in 1932 and was settled only quite recently by Pfister. He showed that only powers of 2 can occur as a minimum and that every such power does occur for some field. Pfister uses results of Cassels on quadratic forms for his proof. The relevant theorems are:

Let F be a field of characteristic $\neq 2$. Let $d \in F$ and x be an indeterminate. Necessary and sufficient for x^2+d to be a sum of $n>1$ squares in $F(x)$ is that either -1 or d is a sum of $n-1$ squares in F.

Let R be the field of real numbers and let $x_1, \cdots, x_n$ be indeterminates over R. Then $x_1^2+ \cdots +x_n^2$ is not a sum of $n-1$ squares in $R(x_1, \cdots, x_n)$.

As the cases 1, 2, 4, 8 show, the result is connected with the composition of sums of squares which will be discussed in the next chapter.

By defining 'positive' in real fields as 'sums of squares' an ordering can be introduced in such fields.

Symmetric matrices over formally real fields have been studied. The set of all their characteristic roots form a field which is real closed. Krakowski and recently also Bender studied symmetric matrices over arbitrary fields with given minimum polynomial.

An application of formally real fields appeared in a very unexpected connection: R. C. Thompson proved (generalizing a theorem by Shoda obtained for algebraically closed fields) that, with the exception of certain 2×2 matrices over GF(2), every unimodular matrix A with elements in a field F is a commutator $B^{-1}C^{-1}BC$ in F. He also studied the question: when can the factors B, C themselves be chosen unimodular? For the case that A is a scalar matrix, this depends, among other things, on whether -1 is a sum of two squares in F. Later Thompson examined the case when the factors B, C have given determinants b, c and then the representation of -1 in the form bx^2+cy^2 becomes critical.

6. Composition of sums of squares, anticommuting matrices, composition algebras. Hurwitz showed that $n=1, 2, 4, 8$ are the only values of n for which identities of the following type hold:

$$\sum_1^n x_j^2 \sum_1^n y_k^2 = \sum_1^n [l_i(x, y)]^2,$$

where l_i are bilinear forms in the x_j, y_k. Pfister's results concerning -1 as a sum of squares are linked with an extension of this problem: he allows the $l_i(x, y)$ to be *rational* functions. In this way he obtains an identity for any n which is a power of 2. For $n=8$ such an identity had been obtained independently by Taussky by a different method and this result was extended to $n=16$ by Eichhorn and Zassenhaus:

The usual way to obtain the above mentioned identities is from the product $\alpha\beta$ of two complex numbers, respectively quaternions, or Cayley numbers, and using the fact that norm $\alpha\beta$ = norm α norm β in all these cases. The identities found by Taussky and Eichhorn and Zassenhaus were, however, derived for $n=2$ from the reals, for $n=4$ from the complex field, for $n=8$ from quaternions, for $n=16$ from Cayley numbers. The method is based on the generalization of the relation $\det X \overline{\det X} = \det XX^*$ when X^* means the transpose conjugate.

The identities are special cases of Gauss' concept of composition of quadratic forms: two n-ary quadratic forms $f(x_i)$, $g(x_i)$ are said to permit composition if the product fg can again be expressed as a quadratic form under a bilinear transformation of the indeterminates.

Hurwitz' proof is based on matrix theory and leads to the enumeration of skew symmetric $n \times n$ matrix pairs which are anticommuting. Such pairs are of interest in many connections and will be mentioned again in Chapter 7. Using an idea of Jordan, von Neumann, and Wigner the theory of group representations was employed by Eckmann (after he replaced Hurwitz' matrices by abstract elements and -1 by an element of order 2) to give a proof of Hurwitz' theorem.

Freudenthal uses a projective geometry over the field of two elements and Desargues' theorem to prove Hurwitz' theorem and Chevalley uses Clifford algebras. An account of this and connected ideas are in a paper by van der Blij.

Later Albert, Kaplansky, and Jacobson imbedded the problem into the study of composition algebras. We start by defining a normed algebra. Let $e_1, \cdots, e_n$ be a basis for the algebra. Let

$$a = a_1 e_1 + \cdots + a_n e_n$$

be an element of the algebra. Define

$$\text{norm } a = |a| = \sum a_i^2.$$

The algebra will be called normed, if

$$|ab| = |a||b|.$$

If the algebra is over the reals, has an identity and is normed then it can be shown that it is either the reals, the complex field, the quaternions or the Cayley numbers. This gives a proof for the Hurwitz theorem. Jacobson treated a more general situation. He starts with a quadratic form $N(x)$ defined on a vector space V over a field of characteristic $\neq 2$. He assumes that $N(x)$ permits composition, i.e., there exists a bilinear composition xy in V such that

$$N(x)N(y) = N(xy), \qquad x, y \in V.$$

The product xy makes an algebra out of V which is then called a composition algebra. However, Albert had shown that forms of dimension 2^n permit composition even for fields of characteristic 2.

The Hilbert identities

$$\left(\sum_1^r x_i^2\right)^m = \sum \rho_k(a_{1k}x_1 + \cdots + a_{rk}x_k)^{2m}$$

with ρ_k positive rationals and a_{ik} integers were used in Hilbert's original solution of the Waring problem, i.e., the proof of the following assertion: every positive integer is a sum of n-th powers of integers and the number of these is determined solely by n. A simpler proof of the identities goes back to Hausdorff. They have the flavour of a composition identity and could possibly be generalized for products

$$\sum x_i^2 \sum y_i^2 \cdots.$$

The associative algebras among the above-mentioned four algebras over the reals are characterized by other properties, e.g., Pontryjagin obtains them from topological properties of fields in which addition and multiplication are continuous functions under some topology. The Gelfand-Mazur theorem states that they are the only normed algebras which are also fields. They will be discussed further in the next chapter.

7. Division algebras over the reals, n-dimensional spheres, n-dimensional Laplace differential equations. Frobenius proved that the reals, the complex numbers and the quaternions are the only division algebras over the reals if commutativity is not required, but associativity retained. In spite of arduous attempts there is still no algebraic proof available for the fact that $n=1, 2, 4, 8$ are the only numbers of base elements for which real division algebras exist, if associativity is not required any longer. The only proofs available so far rely heavily on deep algebraic topology. This approach was initiated by H. Hopf and continued by Stiefel and others with a final break-through by Bott, Kervaire, Milnor, Adams.

The norm in the case of the best known division algebras, the complex numbers, quaternions, Cayley numbers, is the sum of the squares of the coordinates.

The role of these hypercomplex systems in almost any part of mathematics is well known. Quaternions have applications in most abstract parts of mathematics, but also in concrete ones from where they stem. They were introduced by Hamilton and are of great use in theoretical physics, as well as in ring theory, group theory and number theory. If the coordinates are permitted to be complex then the system is no longer a division algebra, but it has other applications. Generalized quaternions are in much use too, they are again algebras with four base elements, but the role of -1 in the products is taken over by other scalars.

In abstract group theory the quaternions are linked with the quaternion group which is the hamiltonian group of lowest order. A hamiltonian group is a non-abelian group, all of whose subgroups are normal. The real group algebra of the quaternion group is homomorphic with the real quaternions.

The norms of these hypercomplex systems link them immediately with the n-dimensional spheres. In particular the fact that no division algebra over the reals with three base elements exists is connected with the fact that the 3-dimensional sphere, the set of triples x_1, x_2, x_3 with $x_1^2+x_2^2+x_3^2=1$, cannot be made into a topological group. The latter result follows easily from Brouwer's fixed point theorem. E. Cartan proved that the n-dimensional sphere is a group space only for $n=1, 2, 4$. It was the study of the topological properties of the higher dimensional spheres which led to the success in the investigation of the real division algebras. That the associative case follows easily from Cartan's result was noticed by Taussky. Recently a new proof of Cartan's result was given by M. Curtis and Dugundji. An earlier proof came from Samelson.

The ring $R[x_1, \cdots, x_n]$ of polynomials in n variables with real coefficients 'modulo the unit sphere', i.e. the ring

$$R[x_1, \cdots, x_n]/(x_1^2 + \cdots + x_n^2 - 1)$$

has been studied in several connections:

Swan showed that a 'unimodular' vector $(x_1, \cdots, x_n)$ over this ring cannot always be completed to a unimodular matrix, unless $n=1, 2, 4$, or 8.

Estes and Butts showed that for $n=3$ composition of quadratic forms is not possible in this ring.

Another subject connected with the classical division algebras is the Laplace differential equation

$$\left(\frac{\partial^2}{\partial x_1^2} + \frac{\partial^2}{\partial x_2^2} + \cdots + \frac{\partial^2}{\partial x_n^2}\right) u = 0.$$

For $n=2$ the Laplace equations for two functions of two real variables are a consequence of the Cauchy-Riemann equations. The question was raised (and answered) by Taussky as to whether an analogous situation exists for other values of n. It turns out this can happen only for $n=2, 4, 8$ and in fact it is a rather simple consequence of the result concerning the division algebras over the reals. However, a purely algebraic proof was obtained by Stiefel subsequently. The 4-dimensional case leads to a set of generalized Cauchy-Riemann equations which were much studied by Fueter and his school for the purpose of generalizing complex function theory. In slightly changed form they appear in theoretical physics as the Dirac equations.

Eichhorn, who had previously contributed to the study of generalized Cauchy-Riemann equations, considered the following more general problem recently: Let X be a vector space over a field F of characteristic $\neq 2$. Let $x \in X$. Consider linear mappings $L(x)$ of X into $\mathrm{Hom}(X, X)$ such that another such mapping exists so that

$$M(x)L(x) = \mu(x)I$$

when I is the identity mapping and $\mu(x) \not\equiv 0$ is a mapping of X into F. The function $\mu(x)$ can be interpreted as a quadratic form and if this form is positive definite and of rank n and X an n-dimensional vector space we are back in the problem of the generalized Cauchy-Riemann equations. However, it is now shown that for any form, whether p.d. or not, as long as it has full rank n, the only possible values of n are 1, 2, 4, 8. But also the case of lower rank $r<n$ is considered. For $n=p2^g$, p odd, Eichhorn obtains $r \leqq 2g+2$. The results are based on the following result: Let $n=p2^g$, p odd. Let F be an arbitrary field of characteristic $\neq 2$. Let A_i, $i=1, \cdots, r-1$, be a set of $n \times n$ matrices with elements in F with the properties: $A_i^2=\alpha_i I$, $\alpha_i \in F$, $\alpha_i \neq 0$ and $A_iA_k+A_kA_i=0$, $i \neq k$. Then $r \leqq 2g+2$.

Connected with this fact is a result by Adams, Lax and Phillips: Let $A_1, \cdots, A_k$ be a set of real $n \times n$ matrices such that $\sum \lambda_i A_i$ is non-singular for all real λ_i, except $\lambda_1 = \cdots = \lambda_n = 0$. Let $n=2^{b+4c}(2a+1)$, $0 \leqq b \leqq 3$. Then $k \leqq 2^b+8c$.

Anticommuting matrices are much connected with the various problems studied in this and the preceding chapter.

Dieudonné studied the following generalization of results by Eddington and by M. H. A. Newman. Let F be a not necessarily commutative field of characteristic $\neq 2$, V an n-dimensional right vector space over F, let σ be an automorphism of F and γ an element of F such that $\gamma^\sigma = \gamma$ and $\xi^\sigma = \gamma^{-1}\xi\gamma$ for $\xi \in F$. Determine the maximal number of semi-linear transformations u_k of V, relative to the automorphism σ, satisfying the relations

$$u_k^2(x) = x\gamma \quad \text{for } x \in V,\ \forall k$$
$$u_h u_k = -u_k u_h \qquad h \neq k.$$

Quaternions and Cayley numbers have many applications in number theory. Lipschitz had studied the ring of quaternions with rational integral coordinates, but Hurwitz later noticed that they do not form a maximal order. By order we understand a subring containing 1 and a basis for the algebra. He constructed the following basis for the maximal order: $(1+i+j+k)/2, i, j, k$. He then studied the factorization of rational primes in this ring. Similar problems have been studied for Cayley numbers by various authors, e.g. Coxeter, Lamont, Linnik, Mahler, Pall, Pall and Taussky, and Rankin.

If $e_1, \cdots, e_n$ ($n=4$, respectively 8) is a basis for an order and $x_1, \cdots, x_n$ indeterminates then

$$(x_1e_1 + \cdots + x_ne_n)(x_1\bar{e}_1 + \cdots + x_n\bar{e}_n)$$

is the norm form of the order if $\bar{e}_i$ is the conjugate of e_i. The problems associated with these forms can then be studied via the associated orders. Here the work of Brandt was basic and recently Kaplansky, Estes and Pall have made contributions.

8. Positive definite polynomials. This chapter is closely linked with the earlier chapters on number theory and on formally real fields. A positive rational number can be expressed as a sum of squares and Hilbert asked whether a real positive definite polynomial, i.e., one which assumes positive values only, can be expressed as a sum of squares of polynomials or, failing this, whether it can be expressed as a sum of squares of rational functions. He gave an example of a positive definite polynomial which cannot be expressed as a sum of squares of polynomials. Recently Motzkin gave a rather simple example of such a polynomial, namely $(x_1^2+x_2^2-3x_3^2)x_1^2x_2^2+x_3^6$. He found this in connection with a study of inequalities in which he expresses the difference of the two sides as sums of squares. Other examples were found recently by R. Robinson:

$$x^2(x^2-1)^2 + y^2(y^2-1)^2 - (x^2-1)(y^2-1)(x^2+y^2-1)$$
$$x^2(x-1)^2 + y^2(y-1)^2 + z^2(z^2-1)^2 + 2xyz(x+y+z-2).$$

Artin solved Hilbert's question completely, showing that an expression by rational functions does indeed exist. Recently the subject was reactivated by asking for a quantitative result, namely the minimum number of terms and explicit representations. It was shown by Pfister that a definite rational function of n variables in a real closed field is a sum of 2^n squares in this field. This is only an upper bound, but for $n=2$ a smaller number will not suffice. Quite recently Cassels, Ellison and Pfister showed that the Motzkin polynomial is not a sum of 3 squares. The lower bound $n+1$ follows from Cassels' Theorem showing that $1+x_1^2+\cdots+x_n^2$ is not a sum of n squares. Ax had shown earlier that Artin's own work can be used to imply the bound 2^n if a further condition is fulfilled which he showed was in fact true for $n=3$.

For polynomials with integral coefficients the following facts have been studied: If the polynomial assumes square values for all integral arguments then $f(x)$ is itself the square of an integral polynomial. A much deeper result was obtained in a diophantine formulation by Siegel: An integral polynomial, not a square, can attain a square value for a finite number of integral values only. Recently LeVeque generalized the question: if $f(x)$ assumes only values which are sums of two squares, is $f(x)$ a sum of two squared polynomials? This was answered affirmatively by several authors.

9. Sums of squares in Galois theory. The converse problem of Galois theory, namely to find a normal algebraic extension with a given Galois group, leads in some special cases connected with the number 2 to sums of squares. The two following cases will be discussed:

(i) The cyclic group of order 4.

(ii) The quaternion group.

Let k be a given field and F a separable normal extension. In case (i) there exists exactly one quadratic field F_0 between k and F. This field is generated by the square root of an element $\mu \in k$. It can be shown that μ is a sum of two squares in k. Conversely, any sum of two squares occurs in this connection. There are various proofs for this. If μ is a sum of squares in k, say $\mu = \mu_1^2 + \mu_2^2$, $\mu_i \in k$, then $F_0 = k(\sqrt{\mu_1^2+\mu_2^2})$ has the property that $-1 = \text{norm}_{F_0/k}(\rho)$ where $\rho \in F_0$. For, every element in F_0 is of the form $\alpha + \beta\sqrt{\mu_1^2+\mu_2^2}$, $\alpha, \beta \in k$ and the norm of this element is $\alpha^2 - \beta^2(\mu_1^2+\mu_2^2)$. This can be made equal to -1 for $\alpha^2 = \mu_1^2/\mu_2^2$ and $\beta^2 = 1/\mu_2^2$. Conversely, if $-1 = \text{norm}_{F_0/k}(\rho)$, $\rho \in F_0$, then F_0 is generated by the square root of a sum of two squares. For $k = Q$ it can happen that -1 is even the norm of a unit.

In case (ii) the field F contains three quadratic fields F_1, F_2, F_3 between k and F. Let F_i be generated by $\sqrt{\mu_i}$, $\mu_i \in k$. Then $\mu_1\mu_2 = \mu^2\mu_3$ where $\mu \in k$. It can be shown by elementary field theory that each μ_i can be represented as a sum of three squares in k. However, two expressions which are sums of three squares do not in general have a product with the same property. Hence μ_1, μ_2 cannot be arbitrary sums of three squares. Pairs of sums of three integers whose product is of the same type can be characterized easily from the known characterization of such integers. Not all such pairs, however, qualify for quaternion fields. A parametric representation for μ_1, μ_2 was given by G. Bucht. The 'sum of three squares' character of such a representation will now be explained (The following treatment is due to Cassels arising out of a discussion with the author.):

The elements μ_1, μ_2 can be obtained as:

$$\mu_1 = \frac{u^2l^2 + s^2u^2 + s^2m^2}{l^2r^2 + l^2v^2 + s^2v^2} = \frac{u^2(l^2 + s^2) + s^2m^2}{v^2(l^2 + s^2) + l^2r^2}$$

$$\mu_2 = \frac{u^2r^2 + m^2r^2 + m^2v^2}{l^2r^2 + l^2v^2 + s^2v^2} = \frac{m^2(r^2 + v^2) + u^2r^2}{l^2(r^2 + v^2) + s^2v^2}.$$

It can be shown, conversely, that any pair of elements $\mu_1, \mu_2 \in K$ for which these relations hold have the property that $k(\sqrt{\mu_1}, \sqrt{\mu_2})$ can be extended to a quaternion field. That μ_1 is a sum of three squares follows from the fact that both denominator and numerator are norms in the field generated by $i\sqrt{l^2+s^2}$, hence their quotient is again such a norm, hence clearly a sum of 3-squares. An analogous fact is true for μ_2. Next we show that $\mu_1\mu_2$ is a sum of three squares. Since μ_1, μ_2 have the same denominator we need only worry about the numerators. Both of them are norms of the field generated by $i\sqrt{u^2+m^2}$; hence their product is a norm in this field. Finally, we give an example showing that not every set μ_1, μ_2, $\mu_1\mu_2$, all sums of three squares, comes in question. Take $\mu_1=3\cdot 73$, $\mu_2=3\cdot 37$. Then

$$\mu_1 v^2 + \mu_2 l^2 = u^2 + m^2,$$

hence

$$(*) \qquad 3(73v^2 + 37l^2) = u^2 + m^2.$$

This implies that $u \equiv m = 0(3)$, hence $v^2+l^2 \equiv 0(3)$, hence $v \equiv l \equiv 0(3)$. Remove a factor 3 from u, m, v, l in (*) and repeat the process. This will finally lead to a contradiction.

A proof for the fact that the μ_i are sums of three squares by non-elementary methods was given by Reichardt.

10. Rational arctangents.*

We now return to a problem concerning the sums of two squares, which, in principle, goes back to Gauss.

In elementary trigonometry one encounters relations of the following form:

$$\arctan 239 = 4 \arctan 5 - 5\pi/4$$

$$\arctan 99 = \arctan 12 - 2 \arctan 5 - \arctan 2 + 5\pi/4.$$

We take up the question of generating all such relations or rather a basis for them. (One of the reasons for the study of those relations is to find convenient methods of calculating π.)

We shall write (x) for that value of arctan x between 0 and $\frac{1}{2}\pi$, so that, in particular, $(1) = \pi/4$. We ask, to begin with, can we have relations of the form

$$(2) = r(1)$$

$$(3) = s(1) + t(2),$$

where r, s, t are integers?

Suppose the first relation holds. Then the complex numbers $1+2i$ and $(1+i)^r$ necessarily have the same argument so that their ratio

$$(1 + 2i)/(1 + i)^r$$

is necessarily real. Since the real part of the denominator is an integer, m say, it follows that this ratio must be $1/m$. If we take the squares of the absolute values of each side of the equation

$$m(1 + 2i) = (1 + i)^r$$

* This chapter was written by John Todd.

we obtain the equation

$$5m^2 = 2^r$$

which manifestly has no solutions.

Similar considerations applied to the second relation lead to the equation

$$10\,m^2 = 2^s 5^t$$

which has a solution $s=3, t=1, m=2$. This gives us the relation

$$(3) = 3(1) - (2).$$

We now introduce the formal definition: (n) is called reducible if it can be expressed in the form

$$(n) = \sum f_r \cdot (n_r),$$

where the n_r are positive integers less than n and the f_r are integers (it can be shown that no change occurs if we allow the f_r to be rational); if no such relation exists we call (n) irreducible. Thus (2) is irreducible, while (3) is reducible. We find that (4), (5), (6) are irreducible but that

$$(7) = -(1) + 2(2)$$
$$(8) = 5(1) - (2) - (5).$$

Consideration of these examples suggested the following theorem:

THEOREM. *A condition necessary and sufficient for the reducibility of* (m) *is that the largest prime factor* $l(m)$ *of* $1+m^2$ *should be less than* $2m$.

We can verify this in the early cases quoted:

n	$1+n^2$	$l(n)$		$2n$
2	5	5	>	4
3	10	5	<	6
4	17	17	>	8
5	26	13	>	10
6	37	37	>	12
7	50	5	<	14
8	65	13	<	16
9	82	41	>	18

This theorem can be established constructively by elementary methods: an algorithm for carrying out the reduction of (n) when it is possible can be given, granted that the factorization of $1+r^2$ is known for $r \leqq n$. A listing of the reductions of (n) for $n \leqq 2089$ is available.

The irreducible (n) are analogous to the ordinary prime numbers and it is possible to ask whether there are theorems about them similar to theorems about prime numbers. In the first place there is an analog of Euclid's theorem: there *is* an infinite number of irreducible arctangents. We can also ask whether there is an infinite number of reducible arctangents—this is also true. Both these results are elementary. Gauss conjectured that the number of ordinary prime numbers $<n$ is approximately $n/\log n$ and this "Prime Number Theorem" was proved much later. Observation of the density of the irreducible arctangents suggests that their density is $\log 2 = .6931$—some theoretical evidence in support of this is available, but the result seems very difficult to prove.

It can also be shown that the arctangent of any rational number can be expressed in terms of the irreducible integral arctangents. For instance

$$(100/17) = (6) + (290) - (4836).$$

This work was carried out (in part) under NSF grant 3909 and 11236. The author is indebted for some references to A. A. Albert, D. Estes, I. Kaplansky, G. Pall, J. Robinson.

This is an account of part of a course with the same title given at California Institute of Technology in 1966. It is an extended version of a lecture given at University of California at Santa Barbara and Riverside, at University of Miami and Rice University.

References to 1

1. L. Henkin, Sums of squares, in Summer Institute for symbolic logic, Cornell University 1957, 2nd edition, IDA, (1962) 284–291.

2. I. Kaplansky, Theory of fields, unpublished manuscript, 1970.

3. G. Kreisel, Sums of squares, in Summer Institute for symbolic logic, Cornell University 1957, 2nd edition, IDA, (1962) 313–320.

4. A. Robinson, On ordered fields and definite functions, Math. Ann., **130** (1955) 257–271; 405–409.

5. ———, Introduction to model theory and to the metamathematics of algebra, North Holland, Amsterdam, 1965.

6. J. Robinson, Existential definiability in arithmetic Trans. Amer. Math. Soc. 72, (1952) 437–444.

7. ———, The decision problem for fields, in Symposium on the theory of models, North Holland, Amsterdam, 1965.

8. R. M. Robinson, The undecidability of pure transcendental extensions of real fields, Z. Math. Logik Gunrdlagen Math., 10 (1964) 275–282.

9. A. Tarski and J. C. C. McKinsey, A decision method for elementary algebra and geometry, University of Calif. Press, Berkeley, 1951.

10. O. Taussky, Sums of Squares, Matematička Biblioteka, Beograd, 41 (1969) 19–27.

11. F. van der Blij, History of the octaves, Simon Stevin, 34 (1960/61) 106–125.

References to 2

1. J. P. Ballantine and D. E. Brown, Pythagorean sets of numbers, this MONTHLY, 45 (1938) 298–301.

2. F. J. M. Barning, On Pythagorean and quasi-Pythagorean triangles and a generation process with the help of unimodular matrices, Math. Centrum Amsterdam, 1963.

3. F. Gross, On the equation $f^n + g^n = h^n$, this MONTHLY, 73 (1966) 1093–1096.

4. G. Iyer, On certain functional equations, J. Indian Math. Soc., 3 (1939) 312–315.

5. J. Mariani, The group of the Pythagorean numbers, this MONTHLY, 69 (1962) 125–128.

6. O. Neugebauer and A. Sachs, Mathematical cuneiform texts, New Haven, 1945.

7. N. E. Sexauer, Pythagorean triples over Gaussian domains, this MONTHLY, 73 (1966) 829–834.

8. W. Sierpiński, Pythagorean triangles, Yeshiva University, New York, 1962.

9. O. Taussky, A generalization of the Pythagorean forms, to be published.

10. H. Zassenhaus, What is an angle?, this MONTHLY, 61 (1954) 369–378.

References to 3

1. P. Bateman, Problem E2051, this MONTHLY, 76 (1969) 190–191.

2. E. Bender, Classes of matrices over an integral domain, Illinois J. Math., 11 (1967) 697–702.

3. ———, Classes of matrices and quadratic fields, Linear Algebra and Appl., 1 (1968) 195–201.

4. ———, Characteristic polynomials of symmetric matrices, Pacific J. Math., 25 (1968) 433–441.

5. H. Cohn, Decomposition into four integral squares in the fields of $\sqrt{2}$, $\sqrt{3}$, Amer. J. Math., 82 (1960) 301–322.

6. C. W. Curtis and I. Reiner, Representation theory of finite groups and associative algebras, Interscience, New York, 1962.

7. L. E. Dickson, Integers represented by positive ternary quadratic forms, Bull. Amer. Math. Soc., 33 (1927) 63–70.

8. J. D. Dixon, Another proof of Lagrange's four square theorem, this MONTHLY, 71 (1964) 286–88.

9. J. Dzewas, Quadratsummen in reell-quadratischen Zahlkörpern, Math. Nachr., 21 (1960) 233–284.

10. D. K. Faddeev, On the characteristic equations of rational symmetric matrices, Doklady Akad. Nauk. SSSR, 58 (1947) 753–754.

11. O. Fraser and B. Gordon, On representing a square as the sum of three squares, this MONTHLY, 76 (1969) 922–923.

12. P. R. Halmos, Note on almost-universal forms, Bull. Amer. Math. Soc., 44 (1938) 141–144.

13. H. Heilbronn, On the representation of a rational as a sum of four squares by means of regular functions, J. London Math. Soc., 39 (1964) 72–76.

14. H. Maass, Über die Darstellung totalpositiver Zahlen des Körpers $R(\sqrt{5})$ als Summe von drei Quadraten, Abh. Math. Sem. Univ. Hamburg, 14 (1941) 185–191.

15. ———, Modulformen und quadratische Formen über dem quadratischen Zahlkörper $R(\sqrt{5})$, Math. Ann., 118 (1941) 65–84.

16. M. Newman, Subgroups of the modular group and sums of squares, Amer. J. Math., 82 (1960) 761–778.

17. G. Pall, On sums of squares, this MONTHLY, 40 (1933) 10–18.

18. S. Ramanujan, On the expression of a number in the form $ax^2+by^2+cz^2+du^2$, Proc. Cambridge Phil. Soc., 19 (1917) 11–21.

19. R. A. Rankin, On the representation of a number as the sum of any number of squares, and in particular of twenty, Acta Arith., 7 (1962) 399–407.

20. A. P. Sapiro, Characteristic polynomials of symmetric matrices, Sibirsk Mat. Z., 3 (1962) 280–291.

21. C. L. Siegel, Darstellung total positiver Zahlen durch Quadrate, Math. Z., 11 (1921) 246–275.

22. O. Taussky, On matrix classes corresponding to an ideal and its inverse, Illinois J. Math., 1 (1957) 108–113.

23. ———, Classes of matrices and quadratic fields, Pacific J. Math., 1 (1951) 127–132.

24. ———, Classes of matrices and quadratic fields II, J. London Math. Soc., 27 (1952) 237–239.

25. R. C. Thompson, Problem E1814, this Monthly, 72 (1965) 782; (1967) 200.

References to 4

1. C. S. Ballantine, Products of positive definite matrices III, J. Algebra, 10 (1968) 174–182.

2. ———, A note on the matrix equation $H=AP+PA^*$, Linear Algebra and Appl., 2 (1969) 37–47.

3. ———, Products of positive definite matrices II, to appear.

4. K. Barner, Über die quaternäre Einheitsform in total reellen algebraischen Zahlkörpern, J. Reine Angew. Math., 229 (1968) 194–208.

5. G. Birkhoff, Linear transformations and invariant cones, this MONTHLY, 74 (1967) 274–276.

6. E. Calabi, Linear systems of real quadratic forms, Proc. Amer. Math. Soc., 15 (1964) 844–846.

7. D. H. Carlson, On real eigenvalues of complex matrices, Pacific J. Math., 15 (1965) 1119–1129.

8. P. Erdös and Ch. Ko, On definite quadratic forms, which are not the sum of two definite or semidefinite forms, Acta Arith., 3 (1939) 102–122.

9. D. Estes and G. Pall, The definite octonary quadratic forms of determinant 1, Illinois J. Math., 14(1970) 159–163.

10. M. Fiedler and V. Pták, Some generalizations of positive definiteness and monotonicity, Numer. Math., 9 (1966) 163–172.

11. W. Givens, Fields of values of a matrix, Proc. Amer. Math. Soc., 3 (1952) 206–209.

12. C. Hermite, Lettres de M. Hermite à M. Jacobi, 2nd letter, in Oeuvres 1, Gauthier-Villars, Paris, 1905, 122–135.

13. M. R. Hestenes, Pairs of quadratic forms, Linear Algebra and Appl., 1(1968) 397–407.

14. M. Kneser, Klassenzahlen definiter quadratischer Formen, Archiv Math., 8 (1957) 241–250.

15. ———, unpublished.

16. Ch. Ko, On the representation of a quadratic form as a sum of squares of linear forms, Quart. J. Math. Oxford, 8 (1937) 81–98.

17. ———, Determination of the class number of positive quadratic forms in nine variables with determinant unity, J. London Math. Soc., 13 (1938) 102–110.

18. ———, On the positive definite quadratic forms with determinant unity, Acta Arith., 3 (1939) 75–85.

19. M. Koecher, Positivitätsbereiche im R^n, Amer. J. Math., 53 (1957) 575–596.

20. A. Korkine and G. Zolotareff, Sur les formes quadratiques positives, Math. Ann., 11 (1877) 242–292.

21. A. Lyapunov, Problème général de la stabilité du mouvement, Commun. Soc. Math., Kharkov, (1892), (1893).

22. W. Magnus, Über die Anzahl der in einem Geschlecht enthaltenen Klassen von positiv definiten quadratischen Formen, Math. Ann., 114 (1937) 465–475; Berichtigung, Math. Ann., 115 (1938) 643–644.

23. J. Milnor, in W. H. Greub, Linear Algebra, Academic Press, New York, 1965.

24. H. Minkowski, Mémoire sur la théorie des formes quadratiques à coefficients entiers, Ges. Abh. 1, Chelsea, New York, 1967. 1–144.

25. L. J. Mordell, A new Waring problem with squares of linear forms, Quart. J. Math. Oxford, 4 (1930) 276–280.

26. ———, The definite quadratic forms in eight variables with determinant unity, J. Math. Pure Appl., 17 (1938) 41–46.

27. M. Newman and O. Taussky, On a generalization of the normal basis in abelian algebraic number fields, Commun. Pure Appl. Math., 9 (1956) 85–91.

28. ——— and ———, Classes of definite unimodular circulants, Canadian Math. J., 9 (1956) 71–73.

29. D. G. Quillen, On the representation of hermitian forms as sums of squares, Invent. Math., 5 (1968) 237–242.

30. R. Redheffer, Remarks on a paper of Taussky, J. Algebra, 2 (1965) 42–47.

31. H. Schneider, Positive operators and an inertia theorem, Numer. Math., 7 (1965) 11–15.

32. P. Stein, Some general theorems on iterants, J. Res. Nat. Bur. Standards, 48 (1952) 82–83.

33. ———, On the range of two functions of positive definite matrices, J. Algebra, 2 (1965) 350–353.

34. ——— and A. Pfeffer, On the ranges of two functions II, ICC Bull., 6 (1967) 81–86.

35. O. Taussky, Automorphs and generalized automorphs of quadratic forms treated as characteristic value relations, Linear Algebra and Appl., 1 (1968) 349–356.

36. ———, Unimodular integral circulants, Math. Z., 63 (1955) 286–289.

37. ———, Problem 4846, this MONTHLY, 66 (1959) 427.

38. ———, Positive definite matrices and their role in the study of the characteristic roots of general matrices, Advances in Math., 2 (1968) 175–186.

39. ———, Matrices C with $C^n \to 0$, J. Algebra, 1 (1964) 5–10.

40. ———, Matrix Theory Research Problem, Bull. Amer. Math. Soc., 71 (1965) 711.

41. ———, Positive definite matrices in 'Inequalities', O. Shisha, Ed., Academic Press, 1967.

42. ———, Stable Matrices in 'Programmation en Analyse Numérique,' J. L. Rigal, Ed., Cahiers Centre Math. Rech. Sci. 1968.

43. R. C. Thompson, Unimodular group matrices with rational integers as elements, Pacific J. Math., 14 (1964) 719–726.

44. ———, Classes of definite group matrices, Pacific J. Math., 17 (1966) 175–190.

45. B. L. van der Waerden, Die Reduktionstheorie der positiven quadratischen Formen, Acta Math., 96 (1956) 265–309.

46. E. P. Wigner, On weakly positive matrices, Canadian J. Math., 15 (1965) 313–317.

47. ——— and M. M. Yanase, On the positive semidefinite nature of a certain matrix expression, Canadian J. Math., 16 (1964) 397–406.

48. M. Wonenburger, Simultaneous diagonalization of symmetric bilinear forms, J. Math. Mech., 15 (1966) 617–622.

49. Yik-Hoi Au Yeung, A theorem on a mapping from a sphere to the circle and the simultaneous diagonalization of two hermitian matrices, Proc. Amer. Math. Soc., 20 (1969) 545–548.

50. ———, Some theorems on the real pencil and diagonalization of two hermitian bilinear functions, Proc. Amer. Math. Soc., 23 (1969) 246–254.

References to 5

1. E. Artin and O. Schreier, Algebraische Konstruktion reeller Körper, Abh. Math. Sem. Univ. Hamburg, 5 (1926) 83–115.

2. E. Bender, The dimensions of symmetric matrices with a given minimum polynomial, Linear Algebra and Appl. To appear.

3. J. W. S. Cassels, On the representation of rational functions as sums of squares, Acta Arith., 9 (1964) 79–82.

4. ———, Représentations comme somme de carrés, Les tendances géométriques en algèbre et théorie des nombres, 55–65, Paris, 1966.

5. P. Chowla, On the representation of -1 as a sum of squares in a cyclotomic field, J. Number Theory, 1 (1969) 208–210.

6. B. Fein and B. Gordon, On the representation of -1 as a sum of two squares in an algebraic number field, J. Number Theory, to appear.

7. H. Kneser, Verschwindende Quadratsummen in Körpern, Jber. Deutsch. Math.-Verein, 44 (1934) 143–146.

8. F. Krakowski, Eigenwerte und Minimalpolynome symmetrischer Matrizen in kommutativen Körpern, Comment. Math. Helv., 32 (1958) 224–240.

9. A. Pfister, Zur Darstellung von -1 als Summe von Quadraten in einem Körper, J. London Math. Soc., 40 (1965) 159–165.

10. J.-P. Serre, Extension de corps ordonnes, C. R. Acad. Sci. Paris, 229 (1949) 576–577.

11. J. Smith, The equation $-1=x^2+y^2$ in certain number fields, J. Number Theory, to appear.

12. R. C. Thompson, Commutators in the special and general linear groups, Trans. Amer. Math. Soc., 101 (1961) 16–33.

13. B. L. van der Waerden, Problem, Jber. Deutsch. Math-Verein, 42 (1932) 71.

14. ———, Modern Algebra I, Frederick Ungar, New York, 1966.

References to 6

1. A. A. Albert, Quadratic forms permitting composition, Ann. Math. (2), 43 (1942) 161–177.

2. ———, Quadratic forms permitting composition, unpublished manuscript, 1962.

3. H. Brandt, Der Kompositionsbegriff bei den quaternären quadratischen Formen, Math. Ann., 91 (1924) 300–315.

4. C. Chevalley, The algebraic theory of spinors, Columbia University Press, New York, 1954.

5. C. W. Curtis, The four and eight square problem and division algebras, in Studies in Modern Algebra 2, A. A. Albert, ed., (1963) 100–125.

6. J. Dieudonné, A problem of Hurwitz and Newman, Duke Math. J., 20 (1953) 381–389.

7. B. Eckman, Gruppentheoretischer Beweis der Satzes von Hurwitz-Radon über die Komposition der quadratischen Formen, Comment. Math. Helv., 15 (1942/3), 358–366.

8. H. Freudenthal, Oktaven, Ausnahmegruppen und Oktavengeometrie, Utrecht, 1951.

9. M. Gerstenhaber, On semicommuting matrices, Math. Z., 83 (1964) 250–260.

10. F. Hausdorff, Zur Hilbertschen Lösung des Waringschen Problems, Math. Ann., 67 (1909) 301–305.

11. A. Hurwitz, Über die Komposition der quadratischen Formen, Math. Ann., 88 (1923) 1–25.

12. N. Jacobson, Composition algebras and their automorphisms, Rend. Circ. Mat. Palermo 7, (1958) 55–80.

13. P. Jordan, J. von Neumann and E. Wigner, On an algebraic generalization of the quantum mechanical formalism, Ann. Math., 2 (1934) 29–64.

14. I. Kaplansky, Composition of binary quadratic forms, Studia Math., 31 (1968) 85–92.

15. ———, Infinite-dimensional quadratic forms permitting composition, Proc. Amer. Math. Soc., 4 (1953) 956–96.

16. P. Kustaanheimo and E. Stiefel, Perturbation theory of Kepler motion based on spinor regularization, J. Reine Angew. Math., 218 (1965) 204–219.

17. K. McCrimmon, A proof of Schafer's conjecture for infinite dimensional forms admitting composition, J. Algebra, 5 (1967) 72–83.

18. M. H. A. Newman, Note on an algebraic theorem of Eddington, J. London Math. Soc., 7 (1932) 93–99.

19. A. Pfister, Multiplikative quadratische Formen, Archiv Math., 16 (1965) 363–370.

20. J. Putter, Maximal sets of anti-commuting skew-symmetric matrices, J. London Math. Soc., 42 (1967) 303–308.

21. R. D. Schafer, Forms permitting composition, Advances in Math., 4 (1970) 127–148.

22. W. Scharlau, Quadratische Formen und Galois-Cohomologie, Invent. Math., 4 (1967) 238–264.

23. O. Taussky, A determinantal identity for quaternions and a new eight square identity, J. Math. Anal. Appl., 15 (1966) 162–164.

24. B. Walsh, The scarcity of cross products on euclidean spaces, this MONTHLY, 74 (1967) 188–194.

25. H. Zassenhaus and W. Eichhorn, Herleitung von acht- und sechzehn-Quadrate–Identitäten mit Hilfe von Eigenschaften der verallgemeinerten Quaternionen und der Cayley-Dicksonschen Zahlen, Archiv Math., 17 (1966) 492–496.

References to 7

1. J. F. Adams, Vector fields on spheres, Ann. Math., 75 (1962) 603–632.

2. ———, P. D. Lax and R. S. Phillips, On matrices whose real linear combinations are

non-singular, Proc. Amer. Math. Soc., 16 (1965) 318–322; Proc. Amer. Math. Soc., 17 (1966) 945–947.

3. A. A. Albert, Absolute-valued real algebras, Ann. Math., 48 (1947) 495–501.

4. ———, Absolute-valued real algebras, Bull. Amer. Math. Soc., 55 (1949) 763–768.

5. R. Arens, Linear topological division algebras, Bull. Amer. Math. Soc., 53 (1947) 623–630.

6. M. F. Atiyah, The role of algebraic topology in mathematics, J. London Math. Soc., 41 (1966) 63–69.

7. F. A. Behrens, Über Systeme reeller algebraischer Gleichungen, Composito Math., 7 (1939) 1–19.

8. G. Benneton, Sur l'arithmétique des quaternions et des biquaternions, Ann. Sci. École Norm. Sup., (3), 60 (1943) 173–214.

9. R. Bott and J. Milnor, On the parallelizability of the spheres, Bull. Amer. Math. Soc., 64 (1958) 87–89.

10. H. Butts and D. Estes, Modules and binary quadratic forms over integral domains, Linear Algebra and Appl., 1 (1968) 153–180.

11. H. Cohn and G. Pall, Sums of four squares in a quadratic ring, Trans. Amer. Math. Soc., 105 (1962) 536–556.

12. H. S. M. Coxeter, Integral Cayley numbers, Duke Math. J., 13 (1946) 561–578.

13. M. L. Curtis and J. Dugundji, Groups which are cogroups, manuscript.

14. L. E. Dickson, On quaternions and their generalizations and the history of the eight square theorem, Ann. Math., 20 (1919) 155–171.

15. W. Eichhorn, Funktionalgleichungen in Vektorräumen, Kompositionsalgebren und Systeme partieller Differentialgleichungen, Aequationes Math., 2 (1969) 287–303.

16. D. Estes and G. Pall, Modules and rings in the Cayley algebra, J. Number Theory, 1 (1969) 163–178.

17. R. Fueter, Die Theorie der regulären Funktionen einer Quaternionenvariablen, Comptes Rendus du congrès int. des mathématiciens, Oslo, 1936, 75–91.

18. I. Gelfand. Normierte Ringe, Mat. Sbornik N. S. 9 (51) (1941) 3–24.

19. J. W. Givens, Tensor coordinates of linear spaces, Ann. Math., 38 (1937) 355–385.

20. W. R. Hamilton, Mathematical papers, Vol. III, Algebra, Cambridge, 1967.

21. H. Hopf, Ein topologischer Beitrag zur reellen Algebra, Comm. Math. Helv., 13 (1941) 219–239.

22. A. Hurwitz, Über die Zahlentheorie der Quaternionen, Ges. Abh., 2, Birkhäuser, 1933, 303–330.

23. N. Jacobson and O. Taussky, Locally compact rings, Proc. Nat. Acad. Sci. USA, 21 (1935) 106–108.

24. I. Kaplansky, Submodules of quaternion algebras, Proc. London Math. Soc., 19 (1969) 219–232.

25. M. Kervaire, Non-parallelizability of the n-sphere for $n>7$, Proc. Nat. Acad. Sci. USA, 14 (1958) 280–283.

26. H. Kestelman, Anticommuting linear transformations, Canadian J. Math., 13 (1961) 614–624.

27. P. J. C. Lamont, Arithmetics in Cayley's algebra, Proc. Glasgow Math. Assoc., 6 (1963) 99–106.

28. Yu. U. Linnik, Quaternions and Cayley numbers; some applications of the arithmetic of quaternions, Uspehi Mat. Nauk (N.S.) 4 (33) (1949) 49–98.

29. ———, Quaternions and Cayley numbers, Math. Centrum Amsterdam, Rapport ZW-1951-002, 1951.

30. R. Lipschitz, Recherches sur les transformations, par des substitutions réelles d'une somme de deux ou de trois carrés en elle-même, J. Math. Pures Appl., 4, 2 (1886) 373–439.

31. D. Lissner, Outer product rings, Trans. Amer. Math. Soc., 116 (1965) 526–535.

32. K. Mahler, On ideals in the Cayley-Dickson algebra, Proc. Roy. Irish Acad. (A), 48 (1943) 123–133.

33. S. Mazur, Sur les anneaux linéaires, C. R. Acad. Sci. Paris, 207 (1938) 1025–1027.

34. G. Pall, On generalized quaternions, Trans. Amer. Math. Soc., 59 (1946) 280–332.

35. G. Pall and O. Taussky, Applications of quaternions to the representations of binary quadratic form as a sum of four squares, Proc. Roy. Irish Acad. (A), 58 (1957), 23–38.

36. ——— and O. Taussky, Factorization of Cayley numbers, J. Number theory, 2 (1970) 74–90.

37. L. S. Pontryagin, Topological groups; Topological groups, Transl. by A. Brown, Gordon and Breach, New York, 1966.

38. J. Radon, Lineare Schären orthogonaler Matrizen, Abh. Math. Sem. Univ. Hamburg, 1 (1922) 1–14.

39. R. A. Rankin, A certain class of multiplicative functions, Duke Math. J., 13 (1946) 281–306.

40. H. Samelson, Über die Sphären, die als Gruppenräume auftreten, Comm. Math. Helv., 13 (1940/41) 149–155.

41. E. Stiefel, Richtungsfelder und Fernparallelismus in n-dimensionalen Mannigfaltigkeiten, Comm. Math. Helv., 8 (1935/6) 3–51.

42. ———, On Cauchy-Riemann equations in higher dimensions, J. Res. Nat. Bur. Standards, 48 (1952) 395–398.

43. M. H. Stone, On the theorem of Gelfand-Mazur, Ann. Polon. Math., 24 (1952) 238–240.

44. R. G. Swan, Vector bundles and projective modules, Trans. Amer. Math. Soc., 105 (1962) 264–277.

45. O. Taussky, Analytical methods in hypercomplex systems, Compositio Math., 3 (1936) 399–407.

46. ———, An algebraic property of Laplace's differential equation, Quart. J. Math. Oxford, 10 (1939) 99–103.

47. ———, (1, 2, 4, 8)-sums of squares and Hadamard matrices, Proc. Symp. on Combinatorics, Los Angeles, 1968, to appear.

48. L. Tornheim, Normed fields over the real and complex numbers, Michigan Math. J., 1 (1952) 61–68.

49. F. van der Blij and T. A. Springer, The arithmetics of octaves of the group G_2, Nederl. Akad. Wetensch. Proc., 62A (1959) 406–418.

50. F. Wright, Absolute valued algebras, Proc. Nat. Acad. Sci. USA, 39 (1953) 330–332.

51. ———, Absolute valued algebras, Proc. Amer. Math. Soc., 11 (1960) 861–866.

References to 8

1. E. Artin, Über die Zerlegung definiter Funktionen in Quadrate, Abh. Math. Sem. Univ. Hamburg, 5 (1926) 100–115.

2. J. Ax, On ternary definite rational functions, Proc. London Math. Soc., to appear.

3. L. Carlitz, Sums of squares of polynomials, Duke Math. J., 3 (1937), 1–7.

4. D. W. Dubois, Note on Artin's solution of Hilbert's 17th problem, Bull. Amer. Math. Soc., 73 (1967) 540–541.

5. W. Habicht, Über die Zerlegung strikt definiter Formen in Quadrate, Comment. Math. Helv., 12 (1946) 317–322.

6. ———, Zerlegung strikte definiter Formen in Quadrate, Comment. Math. Helv., 12 (1940) 317–322.

7. D. Hilbert, Über die Darstellung definiter Formen als Summe von Formenquadraten, Ges. Abh. 2, Springer, Berlin, 1933, 154–161.

8. ———, Mathematische Probleme, in particular problem 17, Ges. Abhandlungen III, Springer, Berlin, 1935, 290–329.

9. E. Landau, Über die Darstellung definiter Funktionen als Summe von Quadraten, Math. Ann., 62 (1906) 272–285.

10. T. S. Motzkin, Algebraic inequalities, in Inequalities, Ed. O. Shisha, Academic Press, New York, 1967, 199–203.

11. A. Pfister, Zur Darstellung definiter Funktionen als Summe von Quadraten, Invent. Math., 4 (1967), 229–237.

12. R. M. Robinson, Some definite polynomials which are not sums of squares of real polynomials (abstract), Notices, Amer. Math. Soc., 16 (1969) 554.

References to 9

1. Gösta Bucht, Arkiv. Math. Astron. and Physik. 6, No. 30.
2. H. Reichardt, Über Normalkörper mit Quaternionengruppe, Math. Z., 41 (1936) 218–221.
3. B. L. van der Waerden, Problem, Jber. Deutsch. Math.-Verein., 43 (1933) 61.

References to 10

1. S. D. Chowla and John Todd, The density of reducible integers, Canadian J. Math., 1 (1949) 297–299.
2. D. H. Lehmer, On arccotangent relations for π, this MONTHLY, 45 (1938) 657–664.
3. John Todd, A problem on arctangent relations, this MONTHLY, 56 (1949) 517–528.
4. ———, Table of arctangents of rational numbers, National Bureau of Standards, Applied Math. Series, No. 11 (1951–1965), U. S. Government Printing Office, Washington, D. C.

This list is not intended to be complete in any respect.

FORMAL POWER SERIES*

IVAN NIVEN, University of Oregon

1. Introduction. Our purpose is to develop a systematic theory of formal power series. Such a theory is known, or at least presumed, by many writers on mathematics, who use it to avoid questions of convergence in infinite series. What is done here is to formulate the theory on a proper logical basis and thus to reveal the absence of the convergence question. Thus "hard" analysis can be replaced by "soft" analysis in many applications.

John Riordan [4] has discussed these matters in a chapter on generating functions, but his interest is in the applications to combinatorial problems. A more abstract discussion is given by de Branges and Rovnyak [1]. Many examples of the use of formal power series could be cited from the literature; we mention only two, one by John Riordan [5] the other by David Zeitlin [6].

The scheme of the paper is as follows. The theory of formal power series is developed in Sections 3, 4, 5, 6, 7, 11, and 12. Applications to number theory and combinatorial analysis are discussed in Sections 2, 8, 9, 10, and in the last part of 11.

The paper is self-contained insofar as it pertains to the theory of formal power series. However, in the applications of this theory, especially in the ap-

* From AMERICAN MATHEMATICAL MONTHLY, vol. 76 (1969), pp. 871–889.

Professor Niven has worked in number theory since his Chicago dissertation on the Waring problem under L. E. Dickson. He was a fellow at Pennsylvania, working with H. Rademacher, and since has been at Illinois, Purdue, and Oregon, interrupted by leaves twice to Berkeley. He was the MAA Hedrick lecturer in 1960 and has long been an active participant in the MAA. His six books include the Carus Monograph *Irrational Numbers* and (with H. S. Zuckerman) *Introduction to the Theory of Numbers*. *Editor.*

plication to partitions in Section 9, we do not repeat here the fundamental results needed from number theory. Thus Sections 9 and 10 may be difficult for a reader who is not too familiar with the basic theory of partitions and the sum of divisors function. This difficulty can be removed by use of the specific references given in these sections; only a few pages of fairly straightforward material are needed as background. In Section 11 on the other hand, the background material is set forth in detail because the source is not too readily available.

2. An example from algebra. To motivate the theory we begin with an illustration from algebra, to be found in Jacobson [2, p. 19]. Let q_n denote the number of ways of associating an n-product $a_1a_2a_3 \cdots a_n$ in a nonassociative system. For example $q_3=2$ because $a_1(a_2a_3)$ and $(a_1a_2)a_3$ are the only possibilities. Similarly $q_4=5$ because of the cases $a_1(a_2(a_3a_4))$, $a_1((a_2a_3)a_4)$, $(a_1a_2)(a_3a_4)$, $(a_1(a_2a_3))a_4$, $((a_1a_2)a_3)a_4$. For $n \geqq 2$ it is easy to establish the recursive formula

$$q_n = \sum_{j=1}^{n-1} q_j q_{n-j}, \tag{1}$$

by the following argument. In imposing a system of parentheses on $a_1a_2a_3 \cdots a_n$ to make it a well-defined n-product, we can begin by writing

$$(a_1a_2 \cdots a_j)(a_{j+1}a_{j+2} \cdots a_n). \tag{2}$$

Now the number of ways of associating the product $a_1a_2 \cdots a_j$ is q_j by definition, and likewise the second factor in (2) can be associated in q_{n-j} ways. Hence (2) can be associated in q_jq_{n-j} ways, and formula (1) follows by considering the possible values for j. Now define the power series

$$f(x) = \sum_{j=1}^{\infty} q_j x^j. \tag{3}$$

Taking for granted (for the moment) the multiplication of power series, we see that for $n \geqq 2$ the coefficient of x^n in $\{f(x)\}^2$ is

$$q_1q_{n-1} + q_2q_{n-2} + q_3q_{n-3} + \cdots + q_{n-1}q_1.$$

But this is q_n by (1), and so we see that $\{f(x)\}^2 = f(x) - x$ or $f^2 - f + x = 0$.

Solving this quadratic equation for f we get

$$f(x) = f = \tfrac{1}{2}\{1 \pm (1-4x)^{1/2}\}. \tag{4}$$

The binomial theorem gives

$$(1-4x)^{1/2} = 1 + \frac{1}{2}(-4x) + \frac{\frac{1}{2}(\frac{1}{2}-1)}{2!}(-4x)^2 + \cdots$$
$$+ \frac{\frac{1}{2}(\frac{1}{2}-1)(\frac{1}{2}-2)\cdots(\frac{1}{2}-n+1)}{n!}(-4x)^n + \cdots.$$

The coefficient of x^n here can be simplified by multiplying numerator and denominator by 2^n to give

$$\frac{(1)(-1)(-3)(-5)\cdots(-2n+3)}{2^n\cdot n!}(-4)^n = -\frac{1\cdot 3\cdot 5\cdots(2n-3)}{n!}\cdot 2^n$$

$$= -\frac{(2n-2)!}{(n!)2^{n-1}(n-1)!}2^n$$

$$= -2\frac{(2n-2)!}{n!(n-1)!}.$$

In view of the minus sign here we see that (4) holds with the minus sign and not the plus sign. Comparing coefficients of x^n in (4) we get the simple formula for q_n,

$$q_n = \frac{(2n-2)!}{n!(n-1)!}. \tag{5}$$

This analysis, however, leaves a number of questions unanswered. Why can we solve the quadratic to derive (4)? Why can we equate coefficients on the two sides of (4) to obtain (5)? To avoid hard analysis in answering such questions, we now develop a theory of formal power series that involves no questions of convergence or divergence. At the end of Section 5 we shall return to the question of the validity of the procedure leading to formula (5).

3. Formal power series. Define α to be an infinite sequence of complex numbers

$$\alpha = [a_0, a_1, a_2, a_3, \cdots]. \tag{6}$$

By P we denote the class of all such infinite sequences α, and these are the formal power series. There are three subsets of P that play a significant role:

P_r: those sequences α all of whose components a_j are real numbers;
P_1: those sequences α with $a_0=1$;
P_0: those sequences α with $a_0=0$.

Although we have specified that the components a_j in the elements of P are complex numbers, the theory could be developed with the a_j in any integral domain.

If $\beta\in P$, say $\beta=[b_0, b_1, b_2, b_3, \cdots]$, define addition by

$$\alpha+\beta = [a_0+b_0, a_1+b_1, a_2+b_2, \cdots].$$

Define multiplication by

$$\alpha\beta = \left[a_0b_0,\ a_1b_0+a_0b_1,\ a_2b_0+a_1b_1+a_0b_2,\ \cdots,\ \sum_{j=0}^{n} a_jb_{n-j},\ \cdots\right].$$

The definition of equality is that $\alpha=\beta$ if and only if $a_j=b_j$ for all j, i.e., $j=0$, 1, 2, 3, $\cdots$.

It is not difficult to establish that the set P is a commutative ring with a unit. The zero element and the unit element are

$$z = [0, 0, 0, 0, \cdots] \quad \text{and} \quad u = [1, 0, 0, 0, \cdots].$$

Given any $\alpha = [a_0, a_1, a_2, a_3, \cdots]$ the additive inverse of α is $-\alpha = [-a_0, -a_1, -a_2, -a_3, \cdots]$. The verification of the associative property of multiplication is not difficult, and it is the only property of any depth in establishing that P is a commutative ring.

Moreover, $\alpha\beta = z$ if and only if $\alpha = z$ or $\beta = z$. If $\alpha = z$ or $\beta = z$ it is obvious that $\alpha\beta = z$. To establish the converse, suppose that $\alpha\beta = z$ but $\alpha \neq z$ and $\beta \neq z$. Let j be the least nonnegative integer such that $a_j \neq 0$, and similarly let k be the least nonnegative integer such that $b_k \neq 0$. Then the component in the $(j+k+1)$-th position in $\alpha\beta$ is

$$\sum_{r=0}^{j+k} a_r b_{j+k-r} = a_j b_k \neq 0,$$

which contradicts $\alpha\beta = z$.

It follows that if $\alpha\beta = \alpha\gamma$ and $\alpha \neq z$ then $\beta = \gamma$, and P is an integral domain. Given any α in P, there corresponds a multiplicative inverse α^{-1} if there is an element α^{-1} in P such that

$$\alpha \cdot \alpha^{-1} = \alpha^{-1} \cdot \alpha = u = [1, 0, 0, 0, \cdots].$$

THEOREM 1. *If* $\alpha = [a_0, a_1, a_2, \cdots]$, α^{-1} *exists if and only if* $a_0 \neq 0$.

Proof. Denote α^{-1} by $[c_0, c_1, c_2, \cdots]$. We see that $\alpha\, \alpha^{-1} = u$ amounts to an infinite system of equations

$$a_0 c_0 = 1, \ a_1 c_0 + a_0 c_1 = 0, \cdots, \sum_{j=0}^{n} a_j c_{n-j} = 0.$$

These equations can be solved successively for $c_0, c_1, c_2, \cdots$ if and only if $a_0 \neq 0$.

LEMMA 2. *Let* $\beta \in P_1$, *so that* β *is of the form* $[1, b_1, b_2, b_3, \cdots]$. *Then for any positive integer* n *we see that* $\beta^n \in P_1$, *say* $\beta^n = [1, c_1, c_2, c_3, \cdots]$. *Also* $c_1 = n\, b_1$ *and for each* $k \geqq 2$ *we have* $c_k = n\, b_k + f_{n,k}(b_1, b_2, \cdots, b_{k-1})$ *where* $f_{n,k}$ *is an appropriate polynomial in* $b_1, b_2, \cdots, b_{k-1}$.

Proof. This result can be readily established by induction on n.

THEOREM 3. *Let* $\alpha \in P_1$, *say* $\alpha = [1, a_1, a_2, a_3, \cdots]$, *and let* n *be any positive integer. Then there is a unique* $\beta \in P_1$, *say* $\beta = [1, b_1, b_2, b_3, \cdots]$, *such that* $\beta^n = \alpha$. *Define* $\alpha^{1/n} = \beta$.

Proof. Using Lemma 2 we can solve the equations

$$nb_1 = a_1, \ nb_2 + f_{2,n}(b_1) = a_2, \cdots, nb_k + f_{k,n}(b_1, b_2, \cdots, b_{k-1}) = a_k, \cdots,$$

successively for $b_1, b_2, b_3, \cdots$.

THEOREM 4. *For any positive integer n and* $\alpha \in P_1$, *we have* $(\alpha^{-1})^n = (\alpha^n)^{-1}$. *Define* $\alpha^{-n} = (\alpha^n)^{-1}$ *and* $\alpha^0 = u$.

Proof. We see that $\alpha^n(\alpha^{-1})^n = \alpha \cdot \alpha \cdots \alpha \cdot \alpha^{-1} \cdot \alpha^{-1} \cdots \alpha^{-1} = u$. (Another way of establishing Theorem 4 is to observe that P_1 is a multiplicative group.)

THEOREM 5. *Let m and n be any integers,* $n > 0$. *To any* $\alpha \in P_1$ *there corresponds a unique* $\beta \in P_1$ *such that* $\alpha^m = \beta^n$, *i.e.,* $\beta = \alpha^{m/n}$.

Proof. This is a corollary of Theorem 3 with α in that theorem replaced by α^m.

4. A power series notation. Let λ denote the particular element $[0, 1, 0, 0, 0, \cdots]$ of P so that

$$\lambda^2 = [0, 0, 1, 0, 0, \cdots], \quad \lambda^3 = [0, 0, 0, 1, 0, 0, \cdots],$$

and in general λ^{n-1} is the sequence with zeros in all positions except the nth, where 1 occurs. We now introduce the notation

$$\sum_{j=0}^{\infty} a_j \lambda^j = a_0 + a_1\lambda + a_2\lambda^2 + a_3\lambda^3 + \cdots \tag{7}$$

for $\alpha = [a_0, a_1, a_2, \cdots]$. What this amounts to is an agreement that a_j in (7) stands for $[a_j, 0, 0, 0, \cdots]$ and that $\lambda^0 = [1, 0, 0, 0, \cdots]$. Thus we are *not* extending the integral domain P to a vector space by introducing scalar multiplication; this could be done, but all we intend by (7) is an alternative, convenient notation for the elements of P. Thus z and u can now be written simply as 0 and 1. The definitions of addition, multiplication, and equality of elements of P can be rewritten as follows. With α as in (7) and

$$\beta = [b_0, b_1, b_2, \cdots] = \sum_{j=0}^{\infty} b_j \lambda^j,$$

then

$$\alpha + \beta = \sum_{j=0}^{\infty} (a_j + b_j)\lambda^j, \quad \alpha\beta = \sum_{j=0}^{\infty} \left(\sum_{k=0}^{j} a_k b_{j-k} \right) \lambda^j,$$

and $\alpha = \beta$ if and only if $a_j = b_j$ for all $j = 0, 1, 2, 3, \cdots$.

For example, in the earlier notation we could write

$$[1, -1, 0, 0, 0, \cdots] \cdot [1, 1, 1, 1, 1, \cdots] = [1, 0, 0, 0, 0, \cdots].$$

This can now be written as $(1-\lambda)(1+\lambda+\lambda^2+\lambda^3+\cdots) = 1$, or $(1-\lambda)^{-1} = 1+\lambda+\lambda^2+\lambda^3+\cdots$. A general binomial theorem is established later, in Theorems 11 and 17.

THEOREM 6. *Let n be any positive integer, let* $\alpha \in P_r$ *and* $\beta \in P_r$, *so that* α *and* β *are real sequences. If n is odd,* $\alpha^n = \beta^n$ *implies* $\alpha = \beta$. *If n is even,* $\alpha^n = \beta^n$ *implies* $\alpha = \beta$ *or* $\alpha = -\beta$.

Proof. We may presume $\alpha \neq 0$ and $\beta \neq 0$. For if $\alpha = 0$, for example, then $\alpha^n = 0$, $\beta^n = 0$ and so $\beta = 0$, $\alpha = \beta$. Let ω denote the nth root of unity

$$\omega = e^{2\pi i/n} = \cos(2\pi/n) + i\sin(2\pi/n).$$

Then $\alpha^n - \beta^n = 0$ can be factored $\alpha^n - \beta^n = \prod_{j=1}^{n}(\alpha - \omega^j\beta) = 0$. If ω^j is not real then $\alpha - \omega^j\beta \neq 0$ because α and β are real sequences with $\alpha \neq 0$ and $\beta \neq 0$. If n is odd, ω^j is real only in the case $j = n$ and hence

$$\alpha - \omega^n\beta = 0, \quad \alpha - \beta = 0, \quad \alpha = \beta.$$

If n is even, ω^j is real in the two cases $j = n$ and $j = n/2$, leading to the conclusion that $\alpha = \beta$ or $\alpha = -\beta$.

Consider an infinite sequence $\alpha_1, \alpha_2, \alpha_3, \cdots$ of elements of P, say

$$\alpha_k = \sum_{j=0}^{\infty} a_{jk}\lambda^j, \qquad k = 1, 2, 3, \cdots. \tag{8}$$

DEFINITION. *A sequence $\alpha_1, \alpha_2, \alpha_3, \cdots$ as in* (8) *is said to be a sequence admitting addition if corresponding to any integer $r \geqq 0$ there is an integer $N = N(r)$ such that for all $n \geqq N$, $a_{0n} = a_{1n} = a_{2n} = \cdots = a_{rn} = 0$.*

If this condition is satisfied we also say that $\sum \alpha_j$ is an *admissible sum*, and we can write

$$\sum_{j=1}^{\infty} \alpha_j = \sum s_r\lambda^r,$$

where for each integer $r \geqq 0$ the coefficient s_r is the coefficient of λ^r in the finite sum $\alpha_1 + \alpha_2 + \cdots + \alpha_N$, i.e.,

$$s_r = a_{r1} + a_{r2} + \cdots + a_{rN}.$$

We note that s_r is the coefficient of λ^r in every finite sum $\alpha_1 + \alpha_2 + \cdots + \alpha_n$ with $n \geqq N$.

LEMMA 7. *Let $\alpha_1, \alpha_2, \alpha_3, \cdots$ be a sequence of elements of P admitting addition. Let $\beta_1, \beta_2, \beta_3, \cdots$ be a rearrangement of the α's in the sense that given any j there exists a unique k such that $\alpha_j = \beta_k$. Then $\beta_1, \beta_2, \beta_3, \cdots$ is also a sequence admitting addition, and*

$$\alpha_1 + \alpha_2 + \alpha_3 + \cdots = \beta_1 + \beta_2 + \beta_3 + \cdots.$$

Proof. Let r be any given nonnegative integer. For n sufficiently large the coefficient of λ^r in $\alpha_1 + \alpha_2 + \alpha_3 + \cdots$ equals the coefficient of λ^r in the finite sum $\alpha_1 + \alpha_2 + \cdots + \alpha_n$. Similarly for n sufficiently large the coefficient of λ^r in $\beta_1 + \beta_2 + \beta_3 + \cdots$ equals the coefficient of λ^r in $\beta_1 + \beta_2 + \cdots + \beta_n$. And clearly $\alpha_1 + \alpha_2 + \cdots + \alpha_n$ and $\beta_1 + \beta_2 + \cdots + \beta_n$ have identical terms in λ^r.

Next we get a result analogous to Lemma 7 for multiplication. Consider an infinite sequence $\gamma_1, \gamma_2, \gamma_3, \cdots$ of elements of P of the form

$$(9) \qquad \gamma_k = \sum_{j=1}^{\infty} c_{jk} \lambda^j, \qquad k = 1, 2, 3, \cdots.$$

Note that the sums begin with $j=1$. If this is a sequence admitting addition, then we say that the related sequence

$$(10) \qquad 1 + \gamma_1,\ 1 + \gamma_2,\ 1 + \gamma_3, \cdots$$

is a sequence admitting multiplication. Furthermore, we write

$$\prod_{k=1}^{\infty} (1 + \gamma_k) = 1 + \sum_{j=1}^{\infty} q_j \lambda^j,$$

where q_r is the coefficient of λ^r in any finite product $\prod_{k=1}^{n} (1+\gamma_k)$ with n sufficiently large that $c_{jk}=0$ for $1 \leqq j \leqq r$ if $k>n$. Then it is clear that we can state a result analogous to Lemma 7 as follows:

LEMMA 7a. *If* (10) *is a sequence admitting multiplication, so is any rearrangement* $1+\delta_1, 1+\delta_2, 1+\delta_3, \cdots$ *of* (10), *and*

$$\prod_{k=1}^{\infty} (1 + \gamma_k) = \prod_{k=1}^{\infty} (1 + \delta_k).$$

5. Formal derivatives. Given any α in P, say $\alpha = \sum_{j=0}^{\infty} a_j \lambda^j$, define the derivative $D(\alpha)$ and the scalar $S(\alpha)$ by

$$(11) \qquad D(\alpha) = \sum_{j=1}^{\infty} j a_j \lambda^{j-1}, \qquad S(\alpha) = a_0.$$

Define $D^2(\alpha) = D(D(\alpha))$, and in general for any positive integer n, the nth derivative is $D^n(\alpha)$. Taking $D^0(\alpha) = \alpha$ for convenience, we can now write a McLaurin series expansion.

THEOREM 8. $\alpha = \sum_{n=0}^{\infty} (1/n!)\, S(D^n(\alpha)) \cdot \lambda^n$.

The proof of this is quite easy.

THEOREM 9. *If* $\alpha \in P$, $\beta \in P$ *then* $D(\alpha+\beta) = D(\alpha) + D(\beta)$ *and* $D(\alpha\beta) = \alpha D(\beta) + \beta D(\alpha)$, *and* $D(\alpha^n) = n\alpha^{n-1} D(\alpha)$ *for any positive integer* n. *Also if* α^{-1} *exists then* $D(\alpha^{-1}) = -\alpha^{-2} D(\alpha)$ *and* $D(\alpha^{-n}) = -n\alpha^{-n-1} D(\alpha)$.

Proof. The formula for $D(\alpha\beta)$ can be established easily by comparing coefficients of λ^n. By using induction on n we get the formula for $D(\alpha^n)$. Next if we differentiate $\alpha\alpha^{-1} = 1$ we get the formula for $D(\alpha^{-1})$. Finally, $\alpha^{-n} = (\alpha^{-1})^n$ can be used to write

$$D(\alpha^{-n}) = D((\alpha^{-1})^n) = n(\alpha^{-1})^{n-1}D(\alpha^{-1}) = -n\alpha^{-n-1}D(\alpha).$$

THEOREM 10. *Let* $\alpha \in P_1$ *so that* $S(\alpha) = 1$. *For any rational number* r, $D(\alpha^r) = r\alpha^{r-1}D(\alpha)$.

Proof. By Theorem 5 there is a unique meaning for α^r. If $r = m/n$ where m and n are integers we can write

$$D((\alpha^r)^n) = n(\alpha^r)^{n-1}D(\alpha^r), \quad D((\alpha^r)^n) = D(\alpha^m) = m\alpha^{m-1}D(\alpha),$$

by Theorem 9. The result follows at once.

A simple version of the binomial theorem can be easily obtained from Theorems 8 and 10, as follows:

THEOREM 11. *For any rational number* r *and any complex number* k,

$$(1 + k\lambda)^r = 1 + r(k\lambda) + \frac{r(r-1)}{2!}(k\lambda)^2 + \cdots$$
$$+ \frac{r(r-1)(r-2)\cdots(r-n+1)}{n!}(k\lambda)^n + \cdots.$$

Proof. First note that $D(1+k\lambda)^r = r(1+k\lambda)^{r-1}D(1+k\lambda) = rk(1+k\lambda)^{r-1}$, and so by induction on n,

$$D^n(1 + k\lambda)^r = r(r-1)(r-2)\cdots(r-n+1)k^n(1 + k\lambda)^{r-n}.$$

Now $(1+k\lambda)^{r-n}$ is a unique element of P_1 by Theorems 3 and 5, and so $S(1+k\lambda)^{r-n} = 1$. It follows that

$$S(D^n(1 + k\lambda)^r) = r(r-1)(r-2)\cdots(r-n+1)k^n.$$

Now use Theorem 8 with α replaced by $(1+k\lambda)^r$, and the result follows.

The form of the binomial theorem just established is sufficient in most applications, for example, to justify the argument given in Section 2. To see this, we replace equation (3) with this definition of α,

$$\alpha = \sum_{j=1}^{\infty} q_j\lambda^j,$$

where the q_j have the same meaning as in Section 2. Then the analysis following equation (3) leads to $\alpha^2 = \alpha - \lambda$. From this we can write $4\alpha^2 - 4\alpha + 1 = 1 - 4\lambda$, or

$$(1 - 2\alpha)^2 = ((1 - 4\lambda)^{1/2})^2.$$

By Theorem 6 it follows that $1 - 2\alpha = (1 - 4\lambda)^{1/2}$, and so by Theorem 11 we conclude that

$$1 - 2q_1\lambda - 2q_2\lambda^2 - 2q_3\lambda^3 - \cdots$$
$$= 1 + \frac{1}{2}(-4\lambda) + \frac{\frac{1}{2}(\frac{1}{2}-1)}{2!}(-4\lambda)^2 + \frac{\frac{1}{2}(\frac{1}{2}-1)(\frac{1}{2}-2)}{3!}(-4\lambda)^3 + \cdots,$$

From the definition of equality in Section 3 we can now equate the coefficients of λ^n to get equation (5).

We want to get a more general form of the binomial theorem, namely the expansion of $(1+\alpha)^r$ where $\alpha \in P_0$, so that $S(\alpha)=0$. To do this we define a formal logarithm. But first we establish one more result about derivatives.

THEOREM 12. *If $\alpha_1+\alpha_2+\alpha_3+\cdots$ is an admissible sum of elements of P in the sense of Section 4, then*

$$D(\alpha_1+\alpha_2+\alpha_3+\cdots) = D(\alpha_1)+D(\alpha_2)+D(\alpha_3)+\cdots.$$

Proof. For any nonnegative integer r the coefficient of λ^r in the infinite sum $\alpha_1+\alpha_2+\alpha_3+\cdots$ equals the coefficient of λ^r in the finite sum $\alpha_1+\alpha_2+\cdots+\alpha_n$ provided $n \geqq N=N(r)$. Hence the coefficients of λ^{r-1} are equal in the equation in Theorem 12. But this holds for all r, so the result follows.

6. Logarithms and the binomial theorem. A formal logarithm is not defined for any element of P, but only for $\alpha \in P_1$, so that $S(\alpha)=1$. For any $\alpha \in P_1$, say $\alpha = 1+\beta$ with $\beta \in P_0$, define

$$L(\alpha) = L(1+\beta) = \beta - \tfrac{1}{2}\beta^2 + \tfrac{1}{3}\beta^3 - \tfrac{1}{4}\beta^4 + \cdots = \sum_{j=1}^{\infty}(-1)^{j+1}\beta^j/j,$$

noting that this is an admissible sum as in Section 4. Thus L is a formal logarithmic function from P_1 to P_0.

THEOREM 13. $D(L(\alpha)) = \alpha^{-1}D(\alpha)$.

Proof. With $\alpha=1+\beta$ we use Theorem 12 to write

$$\begin{aligned} D(L(\alpha)) = D(L(1+\beta)) &= D[\beta - \tfrac{1}{2}\beta^2 + \tfrac{1}{3}\beta^3 - \tfrac{1}{4}\beta^4 + \cdots \\ &= D(\beta) + D(-\tfrac{1}{2}\beta^2) + D(\tfrac{1}{3}\beta^3) + D(-\tfrac{1}{4}\beta^4) + \cdots \\ &= D(\beta) - \beta D(\beta) + \beta^2 D(\beta) - \beta^3 D(\beta) + \cdots \\ &= D(\beta)[1 - \beta + \beta^2 - \beta^3 + \cdots] \\ &= D(\beta)\cdot(1+\beta)^{-1} = D(\alpha)\cdot\alpha^{-1}, \end{aligned}$$

because $D(\alpha)=D(\beta)$ by definition.

THEOREM 14. *If $\alpha \in P_1$ and $\gamma \in P_1$ then $L(\alpha\gamma)=L(\alpha)+L(\gamma)$.*

Proof. We use Theorems 13 and 9 to observe that

$$\begin{aligned} D(L(\alpha\gamma)) = (\alpha\gamma)^{-1}D(\alpha\gamma) &= (\alpha\gamma)^{-1}\{\alpha D(\gamma) + \gamma D(\alpha)\} \\ &= \alpha^{-1}D(\alpha) + \gamma^{-1}D(\gamma) \\ &= D(L(\alpha)) + D(L(\gamma)) \\ &= D(L(\alpha) + L(\gamma)). \end{aligned}$$

Now $L(\alpha\gamma)$ and $L(\alpha)+L(\gamma)$ are elements in P_0, and it is clear from the definition of a derivative that if $\theta_1 \in P_0$ and $\theta_2 \in P_0$ and $D(\theta_1)=D(\theta_2)$, then $\theta_1=\theta_2$.

THEOREM 15. *For any rational number* r, $L(\alpha^r)=rL(\alpha)$.

Proof. By definition $L(1)=0$. Then $\alpha\cdot\alpha^{-1}=1$ implies $L(\alpha)+L(\alpha^{-1})=L(\alpha\cdot\alpha^{-1})$ $=L(1)=0$ and so $L(\alpha^{-1})=-L(\alpha)$. For any integer n we have $L(\alpha^n)=nL(\alpha)$ by induction. If $r=m/n$ where m and n are integers we see that

$$mL(\alpha) = L(\alpha^m) = L((\alpha^r)^n) = nL(\alpha^r).$$

THEOREM 16. $L(\alpha)=0$ *if and only if* $\alpha=1$. *Also if* $L(\alpha)=L(\beta)$ *then* $\alpha=\beta$.

Proof. If $L(\alpha)=0$ then $D(L(\alpha))=D(0)=0$ and so $\alpha^{-1}D(\alpha)=0$. But $\alpha^{-1}\neq 0$ and hence $D(\alpha)=0$ and $\alpha=1$.

THEOREM 17. *If r is rational, if β is an element of P_0 so that $S(\beta)=0$, then*

$$(1+\beta)^r = 1 + r\beta + \frac{r(r-1)}{2!}\beta^2 + \cdots + \frac{r(r-1)(r-2)\cdots(r-n+1)}{n!}\beta^n + \cdots. \tag{12}$$

Proof. For convenience we write

$$\binom{r}{n} = \frac{r(r-1)(r-2)\cdots(r-n+1)}{n!}.$$

Let γ denote the right side of equation (12) so that

$$D(\gamma) = D(\beta)\sum_{j=1}^{\infty} j\binom{r}{j}\beta^{j-1},$$

$$\begin{aligned}
(1+\beta)D(\gamma) &= D(\beta)\sum_{j=1}^{\infty} j\binom{r}{j}\beta^{j-1} + D(\beta)\sum_{j=1}^{\infty} j\binom{r}{j}\beta^{j} \\
&= D(\beta)\sum_{j=1}^{\infty} j\binom{r}{j}\beta^{j-1} + D(\beta)\sum_{j=2}^{\infty}(j-1)\binom{r}{j-1}\beta^{j-1} \\
&= D(\beta)\cdot r + D(\beta)\sum_{j=2}^{\infty}\left\{j\binom{r}{j} + (j-1)\binom{r}{j-1}\right\}\beta^{j-1} \\
&= D(\beta)\cdot r + D(\beta)\sum_{j=2}^{\infty} r\binom{r}{j-1}\beta^{j-1} \\
&= rD(\beta)\left[1 + \sum_{j=1}^{\infty}\binom{r}{j}\beta^{j}\right] = r\gamma D(\beta).
\end{aligned}$$

Multiplying by $\gamma^{-1}(1+\beta)^{-1}$ we get $\gamma^{-1}D(\gamma)=r(1+\beta)^{-1}D(\beta)=r(1+\beta)^{-1}D(1+\beta)$. But $D(L(\gamma))=\gamma^{-1}D(\gamma)$ and $D(L((1+\beta)^r))=D(rL(1+\beta))=r(1+\beta)^{-1}D(1+\beta)$, and so $D(L(\gamma))=D(L((1+\beta)^r))$. Since $L(\gamma)$ and $L(1+\beta)^r$ are in P_0 it follows that $L(\gamma)=L((1+\beta)^r)$, and so $\gamma=(1+\beta)^r$ by Theorem 16.

7. The exponential function. Let β be an element of P_0, so that $S(\beta)=0$. Then we define

$$E(\beta) = 1 + \beta + \frac{\beta^2}{2!} + \frac{\beta^3}{3!} + \cdots = \sum_{n=0}^{\infty} \frac{\beta^n}{n!},$$

so that E is a function from P_0 to P_1. Since $E(\beta)$, as defined, is an admissible sum, we can apply Theorem 12 to get

$$D(E(\beta)) = D(\beta)\left\{1 + \beta + \frac{\beta^2}{2!} + \frac{\beta^3}{3!} + \cdots\right\} = D(\beta)\cdot E(\beta).$$

THEOREM 18. *If $E(\beta)=E(\gamma)$ then $\beta=\gamma$.*

Proof. We observe that $D(E(\beta))=D(E(\gamma))$, so that $D(\beta)\cdot E(\beta)=D(\gamma)\cdot E(\gamma)$. But $E(\beta)\neq 0$ so that $E(\beta)$ and $E(\gamma)$ can be cancelled giving $D(\beta)=D(\gamma)$, and hence $\beta=\gamma$.

THEOREM 19. *If $\beta\in P_0$ then $L(E(\beta))=\beta$. If $\alpha\in P_1$ then $E(L(\alpha))=\alpha$. Thus L and E are inverse functions, L being one-to-one from P_1 onto P_0, and E one-to-one from P_0 onto P_1.*

Proof. By Theorem 13 we see that

$$D(L(E(\beta))) = \{E(\beta)\}^{-1}\cdot D(E(\beta)) = \{E(\beta)\}^{-1}\cdot E(\beta)\cdot D(\beta) = D(\beta).$$

It follows that $L(E(\beta))=\beta$. Next, given any α in P_1 suppose that $E(L(\alpha))=\alpha_1$. Then $L(E(L(\alpha)))=L(\alpha_1)$ and so $L(\alpha)=L(\alpha_1)$. Hence $\alpha=\alpha_1$ by Theorem 16.

THEOREM 20. *Given $\beta\in P_0$, $\gamma\in P_0$, then $E(\beta+\gamma)=E(\beta)\cdot E(\gamma)$.*

Proof. By Theorems 14 and 19 we see that

$$L(E(\beta)\cdot E(\gamma)) = L(E(\beta)) + L(E(\gamma)) = \beta + \gamma.$$

Taking the exponential function of each side, and using Theorem 19 again, we get the result.

By Theorems 15 and 19 we see that $\alpha^r=E(rL(\alpha))$ for any $\alpha\in P_1$ and any rational r. This equation we take as the definition of α^r for any complex number r, so that such properties of exponents as $\alpha^r\cdot\alpha^s=\alpha^{r+s}$ follow at once for complex numbers r and s. Also by use of this definition we note that Theorem 10 can be extended to any complex number r; thus

$$D(\alpha^r) = D(E(rL(\alpha))) = E(rL(\alpha))\cdot D(rL(\alpha)) = \alpha^r\cdot r\alpha^{-1}D(\alpha) = r\alpha^{r-1}D(\alpha).$$

Also Theorem 15 extends to any complex r by use of Theorem 19. Finally, Theorem 17 holds for complex r; in fact the proof of this result needs no alteration for this generalization in view of the extended versions of Theorems 10 and 15 just mentioned.

8. An application to recurrence functions. For any given a, b, x_0, x_1 define a sequence $x_0, x_1, x_2, x_3, \cdots$ by the recurrence relation $x_{n+1}=ax_n+bx_{n-1}$ for $n=1, 2, 3, \cdots$. The Fibonacci sequence is the special case with $a=b=x_0=x_1=1$. The problem is to determine x_n explicitly in terms of a, b, x_0, x_1. If we define $\alpha=x_0+x_1\lambda+x_2\lambda^2+x_3\lambda^3+\cdots$ we see that

$$\alpha - a\lambda\alpha - b\lambda^2\alpha = x_0 + (x_1 - ax_0)\lambda. \tag{13}$$

If k_1 and k_2 are the roots of $k^2-ak-b=0$ we see that (13) can be written as

$$\alpha(1 - k_1\lambda)(1 - k_2\lambda) = x_0 + (x_1 - ax_0)\lambda. \tag{14}$$

CASE 1. Suppose that $k_1=k_2$. Then we see that

$$\alpha = \{x_0 + (x_1 - ax_0)\lambda\}\cdot(1 - k_1\lambda)^{-2}. \tag{15}$$

Now by Theorem 11 or Theorem 17 we have

$$(1 - k_1\lambda)^{-2} = 1 + 2k_1\lambda + 3k_1^2\lambda^2 + 4k_1^3\lambda^3 + 5k_1^4\lambda^4 + \cdots,$$

and so equating coefficients of λ^n in (15) we get

$$\begin{aligned} x_n &= x_0(n + 1)k_1^n + n(x_1 - ax_0)k_1^{n-1} \quad \text{or} \\ x_n &= nx_1k_1^{n-1} - (n - 1)x_0k_1^n. \end{aligned} \tag{16}$$

CASE 2. Suppose that $k_1\neq k_2$. Multiplying the identity

$$k_1 - k_2 = k_1(1 - k_2\lambda) - k_2(1 - k_1\lambda)$$

by $(1-k_1\lambda)^{-1}(1-k_2\lambda)^{-1}$ we get

$$(k_1 - k_2)(1 - k_1\lambda)^{-1}(1 - k_2\lambda)^{-1} = k_1(1 - k_1\lambda)^{-1} - k_2(1 - k_2\lambda)^{-1}.$$

Multiplying this into (14) we have

$$(k_1 - k_2)\alpha = \{x_0 + (x_1 - ax_0)\lambda\}\{k_1(1 - k_1\lambda)^{-1} - k_2(1 - k_2\lambda)^{-1}\}. \tag{17}$$

Also we use $k_1(1-k_1\lambda)^{-1}=k_1+k_1^2\lambda+k_1^3\lambda^2+k_1^4\lambda^3+\cdots+k_1^{n+1}\lambda^n+\cdots$. Equating coefficients of λ^n in (17) we have

$$(k_1-k_2)x_n=x_0(k_1^{n+1}-k_2^{n+1})+(x_1-ax_0)(k_1^n-k_2^n),$$

or

$$x_n = \{x_0(k_1^{n+1} - k_2^{n+1}) + (x_1 - ax_0)(k_1^n - k_2^n)\}/(k_1 - k_2). \tag{18}$$

The results (16) and (18) are well known; an alternative derivation is given in [**3**, page 100]. An entirely different way of treating equation (13) is as follows. We can write

$$\alpha = (1 - a\lambda - b\lambda^2)^{-1}\{x_0 + (x_1 - ax_0)\lambda\}. \tag{19}$$

Now by Theorem 17 we have

$$(1 - a\lambda - b\lambda^2)^{-1} = 1 + (a\lambda + b\lambda^2) + (a\lambda + b\lambda^2)^2 + (a\lambda + b\lambda^2)^3 + \cdots.$$

The coefficient of λ^n here is

$$a^n + \binom{n-1}{1}a^{n-2}b + \binom{n-2}{2}a^{n-4}b^2 + \binom{n-3}{3}a^{n-6}b^3 + \cdots$$
$$= \sum_{j=0}^{[n/2]} \binom{n-j}{j} a^{n-2j}b^j.$$

Equating coefficients of λ^n in (19) gives therefore

$$x_n = x_0 \sum_{j=0}^{[n/2]} \binom{n-j}{j} a^{n-2j}b^j + (x_1 - ax_0) \sum_{j=0}^{[(n-1)/2]} \binom{n-j-1}{j} a^{n-1-2j}b^j.$$

Finally, let us return to the method used for deriving (16) and (18). This method can be used with recurrence relations of higher order. Consider for example any given real (or complex) numbers x_0, x_1, x_2, a, b, c and a recurrence relation

$$x_{n+2} = ax_{n+1} + bx_n + cx_{n-1}, \qquad n = 1, 2, 3, \cdots.$$

If we define $\alpha = x_0 + x_1\lambda + x_2\lambda^2 + x_3\lambda^3 + \cdots$ we note that

$$(20) \qquad \alpha(1 - a\lambda - b\lambda^2 - c\lambda^3) = x_0 + (x_1 - ax_0)\lambda + (x_2 - ax_1 - bx_0)\lambda^2.$$

If the equation $k^3 - ak^2 - bk - c = 0$ has roots k_1, k_2, k_3 say, then (13) can be rewritten as

$$(21) \quad \alpha(1 - k_1\lambda)(1 - k_2\lambda)(1 - k_3\lambda) = x_0 + (x_1 - ax_0)\lambda + (x_2 - ax_1 - bx_0)\lambda^2.$$

There are now three cases depending on the nature of the roots k_1, k_2, k_3: three equal roots, two equal roots, or distinct roots. The case of equal roots follows the pattern of equation (15),

$$\alpha = [x_0 + (x_1 - ax_0)\lambda + (x_2 - ax_1 - bx_0)\lambda^2]\cdot(1 - k_1\lambda)^{-3}.$$

In the other two cases it is a matter of partial fraction expansions, in the sense that constants $q_1, q_2, q_3, q_4, q_5, q_6$ can be found so that

$$(1 - k_1\lambda)^{-2}(1 - k_2\lambda)^{-1} = q_1(1 - k_1\lambda)^{-1} + q_2(1 - k_1\lambda)^{-2} + q_3(1 - k_2\lambda)^{-1},$$
$$(1-k_1\lambda)^{-1}(1-k_2\lambda)^{-1}(1-k_3\lambda)^{-1} = q_4(1-k_1\lambda)^{-1} + q_5(1-k_2\lambda)^{-1} + q_6(1-k_3\lambda)^{-1},$$

in the case of two equal roots or the case of distinct roots, respectively.

For example if $a = 6$, $b = -11$, $c = 6$ then we find that $k_1 = 1$, $k_2 = 2$, $k_3 = 3$, $q_4 = \frac{1}{2}$, $q_5 = -4$, $q_6 = 9/2$. Then (21) implies that

$$\alpha = [x_0 + (x_1 - 6x_0)\lambda + (x_2 - 6x_1 + 11x_0)\lambda^2]$$
$$\cdot[\tfrac{1}{2}(1 - \lambda)^{-1} - 4(1 - 2\lambda)^{-1} + \tfrac{9}{2}(1 - 3\lambda)^{-1}],$$
$$x_n = x_0(\tfrac{1}{2} - 4\cdot 2^n + \tfrac{9}{2}\cdot 3^n) + (x_1 - 6x_0)(\tfrac{1}{2} - 4\cdot 2^{n-1} + \tfrac{9}{2}\cdot 3^{n-1})$$
$$+ (x_2 - 6x_1 + 11x_0)(\tfrac{1}{2} - 4\cdot 2^{n-2} + \tfrac{9}{2}\cdot 3^{n-2}).$$

9. An application to partitions. The notation $p(n)$ represents the number of ways that a positive integer n can be written as a sum of positive integers. Two partitions are not different if they differ only in the order of their summands. As usual, we define $p(0)=1$.

Let α_j denote $1+\lambda^j+\lambda^{2j}+\lambda^{3j}+\cdots$ for every positive integer j. Then $\alpha_1, \alpha_2, \alpha_3, \cdots$ is a sequence admitting multiplication in the sense of (10) in Section 4. By the standard argument, for example in [3, pp. 226, 227], we have

$$\alpha_1 \cdot \alpha_2 \cdot \alpha_3 \cdot \cdots = \prod_{j=1}^{\infty} \alpha_j = \sum_{k=0}^{\infty} p(k)\lambda^k. \tag{22}$$

But also we see that $\alpha_j(1-\lambda^j)=1$ so that $\alpha_j=(1-\lambda^j)^{-1}$, and

$$\prod_{j=1}^{\infty} \alpha_j = \prod_{j=1}^{\infty} (1-\lambda^j)^{-1}. \tag{23}$$

Next let $q^e(n)$ denote the number of partitions of any positive integer n into an even number of distinct summands, and similarly let $q^0(n)$ be the number of partitions of n into an odd number of distinct summands. It is customary to take $q^e(0)=1$ and $q^0(0)=0$. Then the coefficient of λ^n in the expansion of the admissible product

$$(1-\lambda)(1-\lambda^2)(1-\lambda^3)\cdots = \prod_{j=1}^{\infty}(1-\lambda^j)$$

is seen to be $q^e(n)-q^0(n)$ by a simple combinatorial argument. It follows that

$$\prod_{j=1}^{\infty}(1-\lambda^j) = \sum_{n=0}^{\infty}\{q^e(n)-q^0(n)\}\lambda^n. \tag{24}$$

By use of graphs of partitions it can be proved, cf. [3, pp. 224–226], that $q^e(n)-q^0(n)=(-1)^j$ if n is of the form $(3j^2+j)/2$ or $(3j^2-j)/2$ for some nonnegative integer j, and $q^e(n)-q^0(n)=0$ otherwise. It is easy to prove that the sets of positive integers

$$\{(3j^2+j)/2; j=1,2,3,\cdots\}, \{(3j^2-j)/2; j=1,2,3,\cdots\}$$

are distinct, and hence (24) can be written as

$$\begin{aligned}\prod_{j=1}^{\infty}(1-\lambda^j) &= 1+\sum_{j=1}^{\infty}(-1)^j(\lambda^{(3j^2+j)/2}+\lambda^{(3j^2-j)/2}) \\ &= 1-\lambda-\lambda^2+\lambda^5+\lambda^7-\lambda^{12}-\lambda^{15}+\cdots.\end{aligned} \tag{25}$$

This with (22) and (23) implies that

$$\left\{1+\sum_{j=1}^{\infty}(-1)^j(\lambda^{(3j^2+j)/2}+\lambda^{(3j^2-j)/2})\right\}\sum p(k)\lambda^k = 1,$$

$$(1-\lambda-\lambda^2+\lambda^5+\lambda^7-\lambda^{12}-\lambda^{15}+\cdots)\sum p(k)\lambda^k = 1.$$

For any positive integer n, the coefficient of λ^n on the left side of this equation is $p(n)-p(n-1)-p(n-2)+p(n-5)+p(n-7)-p(n-12)-p(n-15)+\cdots$. Thus we have proved the following well-known result of Euler [3, p. 235].

THEOREM 21. *For any positive integers n,*

$$p(n) = p(n-1) + p(n-2) - p(n-5) - p(n-7) + \cdots$$
$$= \sum_{j=1}^{\infty} (-1)^{j+1}\{p(n-(3j^2+j)/2) + p(n-(3j^2-j)/2)\}$$

with $p(t)=0$ if $t<0$, so that the sum is finite.

It should be emphasized that the proof given here of Theorem 21 is not new. The proof above is simply the usual one formulated in terms of the "soft" analysis of formal power series.

10. An application to the sum of divisors function. For any positive integer n let $\sigma(n)$ denote the sum of the positive divisors of n; for example $\sigma(6)=1+2+3+6$. We establish a known recurrence relation [3, p. 236] for $\sigma(n)$, and again the positive integers of the form $(3k^2-k)/2$ and $(3k^2+k)/2$ play a role, namely, the positive integers 1, 2, 5, 7, 12, 15, 22, 26, $\cdots$.

THEOREM 22. *For any positive integer k,*

$$\sigma(k) - \sigma(k-1) - \sigma(k-2) + \sigma(k-5) + \sigma(k-7) - \cdots$$
$$= \begin{cases} (-1)^{j+1}k \text{ if } k = (3j^2+j)/2 \text{ or } k = (3j^2-j)/2, \\ 0 \text{ otherwise.} \end{cases}$$

Proof. Define $\beta = \prod_{j=1}^{k}(1-\lambda^j)$ so that $L(\beta) = \sum_{j=1}^{k} L(1-\lambda^j)$,

$$-D(L(\beta)) = -\beta^{-1}D(\beta) = \sum_{j=1}^{k} j(1-\lambda^j)^{-1}\lambda^{j-1}$$
$$= \sum_{j=1}^{k} \{j\lambda^{j-1} + j\lambda^{2j-1} + j\lambda^{3j-1} + j\lambda^{4j-1} + \cdots\}$$
$$= \sum_{n=1}^{\infty} f(n)\lambda^{n-1},$$

where $f(n)$ is seen to be the sum of all positive divisors of n that do not exceed k. Thus we have $f(n)=\sigma(n)$ if $n \leq k$, and so we can write

$$-\beta^{-1}D(\beta) = \sum_{n=1}^{k} \sigma(n)\lambda^{n-1} + \sum_{n=k+1}^{\infty} f(n)\lambda^{n-1}. \tag{26}$$

Now equation (24) can be written with a finite product

$$\beta = \prod_{j=1}^{k}(1-\lambda^j) = \sum_{n=0}^{\infty}\{q_k^e(n) - q_k^0(n)\}\lambda^n, \tag{27}$$

where $q_k^e(n)$ denotes the number of partitions of n into an even number of distinct summands $\leqq k$, and $q_k^0(n)$ denotes the number of partitions of n into an odd number of distinct summands $\leqq k$. Define $q_k^e(0)=1$ and $q_k^0(0)=0$. If $n\leqq k$ we note that $q_k^e(n)=q^e(n)$ and $q_k^0(n)=q^0(n)$, so (27) can be written as

$$\beta = \sum_{n=0}^{k} \{q^e(n) - q^0(n)\}\lambda^n + \sum_{n=k+1}^{\infty} \{q_k^e(n) - q_k^0(n)\}\lambda^n. \tag{28}$$

We now equate the coefficients of λ^{k-1} in $-D(\beta)$ and in the product $\beta(-\beta^{-1}D(\beta))$. From (28) it is clear that the coefficient of λ^{k-1} in $-D(\beta)$ is

$$-k\{q^e(k) - q^0(k)\} = \begin{cases} -(-1)^j k \text{ if } k = (3j^2 \pm j)/2, \\ 0 \text{ otherwise.} \end{cases}$$

From (28) and (26) the coefficient of λ^{k-1} in $\beta(-\beta^{-1}D(\beta))$ is

$$\begin{aligned}\sigma(k)\{q^e(0) - q^0(0)\} + \sigma(k-1)\{q^e(1) - q^0(1)\} + \sigma(k-2)\{q^e(2) - q^0(2)\} + \cdots \\ = \sigma(k) - \sigma(k-1) - \sigma(k-2) + \sigma(k-5) + \sigma(k-7) - \cdots,\end{aligned}$$

and so the theorem is proved.

11. Trigonometric functions and differential equations. We now return to the general theory of formal power series and make the definitions

$$\sin\alpha = \{E(i\alpha) - E(-i\alpha)\}/2i = \sum_{k=0}^{\infty} \{(-1)^k\alpha^{2k+1}\}/(2k+1)!$$

$$\cos\alpha = \{E(i\alpha) + E(-i\alpha)\}/2 = \sum_{k=0}^{\infty} \{(-1)^k\alpha^{2k}\}/(2k)!,$$

where α is any element in P_0. Thus $\sin\alpha$ is in P_0, but $\cos\alpha$ is in P_1, so we can define $\sec\alpha=(\cos\alpha)^{-1}$ and $\tan\alpha=(\sin\alpha)(\cos\alpha)^{-1}$. However, we cannot now define cosec α and cot α, but in the next section we extend the theory, to encompass these two functions. All the rules of differentiation now apply, such as $D(\sin\alpha)=(\cos\alpha)D(\alpha)$.

The standard theory of homogeneous linear differential equations with constant coefficients is valid. For example, in the second order case, let a and b be any complex numbers, and let r_1 and r_2 be the roots of $x^2+ax+b=0$. Then a solution for ρ in P of the equation $D^2(\rho)+aD(\rho)+b\rho=0$ is

$$\rho = c_1E(r_1\lambda) + c_2E(r_2\lambda)$$

with arbitrary constants c_1 and c_2. It is easy to prove that this is the general solution if $r_1\neq r_2$. If $r_1=r_2$ the general solution is of course $\rho=c_1E(r_1\lambda)+c_2\lambda E(r_1\lambda)$.

We now give a brief sketch of the use of a differential equation to solve a combinatorial problem, as in André [7, p. 172]. Our approach differs from that of André in that we treat the differential equation in a purely formal sense, which he did not. For $n\geqq 2$ let b_n be the number of permutations $a_1, a_2, \cdots,$

a_n of $1, 2, \cdots, n$ such that $a_j > a_{j-1}$ if j is even, and $a_j < a_{j-1}$ if j is odd. Call such a permutation an E-permutation. Similarly, say that $a_1, a_2, \cdots, a_n$ is an O-permutation of $1, 2, \cdots, n$ if $a_j > a_{j-1}$ if j is odd, and $a_j < a_{j-1}$ if j is even. Note that if $a_1, a_2, \cdots, a_n$ is an O-permutation then $n+1-a_1, n+1-a_2, \cdots, n+1-a_n$ is an E-permutation, and conversely. Thus there is a one-to-one correspondence between E-permutations and O-permutations; there are b_n of each type. Define $b_0 = 1$ and $b_1 = 1$.

Next, consider the number of O-permutations with $a_1 = n$. It is not difficult to see that there are b_{n-1} of these. Also, there are no E-permutations with $a_1 = n$. Turning to permutations with $a_2 = n$, there are no O-permutations of this type. However, the number of E-permutations with $a_2 = n$ is $(n-1)b_{n-2}$, or what is the same thing $(n-1)b_1 b_{n-2}$; the reason for this is that a_1 can be any element among $1, 2, \cdots, n-1$ and the rest can be set up as $a_3, a_4, \cdots, a_n$ in b_{n-2} ways. A similar argument shows that there are no E-permutations with $a_3 = n$, whereas the number of O-permutations with $a_3 = n$ is $\binom{n}{2} b_2 b_{n-3}$. Thus by considering all E-permutations and all O-permutations with successively $a_1 = n$, then $a_2 = n$, then $a_3 = n, \cdots$, and finally $a_n = n$, we are led to the recurrence relation

$$2b_n = \sum_{j=0}^{n-1} \binom{n-1}{j} b_j b_{n-j-1} \quad \text{or} \quad 2nc_n = \sum_{j=0}^{n-1} c_j c_{n-j-1},$$

where c_n is defined as $b_n/n!$ for all nonnegative integers n. Taking α to be the formal power series

$$\alpha = \sum_{n=0}^{\infty} c_n \lambda^n$$

we can readily verify that the differential equation $2D(\alpha) = \alpha^2 + 1$ holds. Now it is easy to verify from the definitions of the formal trigonometric functions that $\sin^2\lambda + \cos^2\lambda = 1$, $\sec^2\lambda = 1 + \tan^2\lambda$, $D(\tan\lambda) = \sec^2\lambda$, $D(\sec\lambda) = \sec\lambda\tan\lambda$. Thus the unique formal solution of the differential equation is $\alpha = \tan\lambda + \sec\lambda$. (André gives the solution of the differential equation as $\alpha = \tan(\lambda/2 + \pi/4)$ which has no meaning in our formal definition of the trigonometric functions. The usual formula for $\tan(\alpha+\beta)$ in terms of $\tan\alpha$ and $\tan\beta$ is valid, but $\tan\pi/4 = 1$ cannot be established in the formal theory. In fact $\tan\pi/4$ is not even defined because $\pi/4$ is not an element of P_0, although it is an element of P.) Thus we have

$$\alpha = \sum b_n \lambda^n/n! = \tan\lambda + \sec\lambda.$$

Now the power series for $\tan\lambda$ has odd powers of λ only, with coefficients closely connected with the Bernoulli numbers [8, p. 268]. Similarly the power series for $\sec\lambda$ has even powers of λ only, with coefficients related to the Euler numbers [8, p. 269]. Thus André was able to relate the combinatorial numbers b_n to the Bernoulli numbers for odd n, and to the Euler numbers for even n. (A different approach to this problem has been given recently by R. C. Entringer [9].)

From our point of view in this paper, the important aspect of this is that André's conclusions can be drawn with only a formal use of calculus and differential equations and without any convergence questions in the use of α^2, the square of a power series, in the differential equation. The series expansions for $\tan \lambda$ and $\sec \lambda$ come from those for $\sin \lambda$ and $\cos \lambda$, and these are defined in terms of the exponential functions $E(i\lambda)$ and $E(-i\lambda)$. The formal structure carries the entire argument, with no need for the classical infinitesimal calculus. Of course, such relations as $\sin^2 \lambda + \cos^2 \lambda = 1$ have meaning only in terms of formal power series in this context and not in terms of the geometry of right-angled triangles.

12. Extension to a field. Since the set of formal power series P is a commutative integral domain, it can be imbedded in a field P^* in the classical manner by use of pairs of elements, cf. [2, pp. 87–92]. This construction is very well known in the extension of the integers to the rational numbers. Thus P^* is the field of all pairs (α, β) with $\alpha \in P$, $\beta \in P$ and $\beta \neq 0$. Addition and multiplication are defined by

$$(\alpha_1, \beta_1) + (\alpha_2, \beta_2) = (\alpha_1\beta_2 + \alpha_2\beta_1, \beta_1\beta_2),$$

$$(\alpha_1, \beta_1) \cdot (\alpha_2, \beta_2) = (\alpha_1\alpha_2, \beta_1\beta_2).$$

Two elements (α_1, β_1) and (α_2, β_2) are said to be equal if and only if $\alpha_1\beta_2 = \alpha_2\beta_1$.

If $\beta = 1$ we agree to write α for $(\alpha, \beta) = (\alpha, 1)$, so that P is a subset of P^*. Similarly we agree to write

$$\left(\sum_{j=0}^{\infty} a_j\lambda^j, \lambda^r\right) \quad \text{as} \quad \sum_{j=0}^{\infty} a_j\lambda^{j-r}, \tag{29}$$

where r is a positive integer. We prove in Theorem 23 that every element of P can be written in this way, so that P^* can be thought of as the class of Laurent power series expansions, with a finite number of negative exponents allowed.

To do this we first define the degree of α for any α in P, $\alpha \neq 0$. If $\alpha = \Sigma a_j\lambda^j$ then the degree of α, written $\deg(\alpha)$, is the subscript of the first nonzero coefficient in the sequence of coefficients $a_0, a_1, a_2, \cdots$. If α_1 and α_2 are nonzero elements of P it follows that $\deg(\alpha_1\alpha_2) = \deg(\alpha_1) + \deg(\alpha_2)$. This definition is extended to P^* as follows: if $(\alpha, \beta) \in P^*$ with $\alpha \neq 0$ then $\deg(\alpha, \beta) = \deg(\alpha) - \deg(\beta)$. Degree is well-defined, because if $(\alpha_1, \beta_1) = (\alpha_2, \beta_2)$ then $\alpha_1\beta_2 = \alpha_2\beta_1$ and so we have

$$\deg(\alpha_1) + \deg(\beta_2) = \deg(\alpha_2) + \deg(\beta_1),$$

$$\deg(\alpha_1) - \deg(\beta_1) = \deg(\alpha_2) - \deg(\beta_2).$$

Next for any (α, β) in P^* with $\alpha \neq 0$, let $\deg(\alpha) = m$, $\deg(\beta) = n$ so that $\deg(\alpha, \beta) = m - n$. Then we see that $\beta = \lambda^n\beta_1$ where β_1 has degree 0, so that β_1

has an inverse. It follows that $(\alpha, \beta)=(\alpha, \lambda^n\beta_1)=(\alpha\beta_1^{-1}, \lambda^n)$. Now $\alpha\beta_1^{-1}$ has degree m, so it can be written in the form

$$\alpha\beta_1^{-1} = \sum_{j=m}^{\infty} a_j\lambda^j, \qquad a_m \neq 0.$$

Thus we have

$$(\alpha, \beta) = \sum_{j=m}^{\infty} a_j\lambda^{j-n}, \qquad a_m \neq 0, \tag{30}$$

by virtue of (29).

THEOREM 23. *The representation* (30) *of any nonzero element* (α, β) *of* P^* *is unique.*

Proof. Suppose that (α, β) can also be written as

$$(\alpha, \beta) = \sum_{j=h}^{\infty} c_j\lambda^{j-n}, \qquad c_h \neq 0.$$

By the invariance of degree under different representations we see that $m-n=h-n$ and $m=h$. Also we have

$$(\alpha, \beta) = \left(\sum_{j=m} a_j\lambda^j, \lambda^n\right) = \left(\sum_{j=m}^{\infty} c_j\lambda^j, \lambda^n\right),$$

and so by the definition of equality in P^*,

$$\sum_{j=m}^{\infty} a_j\lambda^{j+n} = \sum_{j=m}^{\infty} c_j\lambda^{j+n}.$$

The theorem follows by the definition of equality in P.

In the preceding section we saw that the trigonometric functions $\sin\alpha$, $\cos\alpha$, $\tan\alpha$, and $\sec\alpha$ could be defined for any element α of P, but not $\operatorname{cosec}\alpha$ and $\cot\alpha$. If $\alpha\neq 0$ we can define the latter two functions from P to P^*; thus

$$\operatorname{cosec}\alpha = (1, \sin\alpha), \quad \cot\alpha = (\cos\alpha, \sin\alpha).$$

A simple calculation shows that

$$\operatorname{cosec}\lambda = \lambda^{-1} + (\lambda/6) + (7\lambda^3/360) + \cdots.$$

Finally, we note that the theory of formal power series, developed here in analogy to power series in a single variable, can be extended in a similar way to the multiple variable case.

Work supported by NSF Grant GP 6510.

References

1. Louis de Branges and James Rovnyak, Square Summable Power Series, Holt, Rinehart and Winston, New York, 1966.

2. Nathan Jacobson, Lectures in Abstract Algebra, vol. 1, Van Nostrand, Princeton, N. J., 1951, p. 19.

3. Ivan Niven and H. S. Zuckerman, An Introduction to the Theory of Numbers, 2nd ed., Wiley, New York, 1966.

4. John Riordan, Generating Functions, Chap. 3 in Applied Combinatorial Mathematics, ed. by E. F. Beckenbach, Wiley, New York, 1964.

5. ———, Combinatorial Identities, Wiley, New York, 1968.

6. David Zeitlin, On convoluted numbers and sums, this MONTHLY, 74 (1967) 235–246.

7. D. André, Sur les permutations alternées, J. Math. (3), 7 (1881) 167–184.

8. J. V. Uspensky and M. A. Heaslet, Elementary Number Theory, McGraw-Hill, New York, 1939.

9. R. C. Entringer, A combinatorial interpretation of the Euler and Bernoulli numbers, Nieuw Arch. Wisk. (3), 14 (1966) 241–246.

10. E. T. Bell, Euler algebra, Trans. Amer. Math. Soc., 25 (1923) 135–154.

11. ———, Algebraic Arithmetic, A.M.S. Coll. Publ. VII, New York, 1927, esp. 124–134.

12. ———, Postulational bases for the umbral calculus, Amer. J. Math., 62 (1940) 717–724.

13. S. Bochner and W. T. Martin, Several Complex Variables, Princeton Univ. Press, Princeton, N. J. 1948, Chap. 1.

14. E. D. Cashwell and C. J. Everett, Formal power series, Pacific J. Math., 13 (1963) 45–64.

15. S. A. Jennings, Substitution groups of formal power series, Canadian J. Math., 6 (1954) 325–340.

16. D. A. Klarner, A ring of sequences generated by rational functions, this MONTHLY, 74 (1967) 813–816.

17. R. Vaidyanathaswamy, Multiplicative arithmetic functions, Trans. Amer. Math. Soc., 33 (1931) 579–662.

18. O. Zariski and P. Samuel, Commutative Algebra, Van Nostrand, Princeton, N. J. 1960, vol. II, Chap. 7.

BIBLIOGRAPHIC ENTRIES: FOUNDATIONS

The references below are to the AMERICAN MATHEMATICAL MONTHLY.

1. Leon Henkin, On mathematical induction, 1960, 323–338.

A rigorous study of induction from the Peano postulates, leading to induction models.

2. F. Cunningham, Jr., A construction of the rational numbers, 1959, 769–777.

Uses algebraic methods, including free cyclic groups.

3. J. A. Dyer, A note on redundancies in the axiom system for the real numbers, 1967, 1244–1246.

Shows that commutativity of addition and multiplication follow from the other axioms.

BIBLIOGRAPHIC ENTRIES: BOOLEAN ALGEBRAS

Except for the two entries labeled MATHEMATICS MAGAZINE, the references below are to the AMERICAN MATHEMATICAL MONTHLY.

1. Paul R. Halmos, The basic concepts of algebraic logic, 1956, 363–387.

An exposition of Boolean algebras and the propositional calculus.

2. E. R. Stabler, Boolean representation theory, 1944, 129–132.

Gives a proof of the representation theorem for Boolean rings.

3. R. E. Smithson, A note on finite Boolean rings, MATHEMATICS MAGAZINE, 1964, 325–327.

Gives an elementary proof of the representation theorem for finite Boolean rings by producing an "orthogonal basis".

4. Franz Hohn, Some mathematical aspects of switching, 1955, 75–90.

Demonstrates how Boolean algebras arise in the analysis of combinatorial circuits.

5. Wai-Kai Chen, Boolean matrices and switching nets, MATHEMATICS MAGAZINE, 1966, 1–8.

Discusses properties of matrices over Boolean algebras and applications to switching. Closely related to the previous reference.

BIBLIOGRAPHIC ENTRY: UNIVERSAL ALGEBRA

1. Steven Feigelstock, A universal subalgebra theorem, AMERICAN MATHEMATICAL MONTHLY, 1965, 884–888

Proves that subalgebras of free universal algebras are free.

BIBLIOGRAPHIC ENTRY: NEAR-RINGS

1. Gerald Berman and Robert J. Silverman, Near-rings, AMERICAN MATHEMATICAL MONTHLY, 1959, 23–34.

An exposition and discussion of some elementary properties of near rings.

BIBLIOGRAPHIC ENTRY: FORMAL POWER SERIES AND APPLICATIONS

1. D. A. Klarner, Algebraic theory for difference and differential equations, AMERICAN MATHEMATICAL MONTHLY, 1969, 366–373.

Using formal power series, gives an algebraic development of the solution of linear differential equations.

AUTHOR INDEX

Numbers in italic type refer to bibliographic entries.